Ecology and Conservation of Parrots in Their Native and Non-Native Ranges

Editors

José L. Tella
Guillermo Blanco
Martina Carrete

MDPI • Basel • Beijing • Wuhan • Barcelona • Belgrade • Manchester • Tokyo • Cluj • Tianjin

Editors

José L. Tella
CSIC- Estación Biológica de Doñana EBD
Spain

Guillermo Blanco
National Museum of Natural Sciences
Spain

Martina Carrete
Universidad Pablo de Olavide
Spain

Editorial Office
MDPI
St. Alban-Anlage 66
4052 Basel, Switzerland

This is a reprint of articles from the Special Issue published online in the open access journal *Diversity* (ISSN 1424-2818) (available at: https://www.mdpi.com/journal/diversity/special_issues/Parrots).

For citation purposes, cite each article independently as indicated on the article page online and as indicated below:

LastName, A.A.; LastName, B.B.; LastName, C.C. Article Title. *Journal Name* **Year**, *Volume Number*, Page Range.

ISBN 978-3-0365-5009-1 (Hbk)
ISBN 978-3-0365-5010-7 (PDF)

Cover image courtesy of Rocco Mosconi.

Contents

MDPI

Editorial

Recent Advances in Parrot Research and Conservation

José L. Tella [1,*], Guillermo Blanco [2] and Martina Carrete [3]

[1] Department of Conservation Biology, Doñana Biological Station CSIC, 41092 Sevilla, Spain
[2] Department of Evolutionary Ecology, Museo Nacional de Ciencias Naturales CSIC, 28006 Madrid, Spain; gblanco@mncn.csic.es
[3] Department of Physical, Chemical and Natural Systems, Universidad Pablo de Olavide, 41013 Sevilla, Spain; mcarrete@upo.es
* Correspondence: tella@ebd.csic.es

Citation: Tella, J.L.; Blanco, G.; Carrete, M. Recent Advances in Parrot Research and Conservation. *Diversity* **2022**, *14*, 419. https://doi.org/10.3390/d14060419

Received: 23 May 2022
Accepted: 23 May 2022
Published: 25 May 2022

Publisher's Note: MDPI stays neutral with regard to jurisdictional claims in published maps and institutional affiliations.

Parrots (Psittaciformes), with about 400 species widely distributed across continents and oceanic islands, stand out among birds for their poor conservation status [1]. According to the IUCN Red List [2], almost 30% of parrot species are threatened with extinction and c. 15% are classified as near threatened. Moreover, almost 60% of all species are experiencing global population declines. Several threats to parrots, such as habitat loss, persecution, and wildlife trade, have recently been addressed globally [3–5]. However, detailed studies on distribution, biology, ecology, population dynamics, population genetics, and specific conservation threats are lacking for most species. The need for further research is exemplified by recent splits of species and descriptions of new species (e.g., [6]) and by the ecological functions of parrots, such as seed dispersal, which have been overlooked until recent years (e.g., [7]). Given the ecosystem services they can provide, the conservation of parrot populations contributes to the health of the habitats in which they live.

The attractiveness of parrots has led to their intensive transport to foreign pet markets around the world [5]. In addition, international trade has caused several parrot species to establish populations outside their native ranges, often resulting in flourishing populations that contrast with the poor conservation status of many native populations. However, studies on non-native populations have been limited mainly to two species of parakeets (the ring-necked parakeets *Psittacula krameri* and the monk parakeet *Myiopsitta monachus*). Much more research is needed on these non-native parrot populations, including aspects such as their establishment and spread processes, population dynamics, potential impacts (negative and positive [8]) on their recipient habitats and communities, the need (or not) for control and/or eradication, or their ecological functions in their invaded regions.

The Special Issue 'Ecology and Conservation of Parrots in Their Native and Non-Native Ranges' offers 23 new research studies and four reviews, thanks to the contribution of 123 authors working in different academic institutions and NGOs in 22 countries. Overall, it combines and synthesizes recent research on native and non-native parrot populations, filling gaps in several research areas, compiling state-of-the-art methodological aspects, and advancing the conservation of threatened species.

This volume progresses the study of parrot distribution and abundance. Along with a review of approaches to modeling parrot distributions [9], other studies advance the prediction of future parrot distributions by taking into account their food plant distributions in Bolivia [10], or show the combination of site-occupancy modeling and citizen science to improve range distributions, and roost-counts to estimate parrot populations in Brazil [11]. Roost counts have also allowed estimations of the global population of a Neotropical parrot species [12]. However, this is not feasible for most parrot species, so roadside surveys are proposed to estimate the relative abundances of entire parrot communities in different biomes around the world [13].

Another group of papers deals with little-known aspects of parrot ecology, such as diseases, movements, or ecological functions. A study on selected bacteria and viruses found

Chlamydia but not beak and feather disease virus (BFDV) in Brazil [14]. A retrospective study also failed to find BFDV in Australia, Argentina, and New Zealand [15], while a new BFDV genotype has recently been found in non-native Spanish parakeet populations [16], and a three-decade study investigated the diversification of this virus and the subsequent waves of infection in Mauritius [17]. Moreover, satellite telemetry of even a few individuals revealed information highly relevant to the conservation of a macaw species in Bolivia [18], and a study of the foraging ecology of another species in Argentina revealed once again the important seed dispersal role of parrots for key plants in the ecosystems where they live [19].

A review compiles the different genetic tools available for the study of parrot evolution, biology, and conservation [20]. Examples of the useful application of these molecular approaches are the demonstration of genetic distinctiveness of isolated parrot populations in Brazil [21], the study of population genetics of wild and captive populations in Mexico and Bolivia [22], and the identification of the geographic origin of traded individuals in Mexico [23].

The wildlife trade is one of the main threats to parrots, and a literature review summarizes actions taken to tackle the illegal parrot trade, making recommendations for improving future efforts [24]. Also related to illegal trade, other papers developed a capture pressure index in Ecuador [25], assessed peoples' perception of poaching to improve conservation programs in Venezuela [26], and demonstrated that parrot poaching is not random but selected for the most attractive species in Colombia [27], and that selective parrot poaching affects parrot populations in Indonesia [28].

Several other papers deal with in situ and ex situ management for parrot conservation. These range from techniques to increase chick survival in the wild [29], to the study of stress physiology in relation to the breeding success of captive individuals destined for reintroductions [30], the challenges faced in establishing reintroduced populations [31,32], and the use of a new technique that could allow the establishment of released individuals in places where parrots are absent [33].

Finally, an updated review has identified 166 introduced parrot species in 120 countries worldwide, of which 60 species have naturalized populations, and 11 species have bred outside their native ranges [34]. The study of naturalized parakeets' home ranges in Spain [35] adds to the scarce information available on the ecology of introduced parrot populations.

We hope that this Special Issue will encourage further research on this fascinating and endangered group of birds.

Author Contributions: J.L.T. wrote the first draft of the manuscript, and G.B. and M.C. edited and contributed additional sections to the manuscript. All authors have read and agreed to the published version of the manuscript.

Funding: Authors were supported by Loro Parque Fundacion (grant reference: PP-146-2018-1).

Acknowledgments: We thank all authors, reviewers, and Academic Editors for their contributions to produce this large special issue.

Conflicts of Interest: The authors declare no conflict of interest.

References

1. McClure, C.J.; Rolek, B.W. Relative conservation status of bird orders with special attention to raptors. *Front. Ecol. Evol.* **2020**, *8*, 420. [CrossRef]
2. IUCN. The IUCN Red List of Threatened Species. Version 2021-3. 2021. Available online: https://www.iucnredlist.org (accessed on 15 May 2022).
3. Vergara-Tabares, D.L.; Cordier, J.M.; Landi, M.A.; Olah, G.; Nori, J. Global trends of habitat destruction and consequences for parrot conservation. *Glob. Change Biol.* **2020**, *26*, 4251–4262. [CrossRef] [PubMed]
4. Barbosa, J.M.; Hiraldo, F.; Romero, M.Á.; Tella, J.L. When does agriculture enter into conflict with wildlife? A global assessment of parrot–agriculture conflicts and their conservation effects. *Divers. Distrib.* **2021**, *27*, 4–17. [CrossRef]
5. Chan, D.T.C.; Poon, E.S.K.; Wong, A.T.C.; Sin, S.Y.W. Global trade in parrots–influential factors of trade and implications for conservation. *Glob. Ecol. Conserv.* **2021**, *30*, e01784. [CrossRef]

6. Braun, M.P.; Reinschmidt, M.; Datzmann, T.; Zamora, R.; Neves, L.; Arndt, T. Influences of oceanic islands & the Pleistocene on the biogeography & evolution of two groups of Australasian parrots (Aves: Psittaciformes: *Eclectus roratus*, *Trichoglossus haematodus* complex). Rapid evolution & implications for taxonomy & conservation. *Eur. J. Ecol.* **2017**, *3*, 47–66.
7. Hernández-Brito, D.; Romero-Vidal, P.; Hiraldo, F.; Blanco, G.; Díaz-Luque, J.A.; Barbosa, J.M.; Symes, C.T.; White, T.H.; Pacífico, E.C.; Sebastián-González, E.; et al. Epizoochory in parrots as an overlooked yet widespread plant–animal mutualism. *Plants* **2021**, *10*, 760. [CrossRef]
8. Hernández-Brito, D.; Carrete, M.; Blanco, G.; Romero-Vidal, P.; Senar, J.C.; Mori, E.; White, T.H.; Luna, Á.; Tella, J.L. The role of monk parakeets as nest-site facilitators in their native and invaded areas. *Biology* **2021**, *10*, 683. [CrossRef]
9. Ferrer-Paris, J.R.; Sánchez-Mercado, A. Contributions of Distribution Modelling to the Ecological Study of Psittaciformes. *Diversity* **2021**, *13*, 611. [CrossRef]
10. Hambuckers, A.; de Harenne, S.; Ledezma, E.R.; Zeballos, L.Z.; François, L. Predicting the Future Distribution of Ara rubrogenys, an Endemic Endangered Bird Species of the Andes, Taking into Account Trophic Interactions. *Diversity* **2021**, *13*, 94. [CrossRef]
11. Zulian, V.; Miller, D.A.W.; Ferraz, G. Endemic and Threatened Amazona Parrots of the Atlantic Forest: An Overview of Their Geographic Range and Population Size. *Diversity* **2021**, *13*, 416. [CrossRef]
12. Dupin, M.K.; Dahlin, C.R.; Wright, T.F. Range-Wide Population Assessment of the Endangered Yellow-Naped Amazon (*Amazona auropalliata*). *Diversity* **2020**, *12*, 377. [CrossRef]
13. Tella, J.L.; Romero-Vidal, P.; Dénes, F.V.; Hiraldo, F.; Toledo, B.; Rossetto, F.; Blanco, G.; Hernández-Brito, D.; Pacífico, E.; Díaz-Luque, J.A.; et al. Roadside Car Surveys: Methodological Constraints and Solutions for Estimating Parrot Abundances across the World. *Diversity* **2021**, *13*, 300. [CrossRef]
14. Vaz, F.F.; Sipinski, E.A.B.; Seixas, G.H.F.; Prestes, N.P.; Martinez, J.; Raso, T.F. Molecular Survey of Pathogens in Wild Amazon Parrot Nestlings: Implications for Conservation. *Diversity* **2021**, *13*, 272. [CrossRef]
15. Ortiz-Catedral, L.; Wallace, C.J.; Heinsohn, R.; Krebs, E.A.; Langmore, N.E.; Vukelic, D.; Bucher, E.H.; Varsani, A.; Masello, J.F. A PCR-Based Retrospective Study for Beak and Feather Disease Virus (BFDV) in Five Wild Populations of Parrots from Australia, Argentina and New Zealand. *Diversity* **2022**, *14*, 148. [CrossRef]
16. Morinha, F.; Carrete, M.; Tella, J.L.; Blanco, G. High Prevalence of Novel Beak and Feather Disease Virus in Sympatric Invasive Parakeets Introduced to Spain from Asia and South America. *Diversity* **2020**, *12*, 192. [CrossRef]
17. Fogell, D.J.; Tollington, S.; Tatayah, V.; Henshaw, S.; Naujeer, H.; Jones, C.; Raisin, C.; Greenwood, A.; Groombridge, J.J. Evolution of Beak and Feather Disease Virus across Three Decades of Conservation Intervention for Population Recovery of the Mauritius Parakeet. *Diversity* **2021**, *13*, 584. [CrossRef]
18. Davenport, L.C.; Boorsma, T.; Carrara, L.; Antas, P.d.T.Z.; Faria, L.; Brightsmith, D.J.; Herzog, S.K.; Soria-Auza, R.W.; Hennessey, A.B. Satellite Telemetry of Blue-Throated Macaws in Barba Azul Nature Reserve (Beni, Bolivia) Reveals Likely Breeding Areas. *Diversity* **2021**, *13*, 564. [CrossRef]
19. Blanco, G.; Romero-Vidal, P.; Carrete, M.; Chamorro, D.; Bravo, C.; Hiraldo, F.; Tella, J.L. Burrowing Parrots Cyanoliseus patagonus as Long-Distance Seed Dispersers of Keystone Algarrobos, Genus Prosopis, in the Monte Desert. *Diversity* **2021**, *13*, 204. [CrossRef]
20. Olah, G.; Smith, B.T.; Joseph, L.; Banks, S.C.; Heinsohn, R. Advancing Genetic Methods in the Study of Parrot Biology and Conservation. *Diversity* **2021**, *13*, 521. [CrossRef]
21. Hellmich, D.L.; Saidenberg, A.B.S.; Wright, T.F. Genetic, but Not Behavioral, Evidence Supports the Distinctiveness of the Mealy Amazon Parrot in the Brazilian Atlantic Forest. *Diversity* **2021**, *13*, 273. [CrossRef]
22. Campos, C.I.; Martinez, M.A.; Acosta, D.; Diaz-Luque, J.A.; Berkunsky, I.; Lamberski, N.L.; Cruz-Nieto, J.; Russello, M.A.; Wright, T.F. Genetic Diversity and Population Structure of Two Endangered Neotropical Parrots Inform In Situ and Ex Situ Conservation Strategies. *Diversity* **2021**, *13*, 386. [CrossRef]
23. Rivera-Ortíz, F.A.; Juan-Espinosa, J.; Solórzano, S.; Contreras-González, A.M.; Arizmendi, M.d.C. Genetic Assignment Tests to Identify the Probable Geographic Origin of a Captive Specimen of Military Macaw (*Ara militaris*) in Mexico: Implications for Conservation. *Diversity* **2021**, *13*, 245. [CrossRef]
24. Sánchez-Mercado, A.; Ferrer-Paris, J.R.; Rodríguez, J.P.; Tella, J.L. A Literature Synthesis of Actions to Tackle Illegal Parrot Trade. *Diversity* **2021**, *13*, 191. [CrossRef]
25. Biddle, R.; Solis-Ponce, I.; Jones, M.; Pilgrim, M.; Marsden, S. Parrot Ownership and Capture in Coastal Ecuador: Developing a Trapping Pressure Index. *Diversity* **2021**, *13*, 15. [CrossRef]
26. Sánchez-Mercado, A.; Blanco, O.; Sucre-Smith, B.; Briceño-Linares, J.M.; Peláez, C.; Rodríguez, J.P. Using Peoples' Perceptions to Improve Conservation Programs: The Yellow-Shouldered Amazon in Venezuela. *Diversity* **2020**, *12*, 342. [CrossRef]
27. Romero-Vidal, P.; Hiraldo, F.; Rosseto, F.; Blanco, G.; Carrete, M.; Tella, J.L. Opportunistic or Non-Random Wildlife Crime? Attractiveness Rather Than Abundance in the Wild Leads to Selective Parrot Poaching. *Diversity* **2020**, *12*, 314. [CrossRef]
28. Nandika, D.; Agustina, D.; Heinsohn, R.; Olah, G. Wildlife Trade Influencing Natural Parrot Populations on a Biodiverse Indonesian Island. *Diversity* **2021**, *13*, 483. [CrossRef]
29. Vigo-Trauco, G.; Garcia-Anleu, R.; Brightsmith, D.J. Increasing Survival of Wild Macaw Chicks Using Foster Parents and Supplemental Feeding. *Diversity* **2021**, *13*, 121. [CrossRef]
30. Ramos-Güivas, B.; Jawor, J.M.; Wright, T.F. Seasonal Variation in Fecal Glucocorticoid Levels and Their Relationship to Reproductive Success in Captive Populations of an Endangered Parrot. *Diversity* **2021**, *13*, 617. [CrossRef]

31. Vilarta, M.R.; Wittkoff, W.; Lobato, C.; Oliveira, R.d.A.; Pereira, N.G.P.; Silveira, L.F. Reintroduction of the Golden Conure (*Guaruba guarouba*) in Northern Brazil: Establishing a Population in a Protected Area. *Diversity* **2021**, *13*, 198. [CrossRef]
32. White, T.H., Jr.; Abreu, W.; Benitez, G.; Jhonson, A.; Lopez, M.; Ramirez, L.; Rodriguez, I.; Toledo, M.; Torres, P.; Velez, J. Minimizing Potential Allee Effects in Psittacine Reintroductions: An Example from Puerto Rico. *Diversity* **2021**, *13*, 13. [CrossRef]
33. Woodman, C.; Biro, C.; Brightsmith, D.J. Parrot Free-Flight as a Conservation Tool. *Diversity* **2021**, *13*, 254. [CrossRef]
34. Preston, C.E.C.; Pruett-Jones, S. The Number and Distribution of Introduced and Naturalized Parrots. *Diversity* **2021**, *13*, 412. [CrossRef]
35. Senar, J.C.; Moyà, A.; Pujol, J.; Tomas, X.; Hatchwell, B.J. Sex and Age Effects on Monk Parakeet Home-Range Variation in the Urban Habitat. *Diversity* **2021**, *13*, 648. [CrossRef]

MDPI

Review

Contributions of Distribution Modelling to the Ecological Study of Psittaciformes

José R. Ferrer-Paris [1,2] and Ada Sánchez-Mercado [1,3,*]

[1] School of Biological, Earth and Environmental Sciences, University of New South Wales, Kensington, NSW 2052, Australia; j.ferrer@unsw.edu.au
[2] Data Science Hub, University of New South Wales, Kensington, NSW 2052, Australia
[3] Escuela de Ciencias Ambientales, Universidad Espiritu Santo, Samborondon 092301, Ecuador
* Correspondence: a.sanchez@unsw.edu.au; Tel.: +61-48-120-3171

Abstract: We provide an overview of the use of species distribution modeling to address research questions related to parrot ecology and conservation at a global scale. We conducted a literature search and applied filters to select the 82 most relevant studies to discuss. The study of parrot species distribution has increased steadily in the past 30 years, with methods and computing development maturing and facilitating their application for a wide range of research and applied questions. Conservation topics was the most popular topic (37%), followed by ecology (34%) and invasion ecology (20%). The role of abiotic factors explaining parrot distribution is the most frequent ecological application. The high prevalence of studies supporting on-ground conservation problems is a remarkable example of reduction in the research–action gap. Prediction of invasion risk and assessment of invasion effect were more prevalent than examples evaluating the environmental or economic impact of these invasions. The integration of species distribution models with other tools in the decision-making process and other data (e.g., landscape metrics, genetic, behavior) could even further expand the range of applications and provide a more nuanced understanding of how parrot species are responding to their even more changing landscape and threats.

Keywords: distribution; conservation; ecology; environmental niche modelling; research selection function; parrots; psittacids; species distribution models; state observation models

Citation: Ferrer-Paris, J.R.; Sánchez-Mercado, A. Contributions of Distribution Modelling to the Ecological Study of Psittaciformes. *Diversity* **2021**, *13*, 611. https://doi.org/10.3390/d13120611

Academic Editors: Guillermo Blanco, José L. Tella and Martina Carrete

Received: 12 October 2021
Accepted: 21 November 2021
Published: 24 November 2021

1. Parrots and Their Important Ecological Role

The order Psittaciformes (including parakeets, macaws, cockatoos, and allies, hereafter parrots) is a diverse order of birds with a wide range of morphological variations and foraging behaviors (~420 spp) [1]. Parrots can reach high density and biomass in many tropical and subtropical regions across the Americas, Africa, Europe, Asia and Oceanian regions [2]. The study of parrots' distribution patterns and factors driving them allows us to improve our understanding of their ecological role. With a wide diversity of biotic requirements (from generalist to specialist apex frugivores) and high prevalence within the bird community, parrots may have a broad influence on the structure of animal and plant communities and ecosystem functions [2–4].

Monitoring the distribution of parrot populations is also an important task for effective management and conservation of both threatened and problematic species [5]. Parrots are among the most threatened avian orders, with 46% of their species under threat and 56% of their populations experiencing population declines [6]. Abundance and distribution declines are driven by modification of their natural habitat and environment, in addition to nest poaching for the illegal market [7,8]. Human-modified environments are quickly encroaching on the most important areas for parrots in the Americas and Oceanian regions [4]; in Australia alone, parrots have lost at least 38% of their potential natural habitat [9].

Reduction in native parrot distribution is, however, only one side of the conservation problem. Parrots are among the most common companion animals, and intentional and unintentional birds released from captivity have been related to the establishment of invasive parrot populations beyond their native distributional limits, causing damage to agriculture and natural environments [9,10].

Species distributions are complex biological phenomena, and many factors interact to determine a species' geographical range [11]. Due to the lack of extensive spatial records of occurrence, it is usually necessary to apply statistical methods to describe and predict species distribution. The key assumption of spatial analysis of wildlife populations is that spatial and temporal patterns in population state variables (i.e., occurrence, abundance or density, richness) represent the response of the species to underlying heterogeneity in external factors such as environment conditions and resource availability [12]. Interpretation of these patterns is scale dependent: at large spatial scales they reflect the overall constraints and conditions influencing species distribution, at intermediate scales they are related to population responses (including meta- and sub-populations), and at small spatial scales they can reflect individual behavioral responses [12].

Species distribution models have become an essential part of the analytical toolkit for ecological studies of birds. This is mainly due to the increasing accumulation and aggregation of basic biodiversity data (species location records), availability of worldwide abiotic environmental variables, and development of Geographic Information Systems [13]. Species distribution models use different algorithms and methods (e.g., MaxEnt, regressions, and occupancy models) to link field observations with spatially explicit explanatory or predictive variables. These variables can then be used to make spatial predictions that can be scaled up to whole landscapes or geographical regions.

Here, we provide an overview of the use of species distribution modeling to understand parrot distribution and place them in the broad conceptual context of the ecological scale at which spatial and temporal patterns are evaluated. For this, we combined quantitative methods of selection and analysis of scientific literature and a narrative discussion of the more relevant studies found. In Section 2, we use a structured search protocol to select relevant scientific literature and classify this sample of publications into a set of topics and general and specific applications. We quantify trends in publication rates and taxonomic and geographical coverage of these topics. In Section 3, we discuss how these methods have been applied to address research questions related to parrot ecology, conservation and biogeography, and in Section 4 we appraise to the extent to which emerging analytical tools have been implemented or could be exploited to explore new research questions in the future.

2. Literature Review of Distribution Modelling in Parrot Species

2.1. Sample of Scientific Literature

Our main objective was to provide a broad overview of the different topics of research. We limited our search to one search engine (Web of Science, WoS) and one language (English), and used a workflow to apply automatic and manual filters to detect the most relevant publications within the extracted list of references. Thus, although quantitative analyses were limited to a single sample of the literature, we used them to illustrate general trends and acknowledge their inherent biases and limitations. Although we do not explicitly compare this sample with other sources, we are confident that this search is representative of the overall trends in the scientific literature. Recent reviews have shown that among several academic search engines, WoS is more selective than Dimensions and GoogleScholar, and has a high degree of overlap with Scopus [14,15]. We are aware that the contribution of publications in non-English languages is high and by focusing our search on only English published papers we obtained a biased sample [16]. However, because several non-English journals include abstracts in English [17] we are confident that we were able to obtain a good representation of topics and applications published on parrot distribution research.

We conducted a literature search on the database Web of Science (WoS) using terms in English related with the focal taxonomic group (Psittaci*, parrot*, macaw*, parakeet*, amazon*, cockato*) in the themes section. This resulted in a total sample of 12,699 documents for the period 1900 to March 2020. At least 88% of the documents were originally published in English, 7.1% in 21 other languages, and 5% did not have information on the original language. Although our sample contains more than 12,000 scientific articles published in the last 100 years, there is a clear difference in the rate of documents per year for the periods before and after 1990 (Figure 1a). This may be an artifact of uneven coverage of the global literature in the WoS database; for example, a lack of digitalization of pre-1990 documents, or an increase in the number of sources included after 1990.

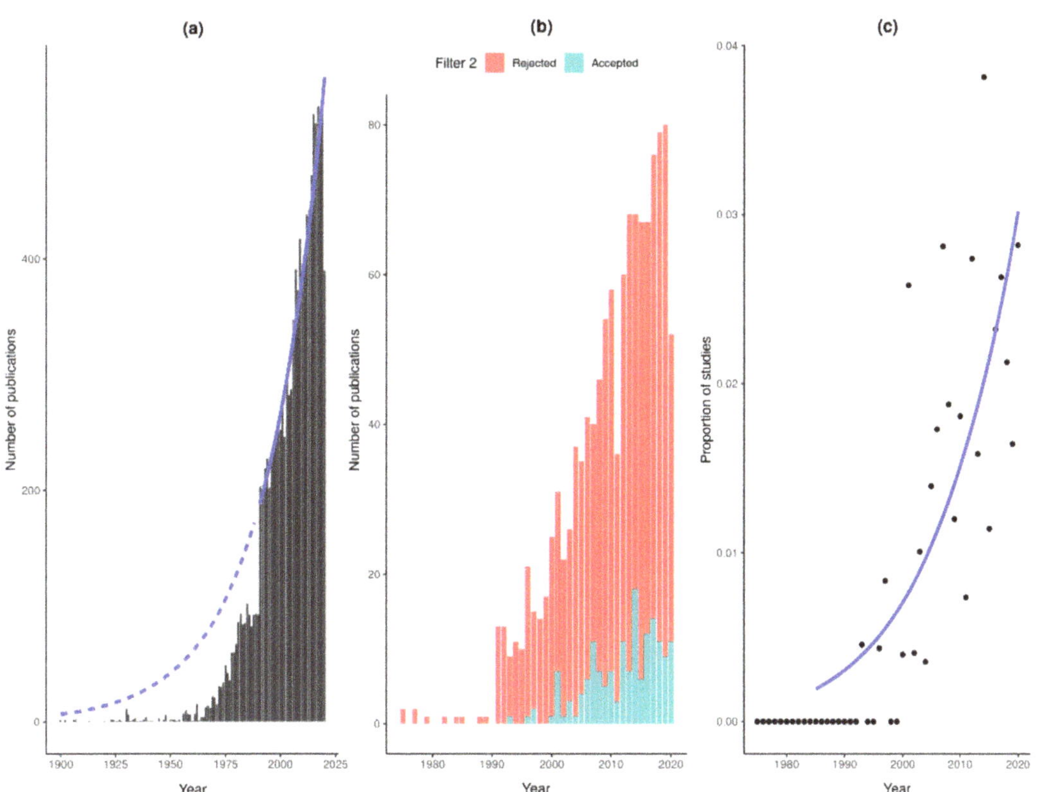

Figure 1. Trends in total number and proportion of publications per year. (**a**) Total sample of 12,699 documents with keywords related to parrots from Web of Science; the blue line indicates the modeled exponential increase in total publications per year using a GLM with a Gaussian distribution and log-link fitted to the years 1990–2020 (solid line) and extrapolated to the previous period (dashed line). (**b**) Number of publications filtered by keywords related to species distribution and manually rejected or accepted. (**c**) Proportion of studies on species distributions in relation to the total sample of publications per year (black dots), the blue line represents the modeled increase in proportion using a GLM with binomial distribution and probit link function.

We then applied successive filters to select the most relevant studies to discuss. In the first filter, we performed an automatic screening of the title, abstract, and authors' key words looking for topic specific words: 'distribution', 'change in distribution', 'range assessment', 'niche model*', 'distribution reduction', 'occupancy model*', 'resource selection', or 'species distribution model*'. With this filter we selected 1210 documents (9.5% of the total sample) that likely had information related to parrot distribution. In the second filter, we reviewed title and abstracts, and, if necessary, also the full text, and rejected 939 documents that were evidently off-topic (e.g., different taxonomic groups, not related to species distribution), in addition to opinion articles, and reviews or overviews with no original data or analysis. The remaining 161 documents (1.26% of the total sample) represent the subset of studies that are directly related to distribution of parrot species. The number of documents selected in the first filter rose sharply in 1990, and the first document included in the second filter is from 1993 (Figure 1b). This gap is partly related to the uneven temporal coverage of the WoS database mentioned above, but even focusing on the publications after 1990 we see a significant positive trend in the proportion of publications related to the study of distribution in parrots (Figure 1c). As we discuss below, the onset of this rising trend coincides with first applications of species distribution models to parrot species around the years 2000–2005.

2.2. Document Classification

For these 161 documents we made a more detailed assessment of the abstract or full text, evaluating whether they met the following criteria: (a) original analysis (no reviews) of species distribution or related state variables; (b) using statistical modeling approach of any kind; (c) using spatial data (location of records and/or spatial cover of explanatory variables); and (d) making explicit spatial predictions of the state variable (Table S1). Based on these criteria, we found that almost half of the documents (79) were focused on reports of species occurrence records without using any modeling approach (i.e., calculation of extent of occurrence and area of occupancy) or were statistical comparisons of naive occurrence estimates, abundance or resource use between habitat types, sampling areas or discrete regions. For the remaining 82 documents that did apply some methods of species distribution modeling, we summarized information on geographical location, target taxa, topics of research and general application (Table 1).

We aggregated information about the countries where the studies were conducted into five main regions following ISO classification: Africa (Eastern, Northern, Southern and Western Africa), the Americas (North America, Latin America and the Caribbean), Asia, Europe, and Oceania [18]. The list of parrot species reported was normalized using the species list of BirdLife International [19] to unify the species scientific names across documents. We identified whether the focal taxon or taxa were native or non-native parrot species. We classified the main research topic addressed (Behavior, Conservation, Ecology, Evolution, Invasion ecology, and Methodological issues), and split these research topics into general and specific applications (Table 1).

Table 1. Major topics addressed in the parrot distribution modelling literature, with general and specific applications and relative examples taken from published case studies. The modelling paradigm and type of data used are shown. ENM = Environmental (or ecological) niche modelling. RFS = Resource Selection Function. SOM = State Observation Models.

Topics	General Application	Specific Issues	Number of Publications	Paradigm Used	Type of Data	Examples
Behavior	Habitat use related to behavior types	Occurrence of behavior types	1	ENM	Literature	[20]
	Temporal distribution patterns	Movement related to environmental factors	1	ENM	Open access databases	[21]
Conservation	Climate change	Change in distribution driven by climate change	6	ENM, SOM	Field work; Open access databases; Citizen science project; Literature; Museum collections	[22–24]
		Combined effects of climate and habitat changes	1	ENM	Literature; Open access databases	[25]
		Evaluating or forecasting the effect of environmental changes	1	ENM	Field work	[26]
	Spatial prediction	Effect of conservation actions	1	ENM	Field work	[27]
		Identification of priority areas for conservation	1	ENM	Literature; Open access databases	[28]
		Resource distribution	2	ENM	Field work; Open access databases	[29,30]
	Threats monitoring	Change in distribution driven by habitat loss	7	ENM, SOM	Field work; Open access databases	[31–36]
		Effect of conservation actions	1	RSF	Field work	[37]
		Fragmentation effect	2	ENM	Field work	[38,39]
		Input for population models/population viability analysis	1	ENM	Field work	[40]
		Threat distribution	2	ENM	Field work	[40,41]
		Threat effect on distribution/occupancy	1	ENM	Field work	[42]
Ecology	Macroecology	Abundance-occupancy relationship	1	SOM	Field work	[43]
		Effect of biotic interactions on distribution	1	ENM	Open access databases; Museum collections	[11]
		Global distribution patterns of diet type	1	ENM	Open access databases	[44]
	Relation with environmental variables	Determining areas for survey	2	ENM	Field work	[20]
		Identifying breeding habitat	5	ENM, RSF	Field work	[45,46]
		Identifying potential habitat	1	ENM	Open access databases	[47]
		Inter-annual variability in distribution	2	ENM, RSF	Field work	[48]
		Variables affecting distribution	8	ENM, SOM, RSF	Field work; Open access databases	[49–53]
	Ecological communities	Richness and alpha-diversity	2	SOM, RSF	Field work; Open access databases	[54,55]

Table 1. *Cont.*

Topics	General Application	Specific Issues	Number of Publications	Paradigm Used	Type of Data	Examples
Evolution	Biogeographic patterns	Change in historical distribution	2	ENM	Open access databases; Museum collections; Field work;	[56,57]
		Understanding distribution of extincted species	1	ENM	Literature; Museum collections	[58]
Invasion ecology	Invasion effect	Impacts on native species	2	ENM, RSF	Field work; Open access databases	[59,60]
	Predictions of invasion risk	Establishment of non-native specie	3	ENM	Open access databases	[27,61,62]
		Limitations into invasion risk	1	RSF	Field work	[63]
		Niche shift	3	ENM	Field work	[21]
		Potential range of invasive species	5	ENM	Literature; Field work; Open access databases	[64,65]
Methodological issues	Improving estimation	Factors affecting distribution estimation	1	SOM	Open access databases; Field work	[66]
	Survey methodology biases	Evaluating citizen science data	1	ENM	Literature; Open access databases; Citizen science; Field work	[65]

We evaluated temporal patterns in parrot distribution modeling publications by aggregating the number of published documents by year and by research topic (Figure 2a). Parrots have been recognized as a model group for global macroecology analysis of species distribution [55] and the first application of distribution modelling techniques for any parrot species focused on the ecology of an endangered species [47]. Ecological questions remained the predominant topic between 2005 and 2011, but the diversification of studies led to a balance between more theoretical and applied research questions. Between 2012 to 2015, Ecological studies had a similar cumulative output as Conservation and Invasion ecology combined, but after 2016 Conservation became the most popular topic (36.6% of all studies up to 2020), followed by Ecology (34%) and Invasion ecology (19.5%; Figure 2a).

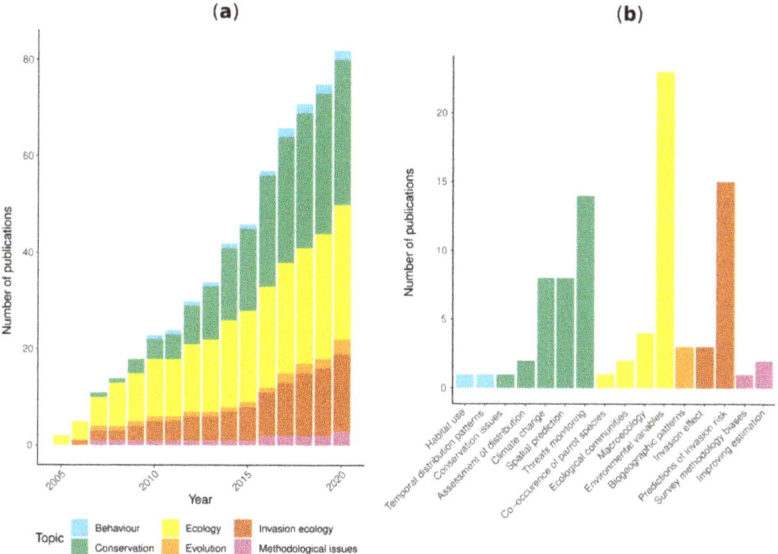

Figure 2. Temporal pattern in published literature in parrots' distribution modelling. (**a**) The accumulated number of publications across years is shown for each topic. (**b**) The total number of published documents by general application within each topic.

Within Conservation topics, threat monitoring and climate change were the most frequent applications, whereas Ecological topics were dominated by applications focused on evaluating the relationship between occurrence and environmental variables (Figure 2b, Table 1). Invasion ecology was mainly focused on predicting invasion risk (Figure 2b, Table 1).

To visualize taxonomic patterns, we aggregated the number of documents by genera, research topic and region, and represented these relationships with a bar and bubble plot. The majority of the reviewed literature was focused on species within their native range (86%; Figure 3). The Americas was the region with the highest diversity in applications, but noticeably studies in invasion effects were almost absent. Oceania was the second most diverse region regarding model's applications, but in this region, studies focused on evolution topics were absent. Africa only had a small number of studies in ecology, conservation, and invasion ecology topics. The Americas, Oceania and African regions had studies in both native and non-native species. In contrast, studies from Asia and Europe have been focused exclusively on predicting invasion risk of non-native parrots (Figure 3).

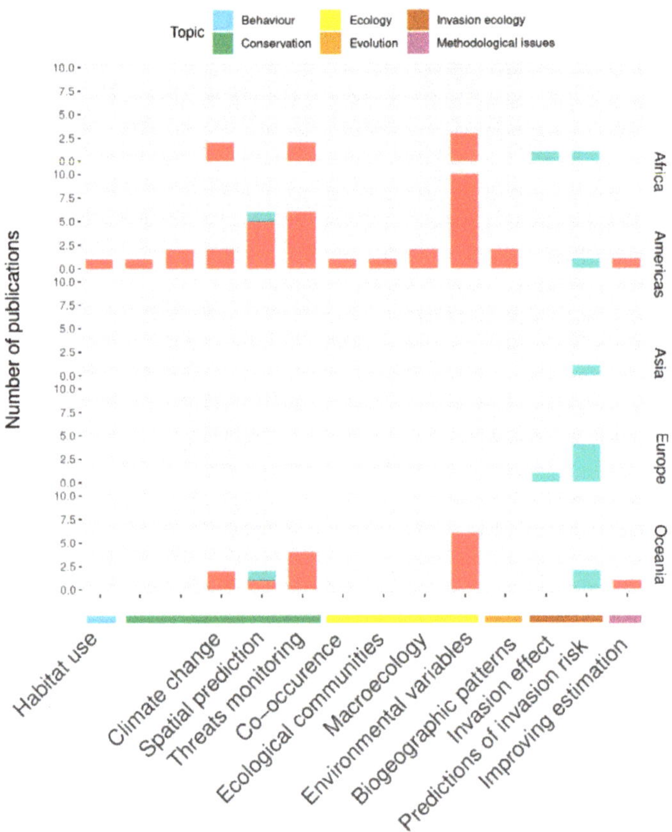

Figure 3. Geographical distribution of applications of the parrots' distribution modelling. Number of published documents by region. Documents focused on native species are in red, whereas those focused on non-native species are in blue. Applications are grouped according to the main topics.

We recorded 52 parrot species in the distribution modeling literature. As in other conservation topics [8], parrots' distribution research shows a clear bias toward widespread species such as *Psittacula krameri*, *Myiopsitta monachus*, and *Amazona oratrix*. At the genus level, *Ara* had the highest diversity of applications, whereas *Psittacula* and *Myiopsitta* only have studies focused on invasion ecology (Figure 4). Taxonomic bias can be explained in part by higher availability of occurrence records for species with wide ranges and/or high abundance. However, although *Amazona*, *Psittacula* and *Myiopsitta* genera are among the top ten species with the largest number of occurrence records in GBIF, altogether they only account for 18% of the parrots' occurrence records (9,880,043 records), with other genera such as *Platycercus*, *Cacatua* and *Trichoglossus*, being better represented in the GBIF database [67]. The high impact of invasive species on socio-economic and environmental contexts likely trigger higher interest in describing the establishment and invasion risk of *Psittacula krameri* and *Myiopsitta monachus* (Figure 4). Conversely, the high prevalence of distribution models for *Amazona oratrix* likely results from a combination of the interest raised by their conservation status and a strong and prolific research team in Mexico, where the three main subpopulations of this species occur [39].

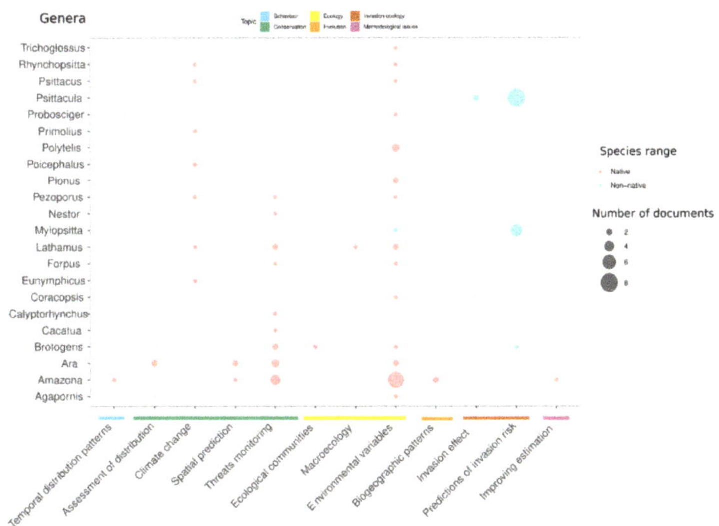

Figure 4. Taxonomic patterns of the published literature in parrots' distribution modelling. The number of species by genus and application are shown. The circle size is proportional to the number of documents. Genera are in alphabetical order from bottom to top. Applications are grouped according to the main topics.

In addition to taxonomic bias, we also identified intrinsic geographic biases in the parrots' distribution research, with publications from the America and Oceania regions dominating the research map. This pattern may represent a combination of: (1) a higher diversity of American parrots compared to other regions (233 spp. in the Americas versus 128 spp. in Asia and 129 in Oceania) [1]; and (2) higher scientific capacity in the Americas and Oceania (65% of detected documents; Figure 4). However, our search strategy, which was focused on English and Spanish keywords, likely resulted in an under-representation of literature published in Asian languages. Future efforts should include a wider range of Asian languages to discern whether the observed spatial pattern responds to a detectability problem or to lower publication rate in Asian countries.

3. Contributions of Species Distribution Models to Parrots Research

The review of literature showed that the study of parrot species distribution has increased steadily in the past 30 years and has likely been boosted by the widespread use of species distribution models in the past 15 years (Figures 1 and 2). These methods have matured alongside the developments of computer capacity for spatial and statistical analysis, and have become part of a standard toolkits, facilitating their application for a wide range of research questions [68].

In the following sections we discuss the many contributions of species distribution models to the study of ecology, conservation and management of parrot species using illustrative examples identified during our review. All literature reviews are inevitably limited by any biases in the initial selection (search engine, languages and keywords) and the involuntary omissions in the subsequent steps of this process. We have highlighted these biases whenever possible, and stress that we do not attempt to offer an exhaustive account of all subjects.

3.1. Distribution Models to Study the Ecology of Parrots

The role of abiotic factors explaining parrot distribution have been to date the most frequent ecological application in parrots' distribution research (Figure 2b, Table 1). This research conceptually aligns with the environmental (or ecological) niche modelling (ENM) paradigm which focuses on estimating the fundamental niches of species, or ecological requirements of species by relating their known geographic distributions (i.e., occurrence records) to a set of environmental or abiotic variables [12]. Niche models predict habitat suitability or potential distribution, but species may not be using their entire potential habitat due to a range of constraining factors. The two most important natural, non-environmental constraints are biotic relationships and accessibility (Figure 5a) [12]. However, large scale patterns in distribution are often the result of geographic variation in the use of resources at the scale of populations or individual movements (Figure 5b,c). The Resource Selection Function (RSF) paradigm compares the frequency in the use of resources (preference) with the overall distribution of resources in the landscape (availability) [69]. RSF are mostly applied to determine preferences at the scale of individual movements, but for long-lived and highly mobile species this can represent very large geographical areas. Thus, ENM and RSF models may overlap in conceptual and practical terms. For example, an early study of the Superb Parrot (*Polytelis swainsonii*) in Australia focused on describing the bioclimatic envelope of the species as a fixed factor influencing its distribution [70]. In a second study, these authors included plant productivity as a covariate related to the availability of resources and were able to explain seasonal and year-to-year variability in abundance and distribution that was not accounted for by the previous static environmental model [71].

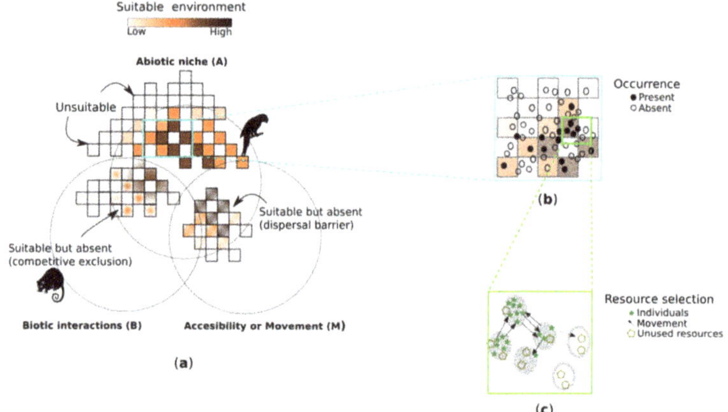

Figure 5. The main paradigms for modelling species distribution operate at different hierarchical levels, requiring different types of data and algorithms. (**a**) The niche modelling (ENM) paradigm predicts suitable environmental conditions where species might occur, but although suitable, this habitat cannot be used due to biotic relationships and dispersal barriers constraints. ENM can use presence-only datasets if algorithms such as Maxent, Random Forest, BIOMOD, and GARP are used, but will require "pseudo-absence" data if GLM methods are used instead. (**b**) State-observation models (SOM) work at the population level and predict the occupancy probability conditioned to the probability of detecting the species given it is present. SOM typically requires data from repeated sampling visits (occasions) to a single site during a time frame over which the population is *closed* (e.g., no changes in occupancy between surveys). (**c**) The Resource Selection Function (RSF) paradigm works at individual and species scales and predicts the probability of an animal or species using a certain resource, proportional to the availability of that resource in the environment. RSF requires two types of data: presence records of the focal species (either at individual or species level) and data on the resources available across the study area.

Incorporation of biotic and biogeographic elements in species distribution models allows their application to be extended to a wider range of research questions in biogeography and evolution, community ecology, etc. [13]. Kissling et al. [42] evaluated the roles of climate and productivity on broad-scale geographical patterns of parrots' richness and whether they show distinct regional differences compared to other frugivore species due to regional patterns of diversification of food plants, niche conservatism and past climate change. Often these broad biogeographical and macroecological patterns can mask more nuanced relationships between species distribution and landscape features at the scale of populations and individuals (Figure 5b). Keighley et al. [49], combined distribution, behavior and genetic information to describe dispersal pathways and barriers for the Palm Cockatoo (*Probosciger aterrimus*) in Australia, and to test hypotheses about key landscape features influencing movement of palm cockatoos throughout their range.

In most cases, local studies rely on field observations of species occurrence to better understand limiting factors and describe temporal changes in occupancy. State-observation models (SOM) use several sampling approaches (multiple visits, multiple observers, distance sampling, etc.) to make joint estimates of a state variable of interest (usually occupancy or abundance) and the observation or detection process (detectability or probability of detection). Normally the SOM paradigm assumes closed populations at each sampling site during the primary sampling periods and explores how species behavior, sampling characteristics and environmental variables might affect detectability across time and geographic space [72]. However, some of these models can be extended to multiple seasons and allow population dynamics between seasons to be studied. Kalle et al. [24] applied dynamic occupancy models to a decade of citizen science-driven presence/absence data on the Cape Parrot (*Poicephalus robustus*) and were able to model recent range dynamics as a function of changing climate conditions and the availability of resources.

3.2. The Biotic Component of Parrots' Distribution

Biotic interactions are undoubtedly an important component of species distributions, and these can determine the relationship between parrot species and their habitat. In particular, the relationship between parrots and their food plants works in both ways; the distribution of diet resources contributes to explain the distribution of 11 parrot species in the Cerrado in Brazil [73], and similarly the distribution and density of three large macaws influences the spatial distribution of motacú palm (*Attalea princeps*) in the Bolivian Amazonas savannas. Parrots have a great behavioral plasticity and different species use several strategies to respond to fluctuations in food availability: switch in diet, shift in habitat use and seasonal movements [74]. This plasticity can lead to seasonal (intra-annual) and inter-annual variability in distribution or resource use correlated with changes in spatial indicators. The identification of these patterns requires not only large-scale, but also local scale variables to better describe landscape elements important for parrots' occurrence. For example, key landscape drivers (e.g., woodland structure) determine the occurrence of *Agapornis lilianae* in the mopane woodlands of Zambia [20]; time since fire influences food resources for the Carnaby's Cockatoo (*Calyptorhynchus latirostris*) in fire-prone landscapes in Australia [67]; whereas seasonal use of foraging habitats explains the dynamics of the Swift Parrot (*Lathamus discolor*) in Australia [45,73] and the Maroon-fronted Parrot (*Rhynchopsitta terrisi*) in Mexico [48], in addition to the aforementioned examples of the Cape Parrot and the Superb Parrot in South Africa and Australia [24,70].

Antagonistic relationships can also have a strong influence on species distribution. For example, the study of Engeman et al. [75] suggests that nest site selection in the Puerto Rican Amazon (*Amazona vittata*) is an adaptive response to predation pressure; parrots select nest sites that allow increased avoidance and detection of predators. Moreover, the high occupancy of introduced mammal predators may represent an additional threat to the endangered the Swift Parrot (*Lathamus discolor*) in Australia [41]. Finally, competition and coexistence of parrot species has also been a subject of research in studies using distribution models in areas of high species diversity in the Neotropics [73,76].

3.3. Applications of Distribution Models to Conservation Problems

The second most widespread application of species distribution models in parrots is related to threat monitoring for conservation (Figure 2b, Table 1). Monitoring of populations is a basic step of conservation planning and management. Distribution models (specially SOM methods) are used to improve the design of sampling protocols to select sampling areas, optimize probability of detection and reduce uncertainty in estimates of probability of occurrence or other state variables [20,66,77]. For example, some less-conspicuous parrots such as the Blue-fronted Amazon (*Amazona aestiva*) and the Peach-fronted Parakeet (*Eupsittula aurea*) require longer observation times in order to improve detectability [73]. This approach is particularly useful when combining robust sampling designs, automated methods such as camera traps and automatic sound recording, and modeling methods for spatial data analysis [29,42,60].

Distribution models are used extensively to evaluate changes in distribution due to habitat loss [31,32,35,78,79]. Habitat fragmentation is often considered a direct threat to the persistence of species in modified landscapes [38,53]. Plasencia-Vazquez et al. [80] combined spatial prediction with different metrics of landscape fragmentation to explore the relationship between forest fragmentation and the geographic potential distribution of different parrot species in the Yucatan Peninsula, Mexico. The combination of current and historic datasets and appropriate modelling methods for each dataset can be useful to make more explicit tests of changes in distribution. For example, Ferrer-Paris et al. [36] compared historical and contemporary distribution of eight species of Amazon parrots (*Amazona*) in Venezuela and found negative changes in widespread species such as *Amazona amazonica*, and *Amazona ochrocephala*, and rare and patchily-distributed species such as *Amazona barbadensis*.

Climate change was the fourth most frequent application in parrot distribution research (Figure 2b, Table 1). Several studies have focused on predicting changes in distribution driven by climate change [22–24,81]. Assessment of impacts was less studied, but Legault et al. [26] presented a new method for assessing how the population size of the New Caledonian Parakeet (*Cyanoramphus saisseti*), the Horned Parakeet (*Eunymphicus cornutus*), and the Ouvéa Parakeet (*Eunymphicus uvaeensis*) in New Caledonia will change in the future based on the relationship between local abundance and modeled habitat suitability obtained using ecological niche models.

Spatial analysis often reveals unexpected inter-species or species–habitat interactions in modified landscapes that may affect already threatened species. Such is the case of novel predators of the Swift Parrot [40,41], in addition to the relationship between modified fire regimes and habitat use in ground parrots, which can inform management actions [30,42]. Species distribution models are also useful for tracking species introduction and recovery of populations. Recio et al. [37] used GPS tracking to evaluate how a forest-dwelling species, *Nestor meridionalis,* selected habitat within its home ranges, showing that this species moved beyond the predator exclusion fence into urban suburbs. In this example the native forest patches throughout the city facilitated dispersal of individuals between refugee and food sources, and long-term survival will require careful urban planning and management to provide the necessary balance of different elements.

3.4. Applications of Distribution Models to Invasion Ecology

Applications related to invasion ecology such as prediction of invasion risk and assessment of invasion effect were also widely detected in the parrot distribution research (Figure 2b, Table 1). Species distributions are dynamic and many species of parrots show a recent natural expansion of their range [24], but in some cases these changes may be confounded by the intervention of humans. Mota-Vargas et al. [57] used ENM methods to compare environmental conditions between historical and recent records of the White-fronted Amazon (*Amazona albifrons*) and discriminate between introductions by humans and natural expansion of its distribution range.

Invasive species and some native species show great adaptability to novel environments, such as urban areas or modified landscapes. Shokuroglou and McCarthy [50] used bird atlas data and Bayesian logistic regression to predict the distribution of the Rainbow Lorikeet (*Trichoglossus moluccanus*) in Melbourne, Australia. Le Louarn et al. [82] compared the use of the urban landscape by a native range-shifting bird (*Corvus monedula*) and an invasive parrot species (*Psittacula krameri*) and found that expansion of the latter is likely driven by its effective ability to exploit urban resources which native species do not exploit. Some tropical islands can become hotspots of exotic species, but not all exotics have the same success as invaders. Falcon and Tremblay [65] analyzed the distribution of parrots in Puerto Rico and found 11 species present only as pets, and at least 29 species in the wild, of which at least 12 were breeding, but most persisted in localized areas and small populations. Only *Brotogeris versicolurus* and *Myiopsitta monachus* showed clear evidence of range expansion.

In most applications the potential risk of invasion or potential distribution of invasive species is predicted from current occurrence records and environmental data layers [64,65,83–85]. Few studies combine these spatial predictions with information about invasion process (i.e., trade, introduction effort, and breeding origin) to improve predictions of environmental suitability and potential niche shifts in the introduced parrots [61,62]. Less prevalent was the evaluation of impacts related to invasion. However, there are some examples of evaluation of economic impacts on agricultural production and human infrastructure [64,83] in addition to some examples of environmental impact through the effects on other animals and competition, and even measurement of the Generic Impact Scoring Scheme [63,83]. Notably, one study goes beyond evaluating the impact of the problem and evaluates the impact of conservation actions such as removing invasive *Trichoglossus moluccanus* [27].

4. Challenges and Opportunities for New Research on Parrots' Distribution

The previous examples of applications of species distribution models reflect how research has adapted to address the dynamic and complexity of species distributions. Here we summarize some of the challenges and opportunities for future research.

4.1. Social Behavior and Geographic Variability

Linking the distribution of species with social structure of parrot populations can provide better insights into intra-specific variability [86]. Most applications of species distribution models have focused on abiotic covariates or combinations of abiotic and inter-specific interactions (use of resources, predator avoidance, competition), but intra-specific interactions (e.g., dispersal and aggregations of individuals) can influence the connectivity of populations and phenotypical or genotypical variability [87,88].

Many aspects of the ecology of parrots are influenced by their social behavior [86]. Most species exhibit long-lasting pair bonds, occupy large home ranges, and congregate in more or less stable foraging and roosting groups. Active defense of year-long territories is rare, and patterns of ranging and dispersal are often seasonal. Geographical variability of behavioral traits such as vocalizations can serve as an indicator of social structure across a range of scales from individual, pairs and flocks, to populations. For example, landscape resistance models revealed strong effects of isolation by elevation on genetic, repertoire and structural call differentiation in Palm Cockatoos [49]. However, if species are highly aggregated, it is likely that standard species distribution modeling (e.g., Maxent) will not provide adequate predictions. In those cases, regression-based methods that account for spatial structure can be used [88].

4.2. From Global to Local

Although many early applications of species distribution models, and particularly ENM, have focused on broad biogeographical or macroecological patterns, current conservation and management applications need more detailed information on changes occurring at the time scales of one or few generations [89]. RSF implicitly account for movement of individuals and are often linked to seasonality in resource distribution by relying on time series of covariates [90]. Multi-season SOM have explicit means to parametrize changes in state variables (e.g., colonization and extinctions). We expect that future applications will continue to explore the links between static and dynamic components of species distribution, for example, by incorporating sink–source dynamics, connectivity and barriers to explain range contraction, shift and expansion of the distribution in the face of climate and land cover change, or invasion processes in new environments [52].

4.3. Automatic Data Collection

Arrays of passive detectors such as camera traps and sound recordings, and the use of unmanned aerial vehicles, have the potential to provide massive streams of data on species presence, abundance and behavior [91–93]. Automatic data collection coupled with machine learning methods to identify species or individuals (image recognition and vocal profiles) have been used to study many emblematic species and to document species diversity [94]. These methods have the great advantage of providing detection histories and allow standardization of sampling protocols and sampling effort across local to regional scales. For example, automated sound recordings were used to model habitat occupancy and post-fire response of the Ground Parrot (*Pezoporus wallicus*) in heathland sites in Australia [42]. An outstanding challenge in this area is the development of virtual platforms for sharing standardized data records that could allow collaboration between research groups and large-scale analysis of spatial and temporal trends [95,96].

4.4. Citizen-Science and Socially-Derived Data Collection

Citizen-science has become the main source of species distribution records for many species, especially for birds [97]. User networks such as eBird [98] and iNaturalist [99] provide large platforms for accessing a great volume of data contributed by enthusiastic ornithologists, photographers and other volunteers. Outstanding challenges are the inherent bias in the distribution of observers and the reliability and accuracy of records [100,101].

Less specific social-media platforms can also be used as a source of additional information on species distribution, but they require more active search and filtering of records. Here some records may come from engaged citizen groups and organizations, but many records are accidental or opportunistic. This is, however, a key source for understanding human interaction with wildlife, for example, legal and illegal pet trade, invasion of exotic species, and human–wildlife conflict [102,103]. For example, Mori et al. [61] used several data sources (from eBird to YouTube) to study worldwide patterns of trade and establishment of exotic *Agapornis* parrots.

4.5. Cyber Infrastructure for Research

Species distribution models, and especially ENM, have become a mainstream tool for ecological analysis and fully integrated into several analytical workflows. For example, virtual laboratories and cloud applications enable new users to learn these workflows and existing users to explore their potential applications and share their results [104,105]. These platforms offer a seamless integration of data into the analytical workflow. For example, biodiversity records from natural history collections (Global Biodiversity Information Facility, Atlas of Living Australia, etc.) and the main citizen-science networks (eBird, iNaturalist) are imported directly. The focus of these applications is on presence only data, and they are compatible with the ENM paradigm and, to a lesser extent, the RSF paradigm. Similar applications for SOM are missing, probably due to the challenge of dealing with different data structures (detection histories, distance sampling, double observer, etc.) and the more complex statistical and computational context.

4.6. Supporting Decision Making

Species distribution models have become an important tool for supporting solutions for on-ground conservation problems. A dialogue between modelers and decision makers increases the opportunities for integration of research outputs into the decision-making process, and will contribute to improve both scientific knowledge and conservation or management outcomes [106].

There are several explicit applications of spatial distribution modeling to support spatial conservation decisions in parrots. Studies in Argentina and Brazil have analyzed the percentage of key parrot habitat covered by protected areas [107–109], and Botero-Delgadillo et al. [65] used models to identify independent conservation units for *Pyrrhura* parakeets in Colombia. The relatively high prevalence of predictive distribution modeling supporting on-ground conservation problems of parrots is particularly encouraging and contrasts with the unclear and less prevalent examples in other conservation contexts [79,106].

5. Conclusions

Parrots are a very attractive and interesting taxonomic group for ecological studies, and our review revealed a large variety of studies related to modeling of species distribution. Since the advent of distribution model paradigms, parrots have been used as model taxa to answer macroecological questions. The complexity of parrots' ecology and behavior likely make it an object of research itself, with an increased interest to address more complex questions at different geographical and ecological scales: from environmental niches to biogeographic and biotic interactions [110]. The flexibility of species distribution models and the maturation process throughout the development of algorithms and analytical approaches have allowed their application to applied research questions. Particularly in parrot distribution research, these models have been used to inform conservation action and management for both threatened and invasive species, becoming a remarkable example of how the research–action gap can be reduced and translated into insightful, science-based conservation actions. The integration of species distribution models with other tools in the decision-making process and other data (e.g., landscape metrics, genetic, behavior) could further expand the range of applications and provide a more nuanced understanding of how parrot species are responding to their even more changing landscape and threats.

Supplementary Materials: The following are available online at https://www.mdpi.com/article/10.3390/d13120611/s1, Table S1: documents used in the review.

Author Contributions: Both authors contributed at the same level in all the stages of manuscript preparation. Conceptualization, J.R.F.-P. and A.S.-M.; methodology, J.R.F.-P. and A.S.-M.; formal analysis, J.R.F.-P. and A.S.-M.; data curation, J.R.F.-P. and A.S.-M.; writing—original draft preparation, J.R.F.-P. and A.S.-M.; writing—review and editing, J.R.F.-P. and A.S.-M. All authors have read and agreed to the published version of the manuscript.

Funding: This research received no external funding.

Conflicts of Interest: The authors declare no conflict of interest.

References

1. IUCN The International Union for Conservation of Nature's Red List of Threatened Species. Available online: https://www.iucnredlist.org/ (accessed on 20 May 2021).
2. Blanco, G.; Hiraldo, F.; Rojas, A.; Dénes, F.v.; Tella, J.L. Parrots as Key Multilinkers in Ecosystem Structure and Functioning. *Ecol. Evol.* **2015**, *5*, 4141–4160. [CrossRef] [PubMed]
3. Blanco, G.; Bravo, C.; Pacifico, E.C.; Chamorro, D.; Speziale, K.L.; Lambertucci, S.A.; Hiraldo, F.; Tella, J.L. Internal Seed Dispersal by Parrots: An Overview of a Neglected Mutualism. *PeerJ* **2016**, *2016*, e1688. [CrossRef] [PubMed]
4. Vergara-Tabares, D.L.; Cordier, J.M.; Landi, M.A.; Olah, G.; Nori, J. Global Trends of Habitat Destruction and Consequences for Parrot Conservation. *Glob. Chang. Biol.* **2020**, *26*, 4251–4262. [CrossRef] [PubMed]
5. Devarajan, K.; Morelli, T.L.; Tenan, S. Multi-Species Occupancy Models: Review, Roadmap, and Recommendations. *Ecography* **2020**, *43*, 1612–1624. [CrossRef]

6. Berkunsky, I.; Quillfeldt, P.; Brightsmith, D.; Abbud, M.; Aguilar, J.; Alemán-Zelaya, U.; Aramburú, R.; Arce Arias, A.; Balas McNab, R.; Balsby, T.; et al. Current Threats Faced by Neotropical Parrot Populations. *Biol. Conserv.* **2017**, *214*, 278–287. [CrossRef]
7. BirdLife International IUCN Red List for Birds. Available online: http://www.birdlife.org (accessed on 10 October 2019).
8. Sánchez-Mercado, A.; Ferrer-Paris, J.R.; Rodríguez, J.P.; Tella, J.L. A Literature Synthesis of Actions to Tackle Illegal Parrot Trade. *Diversity* **2021**, *13*, 191. [CrossRef]
9. Simmonds, J.S.; Watson, J.E.M.; Salazar, A.; Maron, M. A Composite Measure of Habitat Loss for Entire Assemblages of Species. *Conserv. Biol.* **2019**, *33*, 1438–1447. [CrossRef]
10. Carrete, M.; Tella, J.L. Wild-Bird Trade and Exotic Invasions: A New Link of Conservation Concern? *Front. Ecol. Environ.* **2008**, *6*, 207–211. [CrossRef]
11. De Araújo, C.B.; Marcondes-Machado, L.O.; Costa, G.C. The Importance of Biotic Interactions in Species Distribution Models: A Test of the Eltonian Noise Hypothesis Using Parrots. *J. Biogeogr.* **2014**, *41*, 513–523. [CrossRef]
12. Peterson, A.T.; Soberón, J.; Pearson, R.G.; Anderson, R.P.; Martínez-Meyer, E.; Nakamura, M.; Araújo, M.B. *Ecological Niches and Geographic Distributions*; Princeton University Press: Princeton, NJ, USA, 2011.
13. Engler, J.O.; Stiels, D.; Schidelko, K.; Strubbe, D.; Quillfeldt, P.; Brambilla, M. Avian SDMs: Current State, Challenges, and Opportunities. *J. Avian Biol.* **2017**, *48*, 1483–1504. [CrossRef]
14. Martín-Martín, A.; Orduna-Malea, E.; Delgado López-Cózar, E. Coverage of Highly-Cited Documents in Google Scholar, Web of Science, and Scopus: A Multidisciplinary Comparison. *Scientometrics* **2018**, *116*, 2175–2188. [CrossRef]
15. Kumar, V.S.; Singh, P.; Karmakar, M.; Leta, J.; Mayr, P. The Journal Coverage of Web of Science, Scopus and Dimensions: A Comparative Analysis. *Scienciometric* **2021**, 5113–5142. [CrossRef]
16. Angulo, E.; Diagne, C.; Ballesteros-Mejia, L.; Adamjy, T.; Ahmed, D.A.; Akulov, E.; Banerjee, A.K.; Capinha, C.; Dia, C.A.K.M.; Dobigny, G.; et al. Non-English Languages Enrich Scientific Knowledge: The Example of Economic Costs of Biological Invasions. *Sci. Total Environ.* **2021**, *775*, 144441. [CrossRef]
17. O'Neil, D. English as the Lingua Franca of International Publishing. *World Engl.* **2018**, *37*, 146–165. [CrossRef]
18. ISO International Organization for Standardization—ISO 3166 Country Codes. Available online: https://www.iso.org/iso-3166-country-codes.html (accessed on 2 January 2021).
19. BirdLife International Data Zone. Available online: http://datazone.birdlife.org/species/search (accessed on 3 May 2020).
20. Mzumara, T.I.; Martin, R.O.; Tripathi, H.; Phiri, C.; Amar, A. Distribution of a Habitat Specialist: Mopane Woodland Structure Determines Occurrence of near Threatened Lilian's Lovebird Agapornis Lilianae. *Bird Conserv. Int.* **2019**, *29*, 413–422. [CrossRef]
21. Sánchez-Barradas, A.; Santiago-Jiménez, Q.J.; Rojas-Soto, O. Variación Temporal en la Distribución Geográfica y Ecológica de Amazona Finschi, (Psittaciformes: Psittacidae). *Rev. Biol. Trop.* **2017**, *65*, 1194–1207. [CrossRef]
22. Porfirio, L.L.; Harris, R.M.B.; Stojanovic, D.; Webb, M.H.; Mackey, B. Projected Direct and Indirect Effects of Climate Change on the Swift Parrot, an Endangered Migratory Species. *Emu* **2016**, *116*, 273–283. [CrossRef]
23. Monterrubio-Rico, T.C.; Charre-Medellin, J.F.; Sáenz-Romero, C. Current and Future Habitat Availability for Thick-Billed and Maroon-Fronted Parrots in Northern Mexican Forests. *J. Field Ornithol.* **2015**, *86*, 1–16. [CrossRef]
24. Kalle, R.; Ramesh, T.; Downs, C.T. When and Where to Move: Dynamic Occupancy Models Explain the Range Dynamics of a Food Nomadic Bird under Climate and Land Cover Change. *Glob. Chang. Biol.* **2018**, *24*, e27–e39. [CrossRef]
25. Peers, M.J.L.; Thornton, D.H.; Majchrzak, Y.N.; Bastille-Rousseau, G.; Murray, D.L. De-Extinction Potential under Climate Change: Extensive Mismatch between Historic and Future Habitat Suitability for Three Candidate Birds. *Biol. Conserv.* **2016**, *197*, 164–170. [CrossRef]
26. Legault, A.; Theuerkauf, J.; Chartendrault, V.; Rouys, S.; Saoumoé, M.; Verfaille, L.; Desmoulins, F.; Barré, N.; Gula, R. Using Ecological Niche Models to Infer the Distribution and Population Size of Parakeets in New Caledonia. *Biol. Conserv.* **2013**, *167*, 149–160. [CrossRef]
27. Cobden, M.; Alves, F.; Robinson, S.; Heinsohn, R.; Stojanovic, D. Impact of Removal on Occupancy Patterns of the Invasive Rainbow Lorikeet (*Trichoglossus Moluccanus*) in Tasmania. *Austral Ecol.* **2021**, *46*, 31–38. [CrossRef]
28. Botero-Delgadillo, E.; Páez, C.A.; Bayly, N. Biogeography and Conservation of Andean and Trans-Andean Populations of Pyrrhura Parakeets in Colombia: Modelling Geographic Distributions to Identify Independent Conservation Units. *Bird Conserv. Int.* **2012**, *22*, 445–461. [CrossRef]
29. Baños-Villalba, A.; Blanco, G.; Díaz-Luque, J.A.; Dénes, F.v.; Hiraldo, F.; Tella, J.L. Seed Dispersal by Macaws Shapes the Landscape of an Amazonian Ecosystem. *Sci. Rep.* **2017**, *7*, 1–12. [CrossRef]
30. Valentine, L.E.; Fisher, R.; Wilson, B.A.; Sonneman, T.; Stock, W.D.; Fleming, P.A.; Hobbs, R.J. Time since Fire Influences Food Resources for an Endangered Species, Carnaby's Cockatoo, in a Fire-Prone Landscape. *Biol. Conserv.* **2014**, *175*, 1–9. [CrossRef]
31. Marín-Togo, M.C.; Monterrubio-Rico, T.C.; Renton, K.; Rubio-Rocha, Y.; Macías-Caballero, C.; Ortega-Rodríguez, J.M.; Cancino-Murillo, R. Reduced Current Distribution of Psittacidae on the Mexican Pacific Coast: Potential Impacts of Habitat Loss and Capture for Trade. *Biodivers. Conserv.* **2011**, *21*, 451–473. [CrossRef]
32. Monterrubio-Rico, T.C.; Renton, K.; Ortega-Rodrguez, J.M.; Pérez-Arteaga, A.; Cancino-Murillo, R. The Endangered Yellow-Headed Parrot Amazona Oratrix along the Pacific Coast of Mexico. *ORYX* **2010**, *44*, 602–609. [CrossRef]
33. Gibson, L.; Barrett, B.; Burbidge, A. Dealing with Uncertain Absences in Habitat Modelling: A Case Study of a Rare Ground-Dwelling Parrot. *Divers. Distrib.* **2007**, *13*, 704–713. [CrossRef]

34. Cooper, N.W.; Hallworth, M.T.; Marra, P.P. Light-Level Geolocation Reveals Wintering Distribution, Migration Routes, and Primary Stopover Locations of an Endangered Long-Distance Migratory Songbird. *J. Avian Ecol.* **2017**, *48*, 1–11. [CrossRef]
35. Laranjeiras, T.O.; Cohn-Haft, M. Where Is the Symbol of Brazilian Ornithology? The Geographic Distribution of the Golden Parakeet (Guarouba Guarouba - Psittacidae). *Rev. Bras. Ornitol.* **2009**, *17*, 1–19.
36. Ferrer-Paris, J.R.; Sánchez-Mercado, A.; Rodríguez-Clark, K.M.; Rodríguez, J.P.; Rodríguez, G.A. Using Limited Data to Detect Changes in Species Distributions: Insights from Amazon Parrots in Venezuela. *Biol. Conserv.* **2014**, *173*, 133–143. [CrossRef]
37. Recio, M.R.; Payne, K.; Seddon, P.J. Emblematic Forest Dwellers Reintroduced into Cities: Resource Selection by Translocated Juvenile Kaka. *Curr. Zool.* **2016**, *62*, 15–22. [CrossRef]
38. Berkunsky, I.; Simoy, M.v.; Cepeda, R.E.; Marinelli, C.; Kacoliris, F.P.; Daniele, G.; Cortelezzi, A.; Díaz-Luque, J.A.; Friedman, J.M.; Aramburú, R.M. Assessing the Use of Forest Islands by Parrot Species in a Neotropical Savanna. *Avian Conserv. Ecol.* **2015**, *10*, 11–15. [CrossRef]
39. Plasencia-Vázquez, A.H.; Escalona Segura, G. Characterization of the Potential Geographical Distribution Area of Parrot Species in Yucatan Peninsula, Mexico. *Rev. Biol. Trop.* **2014**, *62*, 1509–1522.
40. Stojanovic, D.; Webb, M.H.; Alderman, R.; Porfirio, L.L.; Heinsohn, R. Discovery of a Novel Predator Reveals Extreme but Highly Variable Mortality for an Endangered Migratory Bird. *Divers. Distrib.* **2014**, *20*, 1200–1207. [CrossRef]
41. Allen, M.; Webb, M.H.; Alves, F.; Heinsohn, R.; Stojanovic, D. Occupancy Patterns of the Introduced, Predatory Sugar Glider in Tasmanian Forests. *Austral Ecol.* **2018**, *43*, 470–475. [CrossRef]
42. Bluff, L.A. Ground Parrots and Fire in East Gippsland, Victoria: Habitat Occupancy Modelling from Automated Sound Recordings. *Emu* **2016**, *116*, 402–410. [CrossRef]
43. Webb, M.H.; Heinsohn, R.; Sutherland, W.J.; Stojanovic, D.; Terauds, A. An Empirical and Mechanistic Explanation of Abundance-Occupancy Relationships for a Critically Endangered Nomadic Migrant. *Am. Nat.* **2018**, *193*, 59–69. [CrossRef]
44. Kissling, W.D.; Böhning–Gaese, K.; Jetz, W. The Global Distribution of Frugivory in Birds. *Glob. Ecol. Biogeogr.* **2009**, *18*, 150–162. [CrossRef]
45. Saunders, D.L.; Heinsohn, R. Winter Habitat Use by the Endangered, Migratory Swift Parrot (*Lathamus Discolor*) in New South Wales. *Emu* **2008**, *108*, 81–89. [CrossRef]
46. Webb, M.H.; Terauds, A.; Tulloch, A.; Bell, P.; Stojanovic, D.; Heinsohn, R. The Importance of Incorporating Functional Habitats into Conservation Planning for Highly Mobile Species in Dynamic Systems. *Conserv. Biol.* **2017**, *31*, 1018–1028. [CrossRef] [PubMed]
47. Monterrubio-Rico, T.C.; Álvarez-Jara, M.; Téllez-García, L.; Tena-Morelos, C. Nesting Habitat Characterization for Amazona Oratrix (Psittaciformes: Psittacidae) in the Central Pacific, Mexico. *Rev. Biol. Trop.* **2014**, *62*, 1053–1072. [CrossRef]
48. Ortiz-Maciel, S.G.; Hori-Ochoa, C.; Enkerlin-Hoeflich, E. Maroon-Fronted Parrot (Rhynchopsitta Terrisi) Breeding Home Range and Habitat Selection in the Northern Sierra Madre Oriental, Mexico. *Wilson J. Ornithol.* **2010**, *122*, 513–517. [CrossRef]
49. Keighley, M.v.; Langmore, N.E.; Peñalba, J.v.; Heinsohn, R. Modelling Dispersal in a Large Parrot: A Comparison of Landscape Resistance Models with Population Genetics and Vocal Dialect Patterns. *Landsc. Ecol.* **2020**, *35*, 129–144. [CrossRef]
50. Shukuroglou, P.; McCarthy, M.A. Modelling the Occurrence of Rainbow Lorikeets (*Trichoglossus Haematodus*) in Melbourne. *Austral Ecol.* **2006**, *31*, 240–253. [CrossRef]
51. Rodríguez–Pastor, R.; Senar, J.C.; Ortega, A.; Faus, J.; Uribe, F.; Montalvo, T. Distribution Patterns of Invasive Monk Parakeets (*Myiopsitta Monachus*) in an Urban Habitat. *Anim. Biodivers. Conserv.* **2012**, *35*, 107–117. [CrossRef]
52. Watson, S.J.; Watson, D.M.; Luck, G.W.; Spooner, P.G. Effects of Landscape Composition and Connectivity on the Distribution of an Endangered Parrot in Agricultural Landscapes. *Landsc. Ecol.* **2014**, *29*, 1249–1259. [CrossRef]
53. Plasencia-Vázquez, A.H.; Escalona-Segura, G.; Esparza-Olguín, L.G. Interaction of Landscape Variables on the Potential Geographical Distribution of Parrots in the Yucatan Peninsula, Mexico. *Anim. Biodivers. Conserv.* **2014**, *37*, 191–203. [CrossRef]
54. Kavanagh, R.P.; Stanton, M.A. Bird Population Recovery 22 Years after Intensive Logging near Eden, New South Wales. *Emu* **2003**, *103*, 221–231. [CrossRef]
55. Koleff, P.; Gaston, K.J. Latitudinal Gradients in Diversity: Real Patterns and Random Models. *Ecography* **2001**, *24*, 341–351. [CrossRef]
56. Monterrubio-Rico, T.C.; Villaseñor-Gómez, L.E.; Marin-Togo, M.C.; López-Cordova, E.A.; Fabian-Turja, B.; Sorani-Dalbon, V. Distribución Historica y Actual Del Loro Cabeza Amarilla (*Amazona Oratrix*) en la Costa Central Del Pacífico Mexicano: Ventajas y Limitaciones En El Uso de GARP En Especies Bajo Fuerte Presión de Trafico. *Ornitol. Neotrop.* **2007**, *18*, 263–276.
57. Mota-Vargas, C.; Parra-Noguez, K.P.; Rojas-Soto, O. Análisis Del Conocimiento Histórico de La Distribución Geográfica y Ecológica Del Loro Frente Blanca, Amazona Albifrons, Con Evidencia de Colonización Reciente. *Rev. Mex. Biodivers.* **2020**, *91*, 912708. [CrossRef]
58. Burgio, K.R.; Carlson, C.J.; Tingley, M.W. Lazarus Ecology: Recovering the Distribution and Migratory Patterns of the Extinct Carolina Parakeet. *Ecol. Evol.* **2017**, *7*, 5467–5475. [CrossRef]
59. Strubbe, D.; Matthysen, E.; Graham, C.H. Assessing the Potential Impact of Invasive Ring-Necked Parakeets Psittacula Krameri on Native Nuthatches Sitta Europeae in Belgium. *J. Appl. Ecol.* **2010**, *47*, 549–557. [CrossRef]
60. Mori, E.; Ancillotto, L.; Menchetti, M.; Strubbe, D. 'The Early Bird Catches the Nest': Possible Competition between Scops Owls and Ring-Necked Parakeets. *Anim. Conserv.* **2017**, *20*, 463–470. [CrossRef]

61. Mori, E.; Cardador, L.; Reino, L.; White, R.L.; Hernández-Brito, D.; le Louarn, M.; Mentil, L.; Edelaar, P.; Pârâu, L.G.; Nikolov, B.P.; et al. Lovebirds in the Air: Trade Patterns, Establishment Success and Niche Shifts of Agapornis Parrots within Their Non-Native Range. *Biol. Invasions* **2020**, *22*, 421–435. [CrossRef]
62. Cardador, L.; Carrete, M.; Gallardo, B.; Tella, J.L. Combining Trade Data and Niche Modelling Improves Predictions of the Origin and Distribution of Non-Native European Populations of a Globally Invasive Species. *J. Biogeogr.* **2016**, *43*, 967–978. [CrossRef]
63. Strubbe, D.; Matthysen, E. Invasive Ring-Necked Parakeets Psittacula Krameri in Belgium: Habitat Selection and Impact on Native Birds. *Ecography* **2007**, *30*, 578–588. [CrossRef]
64. Vall-llosera, M.; Woolnough, A.P.; Anderson, D.; Cassey, P. Improved Surveillance for Early Detection of a Potential Invasive Species: The Alien Rose-Ringed Parakeet Psittacula Krameri in Australia. *Biol. Invasions* **2017**, *19*, 1273–1284. [CrossRef]
65. Falcón, W.; Tremblay, R.L. From the Cage to the Wild: Introductions of Psittaciformes to Puerto Rico. *PeerJ* **2018**, *6*, e5669. [CrossRef]
66. Ferrer-Paris, J.R.; Sánchez-Mercado, A. Making Inferences about Non-Detection Observations to Improve Occurrence Predictions in Venezuelan Psittacidae. *Bird Conserv. Int.* **2020**, *30*. [CrossRef]
67. GBIF.org. Psittaciformes Occurrences. Available online: https://www.gbif.org/occurrence/download/0009968-210914110416597 (accessed on 10 September 2021).
68. Thuiller, W. BIOMOD—Optimizing Predictions of Species Distributions and Projecting Potential Future Shifts under Global Change. *Glob. Chang. Biol.* **2003**, *9*, 1353–1362. [CrossRef]
69. Boyce, M.S.; Vernier, P.R.; Nielsen, S.E.; Schmiegelow, F.K.A. Evaluating Resource Selection Functions. *Ecol. Model.* **2002**, *157*, 281–300. [CrossRef]
70. Manning, A.D.; Lindenmayer, D.B.; Nix, H.A.; Barry, S.C. A Bioclimatic Analysis for the Highly Mobile Superb Parrot of South-Eastern Australia. *Emu* **2005**, *105*, 193–201. [CrossRef]
71. Manning, A.D.; Lindenmayer, D.B.; Barry, S.C.; Nix, H.A. Multi-Scale Site and Landscape Effects on the Vulnerable Superb Parrot of South-Eastern Australia during the Breeding Season. *Landsc. Ecol.* **2006**, *21*, 1119–1133. [CrossRef]
72. MacKenzie, D.I.; Nichols, J.D.; Royle, J.A.; Pollock, K.H.; Bailey, L.L.; Heines, J.E. *Occupancy Estimation and Modeling. Inferring Patterns and Dynamics of Species Occurence*; Academic Press: London, UK, 2006.
73. Rodrigues, P.O.; Borges, M.R.; Melo, C. Richness, Composition and Detectability of Psittacidae (Aves) in Three Palm Swamps of the Cerrado Sensu Lato in Central Brazil. *Rev. Chil. Hist. Nat.* **2012**, *85*, 171–178. [CrossRef]
74. Renton, K.; Salinas-Melgoza, A.; de Labra-Hernández, M.Á.; de la Parra-Martínez, S.M. Resource Requirements of Parrots: Nest Site Selectivity and Dietary Plasticity of Psittaciformes. *J. Ornithol.* **2015**, *156*, 73–90. [CrossRef]
75. Engeman, R.; Whisson, D.; Quinn, J.; Cano, F.; Quiñones, P.; White, T.H. Monitoring Invasive Mammalian Predator Populations Sharing Habitat with the Critically Endangered Puerto Rican Parrot Amazona Vittata. *ORYX* **2006**, *40*, 95–102. [CrossRef]
76. Berkunsky, I.; Cepeda, R.E.; Marinelli, C.; Simoy, M.V.; Daniele, G.; Kacoliris, F.P.; Díaz Luque, J.A.; Gandoy, F.; Aramburú, R.M.; Gilardi, J.D. Occupancy and Abundance of Large Macaws in the Beni Savannahs, Bolivia. *Oryx* **2016**, *50*, 113–120. [CrossRef]
77. Wintle, B.A.; Walshe, T.v.; Parris, K.M.; McCarthy, M.A. Designing Occupancy Surveys and Interpreting Non-Detection When Observations Are Imperfect. *Divers. Distrib.* **2011**, *18*, 417–424. [CrossRef]
78. Rivera-Ortíz, F.A.; Oyama, K.; Ríos-Muñoz, C.A.; Solórzano, S.; Navarro-Sigüenza, A.G.; del Coro Arizmendi, M. Habitat Characterization and Modeling of the Potential Distribution of the Military Macaw (*Ara Militaris*) in Mexico. *Rev. Mex. Biodivers.* **2013**, *84*, 1200–1215. [CrossRef]
79. Rios-Muñoz, C.A.; Navarro-Sigüenza, A.G. Efectos Del Cambio de Uso de Suelo En La Disponibilidad Hipotética de Hábitat Para Los Psitácidos de México. *Ornitol. Neotrop.* **2009**, *20*, 491–509.
80. Plasencia Vázquez, A.H.; Escalona Segura, G.; Ferrer Sánchez, Y. The Relationship between Forest Fragmentation and the Potential Geographical Distribution of Psittacids (Psittaciformes: Psittacidae) in the Yucatan Peninsula, Mexico. *Rev. Biol. Trop.* **2017**, *65*, 1470. [CrossRef]
81. Freeman, B.; Sunnarborg, J.; Peterson, A.T. Effects of Climate Change on the Distributional Potential of Three Range-Restricted West African Bird Species. *Condor* **2019**, *121*, 1–10. [CrossRef]
82. Le Louarn, M.; Clergeau, P.; Strubbe, D.; Deschamps-Cottin, M. Dynamic Species Distribution Models Reveal Spatiotemporal Habitat Shifts in Native Range-Expanding versus Non-Native Invasive Birds in an Urban Area. *J. Avian Biol.* **2018**, *49*, jav-01527. [CrossRef]
83. Shivambu, T.C.; Shivambu, N.; Downs, C.T. Impact Assessment of Seven Alien Invasive Bird Species Already Introduced to South Africa. *Biol. Invasions* **2020**, *22*, 1829–1847. [CrossRef]
84. Muñoz, A.R.; Real, R. Assessing the Potential Range Expansion of the Exotic Monk Parakeet in Spain. *Divers. Distrib.* **2006**, *12*, 656–665. [CrossRef]
85. Strubbe, D.; Matthysen, E. Predicting the Potential Distribution of Invasive Ring-Necked Parakeets Psittacula Krameri in Northern Belgium Using an Ecological Niche Modelling Approach. *Biol. Invasions* **2009**, *11*, 497–513. [CrossRef]
86. Salinas-Melgoza, A.; Salinas-Melgoza, V.; Wright, T.F. Behavioral Plasticity of a Threatened Parrot in Human-Modified Landscapes. *Biol. Conserv.* **2013**, *159*, 303–312. [CrossRef]
87. Montoya, D.; Purves, D.W.; Urbieta, I.R.; Zavala, M.A. Do Species Distribution Models Explain Spatial Structure within Tree Species Ranges? *Glob. Ecol. Biogeogr.* **2009**, *18*, 662–673. [CrossRef]

88. Robinson, L.M.; Elith, J.; Hobday, A.J.; Pearson, R.G.; Kendall, B.E.; Possingham, H.P.; Richardson, A.J. Pushing the Limits in Marine Species Distribution Modelling: Lessons from the Land Present Challenges and Opportunities. *Glob. Ecol. Biogeogr.* **2011**, *20*, 789–802. [CrossRef]
89. Ranasinghe, R. On the Need for a New Generation of Coastal Change Models for the 21st Century. *Sci. Rep.* **2020**, *10*, 1–6. [CrossRef] [PubMed]
90. Morato, R.G.; Stabach, J.A.; Fleming, C.H.; Calabrese, J.M.; Paula, C.d.; Ferraz, M.P.M.; Kantek, D.L.Z.; Miyazaki, S.S.; Pereira, T.D.C.; Araujo, G.R.; et al. Space Use and Movement of a Neotropical Top Predator: The Endangered Jaguar. *PLoS ONE* **2016**, *11*, e0168176. [CrossRef] [PubMed]
91. Caravaggi, A.; Banks, P.B.; Burton, A.C.; Finlay, C.M.V.; Haswell, P.M.; Hayward, M.W.; Rowcliffe, M.J.; Wood, M.D. A Review of Camera Trapping for Conservation Behaviour Research. *Remote Sens. Ecol. Conserv.* **2017**, *3*, 109–122. [CrossRef]
92. Sugai, L.S.M.; Silva, T.S.F.; Ribeiro, J.W.; Llusia, D. Terrestrial Passive Acoustic Monitoring: Review and Perspectives. *BioScience* **2019**, *69*, 5–11. [CrossRef]
93. Shonfield, J.; Bayne, E.M. Autonomous Recording Units in Avian Ecological Research: Current Use and Future Applications. *Avian Conserv. Ecol.* **2017**, *12*, 14. [CrossRef]
94. Wäldchen, J.; Mäder, P. Machine Learning for Image Based Species Identification. *Methods Ecol. Evol.* **2018**, *9*, 2216–2225. [CrossRef]
95. Farley, S.S.; Dawson, A.; Goring, S.J.; Williams, J.W. Situating Ecology as a Big-Data Science: Current Advances, Challenges, and Solutions. *BioScience* **2018**, *68*, 563–576. [CrossRef]
96. Thomson, R.; Potgieter, G.C.; Bahaa-el-din, L. Closing the Gap between Camera Trap Software Development and the User Community. *Afr. J. Ecol.* **2018**, *56*, 721–739. [CrossRef]
97. Lee, K.A.; Lee, J.R.; Bell, P. A Review of Citizen Science within the Earth Sciences: Potential Benefits and Obstacles. *Proc. Geol. Assoc.* **2020**, *131*, 605–617. [CrossRef]
98. Sullivan, B.L.; Wood, C.L.; Iliff, M.J.; Bonney, R.E.; Fink, D.; Kelling, S. EBird: A Citizen-Based Bird Observation Network in the Biological Sciences. *Biol. Conserv.* **2009**, *142*, 2282–2292. [CrossRef]
99. iNaturalist INaturalist. Available online: https://www.inaturalist.org/ (accessed on 10 November 2021).
100. Callaghan, C.T.; Rowley, J.J.L.; Cornwell, W.K.; Poore, A.G.B.; Major, R.E. Improving Big Citizen Science Data: Moving beyond Haphazard Sampling. *PLoS Biol.* **2019**, *17*, 1–11. [CrossRef]
101. Ellwood, E.R.; Crimmins, T.M.; Miller-Rushing, A.J. Citizen Science and Conservation: Recommendations for a Rapidly Moving Field. *Biol. Conserv.* **2017**, *208*, 1–4. [CrossRef]
102. Sánchez-Mercado, A.; Asmüssen, M.; Rodríguez-Clark, K.M.; Rodríguez, J.P.; Jedrzejewski, W. Using Spatial Patterns in Illegal Wildlife Uses to Reveal Connections between Subsistence Hunting and Trade. *Conserv. Biol.* **2016**, *30*, 1222–1232. [CrossRef]
103. Sánchez-Mercado, A.; Cardozo-Urdaneta, A.; Moran, L.; Ovalle, L.; Arvelo, M.; Morales-Campo, J.; Coyle, B.; Braun, M.J.; Rodriguez-Clark, K.M. Social Network Analysis Reveals Specialized Trade in an Endangered Songbird. *Anim. Conserv.* **2019**, *23*, 132–144. [CrossRef]
104. Hallgren, W.; Beaumont, L.; Bowness, A.; Chambers, L.; Graham, E.; Holewa, H.; Laffan, S.; Mackey, B.; Nix, H.; Price, J.; et al. The Biodiversity and Climate Change Virtual Laboratory: Where Ecology Meets Big Data. *Environ. Model. Softw.* **2016**, *76*, 182–186. [CrossRef]
105. Velásquez-Tibatá, J.; Olaya-Rodríguez, M.H.; López-Lozano, D.; Gutiérrez, C.; González, I.; Londoño-Murcia, M.C. Biomodelos: A Collaborative Online System to Map Species Distributions. *PLoS ONE* **2019**, *14*, e0214522. [CrossRef]
106. Guisan, A.; Tingley, R.; Baumgartner, J.B.; Naujokaitis-Lewis, I.; Sutcliffe, P.R.; Tulloch, A.I.T.; Regan, T.J.; Brotons, L.; Mcdonald-Madden, E.; Mantyka-Pringle, C.; et al. Predicting Species Distributions for Conservation Decisions. *Ecol. Lett.* **2013**, *16*, 1424–1435. [CrossRef]
107. Monterrubio-Rico, T.C.; Charre-Medellín, J.F.; Pacheco-Figueroa, C.; Arriaga-Weiss, S.; de Dios Valdez-Leal, J.; Cancino-Murillo, R.; Escalona-Segura, G.; Bonilla-Ruz, C.; Rubio-Rocha, Y. Distribución Potencial Histórica y Contemporánea de La Familia Psittacidae En México. *Rev. Mex. Biodivers.* **2016**, *87*, 1103–1117. [CrossRef]
108. Marini, M.Â.; Barbet-Massin, M.; Martinez, J.; Prestes, N.P.; Jiguet, F. Applying Ecological Niche Modelling to Plan Conservation Actions for the Red-Spectacled Amazon (*Amazona Pretrei*). *Biol. Conserv.* **2010**, *143*, 102–112. [CrossRef]
109. Pidgeon, A.M.; Rivera, L.; Martinuzzi, S.; Politi, N.; Bateman, B. Will Representation Targets Based on Area Protect Critical Resources for the Conservation of the Tucuman Parrot? *Condor* **2015**, *117*, 503–517. [CrossRef]
110. Hambuckers, A.; de Harenne, S.; Ledezma, E.R.; Zeballos, L.Z.; François, L. Predicting the Future Distribution of Ara Rubrogenys, an Endemic Endangered Bird Species of the Andes, Taking into Account Trophic Interactions. *Diversity* **2021**, *13*, 94. [CrossRef]

Article

Predicting the Future Distribution of *Ara rubrogenys*, an Endemic Endangered Bird Species of the Andes, Taking into Account Trophic Interactions

Alain Hambuckers [1,*], Simon de Harenne [1,2], Eberth Rocha Ledezma [3], Lilian Zúñiga Zeballos [1,4] and Louis François [2]

[1] Behavioural Biology Lab, Unit of Research SPHERES, University of Liège, Quai Van Beneden 22, , simon.de.harenne@hotmail.com (S.d.H.); lilian.zz510@gmail.com (L.Z.Z.)
[2] Modelling of Climate and Biogeochemical Cycles, Unit of Research SPHERES, University of Liège, Sart Tilman, 4000 Liège, Belgium; louis.francois@uliege.be
[3] Centro de Biodiversidad y Genética, Facultad de Ciencas y Tecnología, Universidad Mayor de San Simón, Cochabamba, Bolivia; e.rocha.ledezma@gmail.com
[4] Museo de Historia Natural Alcide d'Orbigny, Cochabamba, Bolivia
* Correspondence: alain.hambuckers@uliege.be

Citation: Hambuckers, A.; de Harenne, S.; Rocha Ledezma, E.; Zúñiga Zeballos, L.; François, L. Predicting the Future Distribution of *Ara rubrogenys*, an Endemic Endangered Bird Species of the Andes, Taking into Account Trophic Interactions. *Diversity* **2021**, *13*, 94. https://doi.org/10.3390/d13020094

Academic Editor: Miguel Ferrer

Received: 22 January 2021
Accepted: 18 February 2021
Published: 21 February 2021

Abstract: Species distribution models (SDMs) are commonly used with climate only to predict animal distribution changes. This approach however neglects the evolution of other components of the niche, like food resource availability. SDMs are also commonly used with plants. This also suffers limitations, notably an inability to capture the fertilizing effect of the rising CO_2 concentration strengthening resilience to water stress. Alternatively, process-based dynamic vegetation models (DVMs) respond to CO_2 concentration. To test the impact of the plant modelling method to model plant resources of animals, we studied the distribution of a Bolivian macaw, assuming that, under future climate, DVMs produce more conservative results than SDMs. We modelled the bird with an SDM driven by climate. For the plant, we used SDMs or a DVM. Under future climates, the macaw SDM showed increased probabilities of presence over the area of distribution and connected range extensions. For plants, SDMs did not forecast overall response. By contrast, the DVM produced increases of productivity, occupancy and diversity, also towards higher altitudes. The results offered positive perspectives for the macaw, more optimistic with the DVM than with the SDMs, than initially assumed. Nevertheless, major common threats remain, challenging the short-term survival of the macaw.

Keywords: red-fronted macaw; Andes; dynamic vegetation model; biotic interactions; climate change; RCP2.6; RCP8.5

1. Introduction

The problems of habitat destruction and climate change are the main threat to tropical mountain birds. Mountain bird species in the tropics are particularly at risk because they are isolated by hotter lowland zones which often makes them sedentary. In addition, when shifting their distribution up to higher altitudes, the new area of occupancy narrows [1]. The structure of the mountains itself also appears to be a constraining factor limiting the distribution shift with possible decline of habitat quality, for instance, the absence of suitable nesting sites or even vertical gaps between actual and potential future areas of distribution [2].

The slopes of the Andes are recognized as supporting the highest avian diversity in the world combined with high endemism rate but also more than 20% of threatened species [3]. In Bolivia, the red-fronted macaw (*Ara rubrogenys* Lafresnaye, 1847) is one of the 15 endemic species of this country [4]. Less than 30 years ago, *A. rubrogenys* was a little-known species [5]. It lives on the east Andean slopes of south-central Bolivia from

553 m up to 3094 m a.s.l. (Figure 1) and breeds between 1188 and 2696 m [6]. Its natural habitat is mainly semi-deciduous dry forest but this is most often severely degraded by pastoralism and by timber extraction into thorny scrubs with scattered trees [7]. While it was estimated that the threats were of limited extent in the early nineties [8], the status of the species worsens over the course of time with land conversion for agriculture, with poaching for illegal trade, with killing by farmers who consider them a pest and with poisoning with pesticides when they feed on the crops [6,9–11]. The small breeding population (only 67 to 136 pairs) in eight close areas was also pointed out as major risk which increases their extinction risk due to correlated environmental fluctuations [6]. It was found that the birds use agriculture-scrub ecotones more than the forests for foraging, probably because the forests do not offer enough resources. *A. rubrogenys* is now ranked as "Critically Endangered" in Bolivia [11] and on the International Union for Conservation of Nature Red list [12]. In addition, it could be particularly threatened by climate change. In its area of distribution, *A. rubrogenys* uses only terrains along river valleys for roosting, feeding, resting and nesting [9]. Most of the nesting sites are located in steep river cliffs but such environments are not necessarily available at higher altitudes given the magnitude of warming predictions. Climate change is supposed to particularly affect the tropical Andes and notably Bolivia [13]; while warming already averaged 0.1 °C/decade between 1939 and 1998, it accelerated to 0.33 °C/decade between 1980 and 2005. Climate change scenarios suggest warming as high as 7.5 °C by 2080 and important modifications of the precipitation regime with respect to pre-industrial times.

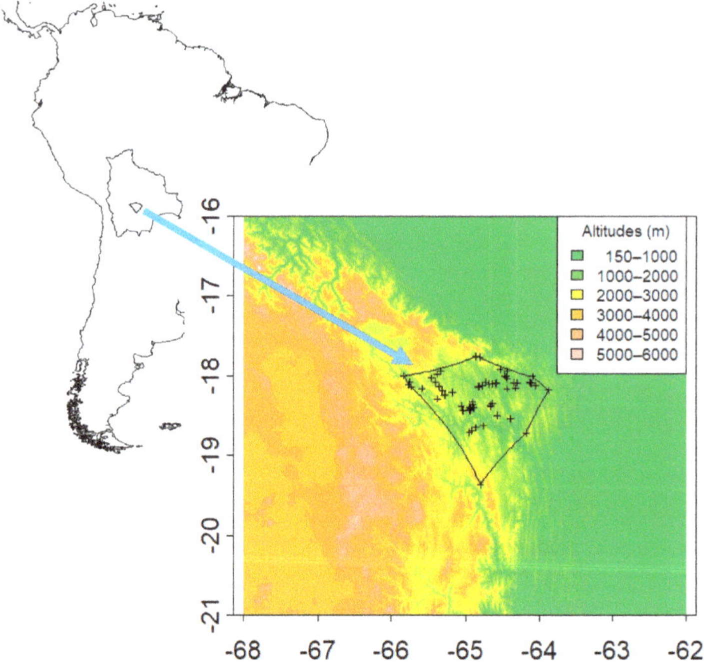

Figure 1. Distribution area (alpha hull) in Bolivia of *Ara rubrogenys* at altitudes with occurrences. 96% of the pixels over the area have an altitude between 1100 and 3000 m.

Species distribution models (SDMs) are based on the computation of an empirical relationship between the presence of a species (a sample of its distribution) and the actual values of the selected explaining factors. Range projections are obtained by computing a probability of presence over the study area using the relationship. The methodology has been applied to parrots several times with different objectives. For instance, comparing the projection of models fitted to historical data with actual ranges allowed to study the conservation status of Andean *Pyrrhura* in Colombia [14] or of *Amazona* in Venezuela [15]. Thanks to climate driven SDMs, the substitution of the actual climate with future climate allows projection of decreasing and shifting ranges in *Amazona pretrei* [16]. Climate driven SDMs also permitted computing of habitat suitability of 13 parrot species in invaded countries and to test the consequences of two successive trade bans in the US and the EU on the invasion success [17]. In some situations, however, climate factors are not sufficient explaining variables, they only improve the fit of models describing the niches in combination with other factors reflecting species requirements like habitat characteristics. Authors reached this conclusion for modelling Bonelli's eagle nesting sites with topography, disturbance, land-use or climate variable at several geographic scales [18]. Bad results were obtained with climate variables only, but climate significantly improves the quality of the prediction offered by the other sets of variables. In light of their results, the authors suggest that snow and low winter temperature may cause physiological stress hampering breeding success; however, they underlined that with more complex models, the interpretation of the effect of each explaining climate factor could become hard to find. Another interesting approach to model bird distribution consists of including biotic interactions as well as abiotic factors. For instance, in [19] SDMs driven by climate were used to simulate several shrub species making up the habitat of a bird species and the outputs of these models were set up as input variables to model animal presence with climate and topography. The authors found that this approach outperforms climate only models, which stressed the importance of taking into account, as far as possible, the different types of niche components to produce consistent simulations. This SDM approach was applied to refine the mapping of the suitability area of *Amazona tucumana* in Argentina and Bolivia. Here, the niche was defined with climate, land-use and the output of another SDM forecasting the distribution of a key plant resource for nestlings, *Podocarpus parlatorei*, providing niche cavities and food [20].

However, while SDMs driven by climate variables are now considered as a standard method to predict plant species distribution under future climate, this approach fails to consider the effect of the increasing CO_2 concentration in air on plant physiology. Indeed, it is well established that increased air CO_2 concentration improves the capacity of plants to resist water stress because the plants can minimize transpiration while still satisfying their CO_2 requirements [21]. Contrary to SDMs, dynamic vegetation models (DVMs) are commonly able to reproduce this effect. Although, questions remain about the acclimation (organism trait responses occurring in days to weeks) and adaptation processes of plants (response occurring through evolutionary processes) to new climates which could lower future changes [22]. Forecasting under future climates with these types of models, forcing, or not, the increasing air CO_2 concentration, gives contrasting results and the projected distributions of plant species or habitats under future climate with increased CO_2 concentrations appear better conserved than with SDMs (e.g., [23,24]).

The objectives of this study are to evaluate the potential impact of climate change on the distribution of *A. rubrogenys*. We compare the results produced by SDMs driven by climate variables for *A. rubrogenys* and for 17 resource plant species over the area of *A. rubrogenys* and the results produced by a DVM for the same plant species, under present conditions and future climates (2070–2100) under RCP2.6 and RCP8.5 forcing. We predict that for the future, approaches with SDMs should conserve less of the original distribution area of *A. rubrogenys* than obtained with DVM.

2. Materials and Methods

2.1. Species Occurrences

Censuses of *A. rubrogenys* were conducted between 2008 and 2010 by two of us (E. Rocha Ledezma and L. Zúñiga Zeballos). The birds were observed between 5 and 11 a.m., in the course of linear surveys and near known nesting, feeding and roosting sites, while prospections were also realized in fields, forests and scrubs. Observations of *A. rubrogenys* in its area of distribution indicated that it mainly feeds on the fruits and seeds of the following wild species [25,26]: *Anadenanthera colubrina* (Vell.) Brenan, *Anisocapparis speciosa* (Griseb.) Cornejo and Iltis, *Aspidosperma quebracho-blanco* Schltdl., *Celtis ehrenbergiana* (Klotzsch) Liebm., *Cenchrus* (i.e., the species occurring in Bolivia in GBIF: *C. alopecuroides* J. Presl., *C. bambusiformis* (E. Fourn.) Morrone, *C. brevisetus* E. Fourn., *C. brownii* Roem. and Schult., *C. chilensis* (É. Desv.) Morrone, *C. ciliaris* L., *C. echinatus* L., *C. insularis* Scribn., *C. latifolius* (Spreng.) Morrone, *C. longisetus* M.C. Johnst, *C. myosuroides* Kunth, *C. nervosus* (Nees) Kuntze, *C. polystachios* (L.) Morrone, *C. purpureus* (Schumach.) Morrone, *C. setosus* Sw., *C. viridis* Spreng.), *Cnidoscolus tubulosus* I.M. Johnst., *Jatropha hieronymi* Kuntze, *Lithraea molleoides* Engl. (only occurrences in South America), *Loxopterygium grisebachii* Hiern ex Griseb., *Neoraimondia herzogiana* (Backeb.) Buxb. and Krainz, *Parasenegalia visco* (Lorentz ex Griseb.) Seigler and Ebinger, *Parkinsonia praecox* (Ruiz and Pav.) Hawkins, *Prosopis chilensis* Stuntz, *Prosopis kuntzei* Harms ex Kuntze, *Sarcomphalus mistol* (Griseb.) Hauenschild, *Schinopsis marginata* Engl. and *Selaginella sellowii* Hieron. We obtained the plant species coordinates of occurrences by querying the Global Diversity Information Facility site database (GBIF) in January 2017 and in August 2020, further checking for duplicates (Table 1). A second check for duplicates was conducted after combining with climate factors for coordinates belonging to the same pixels. This had limited consequences on plant sample size but not for *A. rubrogenys* sample size which dropped to 63. *A. rubrogenys'* area of distribution was defined with the alpha-hull polygon method [27] using the ashape function of the R package "alphahull" [28].

2.2. Climate Data

For the current climate (monthly values for temperature, difference between maximum and minimum daily temperatures, precipitation, relative humidity, sunshine hours, wind speed), we used Worldclim version 2 for the time period between 1970 and 2000 at 2.5 arc-minutes [29]. For the future climate, we used the CMIP5 projections of the HadGEM2-AO global circulation model [30] under the representative concentration pathways of greenhouse gases corresponding to end of 21st century radiative forcings of 2.6 and 8.5 W/m^2 (RCP2.6 and RCP8.5, [31]). RCP2.6 would require a decline of greenhouse gas to reach no emission after 2072. With respect to the pre-industrial period, this scenario would keep global temperature rise below 2 °C in 2100. RCP8.5 is the worst hypothesis with emission of greenhouse gases continuing to increase throughout the 21st century producing global temperature increase between 2.6 and 4.8 °C for the period 2081–2100 [32].

2.3. Dynamic Vegetation Modelling

The DVM CARAIB (CARbon Assimilation In the Biosphere) was mainly described in [23,33–36]. This model was initially conceived to simulate vegetation at global or continental scale and its response to climate change in the future or in the past [37–43]. The model was also applied to agricultural systems [44–46] or to tree species [24,47–49]. It is a grid point model composed of 5 main interacting modules (hydrology budget, photosynthesis and stomatal regulation, carbon allocation and growth, heterotrophic respiration and carbon dynamics in the soil, competition between ecosystem strata and biogeography) and it is also possible to activate natural fire and migration modules. Input data are spatial monthly climates (minimal and maximal temperature, precipitation, relative humidity, sunshine hours and wind speed), CO_2 air concentration, soil texture and color, elevation and a set of information describing the morpho-physiological characteristics of the plant species (traits), like the specific leaf area, leaf and sapwood C:N, plant height, deciduousness nature, et cetera, and climatic thresholds extracted from the distribution samples. Since trait information for the species was lacking, it was replaced by the values of the plant functional type to which the species belongs (Table 4 of Electronic Supplementary Material of [49]). Threshold values controlling germination and mortality under stress conditions are extracted from prescribed percentiles in their actual climate distribution extracted from the occurrence samples [36,40]. For computing the fitness statistics, we also ran the simulation for the present on the coordinates of the sets of pseudo-absences drawn for SDM modelling (see below). As output of the model, we used the net primary productivities of the species fractions (fNPP, $gC/m^2/y$) and we selected thresholds of presence maximizing the true skill statistic (TSS, [50]). The threshold of presence allows us to compute the sensitivity (proportion of presences correctly predicted) and specificity (proportion of absences correctly predicted). We also computed the receiver operating characteristic curve (AUC). Positive TSS values and AUC larger than 0.7 indicate better agreement than random.

Table 1. Species occurrences and data sources (Global Diversity Information Facility site database: GBIF).

Species	Occurrences	Sources
Anadenanthera colubrina	3272	GBIF.org (12 August 2020) GBIF Occurrence Download https://doi.org/10.15468/dl.xccvu7
Anisocapparis speciosa	218	GBIF.org (12 August 2020) GBIF Occurrence Download https://doi.org/10.15468/dl.xxw4c9
Ara rubrogenys	159	This study
Aspidosperma quebracho-blanco	394	GBIF.org (12 August 2020) GBIF Occurrence Download https://doi.org/10.15468/dl.xccvu7
Celtis ehrenbergiana	2528	GBIF.org (11 August 2020) GBIF Occurrence Download https://doi.org/10.15468/dl.ph4g3s
Cenchrus	8069	GBIF.org (12 August 2020) GBIF Occurrence Download https://doi.org/10.15468/dl.3s3af9
Cnidoscolus tubulosus.	159	GBIF.org (6 January 2017) GBIF Occurrence Download http://doi.org/10.15468/dl.wwofy0
Jatropha hieronymi	96	GBIF.org (6 January 2017) GBIF Occurrence Download http://doi.org/10.15468/dl.2ziu73,
Lithraea molleoides	13,271	GBIF.org (12 August 2020) GBIF Occurrence Download https://doi.org/10.15468/dl.m45p2c
Loxopterygium grisebachii	86	GBIF.org (11 August 2020) GBIF Occurrence Download https://doi.org/10.15468/dl.3zkvc4
Neoraimondia herzogiana	29	GBIF.org (12 August 2020) GBIF Occurrence Download https://doi.org/10.15468/dl.xag67z
Parasenegalia visco	34	GBIF.org (12 August 2020) GBIF Occurrence Download https://doi.org/10.15468/dl.mnw4qm
Parkinsonia praecox	1123	GBIF.org (12 August 2020) GBIF Occurrence Download https://doi.org/10.15468/dl.fk3as9
Prosopis chilensis	81	GBIF.org (6 January 2017) GBIF Occurrence Download http://doi.org/10.15468/dl.dh9ski
Prosopis kuntzei Kuntze	68	GBIF.org (6 January 2017) GBIF Occurrence Download http://doi.org/10.15468/dl.dh9ski
Sarcomphalus mistol	217	GBIF.org (11 August 2020) GBIF Occurrence Download https://doi.org/10.15468/dl.mk3r74
Schinopsis marginata Engl.	41	GBIF.org (6 January 2017) GBIF Occurrence Download http://doi.org/10.15468/dl.ngmuf4
Selaginella sellowii	924	GBIF.org (21 August 2020) GBIF Occurrence Download https://doi.org/10.15468/dl.5d5hwt

2.4. *A. rubrogenys* and Plant Species SDM

We used multiple logistic regressions, also called logit models. Studies have showed that it is difficult to rank the SDM methods according to their performances because they vary with dataset properties [51–53], but logistic regression has the best theoretical background and it is used here as a reference methodology for SDM. The models were computed in R with the glm function of the R package "stat" (The R Core Team) and we interpreted the output as a probability of presence (Px). Since absence data are also needed, we generated pseudo-absences for each species (*A. rubrogenys* and plant species). The pseudo-absences were sets of points randomly drawn around the occurrences containing twice as many points as occurrences. Our strategy is based on the fact that for rare events, rather than randomly sampling, it is more efficient to collect positive cases (presences), as much as possible, and to complete them with a limited number of random cases as absences [54]. There is no common rule to select the ratio of absences to presences and this varied considerably in ecological studies (see [55]). For each plant species, we built 10 datasets containing the occurrence coordinates and pseudo-absence coordinates drawn in radius buffers comprised of between 200 and 2000 km around presences. Furthermore, drawing pseudo-absences within the presences is justified by the fact that the species do not occupy their entire range. For each dataset, we identified the best combination of climate factor effects, in other words, a maximum of 6 (only 4 for *N. herzogiana*, due to smaller sample size) linear or quadratic effects, after exhaustive screening based on the Akaike information criterion (AIC), using the glmulti function of the R package "glmulti" [56]. This procedure allowed us to select the shortest distance for the pseudo-absences providing evident AUC increase, which may give more accuracy and meaningful fit of the models [57]. We finally used the datasets giving the highest AUC and the model selected in the above procedure with the lowest AIC. Possibly, we chose a model with slightly higher AIC so that most of the model coefficients may have Z-test p-values lower than 0.05. For validation we tested the selected models against their null models using the likelihood ratio test (with critical p-value = 0.05) and the AUC (auc function of "SDMTools" [58]). The cutting thresholds were also selected thanks to TSS. A linear effect alone reveals a strictly positive or negative ecological response to the considered factor. The combination of a linear effect with the quadratic effect of the same factor produces a bell-shaped ecological response curve. Owing to the logit link function, a quadratic effect alone acts as a threshold. As climate variables, we first selected annual mean temperature, temperature seasonality, maximum temperature of the warmest month, minimum temperature of the coldest month, annual precipitation, precipitation seasonality, precipitation of the wettest quarter and precipitation of the driest quarter. Then, to minimize collinearity, we computed the matrix of Pearson correlation coefficients in the species datasets and pointed the couple of variables with coefficients >0.7 [59]. Thus, we had to drop the maximum temperature of the warmest month, minimum temperature of the coldest month and annual precipitation for the plant species, and additionally, precipitation of the driest quarter for *A. rubrogenys*. For projections, we set missing value pixels with at least one of the explanatory factors outside the range encountered in the calibration datasets.

It should be noted that we tested the approach consisting of including outputs of the plant models to drive *A. rubrogenys* SDM using restricted estimation of the linear coefficients of the plant model output to impose positive coefficients. However, this approach failed mainly due to the fact that the productivity (DVM outputs) or the probability of presence (SDM outputs) were not uniformly higher over the *A. rubrogenys* area than outside. We understood this negative result as that animals are able to rely on resources made by plants growing in sub-optimal conditions, which seemed evident.

3. Results

Climate over the *A. rubrogenys* area varied a lot, exemplified by mean annual temperatures which fall between 12.31 and 23.00 °C and precipitation of the driest quarter which ranged between 6 and 62 mm (Table S1). Under RCP2.6 forcing, temperature and precipitation over the *A. rubrogenys* area increased while seasonality decreased with nevertheless increases of the ranges between extreme higher and lower values. Under RCP8.5 forcing, temperature and precipitation followed the same direction but more intensively than under RCP2.6 forcing.

3.1. Dynamic Vegetation Modelling

Fitness statistics (Table 2) suggested limited to acceptable agreements for plant species DVM modelling. It is normal that the mechanistic model had significantly lower AUC than those obtained with the SDM. Indeed, first there is no training step in the DVM computation and, second, the AUC of the SDM has been maximized in the generation of pseudo-absences by varying the distance to the points of presence. Thus, comparing the performances of the two models in terms of AUC, or any other evaluator using pseudo-absences, cannot be made without bias. In this respect, it should be noted that the performance of the DVM in terms of sensitivity (Se) (which do not use pseudo-absences) is generally quite high, since they are generally larger than 0.8 or even 0.9. Otherwise, owing to excessive computing time, it was not possible to compute maps over the entire areas of distribution of the plant species. All the species but *C. tubulosus* were simulated as occurring over the *A. rubrogenys* presence area (Figures 2 and 3 and Figure S1). The range of fNPP also varied considerably between the species and the climate conditions.

Table 2. Dynamic vegetation modelling (DVM) information: sample size for statistics (*N*), area under the receiver operating characteristic curve (AUC), true skill statistic (TSS), threshold for TSS (net primary productivity of the species fraction: fNPP, $gC/m^2/y$), maximal net primary productivity of the species fraction over the area (Max fNPP, $gC/m^2/y$), sensitivity (Se), specificity (Sp).

Plant Species	*N*	AUC	TSS	Threshold	Max fNPP	Se	Sp
Anadenanthera colubrina	1643	0.67569	0.41083	0.0015	318.03	0.9434	0.46744
Anisocapparis speciosa	147	0.60454	0.40816	4.5738	1131.69	0.95918	0.44898
Aspidosperma quebracho-blanco	218	0.6508	0.38991	0.0706	1064.81	0.94495	0.44495
Celtis ehrenbergiana	1444	0.63466	0.47715	0.1982	1253.74	0.96676	0.51039
Cenchrus	4449	0.66834	0.35042	0.4236	930.89	0.82513	0.52529
Cnidoscolus tubulosus	145	0.68628	0.4	19.5846	1071.89	0.76552	0.63448
Jatropha hieronymi	54	0.73131	0.61111	7.0216	1049.12	0.92593	0.68519
Lithraea molleoides	1870	0.75729	0.61176	0.3059	1184.87	0.94759	0.66417
Loxopterygium grisebachii	49	0.75052	0.57143	1.6223	1112.63	0.91837	0.65306
Neoraimondia herzogiana	25	0.4784	0.4	0.0053	1111.65	0.9600	0.4400
Parasenegalia visco	29	0.6629	0.44828	47.2607	1417.99	0.7931	0.65517
Parkinsonia praecox	550	0.53771	0.38	12.7344	851.75	0.95818	0.42182
Prosopis chilensis	76	0.47152	0.17105	0.0238	1067.33	0.89474	0.27632
Prosopis kuntzei	67	0.62364	0.44776	1.4721	1009.61	0.91045	0.53731
Sarcomphalus mistol	149	0.63799	0.41611	0.7605	1206.76	0.95973	0.45638
Schinopsis marginata	38	0.76939	0.57895	10.2059	1246.94	0.81579	0.76316
Selaginella sellowii	277	0.72557	0.38628	274.332	910.91	0.66065	0.72563

3.2. SDM Modelling

It was possible to obtain models with acceptable to very good agreement for each of the plant species and for *A. rubrogenys*, with fitness indicators generally higher than those obtained for the DVM results (Table 3), except for the sensitivity (Se). For plants, projection maps revealed large areas of potential presence without any presence point which could be due to uneven sampling but also to natural barriers to migration, to competition or to speciation (Figures 4 and 5, Figures S2 and S3). Ten of the plant species were largely simulated as occurring over the *A. rubrogenys'* present area (Figure 6). The simulated distributions over Bolivia sometimes showed large similarities with those produced by the DVM. For *A. rubrogenys*, the model defined a close region over the alpha hull defining the present area (Figure 5). The original range was conserved under the future climates, the

Px increased (Figure 6), while new suitable areas appeared connected with the former one (Figure 5).

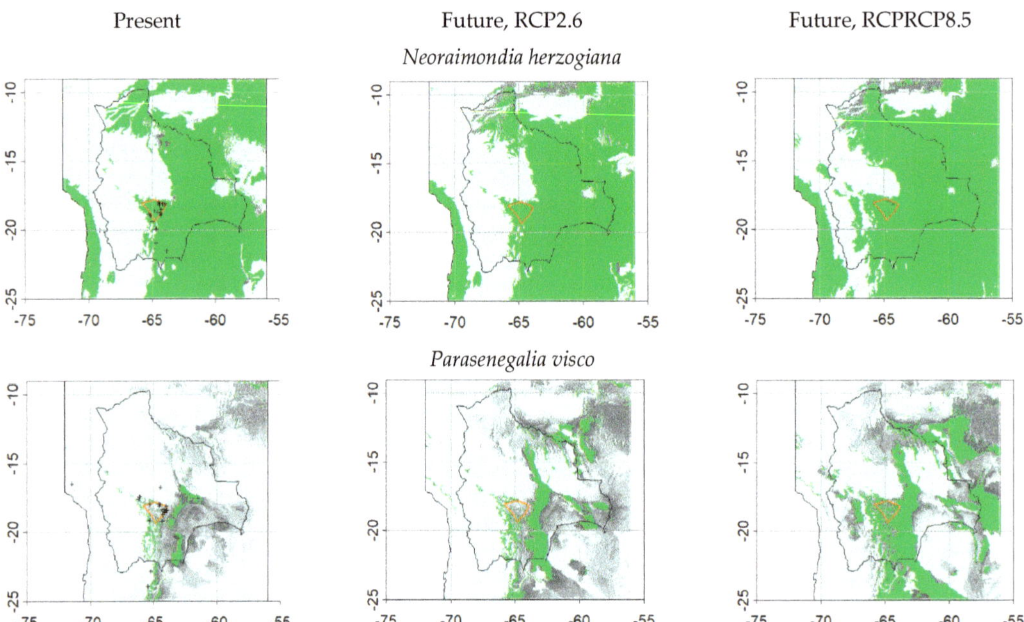

Figure 2. Examples of plant species distributions over Bolivia (*Ara rubrogenys* area delimited by orange lines) predicted by the DVM CARAIB (CARbon Assimilation In the Biosphere) for present and future climates under RCP2.6 andRCP8.5 forcings (+: occurrences of the species, light to dark color variation shows plant net productivity of the species fraction, in blue-gray between 0 and threshold of presence, in green between threshold and maximal value; the other plant species are shown in Figure S1).

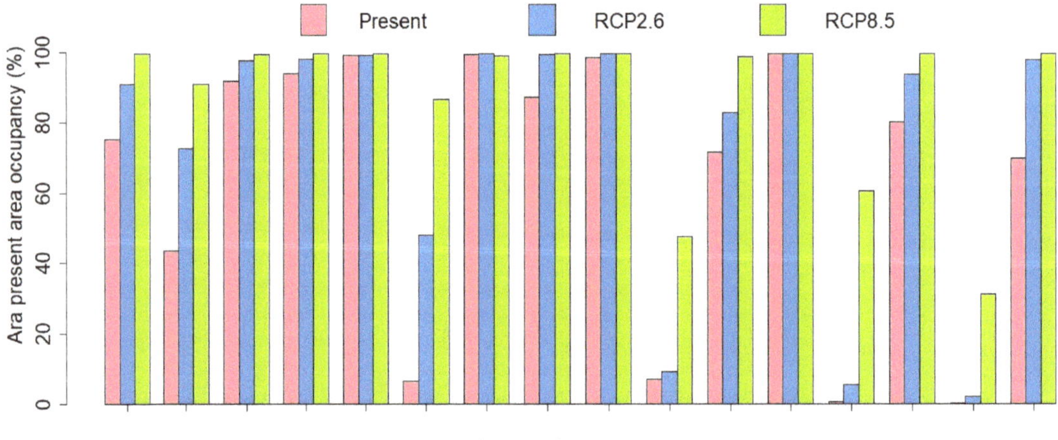

Figure 3. *Cont.*

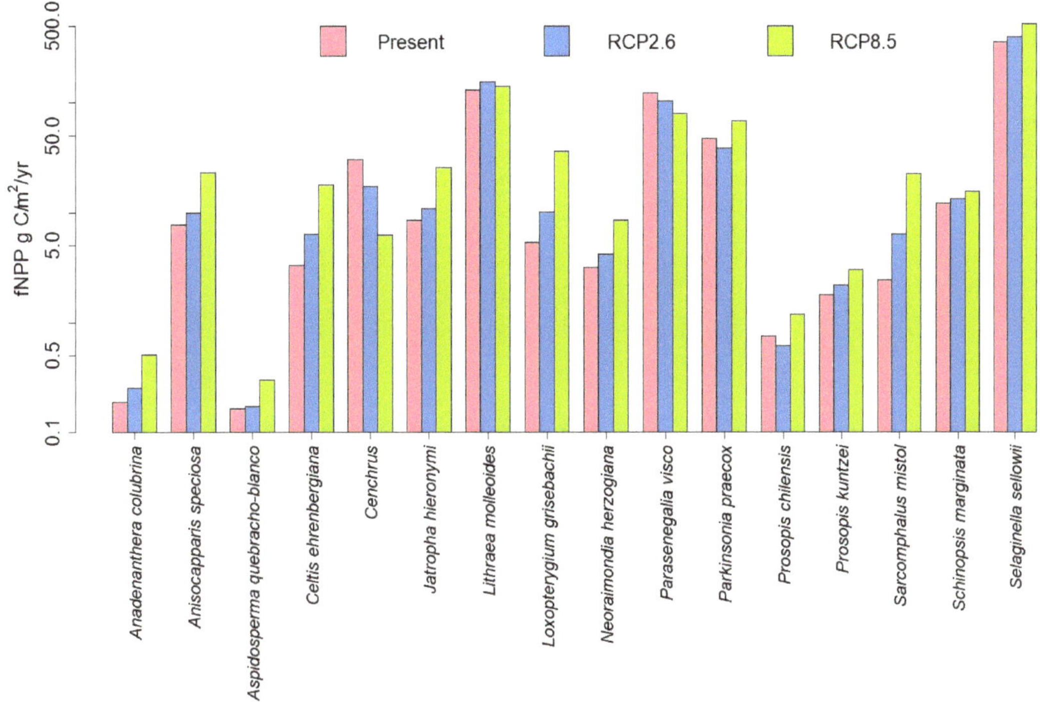

Figure 3. Area of *Ara rubrogenys* occupancy and fraction of the net primary productivity (fNPP) of the resource plant species predicted by the DVM CARAIB for present and future climates under the RCP2.6 andRCP8.5 forcings (*Cnidoscolus tubulosus* was not predicted on this area).

Table 3. Species distribution models (SDM) information: sample size for model estimation and statistics (N), maximal distance from presence for pseudo-absences (D, km), area under the receiver operating characteristic curve (AUC), true skill statistic (TSS), selected threshold (Thresh.), sensitivity (Se), specificity (Sp). p-value of likelihood ratio test was lower than 10^{-5} for each model; formula, i.e., polynomial parts of the logistic models are given in Table S2).

Species	*N*	D	AUC	TSS	Thresh.	Se	Sp
Ara rubrogenys	367	320	0.9775	0.943	0.2756	0.9623	0.9557
Anadenanthera colubrina	8173	1800	0.8584	0.6016	0.1685	0.7823	0.7830
Anisocapparis speciosa	583	1800	0.9445	0.7987	0.2124	0.8776	0.8807
Aspidosperma quebracho-blanco	1018	1800	0.9195	0.733	0.2113	0.8609	0.8591
Celtis ehrenbergiana	6495	1800	0.9008	0.6889	0.2363	0.8322	0.8322
Cenchrus	20,713	2000	0.7794	0.4257	0.2040	0.6982	0.6974
Cnidoscolus tubulosus	463	1400	0.8143	0.4527	0.3983	0.7172	0.7179
Jatropha hieronymi	246	1200	0.9292	0.7876	0.2605	0.8889	0.8802
Lithraea molleoides	4758	1600	0.9338	0.7401	0.5555	0.8698	0.8696
Loxopterygium grisebachii	221	1800	0.9762	0.872	0.2044	0.9388	0.9302
Neoraimondia herzogiana	83	1400	0.9903	0.9483	0.3278	0.9600	0.9655
Parasenegalia visco	97	800	0.9113	0.6805	0.2914	0.8276	0.8235
Parkinsonia praecox	2807	1800	0.9203	0.6987	0.1788	0.8419	0.8418
Prosopis chilensis	238	2000	0.8530	0.694	0.4665	0.8421	0.8395
Prosopis kuntzei	203	1800	0.9231	0.7857	0.3598	0.8955	0.8897
Sarcomphalus mistol	583	1600	0.9367	0.7763	0.2489	0.8800	0.8799
Schinopsis marginata	120	1400	0.9570	0.8723	0.3762	0.9211	0.9268
Selaginella sellowii	1071	1400	0.7794	0.6669	0.3496	0.8267	0.8262

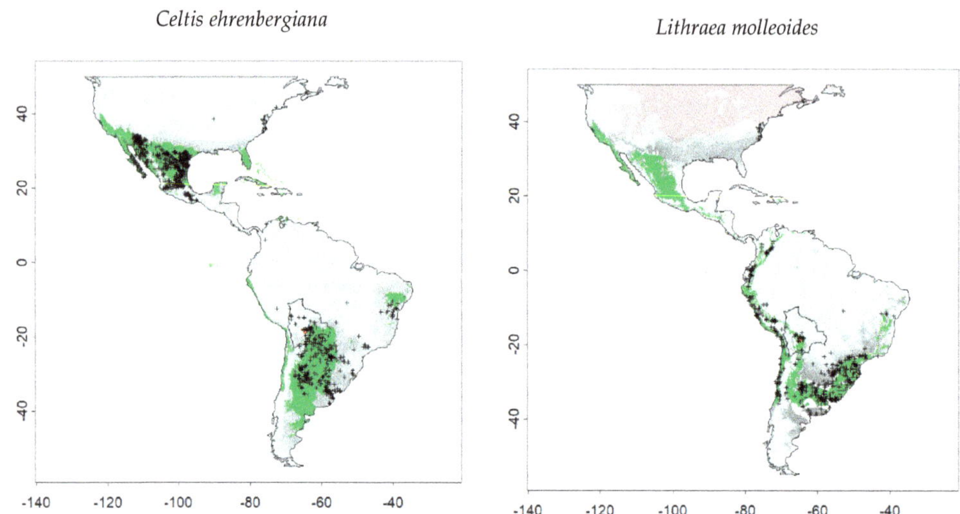

Figure 4. Examples of continental distribution of resource plant species predicted by the SDMs for the present (+: present occurrences of the species, light to dark color variation shows the probability of presence, in blue-grey between 0 and threshold of presence, in green between threshold and maximal value, pink color masks the area where at least one of the climate factors was out of the range of the model computation dataset, *Ara rubrogenys* area in orange; other plant species shown in Figure S2).

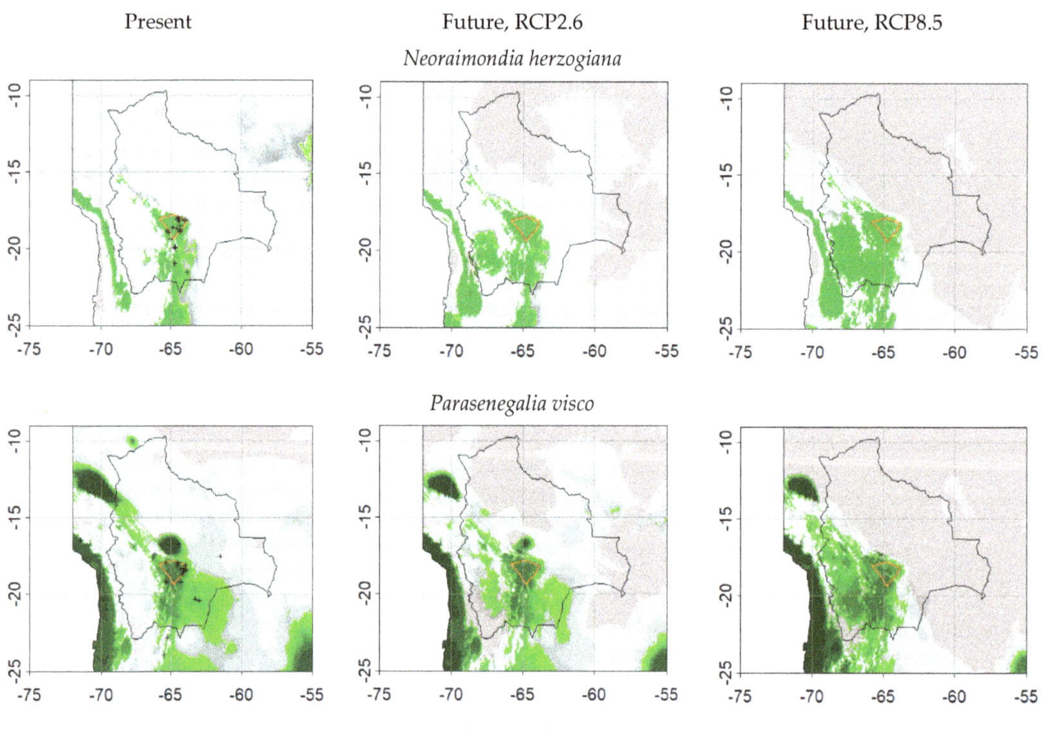

Figure 5. *Cont.*

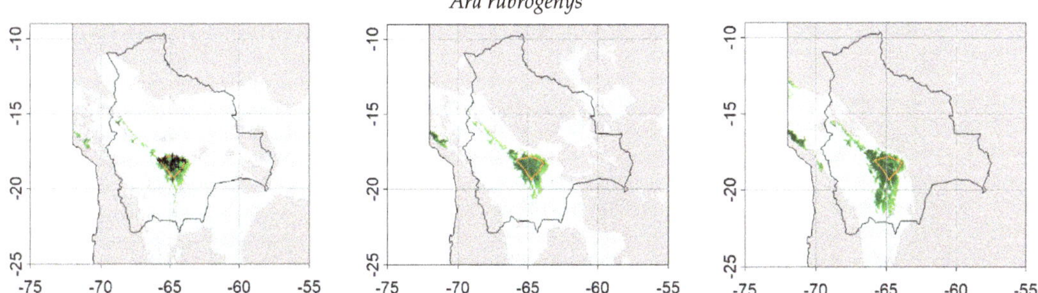

Figure 5. Examples of distributions of plant species and distribution of *Ara rubrogenys* over Bolivia predicted by SDMs for present and future climates under 2.6 and RCP8.5 forcings (+: present occurrences of the species, light to dark color variation shows the probability of presence, in blue-grey between 0 and threshold of presence, in green between threshold and maximal value, pink color masks the area where at least one of the climate factors was out of the range of the model computation dataset, *Ara rubrogenys* area delimited by orange lines; other plant species shown in Figure S3).

3.3. Comparisons of Plant Model Predictions for the Future

Under future conditions, changes predicted by the DVM over Bolivia were limited but plant presences tended rather to spread (Figures 2 and 3 and Figure S1). The trend was more pronounced over *A. rubrogenys*' area, particularly under RCP8.5 forcings while the fNPP of the pixel presence over the *A. rubrogenys* area behaved more or less similarly but with some exceptions (*Cenchrus*, *P. visco*, Figure 3). The mean number of plant species per pixel (plant diversity, Figure S4) computed for present climate was 10.3. It increased to 12.0 under RCP2.6 and 14.2 under RCP8.5 forcings. Furthermore, 80% (RCP2.6) to 96% (RCP8.5) of the pixels lost no species while a majority of them gained up to 11 new species (Figure 7). In addition, the maximal altitude increased for the majority of the species under the future climates (Figure 8).

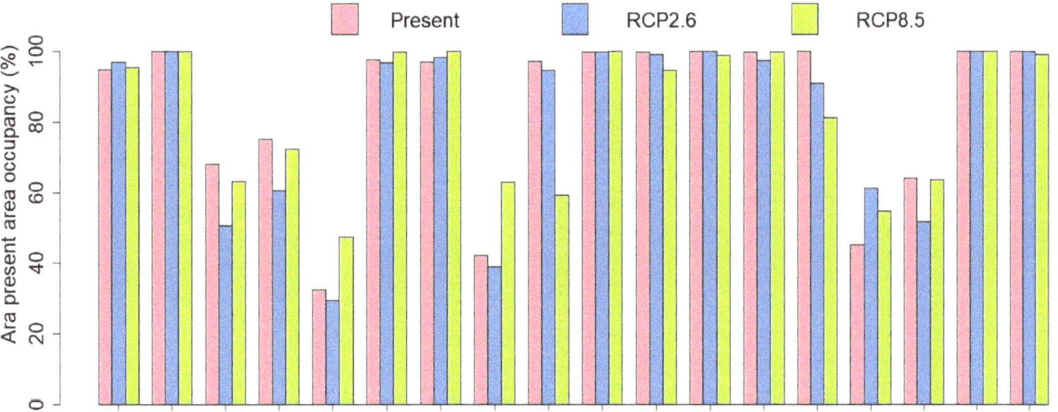

Figure 6. *Cont.*

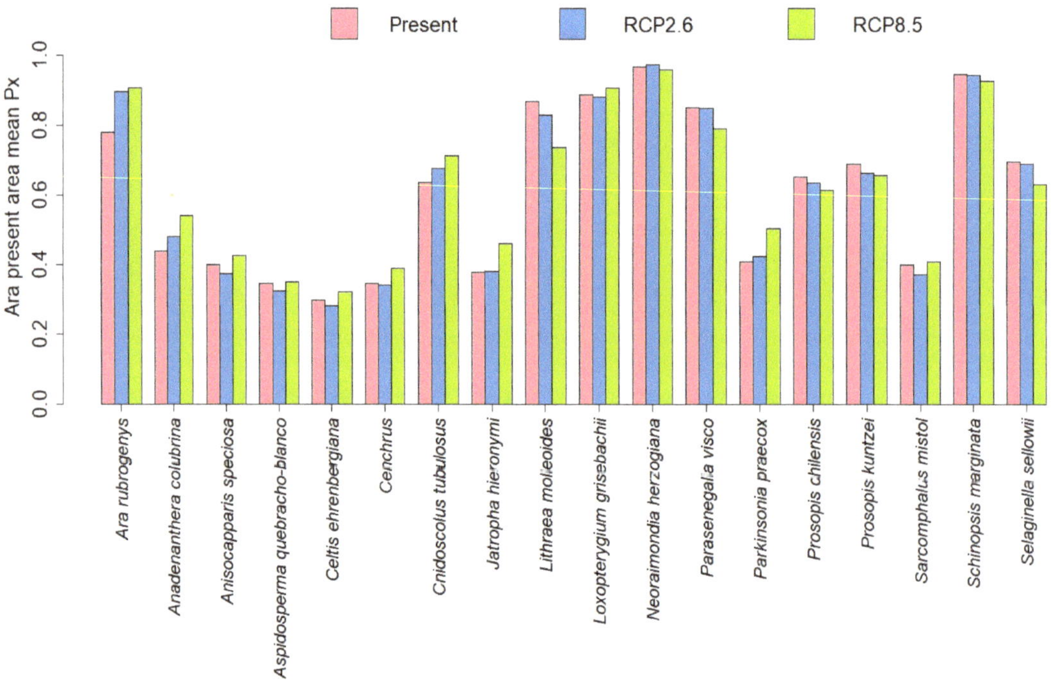

Figure 6. Area of *Ara rubrogenys* occupancy and mean probability of presence (Px) of *A. rubrogenys* and of the 18 resource plant species predicted with SDM for present and future climates under the RCP2.6 andRCP8.5 forcings.

With the SDMs, changes predicted over Bolivia were more significant than with the DVM (Figures 5 and 6 and Figure S3) but generally the presence also tended to spread. Over *A. rubrogenys*' area, no clear trend emerged for occupancy or Px. Diversity over pixels (Figure S4) for the present was higher than with the DVM (mean: 13.2) but decreased to 12.6 under RCP2.6 and to 13 under RCP8.5 forcings. Indeed, more pixels lost species while the gains were more modest (Figure 7). Furthermore, less species than with the SDM had maximal altitude increase under the future climates (Figure 8).With the SDMs, changes predicted over Bolivia were more significant than with the DVM (Figures 5 and 6 and Figure S3) but generally the presence also tended to spread. Over *A. rubrogenys*' area, no clear trend emerged for occupancy or Px. Diversity over pixels (Figure S4) for the present was higher than with the DVM (mean: 13.2) but decreased to 12.6 under RCP2.6 and to 13 under RCP8.5 forcings. Indeed, more pixels lost species while the gains were more modest (Figure 7). Furthermore, less species than with the SDM had maximal altitude increase under the future climates (Figure 8).

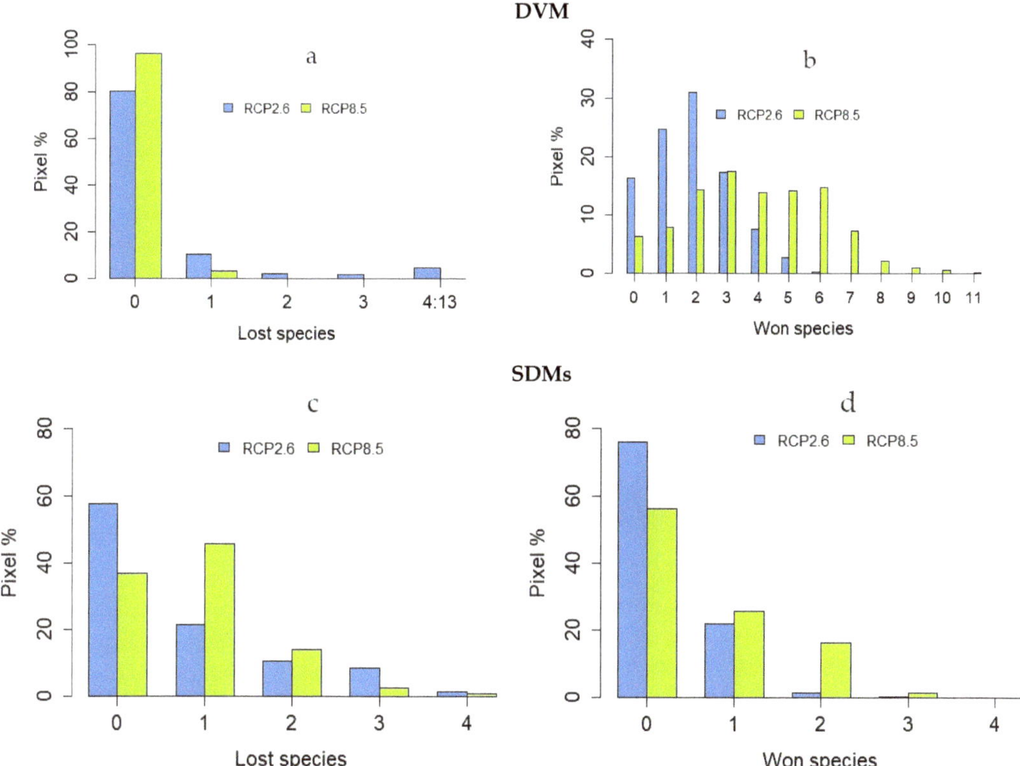

Figure 7. Proportion of pixels exhibiting a given number (0, 1, 2, etc.) of species lost (**a**,**c**) or gained (**c**,**d**) over the area of *Ara rubrogenys* under the RCP2.6 andRCP8.5 forcings compared to the present, for studies with the dynamic vegetation model (DVM: **a**,**b**) and the species distribution model (SDM: **c**,**d**).

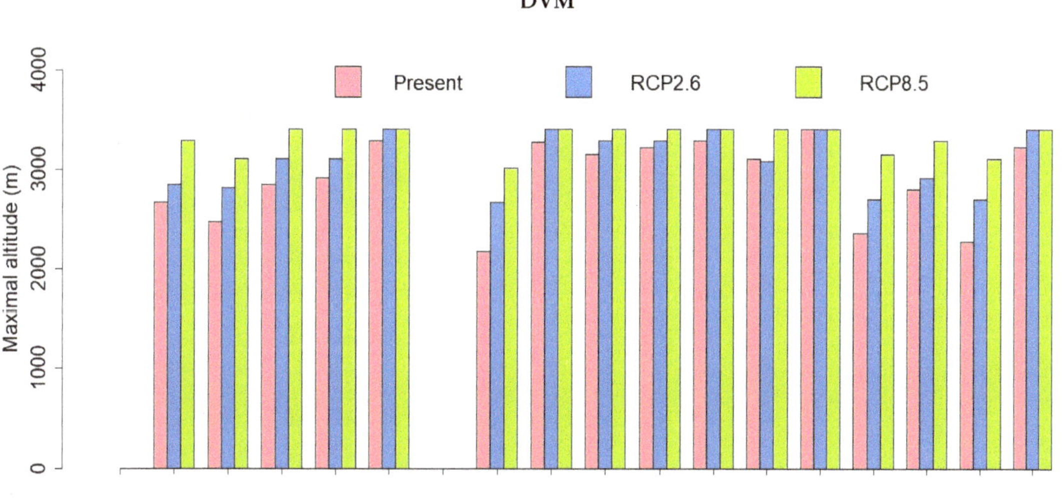

Figure 8. *Cont.*

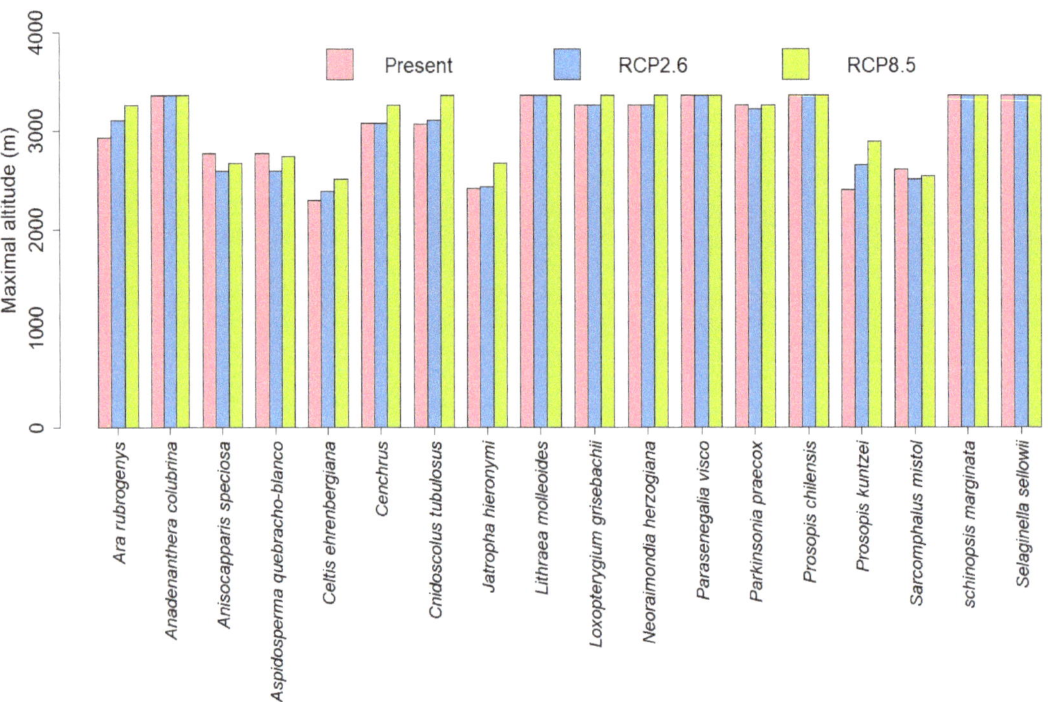

Figure 8. Maximal altitudes of the species over the area of *Ara rubrogenys* for present and future climates under the RCP2.6 andRCP8.5 forcings predicted by the dynamic vegetation model (DVM) and the species distribution model (SDM).

4. Discussion

The SDM allowed accurately modelling of *A. rubrogenys* presence with climate. Under future climate, occupancy, probability of presence and maximal altitude increased and new areas of occupancy appeared, connected with its present area. For the plants, DVM and SDM modelling allowed us to successfully compute the distribution of 17 plant species which are feeding resources of *A. rubrogenys.* Under present conditions, only one species, *C. tubulosus*, was not simulated as present over the area of *A. rubrogenys* by the DVM. SDMs gave better fits than the DVM, except for sensitivity (Se). The DVM mainly produced fNPP, and occupancy increases over the area of *A. rubrogenys*, also towards higher altitudes, under the RCP2.6 and RCP8.5 forcings. With DVM, we also obtained pixel diversity increases resulting from few species loss and more species gains. By contrast, the plant SDMs had no overall response under the future climate conditions for occupancy, probability of presence, altitude or diversity.

The lowest agreement of the DVM simulations compared to the plant SDMs were the result of lower specificity of the DVM. Here, we added an optimization step in which we selected a threshold for each species thanks to TSS. However, these thresholds were low compared to commonly fixed values. They probably reflected the fact that we used fNPP and thus the competition between species for water and light. Lowering the thresholds increases sensitivity at the expense of specificity. It was already noted in previous studies that the DVM may tend to simulate the fundamental niche of the species rather than the realized niche, in the absence of some critical biotic interactions. This could ultimately produce wider distributions compared to those produced by the SDMs as observed elsewhere [49,60]. Nevertheless, the results of both approaches, DVM and SDMs, were rather congruent for present conditions. For the future, the results were mostly in accordance with literature, in other words, the SDMs generally predict more widespread distribution shifts than DVM [24,49,61–63] while this is not absolute. Therefore, it is not surprising that in limited parts of their present area, the Px of some species computed with SDM increased under the RCP forcings. In the Andes, SDMs predicted with dramatic reductions of the Andean vascular plant diversity for the future without excluding some small areas of better climatic stability where species numbers could increase [64]. Otherwise, the DVM results showing increases in diversity and the migrations towards higher altitudes seemed in accordance with the observations from the recent past. Permanent plots in the Andes showed thermophilization, in other words, the increasing number of tropical and subtropical tree species from lower altitudes showing shifts to higher altitudes [65]. This phenomenon, was also observed on European mountain summits. However, it should finally threaten the mountains species of the former communities because they are composed of slower growing species more adapted to harsher conditions compared to species from lower altitudes [66].

It is established that birds are directly sensitive to climate factors. While one of the main advantages of body temperature regulation in homeotherms is to allow an optimal functioning of the organisms under a wide temperature range, the limit of the mechanism is clearly the dissipation of heat in excess during exercise [67]. Thus, air temperature has to be considered as an effective component of the bird's niche. Homeotherms could also be directly sensitive to other climate factors than temperature, for instance, evaporation can limit the survival of birds in deserts during heat waves [68]. Furthermore, it was showed that birds have following their realized climate niche during the course of the last century [69]. For these reasons, the use of species distribution models (SDMs) with climate factors only to predict their future distribution has to be considered as a useful tool. The common response of birds in the Andes to climate change would be a decline [64,70]. Species with restricted distribution, like endemic species, are generally characterized by narrow climate niche and they seem to resist extinction by relatively high local abundance and good demographic resilience resulting from the accumulation of local adaptions [71]. The study of species, including parrots, naturalized outside their original area of distribution indicates however that the climate tolerance could be higher than estimated in the natural area, particularly for the species occupying narrow ranges of climate conditions or marginal climates in their native region [72–74]. With the climate factor, we obtained a very well-defined geographic range (Figure 5). We supposed that the climate niche of *A. rubrogenys* would be broad compared to the other endemic species on the basis of the range of the climate factors over its present area provoked by the steepness of the climate gradient in the mountains (Table S1). Therefore, it would not be surprising that the SDM is driven by climate predicted occupancy increase of the present area under the RCP forcings and also on areas directly connected, at higher altitudes. However, this kind of situation would be rare. Another possibility would be that the area of *A. rubrogenys* is climatically stable enough to guarantee no species loss or even species gain even if those situations are rare [64]. If so, it would also explain the limited changes found for the plant species under the RCP forcings with both modelling approaches. Thus, as long as the non-climate components of the niche of *A. rubrogenys* remain available, we could venture to forecast limited risk of extinction due to climate change provided that the diversity of the plant resources could increase and that species turnover could satisfy animal needs.

While the DVM approach is more promising than the SDM one for evaluating plant future, DVMs are nevertheless perfectible. DVMs need to elaborate an optimal strategy for parametrization [45,48,75]. It is indeed limited by the knowledge of the species-specific traits but also by the response of those traits to environmental factors and results could be improved by determining those responses and integrating them into the DVM [76–78]. The collection of plant traits with biogeographic information is a fast-growing field of knowledge [79]. Since the DVM simulates competition for light and water resources, another interesting point could be to integrate the species of the former communities with the potential newcomers to compute the emerging communities. The newcomers, in addition, could also constitute potential resources.

5. Conclusions

Under the milder RCP2.6 and the harsher RCP8.5 forcings, the SDM for *A. rubrogenys* and SDMs and the DVM for plant feeding resources of *A. rubrogenys* showed that the current area of suitability would be mostly preserved or even broadened. The actual feeding resources could be more precisely evaluated by taking into account the land-use and the productivity of the agricultural species which constitute important feeding resources for *A. rubrogenys* [6,25,26]. Nevertheless, predictions of the evolution of those resources in the course of time would require additional effort since uncertainties increase with the characteristics of the cultivated varieties and the decisions of the landowners or the farmers. The use of the DVM for predicting the future of plant species distribution and productivity is thus improvable. Our result boded well because *A. rubrogenys* would not be subjected to unprecedented climate conditions in the present area and the feeding resources so far would be preserved.

From topology, it could be possible to examine whether new suitable areas include cliffs appropriate for nesting. However, is it possible that the new suitable areas could be colonized even without appropriate cliffs thanks to behavioral flexibility? For instance, *A. rubrogenys* have been observed nesting in the palm *Parajubaea torallyi* [80] but, this tree species has an extremely small range distribution, and there are no other tree species that could offer large cavities for nesting within the range of the red-fronted macaw. Moreover, recent population genetic analyses show a low capacity for the species to colonize distant areas. Despite its restricted range, the population is structured in genetic clusters with low or null gene flow among them, despite that some colonies are separated by few tens of kilometers, that there are no ecological barriers and that macaws make larger daily and seasonal movements for feeding. Therefore, it is highly improbable that the species could colonize very distant areas in its future niche suitability [81].

Thus, while bird decline at global scale seems mainly provoked by increasing temperature [82], the main concerns about *A. rubrogenys*' future remained the direct human threats, in other words, habitat conversion, poaching and killing, which need to be urgently solved before the species rapidly goes extinct in nature [6,11,12].

Supplementary Materials: The following are available online at https://www.mdpi.com/1424-2818/13/2/94/s1, Table S1: Bioclimate variables over *Ara rubrogenys* area and changes under the RCP2.6 and RCP8.5 forcings, Table S2: Coefficients of the climate logistic models for *Ara rubrogenys* and the 17 plant species, Figure S1: Plant species distribution over Bolivia predicted by the DVM CARAIB for present and future under RCP2.6 andRCP8.5 forcings, Figure S2: Plant species continental distribution predicted by the SDMs for the present, Figure S3: Plant species distribution over predicted by the SDM for present and future under RCP2.6 andRCP8.5 forcings, Figure S4: Number of plant species over the area of *Ara rubrogenys* computed with the DVM or the SDMs for present and future climate under RCP2.6 and RCP8.5 forcings.

Author Contributions: Conceptualization, methodology and investigation, A.H. and L.F.; investigations, E.R.L., S.d.H. and L.Z.Z.; data curation, L.Z.Z., writing—original draft preparation and editing, A.H.; review, L.F. All authors have read and agreed to the published version of the manuscript.

Funding: This research received no external funding.

Institutional Review Board Statement: Not applicable.

Data Availability Statement: Information on data from publicly archived datasets used in this study is given in the Section 2.

Acknowledgments: We thank the reviewers for their comments and suggestions and their sharing of information on the red-fronted macaw which allowed us to improve our study.

Conflicts of Interest: The authors declare no conflict of interest. The funders had no role in the design of the study; in the collection, analyses, or interpretation of data; in the writing of the manuscript, or in the decision to publish the results.

References

1. Şekercioğlu Çağan, H.; Primack, R.B.; Wormworth, J. The effects of climate change on tropical birds. *Biol. Conserv.* **2012**, *148*, 1–18. [CrossRef]
2. La Sorte, F.A.; Jetz, W. Tracking of climatic niche boundaries under recent climate change. *J. Anim. Ecol.* **2012**, *81*, 914–925. [CrossRef]
3. Orme, C.D.L.; Davies, R.G.; Burgess, M.; Eigenbrod, F.; Pickup, N.; Olson, V.A.; Webster, A.J.; Ding, T.-S.; Rasmussen, P.C.; Ridgely, R.S.; et al. Global hotspots of species richness are not congruent with endemism or threat. *Nat. Cell Biol.* **2005**, *436*, 1016–1019. [CrossRef]
4. Herzog, S.K.; Maillard, O.; Embert, D.; Caballero, P.; Quiroga, D. Range size estimates of Bolivian endemic bird species revisited: The importance of environmental data and national expert knowledge. *J. Ornithol.* **2012**, *153*, 1189–1202. [CrossRef]
5. Lanning, D.V. Distribution and breeding biology of the Red-fronted Macaw. *Wilson Bull.* **1991**, *103*, 357–365.
6. Tella, J.; Rojas, A.; Carrete, M.; Hiraldo, F. Simple assessments of age and spatial population structure can aid conservation of poorly known species. *Biol. Conserv.* **2013**, *167*, 425–434. [CrossRef]
7. Herzog, S.K.; Kessler, M.; Maijer, S.; Hohnwald, S. Distributional notes on birds of Andean dry forests in Bolivia. *Bull. Br. Ornithol. Club* **1997**, *117*, 223–235.
8. Boussekey, M.; Saint-Pie, J.; Morvan, O. Observations on a population of Red-fronted Macaws Ara rubrogenys in the Río Caine valley, central Bolivia. *Bird Conserv. Int.* **1991**, *1*, 335–350. [CrossRef]
9. Pitter, E.; Christiansen, M.B. Ecology, status and conservation of the Red-fronted Macaw Ara rubrogenys. *Bird Conserv. Int.* **1995**, *5*, 61–78. [CrossRef]
10. Herrera, M.; Hennessey, B. Quantifying the illegal parrot trade in Santa Cruz de la Sierra, Bolivia, with emphasis on threatened species. *Bird Conserv. Int.* **2007**, *17*, 295–300. [CrossRef]
11. Rojas, A.; Zeballos, A.; Rocha, E.; Balderrama, J.A. *Ara rubrogenys Lafrenaye 1847*; Ministerio de Medio Ambiente y Agua: La Paz, Bolivia, 2009; pp. 332–334.
12. BirdLife International. *Ara rubrogenys. The IUCN Red List of Threatened Species*. 2018. Available online: https://dx.doi.org/10.2305/IUCN.UK.2018-2.RLTS.T22685572A131382876.en (accessed on 21 September 2020).
13. Rangecroft, S.; Harrison, S.; Anderson, K.; Magrath, J.; Castel, A.P.; Pacheco, P. Climate Change and Water Resources in Arid Mountains: An Example from the Bolivian Andes. *Ambio* **2013**, *42*, 852–863. [CrossRef]
14. Botero-Delgadillo, E.; Páez, C.A.; Bayly, N. Biogeography and conservation of Andean and Trans-Andean populations of Pyrrhura parakeets in Colombia: Modelling geographic distributions to identify independent conservation units. *Bird Conserv. Int.* **2012**, *22*, 445–461. [CrossRef]
15. Ferrer-Paris, J.R.; Sánchez-Mercado, A.; Rodríguez-Clark, K.M.; Rodríguez, J.P.; Rodríguez, G.A. Using limited data to detect changes in species distributions: Insights from Amazon parrots in Venezuela. *Biol. Conserv.* **2014**, *173*, 133–143. [CrossRef]
16. Marini, M.Â.; Barbet-Massin, M.; Martinez, J.; Prestes, N.P.; Jiguet, F. Applying ecological niche modelling to plan conservation actions for the Red-spectacled Amazon (*Amazona pretrei*). *Biol. Conserv.* **2010**, *143*, 102–112. [CrossRef]
17. Cardador, L.; Lattuada, M.; Strubbe, D.; Tella, J.L.; Reino, L.; Figueira, R.; Carrete, M. Regional Bans on Wild-Bird Trade Modify Invasion Risks at a Global Scale. *Conserv. Lett.* **2017**, *10*, 717–725. [CrossRef]
18. López-López, P.; García-Ripollés, C.; Aguilar, J.M.; Garcia-López, F.; Verdejo, J. Modelling breeding habitat preferences of Bonelli's eagle (Hieraaetus fasciatus) in relation to topography, disturbance, climate and land use at different spatial scales. *J. Ornithol.* **2005**, *147*, 97–106. [CrossRef]
19. Preston, K.L.; Rotenberry, J.T.; Redak, R.A.; Allen, M.F. Habitat shifts of endangered species under altered climate conditions: Importance of biotic interactions. *Glob. Chang. Biol.* **2008**, *14*, 2501–2515. [CrossRef]
20. Pidgeon, A.M.; Rivera, L.; Martinuzzi, S.; Politi, N.; Bateman, B.L. Will representation targets based on area protect critical resources for the conservation of the Tucuman Parrot? *Condor* **2015**, *117*, 503–517. [CrossRef]
21. Fitter, A.; Hay, R. *Environmental Physiology of Plants*; Elsevier: Amsterdam, The Netherlands, 2002; p. 367.
22. Smith, N.G.; Dukes, J.S. Plant respiration and photosynthesis in global-scale models: Incorporating acclimation to temperature and CO. *Glob. Chang. Biol.* **2012**, *19*, 45–63. [CrossRef]
23. Dury, M.; Hambuckers, A.; Warnant, P.; Henrot, A.; Favre, E.; Ouberdous, M.; François, L. Responses of European forest ecosystems to 21st century climate: Assessing changes in interannual variability and fire intensity. *iForest Biogeosci. For.* **2011**, *4*, 82–99. [CrossRef]
24. Raghunathan, N.; François, L.; Huynen, M.-C.; Oliveira, L.C.; Hambuckers, A. Modelling the distribution of key tree species used by lion tamarins in the Brazilian Atlantic forest under a scenario of future climate change. *Reg. Environ. Chang.* **2014**, *15*, 683–693. [CrossRef]
25. Blanco, G.; Hiraldo, F.; Rojas, A.; Dénes, F.V.; Tella, J.L. Parrots as key multilinkers in ecosystem structure and functioning. *Ecol. Evol.* **2015**, *5*, 4141–4160. [CrossRef]

26. Montesinos-Navarro, A.; Hiraldo, F.; Tella, J.L.; Blanco, G. Network structure embracing mutualism–antagonism continuums increases community robustness. *Nat. Ecol. Evol.* **2017**, *1*, 1661–1669. [CrossRef]
27. Burgman, M.A.; Fox, J.C. Bias in species range estimates from minimum convex polygons: Implications for conservation and options for improved planning. *Anim. Conserv.* **2003**, *6*, 19–28. [CrossRef]
28. Pateiro-López, B.; Rodríguez-Casal, A. Generalizing the Convex Hull of a Sample: TheRPackagealphahull. *J. Stat. Softw.* **2010**, *34*, 1–28. [CrossRef]
29. Fick, S.E.; Hijmans, R.J. WorldClim 2: New 1-km spatial resolution climate surfaces for global land areas. *Int. J. Climatol.* **2017**, *37*, 4302–4315. [CrossRef]
30. The HadGEM2 Development Team; Martin, G.M.; Bellouin, N.; Collins, W.J.; Culverwell, I.D.; Halloran, P.R.; Hardiman, S.C.; Hinton, T.J.; Jones, C.D.; McDonald, R.E.; et al. The HadGEM2 family of Met Office Unified Model climate configurations. *Geosci. Model Dev.* **2011**, *4*, 723–757. [CrossRef]
31. Moss, R.H.; Edmonds, J.A.; Hibbard, K.A.; Manning, M.R.; Rose, S.K.; Van Vuuren, D.P.; Carter, T.R.; Emori, S.; Kainuma, M.; Kram, T.; et al. The next generation of scenarios for climate change research and assessment. *Nature* **2010**, *463*, 747–756. [CrossRef] [PubMed]
32. Stocker, T.F.; Qin, D.; Plattner, G.-K.; Tignor, M.; Allen, S.K.; Boschung, J.; Nauels, A.; Xia, Y.; Bex, V.; Midgley, P.M. *Climate Change 2013: The Physical Science Basis; Working Group I Contribution to the Fifth Assessment Report of the Intergovernmental Panel on Climate Change;* Cambridge University Press: Cambridge, UK; New York, NY, USA, 2013. [CrossRef]
33. Warnant, P.; François, L.; Strivay, D.; Gérard, J.-C. CARAIB: A global model of terrestrial biological productivity. *Glob. Biogeochem. Cycles* **1994**, *8*, 255–270. [CrossRef]
34. Gérard, J.C.; Nemry, B.; François, L.M.; Warnant, P. The interannual change of atmospheric CO2: Contribution of subtropical ecosystems? *Geophys. Res. Lett.* **1999**, *26*, 243–246. [CrossRef]
35. Otto, D.; Rasse, D.; Kaplan, J.; Warnant, P.; François, L. Biospheric carbon stocks reconstructed at the Last Glacial Maximum: Comparison between general circulation models using prescribed and computed sea surface temperatures. *Glob. Planet. Chang.* **2002**, *33*, 117–138. [CrossRef]
36. Laurent, J.; Francois, L.; Barhen, A.; Bel, L.; Cheddadi, R. European bioclimatic affinity groups: Data-model comparisons. *Glob. Planet. Chang.* **2008**, *61*, 28–40. [CrossRef]
37. François, L.; Goddéris, Y.; Munhoven, G. Modelling the interactions between biospheric and weathering processes: Towards a mechanistic description of the land environment. *Mineral. Mag.* **1998**, *62A*, 468–469. [CrossRef]
38. François, L.M.; Goddéris, Y.; Warnant, P.; Ramstein, G.; De Noblet, N.; Lorenz, S. Carbon stocks and isotopic budgets of the terrestrial biosphere at mid-Holocene and last glacial maximum times. *Chem. Geol.* **1999**, *159*, 163–189. [CrossRef]
39. François, L.; Ghislain, M.; Otto, D.; Micheels, A. Late Miocene vegetation reconstruction with the CARAIB model. *Palaeogeogr. Palaeoclim. Palaeoecol.* **2006**, *238*, 302–320. [CrossRef]
40. François, L.; Utescher, T.; Favre, E.; Henrot, A.-J.; Warnant, P.; Micheels, A.; Erdei, B.; Suc, J.-P.; Cheddadi, R.; Mosbrugger, V. Modelling Late Miocene vegetation in Europe: Results of the CARAIB model and comparison with palaeovegetation data. *Palaeogeogr. Palaeoclim. Palaeoecol.* **2011**, *304*, 359–378. [CrossRef]
41. Henrot, A.-J.; François, L.; Favre, E.; Butzin, M.; Ouberdous, M.; Munhoven, G. Effects of CO2, continental distribution, topography and vegetation changes on the climate at the Middle Miocene: A model study. *Clim. Past* **2010**, *6*, 675–694. [CrossRef]
42. Henrot, A.-J.; Utescher, T.; Erdei, B.; Dury, M.; Hamon, N.; Ramstein, G.; Krapp, M.; Herold, N.; Goldner, A.; Favre, E.; et al. Middle Miocene climate and vegetation models and their validation with proxy data. *Palaeogeogr. Palaeoclim. Palaeoecol.* **2017**, *467*, 95–119. [CrossRef]
43. Chang, J.; Ciais, P.; Wang, X.; Piao, S.; Asrar, G.; Betts, R.; Chevallier, F.; Dury, M.; François, L.; Frieler, K.; et al. Benchmarking carbon fluxes of the ISIMIP2a biome models. *Environ. Res. Lett.* **2017**, *12*, 045002. [CrossRef]
44. Fontaine, C.M.; Dendoncker, N.; De Vreese, R.; Jacquemin, I.; Marek, A.; Van Herzeleg, A.; Devillet, G.; Mortelmans, D.; François, L. Towards participatory integrated valuation and modelling of ecosystem services under land-use change. *J. Land Use Sci.* **2014**, *9*, 278–303. [CrossRef]
45. Minet, J.; Laloy, E.; Tychon, B.; Francois, L. Bayesian inversions of a dynamic vegetation model at four European grassland sites. *Biogeosciences* **2015**, *12*, 2809–2829. [CrossRef]
46. Fronzek, S.; Pirttioja, N.; Carter, T.R.; Bindi, M.; Hoffmann, H.; Palosuo, T.; Ruiz-Ramos, M.; Tao, F.; Trnka, M.; Acutis, M.; et al. Classifying multi-model wheat yield impact response surfaces showing sensitivity to temperature and precipitation change. *Agric. Syst.* **2018**, *159*, 209–224. [CrossRef]
47. Cheddadi, R.; Henrot, A.-J.; François, L.; Boyer, F.; Bush, M.; Carré, M.; Coissac, E.; De Oliveira, P.E.; Ficetola, F.; Hambuckers, A.; et al. Microrefugia, Climate Change, and Conservation of Cedrus atlantica in the Rif Mountains, Morocco. *Front. Ecol. Evol.* **2017**, *5*. [CrossRef]
48. Dury, M.; Mertens, L.; Fayolle, A.; Verbeeck, H.; Hambuckers, A.; François, L. Refining Species Traits in a Dynamic Vegetation Model to Project the Impacts of Climate Change on Tropical Trees in Central Africa. *Forests* **2018**, *9*, 722. [CrossRef]
49. Raghunathan, N.; François, L.; Dury, M.; Hambuckers, A. Contrasting climate risks predicted by dynamic vegetation and ecological niche-based models applied to tree species in the Brazilian Atlantic Forest. *Reg. Environ. Chang.* **2018**, *19*, 219–232. [CrossRef]

50. Allouche, O.; Tsoar, A.; Kadmon, R. Assessing the accuracy of species distribution models: Prevalence, kappa and the true skill statistic (TSS). *J. Appl. Ecol.* **2006**, *43*, 1223–1232. [CrossRef]
51. Segurado, P.; Araújo, M.B. An evaluation of methods for modelling species distributions. *J. Biogeogr.* **2004**, *31*, 1555–1568. [CrossRef]
52. Meynard, C.N.; Quinn, J.F. Predicting species distributions: A critical comparison of the most common statistical models using artificial species. *J. Biogeogr.* **2007**, *34*, 1455–1469. [CrossRef]
53. Bedia, J.; Busqué, J.; Gutiérrez, J.M. Predicting plant species distribution across an alpine rangeland in northern Spain. A comparison of probabilistic methods. *Appl. Veg. Sci.* **2011**, *14*, 415–432. [CrossRef]
54. King, G.; Zeng, L.; Tomz, M. Logistic Regression in Rare Events Data. *J. Stat. Softw.* **2003**, *8*, 1–27. [CrossRef]
55. Elith, J.; Graham, C.H.; Anderson, R.P.; Dudík, M.; Ferrier, S.; Guisan, A.; Hijmans, R.J.; Huettmann, F.; Leathwick, J.R.; Lehmann, A.; et al. Novel methods improve prediction of species' distributions from occurrence data. *Ecography* **2006**, *29*, 129–151. [CrossRef]
56. Calcagno, V.; De Mazancourt, C. glmulti: AnRPackage for Easy Automated Model Selection with (Generalized) Linear Models. *J. Stat. Softw.* **2010**, *34*, 1–29. [CrossRef]
57. Vanderwal, J.; Shoo, L.P.; Graham, C.; Williams, S.E. Selecting pseudo-absence data for presence-only distribution modeling: How far should you stray from what you know? *Ecol. Model.* **2009**, *220*, 589–594. [CrossRef]
58. VanDerWal, J.; Lorena Falconi, L.; Januchowski, S.; Shoo, L.; Storlie, C. Species Distribution Modelling Tools: Tools for Pro-Cessing Data Associated with Species Distribution Modelling Exercises. 2014. Available online: https://www.rforge.net/SDMTools/ (accessed on 26 November 2020).
59. Dormann, C.F.; Elith, J.; Bacher, S.; Buchmann, C.; Carl, G.; Carré, G.; Marquéz, J.R.G.; Gruber, B.; Lafourcade, B.; Leitão, P.J.; et al. Collinearity: A review of methods to deal with it and a simulation study evaluating their performance. *Ecography* **2012**, *36*, 27–46. [CrossRef]
60. Bolliger, J.; Kienast, F.; Bugmann, H. Comparing models for tree distributions: Concept, structures, and behavior. *Ecol. Model.* **2000**, *134*, 89–102. [CrossRef]
61. Morin, X.; Thuiller, W. Comparing niche- and process-based models to reduce prediction uncertainty in species range shifts under climate change. *Ecolology* **2009**, *90*, 1301–1313. [CrossRef]
62. Cheaib, A.; Badeau, V.; Boe, J.; Chuine, I.; Delire, C.; Dufrêne, E.; François, C.; Gritti, E.S.; Legay, M.; Pagé, C.; et al. Climate change impacts on tree ranges: Model intercomparison facilitates understanding and quantification of uncertainty. *Ecol. Lett.* **2012**, *15*, 533–544. [CrossRef]
63. Takolander, A.; Hickler, T.; Meller, L.; Cabeza, M. Comparing future shifts in tree species distributions across Europe projected by statistical and dynamic process-based models. *Reg. Environ. Chang.* **2018**, *19*, 251–266. [CrossRef]
64. Ramirez-Villegas, J.; Cuesta, F.; Devenish, C.; Peralvo, M.; Jarvis, A.; Arnillas, C.A. Using species distributions models for designing conservation strategies of Tropical Andean biodiversity under climate change. *J. Nat. Conserv.* **2014**, *22*, 391–404. [CrossRef]
65. Fadrique, B.; Báez, S.; Duque, Á.; Malizia, A.; Blundo, C.; Carilla, J.; Osinaga-Acosta, O.; Malizia, L.; Silman, M.; Farfán-Ríos, W.; et al. Widespread but heterogeneous responses of Andean forests to climate change. *Nat. Cell Biol.* **2018**, *564*, 207–212. [CrossRef]
66. Steinbauer, M.J.; Grytnes, J.-A.; Jurasinski, G.; Kulonen, A.; Lenoir, J.; Pauli, H.; Rixen, C.; Winkler, M.; Bardy-Durchhalter, M.; Barni, E.; et al. Accelerated increase in plant species richness on mountain summits is linked to warming. *Nat. Cell Biol.* **2018**, *556*, 231–234. [CrossRef]
67. Heinrich, B. Why Have Some Animals Evolved to Regulate a High Body Temperature? *Am. Nat.* **1977**, *111*, 623–640. [CrossRef]
68. McKechnie, A.E.; Wolf, B.O. Climate change increases the likelihood of catastrophic avian mortality events during extreme heat waves. *Biol. Lett.* **2009**, *6*, 253–256. [CrossRef]
69. Tingley, M.W.; Monahan, W.B.; Beissinger, S.R.; Moritz, C. Birds track their Grinnellian niche through a century of climate change. *Proc. Natl. Acad. Sci. USA* **2009**, *106*, 19637–19643. [CrossRef]
70. Avalos, V.D.R.; Hernández, J. Projected distribution shifts and protected area coverage of range-restricted Andean birds under climate change. *Glob. Ecol. Conserv.* **2015**, *4*, 459–469. [CrossRef]
71. Williams, S.; Williams, Y.M.; Vanderwal, J.; Isaac, J.L.; Shoo, L.P.; Johnson, C.N. Ecological specialization and population size in a biodiversity hotspot: How rare species avoid extinction. *Proc. Natl. Acad. Sci. USA* **2009**, *106*, 19737–19741. [CrossRef] [PubMed]
72. Early, R.; Sax, D.F. Climatic niche shifts between species' native and naturalized ranges raise concern for ecological forecasts during invasions and climate change. *Glob. Ecol. Biogeogr.* **2014**, *23*, 1356–1365. [CrossRef]
73. Bocsi, T.; Allen, J.M.; Bellemare, J.; Kartesz, J.; Nishino, M.; Bradley, B.A. Plants' native distributions do not reflect climatic tolerance. *Divers. Distrib.* **2016**, *22*, 615–624. [CrossRef]
74. Abellán, P.; Tella, J.L.; Carrete, M.; Cardador, L.; Anadón, J.D. Climate matching drives spread rate but not establishment success in recent unintentional bird introductions. *Proc. Natl. Acad. Sci. USA* **2017**, *114*, 9385–9390. [CrossRef]
75. Hartig, F.; Dyke, J.G.; Hickler, T.; Higgins, S.I.; O'Hara, R.B.; Scheiter, S.; Huth, A. Connecting dynamic vegetation models to data—An inverse perspective. *J. Biogeogr.* **2012**, *39*, 2240–2252. [CrossRef]
76. Bloomfield, K.J.; Cernusak, L.A.; Eamus, D.; Ellsworth, D.S.; Prentice, I.C.; Wright, I.J.; Boer, M.M.; Bradford, M.G.; Cale, P.; Cleverly, J.; et al. A continental-scale assessment of variability in leaf traits: Within species, across sites and between seasons. *Funct. Ecol.* **2018**, *32*, 1492–1506. [CrossRef]

77. Gillison, A.N. Plant functional indicators of vegetation response to climate change, past present and future: I. Trends, emerging hypotheses and plant functional modality. *Flora Morphol. Distrib. Funct. Ecol. Plants* **2019**, *254*, 12–30. [CrossRef]
78. Modeling past plant species' distributions in mountainous areas: A way to improve our knowledge of future climate change impacts? *Past Glob. Chang. Mag.* **2020**, *28*, 16–17. [CrossRef]
79. Kattge, J.; Bönisch, G.; Díaz, S.; Lavorel, S.; Prentice, I.C.; Leadley, P.; Tautenhahn, S.; Werner, G.D.A.; Aakala, T.; Abedi, M.; et al. TRY plant trait database – enhanced coverage and open access. *Glob. Chang. Biol.* **2019**, *26*, 119–188. [CrossRef]
80. Rojas, A.; Yucra, E.; Vera, I.; Requejo, A.; Tella, J. A new population of the globally Endangered Red-fronted Macaw Ara rubrogenys unusually breeding in palms. *Bird Conserv. Int.* **2012**, *24*, 389–392. [CrossRef]
81. Blanco, G.; Morinha, F.; Roques, S.; Hiraldo, F.; Rojas, A.; Tella, J.L. Fine-scale genetic structure in the critically endangered red-fronted macaw in the absence of geographic and ecological barriers. *Sci. Rep.* **2021**, *11*, 1–17. [CrossRef] [PubMed]
82. Spooner, F.E.B.; Pearson, R.G.; Freeman, R. Rapid warming is associated with population decline among terrestrial birds and mammals globally. *Glob. Chang. Biol.* **2018**, *24*, 4521–4531. [CrossRef]

Article

Endemic and Threatened *Amazona* Parrots of the Atlantic Forest: An Overview of Their Geographic Range and Population Size

Viviane Zulian [1,*], David A. W. Miller [2] and Gonçalo Ferraz [1]

[1] Programa de Pós-Graduação em Ecologia, Instituto de Biociências, Universidade Federal do Rio Grande do Sul, CP 15007, Porto Alegre 91501-970, RS, Brazil; goncalo.ferraz@ufrgs.br
[2] Department of Ecosystem Science and Management, Pennsylvania State University, 411 Forest Resources Building, University Park, State College, PA 16802, USA; dxm84@psu.edu
* Correspondence: zulian.vi@gmail.com

Abstract: *Amazona* is the largest genus of the Psittacidae, one of the most threatened bird families. Here, we study four species of *Amazona* (*Amazona brasiliensis*, *A. pretrei*, *A. vinacea*, and *A. rhodocorytha*) that are dependent on a highly vulnerable biome: the Brazilian Atlantic Forest. To examine their distribution and abundance, we compile abundance estimates and counts, and develop site-occupancy models of their geographic range. These models integrate data from formal research and citizen science platforms to estimate probabilistic maps of the species' occurrence throughout their range. Estimated range areas varied from 15,000 km^2 for *A. brasiliensis* to more than 400,000 km^2 for *A. vinacea*. While *A. vinacea* is the only species with a statistical estimate of abundance (~8000 individuals), *A. pretrei* has the longest time series of roost counts, and *A. rhodocorytha* has the least information about population size. The highest number of individuals counted in one year was for *A. pretrei* (~20,000), followed by *A. brasiliensis* (~9000). Continued modeling of research and citizen science data, matched with collaborative designed surveys that count parrots at their non-breeding roosts, are essential for an appropriate assessment of the species' status, as well as for examining the outcome of conservation actions.

Keywords: *Amazona*; Psittacidae; species distribution models; data integration models; occupancy models; citizen-science; population size; count data

Citation: Zulian, V.; Miller, D.A.W.; Ferraz, G. Endemic and Threatened *Amazona* Parrots of the Atlantic Forest: An Overview of Their Geographic Range and Population Size. *Diversity* **2021**, *13*, 416. https://doi.org/10.3390/d13090416

Academic Editor: Juan Carlos Illera

Received: 4 June 2021
Accepted: 23 August 2021
Published: 30 August 2021

1. Introduction

Three hundred and ninety-five species of parrots, macaws, and parakeets constitute the Psittacidae family, the largest non-passerine bird family in the world [1]. With 27% (108) of its species threatened with extinction [1], the Psittacidae is the bird family with the highest absolute number of threatened species, that is, species classified as 'vulnerable', 'endangered', 'critically endangered', or 'extinct in the wild', by the International Union for Conservation of Nature and Natural Resources (IUCN). In proportional terms, the Psittacidae come only after the much smaller families of albatrosses and cranes with, respectively, 68% and 66% of their species threatened. Habitat loss and nest poaching are two key factors endangering Psittacidae populations [2,3]. Being dependent on forest habitats, most Psittacidae species require natural cavities to nest [3] and are thus directly impacted by forest clearance [2] and selective logging [4], caused primarily by agro-industrial expansion [5,6]. Nest poaching disproportionately affects species that are colorful, with large body size, relative ease of capture, and that sell for the highest prices [7,8].

The most diverse genus among the Psittacidae is the neotropical genus *Amazona*, or Amazon parrots, with 36 species distributed from northern Argentina to northern Mexico [1]. One half (18) of the *Amazona* species are globally threatened, and 25 species have decreasing population sizes, according to the IUCN Red List [1]. Nest poaching has been reported by Wright et al. [8] as the main cause of mortality in four species: *A. vinacea*, *A. kawalli*, *A. ochrocephala*, and *A. auropalliata*. Habitat loss is also a threat to the genus, especially in those biomes that have been more subjected to deforestation, such as the Atlantic Forest of Brazil. Home to seven *Amazona* species [9], the Atlantic Forest is the second largest rainforest in South America [10,11] and is a global biodiversity hotspot [12]. The biome has lost almost 90% of its forest cover since the onset of European colonization [12], and only 1% of its original extent is presently included in protected areas [10]. According to one projection to 2070 [13], the Atlantic Forest region will lose bird habitat at the rate of 1.2% to 3.3% per decade—the highest rate of loss estimated by that study for any region of the world. Realizing the potential impact of land use in the Atlantic Forest on parrot populations [14], as well as the relative importance of the genus *Amazona* among the Psittacidae, we direct our attention here to what we consider to be the most emblematic *Amazona* species of the Atlantic Forest biome: *A. brasiliensis*, *A. rhodocorytha*, *A. vinacea*, and *A. pretrei*. They are endemic to the Atlantic Forest [15] and classified by the IUCN, respectively, as Near-Threatened, Vulnerable, Endangered, and Vulnerable.

Geographic range and population size are two key descriptors of the state of any living species. Since their temporal trajectories offer evidence of population trends, these two variables inform four out of the five criteria used by the IUCN in assigning species to threat categories [16]. Notwithstanding, the IUCN Red List profiles of these four species in this study reveal substantial uncertainty about their geographic ranges and limited information about how the estimated population sizes were obtained. Our goal here is to fill this knowledge gap to the extent that is possible by compiling information from the ornithological literature and citizen-science platforms. We review information on population sizes based on published abundance estimates and counts of all species. To address geographic ranges, we draw new maps for the four species. Our maps express the species' distribution as occupancy probability per municipality. The statistical models used for producing the new maps integrate data from three different citizen-science platforms (eBird, Wikiaves, and Xeno-Canto) as well as from formal research databases, where available. We hope that improved knowledge about abundance and distribution of *Amazona* species in the Atlantic Forest will help direct future monitoring and conservation efforts, as well as strengthen the basis for threat assessments.

2. Materials and Methods

2.1. Study Area and Data Collection

We organized information about the population size and the geographic range of *Amazona brasiliensis*, *A. pretrei*, *A. rhodocorytha*, and *A. vinacea* following two different approaches. For population size, we compiled all the information about counts or abundance estimates that we could find for each species, including results from peer-reviewed papers, reports, books, and academic theses (Supplementary Tables S1–S3). Count data were collected by four different research teams, during scientific research or monitoring programs. The counts were performed at regularly used roosts or near points of frequent flyover by parrots, at dawn and dusk. For geographic range mapping, we compiled detection–non-detection data from citizen-science platforms and research project databases. Such data were analyzed separately for each species, with a site-occupancy, data-integration model following Zulian, Miller, and Ferraz [17]. We varied the geographic extent, or focal area, used to fit each species' model (Figure 1). Focal areas included either all the states or provinces where the species were detected (*A. rhodocorytha* and *A. vinacea*) or all the municipalities within 150 km of the closest detection (*A. brasiliensis* and *A. pretrei*). These areas ranged from a little over 160,000 km^2, for *A. brasiliensis*, to more than 1.5 million km^2, in the case of *A. vinacea* (Figure 1). We are confident that the extent for each species covers the entirety of each species' potential area of occurrence. The *A. vinacea* range map that we present here is the only map in this paper that combines formal research and citizen-science data. This map is identical to that shown by Zulian, Miller, and Ferraz [17], in a study focused on devising optimal methods for fitting distribution models to multiple data streams, which informs the approach that we used here. Geographic range analyses for the other three species are based uniquely on citizen-science data, as explained below.

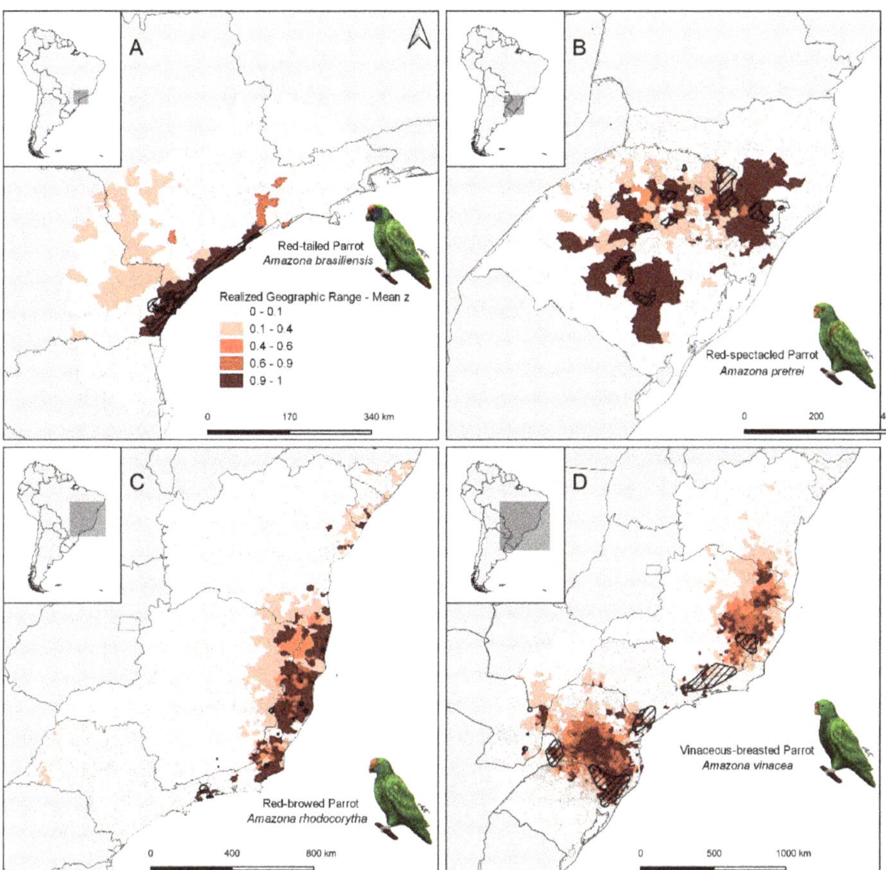

Figure 1. Geographic ranges of the four study species represented by the mean of the true occupancy state (z) estimated for each municipality. Intermediate values—of $z \sim 0.5$—indicate the highest uncertainty about occupancy by each species. Black dashed polygons are the Extant range of each species according to the IUCN Red List of Threatened Species [1].

We obtained records of *A. vinacea*, *A. brasiliensis*, *A. rhodocorytha*, and *A. pretrei* from citizen-science platforms eBird [18], WikiAves [19], and Xeno-canto [20], corresponding to the period between 1 January 2008 and 31 December 2018. For *A. vinacea*, we also included a formal research dataset derived from roost counts and described by Zulian et al. [21]. Our sampling unit is the municipality, where the number varied from 3701 to 405 depending on the species. Citizen-science platforms store data resulting from field visits with highly variable duration, distance covered, observation technique, and observer experience. This lack of standardization requires platform-specific data processing solutions. In particular, eBird data come in the form of checklists, which contain information about observation effort per list. The number of lists per municipality varied from 1 to 3245, with a mean of 33 lists, collected at different times of the year by different observers. WikiAves and Xeno-canto, on the other hand, gather records for a municipality in the form of individual species observations that are not aggregated in any form of observation 'session' per municipality and observer. As a result, we have the equivalent of replicate visits for eBird, but not for the other two platforms, where each municipality has only one 'visit'.

Data processing consisted of some filtering, formatting data matrices, and obtaining effort covariates for all platforms. Starting with eBird, we excluded incomplete checklists, checklists without location information, and checklists that potentially spanned more than one municipality due to long distance (>12 km) or long time traveled (>360 min). We set up a matrix of detection–non-detection histories based on eBird data for each parrot species. In this matrix, municipalities appeared in rows and consecutive checklists of each municipality in columns. Matrix elements were '1', for municipalities and checklists where the parrot species were detected, or '0' where not detected. We calculated three covariates of sampling effort for each eBird checklist and municipality: the total number of species recorded, the number of minutes spent observing, and the number of kilometers traveled. For WikiAves and Xeno-canto, data filtering consisted of deleting sightings of individuals reported as escaped from captivity. Since WikiAves receives photographic and audio records of species, we organized data into two vectors per parrot species, one for the number of photographs and one for the number of audio recordings of that species in each unique municipality. For the WikiAves data, we calculated two covariates of effort: the number of photos and the number of audio recordings of all species, per municipality. Finally, the Xeno-canto platform hosts only audio recordings of bird sounds, so its detection data were easily organized into a single vector per parrot species, holding that species' number of audio recordings per municipality. We also collected the total number of recordings of any species uploaded for each municipality for use as a covariate of Xeno-canto sampling effort. For the *A. vinacea* research data, we created a detection–non-detection matrix with municipalities as rows and counts as columns. Matrix cells corresponding to counts with at least one parrot received a detection ('1'), and those with no parrots received a non-detection ('0'). Here, we used the count's duration, in minutes, as a covariate of sampling effort (see Zulian, Miller, and Ferraz [17] for details).

2.2. Data Analysis

We drew range maps representing the estimated probability of site (or municipality) occupancy by each species during the eleven-year study period. We follow a static approach as originally described by MacKenzie et al. [22] and define 'occupancy' as the probability that a site was occupied by the given species at any point during the whole eleven-year study period. One of the species—*A. pretrei*—is known for its within-year shifts in distribution, which result in exceptionally large concentrations of individuals during the non-breeding season. Therefore, for this species alone, we estimated both the full-year distribution for the species and seasonal range maps. Seasonal distributions were obtained with the same modeling approach applied to four non-overlapping temporal subsets of the data, each corresponding to one trimester of the year and including information from all years. At the core of our statistical approach to site occupancy, there is a process model of the true occupancy state, z_i, of each municipality, i, which takes the value of '1' for those municipalities that are occupied by the species of interest, and '0' for those that are not. This state follows a Bernoulli distribution with a mean ψ_i:

$$z_i \sim Bernoulli\ (\psi_i). \tag{1}$$

The occupancy probability in each municipality i, ψ_i, varies according to n environmental covariates, $X_{n,i}$, according to a generalized linear model with a logit link function. Since the four species of parrots are associated with Atlantic Forest and altitude [23–25], we included the Atlantic Forest cover and average altitude as covariates of municipality i occupancy. We also included the Araucaria Forest cover as a covariate of occupancy by *A. pretrei* and *A. vinacea*, since they rely heavily on Araucaria seeds for food during the winter [23,26,27], and a Dense Forest cover as a covariate of occupancy by *A. brasiliensis*, because this species is apparently associated with dense, lowland coastal forest [25,28,29]. We obtained Atlantic Forest cover data from Ribeiro et al. (in prep.), and Dense Forest cover data from the Brazilian Instituto Brasileiro de Geografia e Estatística (https://www.ibge.gov.br/ (accessed on 30 June 2021)) [30]. Average municipality altitude, x, in meters, is from DIVA-

GIS (https://www.diva-gis.org/ (accessed on 18 November 2019)) [31], log-transformed as $log(x+1)$. Our linear model of occupancy also included a spatial random effect to account for unexplained spatial autocorrelated variation (δ_i):

$$logit(\psi_i) = \beta_0 + \beta_1 * X_{1,i} + \beta_2 * X_{2,i} + \ldots + \delta_i. \quad (2)$$

In this model, occupancy covariates measured at municipality i are given by $X_{1,i}$, $X_{2,i}$ $\ldots$, and β_0, β_1, β_2 $\ldots$ are estimated coefficients. The spatial component of our model follows a conditional auto-regressive (CAR) distribution [17] and was used to estimate correlated spatial variation in the data that is not explained by our covariates. To avoid confounding effects of municipality size variability and to gain replication within spatial units in the CAR analysis, we represented the spatial random effect using a hexagonal lattice overlaid on the study area, with municipalities assigned to the lattice cell that matches their centroid. Hexagonal cells measured 0.5° latitude across, and all the first-order neighbors of each cell were given a weight of 1 when fitting the CAR model.

Ours is a data-integration approach because it models detections from different databases with a joint-likelihood that shares the same occupancy process described above [32,33], for each parrot species. Within each database, detection was expressed as a conditional probability, p_j^*, of detecting the species as a function of an estimated amount of sampling effort, E_j, for visit j [17,33,34]:

$$p_j^* = 1 - (1-p)^{Ej}, \quad (3)$$

where p is the probability of detection per unit of effort. Since we are using indirect, and sometimes several metrics of effort for each data source (our effort covariates), we estimated the parameter E_j for each sample j as a linear function of the covariates. Thus, for each dataset, DS_n, with n varying between one and four (roost counts, eBird, WikiAves, and Xeno-canto), we have:

$$E_j^{DS_n} = \alpha_1 * X1_j + \alpha_2 * X2_j + \alpha_3 * X3_j, \quad (4)$$

were $X1_j$, $X2_j$, and $X3_j$ are effort covariates measured on visit j. We used one to three covariates depending on data type. We fixed p at a value of 0.5, so that the $\alpha_1 \alpha_3$ coefficients express the relationship between covariates and the effort necessary to reach a detection probability of 0.5 per unit of effort. Without fixing p, Equation (3) becomes over-parameterized. Having modeled a conditional probability of detection, p_j^*, we can represent the detection–non-detection data, Y_{ij}, as the outcome of a Bernoulli distribution, that accounts for the true state of each municipality, z_i, and the conditional probability of detection, as follows:

$$Y_{ij} \sim Bernoulli\left(z_i \times p_j^*\right). \quad (5)$$

We fitted all the models using a Bayesian estimator coded in the BUGS language and implemented on WinBugs software [35]. Inference was based on draws from the posterior distribution of model parameters using a Markov Chain Monte Carlo (MCMC) algorithm with three chains, 200,000 iterations, and a burn-in phase of 150,000 (see code in Supplementary S1 in the Supplementary Materials). All results presented here correspond to chains that converged to an R-hat lower than 1.1. We draw maps of 'realized occupancy' given by the mean of the estimated z_i for each municipality and estimated the area of each species' geographic range as the sum of all municipality areas weighted by each municipality's predicted occupancy, ψ_i, estimate.

3. Results

We used a total of 100,289 samples, collected across 3701 municipalities, to inform the estimation of geographic ranges of the four parrot species that we studied. The datasets showed a wide coverage, with more than 90% of the municipalities in each species' study area having at least one sample (Table 1). Alone, the *A. vinacea* dataset accounted for almost 50% of the samples and 58% of the detections, while *A. pretrei* had the smallest dataset, with 18% of the samples and 10% of the detections (Table 1). *A. brasiliensis* had the third highest number of samples, but only 15 municipalities with at least 1 detection. *A. rhodocorytha* had the second smallest sample size, but the second largest number of detections (Table 1).

Our estimated geographic ranges differed from the Extant area calculated from the range maps reported by the IUCN for all four species (Table 1). *A. vinacea* had the largest estimated range, encompassing more than 400,000 square kilometers [17], followed by *A. rhodocorytha*, with approximately 134,000 square kilometers (Figure 1). The discrepancies between the IUCN Extant area and our estimates are not negligible: while our geographic range estimate is three times larger than the IUCN value for *A. vinacea*, it is six times larger for *A. pretrei*. The biggest discrepancy is for *A. rhodocorytha*, for whom the IUCN reports a range 50 times smaller than our estimate. Geographic ranges are an outcome of history and environmental constraints. Our results show how the environmental covariates of Atlantic Forest cover, Araucaria Forest cover, and Altitude help explain the distribution of *A. vinacea*, with all three having strong and positive effects on site-occupancy probability (Table 2). Based on our models, species' detection data, and environmental covariate information, there is no evidence of other statistically distinguishable effects of environmental factors on site occupancy by any of the four species of parrots (i.e., the 95% credible intervals of other coefficients in Table 2 are nearly centered on zero).

Table 1. Sample size, spatial coverage, and number of detections for the four parrot species. Sample size is the number of samples collected form the citizen-science and research datasets, as defined in the text. Coverage is the proportion of municipalities in each study area with at least one sample. The labels n_{det} and n_{muni} show, respectively, the number of parrot detections and the number of municipalities with at least one detection. The last two columns show geographic ranges sizes: the IUCN Extant area is given in each species' online entry to the IUCN Red List of Threatened Species. The estimated geographic range is the sum of municipality areas weighted by the estimated probability the species occurred in each municipality (given here by the mean ± standard deviation of the a posteriori distribution of range size, followed by its 95% credible interval (in parentheses)).

Species	Sample Size	Coverage %	n_{det}	n_{muni}	IUCN Extant Area (km^2)	Estimated Geographic Range (km^2)
Amazona brasiliensis (Red-tailed Parrot)	16,705	99.7	192	15	4750	15,627 ± 8843 (3127–31,414)
Amazona pretrei (Red-spectacled Parrot)	5477	92.7	187	73	10,430	66,203 ± 11,425 (45,727–90,367)
Amazona rhodocorytha (Red-browed Parrot)	30,867	94.2	346	86	2672	134,355 ± 13,922 (109,288–162,828)
Amazona vinacea (Vinaceous-breasted Parrot)	47,240	91.9	1007	339	145,700	434,670 ± 28,911 (382,887–496,550)

Table 2. Coefficients of the occupancy equation in each species model. The numbers show the mean ± standard deviation and 95% credible intervals (in parentheses) of the a posteriori distribution of each parameter.

Species	Atlantic Forest	Dense Forest	Araucaria Forest	Altitude
Amazona brasiliensis	−0.63 ± 2.23 (−5.39–3.48)	−1.33 ± 3.14 (−7.71–4.77)	—	0.23 ± 0.44 (−0.72–0.94)
Amazona pretrei	−0.55 ± 1.08 (−2.70–1.56)	—	0.47 ± 1.05 (−1.55–2.53)	0.15 ± 0.18 (−0.21–0.52)
Amazona rhodocorytha	0.84 ± 0.91 (−1.70–1.84)	—	—	−0.14 ± 0.20 (−0.51–0.25)
Amazona vinacea	2.11 ± 0.86 (0.37–3.79)	—	2.13 ± 0.98 (0.29–4.10)	0.85 ± 0.12 (0.58–1.05)

The subdivision of *A. pretrei* data into trimesters generates four substantially different geographic range maps (Figure 2). During the early breeding season months of July to September, the species is at its most dispersed (Figure 2A). During this period, 39 municipalities throughout the focal area have realized occupancy greater than 0.9 (i.e., mean z > 0.9), even though almost all of them are in the state of Rio Grande do Sul. During the Fall months of April to June, however, *A pretrei* individuals appear aggregated in only 12 municipalities that have realized occupancy greater than 0.9 (Figure 2D). These municipalities form four disjunct clusters in the Rio Grande do Sul and Santa Catarina states. The transition from the aggregated to the dispersed state is faster than the transition from dispersed to aggregated, which takes place from October to March and is represented by the intermediate ranges in panels B and C, of Figure 2.

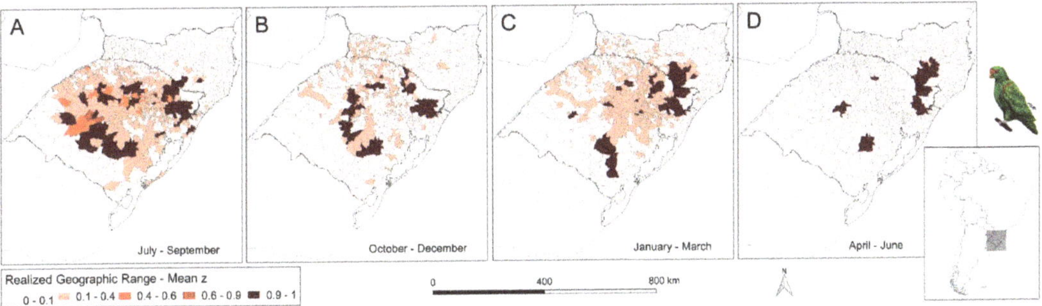

Figure 2. Seasonal variation of the geographic range of the Red-Spectacled Parrot (*Amazona pretrei*) as shown by the mean of the true occupancy state (*z*) estimated for each municipality. Each panel represents a trimester, the sequence starting with July–September (**A**), when the species is most dispersed, and proceeds in three-month intervals to October-December (**B**), January-March (**C**) until April–June (**D**), when it aggregates in only a few municipalities. Darker tones of red indicate higher mean *z*; intermediate tones indicate the highest uncertainty about species presence.

There is much less information about the abundance of Atlantic Forest *Amazona* species than about their geographic range. The species for which we could assemble the longest time series of roost counts was *A. pretrei*, which has a long-term monitoring program led by the same team of researchers since the mid-1990s (Figure 3B). *A. pretrei* is also the species with highest counted number of individuals. Its earliest counts, performed in 1971 by Forshaw and Cooper (ref [36] cited by [37]), returned between 10,000 and 30,000 individuals (Figure 3B). Later, during the 1970s and 1980s, Belton [38] and Varty et al. [37] reported a decline in the number of individuals counted, with recovery during the 1990s. Since 1997, the yearly sum of *A. pretrei* counts has varied around 20,000 individuals (Figure 3B) [39–42]. The second longest time series of roost counts is that of *A. brasiliensis*. This species also has an ongoing monitoring program, coordinated by the same team throughout the last two decades [43]. The sum of *A. brasiliensis* counts has varied, always below 10,000 individuals, over the last three decades [28,29,43–61]. Figure 3A shows a tendency towards increasing counts, but one should not rush to interpret this as evidence of population growth because the count reports do not incorporate corrections for variation in effort through time. *A. vinacea* has the shortest time series of roost counts [21,27,42,62–64] (Figure 3C) but is the only species with a published statistical estimate of population size, which does account for temporal changes in sampling effort, as well as for detection errors [21]. There are two estimates, for the non-breeding seasons of 2016 and 2017, both in the vicinity of 8500 individuals and with 95% credible intervals entirely below the 10,000-individual mark.

We could not assemble a time series of *A. rhodocorytha* counts, as the few published count results were obtained in sparse locations that were not revisited in different years. In 1998, Marsden et al. [65] searched for the species in two separate sites covering 427 km^2 of Bahia and Espírito Santo states, reporting distance-sampling estimates of, respectively, 238 ± 174 and 5990 ± 1680 (mean ± standard error) individuals. Later, in 2008, Klemann-Júnior et al. [66] counted 2295 individuals for all of Espírito Santo state. The *Plano de Ação Nacional para a Conservação dos Papagaios da Mata Atlântica* considers that the *A. rhodocorytha* population size is around 10,000 individuals, based on expert opinion [41], but no more demographic information is available.

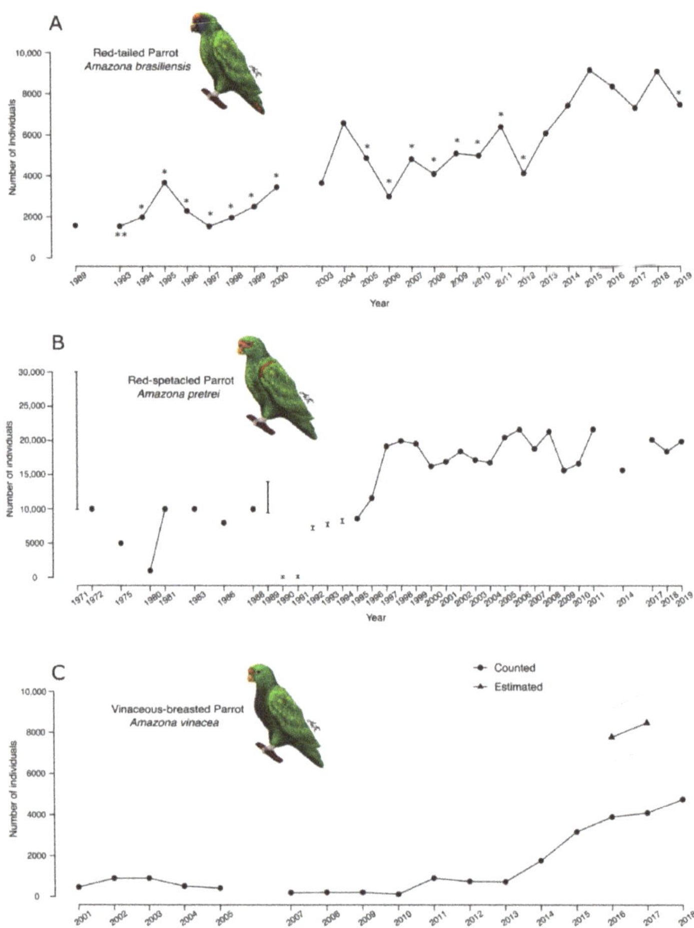

Figure 3. Number of Red-Tailed (**A**), Red-Spectacled (**B**), and Vinaceous-Breasted (**C**) Parrots counted by research teams throughout the last fifty years. Panel (**C**) also includes two estimates of the Vinaceous-Breasted Parrot population size, with gray lines showing bounds of the 95% credible interval of the a posteriori distribution of population size. These are the only statistical estimates of population size available in the literature for any of the study species. There is no plot for the Red-Browed Parrot because we could not find published records of count results for this species. Some of the Red-Spectacled Parrot counts were reported as intervals and appear as vertical lines in panel (**B**). Variations in the number of counted individuals may be due to variation in sampling effort or to real change in population size. Asterisks in panel (**A**) show differences in sampling effort: * corresponds to years that the counts were performed only in Paraná, and ** only in São Paulo. Sources for the numbers shown are [28,29,43–61] (**A**), [36–42,67] (**B**), and [21,27,42,62–64] (**C**).

4. Discussion

4.1. Geographic Range

The four *Amazona* species we studied showed marked differences in geographic range and, most likely, also in population size. The estimated areas of the geographic ranges varied over two orders of magnitude, from the approximately fifteen thousand square kilometers in *A. brasiliensis* to more than four hundred thousand square kilometers in *A. vinacea*. The mean estimated range was larger than the IUCN Extant area for all species, with 95% credible intervals including the IUCN Extant area for only one of them, *A. brasiliensis*. Both our estimated range area and the IUCN Extant areas approximate areas of occupancy as defined by Gaston and Fuller [68]. The marked disparity is likely a reflection of conservative caution in the definition of IUCN Extant areas and of extraordinary sampling coverage afforded by the use of citizen-science data in our estimates.

Geographic-range differences across species partially reflect environmental factors that limit their distribution. The range of *A. brasiliensis* appears to be limited by the highlands of the Serra do Mar [28], which also have high occupancy by *A. vinacea*. Indeed, *A. vinacea* is the only species to show evidence of a positive association between altitude and occupancy probability. Occupancy by *A. vinacea* is also positively associated with Atlantic and Araucaria Forest covers, even though the parrot's range extends beyond that of *Araucaria angustifolia* [17]. None of these associations—with altitude or with any type of forest cover—were evident from the analyses of the other three species—*A. brasiliensis*, *A. pretrei*, or *A. rhodocorytha*. Such lack of statistical association does not mean that they are biologically indifferent to forest cover. They are all cavity-nesters, and will not reproduce without access to tree holes, which are predominantly found in old-growth forests [28,29,66,69–71]. Instead, the focal areas of all three include extensive regions of forest (or of high or low altitude) that happen to not be occupied. This weakens the statistical association with occupancy covariates, not because they do not facilitate occupancy, but because unknown factors not included in our models may be further restricting the parrot distributions.

4.2. Population Size

Of the four species in this study, we only have a statistical estimate of global population size for *A. vinacea*. At around 8500 individuals [21], this estimate is nearly three times the number reported by the IUCN [72]. The local estimate of ~6000 *A. rhodocorytha* individuals for one 461 km^2 site in Espírito Santo, reported by Marsden et al. in 1998 [62], appears too high. This number, which implies a homogeneous density of 13 individuals per km^2 throughout the study site, is more than twice the number counted for the whole state of Espírito Santo by a different team ten years later [63]. There was either a dramatic population reduction in the state or these *A. rhodocorytha* numbers need reconsideration. There are no published estimates or counts of *A. rhodocorytha* for five of the states covered by the range map in Figure 1C. The species' global population size of 10,000 individuals reported by the IUCN [73] and the Brazilian Red List [74] may be reasonable, but neither source provides an explanation of how that number was obtained.

Any considerations about population sizes of *A. pretrei* and *A. brasiliensis* must be based solely on raw counts, as there are no published statistical estimates of population size for these species. Counts are difficult to compare because they do not quantify uncertainty about their values. They are also likely to underestimate real population size because they do not account for detection errors. In the absence of statistical estimates, however, counts offer a reasonable lower bound for population size. *A. pretrei* is the species with the largest counts, exceeding 20,000 individuals in 2006, 2008, and 2011, a number that is also greater than the 16,000 individuals cited by the species' IUCN Red List profile [75,76]. This species' well-known tendency to concentrate in only a few municipalities during the non-breeding season [39] reduces the probability that observers overlook large flocks and makes us relatively more confident of the accuracy of *A. pretrei*'s counts than of the others. Counts of *A. brasiliensis* reached more than 9000 individuals in 2018 [61,77], making it, possibly, the species with the smallest geographic range but the second highest population size in this study. Future research could be aimed at the question of whether *A. brasiliensis* presents an exception to the well-supported positive relationship between area of occupancy and local abundance [78].

4.3. Seasonal Change in Geographic Range

Seasonal movements of aggregation and dispersion, influenced by the reproductive cycle and changes in food availability, are well-documented for *A. brasiliensis* [29], *A. pretrei* [39], and *A. vinacea* [21,27]. Dispersion occurs in the beginning of the breeding season (August–September), when breeding pairs abandon collective roosts to start spending the nights near the nest. By the end of the breeding season—December to March depending on the species—parrots aggregate again in collective roosts, which vary in size from dozens to thousands of individuals [21,27,29,39,79]. Aggregation and dispersal phases of *A. pretrei* occur in nearly non-overlapping parts of the species' range. By early Autumn, individuals concentrate in southeast Santa Catarina [39,80], and they spend the coolest months of the year in this region, feeding on abundant Paraná pine (*Araucaria angustifolia*) seeds [39] while other food resources are scarce [26]. Even though some individuals may overwinter in Rio Grande do Sul, the majority of the *A. pretrei* population spends this period in Santa Catarina, forming groups with thousands of individuals, in the municipalities of Painel, Urupema, Lages, and São Joaquim [39]. Between July and September, *A. pretrei* individuals disperse back to breed in Rio Grande do Sul, reaching at this point their largest geographic range and smallest group sizes [39]. Providing evidence of range dynamics at a larger temporal scale, *A. pretrei*'s center of aggregation has not always been in southeast Santa Catarina. Reports from the 1970s show large wintering aggregations of more than 10,000 individuals in the municipality of Muitos Capões, northern Rio Grande do Sul [3,37,38,67]. By the early 1990s, however, this number had decreased to only a few tens of individuals [39], and larger groups began appearing in Southeast Santa Catarina [39,81]. This shift of more than 100 km to the north follows decades of intense exploration and widespread destruction of Paraná pine forests in RS, which peaked between the 1920s and 1950s [82]. Most likely, scarcity of their most important winter fallback food forced *A. pretrei* into the colder but still relatively abundant Araucaria forests of the new wintering grounds in Santa Catarina [81].

4.4. Long-Term Changes in Geographic Range

Among the four species in our study, *A. vinacea* shows the strongest evidence of range contraction, with local extinctions in parts of Argentina and Paraguay since the 1970s [63,64]. With a historic range that covered southern Paraguay west of Misiones and all the way into central Paraguay to the northwest [63], the occurrence of *A. vinacea* outside Brazil is now restricted to three localities in Argentina [21,63,64] and two in Paraguay [21,63]. Both *A. vinacea* and *A. pretrei* are classified as critically endangered in Argentina [83], which may have had a historical population of the latter [84,85] *A. pretrei* is rarely observed in Paraguay [23,86], where it is also classified as threatened [87]. Belton [38] mentions the possible past occurrence of *A. pretrei* as far north as São Paulo state, in Brazil, but the validity of XIX century records that could backup such possibility is disputed [3,75]. Reviewing information about *A. brasiliensis*, Scherer-Neto [28] cites reports of XIX century sightings in northern Rio Grande do Sul and northeast Santa Catarina (see also [77]), but the validity of these reports, too, is questionable [3]. Even with reliable identification, though, past observation of any species far outside the present range is no firm evidence of range contraction. Individuals may wander away from their species' ranges, sometimes across oceans [88], with sightings in unexpected locations inevitably getting more attention than within a known range, even if they bear no consequence to population dynamics. Parrots introduce the additional complication of having been kept as pets for a long time, so that past sightings in odd places could also be of individuals escaped from captivity. To conclude, *A. rhodocorytha* has the least historical information of the four species, with perhaps one observation deserving special attention: one recent record in the state of Alagoas [89] dispels a previous suggestion of local extinction [3] and confirms the existence of a disjunct population in the extreme north of the distribution.

4.5. Long-Term Change in Population Size

The time series of counts that we report for *A. vinacea* and *A. brasiliensis* show increasing numbers very likely due to an increase in sampling effort. The time series for *A. pretrei* shows relatively small variation for the last two and a half decades. After an apparent decline during the 1970s [37,38], *A. pretrei* counts increased to around 20,000 individuals in 1997. Such increase coincides with the period when *A. pretrei* was shifting its wintering aggregation to Southeast Santa Catarina, where counts have been carried out by the same research team since 1995. Counts of *A. brasiliensis* and *A. vinacea*, on the other hand, have been carried out by different research groups in different locations, with variable degrees of coordination. The highest counts of *A. brasiliensis*, for example, were obtained in 2015 (9176 individuals), and in 2018 (9112 individuals), when research teams visited all known roosts in São Paulo and Paraná. In 2019, however, when only Paraná roosts were visited, approximately 2000 fewer individuals were counted. Similar, effort-driven variation is evident in the *A. vinacea* time series, which had fewer than one thousand individuals counted annually from 2001 to 2013. *A. vinacea* counts have increased since 2014, with the implementation of annual coordinated counts performed at a number of sites, that increased gradually from 20, in 2014, to 67, in 2017. The only period for which we can draw statistical inference about temporal change in the *A. vinacea* population is the transition from 2016 to 2017 [21]. The estimates shown in Figure 3C account for detection error and for variation in effort between the two years. The credible intervals of the abundance estimates, broadly overlapping between the two years, provide no evidence of a substantial change. Future analysis of population trends will require more coordination and replication of counts. This will facilitate statistical analysis of count results and investigation of real trends in population size.

4.6. Concluding Remarks

The future of the four parrot species analyzed in this study is threatened by two key environmental hazards: habitat loss and human exploitation [2]. *A. brasiliensis*, *A. pretrei*, *A. rhodocorytha*, and *A. vinacea* are all impacted by the destruction of the Atlantic Forest, especially because they nest in tree cavities that are much more common in old growth than in secondary forests [90]. Since the arrival of Europeans in South America, almost 90% of the original Atlantic Forest cover was lost [12]. The remaining forest is highly fragmented, with only 20% of its area contained in patches larger than 100 km^2, and 83% of the patches being smaller than 50 hectares [12]. When not replaced by pasture or farmland, cleared forest gives way to exotic tree monocultures, such as *Pinus* and *Eucalyptus* [91]. In coastal areas intensely used by the tourism industry, cleared forests may also give way to urban expansion, which disproportionally affects *A. brasiliensis* [29]. The other primary threat to all four species, human exploitation, comes in the form of nest poaching [8,69,74,92–96]. According to one study [8], nest poaching is the principal cause of nest failure for *A. vinacea* —with more than 80% of 25 monitored nests poached—and *A. brasiliensis*—with 50% of 78 monitored nests poached. Four initiatives have been promoting conservation, as well as research and monitoring of the four species throughout the last three decades: Projeto Charão (for *A. pretrei*, since 1991), Projeto para Conservação do Papagaio-de-cara-roxa (for *A. brasiliensis*, since 1997), Projeto Chauá (for *A. rhodocorytha*, since 2014), and the Programa Nacional para a Conservação do Papagaio-de-peito-roxo (for *A. vinacea*, since 2015). To improve knowledge about population dynamics and manage a response to environmental threats, it is essential that these and similar initiatives expand their reach. Continued tapping of citizen-science data will help to update knowledge about species' ranges. The estimation of abundance and validation of range maps, however, require observers on the ground. Much can be achieved just by sending observers to municipalities with higher uncertainty about species' occurrence, but one can go much further by practicing integration of citizen-science and professional research work on a routine basis. The combination of range mapping based on citizen-science and coordinated observation by research teams throughout the species' ranges offers a powerful tool for accurately monitoring the species' status and for assessing the consequences of management decisions.

Supplementary Materials: The following are available online at https://www.mdpi.com/article/10.3390/d13090416/s1, Supplementary S1: The R and BUGS code for the models used in estimating the parrot's geographic range is available. Table S1: Compilation of the available counts for Amazona brasiliensis with the respective reference. Table S2: Compilation of the available counts for Amazona pretrei with the respective reference. Table S3: Compilation of the available counts and abundance estimates for Amazona vinacea with the respective reference.

Author Contributions: Conceptualization, V.Z. and G.F.; methodology, D.A.W.M.; software, D.A.W.M. and V.Z.; formal analysis, V.Z. and D.A.W.M.; investigation, V.Z. and G.F.; resources, V.Z., D.A.W.M. and G.F.; data curation, V.Z.; writing—original draft preparation, V.Z., D.A.W.M. and G.F.; supervision, G.F.; project administration, V.Z. and G.F.; funding acquisition, V.Z. and G.F. All authors have read and agreed to the published version of the manuscript.

Funding: V.Z. received fellowships from the Brazilian government's CAPES and PrInt/CAPES programs, as well as research funding from Funbio, Instituto Humanize, and The Rufford Foundation (19835-1 and 23709-2).

Institutional Review Board Statement: Not applicable.

Informed Consent Statement: Not applicable.

Data Availability Statement: The R codes and datasets used for the analyses are openly available at GitHub at: https://github.com/vivizulian/DataIntegrationModels (accessed on 3 April 2021).

Acknowledgments: We are grateful to Reinaldo Guedes who volunteered his free time over the last twelve years to developing and maintaining WikiAves, the most successful citizen-science initiative in Brazil. This paper would not be possible without the help of thousands of bird observers who uploaded photos, audio recordings, and birding lists to eBird, WikiAves, and Xeno-canto. Glayson Bencke kindly shared his time and knowledge about parrot biology with us in several insightful conversations that contributed to shaping the best part of this paper. We also thank the three reviewers that helped improve the manuscript.

Conflicts of Interest: The authors declare no conflict of interest.

References

1. BirdLife International. The IUCN Red List of Threatened Species. 2021. Available online: https://www.iucnredlist.org/ (accessed on 2 February 2020).
2. Collar, N.J.; Juniper, A.T. Dimensions and Causes of the Parrot Conservation Crisis. In *New World Parrots in Crisis: Solutions from Conservation Biology*; Smithsonian Institution Press: Washington, DC, USA, 1992.
3. Juniper, T.; Parr, M. *Parrots: A Guide to the Parrots of the World*; Yale University Press: New Haven, CT, USA; London, UK, 1998.
4. Cockle, K.L.; Martin, K.; Drever, M.C. Supply of Tree-Holes Limits Nest Density of Cavity-Nesting Birds in Primary and Logged Subtropical Atlantic Forest. *Biol. Conserv.* **2010**, *143*, 2851–2857. [CrossRef]
5. Foley, J.A. Global Consequences of Land Use. *Science* **2005**, *309*, 570–574. [CrossRef]
6. Berkunsky, I.; Quillfeldt, P.; Brightsmith, D.J.; Abbud, M.C.; Aguilar, J.M.R.E.; Alemán-Zelaya, U.; Aramburú, R.M.; Arce Arias, A.; Balas McNab, R.; Balsby, T.J.S.; et al. Current Threats Faced by Neotropical Parrot Populations. *Biol. Conserv.* **2017**, *214*, 278–287. [CrossRef]
7. Tella, J.L.; Hiraldo, F. Illegal and Legal Parrot Trade Shows a Long-Term, Cross-Cultural Preference for the Most Attractive Species Increasing Their Risk of Extinction. *PLoS ONE* **2014**, *9*, e107546. [CrossRef] [PubMed]
8. Wright, T.F.; Toft, C.A.; Enkerlin-Hoeflich, E.; Gonzalez-Elizondo, J.; Albornoz, M.; Rodríguez-Ferraro, A.; Rojas-Suárez, F.; Sanz, V.; Trujillo, A.; Beissinger, S.R.; et al. Nest Poaching in Neotropical Parrots. *Conserv. Biol.* **2001**, *15*, 710–720. [CrossRef]
9. del Hoyo, J.; Elliott, A.; Sargatal, J.; Christie, D.A.; de Juana, E. *Handbook of the Birds of the World Alive*; Lynx Edicions: Barcelona, Spain, 2017.
10. Ribeiro, M.C.; Martensen, A.C.; Metzger, J.P.; Tabarelli, M.; Scarano, F.R.; Fortin, M.-J. The Brazilian Atlantic Forest: A Shrinking Biodiversity Hotspot. In *Biodiversity Hotspots*; Zachos, F.E., Habel, J.C., Eds.; Springer: Berlin/Heidelberg, Germany, 2011; ISBN 978-3-642-20991-8.

11. Tabarelli, M.; Aguiar, A.V.; Ribeiro, M.C.; Metzger, J.P.; Peres, C.A. Prospects for Biodiversity Conservation in the Atlantic Forest: Lessons from Aging Human-Modified Landscapes. *Biol. Conserv.* **2010**, *143*, 2328–2340. [CrossRef]
12. Ribeiro, M.C.; Metzger, J.P.; Martensen, A.C.; Ponzoni, F.J.; Hirota, M.M. The Brazilian Atlantic Forest: How Much Is Left, and How Is the Remaining Forest Distributed? Implications for Conservation. *Biol. Conserv.* **2009**, *142*, 1141–1153. [CrossRef]
13. Powers, R.P.; Jetz, W. Global Habitat Loss and Extinction Risk of Terrestrial Vertebrates under Future Land-Use-Change Scenarios. *Nat. Clim. Chang.* **2019**, *9*, 323–329. [CrossRef]
14. Vergara-Tabares, D.L.; Cordier, J.M.; Landi, M.A.; Olah, G.; Nori, J. Global Trends of Habitat Destruction and Consequences for Parrot Conservation. *Glob. Chang. Biol.* **2020**, *26*, 4251–4262. [CrossRef] [PubMed]
15. Vale, M.M.; Tourinho, L.; Lorini, M.L.; Rajão, H.; Figueiredo, M.S.L. Endemic Birds of the Atlantic Forest: Traits, Conservation Status, and Patterns of Biodiversity. *J. Field Ornithol.* **2018**, *89*, 193–206. [CrossRef]
16. Mace, G.M.; Collar, N.J.; Gaston, K.J.; Hilton-Taylor, C.; AkçAkaya, H.R.; Leader-Williams, N.; Milner-Gulland, E.J.; Stuart, S.N. Quantification of Extinction Risk: IUCN's System for Classifying Threatened Species. *Conserv. Biol.* **2008**, *22*, 1424–1442. [CrossRef] [PubMed]
17. Zulian, V.; Miller, D.A.W.; Ferraz, G. Integrating citizen-science and planned-survey data improves species distribution estimates. *Divers. Distrib.* **2021**. [CrossRef]
18. eBird. Available online: https://ebird.org/home (accessed on 2 February 2020).
19. WikiAves. WikiAves, a Enciclopédia Das Aves Do Brasil. Available online: http://www.wikiaves.com.br (accessed on 8 January 2020).
20. Xeno-Canto. Available online: https://www.xeno-canto.org/ (accessed on 9 January 2019).
21. Zulian, V.; Müller, E.S.; Cockle, K.L.; Lesterhuis, A.; Tomasi Júnior, R.; Prestes, N.P.; Martinez, J.; Kéry, M.; Ferraz, G. Addressing Multiple Sources of Uncertainty in the Estimation of Global Parrot Abundance from Roost Counts: A Case Study with the Vinaceous-Breasted Parrot (*Amazona vinacea*). *Biol. Conserv.* **2020**, *248*, 108672. [CrossRef]
22. MacKenzie, D.I.; Nichols, J.D.; Lachman, G.B.; Droege, S.; Andrew Royle, J.; Langtimm, C.A. Estimating Site Occupancy Rates When Detection Probabilities Are Less than One. *Ecology* **2002**, *83*, 2248–2255. [CrossRef]
23. Collar, N.; Boesman, P.F.D. Red-spectacled Parrot (*Amazona pretrei*). In *Birds of the World*; Billerman, S.M., Keeney, B.K., Rodewald, P.G., Schulenberg, T.S., Eds.; Cornell Lab of Ornithology: Ithaca, NY, USA, 2020.
24. Collar, N.; Boesman, P.F.D.; Sharpe, C. Red-browed Parrot (*Amazona rhodocorytha*). In *Birds of the World*; Billerman, S.M., Keeney, B.K., Rodewald, P.G., Schulenberg, T.S., Eds.; Cornell Lab of Ornithology: Ithaca, NY, USA, 2020.
25. Collar, N.; Boesman, P.F.D.; Sharpe, C. Red-tailed Parrot (*Amazona brasiliensis*). In *Birds of the World*; Billerman, S.M., Keeney, B.K., Rodewald, P.G., Schulenberg, T.S., Eds.; Cornell Lab of Ornithology: Ithaca, NY, USA, 2020.
26. Prestes, N.P.; Martinez, J.; Peres, A.R. Dieta alimentar do papagaio-charão (*Amazona pretrei*). In *Biologia Da Conservação: Estudo de Caso com Papagaio-Charão e Outros Papagaios Brasileiros*; Martinez, J., Prestes, N.P., Eds.; UPF Editora: Passo Fundo, RS, Brazil, 2008.
27. Prestes, N.P.; Martinez, J.; Kilpp, J.C.; Batistela, T.; Turkievicz, A.; Rezende, É.; Gaboardi, V.T.R. Ecologia e Conservação de *Amazona vinacea* Em Áreas Simpátricas Com *Amazona pretrei*. *Ornithologia* **2014**, *6*, 109–120.
28. Scherer-Neto, P. Contribuição à Biologia do Papagaio-de-Cara-Roxa *Amazona Brasiliensis* (Linnaeus, 1758) (Psittacidae, Aves). Master's Thesis, Universidade Federal do Paraná, Curitiba, PR, Brazil, 1989.
29. Martuscelli, P. Ecology and Conservation of the Red-Tailed Amazon *Amazona brasiliensis* in South-Eastern Brazil. *Bird Conserv. Int.* **1995**, *5*, 405–420. [CrossRef]
30. IBGE Mapa de Vegetação. 2012. Available online: https://www.ibge.gov.br (accessed on 30 June 2021).
31. Hijmans, R.J.; Guarino, L.; Cruz, M.; Rojas, E. Computer Tools for Spatial Analysis of Plant Genetic Resources Data: 1. DIVA-GIS. *Plant Genet. Resour. Newsl.* **2001**, *127*, 15–19.
32. Pacifici, K.; Reich, B.J.; Miller, D.A.W.; Gardner, B.; Stauffer, G.; Singh, S.; McKerrow, A.; Collazo, J.A. Integrating Multiple Data Sources in Species Distribution Modeling: A Framework for Data Fusion. *Ecology* **2017**, *98*, 840–850. [CrossRef]
33. Miller, D.A.W.; Pacifici, K.; Sanderlin, J.S.; Reich, B.J. The Recent Past and Promising Future for Data Integration Methods to Estimate Species' Distributions. *Methods Ecol. Evol.* **2019**, *10*, 22–37. [CrossRef]
34. Stauffer, G.E.; Miller, D.A.W.; Williams, L.M.; Brown, J. Ruffed Grouse Population Declines after Introduction of West Nile Virus. *Jour. Wild. Mgmt.* **2018**, *82*, 165–172. [CrossRef]
35. Lunn, D.J.; Thomas, A.; Best, N.; Spiegelhalter, D. WinBUGS-a Bayesian Modelling Framework: Concepts, Structure, and Extensibility. *Stat. Comput.* **2000**, *10*, 325–337. [CrossRef]
36. Forshaw, J.M.; Cooper, W.T. *Parrots of the World*, 2nd ed.; Lansdowne Press: Melbourne, Australia, 1978.
37. Varty, N.; Bencke, G.A.; Bernardini, L.M.; Cunha, A.S.; Dias, E.V.; Fontana, C.S.; Guadagnin, D.L.; Kindel, A.; Kindel, E.; Raymundo, M.M.; et al. *Conservação Do Papagaio-Charão Amazona pretrei No Sul Do Brasil: Um Plano de Ação Preliminar*; Divulgações do Museu de Ciências e Tecnologia UBEA/PUCRS: EDIPUCRS: Porto Alegre, RS, Brazil, 1994.
38. Belton, W. *Birds of Rio Grande Do Sul, Brazil. Part 1. Rheidae through Furnariidae*; Bulletin of the American Museum of Natural History: New York, NY, USA, 1984; Volume 178.
39. Martinez, J.; Prestes, N.P. Tamanho populacional, tamanho médio de bando e outros aspectos demográficos do papagaio-charão (*Amazona pretrei*). In *Biologia da Conservação: Estudo de Caso Com Papagaio-Charão e Outros Papagaios Brasileiros*; Martinez, J., Prestes, N.P., Eds.; UPF Editora: Passo Fundo, RS, Brazil, 2008.
40. Tella, J.L.; Dénes, F.V.; Zulian, V.; Prestes, N.P.; Martínez, J.; Blanco, G.; Hiraldo, F. Endangered Plant-Parrot Mutualisms: Seed Tolerance to Predation Makes Parrots Pervasive Dispersers of the Parana Pine. *Sci. Rep.* **2016**, *6*, 31709. [CrossRef]

41. Schunck, F.; Somenzari, M.; Lugarini, C.; Soares, E.S. *Plano de Ação Nacional Para a Conservação Dos Papagaios Da Mata Atlântica*; Instituto Chico Mendes de Conservação da Biodiversidade-ICMBio: Brasília, DF, Brazil, 2011.
42. ICMBio. Plano de Ação Nacional Para a Conservação Dos Papagaios—PAN Papagaios: Matriz de Monitoria. 2020. Available online: https://www.icmbio.gov.br/portal/faunabrasileira/plano-de-acao-nacional-lista/837-plano-de-acao-nacional-para-conservacao-dos-papagaios-da-mata-atlantica (accessed on 2 April 2020).
43. *SPVS Relatório Anual 2019*; Sociedade de Pesquisa em Vida Selvagem e Educação Ambiental: Curitiba, Brazil, 2019; p. 52.
44. Scherer-Neto, P.; Toledo, M.C. Avaliação Populacional Do Papagaio-de-Cara-Roxa (*Amazona brasiliensis*) (Psittacidae) No Estado Do Paraná, Brasil. *Ornitol. Neotrop.* **2007**, *18*, 379–393.
45. Galetti, M.; Schunck, F.; Ribeiro, M.; Paiva, A.A.; Toledo, R.; Fonseca, L. Distribuição e Tamanho Populacional Do Papagaio-de-Cara-Roxa *Amazona brasiliensis* No Estado de São Paulo. *Rev. Bras. Ornitol.* **2006**, *14*, 239–247.
46. *SPVS Relatório Anual 2003*; Sociedade de Pesquisa em Vida Selvagem e Educação Ambiental: Curitiba, PR, Brazil, 2003; p. 20.
47. *SPVS Relatório Anual 2004*; Sociedade de Pesquisa em Vida Selvagem e Educação Ambiental: Curitiba, PR, Brazil, 2004; p. 16.
48. *SPVS Relatório Anual 2005*; Sociedade de Pesquisa em Vida Selvagem e Educação Ambiental: Curitiba, PR, Brazil, 2005; p. 8.
49. *SPVS Relatório Anual 2006*; Sociedade de Pesquisa em Vida Selvagem e Educação Ambiental: Curitiba, PR, Brazil, 2006; p. 12.
50. *SPVS Relatório Anual 2007*; Sociedade de Pesquisa em Vida Selvagem e Educação Ambiental: Curitiba, PR, Brazil, 2007; p. 16.
51. *SPVS Relatório Anual 2008*; Sociedade de Pesquisa em Vida Selvagem e Educação Ambiental: Curitiba, PR, Brazil, 2008; p. 12.
52. *SPVS Relatório Anual 2009*; Sociedade de Pesquisa em Vida Selvagem e Educação Ambiental: Curitiba, PR, Brazil, 2009; p. 6.
53. *SPVS Relatório Anual 2010*; Sociedade de Pesquisa em Vida Selvagem e Educação Ambiental: Curitiba, PR, Brazil, 2010; p. 4.
54. *SPVS Relatório Anual 2011*; Sociedade de Pesquisa em Vida Selvagem e Educação Ambiental: Curitiba, PR, Brazil, 2011; p. 13.
55. *SPVS Relatório Anual 2012*; Sociedade de Pesquisa em Vida Selvagem e Educação Ambiental: Curitiba, PR, Brazil, 2012; p. 38.
56. *SPVS Relatório Anual 2013*; Sociedade de Pesquisa em Vida Selvagem e Educação Ambiental: Curitiba, PR, Brazil, 2013; p. 27.
57. *SPVS Relatório Anual 2014*; Sociedade de Pesquisa em Vida Selvagem e Educação Ambiental: Curitiba, PR, Brazil, 2014; p. 27.
58. *SPVS Relatório Anual 2015*; Sociedade de Pesquisa em Vida Selvagem e Educação Ambiental: Curitiba, PR, Brazil, 2015; p. 35.
59. *SPVS Relatório Anual 2016*; Sociedade de Pesquisa em Vida Selvagem e Educação Ambiental: Curitiba, PR, Brazil, 2016; p. 52.
60. *SPVS Relatório Anual 2017*; Sociedade de Pesquisa em Vida Selvagem e Educação Ambiental: Curitiba, PR, Brazil, 2017; p. 48.
61. *SPVS Relatório Anual 2018*; Sociedade de Pesquisa em Vida Selvagem e Educação Ambiental: Curitiba, PR, Brazil, 2018; p. 44.
62. Abe, L.M. Caracterização do Hábitat do Papagaio-de-Peito-Roxo *Amazona vinacea* (Kuhl, 1820) No Município de Tunas do Paraná, Região Metropolitana de Curitiba, Paraná. Master's Thesis, Universidade Federal do Paraná, Curitiba, PR, Brazil, 2004.
63. Cockle, K.; Capuzzi, G.; Bodrati, A.; Clay, R.; del Castillo, H.; Velázquez, M.; Areta, J.I.; Fariña, N.; Fariña, R. Distribution, Abundance, and Conservation of Vinaceous Amazons (*Amazona vinacea*) in Argentina and Paraguay. *J. Field Ornithol.* **2007**, *78*, 21–39. [CrossRef]
64. Segovia, J.M.; Cockle, K.L. Conservación del Loro Vinoso (*Amazona vinacea*) En Argentina. *El Hornero* **2012**, *27*, 27–37.
65. Marsden, S.J.; Whiffin, M.; Sadgrove, L.; Guimarães, P. Parrot Populations and Habitat Use in and around Two Lowland Atlantic Forest Reserves, Brazil. *Biol. Conserv.* **2000**, *96*, 209–217. [CrossRef]
66. Klemann-Júnior, L.; Neto, P.S.; Monteiro, T.V.; Ramos, M.; de Almeida, R. Mapeamento da distribuição e conservação do chauá (*Amazona rhodocorytha*) no estado do Espírito Santo, Brasil. *Ornitol. Neotrop.* **2008**, *19*, 14.
67. Silva, F. Contribuição Ao Conhecimento Da Biologia Do Papagaio Charão, *Amazona pretrei* (TEMMINCK, 1830) (Psittacidae, Aves). *Iheringia Sér. Zool.* **1981**, *58*, 79–85.
68. Gaston, K.J.; Fuller, R.A. The Sizes of Species' Geographic Ranges. *J. Appl. Ecol.* **2009**, *46*, 1–9. [CrossRef]
69. Prestes, N.P.; Martinez, J.; Meyrer, P.A.; Hansen, L.H.; de Negri Xavier, M. Nest Characteristics of the Red-Spectacled Amazon *Amazona pretrei* Temminck, 1830 (Psittacidae). *Ararajuba* **1997**, *5*, 151–158.
70. Prestes, N.P.; Martinez, J.; Rezende, É. Biologia reprodutiva do papagaio-charão (*Amazona pretrei*). In *Biologia Da Conservação: Estudo de Caso com Papagaio-Charão e Outros Papagaios Brasileiros*; Martinez, J., Prestes, N.P., Eds.; Editora UPF: Passo Fundo, RS, Brazil, 2008; Volume 1.
71. Forshaw, J.M. *Parrots of the World*; Princeton Field Guides; Princeton University Press: Princeton, NJ, USA, 2010; ISBN 978-0-691-14285-2.
72. BirdLife International IUCN Red List of Threatened Species: Amazona Vinacea. 2017. Available online: http://dx.doi.org/10.2305/IUCN.UK.2016-3.RLTS.T22686374A93109194.en (accessed on 18 March 2019).
73. BirdLife International IUCN Red List of Threatened Species: Amazona Rhodocorytha. 2017. Available online: https://dx.doi.org/10.2305/IUCN.UK.2017-3.RLTS.T22686288A118968809.en (accessed on 18 March 2019).
74. ICMBio. Livro Vermelho da Fauna Brasileira Ameaçada de Extinção: Volume III—Aves. In *Livro Vermelho da Fauna Brasileira Ameaçada de Extinção*; Instituto Chico Mendes de Conservação da Biodiversidade. (Org.): Brasília, DF, Brazil, 2018; Volume 3, p. 712.
75. BirdLife International IUCN Red List of Threatened Species: *Amazona Pretrei*. 2016. Available online: https://dx.doi.org/10.2305/IUCN.UK.2016-3.RLTS.T22686251A93104759.en (accessed on 18 March 2019).
76. *Parrots: Status Survey and Conservation Action Plan, 2000–2004*; Snyder, N.F.R.; McGowan, P.; Gilardi, J.; Grajal, A. (Eds.) IUCN/SSC Action Plans for the Conservation of Biological Diversity; IUCN: Gland, Switzerland, 2000; ISBN 978-2-8317-0504-0.
77. BirdLife International IUCN Red List of Threatened Species: *Amazona Brasiliensis*. 2017. Available online: https://dx.doi.org/10.2305/IUCN.UK.2017-3.RLTS.T22686296A118478685.en (accessed on 27 April 2021).

78. Holt, A.R.; Gaston, K.J.; He, F. Occupancy-Abundance Relationships and Spatial Distribution: A Review. *Basic Appl. Ecol.* **2002**, *3*, 1–13. [CrossRef]
79. Sipinski, E.A.B.; Abbud, M.C.; Sezerban, R.M.; Serafini, P.P.; Boçon, R.; Manica, L.T.; de Camargo Guaraldo, A. Tendência Populacional Do Papagaio-de-Cara-Roxa (*Amazona brasiliensis*) No Litoral Do Estado Do Paraná. *Ornithologia* **2014**, *6*, 136–143.
80. Marini, M.Â.; Barbet-Massin, M.; Martinez, J.; Prestes, N.P.; Jiguet, F. Applying Ecological Niche Modelling to Plan Conservation Actions for the Red-Spectacled Amazon (*Amazona pretrei*). *Biol. Conserv.* **2010**, *143*, 102–112. [CrossRef]
81. Martinez, J.; Prestes, N.P. Um pouco da história do papagaio-charão (*Amazona pretrei*). In *Biologia da Conservação: Estudo de Caso Com Papagaio-Charão E Outros Papagaios Brasileiros*; Martinez, J., Prestes, N.P., Eds.; UPF Editora: Passo Fundo, RS, Brazil, 2008.
82. Martinez, J.; Prestes, N.P.; Rezende, É. As ameaças enfrentadas pelo papagaio-charão (*Amazona pretrei*). In *Biologia Da Conservação: Estudo de Caso com Papagaio-Charão e Outros Papagaios Brasileiros.*; Martinez, J., Prestes, N.P., Eds.; Editora UPF: Passo Fundo, RS, Brazil, 2008.
83. Ministerio de Ambiente e Desarollo Sustentable. Aves Argentinas. In *Categorizacion de Las Aves de La Argentina Según Su Estado de Conservación*; Informe del Ministerio de Ambiente y Desarrollo Sustentable de la Nación y de Aves Argentinas: Buenos Aires, Argentina, 2015.
84. BirdLife International Handbook of the Birds of the World *Amazona Pretrei*. 2016. Available online: https://dx.doi.org/10.2305/IUCN.UK.2016-3.RLTS.T22686251A93104759.en (accessed on 18 March 2019).
85. Bodrati, A.; Cockle, K. New Records of Rare and Threatened Birds from the Atlantic Forest of Misiones, Argentina. *Cotinga* **2006**, *26*, 20–24.
86. Brooks, M.; Clay, R.P.; Lowen, C.; Butchart, H.M.; Barnes, R.; Esquivel, E.Z.; Etcheverry, N.I.; Vincent, J.P. New Information on Nine Birds from Paraguay. *Ornitol. Neotrop.* **1995**, *6*, 129–134.
87. Ministerio del Ambiente y Desarrollo Sostenible. *Resolución n 254/19 Por La Cual Se Actualiza El Listado de Las Especies Protegidas de La Vida Silvestre de La Classe Aves*; Gobierno Nacional: Asunción, Paraguay, 2019.
88. Alerstam, T. Ecological Causes and Consequences of Bird Orientation. *Experientia* **1990**, *46*, 405–415. [CrossRef]
89. Roda, S.A. *Distribuição de aves Endêmicas e Ameaçadas em Usinas de Açúcar e Unidades de Conservação do Centro Pernambuco*; Centro de Pesquisas Ambientais do Nordeste: Recife, PE, Brazil, 2005; p. 42.
90. Katayama, M.V.; Zima, P.V.Q.; Perrella, D.F.; Francisco, M.R. Successional Stage Effect on the Availability of Tree Cavities for Cavity-Nesting Birds in an Atlantic Forest Park from the State of São Paulo, Brazil. *Biota Neotrop.* **2017**, *17*. [CrossRef]
91. Baptista, S.R.; Rudel, T.K. A Re-Emerging Atlantic Forest? Urbanization, Industrialization and the Forest Transition in Santa Catarina, Southern Brazil. *Environ. Conserv.* **2006**, *33*, 195. [CrossRef]
92. Bonfanti, T.; Meurer, C.; Martinez, J.; Prestes, N.P. A captura de papagaios: Espécies encontradas em cativeiro no norte e nordeste do Rio Grande do Sul. In *Biologia Da Conservação: Estudo de Caso com Papagaio-Charão e Outros Papagaios Brasileiros*; Martinez, J., Prestes, N.P., Eds.; Editora UPF: Passo Fundo, RS, Brazil, 2008.
93. Martuscelli, P. A Parrot with a Tiny Distribution and a Big Problem: Will Illegal Trade Wipe out the Red-Tailed Amazon? *PsittaScene* **1994**, *6*, 3–7.
94. de Cavalheiro, M.L. Qualidade do Ambiente e Características Fisiológicas do Papagaio-de-Cara-Roxa (*Amazona brasiliensis*) Ilha Comprida-São Paulo. Master's Thesis, Universidade Federal do Paraná UFPR, Curitiba, PR, Brazil, 1999.
95. Carrillo, A.C.; Batista, D.B. A conservação do papagaio-da-cara-roxa (*Amazona brasiliensis*) no estado do Paraná: Uma experiência de Educação Ambiental no ensino formal. *Rev. Árvore* **2007**, *31*, 113–122. [CrossRef]
96. Cordeiro, P.H.C. A fragmentação da Mata Atlântica no sul da Bahia e suas implicaçãoes na conservação dos psitacídeos. In *Ecologia e Conservação de Psitacídeos no Brasil*; Melopsittacus Publicações Científicas: Belo Horizonte, MG, Brazil, 2002; p. 236.

Article

Range-Wide Population Assessment of the Endangered Yellow-Naped Amazon (*Amazona auropalliata*)

Molly K. Dupin [1,*], Christine R. Dahlin [2] and Timothy F. Wright [1]

[1] Biology Department, New Mexico State University, Las Cruces, NM 88003, USA; wright@nmsu.edu
[2] Biology Department, University of Pittsburgh at Johnstown, Johnstown, PA 15904, USA; cdahlin@pitt.edu
* Correspondence: mdupin@nmsu.edu; Tel.: +1-717-887-1577

Received: 7 September 2020; Accepted: 27 September 2020; Published: 30 September 2020

Abstract: Yellow-naped amazons, *Amazona auropalliata*, have experienced a dramatic population decline due to persistent habitat loss and poaching. In 2017, BirdLife International changed the species' status from threatened to endangered and estimated that between 10,000 and 50,000 individuals remained in the wild. An accurate estimate of the number of remaining wild individuals is critical to implementing effective conservation plans. Wright et al. conducted roost count surveys in Costa Rica and Nicaragua during 2016 and published their data in 2019; however, no population data exists for the rest of the range. We conducted roost counts at 28 sites across Mexico, Guatemala, and the Bay Islands in Roatan during 2018 and 2019. We counted 679 birds and combined our data with the published Wright et al. (2019) data for a total of 2361 wild yellow-naped amazons observed across the species' range. There were fewer roosts detected in the northern region of the range than in the southern region. We found that roosts were most likely to occur in built-up rural and pasture habitat, with 71% found within 100 m of human habitation. Our results illustrate the need for immediate conservation action to mitigate decline, such as enforced legal action against poaching, nest guarding, and increased community education efforts.

Keywords: eBird; endangered species; parrot conservation; population survey; roost counts

1. Introduction

The rapid collection of population and demographic data on wild populations of endangered species has become increasingly imperative in the face of the modern mass extinction. It is estimated that in the upcoming decades we will see drastic population declines which could result in the extinction of 54% of all species due to global warming, habitat loss caused by agricultural expansion, deforestation, unregulated grazing, urbanization, and other human activities [1,2]. Human alterations to the environment have already resulted in a substantial proportion of habitat types becoming rarer and more fragmented. Fragmentation of a species' habitat minimizes the opportunity for affected individuals to breed, forage, and interact socially with conspecifics, which can result in population decline and increases the likelihood of local extinction [3].

Seasonally dry tropical forest habitat, which once covered major swaths of Mesoamerica, has suffered significant losses due to deforestation and increased agricultural production [4,5]. This habitat type is home to a substantial proportion of the world's diversity, especially with regard to bats and birds; however, in the year 2000, only 30% of the original extent of this land cover remained in Central America [5]. As tropical dry forests become smaller and more fragmented, remaining patches sustain fewer residents and other individuals are pushed into human-altered habitat such as pastures and rural villages, where they are often exposed to human interaction. This change puts threatened and endangered individuals at a higher risk for events which can exacerbate population decline, such as direct persecution and poaching [4,6,7].

Parrots and cockatoos of the order Psittaciformes have been experiencing an especially rapid decline for the past four decades, largely due to a combination of anthropogenic factors (e.g., poaching for the pet trade, habitat destruction, introduction of exotic species, and direct persecution) and biological ones (e.g., disease, lack of breeding individuals) [7,8]. As a result, they are one of the most threatened avian orders, with over 28% of species classified as vulnerable, endangered, or critically endangered and an additional 14% classified as near threatened by the International Union for the Conservation of Nature (IUCN) [9]. Berkunsky et al. [7] found that 38% of 192 neotropical parrot populations were experiencing decline, primarily due to agricultural activity and poaching for the pet trade. To combat these declines, basic population data are necessary to allow biologists to craft management plans and determine where to most effectively concentrate management actions. However, population data is time-consuming and costly to collect from many parrot species due to their generally high mobility and large ranges [10]. Management actions, including captive breeding, reintroduction programs, or nest protection programs, can cost organizations millions of dollars, and therefore should be targeted to key areas [11].

Yellow-naped amazons, *Amazona auropalliata*, are large, charismatic parrots that are native to Mesoamerica, occupying lowland mangrove and tropical dry forest habitat from southern Mexico to northern Costa Rica along both the Pacific and Caribbean coasts [12] (Figure 1). In 2017 the IUCN declared yellow-naped amazons endangered [12]. This status change was supported in part by data Wright et al. collected in 2016 and published in 2019 [13]. The study showed that populations sampled in Costa Rica had experienced a mean decline of 54% in only 11 years. Additionally, long-term nest monitoring of this species indicated that only 11% of yellow-naped amazon young successfully fledged the nest, and the highest cause of mortality was due to poaching for the pet trade [14,15]. Yellow-naped amazons exhibit fission-fusion flock patterns throughout the day and sleep in temporally stable, communal roosts [16]. Long-term monitoring of populations via roost counts in Costa Rica and Nicaragua have shown that yellow-naped amazons are particularly tolerant of human-disturbed habitat, which means that fragmentation increases the risk of poaching or persecution for this species [13]. An informal survey of Costa Rican inhabitants indicated overall compassion for the species, but also a lack of fear regarding the legal repercussions of illegal poaching [14,17]. While periodic monitoring of populations of yellow-naped amazons in the southern region of the range has demonstrated an ongoing decline, at present there is only fragmentary data available regarding the status of populations in the northern regions of the range [13].

Counting wild parrot populations is challenging but essential for understanding population trends of species on both regional and range-wide scales. Historically, population data is collected via traditional field-based counts which are a reliable method to determine the number of individuals in an area and evaluate population fluctuations due to immigration, emigration, and mortality events [18]. Publications using roost counts have generally been successful in estimating the number of parrots in a particular area or region; however, roost counts also present many challenges due to required time and funding, detectability of animals, and unpredictable roost use in some species [19,20]. Other forms of field counts such as transects and point counts are less commonly used on wild parrots due to their highly mobile lifestyles, sparse distribution, and large home ranges, and their general wariness around humans, which makes them difficult to count [10] (but see Joyner's transect guide [21]). Thus, scientists have been seeking alternative ways to collect population data on various species in the wild.

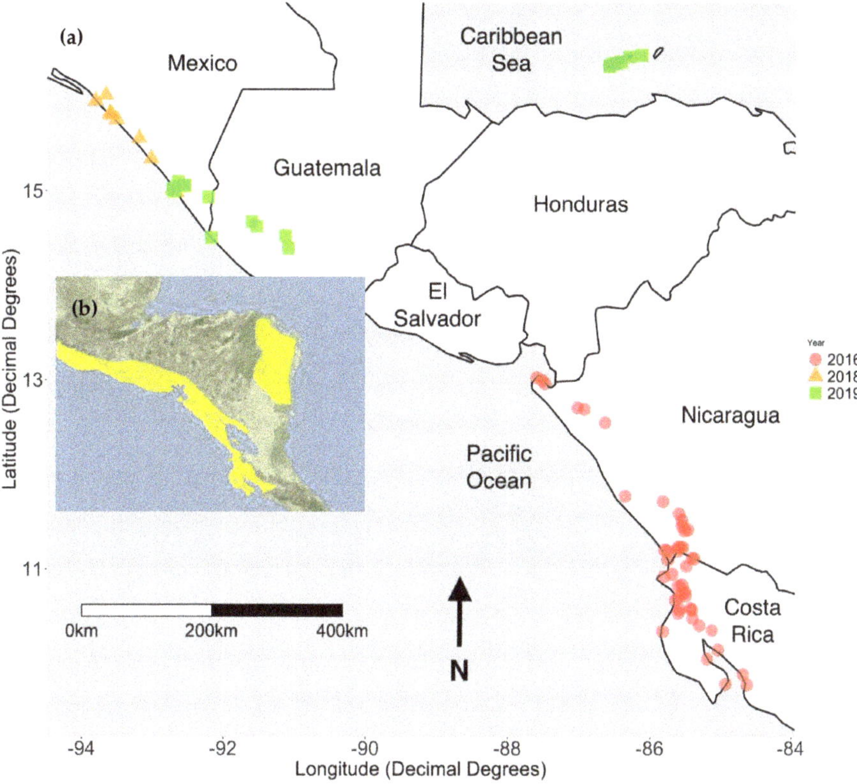

Figure 1. (**a**) A map of all sites sampled during 2018 and 2019. Sites from the Wright et al. study [13] conducted in 2016 are also included. The color and shape of each point corresponds to the year the site was sampled. (**b**) A species range polygon for the yellow-naped amazon provided by BirdLife [22].

One such alternative method is the use of volunteer-based data collection. For example, during the yearly American Breeding Bird Survey (ABBS) hosted by the United States Geological Survey and Canadian Wildlife Service, volunteers across North America collect visual and aural data on bird species encountered in various areas [23]. Another popular example is the online, open-access database eBird (www.eBird.org), which displays observations reported by birdwatchers, including information on species, geographic location, and date of observation. eBird has been used previously to assess species' presence and diversity, migration patterns, distribution, and in some cases, general population trends [23]. However, it should be noted that eBird reports do not follow a standardized protocol and individuals submitting reports are untrained, therefore the data are not standardized in the same way as structured field counts [24]. This presents a dilemma for conservationists who wish to use citizen science as a tool for tracking population trends. In fact, while Walker and Taylor [25] found that eBird reports closely followed those of the ABBS, Kamp et al. [24] found that citizen science databases did not accurately indicate the well-known decline of several common bird species in Denmark. Furthermore, there have been no such comparisons for tropical species like the yellow-naped amazon.

We aimed to estimate the number of remaining yellow-naped amazon individuals in the wild across the species' Mesoamerican range using traditional roost counts and, for comparison, eBird reports from the same months and years. We performed roost counts in the northern range of the yellow-naped amazon during 2018 and 2019 and combined these data with those collected by Wright et al. [13] in 2016 using the same methodology. We also recorded the distance to human habitation, habitat type, and elevation at each site and combined these results with previously collected data published by Wright et al. [13]. We assessed the difference between our field counts of yellow-naped amazons with reports on eBird over the same period of time. Finally, for each region we used habitat type and number of individuals observed at communal roosts to determine locations which should be considered conservation priorities for this species.

2. Materials and Methods

We conducted population counts of the yellow-naped amazon in Mexico in 2018 and in Mexico, Guatemala, and the Bay Islands, Honduras in 2019. Specific sites were chosen using a site selection process, which entailed a detailed examination of unsampled areas within the range to identify locations in which yellow-naped amazons were most likely to occur. Multiple sources were utilized in this process, including detailed consultation with local experts and other conservationists, and historical reports of the species on www.eBird.org. Areas where parrots had been reported within the past two years were selected as priority sites during the field season and were scouted thoroughly. We also explored some historically populated areas where this species was reported to no longer exist.

Roost counts were conducted at each site using the same protocol as Wright et al. [13] to maximize data compatibility between the two studies. Counts took place in the morning beginning before dawn, and in the evening prior to sunset, and were done in both the morning and evening at each site when time and weather permitted. Counts were performed during June and July, outside the species breeding period. No single site was counted more than twice in one year. Observers were stationed within each site in a manner which allowed maximum visibility of the parrots' flight path. Each observer was equipped with binoculars, a notepad, a compass, a GPS device, and a watch. Observers recorded the number of birds flying into or out of the roost, the direction of flight, time of day, and the location and altitude of the roost using GPS. In some cases, roosts were difficult or even impossible to reach, therefore GPS location was taken as close to the roost as possible and a thorough description of the roost itself was recorded. In the event that multiple observers counted at the same site and the number of birds differed between observers, the highest number was used. Roost behavior was noted even if the physical roost could not be directly observed. This includes groups of parrots consistently flying from the same direction during and shortly after dawn, or groups of parrots flying in the same direction or toward the same area during or shortly before dusk. In addition to direct counts, we made estimations of roost size in which we supplemented our visual observations with aural observations; we separately report both counts and estimates for each roost (see Table 1). At each roost site for 2018 and 2019, the dominant habitat type was classified as one of five habitat types: mangrove, tropical dry forest, tropical pine, agricultural, or built-up rural. These types were categorized by: (i) the presence of trees and shrubs flooded with brackish or saltwater along coasts and waterways (mangrove); (ii) dense swaths of tall trees like *Ceiba pentandra* (ceiba), *Hura crepitans* (jabillo), and *Enterolobium cyclocarpum* (guanacaste) (tropical dry forest); (iii) the presence of open fields with some tree stands and low-intensity agricultural areas (pasture/agriculture); (iv) stands dominated by *Pinus caribaea* stands (tropical pine); or (v) human-modified landscape that included features such as homes, small buildings, schools, and roads (built-up rural). We combined the 2018 and 2019 datasets with unpublished roost habitat data from the sites counted by Wright and Dahlin in 2016. We also recorded the proximity of human habitation by recording whether roosts were within 100 m of human infrastructure.

Table 1. A list of data collected from Mexico, Guatemala, and Honduras during 2018 and 2019.

Country	Year	Site Name	Latitude	Longitude	Roost Behavior	Roost Observed	Roost Count	Roost Estimate	eBird Count	Habitat Type
Mexico	2018	Aztlan Site 3	14°58.841′	92°40.158′	Yes	Yes	33	-	24	Pasture/Agriculture
Mexico	2018	Aztlan Site 2	14°59.596′	92°40.731′	Yes	No	38	-	24	Pasture/Agriculture
Mexico	2018	Aztlan Site 1	15°0.045′	92°41.431′	Yes	No	51	-	24	Pasture/Agriculture
Mexico	2018	Las Brisas de Hueyate	15°1.422′	92°43.166′	Yes	No	113	114	0	Mangrove
Mexico	2018	Manguito	15°45.18′	93°30.99′	No	No	0	-	0	Mangrove
Mexico	2018	Rancho el Piñon	16°0.667′	93°40.2′	No	No	2	4	0	Built-up rural
Mexico	2018	Ponte Duro	15°48.485′	93°35.728′	No	No	3	4	3	Built-up rural
Mexico	2018	Ponte Duro Mangroves	15°49.063′	93°36.816′	Yes	No	16	-	8	Mangrove
Mexico	2018	Puerto Arista	15°56.98′	93°48.871′	No	No	0	-	0	Built-up rural
Mexico	2018	Roberto Barrios	15°20.435′	93°1.412′	Yes	Yes	23	-	4	Pasture/Agriculture
Mexico	2018	Salto de Agua	15°33.326′	93°11.633′	Yes	Yes	12	-	33	Pasture/Agriculture
Mexico	2019	Aztlan Town Roost	15°1.003′	92°42.278′	Yes	Yes	33	74	19	Built-up rural
Mexico	2019	Las Palmas	14°59.943′	92°41.455′	Yes	No	39	40	0	Built-up rural
Mexico	2019	Huixtla	15°3.233′	92°32.991′	No	No	0	-	41	Built-up rural
Mexico	2019	Hidalgo	15°5.872′	92°38.266′	No	No	0	-	0	Built-up rural
Mexico	2019	Tapachula	14°55.965′	92°13.33′	No	No	0	-	0	Built-up rural
Guatemala	2019	Tilapa	14°30.18′	92°10.753′	No	No	0	-	8	Built-up rural
Guatemala	2019	San Martín Zapotitlan	14°37.303′	91°32.358′	No	No	0	-	3	Built-up rural
Guatemala	2019	Finca Patrocinio	14°40.221′	91°36.538′	No	No	5	10	2	Tropical dry
Guatemala	2019	Los Tarrales	14°31.328′	91°8.34′	No	No	3	8	9	Tropical dry
Guatemala	2019	Pineapple Plantation	14°23.458′	91°5.732′	Yes	No	8	12	43	Pasture/Agriculture
Honduras	2019	Sandy Bay	16°19.133′	86°34.716′	No	No	22	-	55	Built-up rural
Honduras	2019	Guava Grove	16°18.985′	86°34.701′	Yes	Yes	20	24	4	Pasture/Agriculture
Honduras	2019	Mud Hole	16°20.81′	86°31.843′	No	No	0	-	0	Built-up rural
Honduras	2019	Parrot Tree	16°21.866′	86°24.851′	No	No	0	-	0	Built-up rural
Honduras	2019	Los Fuertes	16°20.913′	86°28.481′	No	No	0	-	0	Built-up rural
Honduras	2019	Undisclosed	–	–	Yes	No	248	266	0	Tropical dry
Honduras	2019	Port Royal National Park	16°25.083′	86°17.8′	Yes	No	10	-	10	Tropical pine forest

The roost count results were summed at both the country and range-wide levels to provide a minimum count of yellow-naped amazons per country and to assess the species population as a whole. This was accomplished by combining survey data from Costa Rica and Nicaragua published by Wright et al. [13] with the data we collected in 2018 and 2019. When making these totals we removed one of two repeat counts at two sites in Mexico during 2018/2019. We compared these data to reports of yellow-naped amazon sightings on www.eBird.org and used a paired t-test to assess whether the difference between the two totals was significant. We then used a Spearman's rank correlation to evaluate the relationship between eBird reports and roost count data. To maximize comparability, we only counted eBird reports that were located within 5 km of the roost GPS location that we recorded in the field.

We tested for differences in mean roost size between countries using a Kruskal-Wallis test, due to the unequal variances present in the data. We compared the sizes of roosts within 100 m and farther than 100 m from human habitat by using Welch's two-sample t-test, which accounts for unequal variances. We tested for an association between roost size and elevation using a Spearman's rank correlation. All means are listed ± SD, and all alpha values of significance were $p < 0.05$.

3. Results

3.1. Roost Count Results

Twenty-eight roost counts were completed within the northern portion of the yellow-naped amazon range during 2018 and 2019 (Figure 1), during which our team counted 679 yellow-naped amazons. Birds were observed at 18 of the 28 sites, and only three of those had more than 50 birds counted (Figure 2). Eleven sites were sampled in Chiapas, Mexico during 2018, and five sites were sampled in Mexico during 2019, with a total of 363 birds observed in Mexico. In 2019 in Guatemala, 16 birds were counted within five sites; on the island of Roatán, Honduras, we counted 52 birds within six sites, and we counted 248 birds at a private, undisclosed location in the Bay Islands (see Table 1). Roosts in Guatemala were no larger than eight birds, and on Roatan, the largest roost was 22 birds. Previously Wright et al. [13] reported surveying 25 sites in Costa Rica, only four of which had no parrot sightings. They counted 990 individuals across the remaining 21 sites. In the same study, Wright et al. surveyed 19 areas in Nicaragua, of which two had no observations of yellow-naped amazons, and 692 parrots were reported [13]. To summarize, in Mexico, 363 yellow-naped amazons were observed, while 16 were seen in Guatemala, and 300 were observed across the Bay Islands. Wright et al. [13] reported 692 parrots in Nicaragua and 990 in Costa Rica. Across these two complementary studies, a total of 2361 yellow-naped amazons were counted across the range. Using the combined dataset, the Kruskal-Wallis test did not detect any significant differences in median roost size between the countries we surveyed in the number of yellow-naped amazons observed (chi-squared = 6.8985, df = 4, $p = 0.141$).

3.2. Estimated Differences between Traditional Roost Counts and eBird Database Reports

We combined our roost count data with previously published roost count data from Costa Rica and Nicaragua collected with the same methodology by Wright et al. [13]. This added 44 sites to our data, for a total of 72 sites sampled across the yellow-naped amazon range. We found a significant relationship between eBird and roost count data ($R_s = 0.369$, $p = 0.00143$). There was, however, a consistent difference between counts obtained with the two approaches, with the number of birds counted using roost counts (33.0 ± 55.3) significantly higher than those from eBird reports (16.5 ± 35.3) for submissions within 5 km the corresponding roost count location (N = 72, t = 2.6061, $p = 0.011$).

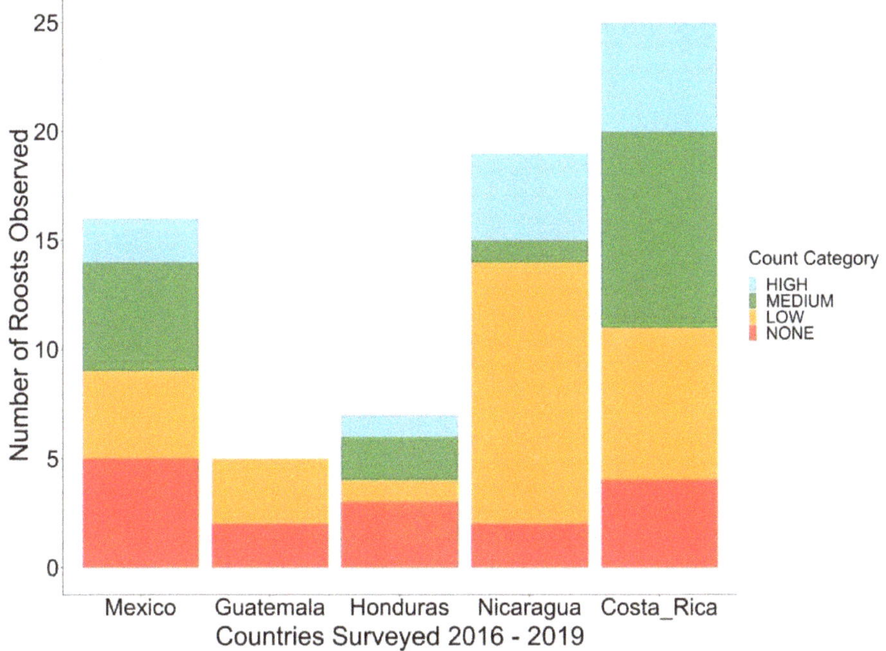

Figure 2. The number of none, low, medium, and high roosts in each country. Roosts were assigned a count category based on the number of birds present. None = 0 birds counted at the roost, low = 1–20 birds, medium = 21–50 birds, and high = 51+ birds.

3.3. *Roost Characteristics*

Across the entire range, we found a significant difference between the size of roosts within 100 m of human habitation and outside 100 m of human habitation ($t = -2.89$, $df = 44.725$, $p = 0.006$). Roosts within 100 m of human habitation were larger on average (47.18 ± 63.31 versus 14.81 ± 16.69); however, using only the 2018 and 2019 data from the northern range, we were not able to find a significant difference between roost size within and outside 100 m of human habitation ($t = -0.217$, $df = 25.075$, $p = 0.83$). We categorized all roost sites by dominant habitat type and found that the preferred types were built-up rural (35% of roosts) and pasture (31% of roosts) (Figure 3). Some roosts were located on the edge of intensive agricultural fields such as pineapple plantations, however no roosts were located within any high-intensity agricultural areas (Supplementary Table S1). Additionally, we found a weak negative association between roost size and elevation of roosts using a Spearman's rank test ($r_s = 0.25$, $N = 72$, $p = 0.036$). With few exceptions, roosts surveyed in 2018 and 2019 consistently occurred below 300 m elevation (5 of 28 roosts, 82% below 300 m).

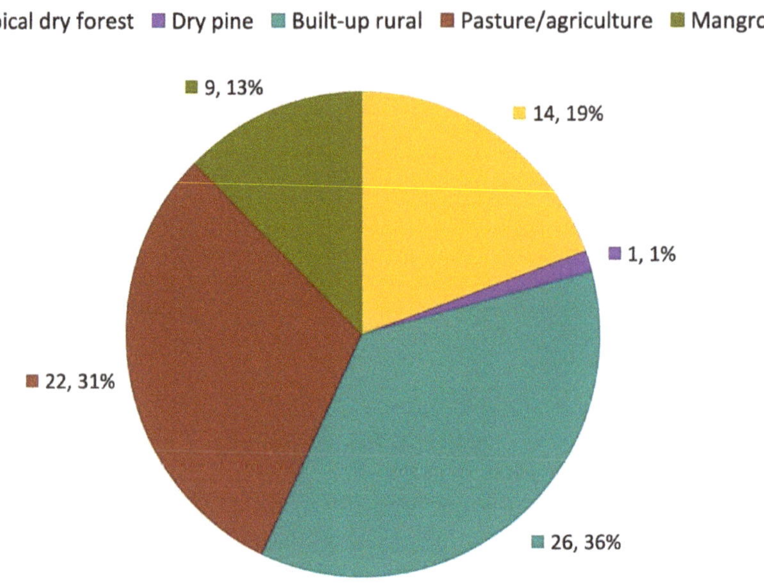

Figure 3. The proportion of yellow-naped amazon roosts in each of six habitat types (N = 72).

4. Discussion

Yellow-naped amazons have experienced a dramatic decline across their range for several decades due to habitat fragmentation, land alteration, and most persistently, poaching for the pet trade [13]. Long-term studies, however, have only been conducted in the southern portion of this species' range, despite evidence that the same threats of unregulated capture of the pet trade and loss of habitat were present in the northern portion [8,26]. In 2017, a report published by BirdLife International on the yellow-naped amazon estimated there may be less than 10,000 remaining individuals in the wild, but acknowledged great uncertainty regarding populations in the northern part of the range. This uncertainty has presented a major challenge to conservationists hoping to implement effective range-wide management strategies.

4.1. Range-Wide Population Estimates for the Yellow-Naped Amazon

Our study is the first of which we are aware to execute a range-wide population estimate for the yellow-naped amazon. We performed roost counts at 28 sites in Mexico, Guatemala, and Honduras, and counted a total of 679 birds between 2018 and 2019. Of the 28 sites we sampled, 10 were recorded as having no parrots at all. For comparison, Wright et al. [13] counted 1682 birds in Costa Rica and Nicaragua in 2016, with only 6 out of 44 sites having no birds. In total, only 2361 yellow-naped amazons have been observed across their Mesoamerican range between these two surveys conducted from 2016 to 2019 with similar methods and overlapping personnel. Our results, combined with those of Wright et al. [13], provide a better estimate of the global population of this iconic species and reinforce the conclusion that it has experienced a drastic population decline.

The results for our range-wide survey have indicated that there exist several regions which should be given special consideration as high conservation priorities. In the northern portion of the species range these include the Reserva de la Biosfera la Encrucijada, Mexico and the Bay Islands in Honduras. The Reserva de la Biosfera la Encrucijada was the most prominently populated region of Mexico, with several roosts located in the southern part of the reserve. Of the 2361 birds counted across the range, approximately 15% of individuals were located in Reserva de la Biosfera la Encrucijada. The Bay Islands contained relatively few roosts, with 83% of birds there concentrated in one roost locale. We identified several locations that should be prioritized in the Southern part of the range, including the Island of Ometepe with roost sites Peña Inculta, Mérida, and Tichana in Nicaragua. Costa Rica as a whole still has 13 sites with more than 30 birds counted or estimated, and thus strong conservation measures should be undertaken throughout this country. Two sites in particular with over 100 birds each should be prioritized; they are Cuajuiniquil and Finca Charlie Red [13].

Logistical constraints prevented us from surveying all of the yellow-naped amazon range. We were unable to survey El Salvador, as well as regions within southeast Honduras and northeast Nicaragua along the Caribbean coast. Anecdotal evidence suggests that El Salvador has fewer than 100 remaining individuals (Nestor Herrera, pers. comm.). Joyner's transect guide details counts from the last several years in Honduras. Her team counted 115 yellow-naped amazons in northern Honduras during 2015, 94 birds in Chismuyu Bay in 2017, and nearly 500 on Guanaja Island in 2018, emphasizing the importance of the Honduras Bay Islands for this species [21]. Wiedenfeld, Molina, Hille, and López conducted counts throughout the Caribbean regions of Nicaragua during 2013 and observed 73 yellow-naped amazons (Martín Lezama, pers. comm.).

4.2. Threats to Populations

Regions within the yellow-naped amazon range in which no birds were observed have typically experienced heavy human modification through agriculture or logging. Increased agricultural production has been influencing regions of Central America for the past several decades, and many areas of suitable habitat in Costa Rica and Nicaragua have been converted into high-intensity crop sites for export products such as sugar cane, rice, oil palm, and pineapples [27,28]. These landscapes lack tree stands used by parrots for foraging and roosting [13]. Logging of trees removes large portions of suitable habitat and creates or exacerbates already-existing fragmentation, which puts species like the yellow-naped amazon at a higher risk of exposure to humans. Humans may encourage contact with parrot populations by providing them with food and water [29]. Poaching of yellow-naped amazons removes individuals with future breeding potential and has extirpated some populations. The surveys our team conducted in the northern part of the range mirror these observations. In the western San Marcos Department of Guatemala, local inhabitants reported historical populations of yellow-napes, stating that at one time suburban areas around Aldea El Chico were more densely vegetated and home to a variety of birds, but as a result of landscape changes for agriculture and housing structures this is no longer the case (Dupin, pers. obs.). Historical submissions for sightings of yellow-naped amazons on eBird support this account, although sightings in these locations were only ever of a few birds (www.eBird.com). North of Retalhuleu in Guatemala, there exists plentiful amounts of suitable habitat, yet our team was unable to observe any birds, with the exception of 5 birds flying overhead near Finca Patrocinio. Our team's conversations with local residents revealed that low population numbers of yellow-naped amazons may be a result of historically high levels of poaching within the region, and that there still exists a market for this species (MKD, unpublished data). Yellow-naped amazons are especially popular because of their remarkable vocal mimicry skills and beautiful color.

4.3. The Utility of Citizen Science Approaches to Monitoring Populations

The perceived usefulness and popularity of citizen science databases in collecting population data [23,30–32] motivated us to examine the utility of the eBird database for conducting a population census of the yellow-naped amazon. We compared our roost count data to the reports of yellow-naped amazon sightings on the citizen science database eBird. While we did detect a statistically significant relationship between these two counts, this relationship was weak and there was often a substantial disparity between the two methods in the number of individuals detected. Our standardized roost counts consistently detected more birds than did the eBird reports for the same sites. It should be noted that eBird reports for yellow-naped amazon sightings did not provide the time of day during which observations took place. Thus, this discrepancy may be attributed to eBird users counting daytime foraging groups, which tend to be smaller, in addition to the occasional roost. Although eBird has been recognized as a useful tool for fine-scale mapping and tracking temporal changes in the distribution patterns of some species [33], our findings highlight the limitations of this approach for population estimates of an endangered species, particularly in developing countries of Central America where reporting intensity may be lower. This problem may be exacerbated by several factors exhibited by the yellow-naped amazon: they are rare, they use dense mangrove and forested habitat, they range widely, they are quiet for most of the day while foraging, and roost in very specific, sometimes inaccessible locations. Thus, knowledge of their roosting locales is important for complete censusing of populations.

An important bias to consider regarding eBird is that reports can be created and submitted by anyone who creates an account through the website, which introduces the potential for reporting error in submissions made by less experienced birders. A 2016 study showed that a citizen science database in Denmark failed to indicate the decline of several common bird species, primarily due to the inexperience of observers in bird identification or lack of standardized protocols [24]. We also noticed a pattern of inaccurate species identification on eBird. For example, when investigating eBird reports of yellow-naped amazons in the field, we would often instead find flocks of the sympatric white-fronted amazon, *Amazona albifrons*, whose similar body coloration and conformation often led to confusion by local inhabitants (Dupin, pers. obs.) and possibly birdwatchers as well. Thus, we do not consider this approach for collecting citizen science data to be an accurate estimator for endangered species such as the yellow-naped amazon.

One alternative citizen science approach which we believe has the potential to aid in the conservation of various parrot species is the long-term monitoring of populations across the range via minimally trained volunteers. The African-Eurasian Waterbird Census and the Portuguese Society for the Study of Birds are just two examples of successful programs which have incorporated the long-term monitoring of one or more species using volunteers [34,35]. In 2019, we developed the Mesoamerican Parrot Census Network for the yellow-naped amazon (https://parrotcensus.com) with the goal of joining together conservationists, researchers, and interested members of the general public in the common goal of preventing species extinction. Through this network, we aim to collect long-term population data on the yellow-naped amazon in the form of regular roost counts using a standardized protocol.

4.4. Implications for Conservation and Management

In 2017, BirdLife International estimated that there remained 10,000–50,000 yellow-naped amazons remaining in the wild, based on available data [12]. Our team counted less than 3000 scattered among fragmented habitat in our range-wide survey, which illustrates the need for immediate action to mitigate this species' decline. Our survey has shown that in order to ensure the success of the yellow-naped amazon, strategic and targeted conservation plans should be implemented immediately and focused in areas where there still remain healthy populations and suitable habitat. It is our opinion that these efforts should focus primarily on in situ conservation efforts such as habitat protection, enforcement of poaching repercussions, nest guarding, and gaining an improved understanding of population distributions and movements throughout the range. The blue-throated macaw, *Ara glaucogularis*, is a recent example of a species facing extinction that has begun to reverse its decline with the implementation and maintenance of in situ conservation measures [36–38]. An additional critical approach is enhancing community education and involvement through local pride [36]. Dahlin et al. [14] also write about their use of education programs as an added measure to teach children about why parrots are important [14,39]; similar programs have been implemented with the scarlet macaw, *Ara macao*, in Costa Rica [37]. More effort to promote eco-tourism would provide economic opportunities to members of the local community while bringing public attention to diverse and endangered wildlife. Ecotourism measures were also suggested as conservation measures for the declining blue-throated macaw populations in the early 2000s [36]. Pires advocates for situational crime prevention, improved policy and legislation with organizations such as CITES, and eco-tourism focusing on endangered and threatened species as the most effective approach to eliminating the threat of poaching [40,41]. Finally, there are substantial numbers of yellow-naped amazons held as pets within the countries of Mesoamerica, and programs aimed at rehabilitation, captive breeding and eventually reintroduction, although resource-intensive, should be considered.

We believe that our results, in conjunction with those of Wright et al. [13], and the recent up-listing of the yellow-naped amazon to endangered status, highlight the need for immediate conservation action for this species. We recommend that conservation funding and planning should be focused on the areas that we have designated as priorities in an effort to focus limited time and funding on healthy, breeding populations. Increased efforts toward habitat and nest protection should also be considered, such as with camera traps and provision of protected artificial nests [38,39]. Surveys should be conducted routinely with the remaining wild populations of this species to monitor and manage its decline. Historically populated regions of eastern Honduras and Nicaragua should also be investigated and routinely monitored. By implementing these strategies and following examples of programs that have aided in the recovery and success of other endangered parrot species, we believe that we can reverse the trend of population decline in the yellow-naped amazon.

Supplementary Materials: The following are available online at http://www.mdpi.com/1424-2818/12/10/377/s1, Table S1: All sites surveyed for yellow-naped amazons with each site's habitat type.

Author Contributions: Conceptualization, M.K.D. and C.R.D. and T.F.W.; methodology, M.K.D. and C.R.D. and T.F.W.; formal analysis, M.K.D.; investigation, M.K.D. and C.R.D. and T.F.W.; resources, T.F.W.; data curation, M.K.D.; writing—original draft preparation, M.K.D.; writing—review and editing, C.R.D. and T.F.W.; visualization, M.K.D.; supervision, T.F.W.; project administration, T.F.W.; funding acquisition, M.K.D. and T.F.W. and C.R.D. All authors have read and agreed to the published version of the manuscript.

Funding: This research was funded by the World Parrot Trust (grant date April 2018), and New Mexico State University's College of Arts and Sciences.

Acknowledgments: We would like to extend our gratitude to Martín Lezama (Universidad Centroamericana), Edith Belen Jimenez-Diaz, María Cristina Contreras-Meda, Candelario Giron Montes, Eric Anderson, Cara Dunbar, and Jamie Gilardi for advice and assistance. Thank you to LoraKim Joyner and One Earth Conservation for their collaboration, to all volunteers working with the Mesoamerican Parrot Census Network, and to our field assistants Lorena Cabada-Gomez and Carlos. We would also like to thank David Wiedenfeld (University of Oklahoma), José Molina (Oficina CITES- Nicaragua), David Hille (University of Oklahoma/Northwest Nazerne Univesrity) for population estimates in the Caribbean regions of Nicaragua. The Reserva la Biosfera de Encrucijada, the Area de Conservación Guanacaste, and the Horizontes Field Station in Costa Rica aided in our data collection by permitting us to conduct our fieldwork.

Conflicts of Interest: The authors declare no conflict of interest.

References

1. Rosenberg, K.V.; Dokter, A.M.; Blancher, P.J.; Sauer, J.R.; Smith, A.C.; Smith, P.A.; Stanton, J.C.; Panjabi, A.; Helft, L.; Parr, M.; et al. Decline of the North American avifauna. *Science* **2019**, *366*, 120–124. [CrossRef] [PubMed]
2. Urban, M.C. Accelerating extinction risk from climate change. *Science* **2015**, *348*, 571–573. [CrossRef] [PubMed]
3. Fahrig, L. Effects of habitat fragmentation on biodiversity. *Annu. Rev. Ecol. Evol. Syst.* **2003**, *34*, 487–515. [CrossRef]
4. Sabogal, C. Regeneration of tropical dry forests in Central-America, with examples from Nicaragua. *J. Veg. Sci.* **1992**, *3*, 407–416. [CrossRef]
5. Chazdon, R.L.; Harvey, C.A.; Martínez-Ramos, M.; Balvanera, P.; Stoner, K.E.; Schondube, J.E.; Avila Cabadilla, L.D.; Flores-Hidalgo, M. Seasonally dry tropical forest biodiversity and conservation value in agricultural landscapes of Mesoamerica. In *Seasonally Dry Tropical Forests*; Dirzo, R., Young, H.S., Mooney, H.A., Ceballos, G., Eds.; Island Press: Washington, DC, USA, 2011; pp. 195–219.

6. Visco, D.M.; Michel, N.L.; Boyle, W.A.; Sigel, B.J.; Woltmann, S.; Sherry, T.W. Patterns and causes of understory bird declines in human-disturbed tropical forest landscapes: A case study from Central America. *Biol. Conserv.* **2015**, *191*, 117–129. [CrossRef]
7. Berkunsky, I.; Quillfeldt, P.; Brightsmith, D.J.; Abbud, M.C.; Aguilar, J.; Aleman-Zelaya, U.; Aramburu, R.M.; Ariash, A.A.; McNab, R.B.; Balsby, T.J.S.; et al. Current threats faced by Neotropical parrot populations. *Biol. Conserv.* **2017**, *214*, 278–287. [CrossRef]
8. Wright, T.F.; Toft, C.A.; Enkerlin-Hoeflich, E.; Gonzalez-Elizondo, J.; Albornoz, M.; Rodriguez-Ferraro, A.; Rojas-Suarez, F.; Sanz, V.; Trujillo, A.; Beissinger, S.R.; et al. Nest poaching in neotropical parrots. *Conserv. Biol.* **2001**, *15*, 710–720. [CrossRef]
9. BirdLife International. Psittaciformes assessment. In *The IUCN Red List of Threatened Species*; The International Union for Conservation of Nature: Cambridge, UK, 2019.
10. Denes, F.V.; Tella, J.L.; Beissinger, S.R. Revisiting methods for estimating parrot abundance and population size. *Emu Austral Ornithol.* **2018**, *118*, 67–79. [CrossRef]
11. Loss, S.R.; Will, T.; Marra, P. Direct mortality of birds from anthropogenic causes. *Annu. Rev. Ecol. Evol. Syst.* **2015**, *46*, 99–120. [CrossRef]
12. BirdLife International. *Amazona auropalliata. The IUCN Red List of Threatened Species 2017*; E.T22686342A118961453; The International Union for Conservation of Nature: Cambridge, UK, 2017.
13. Wright, T.F.; Lewis, T.C.; Lezama-Lopez, M.; Smith-Vidaurre, G.; Dahlin, C.R. Yellow-naped Amazon Amazona auropalliata populations are markedly low and rapidly declining in Costa Rica and Nicaragua. *Bird Conserv. Int.* **2019**, *29*, 291–307. [CrossRef]
14. Dahlin, C.R.; Blake, C.; Rising, J.; Wright, T.F. Long-term monitoring of Yellow-naped Amazons (*Amazona auropalliata*) in Costa Rica: Breeding biology, duetting, and the negative impact of poaching. *J. Field Ornithol.* **2018**, *89*, 1–10. [CrossRef]
15. Wright, T.F.; Dahlin, C.R. Vocal dialects in parrots: Patterns and processes of cultural evolution. *Emu* **2018**, *118*, 50–66. [CrossRef] [PubMed]
16. Wright, T.F. Regional dialects in the contact call of a parrot. *Proc. R. Soc. B Biol. Sci.* **1996**, *263*, 867–872. [CrossRef]
17. Drews, C. The state of wild animals in the minds and households of a neotropical society: The Costa Rican case study. In *The State of the Animals II: 2003*; Rowan, A.N., Salem, D.J., Eds.; Humane Society Press: Washington, DC, USA, 2003; pp. 193–205.
18. Henry, E.H.; Haddad, N.M.; Wilson, J.; Hughes, P.; Gardner, B. Point-count methods to monitor butterfly populations when traditional methods fail: A case study with Miami blue butterfly. *J. Insect Conserv.* **2015**, *19*, 519–529. [CrossRef]
19. Matuzak, G.D.; Brightsmith, D.J. Roosting of Yellow-naped Parrots in Costa Rica: Estimating the size and recruitment of threatened populations. *J. Field Ornithol.* **2007**, *78*, 159–169. [CrossRef]
20. De Moura, L.N.; Vielliard, J.M.E.; Da Silva, M.L. Seasonal fluctuation of the Orange-winged Amazon at a roosting site in Amazonia. (Report). *Wilson J. Ornithol.* **2010**, *122*, 88. [CrossRef]
21. Joyner, L. *Guide to Multiple Point Fixed Transects in Parrot Monitoring*; One Earth Conservation: New York, NY, USA, 2020.
22. BirdLife International and Handbook of the Birds of the World. Bird Species Distribution Maps of the World. Version 2019.1. 2019. Available online: http://datazone.birdlife.org/species/requestdis (accessed on 10 June 2020).
23. Horns, J.J.; Adler, F.R.; Sekercioglu, C.H. Using opportunistic citizen science data to estimate avian population trends. *Biol. Conserv.* **2018**, *221*, 151–159. [CrossRef]
24. Kamp, J.; Oppel, S.; Heldbjerg, H.; Nyegaard, T.; Donald, P.F. Unstructured citizen science data fail to detect long-term population declines of common birds in Denmark. *Divers. Distrib.* **2016**, *22*, 1024–1035. [CrossRef]
25. Walker, J.; Taylor, P.D. Using eBird data to model population change of migratory bird species. *Avian Conserv. Ecol.* **2017**, 12. [CrossRef]
26. Roldán-Clará, B.; López-Medellín, X.; Espejel, I.; Arellano, E. Literature review of the use of birds as pets in Latin-America, with a detailed perspective on Mexico. *Ethnobiol. Conserv.* **2014**, 3. [CrossRef]
27. Harvey, C.A.; Alpízar, F.; Chacón, M.; Madrigal, R. *Assessing Linkages between Agriculture and Biodiversity in Central America: Historical Overview and Future Perspectives*; The Nature Conservancy: San José, Costa Rica, 2004.

28. Aide, T.M.; Clark, M.L.; Grau, H.R.; Lopez-Carr, D.; Levy, M.A.; Redo, D.; Bonilla-Moheno, M.; Riner, G.; Andrade-Nunez, M.J.; Muniz, M. Deforestation and Reforestation of Latin America and the Caribbean (2001–2010). *Biotropica* **2013**, *45*, 262–271. [CrossRef]
29. Tryjanowski, P.; Kosicki, J.Z.; Hromada, M.; Mikula, P. The emergence of tolerance of human disturbance in Neotropical birds. *J. Trop. Ecol.* **2020**, *36*, 1–5. [CrossRef]
30. Dickinson, J.L.; Zuckerberg, B.; Bonter, D.N. Citizen Science as an Ecological Research Tool: Challenges and Benefits. *Annu. Rev. Ecol. Evol. Syst.* **2010**, *41*, 149–172. [CrossRef]
31. Sullivan, B.L.; Aycrigg, J.L.; Barry, J.H.; Bonney, R.E.; Bruns, N.; Cooper, C.B.; Damoulas, T.; Dhondt, A.A.; Dietterich, T.; Farnsworth, A.; et al. The eBird enterprise: An integrated approach to development and application of citizen science. *Biol. Conserv.* **2014**, *169*, 31–40. [CrossRef]
32. Robinson, O.J.; Ruiz-Gutierrez, V.; Fink, D.; Meese, R.J.; Holyoak, M.; Cooch, E.G. Using citizen science data in integrated population models to inform conservation. *Biol. Conserv.* **2018**, *227*, 361–368. [CrossRef]
33. Pimm, S.L.; Alibhai, S.; Bergl, R.; Dehgan, A.; Giri, C.; Jewell, Z.; Joppa, L.; Kays, R.; Loarie, S. Emerging technologies to conserve biodiversity. *Trends Ecol. Evol.* **2015**, *30*, 685–696. [CrossRef] [PubMed]
34. Berglund, P.-A.; Hentati-Sundberg, J. *Arctic Seabirds Breeding in the African-Eurasian Waterbird Agreement (AEWA) Area: Status and Trends 2014*; Conservation of Arctic Flora and Fauna: Akureyri, Iceland, 2015.
35. Lourenço, P.M.; Alonso, H.; Alves, J.A.; Carvalho, A.T.; Catry, T.; Costa, H.; Costa, J.S.; Dias, M.P.; Encarnação, V.; Fernandes, P.; et al. Monitoring waterbird populations in the Tejo estuary, Portugal: Report for the decade 2007–2016. *Airo* **2018**, *25*, 3–31.
36. Hesse, A.J.; Duffield, G.E. The status and conservation of the Blue-Throated Macaw Ara glaucogularis. *Bird Conserv. Int.* **2000**, *10*, 255–275. [CrossRef]
37. Herzog, S.K.; Maillard, Z.O.; Boorsma, T.; Sanchez-Avila, G. First systematic sampling approach to estimating the global population size of the Critically Endangered Blue-throated Macaw *Ara glaucogularis*. *Bird Conserv. Int.* **2020**, 1–19. [CrossRef]
38. Berkunsky, I.; Diaz Luque, J.A.; Kacoliris, F.P.; Daniele, G.; Milpacher, S.; Gilardi, J.D.; Martin, S. Blue-throated Macaw—10 Years. *PsittaScene* **2012**, *24*, 3–5.
39. Vaughan, C.; Nemeth, N.; Marineros, L. Ecology and management of natural and artificial scarlet macaw (*Are macao)* nest cavities in Costa Rica. *Ornitol. Neotrop.* **2003**, *14*, 381–396.
40. Ferreira Pires, S. The Illegal Parrot Trade in the Neo-Tropics. Ph.D. Thesis, Rutgers University, Newark, NJ, USA, 2012.
41. Pires, S.F. The illegal parrot trade: A literature review. *Glob. Crime* **2012**, *13*, 176–190. [CrossRef]

 *diversity*

Article

Roadside Car Surveys: Methodological Constraints and Solutions for Estimating Parrot Abundances across the World

José L. Tella [1,*], Pedro Romero-Vidal [2], Francisco V. Dénes [3,4], Fernando Hiraldo [1], Bernardo Toledo [1], Federica Rossetto [5], Guillermo Blanco [6], Dailos Hernández-Brito [1], Erica Pacífico [1], José A. Díaz-Luque [7], Abraham Rojas [8], Alan Bermúdez-Cavero [1,9], Álvaro Luna [1,10], Jomar M. Barbosa [1,11] and Martina Carrete [2]

1 Department of Conservation Biology and Integrative Biology, Doñana Biological Station CSIC, 41092 Sevilla, Spain; hiraldo@ebd.csic.es (F.H.); ber.toledo@hotmail.com (B.T.); dailoshb@gmail.com (D.H.-B.); ericapacifico81@gmail.com (E.P.); becao82@gmail.com (A.B.-C.); alvaro.luna@universidadeuropea.es (Á.L.); jomarmbarbosa@gmail.com (J.M.B.)
2 Department of Physical, Chemical and Natural Processes, Universidad Pablo de Olavide, 41013 Sevilla, Spain; pedroromerovidal123@gmail.com (P.R.-V.); mcarrete@upo.es (M.C.)
3 Department of Ecology, Biosciences Institute, University of São Paulo, 05508-220 São Paulo, Brazil; francisco.denes@ib.usp.br
4 Department of Migration, Max Planck Institute of Animal Behavior, D-78315 Radolfzell, Germany
5 Biodiversity Research Institute (CSIC, UO, PA), Oviedo University, 33600 Mieres, Spain; federica.rossetto1991@gmail.com
6 Department of Evolutionary Ecology, Museo Nacional de Ciencias Naturales CSIC, 28006 Madrid, Spain; g.blanco@csic.es
7 CLB Foundation, Santa Cruz de la Sierra 5030, Bolivia; jose.diaz@endangeredconservation.org
8 Zoológico Municipal de Fauna Sudamericana, Santa Cruz de la Sierra 1026, Bolivia; mrojas_bo@hotmail.com
9 Laboratorio de Biología y Genética, Universidad Nacional de San Cristóbal de Huamanga, Ayacucho 05001, Peru
10 Department of Health Sciences, Faculty of Biomedical and Health Sciences, Universidad Europea de Madrid, 28670 Madrid, Spain
11 Department of Applied Biology, Universidad Miguel Hernández, Elche, 03202 Alicante, Spain
* Correspondence: tella@ebd.csic.es

Citation: Tella, J.L.; Romero-Vidal, P.; Dénes, F.V.; Hiraldo, F.; Toledo, B.; Rossetto, F.; Blanco, G.; Hernández-Brito, D.; Pacífico, E.; Díaz-Luque, J.A.; et al. Roadside Car Surveys: Methodological Constraints and Solutions for Estimating Parrot Abundances across the World. *Diversity* **2021**, *13*, 300. https://doi.org/10.3390/d13070300

Academic Editor: Michael Wink

Received: 10 May 2021
Accepted: 29 June 2021
Published: 1 July 2021

Abstract: Parrots stand out among birds because of their poor conservation status and the lack of available information on their population sizes and trends. Estimating parrot abundance is complicated by the high mobility, gregariousness, patchy distributions, and rarity of many species. Roadside car surveys can be useful to cover large areas and increase the probability of detecting spatially aggregated species or those occurring at very low densities. However, such surveys may be biased due to their inability to handle differences in detectability among species and habitats. We conducted 98 roadside surveys, covering > 57,000 km across 20 countries and the main world biomes, recording ca. 120,000 parrots from 137 species. We found that larger and more gregarious species are more easily visually detected and at greater distances, with variations among biomes. However, raw estimates of relative parrot abundances (individuals/km) were strongly correlated (r = 0.86–0.93) with parrot densities (individuals/km^2) estimated through distance sampling (DS) models, showing that variability in abundances among species (>40 orders of magnitude) overcomes any potential detectability bias. While both methods provide similar results, DS cannot be used to study parrot communities or monitor the population trends of all parrot species as it requires a minimum of encounters that are not reached for most species (64% in our case), mainly the rarest and more threatened. However, DS may be the most suitable choice for some species-specific studies of common species. We summarize the strengths and weaknesses of both methods to guide researchers in choosing the best-fitting option for their particular research hypotheses, characteristics of the species studied, and logistical constraints.

Keywords: bird abundance; census; bird density; detectability; distance sampling; Psittaciformes

1. Introduction

Parrots (Order Psittaciformes) stand out among birds because of their poor conservation status [1,2] and the lack of knowledge on their population sizes and trends. According to the most recent IUCN evaluation, almost 30% of the 402 extant parrot species are threatened with extinction, while accurate information on their population numbers and changes in abundances is lacking for most species [3]. The paucity of information on population sizes, densities, and changes in the abundance of parrots across the world was highlighted six years ago [4], calling for further development and application of monitoring methods to better understand how parrot populations are responding to the variety of threats they face [1]. In fact, a recent review relating conservation threats to population trends in the Neotropics, the realm with the highest richness of parrot species [1], revealed the scarcity of data on actual abundances and population trends [5]. The situation is similar for the other realms, even for the Afrotropics [6] where parrot species richness is the lowest [1].

Estimating parrot abundance is challenging because many species naturally occur at very low densities [4], while others have heavily patched distributions or very restricted ranges [3]. Moreover, widespread threats such as habitat loss, illegal trade, and persecution [7–9] may be drastically reducing parrot population sizes and ranges, making the design of monitoring programs even more difficult. Moreover, some parrot species are highly gregarious and aggregate in large communal roots, and thus estimates of overall population size can be obtained when all roosts are located and can be properly surveyed [10]. However, this is not feasible for most parrots species, as roost sites may often change [11], they can not be located in large, inaccessible areas, or simply because not all species gather in large communal roosts. Then, researchers are forced to use alternative methodologies such as point counts and line transects, traditionally used for many avian taxa, to obtain estimates of relative abundances and densities [10]. A recent review has compiled different sampling and analytical methods for estimating parrot abundances [10]. Although the efficiency of walk line transects and point counts to estimate parrot abundances may differ among studies [11–13], both methods are constrained by the small geographic scale at which they can be done. Therefore, they may not be logistically affordable for surveying parrot species that are patchily distributed and with very low densities, as a very large number of sampling sites (e.g., up to 2000) are required for surveying uncommon species [12]. Conversely, roadside car surveys allow the coverage of very large areas, thus accounting for the large home ranges and mobility of many common parrot species and increasing the probability of detecting individuals of species occurring at very low densities and/or those that are spatially aggregated [10].

Roadside car surveys have been largely used to survey conspicuous species (mostly raptors, e.g., [14–16]), providing an easy-to-obtain measure of relative abundance (number of individuals recorded/km surveyed). Recently, roadside car surveys have also been used to relate the relative abundances of parrot species to habitat changes [17,18], the role of parrots as seed dispersers [19,20] or their roles in other ecological functions [21], or to assess how parrots are selectively poached for their use as pets [22]. Their gregariousness and especially their loud vocalization behavior [10] makes this method even more appropriate for parrots because vocalizations facilitate their detection compared to other taxa such as raptors, which are mostly only visually detected and thus more difficult to record when perched hidden by the vegetation. The easier aural than visual detection of parrots was revealed by Lee & Marsden [23], showing that only 4% out of 2,681 parrot detections obtained through walk line transects were of silent, seen-only groups. However, as for point counts [24] and walk line transects [23], several parrot encounters correspond to aural-only detections, and thus the number of unobserved individuals cannot be recorded for estimating abundances [22–24]. A proposed solution for this problem, both for point counts, walking and car transects, is to substitute missing count data (i.e., aural-only encounters) with the average flock size obtained for the species during the survey [22–24]. However, there is no evaluation of how this methodological approach may affect the estimates of abundance. Another obvious problem for all three methods is that the probability of detection decreases with the distance of encountered birds from the observer and that this distance-dependent probability of detection may vary among species and habitats [10]. This problem is easily solved through distance sampling (DS) modeling, currently implemented in accessible statistical packages, which allows the calculation of probabilities of detection to estimate densities (individuals/km^2) of the studied species [10]. However, this much more desirable approximation comes with the caveat that robust DS modeling requires a minimum of visual encounters [10], from which distance measurements can be taken to inform models, which in some cases could reach 40–50 contacts [13]. Unfortunately, this analytical constraint makes it impossible to estimate the abundances for rare parrot species occurring at very low densities [20,25] or those relatively abundant but highly gregarious species recorded in high numbers of individuals in a few very large flocks [26], with numbers of encounters that are insufficient for DS modeling. Nonetheless, recent work showed a strong correlation between distance-uncorrected relative parrot abundances obtained through roadside car surveys and distance-corrected densities for a sample of species with enough visual encounters needed for DS modeling [22] (see also unpublished results offered by [10]). These results support the idea that distance-uncorrected relative abundances of parrots obtained through roadside car surveys are good proxies of their actual abundances, especially when the high variability in abundance among species overcomes the main sources of sampling error, i.e., differences in detectability [22]. Nonetheless, further research embracing different parrot communities and biomes is needed before generalizing these conclusions.

Here, we take advantage of an unprecedented data set that compiles our roadside car surveys conducted over 10 years, covering 20 countries and all continents and biomes inhabited by parrots across the world. We first assessed sources of variability related to the percentage of aural-only encounters and the distance at which parrots were detected. We hypothesized that parrot detectability in roadside surveys is a function of species size and gregariousness, and the openness of the surveyed habitat. We predicted that larger and more gregarious species should be more easily detected visually and at greater distances, and that detection should also vary among biomes since they range from very open (e.g., Deserts and Xeric Scrublands) to highly concealing forested habitats (e.g., Tropical and Subtropical Broadleaf Moist Forests). We then correlated distance-uncorrected relative densities (individuals/km) with density estimates (individuals/km^2) obtained through DS modeling, using different thresholds for a minimum of visual contacts. We evaluated how adding an estimation of the number of only heard (unseen) individuals [22–24] affects these correlations. We found a strong correlation between these estimates of parrot abundances and discuss the pros and cons of both methods, including the loss of whole surveys and the traits of species that are excluded when using DS and not reaching the minimum numbers of visual encounters needed for statistical modeling. We aim to guide researchers in choosing the best-suited methodology given their research objectives and study species.

2. Materials and Methods

2.1. Study Areas and Field Work

We selected several countries from the main five parrot-inhabited realms (Neotropic, Afrotropic, Indomalayan, and Australasia). These regions represent the richest to the poorest parrot communities worldwide [1]. This work was embedded within different research projects, having all in common our need to estimate the relative abundance of each species within each parrot community. We used these estimates to answer different questions, such as those related to their relative contribution to ecological functions [19–21], assessing poaching pressure [22], or the effects of habitat transformations on parrot abundances [17,18]. Therefore, for each country, we designed road itineraries to cover the main biomes and ecoregions occupied by parrots (obtained from https://ecoregions2017.appspot.com/; accessed 15 January 2021) and the distribution of as many parrot species as possible (obtained from [3] and a variety of regional bird field guides). Using satellite maps, we selected unpaved and low-transit paved roads that crossed from pristine to highly humanized habitats (e.g., agricultural and urbanized areas), thus maximizing the chances of finding a variety of parrot species, from those intolerant to habitat transformations to those benefitting from anthropogenic changes (e.g., [17,18,27–29]).

Most of the fieldwork was done between 2011 and 2020 (Supplementary S1), through expeditions that typically lasted 3–5 weeks. Some small countries were well surveyed through a single expedition (e.g., Costa Rica), while some of the largest (e.g., Brazil) required many expeditions to cover the greater variety of biomes, ecoregions, and parrot communities. In such cases, results obtained from a single ecoregion/biome/country in different expeditions (usually conducted in different years) were pooled to increase sample sizes (number of km surveyed and number of parrots recorded) and thus better represent the whole parrot community and increase the precision of estimates [12]. Only Australia, Colombia, and India were partially surveyed due to logistical constraints (Supplementary S1). Surveys were conducted in different seasons and across the annual cycles of parrots. However, this should not be problematic for the objectives of this paper, since our analyses compare results of two parrot abundance estimates simultaneously obtained within each ecoregion/biome/country surveyed (see below). Rather, the large geographic and temporal scales of our surveys reinforced and allowed the generalization our results.

2.2. Roadside Surveys

Typically, and similarly to other roadside parrot surveys [17–22], the driver and two experienced observers drove a 4 × 4 vehicle at low speed (10–40 km/h) following previously designed itineraries from dawn to dusk, avoiding rainy and hot middays when parrot activity declines [30,31]. All parrots detected were recorded, briefly stopping when needed to identify species and/or count the number of individuals in flocks. Observers were familiar with the parrot species surveyed, as surveys were combined with behavioral and foraging studies across all study areas (see e.g., [32–35]), so they were able to visually and aurally identify them. Moreover, several authors participated in different surveys, and the first author participated in 91% of all surveys, so each survey included researchers with accumulated experience in identifying parrots. For a subsample of surveys (those conducted since 2018), we also recorded the mode of detection (i.e., whether parrots were first detected aurally, visually, or both) and their behavior at first detection (i.e., resting, feeding, or flying). Following previous recommendations [13] and studies [17–22], we considered both perched and flying individuals for estimating parrot abundances (see Discussion for pros and cons of including flying birds), thus also making distance-corrected and uncorrected estimates (see below) comparable. We paid special attention to the flying direction and group size of parrots in flight to avoid double counting of flocks [13].

Distances of detection (i.e., the perpendicular distance from parrots to the road when they were first detected) were recorded to compare two estimates of parrot abundance (see below). Detection distance was estimated visually for short distances or using a laser rangefinder incorporated into binoculars for large distances (Leica Geovid 10 × 42, range: 8–1500 m), measuring the distance to the closest tree for flying flocks. In the case of loose flocks, we measured the distance to the closest individual in the flock. In many instances, parrots were only heard and the species identified through their vocalizations because they were concealed by vegetation. Therefore, we could not record the distance of detection nor the number of individuals. Thus, we classified detections as aural (only heard) or visual (seen or both seen and heard).

Since 2018, all roadside surveys and parrot counts were recorded using the ObsMapp application for smartphones, which uploads the observations to the citizen science platform Observation (www.observation.org; accessed 15 January 2021). Therefore, all records, exact location, and associated information can be viewed and downloaded from this web platform (searching for the observers Pedro Romero-Vidal, Dailos Hernández-Brito, and José Luis Tella) by any researcher in the future.

2.3. Distance Sampling Modeling

Distance sampling (DS) models were fit for each combination of country, ecoregion, and species (henceforth study case). The maximum detection distance was fixed at 500 m for all species. While this value may not be optimal for some species and/or habitat types, it encompasses most of the detections (see Results Section 3.3.2). More importantly, having a single maximum distance allows straightforward comparisons among study cases. We restricted DS modeling to those study cases with at least 10 visual contacts within 500 m of distance. We conservatively used this encounter threshold as it was the minimum required for DS modeling in a previous whole-parrot community study [22], thus allowing us to include as many species and study cases as possible. In fact, a minimum of 10 contacts of the target species was suggested to obtain useful, if imprecise, parrot density estimates [4]. Nonetheless, we also tested how results could change by gradually increasing the threshold up to 50 visual contacts per species (see below). Because the number of individuals in a group can influence detection, we evaluated the potential correlation between group size and detection distance using Spearman correlation tests. We binned distance data for each study case to facilitate the fitting of detection functions, using breaks every 25 (a), 50 (b), and 100 (c) m (i.e., a: 0–25, 25–50, ... , 475–500 m; b: 0–50, 50–100, ... , 450–500 m; c: 0–100, 100–200, ... , 400–500 m). For each binning setup (a, b, and c), we fitted DS models with a half-normal key function as previously recommended after visual inspection of the histograms of distances [36,37], but also using the hazard rate and the uniform key as alternative functions. We compared models with no adjustment terms and with cosine, Hermite polynomial, and simple polynomial adjustments, up to order 5. For models where group size was correlated with detection distances, we also fitted a DS model with group size as a covariate. Akaike's Information Criterion (AIC) was used to compare models within a distance break set [38], but it cannot be used to compare models fit to data with different binning setups [36]. Thus, we performed chi-square goodness-of-fit tests to compare the best models from each binning setup and identify the best fitting model (highest chi-square test p-value) for each study case. To allow visual inspection of our DS models and chi-square tests, we provide, for each study case, a histogram of detection distances (with Sturges's breaks), the plot of group size x detection distances (with Spearman correlation test p-value), and the estimated detection functions from the best DS models for each binning setup, overlaid on the histogram of detection distances with the respective distance breaks (Supplementary S3).

Detection probability (P) was obtained from the best model for each study case. Then, abundance (N) was calculated by dividing the number of observed individuals by the estimated P within 500 m maximum distance (or a 1 km-wide strip centered on the road). Density was calculated by dividing N by the length (in km) surveyed for each case, providing an estimate of individuals/km^2 (the width surveyed was 1 km). Analyses were done in R using the "Distance" package [39,40].

2.4. Traits of Parrot Species

We obtained two measures of parrot size, body length (in cm) and body mass (in g), from [41]. As a proxy of the gregariousness of a species, we used our own data on flock sizes. For analyses based on study cases, we used the average flock size of the species recorded within each study case. For analyses at the species level, we used the overall average flock size after pooling data when a species was surveyed in more than one study case. Average flock sizes were unrelated to the body length (Spearman correlation, $r_s = -0.02$, $p = 0.84$) and body mass ($r_s = -0.09$, $p = 0.28$) of the 131 species visually recorded in our study. However, body length and body mass were strongly correlated ($r_s = 0.88$, $p < 0.001$), so both variables were alternatively fitted in models accounting for the relationship between detectability and body size (see below). Results were nearly identical but the effect of body mass was always slightly stronger than that of body length, so the later results are not shown for simplicity. The global conservation status of each species was obtained from the 2020 IUCN Red List [3].

2.5. Statistical Analyses

We used Generalized Linear Models (GLM) to assess how the number of parrot encounters and the number of parrot species recorded (negative binomial error distribution, log link function) varied among realms and with the lengths of surveys. Moreover, we evaluated how the percentage of aural encounters, distances of detection, and probabilities of detection (P) (log-transformed; normal error distribution, identity link function) were affected by the body mass and flock size of the species and the biomes they occupied. For the proportion of aural encounters, we restricted analyses to species with at least 15 encounters to reduce error biases in the estimation of proportions [42].

The relationship between relative abundances (individuals observed/km; response variable) and densities of parrots (individuals estimated/km^2) obtained through DS modeling was assessed with non-parametric Spearman correlation and linear regressions on raw and log-transformed data, respectively. As the robustness of DS models and thus the precision of their estimated densities may increase with sample sizes (i.e., number of contacts [12]), we performed five regressions by restricting data to cases with at least 10, 20, 30, 40, and 50 visual encounters at distances $\leq$ 500 m. Following previous recommendations to avoid the underestimation of secretive species [22–24], we also estimated relative abundances by summing to the number of observed individuals the estimation of those not observed (number of aural-only contacts $\times$ average flock size obtained in each study case), divided by the km surveyed, and repeated the same regression on densities obtained from DS models. Finally, we assessed whether these relationships are influenced by body mass, flock size, biome, and the number of visual encounters through a GLM (response variable: log-transformed relative abundance; normal error distribution, identity link function).

The characteristics of case studies and species (body mass, flock size, relative abundance, conservation status) not available for estimating their densities through DS modeling due to the low number of visual contacts were identified using GLMs (response variable: available/not available; binomial error distribution, logistic link function).

Our data set included species that were surveyed in different case studies (mean = 4.2, median = 2 case studies per species), thus providing replicates that allow for the testing of the relative contribution of species traits and biomes on detectability and abundance estimates through the multivariate models described above. These models would require controlling for species identity to account for pseudoreplication. However, models fitting species identity as random or fixed effects together with species traits confounded their individual effects as species had unique values of body size and almost-unique values of flock size. As our research goal was not to simply assess whether species differ among them but to know what species traits explain these differences, we show models including species traits without controlling for their identity. Models did not show data overdispersion, and the percentage of deviance explained by GLMs and adjusted R^2 for linear regressions are provided to show the variability in the data captured by our models. Statistical analyses were performed using SPSS v. 27.

3. Results

3.1. Distribution of Surveys and Species Recorded

We conducted 98 surveys, covering a total of 57,241.44 km across 81 ecoregions, from 11 biomes and 20 countries belonging to the Neotropic (48,612.32 km), Afrotropic (6499.65 km), Indomalayan (1405.72 km), and Australasia (723.75) realms (Figure 1, Supplementary S1).

Surveys averaged 584.1 km in length (range: 35.88–6899.48 km, N = 98), and 75% of them were longer than 150 km (Supplementary S1). The number of parrot encounters varied between 0 and 1263 per survey (mean 162.1 + 199.4 SD, median 98, Supplementary S1), and a GLM revealed it was unrelated to survey length (χ^2 = 1.58, p = 0.21) but varied among realms (χ^2 = 64.44, p < 0.001), with no significant interaction between survey length and realm (χ^2 = 2.07, p = 0.56). The average number of parrot encounters per survey decreased as follows: Neotropic > Indomalayan > Australasia > Afrotropic. The number of parrot species recorded per survey ranged from 0 to 25 (mean 5.87 + 5.27 SD, median 4 species, Supplementary S1). Similarly to the number of encounters, a GLM showed that the number of species recorded was unrelated to survey length ($\chi^2 = 0.36$, $p = 0.85$) and varied among realms ($\chi^2 = 21.71$, $p < 0.001$), with no significant interaction between survey length and realm ($\chi^2 = 0.63$, $p = 0.89$). The average number of species recorded per survey decreased as follows: Australasia > Neotropic > Indomalayan > Afrotropic (Supplementary S1).

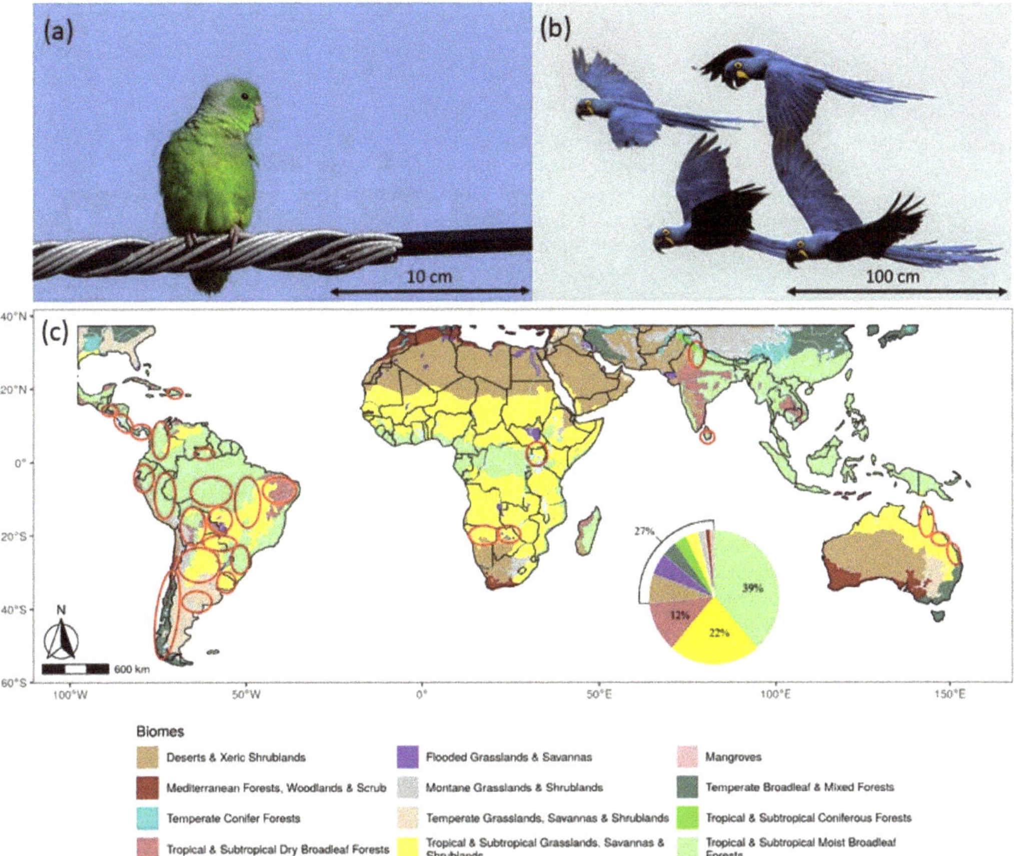

Figure 1. Roadside surveys allowed us to record from (**a**) the smallest (green-rumped parrotlet Forpus passerinus) to (**b**) the largest parrot species (hyacinth macaws Anodorhynchus hyacinthinus) through 98 surveys conducted in 20 countries (**c**). The surveyed areas are roughly depicted with red ellipses over the world biomes. Each area may include several surveys, biomes, and ecoregions. The inserted pie chart shows the percentage of surveys conducted within each biome. Photographs: José L. Tella.

As each of the 98 surveys covered different combinations of biomes, ecoregions, and countries (Supplementary S1), and up to 25 species were recorded per survey, we obtained a total of 575 estimates of species-specific parrot abundances (i.e., study cases).

3.2. Traits of the Species Recorded

Overall, we recorded 137 parrot species from 49 genera distributed among the Neotropic (110 spp), Afrotropic (6 spp), Indomalayan (6 spp), and Australasia (16 spp) realms (Supplementary S2). Species ranged in size from the smallest parrotlets (Forpus passerinus, body length 12.5 cm, body mass 23 g) to the largest macaws (Anodorhynchus hyacinthinus, body length 95 cm, body mass 1565 g; Figure 1). Species also greatly varied in gregariousness, as reflected by their average flock size that ranged from 1 to 106.4 individuals (mean = 8.37 ± 12.35 SD, median = 5). Regarding their conservation status, most of the species recorded were classified as Least Concern (66.4%), while 10.9% were Near Threatened, 12.4% Vulnerable, 7.3% Endangered, and 2.9% Critically Endangered according to the IUCN Red List. As we recorded from the rarest to the commonest species (Supplementary S2), the number of encounters per species ranged from 1 to 2127 (mean = 109.9 ± 278.9 SD, median = 14, N = 15,072).

3.3. Sources of Variation in the Detectability of Species

3.3.1. Aural and Visual Encounter Rates

Considering the smaller data set of parrot encounters in which we recorded the mode of detection (N = 9617 encounters), 15.6% were detected visually, 46% were detected aurally, and 38.4% were simultaneously seen and heard. Parrot detections summing those exclusively heard plus those heard and seen accounted for 84.4% of the encounters.

Using the whole data set, we recorded a total of 15,072 parrot encounters of which 5325 (35.33%) were only aural, thus allowing records of 119,797 observed individuals and an unknown number of unseen individuals identified to the species level through their vocalizations. The proportion of aural encounters differed among species, ranging from 0% to 100% (mean = 23.9%, median = 16.5%; 6 species were only aurally registered, see Supplementary S2). Considering those study cases with at least 15 encounters (N = 191), a GLM showed that the proportion of aural encounters decreased with body mass ($\chi^2 = 69.04$, $p < 0.001$, Figure 2a) and to a lesser extent with average flock size ($\chi^2 = 6.09$, $p < 0.014$, Figure 2b) of the species, meaning that the larger and more gregarious species were more easily recorded visually, with no statistically significant variation among biomes ($\chi^2 = 17.58$, $p = 0.063$, Figure 2c). This model explained 34.48% of the deviance.

As proposed in previous works, a method to avoid underestimating the number of parrots due to aural-only encounters is to multiply them by the average flock size recorded within each species-specific study case and summing this estimate of unseen (but heard) individuals to the number of visually recorded individuals, thus obtaining a more reliable estimate per species. By applying this factor of correction to our whole data set (575 study cases), the total number of parrots recorded increased by 22.6% (i.e., from 119,797 observed individuals to 154,759 estimated individuals). Importantly, this increment largely varied among species, ranging from 0 to 73.8% (mean = 20.4 + 19.8 SD, median = 14.3, N = 131 species; the increment could be not calculated for the six species that were only aurally encountered).

3.3.2. Distance-Dependent Detectability

The distance at which parrots were detected was influenced by several factors. When analyzing the smaller data set in which both the type of detection and behavior of birds were recorded, a GLM showed that distances (range 4–1400 m, mean = 89.8 ± 102.7 SD, median = 60.0, N = 4,783) were lower for aural than for visual detections ($\chi^2 = 82.83$, $p < 0.001$) and for perching than for flying birds ($\chi^2 = 349.68$, $p < 0.0001$), while they were larger for larger flocks ($\chi^2 = 37.39$, $p < 0.001$) and species with larger body mass ($\chi^2 = 449.75$, $p < 0.0001$), with significant variations among biomes ($\chi^2 = 125.42$, $p < 0.0001$) (deviance explained by the model: 22.24%). These and probably other unmeasured sources of variation suggest the need for modeling distance-dependent probabilities of detection for unbiased estimation of parrot abundances.

Using the whole data set, we could calculate distance-dependent probabilities of detection (P) through DS modeling for 208 study cases with at least 10 visual encounters within 500 m of the transect line per species. Distances ranged between 0 and 1498 m (mean = 76.1 ± 95.8 SD, median = 50, N = 8491), while 99.3% of the distances were ≤500 m.

The half-normal key function was the detection function best fitting the data in most of the study cases (51.4%), followed by the hazard rate (42.8%) and the uniform (5.8%) functions. The best-fitted models included different cosine adjustments in 24 (11.5%) of the cases, and only in 10 cases (4.8%) included group size as a covariate. The resulting P ranged from 0.01 to 1 (mean: 0.22 ± 0.15 SD, median = 0.19). It is worth noting that the extremely low values of P (ranging from 0.01 to 0.05, in 18 study cases obtained through the hazard rate and in one case obtained through the half-normal functions) may be attributable to cases where parrots were attracted by feeding/nesting resources available close to the roads, thus violating a key assumption of DS modeling and making these values questionable (see Discussion Section 4.1).

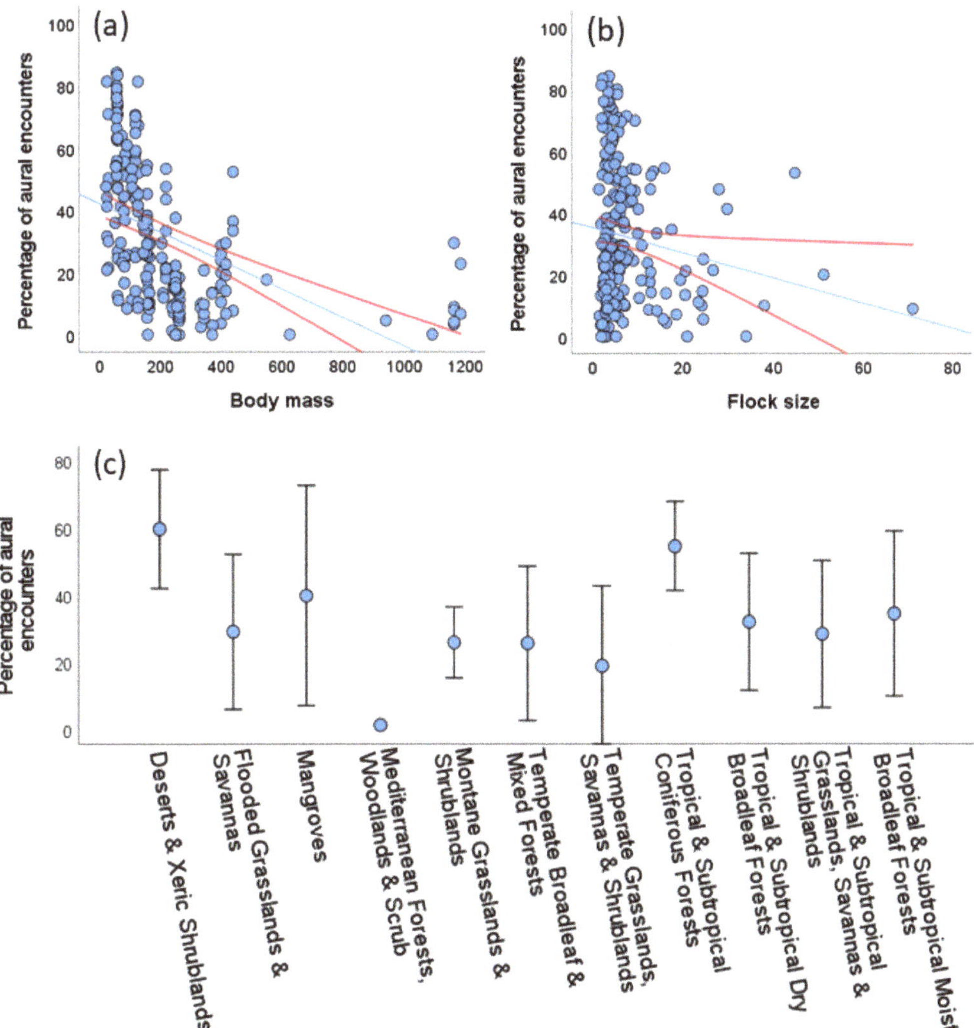

Figure 2. Univariate relationships between the percentage of aural encounters (i.e., when parrots were only heard), and (**a**) their body mass (in g), (**b**) their average flock size (number of individuals/number of visual encounters), and (**c**) the biomes surveyed in 191 study cases with at least 15 encounters per parrot species. Red lines (in **a**,**b**) and bars (in **c**) show 95% confidence intervals. See Results for multivariate analyses.

A GLM showed that P was positively related to the body mass ($\chi^2 = 25.02$, $p < 0.001$, Figure 3a) and the average flock size ($\chi^2 = 25.31$, $p < 0.001$, Figure 3b) of the species, meaning that the larger and more gregarious species were detected farther from the road than the smaller and less gregarious species, with significant differences among biomes ($\chi^2 = 30.38$, $p < 0.001$) despite the large overlap shown by biomes in univariate plots (Figure 3c). This model explained 35.7% of the deviance. When excluding the 19 questionable P values (black dots in Figure 3a,b) from the GLM, the results were similar (body mass: $\chi^2 = 62.68$, $p < 0.001$; flock size: $\chi^2 = 31.83$, $p < 0.001$; biomes: $\chi^2 = 27.19$, $p < 0.001$; deviance explained by the model: 35.13%).

3.4. Relationships between Densities and Relative Abundances

Parrot densities (individuals estimated/km^2) were obtained by correcting the number of individuals observed by their P obtained through DS modeling, for the 208 study cases with at least 10 visual encounters at distances < 500 m per parrot species. Densities ranged from 0.04 to 97.4 individuals/km^2 (mean = 5.1 ± 11.2 SD, median = 1.8). We also calculated the relative abundances (number of observed individuals/km) for the same dataset, which ranged from 0.02 to 7.31 individuals/km (mean = 0.57 ± 0.86 SD, median = 0.30). The relative abundances of the species were uncorrelated to their probabilities of detection (Spearman correlation, $r_s = -0.10$, $p = 0.15$, N = 208).

Despite the large differences in P among case studies, the fact that both densities and relative abundances of parrots varied within >40 orders of magnitude, leads to a strong positive correlation between these two estimates of abundance (Spearman correlation of raw data: $r_s = 0.83$, $p < 0.001$; linear regression of log-transformed values: $r = 0.83$, estimate: 0.659 ± 0.031 SE, $p < 0.001$, adjusted-$R^2 = 0.69$, N = 208; Figure 4a). This correlation becomes stronger when excluding the 19 densities obtained from the extremely low, questionable P values (linear regression of log-transformed values: $r = 0.92$, estimate: 0.799 + 0.025 SE, $p < 0.001$, adjusted-$R^2 = 0.84$, N = 189; Figure 4b).

Nearly identical results were obtained when restricting the dataset to study cases with at least 20, 30, 40, and 50 visual encounters at distances < 500 m per species to increase the robustness of DS modeling ($r = 0.86$–0.91, all $p < 0.001$), despite the fact that study cases were reduced to 120, 74, 65, and 52, respectively. Therefore, estimates of parrot abundances are equivalent whether or not controlling for differences in detectability.

As suggested in previous works, a way to avoid the underestimation of parrot species with varying percentages of aural encounters is to estimate the number of unobserved individuals by multiplying them by the average flock size of the species obtained in the same survey. This estimated relative abundance index (i.e., (number of observed individuals + number of estimated heard individuals)/km) correlates equally well with densities obtained through DS modeling (linear regression of log-transformed values: $r = 0.83$, $p < 0.001$, estimate: 0.725 + 0.033 SE, adjusted-$R^2 = 0.70$, N = 208; Figure 4c); thus, its use is recommended to avoid the underestimation of parrot numbers. As before, the correlation results stronger when excluding the 19 questionable desnities ($r = 0.93$, $p < 0.001$, estimate: 0.881 + 0.026 SE, adjusted-$R^2 = 0.86$, N = 189; Figure 4d). This relationship remains similar in a GLM (estimate: 0.825 + 0.034 SE, $\chi^2 = 565.89$, $p < 0.0001$) when controlling for a much smaller effect of flock size (estimate: 0.006, SE: 0.002, $\chi^2 = 14.28$, $p < 0.001$), with no significant effects of body mass ($\chi^2 = 0.05$, $p = 0.82$), biomes ($\chi^2 = 16.93$, $p = 0.06$), and number of visual encounters ($\chi^2 = 0.07$, $p = 0.79$). This model explained 87.4% of the deviance.

3.5. Characteristics of the Species and Surveys Lost When Using Distance Sampling

From the 575 study cases obtained, in 367 (63.8%) DS modeling was not possible because the number of visual contacts was <10. The number of study cases lost when using DS mostly corresponded to those showing lower relative abundances (individuals/km; $\chi^2 = 82.13$, $p < 0.0001$, Figure 5a), with a smaller positive effect of average flock size ($\chi^2 = 20.80$, $p < 0.001$). This may be explained by the fact that some common species are highly gregarious and thus can be recorded in high numbers (see large data dispersion in Figure 5a) but with a low number of flocks encountered, thus not allowing DS modeling. The loss of cases from DS modeling was unrelated to the body mass of the species ($\chi^2 = 0.45$, $p = 0.50$) (deviance explained by model: 30.35%).

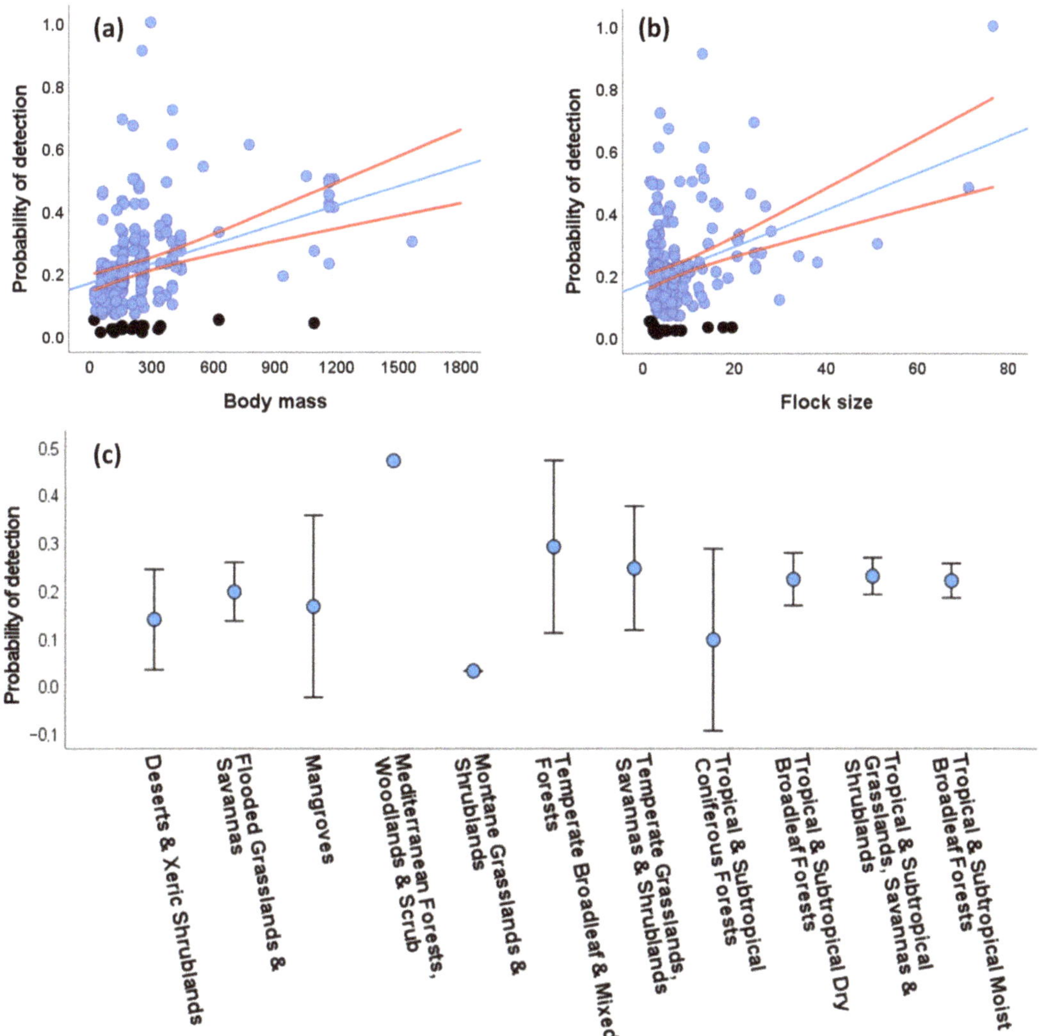

Figure 3. Univariate relationships between the probability of detection (P) of parrots obtained through distance sampling (DS) modeling and (**a**) parrot body mass (in g), (**b**) average flock size (number of individuals/number of visual encounters), and (**c**) the biomes surveyed for 208 study cases with at least 10 visual encounters at distances ≤ 500 m per parrot species. Black dots (in **a**,**b**) correspond to extremely low, questionable P values (see text for more details). Red lines (in **a**,**b**) and bars (in **c**) show 95% confidence intervals. See Results (Section 3.3.2) for multivariate analyses.

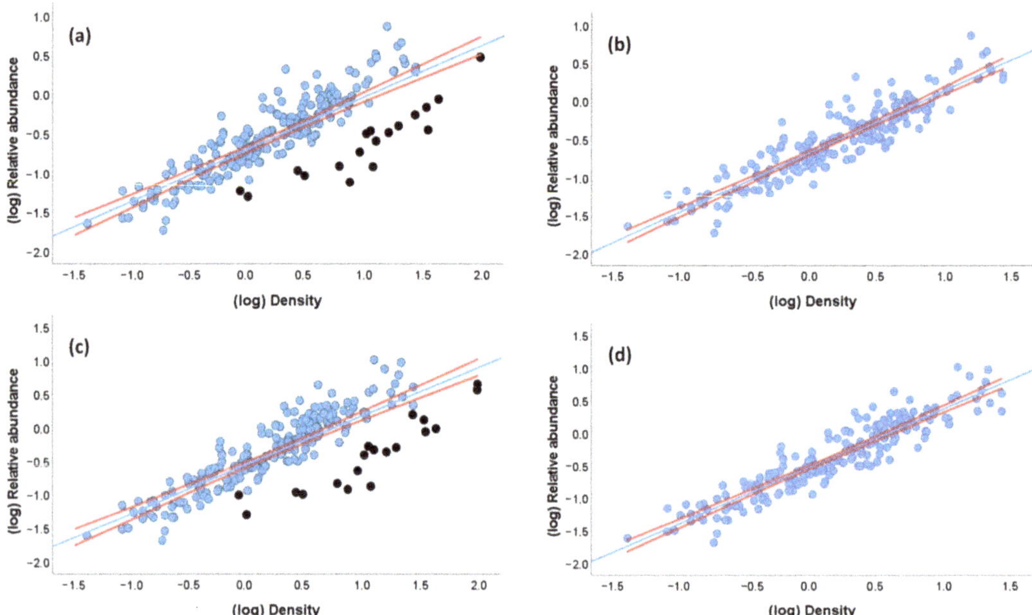

Figure 4. Relationship between (**a**) the relative abundance (individuals/km) and density (individuals/km^2) of parrots when including densities obtained from questionable probabilities of detection (black dots) and (**b**) excluding them, and (**c**) between the estimated relative abundance (i.e., (number of observed individuals + number of estimated heard individuals)/km) and density (individuals/km^2) of parrots when including densities obtained from questionable probabilities of detection (black dots) and (**d**) excluding them. Densities were obtained through distance sampling modeling for 208 study cases with at least 10 visual encounters at distances < 500 m. Red lines represent the 95% CI for the regression lines.

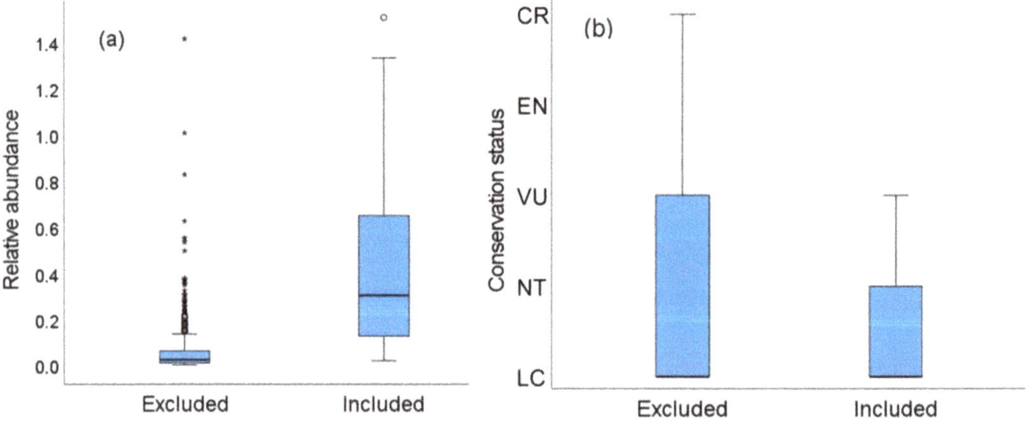

Figure 5. Several study cases and species were excluded from DS modeling because they did not met the minimum number of visual contacts to allow for estimating probabilities of detection and densities. (**a**) Study cases excluded (64% of 575) corresponded to species with lower relative abundances; (**b**) Species excluded (47% of 137) showed a poorer global conservation status (LC: Least Concern, NT: Near Threatened, VU: Vulnerable, EN: Endangered, CR: Critically Endangered).

DS modeling could not be applied to 64 (46.7%) of the 137 species surveyed even when pooling all surveys across world ecoregions together, as they did not reach a minimum of 10 visual encounters. The percentage of species excluded varied among realms, the highest being in the Afrotropics (100%, N = 6 species), followed by the Indomalayan (16.7%, N = 6 spp), Australasia (25%, N = 16 spp), and Neotropic (33.6%, N = 110 spp) realms. The species excluded from DS modeling significantly showed a poorer global conservation status ($\chi^2 = 7.51$, $p < 0.01$; 72.32% of deviance explained, Figure 5b).

DS modeling could be not applied for 34 (34.7%) of the 98 surveys conducted, as they did not include a single species reaching a minimum of 10 visual encounters. The percentage of surveys excluded for modeling varied among realms, with the highest being in the Afrotropics (100%, N = 16 surveys), followed by the Neotropic (23.3%, N = 73), Indomalayan (16.7%, N = 6), and Australasia (0%, N = 3) realms.

4. Discussion

Roadside car surveys have been largely recommended to estimate the abundances of large and conspicuous birds which occur at low densities, such as raptors [43]. Recently, this methodology has been applied to parrots, although there is no proper evaluation of its strengths and weaknesses [10]. After our experience conducting roadside raptor surveys in a variety of tropical biomes [15,16], we considered this method to be even more adequate for parrots given that their frequent and loudly vocal activity makes them more easily detectable than the more silent raptors. In fact, 85% of our parrot encounters were aurally detected. The loud behavior of parrots largely reduces the problems in detecting raptors in forested biomes [15]. Supporting this, we found that the proportion of aural detections was related to the body mass and gregariousness of the species but not to the biomes they inhabit, which included habitats largely differing in openness, from steppes to rainforests. Therefore, through our large-scale roadside surveys, we were able to record c. 35% of the extant parrot species across the world biomes, including the commonest to the rarest and even Critically Endangered species. The former species, as well as those common but highly gregarious or patchily distributed, are difficult to survey through walked line transects and point counts because of their low encounter rates [10]. Moreover, we have demonstrated that distance-uncorrected estimates of parrot abundances are strongly correlated to those obtained when using DS modeling, thus providing a good proxy of the actual relative densities of the species. Nonetheless, roadside parrot surveys have several limitations regarding the design and length of surveys and the detectability of the species, which can be addressed as discussed below.

4.1. Roadside Parrot Surveys: Caveats, Solutions, and Prospects

As for raptors and other avian taxa [15,43], parrot abundances obtained through roadside surveys can be biased by the spatial distribution of roads and the response of the species to them. Recent studies have shown that coexisting bird species may differentially respond to roads, some decreasing but others increasing their abundances close to them, also differing in their responses between major and minor roads [44,45]. As some scavengers and birds of prey may be attracted by roadkills and the larger availability of prey and perching sites (e.g., power lines, poles) close to roads [15,46], some parrots can also be attracted by feeding resources, large trees and perching sites available close to roads. In fact, we could confirm that most of the extremely low probabilities of detection we obtained corresponded to study cases where parrots were attracted by feeding resources most often available in the gutters of the roads, such as fruiting trees (e.g., Burrowing parrots *Cyanoliseus patagonus* in Argentina, [35]) or herb seeds (e.g., Galahs *Eolophus roseicapilla* in Australia, [47]), or by lines of eucalyptus trees and power lines running in parallel to roads in deforested areas of Argentina, Paraguay, Uruguay and Brazil, substrates where Monk parakeets *Myiopsitta monachus* build their large communal nests [48]. In these few cases, extremely low probabilities of detection did not result from parrots being hard to detect at large distances from roads but from the fact that they were aggregated around them.

These particular circumstances violate a key assumption of DS modeling, i.e., that animal locations are independent of the line transect position [38], thus questioning its use as they may lead to the obtention of inflated densities (see Results Section 3.4 and Figure 4a,c).

On the other hand, some parrots may avoid roads because of human disturbance. This so-called "disturbance effect" may even affect bird abundances obtained from point counts because of the presence of observers [49], and thus traffic should also affect the behavior of parrots. We tried to minimize this disturbance effect by selecting *a priori*, using recent satellite images, minor paved roads and unpaved roads with little or no traffic, often only accessible using 4 × 4 vehicles. The fact that the relative abundances (individuals/km) of the species were uncorrelated to their distance-dependent probabilities of detection suggests that the less encountered species are actually uncommon (as is also supported by their IUCN Red List evaluations [3]), rather than their abundances being underestimated because they avoid roads and thus remain undetected. Moreover, through this work we found that parrots, from the smallest to the largest species, were largely undisturbed by the vehicle, allowing us to approach them at short distances, even taking detailed photographs (e.g., [34,35]). This agrees with the perception of high behavioral flexibility of parrots when facing human disturbance (e.g., [18,27,50]). In fact, recent studies have shown that the inter-individual variability of birds in their tolerance to sources of human disturbance such as roads [51] and human presence [52] is related to the relative brain size of the species, and parrots are among the birds with larger brains showing less fear of humans [52]. Nonetheless, further well-designed studies are needed to delve more deeply into these aspects and to evaluate how parrots respond to roads with high traffic intensity.

Another problem of roadside surveys is that habitat composition and configuration near roads may differ from the surrounding areas, thus leading to bird abundance biases [10,53]. The occurrence of these potential biases can be assessed *a posteriori* by comparing habitat composition along the roads surveyed with surrounding areas [54] but, ideally, can be largely avoided by carefully selecting the roads *a priori* using satellite images. In our case, within each survey, we intentionally selected roads crossing both protected and unprotected habitats with different degrees of transformation, as we were interested in surveying whole parrot communities that included habitat-sensitive species but also those that are favored by low-intensive agricultural and urban habitats [17,18,27–29]. In other cases, however, researchers may be interested in surveying a particular species and in such a case they should ensure the selected roads cover and represent the habitats used by this species and not others. Alternatively, they may be interested in species responses to habitat transformation. Road transects can be divided into small sections whose habitats can be measured [43], and thus long surveys crossing fragments of habitats with different degrees of transformation, from pristine to urban areas within the same study area, allow for testing changes in parrot abundances based on changes in land use [17,18]. The length of the section can be used as a proxy of the size of the habitat patch crossed when acquiring large data sets, and thus testing the effects of habitat transformation together with patchiness on single-species parrot abundances [18]. The same approach can be translated to multi-species studies, obtaining estimates of total abundance, diversity, and species richness (by simply recording presence/absence of each species) for each roadside habitat section [15,55]. Another approach is to compare the habitat composition within a buffer centered on each detected parrot with that around random points selected from the same roadside survey, combining field data with remote sensing tools [55]. These approaches have still been little explored and have the potential to increase our knowledge on the responses of different parrot species and communities to very large-scale changes in land use and habitat fragmentation, and are urged given the further habitat loss predicted for parrots worldwide [7].

4.2. Do We Need to Account for Parrot Detectability?

As for other avian taxa, it is widely assumed that detectability varies among parrot species [10]. However, differences in distance-dependent detectability among parrot species have been little reported [13], and even less is known about which parrot traits explain these differences. Observations of flying parrots recorded from Amazonian rainforest canopy points showed that larger-bodied species were detected at greater distances, and that average flock sizes were negatively related to their body mass [56]. Here, analyzing a large data set that includes a variety of species and biomes, we show that not only the distances of detection but also the probabilities of visual detections are positively related to the body mass and gregariousness of a species. Moreover, there are other potential sources of variation in parrot detectability that we could not assess through our large-scale approach. For example, visual (but not aural) detectability may vary within species and biomes due to habitat transformations (it could be higher in agriculture than in forest habitats) and seasonal changes in vegetation structure (it could be higher during the dry season in deciduous tropical dry forests when most trees lose their leaves).

Breeding phenology may also affect parrot detectability since the gregariousness of some species decreases during the nesting period [10,11] and nesting pairs may be more tied to their nesting sites and thus less mobile and detectable. Therefore, it is important to consider potential seasonal changes in parrot behavior and to account for variation in parrot detectability when performing censuses.

Accordingly, our distance-dependent probabilities of detection (P) were positively related to the body mass and gregariousness of the species and varied among biomes. Even though we relied on a minimum of 10 visual encounters, which can lead to useful but imprecise density estimates [4], the densities obtained were within the ranges obtained for the same parrot genera through DS modeling using walked line transects and point counts [4]. As highlighted in the same review, parrot densities obtained through different methods, even including roost counts, are quite similar when looking at differences among species [4]. This is likely due to the fact that differences in natural (and/or human-induced) abundances among parrot species [3] are so high (in our study within >40 orders of magnitude) that any biases due to differential detectability or other methodological biases are overcome in interspecific comparisons. Then, perhaps not surprisingly, our results allow us to confirm and generalize previous findings [22], showing a strong correlation between detectability-corrected and uncorrected estimates of parrot abundance at a global scale. Notably, the same correlation holds when increasing the minimum threshold of encounters to increase the robustness of DS modeling and when including estimates of the number of unseen (only-heard) individuals, while it is not affected by the body mass of the species, biomes, or the number of encounters per species. Therefore, simple estimates of relative parrot abundance (individuals/km) can be used as good proxies of their estimated detectability-corrected densities. This does not mean however that one method is better than the other, nor that distance sampling is not needed for roadside parrot surveys. The choice should be balanced attending to different methodological constraints and research objectives, as further discussed below.

4.3. *Pros and Cons of Distance Sampling*

A major challenge for estimating parrot densities is obtaining enough encounters from all species for DS modeling [4]. For example, density estimates could be obtained for only 9 of 17 parrot species after significant effort conducting walked line transects (accumulating 2,412 km surveyed over 3 years) in two small Amazonian study areas [23]. In our study, 64% of the case studies, 47% of the parrot species, and 35% of all surveys had to be excluded from DS modeling. This occurred despite pooling data from the same ecoregions/countries obtained in different seasons and years, when available, to increase sample sizes, to better represent the whole parrot community, and increase the precision of estimates [12], and even though we used the lowest number of visual encounters required for DS modeling [4]. Concerningly, most of the species excluded are threatened or uncommon in the wild, but there are also some common but highly gregarious species, varying among realms. The extreme case is exemplified by the Afrotropic realm, where all study cases, species, and surveys were excluded despite the high survey effort invested (Supplementary S1).

Obviously, the percentage of exclusions from DS modeling would be much higher if we had separated surveys by years or seasons or split them into habitat-category sections [17], or had increased the minimum number of encounters for obtaining more precise density estimates [13], as many researchers may require for dealing with their research objectives.

Some procedures have been proposed to solve the problem of insufficient detections for parrot DS modeling. One is to use the records of a coexisting common species to model its probability of detection and use it for estimating the density of a congeneric, similar-sized rare species from which insufficient encounters were obtained [20]. However, after our experience, all species from the same genus (e.g., large macaws *Ara*, amazon parrots *Amazona*) are often equally scarce within the same survey, and thus all are unavoidably excluded from DS modeling. Another solution applied is pooling all records from rare species (even from different genera) to estimate a common probability of detection and derived species-specific density estimates [57]. However, these estimates must be taken with caution as the assumption that the detectability of different species is equivalent may be violated [53].

Rather than forcing the obtention of somewhat questionable density estimates when species-specific data are lacking, we recommend relying on simple relative abundances (individuals/km) when roadside surveys focus on whole parrot communities that include uncommon species, as they offer abundance estimates equivalent to detectability-corrected densities. Moreover, not recording distances has some advantages. On the one hand, the calculus of relative abundances is very simple and does not require the statistical skills needed for DS modeling. On the other hand, the field-work time saved by not recording distances (i.e., in surveys of rich and abundant parrot communities, researchers often must stop the car every few minutes to record them) can be invested in conducting longer roadside surveys, thus better representing the areas and parrot communities surveyed. This may be an important advantage, as parrot surveys are often logistically constrained by climatic conditions, and the time and funds available. Contrarily, we recommend DS modeling when researchers focus on one or a few common species, as they can then obtain more precise estimates of abundance by increasing the number of encounters (not paying attention to the rest of the species) and the best-fitting detection functions, as is done with point counts and line transects [13]. Even more importantly, DS modeling allows the calculation of densities that can be carefully extrapolated to the extent of suitable habitat and thus estimate the size of parrot populations, as has been done using point counts on islands [24,58]. A stratified design of large-scale roadside surveys could allow the estimation of population sizes for common parrot species with country- and even continental-level distributions, something that could be logistically unaffordable through point counts and walked line transects.

Finally, as a word of caution, researchers must keep in mind that distance sampling modeling was developed to correct for the imperfect detection of species in census surveys, but that the violation of some assumptions may also generate imperfect results. For the case of parrots, some assumptions of DS modeling are often violated: that all individuals encountered are accurately counted and their distances of detection exactly measured, and that encountered birds do not move while conducting the survey [10,38,53]. We have shown that the first assumption is not only violated in walked line transects and point counts [23,34] but also in roadside car surveys (see also [22]). In our surveys, 24% of the encounters corresponded to aural contacts of an unknown number of unseen individuals. Concerningly, the proportion of aural contacts was not randomly distributed but varied from 0 to 100% among species, being related to their body mass and gregariousness. As a solution following previous works [12,22–24], we estimated the number of unseen birds by substituting aural contacts with the average flock size of the species obtained from the same survey (this is important as average flock sizes may vary among seasons and regions). We used average flock size for consistency with previous works that adopted this solution [22–24] and because it is often reported as a measure of gregariousness (e.g., [13,56]). Given the often right-skewed distribution of flock sizes, researchers could use the median instead of the mean, although results should not markedly differ. In any case, we recommend incorporating this procedure to avoid the underestimation of parrot numbers in roadside surveys (in our case reaching 23% on average), resulting in relative parrot abundances that strongly correlated to distance-corrected densities. However, incorporating these estimates of an unseen number of individuals into DS modeling is challenging given the difficulties of estimating their distances of detection. Some solutions have been proposed when conducting parrot walked line transects and point counts, such as measuring distances to other objects at a similar distance if the heard parrot/flock was not visible [13,24] or categorizing these estimated distances to unseen parrots into intervals [58]. These estimations require expert observer skills and thus, researchers must be careful to do not introduce distance biases that would affect DS density estimates [38].

Regarding bird movements, DS modeling was conceptually developed as a 'snapshot' method in which animals are ideally 'frozen' while the survey is conducted, but in practice animals often make non-responsive movements (i.e., not disturbed by the observer) [38]. Buckland et al. [53] suggested that this assumption must be relaxed to include flying individuals in avian taxa that spend large proportions of their time in flight, such as seabirds and raptors. This is also the case for parrots. Except for a few low-mobility forest species (e.g., genus *Pionites*), most parrot species make long daily trips looking for food and moving between foraging, breeding, and roosting sites [10]. In fact, 36% of our parrot encounters corresponded to birds/flocks detected in non-responsive flights. Excluding these records would underestimate parrot abundances, with non-random biases according to the different flight propensities among species. Using walked line transects, Legault et al. [13] found that excluding flying birds caused an underestimation of parrot densities that varied between 7% and 67% among species. In their review on distance sampling approaches and assumptions, Thomas et al. [38] indicated that, in practice, non-responsive movement in walked line-transect surveys is not problematic provided it is slow relative to the speed of the observer, and thus it should be even less problematic for the faster-speed road car surveys. Therefore, we support the inclusion of flying parrots in roadside car surveys, as for walked line transects [13], but also suggest that researchers record the behavior of parrots (perching, foraging, flying) encountered. This may later allow researchers to decide whether to include flying birds in DS estimates [13] and to assess for example foraging habitat preferences by restricting records to foraging birds [17].

Researchers should be not discouraged by the limitations of DS modeling applied to roadside car surveys. Rather, they should be aware of how and when its application is feasible for their study species. On the other hand, some analytical advances for estimating parrot abundances [10] such as the use of hierarchical (N-mixture) models [59] have been recently applied to parrot roost counts [60], walked transects, and point counts [61], and have the potential to be used in roadside parrot surveys as has been done for raptors [16].

5. Conclusions

While roost counts may allow estimating regional and even global populations sizes of some parrot species [11,60,62,63], they are not affordable for most parrot populations and species and thus estimates of densities are often obtained using point counts or walk line transects [10]. However, these methodologies may fail to record rare and patchily distributed species, a problem that could be solved using large-scale roadside car surveys [10]. Here, compiling roadside car surveys conducted across the world biomes and continents inhabited by parrots, we have assessed how the aural- and distance-dependent probabilities of detection are affected by species traits and biomes as well as the pros and cons of roadside car surveys using or not using DS modeling, providing potential solutions for the problems encountered. We have demonstrated that distance-uncorrected estimates of parrot abundances are strongly correlated to those obtained using DS modeling, thus offering a good proxy for the actual relative densities of the species. This however does not mean that one method is better than the other. While DS modeling generally can not be used when dealing with whole parrot communities, because it results in the exclusion of a high percentage of surveys and species (mostly those uncommon and threatened ones), it may be useful for species-specific studies of common species. As learned from comparisons of other survey methodologies [10,49,59], the choice of the most suitable method is context-dependent. We summarize in Table 1 the strengths and weaknesses of using or not using DS attending to sampling effectiveness, which is understood here as the ability of either method to record birds that are present, to methodological constraints, and to the output variables required to reach different research goals. We hope this comprehensive summary will help guide researchers in choosing the best–fitting option for their particular research hypotheses, characteristics of the species studied, and logistical constraints.

Table 1. Comparison of strengths (+) and weaknesses (-) when using distance sampling modeling (DS Yes) or not (DS No) for estimating parrot abundances through roadside surveys, attending to the shortcomings of both methods and the objectives of studies. Equal signs (=) denote similar performance.

	DS Yes	DS No	Justification
Sampling effectiveness			
Attraction effect	-	+	DS may inflate densities of parrots attracted by roadside resources
Avoidance effect	+	-	DS may account for the potential avoidance of highly transited roads
Aural-only encounters	-	+	Estimating distances for DS from non-visual encounters is challenging
Flying individuals	=	=	Including flying individuals should not affect results from roadside surveys
Uncommon species	-	+	Encounters of naturally scarce and threatened species are not sufficient for DS
Gregarious species	-	+	Encounters of common but highly gregarious species may not be sufficient for DS
Detectability	+	-	DS allows the correction of abundances for distance-dependent detectability and associated covariates
Methodological constraints			
Survey length	-	+	DS requires longer surveys to obtain enough encounters for statistical modelling
Time invested	-	+	Time saved by not recording distances allows for longer surveys
Data analysis	-	+	DS requires statistical modeling instead of simple divisions
Output variables			
Single-species abundance	+	-	More accurate estimates can be obtained through DS for common species
Multi-species abundances	-	+	DS excludes a high percentage of species
Occupancy	-	+	Only presence/absence data are required
Species richness	-	+	DS is not needed
Species diversity	-	+	DS excludes a high percentage of species
Density	+	-	DS allows for calculating densities for species with enough encounters
Population size	+	-	DS allows extrapolating densities to the species distribution and thus estimating population size

Supplementary Materials: The following are available online at https://www.mdpi.com/article/10.3390/d13070300/s1, Supplementary S1: Details of the surveys conducted, Supplementary S2: Details of the species surveyed, Supplementary S3: histogram of detection distances, the plot of group size x detection distances (with Spearman correlation test p-value), and the estimated detection functions from the best DS models for each binning setup (including X2 goodness-of-fit tests), overlaid on the histogram of detection distances with the respective distance breaks, for each case study.

Author Contributions: Conceptualization, J.L.T., P.R.-V., F.R. and M.C.; methodology, J.L.T., F.V.D.; validation, J.L.T., F.H., G.B., F.V.D., and P.R.-V. fieldwork, all authors; formal analysis, J.L.T., F.V.D., and F.R.; resources, P.R.-V., B.T., and D.H.-B.; data curation, P.R.-V., B.T., and J.L.T.; writing—original draft preparation, J.L.T.; writing—review and editing, M.C., F.V.D., B.T., F.R., G.B., F.H., and J.M.B.; visualization, M.C.; supervision, J.L.T. and M.C.; project administration, M.C. and J.L.T.; funding acquisition, J.L.T. and M.C. All authors have read and agreed to the published version of the manuscript.

Funding: This research was funded by Fundación Biodiversidad (Spanish Ministerio de Medio Ambiente, project 52I.CA2109), Fundación Repsol, Spanish Ministerio de Ciencia e Innovación (Project CGL2015-71378-P), and mostly by Loro Parque Fundación (Project SEJI/2018/024).

Institutional Review Board Statement: Not applicable.

Informed Consent Statement: Not applicable.

Data Availability Statement: Main data are provided as Supplementary Materials. Additional information can be requested from the corresponding author.

Acknowledgments: We thank Isabel Afán (LAST-EBD) for helping in the design of road surveys using satellite maps. Logistic and technical support was provided by Doñana ICTS-RBD. We also thank John Griffith for his support in Australia, Julio Rabadán for guiding us in using and improving Observation, and two anonymous reviewers for their valuable suggestions.

Conflicts of Interest: The authors declare no conflict of interest.

References

1. Olah, G.; Butchart, S.H.M.; Symes, A.; Guzmán, I.M.; Cunningham, R.; Brightsmith, D.; Heinsohn, R. Ecological and socio-economic factors affecting extinction risk in parrots. *Biodivers. Conserv.* **2016**, *25*, 205–223. [CrossRef]
2. McClure, C.J.W.; Rolek, B.W. Relative Conservation Status of Bird Orders With Special Attention to Raptors. *Front. Ecol. Evol.* **2020**, *8*, 420. [CrossRef]
3. IUCN. The IUCN Red List of Threatened Species. Available online: https://www.iucnredlist.org (accessed on 17 January 2021).
4. Marsden, S.J.; Royle, K. Abundance and abundance change in the world's parrots. *Ibis* **2015**, *157*, 219–229. [CrossRef]
5. Berkunsky, I.; Quillfeldt, P.; Brightsmith, D.J.; Abbud, M.C.; Aguilar, J.M.R.E.; Alemán-Zelaya, U.; Aramburú, R.M.; Arce Arias, A.; Balas McNab, R.; Balsby, T.J.S.; et al. Current threats faced by Neotropical parrot populations. *Biol. Conserv.* **2017**, *214*, 278–287. [CrossRef]
6. Martin, R.O.; Perrin, M.R.; Boyes, R.S.; Abebe, Y.D.; Annorbah, N.D.; Asamoah, A.; Bizimana, D.; Bobo, K.S.; Bunbury, N.; Brouwer, J.; et al. Research and conser-vation of the larger parrots of Africa and Madagascar: A review of knowledge gaps and opportunities. *Ostrich* **2014**, *85*, 205–233. [CrossRef]
7. Vergara-Tabares, D.L.; Cordier, J.M.; Landi, M.A.; Olah, G.; Nori, J. Global trends of habitat destruction and consequences for parrot conservation. *Glob. Chang. Biol.* **2020**, *26*, 4251–4262. [CrossRef] [PubMed]
8. Barbosa, J.M.; Hiraldo, F.; Romero, M.A.; Tella, J.L. When does agriculture enter into conflict with wildlife? A global assess-ment of parrot–agriculture conflicts and their conservation effects. *Divers. Distrib.* **2021**, *27*, 4–17. [CrossRef]
9. Sánchez-Mercado, A.; Ferrer-Paris, J.; Rodríguez, J.; Tella, J.L. A Literature Synthesis of Actions to Tackle Illegal Parrot Trade. *Diversity* **2021**, *13*, 191. [CrossRef]
10. Dénes, F.V.; Tella, J.L.; Beissinger, S.R. Revisiting methods for estimating parrot abundance and population size. *Emu Austral. Ornithol.* **2017**, *118*, 67–79. [CrossRef]
11. Casagrande, D.G.; Beissinger, S.R. Evaluation of Four Methods for Estimating Parrot Population Size. *Condor* **1997**, *99*, 445–457. [CrossRef]
12. Marsden, S.J. Estimation of parrot and hornbill densities using a point count distance sampling method. *Ibis* **2008**, *141*, 327–390. [CrossRef]
13. Legault, A.; Theuerkauf, J.; Baby, E.; Moutin, L.; Rouys, S.; Saoumoé, M.; Verfaille, L.; Barré, N.; Chartendrault, V.; Gula, R. Standardising distance sampling surveys of parrots in New Caledonia. *J. Ornithol.* **2013**, *154*, 19–33. [CrossRef]
14. Sánchez-Zapata, J.A.; Calvo, J.F. Raptor distribution in relation to landscape composition in semi-arid Mediterranean habitats. *J. Appl. Ecol.* **1999**, *36*, 254–262. [CrossRef]
15. Carrete, M.; Tella, J.L.; Blanco, G.; Bertellotti, M. Effects of habitat degradation on the abundance, richness and diversity of raptors across Neotropical biomes. *Biol. Conserv.* **2009**, *142*, 2002–2011. [CrossRef]
16. Dénes, F.V.; Solymos, P.; Lele, S.; Silveira, L.F.; Beissinger, S.R. Biome-scale signatures of land-use change on raptor abun-dance: Insights from single-visit detection-based models. *J. Appl. Ecol.* **2017**, *54*, 1268–1278. [CrossRef]
17. Tella, J.; Rojas, A.; Carrete, M.; Hiraldo, F. Simple assessments of age and spatial population structure can aid conservation of poorly known species. *Biol. Conserv.* **2013**, *167*, 425–434. [CrossRef]
18. Luna, Á.; Romero-Vidal, P.; Hiraldo, F.; Tella, J.L. Cities may save some threatened species but not their ecological functions. *PeerJ* **2018**, *6*, e4908. [CrossRef]
19. Tella, J.L.; Dénes, F.; Zulian, V.; Prestes, N.P.; Martínez, J.; Blanco, G.; Hiraldo, F. Endangered plant-parrot mutualisms: Seed tolerance to predation makes parrots pervasive dispersers of the Parana pine. *Sci. Rep.* **2016**, *6*, 31709. [CrossRef]
20. Baños-Villalba, A.; Blanco, G.; Díaz-Luque, J.A.; Dénes, F.; Hiraldo, F.; Tella, J.L. Seed dispersal by macaws shapes the landscape of an Amazonian ecosystem. *Sci. Rep.* **2017**, *7*, 1–12. [CrossRef]
21. Blanco, G.; Hiraldo, F.; Rojas, A.; Dénes, F.; Tella, J.L. Parrots as key multilinkers in ecosystem structure and functioning. *Ecol. Evol.* **2015**, *5*, 4141–4160. [CrossRef]
22. Romero-Vidal, P.; Hiraldo, F.; Rosseto, F.; Blanco, G.; Carrete, M.; Tella, J.L. Opportunistic or Non-Random Wildlife Crime? Attractiveness rather than Abundance in the Wild Leads to Selective Parrot Poaching. *Diversity* **2020**, *12*, 314. [CrossRef]
23. Lee, A.T.K.; Marsden, S.J. The Influence of Habitat, Season, and Detectability on Abundance Estimates across an Amazonian Parrot Assemblage. *Biotropica* **2012**, *44*, 537–544. [CrossRef]

24. Reuleaux, A.; A Siregar, B.; Collar, N.J.; Panggur, M.R.; Mardiastuti, A.; Jones, M.J.; Marsden, S.J. Protected by dragons: Density surface modeling confirms large population of the critically endangered Yellow-crested Cockatoo on Komodo Island. *Condor* **2020**, *122*, 042. [CrossRef]
25. Lopes, D.C.; Martin, R.O.; Henriques, M.; Monteiro, H.; Cardoso, P.; Tchantchalam, Q.; Pires, A.J.; Regalla, A.; Catry, P. Combining local knowledge and field surveys to determine status and threats to Timneh Parrots Psittacus timneh in Guinea-Bissau. *Bird Conserv. Int.* **2019**, *29*, 400–412. [CrossRef]
26. Grilli, P.G.; Soave, G.E.; Arellano, M.L.; Masello, J.F. Abundancia relativa del Loro Barranquero (Cyanoliseus patagonus) en la provincia de Buenos Aires y zonas limítrofes de La Pampa y Río Negro, Argentina. *Hornero* **2012**, *27*, 63–71.
27. Salinas-Melgoza, A.; Salinas-Melgoza, V.; Wright, T.F. Behavioral plasticity of a threatened parrot in human-modified land-scapes. *Biol. Conserv.* **2013**, *159*, 303–312. [CrossRef]
28. Figueira, L.; Tella, J.L.; de Camargo, U.M.; Ferraz, G. Autonomous sound monitoring shows higher use of Amazon old growth than secondary forest by parrots. *Biol. Conserv.* **2015**, *184*, 27–35. [CrossRef]
29. Flores-López, E.; Montero-Castro, J.C.; Monterrubio-Rico, T.C.; Ibarra-Manríquez, G.; López-Toledo, L.; Bonilla-Ruz, C. Differential Use of Forest Patches by the Military Macaw Ara militaris (Psittacidae) in Coastal Tropical Forests of Jalisco, Mexico. *Ardeola* **2020**, *67*, 423–432. [CrossRef]
30. Wirminghaus, J.; Downs, C.T.; Perrin, M.; Symes, C. Abundance and activity patterns of the Cape parrot (Poicephalus robustus) in two afromontane forests in South Africa. *Afr. Zool.* **2001**, *36*, 71–77. [CrossRef]
31. Salinas-Melgoza, A.; Renton, K. Seasonal variation in activity patterns of juvenile Lilac-crowned parrots in tropical dry forest. *Wilson Bull.* **2006**, *117*, 291–295. [CrossRef]
32. Montesinos-Navarro, A.; Hiraldo, F.; Tella, J.L.; Blanco, G. Network structure embracing mutualism–antagonism continuums increases community robustness. *Nat. Ecol. Evol.* **2017**, *1*, 1661–1669. [CrossRef]
33. Sebastián-González, E.; Hiraldo, F.; Blanco, G.; Hernández-Brito, D.; Romero-Vidal, P.; Carrete, M.; Gómez-Llanos, E.; Pacífico, E.C.; Díaz-Luque, J.A.; Dénes, F.V.; et al. The extent, frequency and ecological functions of food wasting by parrots. *Sci. Rep.* **2019**, *9*, 1–11. [CrossRef]
34. Hernández-Brito, D.; Romero-Vidal, P.; Hiraldo, F.; Blanco, G.; Díaz-Luque, J.A.; Barbosa, J.M.; Symes, C.T.; White, T.H.; Pacífico, E.C.; Sebastián-González, E.; et al. Epizoochory in parrots as an overlooked yet widespread plant–animal mutualism. *Plants* **2021**, *10*, 760. [CrossRef]
35. Blanco, G.; Romero-Vidal, P.; Carrete, M.; Chamorro, D.; Bravo, C.; Hiraldo, F.; Tella, J. Burrowing Parrots *Cyanoliseus patagonus* as Long-Distance Seed Dispersers of Keystone Algarrobos, Genus *Prosopis*, in the Monte Desert. *Diversity* **2021**, *13*, 204. [CrossRef]
36. Buckland, S.T.; Anderson, D.R.; Burnham, K.P.; Laake, J.L.; Borchers, D.L.; Thomas, L. *Introduction to Distance Sampling: Estimating Abundance of Biological Populations*; Oxford University Press: Oxford, UK, 2001.
37. Buckland, S.T. Point transect surveys for songbirds: Robust methodologies. *Auk* **2006**, *123*, 345–357. [CrossRef]
38. Thomas, L.; Buckland, S.T.; Rexstad, E.; Laake, J.L.; Strindberg, S.; Hedley, S.L.; Bishop, J.R.B.; Marques, T.A.; Burnham, K.P. Distance software: Design and analysis of distance sampling surveys for estimating population size. *J. Appl. Ecol.* **2010**, *47*, 5–14. [CrossRef] [PubMed]
39. Miller, D.L.; Rexstad, E.; Thomas, L.; Marshall, L.; Laake, J.L. Distance Sampling in R. *J. Stat. Softw.* **2019**, *89*, 1–28. [CrossRef]
40. R Core Team. *R: A Language and Environment for Statistical Computing*; R Foundation for Statistical Computing: Vienna, Austria, 2018; Available online: http://www.R-project.org/ (accessed on 6 July 2019).
41. Burgio, K.R.; Davis, K.E.; Dreiss, L.M.; Cisneros, L.M.; Klingbeil, B.; Presley, S.J.; Willig, M.R. Phylogenetic supertree and functional trait database for all extant parrots. *Data Brief.* **2019**, *24*, 103882. [CrossRef]
42. Jovani, R.; Tella, J.L. Parasite prevalence and sample size: Misconceptions and solutions. *Trends Parasitol.* **2006**, *22*, 214–218. [CrossRef]
43. Bibby, C.J.; Burgess, N.D.; Hill, D.A. *Bird Census Techniques*; Elsevier: Amsterdam, The Netherlands, 1992.
44. Mammides, C.; Kounnamas, C.; Goodale, E.; Kadis, C. Do unpaved, low-traffic roads affect bird communities? *Acta Oecologica* **2016**, *71*, 14–21. [CrossRef]
45. Cooke, S.; Balmford, A.; Donald, P.F.; E Newson, S.; Johnston, A. Roads as a contributor to landscape-scale variation in bird communities. *Nat. Commun.* **2020**, *7*, 1–10. [CrossRef]
46. Meunier, F.D.; Verheyden, C.; Jouventin, P. Use of roadsides by diurnal raptors in agricultural landscapes. *Biol. Conserv.* **2000**, *92*, 291–298. [CrossRef]
47. Blanco, G.; Bravo, C.; Chamorro, D.; Lovas-Kiss, A.; Hiraldo, F.; Tella, J.L. Herb endozoochory by cockatoos: Is 'foliage the fruit"? *Austral. Ecol.* **2020**, *45*, 122–126. [CrossRef]
48. Hernández-Brito, D.; Carrete, M.; Blanco, G.; Romero-Vidal, P.; Senar, J.C.; Mori, E.; White, T.H., Jr.; Luna, A.; Tella, J.L. The Role of Monk Parakeets as Nest-Site Facilitators in Their Native and Invaded Areas. *Biology* **2021**. (under review).
49. Darras, K.; Batáry, P.; Furnas, B.J.; Grass, I.; Mulyani, Y.A.; Tscharntke, T. Autonomous sound recording outperforms human observation for sampling birds: A systematic map and user guide. *Ecol. Appl.* **2019**, *29*, e01954. [CrossRef] [PubMed]
50. Tella, J.; Canale, A.; Carrete, M.; Petracci, P.; Zalba, S.M. Anthropogenic Nesting Sites Allow Urban Breeding in Burrowing Parrots Cyanoliseus patagonus. *Ardeola* **2014**, *61*, 311–321. [CrossRef]
51. Carrete, M.; Tella, J.L. Individual consistency in flight initiation distances in burrowing owls: A new hypothesis on disturb-ance-induced habitat selection. *Biol. Lett.* **2010**, *6*, 167–170. [CrossRef]

52. Carrete, M.; Tella, J.L. Inter-individual variability in fear of humans and relative brain size of the species are related to contemporary urban invasion in birds. *PLoS ONE* **2011**, *6*, e18859. [CrossRef]
53. Buckland, S.; Marsden, S.J.; Green, R.E. Estimating bird abundance: Making methods work. *Bird Conserv. Int.* **2008**, *18*, S91–S108. [CrossRef]
54. Keller, C.M.; Scallan, J.T. Potential roadside biases due to habitat changes along breeding bird survey routes. *Condor* **1999**, *101*, 50–57.
55. Romero-Vidal, P.; Barbosa, J.M.; Blanco, G.; Hiraldo, F.; Carrete, M.; Tella, J.L. Deforestation or overharvesting? Pet-keeping cultural burden rather than habitat transformation causes selective parrot defaunation in Costa Rica. 2021. (in preparation)
56. Gilardi, J.D.; Munn, C.A. Patterns of Activity, Flocking, and Habitat Use in Parrots of the Peruvian Amazon. *Condor* **1998**, *100*, 641–653. [CrossRef]
57. Española, C.P.; Collar, N.J.; Marsden, S.J. Are populations of large-bodied avian frugivores on Luzon, Philippines, facing imminent collapse? *Anim. Conserv.* **2013**, *16*, 467–479. [CrossRef]
58. Rivera-Milán, F.F.; Simal, F.; Bertuol, P.; Boomer, G.S. Population monitoring and modelling of yellow-shouldered parrot on Bonaire, Caribbean Netherlands. *Wildl. Biol.* **2018**, *2018*, wlb.00384. [CrossRef]
59. Dénes, F.V.; Silveira, L.F.; Beissinger, S.R. Estimating abundance of unmarked animal populations: Accounting for imperfect detection and other sources of zero inflation. *Methods Ecol. Evol.* **2015**, *6*, 543–556. [CrossRef]
60. Zulian, V.; Müller, E.S.; Cockle, K.L.; Lesterhuis, A.; Júnior, R.T.; Prestes, N.P.; Martinez, J.; Kéry, M.; Ferraz, G. Addressing multiple sources of uncertainty in the estimation of global parrot abundance from roost counts: A case study with the Vina-ceous-breasted Parrot (Amazona vinacea). *Biol. Conserv.* **2020**, *248*, 108672. [CrossRef]
61. Geary, M.; Brailsford, C.J.; Hough, L.I.; Baker, F.; Guerrero, S.; Leon, Y.M.; Collar, N.J.; Marsden, S.J. Street-level green spaces support a key urban population of the threatened Hispaniolan parakeet Psittacara chloropterus. *Urban. Ecosyst.* **2021**, 1–8. [CrossRef]
62. Barbosa, A.E.A.; Tella, J.L. How much does it cost to save a species from extinction? Costs and rewards of conserving the Lear's macaw. *R. Soc. Open Sci.* **2019**, *6*, 190190. [CrossRef] [PubMed]
63. Dupin, M.K.; Dahlin, C.R.; Wright, T.F. Range-Wide Population Assessment of the Endangered Yellow-Naped Amazon (*Amazona auropalliata*). *Diversity* **2020**, *12*, 377. [CrossRef]

Communication

Molecular Survey of Pathogens in Wild Amazon Parrot Nestlings: Implications for Conservation

Frederico Fontanelli Vaz [1], Elenise Angelotti Bastos Sipinski [2], Gláucia Helena Fernandes Seixas [3,4], Nêmora Pauletti Prestes [5], Jaime Martinez [5] and Tânia Freitas Raso [1,*]

[1] Avian Ecopathology Laboratory, School of Veterinary Medicine and Animal Science, University of São Paulo, 05508-270 São Paulo, Brazil; fredfontanelli@yahoo.com.br
[2] Society for Wildlife Research and Environmental Education, 80520-310 Curitiba, Brazil; tise@spvs.org.br
[3] Parque das Aves, 85855-750 Foz do Iguaçu, Brazil; glauciaseixas@hotmail.com
[4] Neotropical Foundation of Brazil, 79290-000 Bonito, Brazil
[5] Institute of Biological Sciences, University of Passo Fundo, 99042-800 Passo Fundo, Brazil; prestes@upf.br (N.P.P.); martinez@upf.br (J.M.)
* Correspondence: tfraso@usp.br

Abstract: South America presents the greatest Psittacidae diversity in the world, but also has the highest numbers of threatened parrot species. Recently, exotic viruses have been detected in captive native psittacine birds in Brazil, however, their impacts on the health of wild parrots are still unknown. We evaluated the presence of *Chlamydia psittaci*, *Psittacid alphaherpesvirus* 1 (PsHV-1), avipoxvirus and beak and feather disease virus (BFDV) in wild *Amazona aestiva*, *A. brasiliensis* and *A. pretrei* nestlings and in wild caught *A. aestiva* nestlings seized from illegal trade. Samples were collected from 205 wild nestlings and 90 nestlings from illegal trade and pathogen-specific PCR was performed for each sample. *Chlamydia* DNA prevalence was 4.7% in *A. aestiva* and 2.5% in *A. brasiliensis* sampled from the wild. Sequencing revealed that the *C. psittaci* sample belonged to the genotype A. PsHV-1, avipoxvirus and BFDV DNA was not detected. These results have conservation implications since they suggest that wild parrot populations have a low prevalence of the selected pathogens and, apparently, they were not reached by the exotic BFDV. Stricter health protocols should be established as condition to reintroduction of birds to the wild to guarantee the protection of Neotropical parrots.

Keywords: wild parrots; *Chlamydia psittaci*; *Psittacid alphaherpesvirus* 1; avipoxvirus; beak and feather disease virus; conservation threats

Citation: Vaz, F.F.; Sipinski, E.A.B.; Seixas, G.H.F.; Prestes, N.P.; Martinez, J.; Raso, T.F. Molecular Survey of Pathogens in Wild Amazon Parrot Nestlings: Implications for Conservation. *Diversity* **2021**, *13*, 272. https://doi.org/10.3390/d13060272

Academic Editors: José L. Tella, Guillermo Blanco and Martina Carrete

Received: 26 April 2021
Accepted: 12 June 2021
Published: 16 June 2021

1. Introduction

Psittacidae diversity in South America is the greatest in the world and Brazil is the country with the largest number of species. Among the 411 known species, 86 occur in the national territory [1]. Unfortunately, Brazil is also in the first position when it comes to threatened species, with Psittaciformes being one of the most threatened, containing 25 native species in the Global International Union for Conservation of Nature (IUCN) Red List [2].

Amazon parrots are prominent among the national species, being the first among the most trafficked psittacine birds in Brazil. *Amazona* genus comprises 12 species in Brazil and these birds are threatened mainly by the illegal trade and loss of their habitat. Currently, one third of the native Amazon species is threatened [2].

Among these species, the red-spectacled Amazon parrot (*Amazona pretrei*) is threatened within the vulnerable category. The red-tailed Amazon parrot (*Amazona brasiliensis*) has left the IUCN Red List, entering the near-threatened status. Both species have a restricted distribution and exist only in Brazilian territory. The blue-fronted Amazon parrot (*Amazona aestiva*) is also in the near-threatened category and has a wide distribution, including Brazil, Argentina, Bolivia, and Paraguay territories. However, there is a special interest in this species because it is the main target of the illegal trade [2,3].

Nonetheless, another current challenge to wildlife conservation efforts is the dissemination of infectious diseases. As parrots are extremely popular pets, the demand created around the world has led to an international movement of over 19 million birds since 1975 [4], which triggers the spread of pathogens. Disease emergence can be triggered by translocation; introduction of infected animals, pathogens or vectors to new geographic regions; human or domestic animals' encroachment, spill-over, ex situ contact and ecological manipulation [5]. Amazon parrots are subjected to at least three of these situations [6], therefore, their health assessment in the wild is an important addition to their conservation efforts.

The illegal trade of wild birds is still a reality in Brazil, and only a small part of the nestlings removed from nature is apprehended by environmental authorities. These birds are mixed in rehabilitation centres with resident or pet birds and are often released into the wild without any health criteria [7]. In addition, national and international cross border movement of birds continues as the result of smuggling and legal trade of domestically raised birds [8], creating the perfect scenario for disease dissemination in wild and captive animals, as trafficked and imported birds are fed with improper diets, housed in crowded unhygienic conditions, and mixed with other species [9,10]. The global spread of diseases has caused a significantly negative conservation impact on captive and wild populations [11,12]. Highly resistant viruses in the environment and persistent subclinical infections make controlling these pathogens a challenge [8].

Chlamydia psittaci and the Psittacid herpesvirus 1 (PsHV-1) are relevant pathogens that affect parrots and have been observed in captive psittacine in Brazil [13,14], including occasional outbreaks [10]. A neglected virus in wild birds, the avipoxvirus, has also caused an outbreak in psittacine species located in a facility in Brazil [15]. In addition, the Psittacine Beak and Feather Disease (PBFD) caused by a circovirus, is an exotic pathogen introduced in the country [16], that has been reported in exotic and native pet birds [17,18].

The results of all the negative anthropogenic actions for the health of wild Amazon parrots in Brazil are unknown and information is scarce in the literature [19–21]. The aim of the present study was to investigate the presence of *C. psittaci* and viral pathogens DNA in wild Amazon parrot nestlings and in wild caught nestlings recently apprehended from illegal trade in Brazil, and to discuss the implications of the results for the conservation of psittacine birds.

2. Materials and Methods

The study comprised three species of parrots in four states of Brazil (Figure 1). *Amazona pretrei* nestlings were sampled in a fragmented area of the southern fields, in the municipality of Pontão, state of Rio Grande do Sul. *Amazona brasiliensis* parrots were studied on three islands (Ilha Rasa, Ilha Gamela and Ilha Chica), in the state of Paraná, located in the Guaraqueçaba Environmental Protection Area, which has an extensive area of Atlantic Forest. This species was also sampled in Comprida Island, state of São Paulo, another Atlantic Forest area within the Ilha Comprida Environmental Protection Area. *Amazona aestiva* parrots were sampled in Miranda, state of Mato Grosso do Sul, located in the Brazilian Pantanal wetlands.

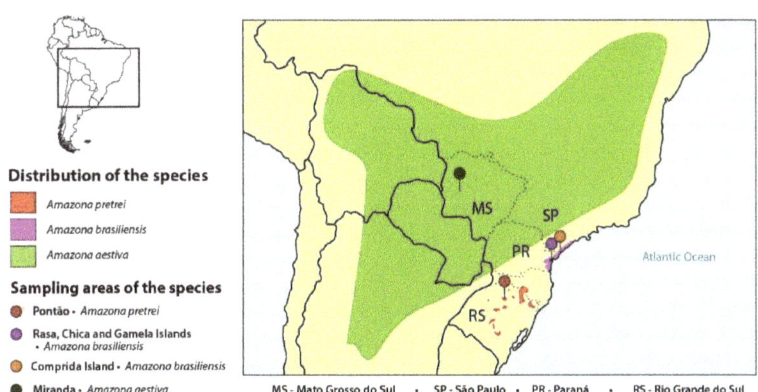

Figure 1. Distribution of *Amazona pretrei, Amazona brasiliensis* and *Amazona aestiva* in Brazil and South America and sampling areas.

Oropharyngeal and cloacal swab samples and/or blood samples were collected from Amazon parrot nestlings in field expeditions during the breeding seasons (October to January) from 2013 to 2018. All samples were collected by trained professionals. Natural and artificial nests were accessed using ladders or climbing equipment. The birds were removed from the nests, examined, sampled, and then placed back in the nests. Swab samples were kept frozen in microtubes containing viral transport media, and blood samples were kept frozen in microtubes until the analyses. One *A. brasiliensis* was found recently dead inside a nest and was necropsied for sample collection. Liver and spleen fragments were collected and kept frozen until laboratory analysis.

Additionally, in 2015, 413 wild caught *A. aestiva* nestlings were seized from the illegal trade by environmental officials in three different locations (A—262 birds, B—116 birds, C—35 birds) in two states of Brazil (Mato Grosso do Sul and Paraná). The nestlings apprehended were submitted to the Wildlife Rehabilitation Center (WRC) located in the city of Campo Grande, Mato Grosso do Sul. Birds were different ages, ranging from recently hatched to fully fledged nestlings (about 5 to 50 days). They were housed inside boxes in a proper room, separated by bird size and by origin. Biological samples were collected in the first 72 h after the birds were seized. Cloacal swab samples were randomly collected from approximately 20% of the nestlings from each box, totalizing 90 nestlings, 21.6% of the nestlings received in that year. Blood was collected from the brachial vein in 30 of these parrots and all samples were kept frozen until analysis.

Genomic DNA extraction was performed using the NucleoSpin Tissue kit® (Macherey-Nagel, Germany) according to manufacturer's instructions. Each sample was digested in 200 μL of lysis buffer and proteinase K (20 mg/mL) at 56 °C for 12 h prior extraction. DNA was extracted from 326 samples (swab/blood samples and a fragment of liver/spleen) from 205 birds. All samples were screened for the presence of chlamydial DNA using a conventional PCR targeting 111bp of the Chlamydiaceae 23S rRNA gene with primers from Enrich et al. [22]. A 25 μL reaction mix containing 3 μL of genomic DNA, 10 pmol of each primer, 12.5 μL of buffer (DreamTaq Green PCR Master Mix, Thermo Fisher Scientific, Waltham, MA, USA) and nuclease-free water qsp was used in the reaction. Target DNA was amplified performing a conventional PCR using an initial denaturation of 60 s at 96 °C, then 40 cycles of 30 s at 94 °C, 60 s at 50 °C and 30 s at 72 °C, followed by a final extension of 4 min at 72 °C. The samples were also screened for the presence of viruses using conventional PCRs for PsHV-1 [23], avipoxvirus [24] and BFDV [25]. Following this, positive samples in the *Chlamydiaceae* PCR were evaluated using a second PCR assay to amplify a fragment of *C. psittaci ompA* gene [26]. The primer sequences used for all agents can be found in the Table S1.

Negative and positive standard controls were used in all PCR reactions for each agent. Nuclease-free water was used as negative control. The *C. psittaci* genotype A, Cpsi/Mm/BR01 DNA was used as positive control (GenBank accession number JQ926183) for *Chlamydiacea* and *omp*A PCR assays. BFDV (strain from a *Psittacus erithacus*), herpesvirus (PsHV-1 genotype 3 from an *Ara ararauna*) and Pox vaccine (Pox pigeon, Biovet, Brazil) DNA were used as positive controls for the other PCR assays. All reactions were carried out using the thermal cycler Axygen® Maxygene (Axygen, Union City, CA, USA). The products were analysed by electrophoresis in a 1.5% agarose gel stained with GelRed® (Biotium, Fremont, CA, USA) nucleic acid stain.

Amplified products from the *omp*A PCR assays were purified from the agarose gel using a commercial kit (NucleoSpin Gel and PCR Clean-Up, Macherey Nagel, Düren, North Rhine-Westphalia, Germany) according to the manufacturer's instructions and sequenced in dual-direction by Sanger sequencing (Genome Research Center, University of São Paulo, São Paulo, Brazil). The chromatograms were analysed for quality using MEGA X software, and sequences were compared with data available on GenBank through a BLAST search. The nucleotide alignment was performed using MAFFT version 7 with the FFT-NS-I algorithm [27]. A neighbor joining tree was constructed using Mega X [28]. The Tamura-Nei model was chosen to create the tree tested by bootstrapping with 1000 replicates.

3. Results

A total of 205 Amazon parrot nestlings from wildlife were sampled as shown in Table 1. All the birds showed no clinical signs that could suggest infection by any of the pathogens here investigated. Liver and spleen fragments were collected from one wild *A. brasiliensis* nestling that was found recently dead inside one nest at Rasa Island, but pathogens DNA were not detected in those samples. None of the nestling samples tested yielded positive PCR results for PsHV-1, avipoxvirus or BFDV.

Table 1. Number of wild Amazon parrot nestlings sampled according to breeding season in the states of Rio Grande do Sul (RS), Paraná (PR), Mato Grosso do Sul (MS) and São Paulo (SP), Brazil.

Amazon Species	State	2013/2014	2015	2016	2017	2018	Total Samples (Swab and/or Blood)	Number of Birds Sampled for Selected Virus **/Positive (%)	Number of *Chlamydia* Positive/Total Birds (%)
A. pretrei	RS	0	0	4	0	0	8	4 (0%)	0/4 (0%)
A. brasiliensis	PR/SP	74 *	0	28 #	15	21	230	138 (0%)	2/80 (2.5%)
A. aestiva	MS	17	17	23	3	0	89	63 (0%)	3/63 (4.8%)
Total							327	205 (0%)	5/147 (3.4%)

* Just 16 swabs were tested for *Chlamydia* in this period. Total number of birds tested for *Chlamydia* = 147. # Liver and spleen fragments sampled from one carcass. ** PsHV-1, avipoxvirus and BFDV.

Chlamydia prevalence found for all the parrots evaluated was 3.4% (5/147); these samples were from parrots sampled between 2015–2018. From the *Chlamydia*-positive samples, 4.8% (3/63) were collected from *A. aestiva* (cloacal swab samples), and 2.5% (2/80) were collected from *A. brasiliensis* (one cloacal/oropharyngeal swab sample and one blood sample). The *A. pretrei* nestlings evaluated were negative (0/4, 0%). *Chlamydia psittaci* nucleotide sequencing was possible only in the blood sample from an *A. brasiliensis* nestling (Cpsi/Ab/BR02; GenBank accession number MT741095). This partial *ompA* gene sequence was analysed and aligned with reference sequences available on GenBank (Table S2). The phylogenetic tree is presented in Figure 2. The sequence obtained was confirmed as *C. psittaci* as it had a high percentage of identity (99.25%) with other *C. psittaci* sequences, clustering within the Genotype A.

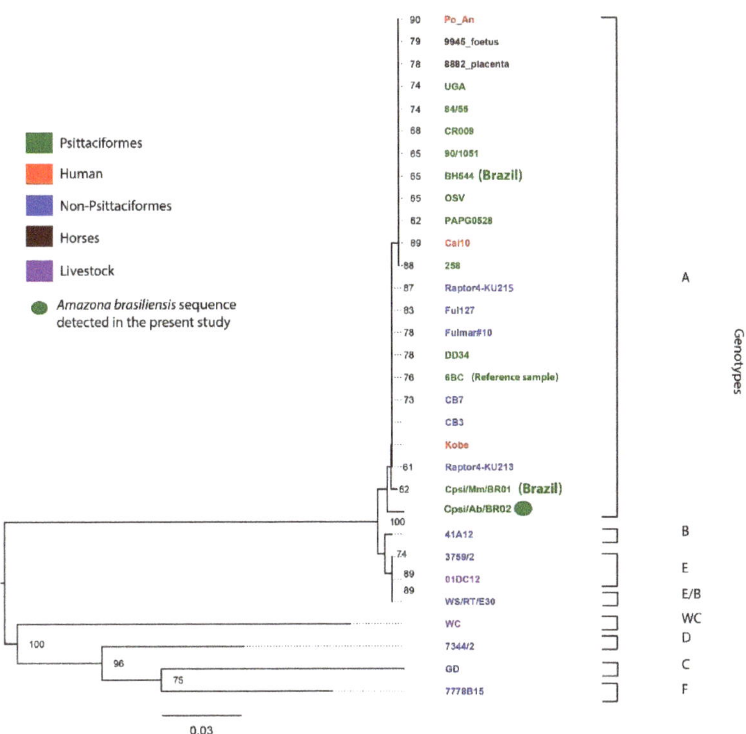

Figure 2. A mid-point rooted, neighbor joining phylogeny of the partial DNA sequences of the *Chlamydia psittaci* outer membrane protein gene alignment (1000 bootstrap replicates). Bootstrap support of nodes is shown if it exceeds 60%. (Cpsi/Ab/BR02, GenBank accession number MT741095).

Regarding the nestlings from illegal trade, all samples (90 swabs and 30 blood samples) collected in the first 72 h after apprehension of the birds were negative for the agents analysed (*Chlamydia psittaci,* PsHV-1, avipoxvirus and BFDV).

4. Discussion

Anthropogenic activities have been a trigger for dissemination of diseases in psittacine birds as shown in previous studies around the world [5], including the international introduction of pathogens to wild and captive naïve populations [11], and outbreaks in wild and captive birds [29]. Nevertheless, the impact of these actions on the health of wild Brazilian parrots is unknown, as a large-scale assessment has never been performed. The results of this study showed low prevalence of *C. psittaci* and no viral detection in the wild Amazon parrot nestlings sampled, which apparently have not yet been reached by the global spread of relevant psittacine pathogens, which is not the case of BFDV in other countries [30].

C. psittaci is a bacterium considered endemic in Brazilian psittacine birds. It was first detected in wild *A. aestiva* (2/32, 6.3%) in Pantanal, Mato Grosso do Sul, in 2006, using a semi-nested PCR and complement fixation test [19]. This prevalence is in accordance with our results (3/63, 4.8%) in *A. aestiva* nestlings evaluated from the same region of Pantanal. Other studies in *A. brasiliensis* nestlings in Rasa Island, Paraná, showed 0.8% [20] and 0% [21] of *C. psittaci* prevalence which are in accordance with our results for this species (2/80, 2.5%). Unfortunately, no *C. psittaci* sequences from these previous studies are available for comparison. Therefore, even though some parrot populations have the bacteria circulating, the overall prevalence seems to be stable over the years and, in some of them, the circulating genotypes are unknown. Conventional PCR is widely used in research of pathogens in captive birds; however, a minor limitation is that this assay can have less sensitivity than real time PCR and maybe low copy number samples can be missed. The sampled birds here were very young fledglings and possibly were not even infected. This fact was demonstrated in a previous study [21], in which the combined serology and PCR results were negative for *Chlamydia*, demonstrating that the nestlings were not infected in the sample collection time. Therefore, we believe that the chlamydial prevalence in these populations is indeed low. In other Latin American countries, only serologic surveys are available for *C. psittaci* in the wild: *A. aestiva* in Bolivia [31] and *Aratinga weddelli* and *Brotogeris sanctithomae* in Peru [32]. However, no antibodies against *C. psittaci* were detected.

The amplified *C. psittaci ompA* fragment clustered with previously described *ompA* Genotype A. Based on BLAST analyses, our sequence had 99% identity with the reference sequence 6BC (NC_017287) and with two Brazilian *C. psittaci* sequences: one from a long-term captive *A. aestiva* (Genbank MH138296) and one from a wild caught monk parakeet (*Myiopsitta monachus*) sampled after being seized from poachers in Southern Brazil (Genbank JQ926183) [33]. Furthermore, the highest percentage of nucleotide identity (99.25%) was observed with *C. psittaci* found in birds and humans from Europe and Asia (Genbank CP033059, KP893667 and AB468956). The sequence obtained here and from previous studies in Brazil suggest that Genotype A can be the main circulating genotype in wild psittacine birds in the country. Moreover, Genotype A is most frequently associated with psittacosis cases in humans. In Brazil, the potential of monk parakeets (*Myiopsitta monachus*) in transmitting *C. psittaci* to humans has already been documented [34].

In the present study, no BFDV, PsHV and avipoxvirus DNA was detected in wild and in smuggled nestlings. Nevertheless, the wild caught nestlings seized from the illegal trade were not sampled later to evaluate housing long term effects on their health.

Pacheco's disease (PD) is caused by PsHV-1 and it was first recognized in parrots in Brazil [35], and only later it was seen in many psittacine birds exported from South America to Europe and North America [36]. Even so, there are only sporadic reports on PD occurrence in parrots in Brazil [14] and there is no information on the genotypes circulating in the country. So far, in captive birds, only Genotype 1 was found in 18 Amazon individuals [37]. Negative results for avipoxvirus were also reported in 29 captive *A. vinacea* using the same primers in a conventional PCR [38]. However, avipoxvirus outbreaks have been reported in captive native [15] and exotic [39] Psittaciformes in Brazil, showing low and high mortality rates, respectively. BFDV has been recently detected in captive exotic and native species [17,18] in Brazil. Based on the initial findings of the present study, it is likely that wild parrot populations can still be unreached by the global spread of BFDV [40]. Because of the high dissemination capacity and the immunosuppressive effects, BFDV has worried avian veterinarians around the world [41]. Hence, further research must be done to provide more detailed data on prevalence, diversity of genotypes and host range of these viruses in wild and captive psittacine birds in the South America.

Wildlife rehabilitation centers are responsible for receiving injured or apprehended wild animals, which are mainly represented by native species seized from illegal trade or illegally maintained as pets by owners or by irregular breeders and, occasionally, exotic species from irregular captivity. Frequently, these places release native birds to the wild after being recovered. This situation is concerning and there is an imminent risk of introduction of pathogens to wild Brazilian psittacine populations, as thousands of birds are released to the wild every year without any health criteria/quarantine. Considering these data and the negative results for BFDV reported here in wild parrots, little has been done in mitigating health threats and to improve the protection of the Brazilian parrot fauna. Therefore, the elaboration of a national health program for relevant pathogens that affects psittacine birds is extremely urgent. Once introduced in a captive or wild population, it is difficult or even impossible to eradicate the BFDV, and many birds would have to be euthanized to achieve this. Thus, there is no doubt that prevention methods are the best approach to control the spread of this virus [41].

5. Conclusions

Our study reveals a longitudinal pathogens assessment of wild psittacine fledglings showing low chlamydial prevalence and no detection of some important parrot viruses. Even though the incidence of exotic viral diseases is increasing in captive psittacine in Brazil, it is still early to assess the real impact in wildlife parrots. Unfortunately, the capture and sampling of adult Amazon parrots is not an easy procedure in natural conditions, which could provide more robust data about the health status of these populations. Further, the infectious diseases control in psittacine from captivity or from illegal trade must be carried out carefully before releasing the birds into the wild avoiding the dissemination of pathogens that have the potential to negatively impact the conservation of Neotropical parrots.

Supplementary Materials: The following are available online at https://www.mdpi.com/article/10.3390/d13060272/s1, Table S1: primer sequences used for detection of viruses and *Chlamydia psittaci* in wild Amazon parrot from Brazil. Table S2: GenBank accessions, host species, origins, and *Chlamydia psittaci* strains used for comparation in this study.

Author Contributions: Conceptualization: T.F.R., F.F.V. Formal analysis: T.F.R., F.F.V. Funding acquisition: T.F.R., F.F.V. Investigation: F.F.V., E.A.B.S., G.H.F.S., N.P.P., J.M., T.F.R. Methodology: F.F.V., E.A.B.S., G.H.F.S., N.P.P., J.M., T.F.R. Project administration: T.F.R. Supervision: T.F.R. Writing—original draft: F.F.V., T.F.R. Writing—Review and Editing: F.F.V., E.A.B.S., G.H.F.S., N.P.P., J.M., T.F.R. All authors have read and agreed to the published version of the manuscript.

Funding: FFV was recipient of PhD fellowship by the FAPESP (2016/03928-2).

Institutional Review Board Statement: Captures and all sampling collection in this study were approved by the Brazilian Ministry of Environment (SISBIO 43876-1, 4993-6, 35621-4) and by the Ethics Committee on Animal Use of the School of Veterinary Medicine and Animal Science of the University of São Paulo (CEUA 9545290116).

Informed Consent Statement: Not applicable.

Data Availability Statement: Not applicable.

Acknowledgments: We thank Nara Pontes and the staff of the Wildlife Rehabilitation Centre (CRAS, IMASUL) in Campo Grande/MS for assistance in the collection of samples in smuggled nestlings.

Conflicts of Interest: The authors declare no conflict of interest.

References

1. Piacentini, V.Q.; Aleixo, A.; Agne, C.E.; Mauricio, G.N.; Pacheco, J.F.; Bravo, G.A.; Brito, G.R.R.; Naka, N.L.; Olmos, F.; Posso, S.; et al. Annotated checklist of the birds of Brazil by the Brazilian Ornithological Records Committee. *Rev. Bras. Ornitol.* **2015**, *23*, 91–298. [CrossRef]
2. IUCN—International Union for Conservation of Nature. *Red List of Threatened Species*, Version 2010-2; 2020. Available online: http://www.iucnredlist.org (accessed on 15 March 2021).
3. Seixas, G.H.F.; Mourão, G. Communal roosts of the Blue-fronted Amazons (*Amazona aestiva*) in a large tropical wetland: Are they of different types? *PLoS ONE.* **2018**, *13*, e0204824. [CrossRef]
4. CITES—Convention on International Trade in Endangered Species of Wild Fauna and Flora. *CITES Trade Database—1975–2018*; 2018. Available online: https://trade.cites.org/ (accessed on 15 March 2021).
5. Daszak, P.; Cunningham, A.A.; Hyatt, A.D. Emerging infectious diseases of wildlife—threats to biodiversity and human health. *Science* **2000**, *28*, 443–449. [CrossRef] [PubMed]
6. Berkunsky, I.; Quillfeldt, P.; Brightsmith, D.J.; Abbud, M.C.; Aguilar, J.M.R.E.; Alemán-Zelaya, U.; Aramburú, R.M.; Arce Arias, A.; Balas McNab, R.; Balsby, T.J.S.; et al. Current threats faced by Neotropical parrot populations. *Biol. Cons.* **2017**, *214*, 278–287. [CrossRef]
7. Marini, M.A.; Garcia, F.I. Bird conservation in Brazil. *Cons. Biol.* **2005**, *19*, 665–671. [CrossRef]
8. Phalen, D.N. The globalization of parrot viruses as the result of the trade in wild caught parrots and its potential conservation implications. In Proceedings of the Wildlife Diseases Association International Conference, Sunshine Coast, Australia, 26–30 July 2015; p. 24.

9. Gaskin, J.M.; Robbins, C.M.; Jacobson, E.R. An explosive outbreak of Pacheco's parrot disease and preliminary experimental findings. In Proceedings of the American Association of Zoo Veterinarians, Knoxville, TN, USA, 7–9 November 1978; pp. 365–374.
10. Raso, T.F.; Godoy, S.N.; Milanelo, L.; Souza, C.A.I.; Matushima, E.R.; Araújo Júnior, J.P.; Pinto, A.A. An outbreak of chlamydiosis in captive Blue-fronted Amazon parrots (*Amazona aestiva*) in Brazil. *J. Zoo Wildl. Med.* **2004**, *35*, 94–96. [CrossRef] [PubMed]
11. Kundu, S.; Faulkes, C.G.; Greenwood, A.G.; Jones, C.G.; Kaiser, P.; Lyne, O.D.; Black, S.A.; Chowrimootoo, A.; Groombridge, J.J. Tracking viral evolution during a disease outbreak: The rapid and complete selective sweep of a circovirus in the endangered Echo parakeet. *J. Virol.* **2012**, *86*, 5221–5229. [CrossRef]
12. Olsen, G.H.; Crosta, L.; Gartrell, B.D.; Marsh, P.M.; Stringfield, C.E. Conservation of avian species. In *Current Therapy in Avian Medicine and Surgery*; Speer, B.L., Ed.; Elsevier: St.Louis, MO, USA, 2016; pp. 719–748.
13. Raso, T.F.; Teixeira, R.H.F.; Carrasco, A.O.T.; Araújo, J.P., Jr.; Pinto, A.A. *Chlamydophila psittaci* infections in hyacinth macaws (*Anodohynchus hyacinthinus*) confiscated in Brazil. *J. Zoo Wildl. Med.* **2013**, *44*, 169–172. [CrossRef]
14. Luppi, M.M.; Luiz, A.P.M.F.; Coelho, F.M.; Malta, M.C.C.; Preis, I.S.; Ecco, R.; Fonseca, F.G.; Resende, M. Identification and isolation of psittacid herpesvirus from psittacids in Brazil. *Vet. Microbiol.* **2011**, *154*, 69–77. [CrossRef] [PubMed]
15. Esteves, F.C.B.; Marín, S.Y.; Resende, M.; Silva, A.S.G.; Coelho, H.L.G.; Barbosa, M.B.; D'Aparecida, N.S.D.; Resende, J.S.; Torres, A.C.D.; Martins, N.R.S. Avian pox in native captive psittacines, Brazil. *Emerg. Infect. Dis.* **2015**, *23*, 154–156. [CrossRef]
16. Werther, K.; Raso, T.F.; Durigon, E.L.; Latimer, K.S.; Campagnoli, R.P. Psittacine Beak and Feather Disease in Brazil: Case report. *Braz. J. Poult. Sci.* **1999**, *1*, 85–88.
17. Araújo, A.V.; Andery, D.A.; Ferreira, F.C., Jr.; Ortiz, M.C.; Marques, M.V.R.; Marin, S.Y.; Vilela, D.A.R.; Resende, J.S.; Resende, M.; Donatti, R.V.; et al. Molecular diagnosis of beak and feather disease in native psittacine species. *Brazil. J. Poult. Sci.* **2015**, *17*, 451–458. [CrossRef]
18. Azevedo, N.P. Detecção de bornavírus, poliomavírus e circovírus em amostras biológicas, utilizando PCR e RT-PCR, de psitacídeos com diferentes aspectos clínicos. Master's Thesis, University of São Paulo, São Paulo, 2014. Available online: https://teses.usp.br/teses/disponiveis/10/10133/tde-07102014-152516/pt-br.php (accessed on 8 January 2021).
19. Raso, T.F.; Seixas, G.H.F.; Guedes, N.M.R.; Pinto, A.A. *Chlamydophila psittaci* in free-living Blue-fronted Amazon parrots (*Amazona aestiva*) and Hyacinth macaws (*Anodorhynchus hyacinthinus*) in the Pantanal of Mato Grosso do Sul, Brazil. *Vet. Microbiol.* **2006**, *117*, 235–241. [CrossRef]
20. Ribas, J.M.; Sipinski, E.A.B.; Serafini, P.P.; Ferreira, V.L.; Raso, T.F.; Pinto, A.A. *Chlamydophila psittaci* assessment in threatened Red-tailed Amazon (*Amazona brasiliensis*) parrots in Paraná, Brazil. *Ornithologia* **2014**, *6*, 144–147.
21. Vaz, F.F.; Serafini, P.P.; Locatelli-Dittrich, R.; Meurer, R.; Durigon, E.L.; Araújo, J.; Thomazelli, L.M.; Ometto, T.; Spinski, E.A.B.; Sezerban, R.M.; et al. Survey of pathogens in threatened wild red-tailed Amazon parrot (*Amazona brasiliensis*) nestlings in Rasa Island, Brazil. *Brazil J. Microbiol.* **2017**, *48*, 747–753. [CrossRef]
22. Ehricht, R.; Slickers, P.; Goellner, S.; Hotzel, H.; Sachse, K. Optimized DNA microarray assay allows detection and genotyping of single PCR-amplifiable target copies. *Mol. Cell Probes.* **2006**, *20*, 60–63. [CrossRef]
23. Tomaszewski, E.K.; Kaleta, E.F.; Phalen, D.N. Molecular phylogeny of the psittacid herpesviruses causing Pacheco's disease: Correlation of genotype and phenotypic expression. *J. Virol.* **2003**, *77*, 11260–11267. [CrossRef] [PubMed]
24. Lee, L.H.; Lee, K.H. Application of the polymerase chain reaction for the diagnosis of fowl poxvirus infection. *J. Virol. Meth.* **1997**, *63*, 113–119. [CrossRef]
25. Ypelaar, I.; Bassami, M.R.; Wilcox, G.E.; Raidal, S.R. A universal polymerase chain reaction for the detection of psittacine beak and feather disease virus. *Vet. Microbiol.* **1999**, *16*, 141–148. [CrossRef]
26. Sachse, K.; Laroucau, K.; Vorimore, F.; Magnino, S.; Feige, J.; Müller, W.; Kube, S.; Hotzel, H.; Schubert, E.; Slickers, S.; et al. DNA microarray-based genotyping of *Chlamydophila psittaci* strains from culture and clinical samples. *Vet. Microbiol.* **2009**, *135*, 22–30. [CrossRef]
27. Katoh, K.; Standley, D.M. MAFFT multiple sequence alignment software version 7: Improvements in performance and usability. *Mol. Biol. Evol.* **2013**, *30*, 772–780. [CrossRef] [PubMed]
28. Tamura, K.; Peterson, D.; Peterson, N.; Stecher, G.; Nei, M.; Kumar, S. MEGA5: Molecular evolutionary genetics analysis using maximum likelihood, evolutionary distance, and maximum parsimony methods. *Mol. Biol. Evol.* **2011**, *28*, 2731–2739. [CrossRef]
29. Regnard, G.L.; Boyes, R.S.; Martin, R.O.; Hitzeroth, I.I.; Rybicki, E.P. Beak and feather disease viruses circulating in Cape parrots (*Poicepahlus robustus*) in South Africa. *Arch. Virol.* **2015**, *160*, 47–54. [CrossRef]
30. Fogell, D.J.; Martin, R.O.; Bunbury, N.; Lawson, B.; Sells, J.; McKeand, A.M. Trade and conservation implications of new beak and feather disease virus detection in native and introduced parrots. *Cons Biol.* **2018**, *32*, 1325–1335. [CrossRef]
31. Deem, S.L.; Noss, A.J.; Cuellar, R.L.; Karesh, W.B. Health evaluation of free-ranging and captive blue-fronted amazon parrots (*Amazona aestiva*) in the Gran Chaco, Bolivia. *J. Zoo Wildl. Med.* **2005**, *36*, 598–605. [CrossRef]
32. Gilardi, K.V.K.; Lowentime, L.J.; Gilardi, J.D.; Munn, C.A.A. A survey for selected viral, chlamydial, and parasitic diseases in wild dusky-headed parakeets (*Aratinga weddelli*) and tui parakeets (*Brotogeris sanctithomae*) in Peru. *J. Wildl. Dis.* **1995**, *31*, 523–528. [CrossRef] [PubMed]
33. Raso, T.F.; Almeida, M.A.B.; Santos, E.; Bercini, M.; Barbosa, M. Genetic characterization of *Chlamydophila psittaci* from monk parakeet (*Myiopsitta monachus*). In Proceedings of the Congresso Latinoamericano de Microbiologia, Santos, Brazil, 28–31 October 2012; p. 152.

34. Raso, T.F.; Ferreira, V.L.; TIMM, L.N.; Abreu, M.F.T. Psittacosis domiciliary outbreak associated with monk parakeets (*Myiopsitta monachus*) in Brazil: Need for surveillance and control. *JMM Case Rep.* **2014**, *1*, e003343. [CrossRef]
35. Pacheco, G.; Bier, O. Epizootie chex les perroquets du Brésil. Relations avec le psittacose. *CR Soc. Biol.* **1930**, *105*, 109–111.
36. Miller, T.D.; Millar, D.L.; Naqi, S.A. Isolation of Pacheco's disease herpesvirus in Texas. *Avian Dis.* **1979**, *23*, 753–756. [CrossRef]
37. Luppi, M.M.; Luiz, A.P.M.F.; Coelho, F.M.; Ecco, R.; Fonseca, F.G.; Resende, M. Genotypic characterization of psittacid herpesvirus isolates from Brazil. *Braz. J. Microbiol.* **2016**, *47*, 217–224. [CrossRef] [PubMed]
38. Saidenberg, A.B.S.; Zuniga, E.; Melville, P.A.; Salaberry, S.; Benites, N.R. Health-screening protocols for vinaceous amazons (*Amazona vinacea*) in a reintroduction project. *J. Zoo Wildl. Med.* **2015**, *46*, 704–712. [CrossRef] [PubMed]
39. Murer, L.; Westenhofen, M.; Kommers, G.D.; Furian, T.Q.; Borges, K.A.; Kunert-Filho, H.C.; Streck, A.F.; Lovato, M. Identification and phylogenetic analysis of clade C *Avipoxvirus* in a fowlpox outbreak in exotic psittacines in southern Brazil. *J. Vet. Diagn. Investig.* **2018**, *30*, 946–950. [CrossRef]
40. Fogell, D.J.; Martin, R.O.; Groombridge, J.J. Beak and feather disease virus in wild and captive parrots: An analysis of geographic and taxonomic distribution and methodological trends. *Arch. Virol.* **2016**, *161*, 2059–2074. [CrossRef]
41. Raidal, S.R.; Peters, A. Psittacine beak and feather disease: Ecology and implications for conservation. *Emu—Austral Ornithol.* **2018**, *118*, 80–93. [CrossRef]

Communication

A PCR-Based Retrospective Study for Beak and Feather Disease Virus (BFDV) in Five Wild Populations of Parrots from Australia, Argentina and New Zealand

Luis Ortiz-Catedral [1,2], Connor J. Wallace [1], Robert Heinsohn [3], Elizabeth A. Krebs [4], Naomi E. Langmore [5], Dusan Vukelic [6], Enrique H. Bucher [7], Arvind Varsani [8,9] and Juan F. Masello [10,*]

1 Ecology and Conservation Lab, School of Natural and Mathematical Sciences, Massey University, Auckland 0745, New Zealand; luis@parrots.org (L.O.-C.); connorw9@me.com (C.J.W.)
2 Oceania Conservation Program, World Parrot Trust, Hayle TR27 4HB, UK
3 Fenner School of Environment and Society, Australian National University, Canberra, ACT 2601, Australia; robert.heinsohn@anu.edu.au
4 Wildlife Research Division, Science and Technology Branch, Environment and Climate Change Canada, Delta, BC V4K 3N2, Canada; elsie.krebs@ec.gc.ca
5 Research School of Biology, Australian National University, Canberra, ACT 2601, Australia; naomi.langmore@anu.edu.au
6 School of Biological Sciences, University of Canterbury, Christchurch 8140, New Zealand; dvukelic77@gmail.com
7 Centro de Zoología Aplicada, Facultad de Ciencias Exactas, Físicas y Naturales, Universidad Nacional de Córdoba, Córdoba 5000, Argentina; buchereh@gmail.com
8 The Biodesign Center for Fundamental and Applied Microbiomics, Center for Evolution and Medicine, School of Life Sciences, Arizona State University, Tempe, AZ 85287, USA; arvind.varsani@asu.edu
9 Structural Biology Research Unit, Department of Integrative Biomedical Sciences, University of Cape Town, Cape Town 7701, South Africa
10 Department of Animal Ecology & Systematics, Justus Liebig University Giessen, Heinrich-Buff-Ring 26, 35392 Giessen, Germany
* Correspondence: juan.f.masello@bio.uni-giessen.de

Citation: Ortiz-Catedral, L.; Wallace, C.J.; Heinsohn, R.; Krebs, E.A.; Langmore, N.E.; Vukelic, D.; Bucher, E.H.; Varsani, A.; Masello, J.F. A PCR-Based Retrospective Study for Beak and Feather Disease Virus (BFDV) in Five Wild Populations of Parrots from Australia, Argentina and New Zealand. *Diversity* **2022**, *14*, 148. https://doi.org/10.3390/d14020148

Academic Editors: Guillermo Blanco, José L. Tella and Martina Carrete

Received: 21 December 2021
Accepted: 13 February 2022
Published: 18 February 2022

Abstract: The beak and feather disease virus (family *Circovirdae*) is a virus of concern in the conservation of wild Psittaciformes globally. We conducted a PCR screening for the beak and feather disease virus (BFDV) using samples collected during previous field studies (1993–2014) in five populations of parrots of the Southern Hemisphere: Eclectus parrots (*Eclectus roratus*) and Crimson rosellas (*Platycercus elegans*) from Australia, Burrowing parrots (*Cyanoliseus patagonus*) and Monk parakeets from Argentina (*Myiopsitta monachus*), and Forbes' parakeet from New Zealand (*Cyanoramphus forbesi*). A total of 612 samples were screened. BFDV was not detected in any of the sampled birds. Our results provide a retrospective screening, covering three different tribes of Old and New World parrots, including two of the most numerous species, and contributing a large set of negative results. Furthermore, our results suggest that geographical and temporal differences in BFDV distribution may exist and merit further research, as a critical component in the efforts to manage the disease and its epidemiological aspects. The results presented here hold the potential to provide a baseline for future studies investigating the temporal evolution and the spread of BFDV.

Keywords: BFDV; *Circoviridae*; infectious disease; Psittaciformes; surveillance; viral infection; vulnerable taxa; wild populations

1. Introduction

Existing and emerging pathogens can drive rapid changes in population numbers and in the genetic diversity of the wild host population [1]. Pathogens have caused declines in previously large populations or even increased the rate of decline in endangered species [2–4]. Moreover, global pet trade and climate changes hold great potential to extend current pathogen distributions and need to be considered as potential risk factors for the introduction of disease to wildlife [5–7]. For this reason, infectious disease has become a major challenge for conservation; thus, knowledge of the extent of infectious diseases in wildlife populations has become increasingly important for conservation work [8,9].

Parrots and cockatoos (Psittaciformes) have long been recognized as one of the most threatened orders of birds globally, with nearly a third of all known species classified as 'at risk of extinction', and a larger number facing population decline [10,11]. There are multiple factors associated with declining parrot populations, however, capture of wild parrots for the pet trade, intensified agriculture, hunting, and logging are the most frequent threats [10,11], with depredation by introduced species being a serious threat on islands [12]. Moreover, susceptibility to diseases substantially threatens some parrots e.g., Philippine cockatoo *Cacatua haematuropygia*, Cape parrots (*Poicepahlus robustus*), blue-headed racquettail *Prioniturus platenae*, orange-bellied parrot *Neophema chrysogaster* [13–15].

The potentially negative effects of diseases for the survival of endangered parrots have been widely acknowledged [11,16,17] and have triggered abundant research. Studies on diseases, health and pathogens of captive parrots are published regularly [15,18,19]. Nevertheless, there is limited information on pathogenic infection in free-living Psittaciformes [20–29]. This paucity of studies on pathogens and diseases among free-living parrots makes it clear that we only partially understand their role as a threatening factor.

The beak and feather disease virus (BFDV) is a small circular single stranded DNA virus in the family *Circoviridae* [30,31], often cited as a pathogen of conservation concern for parrots in the wild, as well as in captivity [6,8,29,32], given its immune-suppressive effect in infected birds [33,34]. Abnormal plumage and morphological development, anaemia, damage of the lymphoid tissue, feather loss and weight loss among infected birds are common symptoms associated with this viral infection [35].

BFDV infects predominantly Psittaciformes [35], and is reported to cause high mortalities in avicultural collections [36] and in at least two free-living populations [37–39]. Recent evidence indicates, however, that BFDV can also infect non-parrot species [40]. In general, the virus has been reported as infecting over 10% of known parrot species, a figure that comes mostly from studies on captive birds [8,18,41,42]. Despite a wealth of information on captive birds (e.g., [18,41,43,44]), the prevalence of the virus in wild populations remains largely unknown for most regions except Australia, Mauritius, New Caledonia and New Zealand [8,26–28,42,45–49].

The advances in molecular techniques to detect the virus (e.g., [28,46,50] open up an opportunity to conduct large scale surveys for BFDV among wild populations of Psittaciformes, and especially to screen large collections of blood samples from long term studies on parrots. Here, we present a retrospective study investigating the presence of BFDV among five wild populations of Psittaciformes belonging to three different tribes: (a) Psittaculini, the Eclectus parrot (*Eclectus roratus*) from tropical Australia, (b) Platycercini, the Crimson rosella (*Platycercus elegans*) from temperate Australia, and the Forbes' parakeet (*Cyanoramphus forbesi*) from the Chatham Islands, New Zealand, and (c) Arini, the Burrowing parrot (*Cyanoliseus patagonus*) from the Patagonian steppes and Monk parakeet (*Myiopsitta monachus*) from Central Argentina.

2. Methods

We used 612 blood samples collected during previous studies (Table 1), to investigate the presence of BFDV. Details on the sample and populations sizes for each species are given in Table 1. Every individual was sampled once.

Table 1. Details on blood samples from five wild populations of Psittaciformes in this study.

Species	Estimation of Population Size	Reference for Population Size	Year of Sample Collection	Blood Samples (n)		Total
				Adult	Nestling	
		Psittaculini				
Eclectus roratus	3000	[51]	1997–2007	24	291	315
		Platycercini				
Platycercuselegans	550	[52]	1993–1995	17	52	69
Cyanoramphus forbesi	1000	[53]	2014	95	–	95
		Arini				
Cyanoliseus patagonus	75,000	[54]	December 1998, December 1999	49t	55	104
Myopsitta monachus	500	[55] and E.H.B. unpubl. data	December 2000	29	–	29

Samples from Eclectus parrots were taken over the course of a long-term study (1997–2007) on Cape York Peninsula in northern Queensland Australia (12°45′ S, 143°17′ E) [56,57]. Most samples were taken from nestlings in nest hollows 15–25 m above the ground in rainforest trees. Adults were also captured using mist nets set at similar heights in the rainforest canopy. Approximately 100 µL of blood was taken from the brachial vein of each captured individual. Eclectus parrot blood was stored in 70% ethanol [57,58].

Samples from Crimson rosellas were collected from adult and nestling birds breeding in Black Mountain Nature Reserve, Australian Capital Territory (35°16′28″ S, 149°05′55″ E) [52]. Birds were captured in nest-boxes between 1993 and 1996; a small blood sample (50 to 100 µL) was taken from the brachial vein of each individual, and preserved in Queen's Buffer (10 mM Tris, 10 mM NaCl, 10 mM disodium EDTA, 1% n-lauroylsarcosine, pH 8.0) [59]. Blood samples were taken from adults on capture and from nestlings between 25 and 30 days of age.

Forbes' parakeets were captured using mist-nets on Mangere Island, Chatham Islands (44°26′ S, 176°29′ W), in March 2014. Blood samples (200 µL) were taken by puncture of the brachial vein immediately after capture and preserved in Queen's Buffer [59]. Only adults were sampled.

Burrowing parrots were captured at its major colony in El Cóndor, north-eastern Patagonia, Argentina (41°04′ S, 62°50′ W) during regular nest inspections in December 1998 and December 1999 [54]. Adults were sampled when found in the nest; nestlings were sampled between the age of 38 and 60 days. Monk parakeet samples were obtained in an area of 600 ha, situated near Jesús María, Córdoba, Argentina (31°05′ S, 64°11′ W) [55]. Monk parakeets were captured in their nests during December 2000. Blood samples (200 µL) of the adult and nestling burrowing parrots, as well as of adult monk parakeets, were taken by puncture of the brachial vein immediately after capture. The blood was stored in 70% ethanol [58].

In 2014, DNA was extracted from 10 µL of blood, which was added to 10 µL of 'lysis solution' from the Extract-n-AmpTM Blood PCR Kit (Sigma-Aldrich, St Louis, MO, USA) and incubated for 10 min at room temperature. Ninety microliters of this kit's 'neutralization solution' was subsequently added to yield crude total DNA. One microliter of the crude extract was used as template in the subsequent PCR [46]. Extracted DNA was stored at −20 °C. In addition, in 2014, as described in previously published studies [18,46,47,60], BFDV specific PCR screening was carried out using KAPA Blood PCR Kit Mix B (KAPA Biosystems, Wilmington, DE, USA) using the primer pair forward 5′-TTAACAACCCTACAGACGGCGA-3′ and reverse 5′-GGCGGAGCATCTCGCAATAAG-3′, which target a 605 bp region of the *rep* gene of BFDV [61]. The reaction volume was 25 µL with 1 µL of 10 µM F/R primer pair, 12.5 µL of the 2xKAPA Blood PCR Kit Mix, 1 µL of DNA templates and 10.5 µL of sterile molecular grade water. The PCR program contained an initial step of 94 °C for 5 min, which was followed by 25 cycles of 94 °C for 30 s, 56 °C for 30 s and 72 °C for 45 s and with a final 1 min extension step at 72 °C and cooling to 4 °C for 10 min. DNA from a BFDV-infected red-fronted parakeet (*Cyanoramphus novaezelandiae*) from Little Barrier Island was used as a positive control [62]. The total DNA used as positive control was extracted from 60 μL of blood using the Qiagen QIAamp DNA minikit (Qiagen, Hilden, Germany) according to the manufacturer's protocols.

3. Results

We did not detect BFDV in any of the blood samples investigated by PCR.

4. Discussion

Surveillance for pathogens is a fundamental element for understanding the temporal and spatial prevalence of wildlife diseases and for understanding transmission pathways and effects on animal populations [63]. We applied a commonly used PCR screen [18,46,47,60] to detect viral DNA in blood samples collected during previous field studies of Eclectus parrots, Crimson rosellas, Forbes' parakeets, Burrowing parrots and Monk parakeets. Our negative results suggest that BFDV was not present in the studied populations at the time of sampling, and show some differences with previous studies, which could be related to temporal, geographical and captive versus wild population differences in BFDV prevalence and distribution. BFDV has previously been reported from captive Eclectus parrots [45,64,65]; however, the wild population here investigated is isolated from large human populations and parrots kept in captivity. Free-ranging Crimson rosellas on Norfolk Island and in Victoria, Australia, have been reported with BFDV [26–28,66], yet the samples in the current study originate from a population within and surrounding the city of Canberra, where a previous BFDV study found a very low number of potentially infected individuals [67]. BFDV has been reported on close relatives of Forbes' parakeets, including red-fronted parakeets and yellow-crowned parakeets (*Cyanoramphus auriceps*) [46], but has not been detected in other *Cyanoramphus* species in the wild. For Monk parakeets, the virus has been found in 37% of sampled individuals belonging to a feral population in Spain [68]. This high prevalence could be related to the origin of the birds, which accidentally escaped from captivity, where BFDV has been reported frequently [8,18,36]. To our knowledge, BFDV infection in Burrowing parrots is unknown for either captive or free-living individuals.

There are an increasing number of field studies with Psittaciformes worldwide; commonly, blood samples are collected. Those samples could be used to increase the range of species screened in the wild, allowing for a better understanding of the geographical distribution of BFDV. Moreover, Fogell et al. [8] pointed out that two biases currently exist in BFDV research, namely, the lack of (1) research in regions of the world such as South America and Southeast and Southern Asia, both characterised by a high parrot diversity, and (2) publications reporting negative results. Recent studies are starting to fill those gaps. Vaz et al. [29] using pathogen-specific PCR, evaluated the presence of BFDV. As in our study, Vaz et al. [29] detected no BFDV DNA in a large sample of 205 wild nestlings and 90 nestlings from the illegal trade. Moreover, we are confident that our study also makes a substantial contribution to BFDV research by providing further screening results for South American parrots, including two of the most numerous species, and by contributing a large screening with negative results, obtained with a methodology thoroughly tested [18,46,47,60]. Furthermore, our results suggest that geographical differences in BFDV distribution may exist and merit further research, as a critical component in the efforts to manage the disease and its epidemiological aspects. Lastly, the results presented here hold the potential to provide a baseline for future studies investigating the temporal evolution and the spreading of BFDV. However, two final cautionary remarks are needed. First, we acknowledge that there is a possibility that the nucleic acid may be damaged in storage and transport; this may impact the amplification of the target virus sequences in some of the samples. Second, the widely applied PCR protocol [18,46,47,60] used in this study has some limitation. BFDV is known for a high genetic diversity [68–70]; it cannot be fully excluded that the primers used in this investigation might have missed some genetic variants. Thus, future studies should evaluate the presence of the virus based on any previous identification BFDV sequences from these hosts in captivity or introduction on new regions. Nonetheless, the primer pair we have used in this study binds with 100% complementarity to a BFDV sequence (GenBank Accession # MT303064) derived from the blood sample of Monk parakeets in Spain [68].

Author Contributions: Conceptualization, L.O.-C., R.H., A.V. and J.F.M.; methodology, L.O.-C., C.J.W., R.H., A.V., D.V., E.H.B. and J.F.M.; validation, L.O.-C. and A.V.; formal analysis, L.O.-C., C.J.W., R.H., A.V., D.V., E.H.B. and J.F.M.; investigation, L.O.-C., C.J.W., R.H., A.V., D.V., E.H.B. and J.F.M.; resources, L.O.-C., R.H., A.V., E.A.K., N.E.L., E.H.B. and J.F.M.; writing—original draft preparation, L.O.-C., C.J.W., R.H., A.V., E.H.B. and J.F.M.; writing—review and editing, L.O.-C., R.H., A.V. and J.F.M.; supervision, A.V. and J.F.M.; funding acquisition, L.O.-C., A.V., E.H.B. and J.F.M. All authors have read and agreed to the published version of the manuscript.

Funding: Funding for data analysis and manuscript preparation was provided by the National Council of Science and Technology of Mexico (CONACYT), Claude McCarthy Fellowship, New Zealand Vice-Chancellor's Committee and Mohamed bin Zayed Conservation Trust (to L.O.-C.). The Eclectus parrot research was funded by the Australian Research Council, National Geographic Society and Winifred Violet Scott Foundation. The molecular analyses were supported by an Early career grant (University of Canterbury) awarded to A.V. J.F.M. research was supported by the City Council of Viedma (Río Negro, Argentina), World Parrot Trust, Liz Claiborne Art Ortenberg Foundation, and Wildlife Conservation Society. E.H.B. research was partially funded by the US Fish and Wildlife Service, International Affairs division, and a co-operation grant between the IB of the BMBF of Germany (ARG 99/020) and the Argentinean SECyT (AL/A99-EXIII/003).

Institutional Review Board Statement: The Eclectus parrot research was conducted under license from the Queensland Government and the ANU Animal Ethics Committee (Permit No: C.R.E.35.04). Crimson Rosellas sampling was conducted in accordance with ANU Animal Ethics guidelines (ANU Animal Ethics Permit J.BTZ.22.93), an ACT Parks Capture and Release Permit LT96023, and an ABBBS banding permit 1778. Forbes' parakeets sampling was approved by the Department of Conservation, 19621-FAU, New Zealand. Burrowing Parrot sampling was carried out under permission of the Dirección de Fauna de la Provincia de Río Negro, Argentina (143089-DF-98). Monk parakeet sampling permit was granted by the Consejo National de Investigaciones Cientificas y Técnicas of Argentina (CONICET) to E.H.B.

Data Availability Statement: All data are available in the main text.

Acknowledgments: We especially thank C. Blackman, P. Huybers, E. Huybers, S. Legge, Anja Quellmalz and Pablo Manzano Baena for assistance with field work, Erio Curto for allowing us to sample nests in his property in Marull, Córdoba, Luke Martin, Florence Gaude and Kevin Parker for conducting sample collection for Forbes' parakeets, Tansy Bliss and Dave Houston for assistance with permits.

Conflicts of Interest: The authors declare no conflict of interest.

References

1. Altizer, S.; Harvell, D.; Friedle, E. Rapid evolutionary dynamics and disease threats to biodiversity. *Trends Ecol. Evol.* **2003**, *18*, 589–596. [CrossRef]
2. Steinmetz, H.W.; Bakonyi, T.; Weissenböck, H.; Hatt, J.-M.; Eulenberger, U.; Robert, N.; Hoop, R.; Nowotny, N. Emergence and establishment of Usutu virus infection in wild and captive avian species in and around Zurich, Switzerland—Genomic and pathologic comparison to other central European outbreaks. *Vet. Microbiol.* **2011**, *148*, 207–212. [CrossRef] [PubMed]
3. Kleyheeg, E.; Slaterus, R.; Bodewes, R.; Rijks, J.; Spierenburg, M.A.H.; Beerens, N.; Kelder, L.; Poen, M.; Stegeman, J.; Fouchier, R.A.M.; et al. Deaths among wild birds during highly pathogenic Avian Influenza A(H5N8) Virus Outbreak, The Netherlands. *Emerg. Infect. Dis. J.* **2017**, *23*, 2050. [CrossRef] [PubMed]
4. Krone, O.; Globig, A.; Ulrich, R.; Harder, T.; Schinköthe, J.; Herrmann, C.; Gerst, S.; Conraths, F.J.; Beer, M. White-Tailed Sea Eagle (*Haliaeetus albicilla*) die-off due to infection with highly pathogenic Avian Influenza Virus, Subtype H5N8, in Germany. *Viruses* **2018**, *10*, 478. [CrossRef] [PubMed]
5. Jackson, H.; Strubbe, D.; Tollington, S.; Prys-Jones, R.; Matthysen, E.; Groombridge, J.J. Ancestral origins and invasion pathways in a globally invasive bird correlate with climate and influences from bird trade. *Mol. Ecol.* **2015**, *24*, 4269–4285. [CrossRef] [PubMed]

6. Fogell, D.J.; Martin, R.O.; Bunbury, N.; Lawson, B.; Sells, J.; McKeand, A.M.; Tatayah, V.; Trung, C.T.; Groombridge, J.J. Trade and conservation implications of new beak and feather disease virus detection in native and introduced parrots. *Conserv. Biol.* **2018**, *32*, 1325–1335. [CrossRef]
7. Ortiz-Catedral, L.; Brunton, D.; Stidworthy, M.F.; Elsheikha, H.M.; Pennycott, T.; Schulze, C.; Braun, M.; Wink, M.; Gerlach, H.; Pendl, H.; et al. *Haemoproteus minutus* is highly virulent for Australasian and South American parrots. *Parasites Vectors* **2019**, *12*, 40. [CrossRef] [PubMed]
8. Fogell, D.J.; Martin, R.O.; Groombridge, J.J. Beak and feather disease virus in wild and captive parrots: An analysis of geographic and taxonomic distribution and methodological trends. *Arch. Virol.* **2016**, *161*, 2059–2074. [CrossRef] [PubMed]
9. Jayasinghe, M.; Midwinter, A.; Roe, W.; Vallee, E.; Bolwell, C.; Gartrell, B. Seabirds as possible reservoirs of *Erysipelothrix rhusiopathiae* on islands used for conservation translocations in New Zealand. *J. Wildl. Dis.* **2021**, *57*, 534–542. [CrossRef]
10. Olah, G.; Butchart, S.H.M.; Symes, A.; Guzmán, I.M.; Cunningham, R.; Brightsmith, D.J.; Heinsohn, R. Ecological and socio-economic factors affecting extinction risk in parrots. *Biodivers. Conserv.* **2016**, *25*, 205–223. [CrossRef]
11. Berkunsky, I.; Quillfeldt, P.; Brightsmith, D.J.; Abbud, M.C.; Aguilar, J.M.R.E.; Alemán-Zelaya, U.; Aramburú, R.M.; Arce Arias, A.; Balas McNab, R.; Balsby, T.J.S.; et al. Current threats faced by Neotropical parrot populations. *Biol. Conserv.* **2017**, *214*, 278–287. [CrossRef]
12. Ortiz-Catedral, L.; Nias, R.; Fitzsimons, J.; Vine, S.; Christian, M. Back from the brink–again: The decline and recovery of the Norfolk Island green parrot. In *Recovering Australian Threatened Species: A Book of Hope*; Garnett, S., Latch, P., Lindenmayer, D., Woinarski, J., Eds.; CSIRO Publishing: Clayton South, Australia, 2018.
13. Snyder, N.; McGowan, P.; Gilardi, J.; Grajal, A. *Parrots. Status Survey and Conservation Action Plan 2000–2004*; IUCN: Gland, Switzerland; Cambridge, UK, 2000.
14. Regnard, G.L.; Boyes, R.S.; Martin, R.; Hitzeroth, I.I.; Rybicki, E.P. Beak and feather disease viruses circulating in Cape parrots (*Poicepahlus robustus*) in South Africa. *Arch. Virol.* **2015**, *160*, 47–54. [CrossRef] [PubMed]
15. Das, S.; Smith, K.; Sarker, S.; Peters, A.; Adriaanse, K.; Eden, P.; Ghorashi, S.A.; Forwood, J.K.; Raidal, S.R. Repeat spillover of beak and feather disease virus into an endangered parrot highlights the risk associated with endemic pathogen loss in endangered species. *J. Wildl. Dis.* **2020**, *56*, 896–906. [CrossRef] [PubMed]
16. Wilson, M.H.; Kepler, C.B.; Snyder, N.F.R.; Derrickson, S.R.; Dein, F.J.; Wiley, J.W.; Wunderle, J.M.; Lugo, A.E.; Graham, D.L.; Toone, W.D. Puerto Rican Parrots and potential limitations of the metapopulation approach to species conservation. *Conserv. Biol.* **1994**, *8*, 114–123. [CrossRef]
17. Gartrell, B.D.; Alley, M.R.; Mack, H.; Donald, J.; McInnes, K.; Jansen, P. Erysipelas in the critically endangered kakapo (*Strigops habroptilus*). *Avian Pathol.* **2005**, *34*, 383–387. [CrossRef] [PubMed]
18. Julian, L.; Piasecki, T.; Chrząstek, K.; Walters, M.; Muhire, B.; Harkins, G.W.; Martin, D.P.; Varsani, A. Extensive recombination detected among beak and feather disease virus isolates from breeding facilities in Poland. *J. Gen. Virol.* **2013**, *94*, 1086–1095. [CrossRef] [PubMed]
19. Piasecki, T.; Harkins, G.W.; Chrząstek, K.; Julian, L.; Martin, D.P.; Varsani, A. Avihepadnavirus diversity in parrots is comparable to that found amongst all other avian species. *Virology* **2013**, *438*, 98–105. [CrossRef] [PubMed]
20. Raidal, S.R.; Mcelnea, C.L.; Cross, G.M. Seroprevalence of psittacine beak and feather disease in wild psittacine birds in New-South-Wales. *Aust. Vet. J.* **1993**, *70*, 137–139. [CrossRef]
21. Gilardi, K.V.; Lowenstine, L.J.; Gilardi, J.D.; Munn, C.A. A survey for selected viral, chlamydial, and parasitic diseases in wild dusky-headed parakeets (*Aratinga weddellii*) and tui parakeets (*Brotogeris sanctithomae*) in Peru. *J. Wildl. Dis.* **1995**, *31*, 523–528. [CrossRef] [PubMed]
22. Deem, S.L.; Noss, A.J.; Cuellar, R.L.; Karesh, W.B. Health evaluation of free-ranging and captive blue-fronted Amazon parrots (*Amazona aestiva*) in the Gran Chaco, Bolivia. *J. Zoo Wildl. Med.* **2005**, *36*, 598–605. [CrossRef]
23. Ha, H.J.; Anderson, I.L.; Alley, M.R.; Springett, B.P.; Gartrell, B.D. The prevalence of beak and feather disease virus infection in wild populations of parrots and cockatoos in New Zealand. *N. Z. Vet. J.* **2007**, *55*, 235–238. [CrossRef] [PubMed]
24. Ortiz-Catedral, L.; Ismar, S.M.H.; Baird, K.; Ewen, J.G.; Hauber, M.E.; Brunton, D.H. No evidence of *Campylobacter*, *Salmonella* and *Yersinia* in free-living populations of the red-crowned parakeet (*Cyanoramphus novaezelandiae*). *N. Z. J. Zool.* **2009**, *36*, 379–383. [CrossRef]
25. Ortiz-Catedral, L.; McInnes, K.; Hauber, M.E.; Brunton, D.H. First report of beak and feather disease virus (BFDV) in wild Red-fronted Parakeets (*Cyanoramphus novaezelandiae*) in New Zealand. *Emu* **2009**, *109*, 244–247. [CrossRef]
26. Martens, J.M.; Stokes, H.S.; Berg, M.L.; Walder, K.; Bennett, A.T.D. Seasonal fluctuation of beak and feather disease virus (BFDV) infection in wild Crimson Rosellas (*Platycercus elegans*). *Sci. Rep.* **2020**, *10*, e7894. [CrossRef] [PubMed]
27. Martens, J.M.; Stokes, H.S.; Berg, M.L.; Walder, K.; Raidal, S.R.; Magrath, M.J.; Bennett, A.T. A non-invasive method to assess environmental contamination with avian pathogens: Beak and feather disease virus (BFDV) detection in nest boxes. *PeerJ* **2020**, *8*, e9211. [CrossRef]
28. Martens, J.M.; Stokes, H.S.; Berg, M.L.; Walder, K.; Raidal, S.R.; Magrath, M.J.L.; Bennett, A.T.D. Beak and feather disease virus (BFDV) prevalence, load and excretion in seven species of wild caught common Australian parrots. *PLoS ONE* **2020**, *15*, e0235406. [CrossRef]
29. Vaz, F.F.; Sipinski, E.A.B.; Seixas, G.H.F.; Prestes, N.P.; Martinez, J.; Raso, T.F. Molecular Survey of Pathogens in Wild Amazon Parrot Nestlings: Implications for Conservation. *Diversity* **2021**, *13*, 272. [CrossRef]

30. Breitbart, M.; Delwart, E.; Rosario, K.; Segalés, J.; Varsani, A.; ICTV Report Consortium. ICTV Virus Taxonomy Profile: Circoviridae. *J. Gen. Virol.* **2017**, *98*, 1997–1998. [CrossRef]
31. Rosario, K.; Breitbart, M.; Harrach, B.; Segalés, J.; Delwart, E.; Biagini, P.; Varsani, A. Revisiting the taxonomy of the family Circoviridae: Establishment of the genus *Cyclovirus* and removal of the genus *Gyrovirus*. *Arch. Virol.* **2017**, *162*, 1447–1463. [CrossRef]
32. Raidal, S.R.; Peters, A. Psittacine beak and feather disease: Ecology and implications for conservation. *Emu* **2018**, *118*, 80–93. [CrossRef]
33. Todd, D. Circoviruses: Immunosuppressive threats to avian species: A review. *Avian Pathol.* **2000**, *29*, 373–394. [CrossRef] [PubMed]
34. Ortiz-Catedral, L. No T-cell-mediated immune response detected in a red-fronted parakeet (*Cyanoramphus novaezelandiae*) infected with the Beak and Feather Disease Virus (BFDV). *Notornis* **2010**, *57*, 81–84.
35. Ritchie, B.W.; Niagro, F.D.; Lukert, P.D.; Latimer, K.S.; Steffens III, W.L.; Pritchard, N. A review of psittacine beak and feather disease: Characteristics of the PBFD virus. *J. Assoc. Avian Vet.* **1989**, *3*, 143–149. [CrossRef]
36. Kock, N.; Hangartner, P.; Lucke, V. Variation in clinical disease and species susceptibility to psittacine beak and feather disease in Zimbabwean lovebirds. *J. Vet. Res.* **1993**, *60*, 159–161.
37. Malham, J.; Kovac, E.; Reuleaux, A.; Linnebjerg, J.; Tollington, S.; Raisin, C.; Marsh, P.; McPherson, S. *Results of Screening Echo and Ringneck Parakeets for Psittacine Beak and Feather Disease in Mauritius March 2008*; Mauritian Wildlife Foundation: Vacoas-Phoenix, Mauritius; National Parks and Conservation Service of Mauritius: Reduit, Mauritius; International Zoo Veterinary Group: Keighley, UK; Durrel Wildlife Conservation Trust: Jersey, Channel Islands; IBL Aviation, Shipping and Other Services: Port Louis, Mauritius; Chester Zoo: Chester, UK; The World Parrot Trust: Hayle, UK, 2008.
38. Peters, A.; Patterson, E.I.; Baker, B.G.B.; Holdsworth, M.; Sarker, S.; Ghorashi, S.A.; Raidal, S.R. Evidence of psittacine beak and feather disease virus spillover into wild critically endangered orange-bellied parrots (*Neophema chrysogaster*). *J. Wildl. Dis.* **2014**, *50*, 288–296. [CrossRef] [PubMed]
39. Olah, G.; Smith, B.T.; Joseph, L.; Banks, S.C.; Heinsohn, R. Advancing Genetic Methods in the Study of Parrot Biology and Conservation. *Diversity* **2021**, *13*, 521. [CrossRef]
40. Amery-Gale, J.; Marenda, M.; Owens, J.; Eden, P.A.; Browning, G.; Devlin, J. A high prevalence of beak and feather disease virus in non-psittacine Australian birds. *J. Med. Microbiol.* **2017**, *66*, 1005–1013. [CrossRef]
41. Varsani, A.; Regnard, G.L.; Bragg, R.; Hitzeroth, I.I.; Rybicki, E.P. Global genetic diversity and geographical and host-species distribution of beak and feather disease virus isolates. *J. Gen. Virol.* **2011**, *92*, 752–767. [CrossRef]
42. Fogell, D.J.; Tollington, S.; Tatayah, V.; Henshaw, S.; Naujeer, H.; Jones, C.; Raisin, C.; Greenwood, A.; Groombridge, J.J. Evolution of Beak and Feather Disease Virus across three decades of conservation intervention for population recovery of the Mauritius parakeet. *Diversity* **2021**, *13*, 584. [CrossRef]
43. Ortiz-Catedral, L.; Kearvell, J.; Brunton, D.H. Re-introduction of captive-bred Malherbe's parakeet to Maud Island, Marlborough Sounds, New Zealand. In *Global Re-Introduction Perspectives: Additional Case-Studies from around the Globe*; Soorae, P.S., Ed.; IUCN/SSC Re-Introduction Specialist Group: Abu Dhabi, United Arab Emirates, 2010; pp. 151–154.
44. Varsani, A.; Villiers, G.; Regnard, G.; Bragg, R.; Kondiah, K.; Hitzeroth, I.; Rybicki, E. A unique isolate of beak and feather disease virus isolated from budgerigars (*Melopsittacus undulatus*) in South Africa. *Arch. Virol.* **2010**, *155*, 435–439. [CrossRef]
45. Julian, L.; Lorenzo, A.; Chenuet, J.-P.; Bonzon, M.; Marchal, C.; Vignon, L.; Collings, D.A.; Walters, M.; Jackson, B.; Varsani, A. Evidence of multiple introductions of beak and feather disease virus into the Pacific islands of Nouvelle-Caledonie (New Caledonia). *J. Gen. Virol.* **2012**, *93*, 2466–2472. [CrossRef] [PubMed]
46. Massaro, M.; Ortiz-Catedral, L.; Julian, L.; Galbraith, J.A.; Kurenbach, B.; Kearvell, J.; Kemp, J.; van Hal, J.; Elkington, S.; Taylor, G.; et al. Molecular characterisation of beak and feather disease virus (BFDV) in New Zealand and its implications for managing an infectious disease. *Arch. Virol.* **2012**, *157*, 1651–1663. [CrossRef] [PubMed]
47. Jackson, B.; Harvey, C.; Galbraith, J.; Robertson, M.; Warren, K.; Holyoake, C.; Julian, L.; Varsani, A. Clinical beak and feather disease virus infection in wild juvenile eastern rosellas of New Zealand; biosecurity implications for wildlife care facilities. *N. Z. Vet. J.* **2014**, *62*, 297–301. [CrossRef] [PubMed]
48. Eastwood, J.R.; Berg, M.L.; Spolding, B.; Buchanan, K.L.; Bennett, A.T.D.; Walder, K. Prevalence of beak and feather disease virus in wild *Platycercus elegans*: Comparison of three tissue types using a probe-based real-time qPCR test. *Aust. J. Zool.* **2015**, *63*, 1–8. [CrossRef]
49. Raidal, S.R.; Sarker, S.; Peters, A. Review of psittacine beak and feather disease and its effect on Australian endangered species. *Aust. Vet. J.* **2015**, *93*, 466–470. [CrossRef]
50. Rahaus, M.; Desloges, N.; Probst, S.; Loebbert, B.; Lantermann, W.; Wolff, M. Detection of beak and feather disease virus DNA in embryonated eggs of psittacine birds. *Vet. Med.* **2008**, *53*, 53–58. [CrossRef]
51. Legge, S.; Heinsohn, R.; Garnett, S. Availability of nest hollows and breeding population size of eclectus parrots, *Eclectus roratus*, on Cape York Peninsula, Australia. *Wildlife Res.* **2004**, *31*, 149–161. [CrossRef]
52. Krebs, E.A. Breeding biology of crimson rosellas (*Platycercus elegans*) on Black Mountain, Australian Capital Territory. *Aust. J. Zool.* **1998**, *46*, 119–136. [CrossRef]
53. Bliss, T. *Forbes' Parakeet (Cyanoramphus forbesi) Report on Field Work during the 2012–2015 Breeding Seasons and Analysis of Monitoring Results 1999 to 2015*; Department of Conservation: Chatham Islands, New Zealand, 2016.
54. Masello, J.F.; Pagnossin, M.L.; Sommer, C.; Quillfeldt, P. Population size, provisioning frequency, flock size and foraging range at the largest known colony of Psittaciformes: The Burrowing Parrots of the north-eastern Patagonian coastal cliffs. *Emu* **2006**, *106*, 69–79. [CrossRef]

55. Bucher, E.H.; Martin, L.F.; Martella, M.B.; Navarro, J.L. Social behaviour and population dynamics of the Monk Parakeet. *Proc. Int. Ornithol. Congr.* **1991**, *20*, 681–689.
56. Heinsohn, R.; Legge, S.; Endler, J.A. Extreme Reversed Sexual Dichromatism in a Bird Without Sex Role Reversal. *Science* **2005**, *309*, 617–619. [CrossRef] [PubMed]
57. Heinsohn, R.; Ebert, D.; Legge, S.; Peakall, R. Genetic evidence for cooperative polyandry in reverse dichromatic *Eclectus* parrots. *Anim. Behav.* **2007**, *74*, 1047–1054. [CrossRef]
58. Arctander, P. Comparative studies of avian DNA by restriction fragment length polymorphism analysis: Convenient procedures based on blood samples from live birds. *J. Ornithol.* **1988**, *129*, 205–216. [CrossRef]
59. Seutin, G.; White, B.N.; Boag, P.T. Preservation of avian blood and tissue samples for DNA analyses. *Can. J. Zool.* **2014**, *69*, 82–90. [CrossRef]
60. Jackson, B.; Varsani, A.; Holyoake, C.; Jakob-Hoff, R.; Robertson, I.; McInnes, K.; Empson, R.; Gray, R.; Nakagawa, K.; Warren, K. Emerging infectious disease or evidence of endemicity? A multi-season study of beak and feather disease virus in wild red-crowned parakeets (*Cyanoramphus novaezelandiae*). *Arch. Virol.* **2015**, *160*, 2283–2292. [CrossRef] [PubMed]
61. Ritchie, P.A.; Anderson, I.L.; Lambert, D.M. Evidence for specificity of psittacine beak and feather disease viruses among avian hosts. *Virology* **2003**, *306*, 109–115. [CrossRef]
62. Ortiz-Catedral, L.; Kurenbach, B.; Massaro, M.; McInnes, K.; Brunton, D.; Hauber, M.; Martin, D.; Varsani, A. A new isolate of beak and feather disease virus from endemic wild red-fronted parakeets (*Cyanoramphus novaezelandiae*) in New Zealand. *Arch. Virol.* **2010**, *155*, 613–620. [CrossRef]
63. Leendertz, F.H.; Pauli, G.; Maetz-Rensing, K.; Boardman, W.; Nunn, C.; Ellerbrok, H.; Jensen, S.A.; Junglen, S.; Christophe, B. Pathogens as drivers of population declines: The importance of systematic monitoring in great apes and other threatened mammals. *Biol. Conserv.* **2006**, *131*, 325–337. [CrossRef]
64. Khalesi, B.; Bonne, N.; Stewart, M.; Sharp, M.; Raidal, S. A comparison of haemagglutination, haemagglutination inhibition and PCR for the detection of psittacine beak and feather disease virus infection and a comparison of isolates obtained from loriids. *J. Gen. Virol.* **2005**, *86*, 3039–3046. [CrossRef]
65. Sariya, L.; Prompiram, P.; Khocharin, W.; Tangsugjai, S.; Phonarknguen, R.; Ratanakorn, P.; Chaichoun, K. Genetic analysis of beak and feather disease virus isolated from captive psittacine birds in Thailand. *Southeast Asian J. Trop. Med. Public Health* **2011**, *42*, 851–858. [PubMed]
66. Hermes, N.; Evans, O.; Evans, B. Norfolk Island birds: A review. *Notornis* **1986**, *33*, 141–149.
67. Peachey, M. Psittacine beak and feather viral disease in parrots in the ACT. *Canberra Bird Notes* **2013**, *38*, 106–118.
68. Morinha, F.; Carrete, M.; Tella, J.L.; Blanco, G. High prevalence of novel beak and feather disease virus in sympatric invasive parakeets introduced to Spain from Asia and South America. *Diversity* **2020**, *12*, 192. [CrossRef]
69. Heath, L.; Martin, D.P.; Warburton, L.; Perrin, M.; Horsfield, W.; Kingsley, C.; Rybicki, E.P.; Williamson, A.-L. Evidence of unique genotypes of psittacine beak and feather disease in southern Africa. *J. Virol.* **2004**, *78*, 9277–9284. [CrossRef] [PubMed]
70. Raue, R.; Johne, R.; Crosta, L.; Burkle, M.; Gerlach, H.; Muller, H. Nucleotide sequence analysis of a C1 gene fragment of psittacine beak and feather disease virus amplified by real-time polymerase chain reaction indicates a possible existence of genotypes. *Avian Pathol.* **2004**, *33*, 41–50. [CrossRef] [PubMed]

Communication

High Prevalence of Novel Beak and Feather Disease Virus in Sympatric Invasive Parakeets Introduced to Spain From Asia and South America

Francisco Morinha [1,*], Martina Carrete [2], José L. Tella [3] and Guillermo Blanco [1]

[1] Department of Evolutionary Ecology, National Museum of Natural Sciences (MNCN), Spanish National Research Council (CSIC), José Gutiérrez Abascal 2, 28006 Madrid, Spain; g.blanco@csic.es

[2] Department of Physical, Chemical and Natural Systems, University Pablo de Olavide, Ctra. de Utrera km. 1, 41013 Sevilla, Spain; mcarrete@upo.es

[3] Department of Conservation Biology, Estación Biológica de Doñana (CSIC), Avda. Américo Vespucio, 41092 Sevilla, Spain; tella@ebd.csic.es

* Correspondence: franciscomorinha@hotmail.com or franciscomorinha@mncn.csic.es

Received: 26 April 2020; Accepted: 11 May 2020; Published: 13 May 2020

Abstract: The psittacine beak and feather disease (PBFD) is a globally widespread infectious bird disease that mainly affects species within the Order Psittaciformes (parrots and allies). The disease is caused by an avian circovirus (the beak and feather disease virus, BFDV), which is highly infectious and can lead to severe consequences in wild and captive populations during an outbreak. Both legal and illegal trading have spread the BFDV around the world, although little is known about its prevalence in invasive parrot populations. Here, we analyze the BFDV prevalence in sympatric invasive populations of rose-ringed (*Psittacula krameri*) and monk parakeets (*Myiopsitta monachus*) in Southern Spain. We PCR-screened 110 blood samples (55 individuals from each species) for BFDV and characterized the genotypes of five positives from each species. About 33% of rose-ringed parakeets and 37% of monk parakeets sampled were positive for BFDV, while neither species showed disease symptoms. The circovirus identified is a novel BFDV genotype common to both species, similar to the BFDV genotypes detected in several parrot species kept in captivity in Saudi Arabia, South Africa and China. Our data evidences the importance of an accurate evaluation of avian diseases in wild populations, since invasive parrots may be bringing BFDV without showing any visually detectable clinical sign. Further research on the BFDV prevalence and transmission (individual–individual, captive–wild and wild–captive) in different bird orders and countries is crucial to understand the dynamics of the viral infection and minimize its impact in captive and wild populations.

Keywords: circovirus; PBFD; BFDV; rose-ringed parakeet; monk parakeet; invasive species

---

1. Introduction

Psittacine Beak and Feather Disease (PBFD) is one of the most relevant infectious diseases affecting wild and captive parrot species [1,2]. This disease is caused by the Beak and Feather Disease Virus (BFDV), which is a highly infectious and mutable single-stranded DNA (ssDNA) circovirus [3]. The main PBFD symptoms include feather lesions (loss of feathers, improper moulting, malformations and colour changes) and abnormal growth of the beak, although infected individuals can be asymptomatic [4]. Although information is scarce, the prevalence of BFDV is high in native parrots of Oceania, Africa and Asia [5], including wild populations of some threatened species [6–9]. To our knowledge, no surveillance of BFDV has been conducted on free-living parrots in South America, the other main stronghold for parrots. A screening conducted on captive individuals seized from illegal trade in Brazil showed evidence of BFDV at low prevalence in two native species [10]. The genomic similarity of the isolates with reference strains from Asia and Oceania suggested an exotic origin of BFDV strains disseminated in captivity in South America [10,11].

BFDV has been detected in most captive parrot species around the world, showing a high genetic variability, with thousands of genotypes and novel ones described each year [5,12,13]. The international legal and illegal trade on captive birds [14,15] has induced a fast spread of BFDV worldwide, a threat that is enhanced in areas where parrots have established invasive populations [16]. Invasive parrot populations can be natural viral hosts, although their role in the genetic diversification of the virus and the spread into native populations of parrots and other avian orders has not yet been explored. However, it has been reported that invasive rose-ringed parakeets (*Psittacula krameri*) could have been involved in the outbreak of BFDV in the threatened population of Echo parakeets (*Psittacula eques*) on the island of Mauritius, with devastating effects due to its high mortality rate [2,9]. Moreover, the presence of BFDV has been recently reported in various non-psittacine species, which may both suffer the disease or act as reservoirs, with different prevalences depending on the order [17]. All these aspects increase the epidemiological complexity of the disease and its diagnosis [18], making urgent a comprehensive understanding of its drivers to prevent epidemic outbreaks impacting common but also threatened species.

The rose-ringed parakeet (*P. krameri*) and the monk parakeet (*Myiopsitta monachus*) were the first and fourth most traded parrots worldwide [14], and have established exotic populations in various countries around the world mostly as a consequence of accidental escapes of individuals kept as pets [19]. Recent studies have shown that the exotic populations of these two species mainly originated from individuals traded from India and an area between Argentina and Uruguay, respectively [20–22]. Governmental and environmental organizations have applied different measures to control these populations [23–25], highlighting that these invasive birds can be natural reservoirs of infectious bacteria, fungi and viruses of zoonotic concern. However, scientific evidence is scarce [26,27], and only available for rose-ringed parakeets introduced in some European countries [28,29] and on the island of Mauritius [9]. Given the BFDV mutagenic potential and unpredictable viral effects, its study in invasive populations is important to prevent dangerous outbreaks in novel hosts among wild species [9].

Here, we evaluate the prevalence of BFDV in sympatric populations of the invasive rose-ringed parakeet and monk parakeet in Spain and then proceed to its genetic characterization. We reviewed the information on BFDV *rep* gene isolated from captive rose-ringed parakeets worldwide and assessed whether specific BFDV variants can infect both parrot species despite their different geographic origin. Results are discussed in the context of the global spread of BFDV through wildlife trade and its potential impact on native species of the recipient communities.

2. Materials and Methods

From 2015 to 2018, parakeets of both species were captured with mist nets in Sevilla (Southern Spain), where they coexist in urban parks [30]. Individuals were banded, examined for lesions in the beak and plumage alterations, measured for several traits, banded and released. A sample of blood (ca. 0.05 mL) was collected from the brachial vein and stored in absolute ethanol for molecular analysis. A random sample of the parakeets (55 individuals from each species) was screened for the presence of BFDV (n = 17, 16, 12, 10 in 2015–2018 for *P. krameri*; n= 25, 13, 17 in 2016–2018 for *M. monachus*). The capture and extraction of blood samples from parakeets were approved by the Ethical Committee of EBD-CSIC (our national research institution), codes: 11_27-Tella and 12_48-Tella.

The DNA was isolated from blood using the Quick-DNA Miniprep Kit (Zymo Research) using an optimized protocol (i.e., samples were digested at 56 °C for 12 h in a solution containing 200 µL of genomic lysis buffer and 20 µL of 20-mg/mL proteinase K). The screening of BFDV was performed on all samples using two primer sets (5′-AACCCTACAGACGGCGAG-3′ and 5′-GTCACAGTCCTCCTTGTACC-3′, [31]; 5′-TTAACAACCCTACAGACGGCGA-3′ and 5′-GGCGGAGCATCTCGCAATAAG-3′, [32]) that amplify a partial sequence of the replication-associated protein (*rep*) gene. PCR was performed in a reaction mixture of 20 µL containing 10 µL of 2x MyTaq HS Mix (Bioline), 250 nM of each primer (final concentration) and ~20 ng of template DNA. The amplification protocol was composed of the following steps: 95 °C for 5 min followed by 40 cycles of 95 °C for 30 s, 58 °C for 1 min, 72 °C for 30 s, and a final extension at 60 °C for 10 min. Negative controls and non-template controls were used in all PCR reactions to exclude contamination issues. Population-prevalence estimates and 95% CI were calculated with Epitools [33] following Fogell et al. [16].

Samples were considered BFDV positive if at least one primer pair PCR had an accurate amplification of the expected size fragment. Five random positive samples of each species were bi-directionally sequenced by Sanger sequencing to assess genotype diversity. Sequences were visualized and edited using the software Geneious v.11.1.5. Then, our sequences were compared with data available in GenBank through a BLAST search [34]. The program BEAST v.2.6.0 [35] was used to construct a Maximum Clade Credibility (MCC) tree using the 100 sequences most related with the genotype characterized in this study. These sequences were obtained through a BLAST search and covered a total of 77 BFDV genotypes (see Supplementary database 1). The model of sequence evolution HKY + I + G4 was selected after careful estimation of this evolutionary parameter with jModelTest v.2.1.7 [36]. The proportion of invariable sites was set to 0.491 and the alpha shape parameter (α) was set to 0.643, using a normal distribution for the rate prior and letting the program to estimate the mutation rates. The strict clock and Yule model priors were used. The software run consisted of 200 million steps, with a sampling of the chains every 20,000 steps and a burn-in of 10%. The adequate convergence and mixing of the chains and sufficient effective sample sizes (ESS) were checked with Tracer v.1.7.1 [37]. The consensus tree was visualized and edited using the software FigTree v.1.4.4 [38]. The same procedures (changing only the model of evolution to HKY + G4 using $\alpha = 0.115$) were used to construct an MCC tree using only *P. krameri rep* gene sequences with a 100% coverage of the query sequence in the BLAST search.

3. Results

A similar prevalence of BFDV was found in both species (rose-ringed parakeet: 18 of 55, 32.7%, 95% CI: 21.8–45.9; monk parakeet: 20 of 55, 36.4%, 95% CI: 24.9–49.6; Fisher's exact test, $p = 0.84$). Individuals positive to BFDV were found in 2016 (50.0%, $n = 16$) and 2017 (83.3%, $n = 12$) but not in 2015 (0.0%, $n = 17$) and 2018 (0.0%, $n = 10$) for *P. krameri*. BFDV-positive samples were found in all sampling years for *M. monachus* (2016, 24.0%, $n = 25$; 2017, 38.5%, $n = 13$; 2018, 52.9%, $n = 17$). No individual showed visible signs of the disease.

The partial sequence of the *rep* gene isolated and characterized for rose-ringed parakeets and monk parakeets revealed a novel and unique BFDV genotype common to both species (GenBank accessions: MT303063 and MT303064). Our analysis indicates that the novel genotype differed by nine to 13 nucleotide substitutions with the most similar genotypes among all BFDV sequences described to date for different psittacine species (Figure 1). These closest variants were found in captive individuals of several parrot species (*P. krameri, Psittacus erithacus, Nymphicus hollandicus, Agapornis fischeri* and *Poicephalus gulielmi*) in Saudi Arabia, China and South Africa (Figure 1). Phylogenetic data suggest that genotypes isolated in Saudi Arabia and South Africa diverged from the novel circovirus genotype found in Spain (Figure 1). The phylogeny of the *rep* sequences isolated from *P. krameri* in various countries supports the proximity of the BFDV genotypes isolated from wild individuals in an invasive population in Spain with captive individuals in Saudi Arabia, which share an ancestor with the two genotypes also detected in captive individuals in Poland (Figure 2).

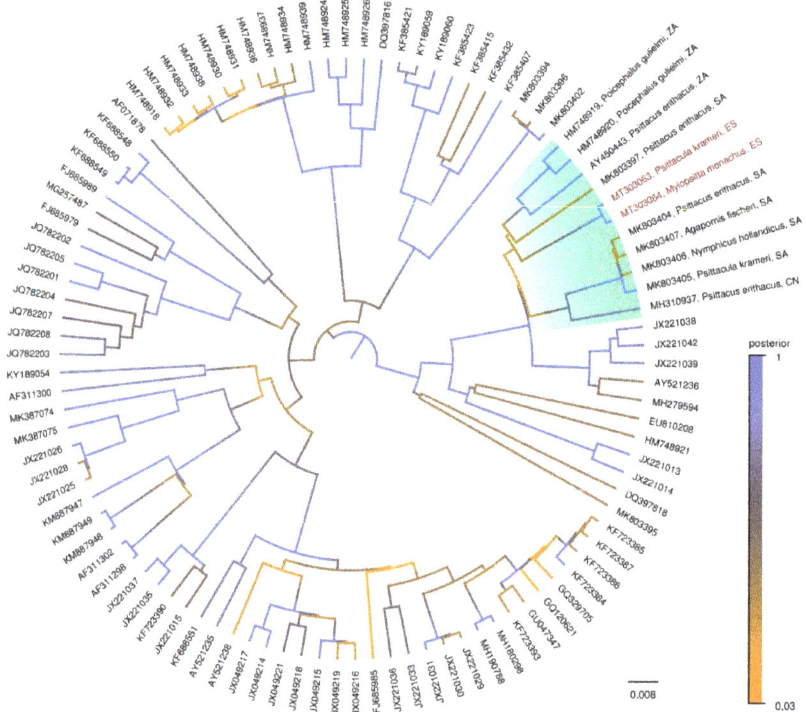

Figure 1. Maximum Clade Credibility tree for the beak and feather disease virus (BFDV) *rep* gene partial nucleotide sequences (603 bases). Posterior probabilities for the nodes are represented with a colour gradient scale. All partial *rep* sequences are identified by GenBank accession number. The clade that includes the novel circovirus genotype is highlighted and also shows species and country codes (ISO 3166-1 alpha-2). The detailed list of host species, country of virus isolation and genotypes are presented as supplementary database 1.

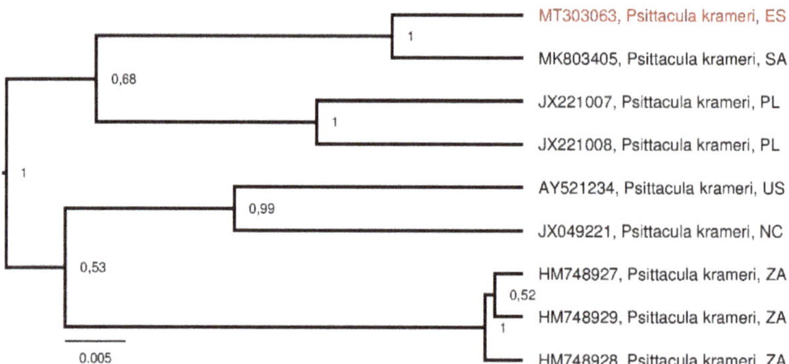

Figure 2. Maximum clade credibility tree inferred using partial nucleotide sequences of the BFDV *rep* gene isolated from rose-ringed parakeets (*Psittacula krameri*). Posterior probabilities are indicated for each node. All partial rep sequences are defined by GenBank accession number, species and country codes (ISO 3166-1 alpha-2). The novel circovirus genotype of Spain (ES) is highlighted.

The partial BFDV sequences isolated from rose-ringed parakeets in Spain and Saudi Arabia differ in 9 (1.5%) nucleotides and 1 (0.5%) amino acid (Table 1). However, the pattern of nucleotide and amino acid variations between isolates from different countries and continents is extreme, ranging from nine (1.5%) to 49 (8.1%) nucleotides and from zero to 16 (8.0%) amino acids (Table 1).

Table 1. Number of nucleotide (below diagonal) and amino acid (above diagonal) variations between partial sequences of the BFDV *rep* gene isolated from rose-ringed parakeets (*Psittacula krameri*). The percentage of variation considering the sequence length (603 nucleotides and 201 amino acids) is noted in parentheses.

	MT303063 (ES)	**MK803405 (SA)**	**JX221007 (PL)**	**JX221008 (PL)**	**AY521234 (US)**	**JX049221 (NC)**	**HM748927 (ZA)**	**HM748929 (ZA)**	**HM748928 (ZA)**
MT303063 (ES)	-	1 (0.5)	11 (5.4)	4 (1.9)	11 (5.4)	7 (3.5)	7 (3.5)	7 (3.5)	8 (4.0)
MK803405 (SA)	9 (1.5)	-	10 (5.0)	3 (1.5)	10 (5.0)	6 (3.0)	8 (4.0)	8 (4.0)	9 (4.5)
JX221007 (PL)	39 (6.5)	33 (5.5)	-	9 (4.5)	16 (8.0)	12 (6.0)	14 (7.0)	14 (7.0)	15 (7.5)
JX221008 (PL)	31 (5.1)	25 (4.1)	20 (3.3)	-	9 (4.5)	5 (2.5)	7 (3.5)	7 (3.5)	8 (4.0)
AY521234 (US)	33 (5.5)	27 (4.4)	46 (7.6)	36 (6.0)	-	7 (3.5)	9 (4.5)	9 (4.5)	10 (5.0)
JX049221 (NC)	28 (4.6)	22 (3.6)	44 (7.3)	32 (5.3)	22 (3.6)	-	6 (3.0)	6 (3.0)	7 (3.5)
HM748927 (ZA)	31 (5.1)	31 (5.1)	47 (7.8)	40 (6.6)	34 (5.6)	33 (5.5)	-	0 (0.0)	1 (0.5)
HM748929 (ZA)	32 (5.3)	31 (5.1)	47 (7.8)	40 (6.6)	33 (5.5)	34 (5.6)	1 (0.2)	-	1 (0.5)
HM748928 (ZA)	33 (5.5)	33 (5.5)	49 (8.1)	42 (7.0)	36 (6.0)	33 (5.5)	2 (0.3)	3 (0.5)	-

4. Discussion

Our study revealed a high prevalence of BFDV in the two sympatric invasive populations of rose-ringed parakeets and monk parakeets sampled in Southern Spain. No individual showed visible signs of the disease, which suggests that most of them were asymptomatic carriers or that ill (i.e., symptomatic) individuals die soon because of the disease [4,8,9] or are rapidly eliminated from the wild by natural selection [39]. The high prevalence of BFDV in the rose-ringed parakeet contrasts with the very low values in blood reported for other invasive populations of this species in Europe, Asia and Africa (from 0.0% in Germany [29] to 16.1% in Mauritius [16]). These values were lower than those reported in native populations in Asia (100% in Bangladesh and 71.4% in Pakistan [16]). It is worth mentioning that prevalence detected in our study area are similar to those found in native species from Australia and New Zealand, where the BFDV is endemic [40–42]. Differences in prevalence can be real among populations, although they can also arise due to the tissue analysed and to the selection of primers [16]. In our case, following previous studies showing high variability in the amplification specificity and sensitivity between different primers sets [43], we used two different sets of primers to search for BFDV. Positive samples for the virus only amplified with the primers reported by Ritchie et al. [32], likely due to variations in viral copy numbers or mutations in the primer binding sites in some BFDV genotypes. However, several studies only focused on one primer set for BFDV screening [10,12,17], and thus may have underestimated the actual prevalence in these populations. Further research using different molecular markers is thus needed to increase the robustness of the BFDV diagnosis test.

The evidence of BFDV infections in monk parakeets is restricted to a presumed case reporting beak lesions in a Greek invasive population, although no genetic, microbiological or other type of validation was provided [44]. No further information is available in its native or invasive range, so this is the first estimate of prevalence of BFDV in this species in the wild.

Genotype variations are frequent in circovirus and there are thousands of BFDV genotypes described and available through genome browsers [5,12,13]. This variability may explain species-specific susceptibility and infection impact, which deserves further research on the potential infectiousness to native species in the invasive range of these parakeet species. The novel genotype characterized in this study and its closest viral variants can colonize psittacids from all continents. This has important implications for the spread of this and other viral variants on native parrots interacting with invasive ones [9] or with individuals escaped from captivity, especially in the Neotropics where many parrot species are of conservation concern [45].

Phylogenetic data suggest that genotypes isolated in Saudi Arabia and South Africa diverged from the novel circovirus genotype found in Spain, which share an ancestor with the two genotypes also detected in captive individuals in Poland. These countries imported and exported thousands of captive-bred and wild parrots (and other birds) from and towards Spain and other European countries in recent decades according to CITES Trade Database [46]. This result highlights the potential role played by the international bird trade in the spread of wildlife infectious diseases and the emergence of zoonosis [8,47,48]. Specifically, the capacity of the same circovirus variant to colonize different parrot species, one a native of Asia and the other from South America, in one area of their European invasive range emphasizes the complex and concerning outcomes of trade-driven biological invasions on the global circulations of pathogens.

Invasive parrot populations with high prevalence of BFDV are not only of concern in areas of coexistence with other parrot species. There is evidence that BFDV can be transmitted from native parrots to several avian species of different orders [17], with some individuals showing the typical beak and plumage alterations of this disease [43,49]. A previous study showed BFDV symptoms and the presence of the causative virus in captive Gouldian finches (*Chloebia gouldiae*), which were suggested to be infected by invasive monk parakeets in Italy [50]. However, no test of the occurrence of the virus in the parakeets was conducted. Thus, the high prevalence of BFDV in the invasive parakeet populations sampled in southern Spain should be carefully considered, as it may have important consequences for the conservation of native birds, particularly those sharing habitats or nests with these invasive species and those predating on them both in urban and rural habitats [30,51,52]. These concerns increase due to the fast spread rates of these two parakeet species in Spain [19,53].

5. Conclusions

We show a high prevalence of a novel BFDV in free-ranging, sympatric invasive populations of two parakeets native to different continents in Southern Spain. Although previous studies have demonstrated the presence of BFDV in rose-ringed parakeets, this is the first genetic evidence of BFDV in monk parakeets. Both legal and illegal trade can contribute considerably to the dissemination of the virus in non-endemic regions, which can have important impacts not yet considered on native birds [54]. Thus, surveillance of invasive populations should be mandatory taking into account the high mutation rate of the virus and the possible cross-transmission to native species. We strongly encourage a strict control or total ban on the international bird trade to avoid the spread of this and other pathogens potentially threatening wildlife and public health.

Supplementary Materials: The following are available online at http://www.mdpi.com/1424-2818/12/5/192/s1, Database 1: detailed list of GenBank accessions, host species, country of virus isolation and genotypes of the BFDV *rep* gene.

Author Contributions: Conceptualization, F.M., M.C., J.L.T., and G.B.; methodology, F.M., M.C., J.L.T., and G.B.; software, F.M.; validation, F.M.; formal analysis, F.M.; investigation, F.M., M.C., J.L.T., and G.B.; resources, F.M., M.C., and J.L.T.; data curation, F.M.; writing—original draft preparation, F.M., and G.B.; writing—review and editing, F.M., M.C., J.L.T., and G.B.; visualization, F.M., M.C., J.L.T., and G.B.; supervision, F.M., M.C., J.L.T., and G.B.; project administration, M.C., and J.L.T.; funding acquisition, M.C., and J.L.T. All authors have read and agreed to the published version of the manuscript.

Funding: F.M. was supported by a Juan de la Cierva postdoctoral fellowship from Spain's Ministry of Science and Innovation (FJCI-2017-32055). This work was partially supported by Loro Parque Fundación.

Acknowledgments: Logistic and technical support for laboratory and field work was provided by Doñana ICTS-RBD. We thank Centro Andaluz de Arte Contemporáneo for allowing us to capture parakeets within its urban gardens, and to Simon Young for editing the English.

Conflicts of Interest: The authors declare no conflict of interest.

References

1. Fogell, D.J.; Martin, R.O.; Groombridge, J.J. Beak and feather disease virus in wild and captive parrots: An analysis of geographic and taxonomic distribution and methodological trends. *Arch. Virol.* **2016**, *161*, 2059–2074. [CrossRef]
2. Raidal, S.R.; Peters, A. Psittacine beak and feather disease: Ecology and implications for conservation. *Emu* **2018**, *118*, 80–93. [CrossRef]
3. Sarker, S.; Patterson, E.I.; Peters, A.; Baker, G.B.; Forwood, J.K.; Ghorashi, S.A.; Holdsworth, M.; Baker, R.; Murray, N.; Raidal, S.R. Mutability dynamics of an emergent single stranded DNA virus in a naïve host. *PLoS ONE* **2014**, *9*, e85370. [CrossRef] [PubMed]
4. Pass, D.A.; Perry, R.A. The pathology of psittacine beak and feather disease. *Aust. Vet. J.* **1984**, *61*, 69–74. [CrossRef]
5. Harkins, G.W.; Martin, D.P.; Christoffels, A.; Varsani, A. Towards inferring the global movement of beak and feather disease virus. *Virology* **2014**, *450*, 24–33. [CrossRef]
6. Warburton, L.; Perrin, M. Evidence of psittacine beak and feather disease in wild Black-cheeked Lovebirds in Zambia. *Papageien* **2002**, *5*, 166–169.
7. Regnard, G.L.; Boyes, R.S.; Martin, R.O.; Hitzeroth, I.I.; Rybicki, E.P. Beak and feather disease viruses circulating in Cape parrots (*Poicepahlus robustus*) in South Africa. *Arch. Virol.* **2015**, *160*, 47–54. [CrossRef] [PubMed]
8. Fogell, D.J.; Groombridge, J.J.; Tollington, S.; Canessa, S.; Henshaw, S.; Zuel, N.; Jones, C.G.; Andrew Greenwood, A.; Ewen, J.G. Hygiene and biosecurity protocols reduce infection prevalence but do not improve fledging success in an endangered parrot. *Sci. Rep.* **2019**, *9*, 1–10. [CrossRef] [PubMed]
9. Kundu, S.; Faulkes, C.G.; Greenwood, A.G.; Jones, C.G.; Kaiser, P.; Lyne, O.D.; Black, S.A.; Aurelie Chowrimootoo, A.; Groombridge, J.J. Tracking viral evolution during a disease outbreak: The rapid and complete selective sweep of a circovirus in the endangered Echo parakeet. *J. Virol.* **2012**, *86*, 5221–5229. [CrossRef] [PubMed]
10. Araújo, A.V.; Andery, D.A.; Ferreira, F.C., Jr.; Ortiz, M.C.; Marques, M.V.R.; Marin, S.Y.; Vilela, D.A.R.; Resende, J.S.; Resende, M.; Donatti, R.V.; et al. Molecular diagnosis of beak and feather disease in native Brazilian psittacines. *Braz. J. Poult. Sci.* **2015**, *17*, 451–458. [CrossRef]
11. González-Hein, G.; Gil, I.A.; Sanchez, R.; Huaracan, B. Prevalence of Aves Polyomavirus 1 and Beak and Feather Disease Virus From Exotic Captive Psittacine Birds in Chile. *J. Avian Med. Surg.* **2019**, *33*, 141–149. [CrossRef] [PubMed]
12. Massaro, M.; Ortiz-Catedral, L.; Julian, L.; Galbraith, J.; Kurenbach, B.; Kearvell, J.; Kemp, J.; Hal, J.; Elkington, S.; Taylor, G.; et al. Molecular characterisation of beak and feather disease virus (BFDV) in New Zealand and its implications for managing an infectious disease. *Arch. Virol.* **2012**, *157*, 1651–1663. [CrossRef] [PubMed]
13. Varsani, A.; Regnard, G.L.; Bragg, R.; Hitzeroth, I.I.; Rybicki, E.P. Global genetic diversity and geographical and host-species distribution of beak and feather disease virus isolates. *J. Gen. Virol.* **2011**, *92*, 752–767. [CrossRef] [PubMed]
14. Cardador, L.; Lattuada, M.; Strubbe, D.; Tella, J.L.; Reino, L.; Figueira, R.; Carrete, M. Regional bans on wild-bird trade modify invasion risks at a global scale. *Conserv. Lett.* **2017**, *10*, 717–725. [CrossRef]
15. Ribeiro, J.; Reino, L.; Schindler, S.; Strubbe, D.; Vall-llosera, M.; Bastos Araújo, M.; Capinha, C.; Carrete, M.; Mazzoni, S.; Monteiro, M.; et al. Trends in legal and illegal trade of wild birds: A global assessment based on expert knowledge. *Biodivers. Conserv.* **2019**, *28*, 3343–3369. [CrossRef]
16. Fogell, D.J.; Martin, R.O.; Bunbury, N.; Lawson, B.; Sells, J.; McKeand, A.M.; Tatayah, V.; Trung, C.T.; Groombridge, J.J. Trade and conservation implications of new beak and feather disease virus detection in native and introduced parrots. *Conserv. Biol.* **2018**, *32*, 1325–1335. [CrossRef]
17. Amery-Gale, J.; Marenda, M.S.; Owens, J.; Eden, P.A.; Browning, G.F.; Devlin, J.M. A high prevalence of beak and feather disease virus in non-psittacine Australian birds. *J. Med. Microbiol.* **2017**, *66*, 1005–1013. [CrossRef]
18. Rahaus, M.; Wolff, M.H. Psittacine beak and feather disease: A first survey of the distribution of beak and feather disease virus inside the population of captive psittacine birds in Germany. *J. Vet. Med. B* **2003**, *50*, 368–371. [CrossRef]

19. Abellán, P.; Tella, J.L.; Carrete, M.; Cardador, L.; Anadón, J.D. Climatic matching drives spread rate but not establishment success in recent unintentional bird introductions. *PNAS* **2017**, *114*, 9385–9390. [CrossRef]
20. Jackson, H.; Strubbe, D.; Tollington, S.; Prys-Jones, R.; Matthysen, E.; Groombridge, J.J. Ancestral origins and invasion pathways in a globally invasive bird correlate with climate and influences from bird trade. *Mol. Ecol.* **2015**, *24*, 4269–4285. [CrossRef] [PubMed]
21. Cardador, L.; Carrete, M.; Gallardo, B.; Tella, J.L. Combining trade data and niche modelling improves predictions of the origin and distribution of non-native European populations of a globally invasive species. *J. Biogeogr.* **2016**, *43*, 967–978. [CrossRef]
22. Edelaar, P.; Roques, S.; Hobson, E.A.; Gonçalves da Silva, A.; Avery, M.L.; Russello, M.A.; Senar, J.C.; Wright, T.F.; Carrete, M.; Tella, J.L. Shared genetic diversity across the global invasive range of the Monk Parakeet suggests a common restricted geographic origin and the possibility of convergent selection. *Mol. Ecol.* **2015**, *24*, 2164–2176. [CrossRef] [PubMed]
23. Kumschick, S.; Blackburn, T.M.; Richardson, D.M. Managing alien bird species: Time to move beyond "100 of the worst" lists? *Bird Conserv. Int.* **2016**, *26*, 154–163. [CrossRef]
24. Avery, M.L.; Shiels, A.B. Monk and rose-ringed parakeets. In *Ecology and Management of Terrestrial Vertebrate Invasive Species in the United States*, 1st ed.; Pitt, W.C., Beasley, J.C., Witmer, G.W., Eds.; CRC Press: Boca Raton, FL, USA, 2017; pp. 333–358.
25. Nentwig, W.; Bacher, S.; Kumschick, S.; Pyšek, P.; Vilà, M. More than "100 worst" alien species in Europe. *Biol. Invasions* **2018**, *20*, 1611–1621. [CrossRef]
26. Mazza, G.; Tricarico, E.; Genovesi, P.; Gherardi, F. Biological invaders are threats to human health: An overview. *Ethol. Ecol. Evol.* **2014**, *26*, 112–129. [CrossRef]
27. Mori, E.; Meini, S.; Strubbe, D.; Ancillotto, L.; Sposimo, P.; Menchetti, M. Do alien free-ranging birds affect human health? A global summary of known zoonoses. In *Invasive Species and Human Health*, 1st ed.; Mazza, G., Tricarcio, E., Eds.; CABI Editions: Wallingford, UK, 2018; pp. 120–129.
28. Sa, R.C.; Cunningham, A.A.; Dagleish, M.P.; Wheelhouse, N.; Pocknell, A.; Borel, N.; Hannah, L.; Peck, H.L.; Lawson, B. Psittacine beak and feather disease in a free-living ring-necked parakeet (*Psittacula krameri*) in Great Britain. *Eur. J. Wildl. Res.* **2014**, *60*, 395–398. [CrossRef]
29. Kessler, S.; Heenemann, K.; Krause, T.; Twietmeyer, S.; Fuchs, J.; Lierz, M.; Corman, V.M.; Vahlenkamp, T.M.; Rubbenstroth, D. Monitoring of free-ranging and captive Psittacula populations in Western Europe for avian bornaviruses, circoviruses and polyomaviruses. *Avian Pathol.* **2020**, *49*, 119–130. [CrossRef]
30. Hernández-Brito, D.; Carrete, M.; Popa-Lisseanu, A.; Ibáñez, C.; Tella, J.L. Crowding in the city: Losing and winning competitors of an invasive bird. *PLoS ONE* **2014**, *9*, e100593. [CrossRef]
31. Ypelaar, I.; Bassami, M.R.; Wilcox, G.E.; Raidal, S.R. A universal polymerase chain reaction for the detection of psittacine beak and feather disease virus. *Vet. Microbiol.* **1999**, *68*, 141–148. [CrossRef]
32. Ritchie, P.A.; Anderson, I.L.; Lambert, D.M. Evidence for specificity of psittacine beak and feather disease viruses among avian hosts. *Virology* **2003**, *306*, 109–115. [CrossRef]
33. Sergeant, E. 2018. Epitools Epidemiological Calculators. Ausvet Pty Ltd. Available online: http://epitools.ausvet.com.au/ (accessed on 15 April 2020).
34. Boratyn, G.M.; Camacho, C.; Cooper, P.S.; Coulouris, G.; Fong, A.; Ma, N.; Madden, T.L.; Matten, W.T.; McGinnis, S.D.; Merezhuk, Y. BLAST: A more efficient report with usability improvements. *Nucleic Acids Res.* **2013**, *41*, W29–W33. [CrossRef] [PubMed]
35. Bouckaert, R.; Heled, J.; Kühnert, D.; Vaughan, T.; Wu, C.H.; Xie, D.; Suchard, M.A.; Rambaut, A.; Drummond, A.J. BEAST 2: A software platform for Bayesian evolutionary analysis. *PLoS Comput. Biol.* **2014**, *10*, e1003537. [CrossRef] [PubMed]
36. Darriba, D.; Taboada, G.L.; Doallo, R.; Posada, D. jModelTest 2: More models, new heuristics and parallel computing. *Nat. Methods* **2012**, *9*, 772. [CrossRef] [PubMed]
37. Rambaut, A.; Drummond, A.J.; Xie, D.; Baele, G.; Suchard, M.A. Posterior summarization in Bayesian phylogenetics using Tracer 1.7. *Syst. Biol.* **2018**, *67*, 901–904. [CrossRef] [PubMed]
38. Rambaut, A. Figtree Software (v.1.4.4). 2018. Available online: http://tree.bio.ed.ac.uk/software/figtree/ (accessed on 20 February 2020).
39. Genovart, M.; Negre, N.; Tavecchia, G.; Bistuer, A.; Parpal, L.; Oro, D. The young, the weak and the sick: Evidence of natural selection by predation. *PLoS ONE* **2010**, *5*, e9774. [CrossRef] [PubMed]

40. Ha, H.J.; Anderson, I.L.; Alley, M.R.; Springett, B.P.; Gartrell, B.D. The prevalence of beak and feather disease virus infection in wild populations of parrots and cockatoos in New Zealand. *N. Z. Vet. J.* **2007**, *55*, 235–238. [CrossRef]
41. Ortiz-Catedral, L.; McInnes, K.; Hauber, M.E.; Brunton, D.H. First report of beak and feather disease virus (BFDV) in wild Red-fronted Parakeets (*Cyanoramphus novaezelandiae*) in New Zealand. *Emu* **2009**, *109*, 244–247. [CrossRef]
42. Eastwood, J.R.; Berg, M.L.; Ribot, R.F.; Buchanan, K.L.; Walder, K.; Bennett, A.T. Prevalence of BFDV in wild breeding Platycercus elegans. *J. Ornithol.* **2019**, *160*, 557–565. [CrossRef]
43. Sarker, S.; Moylan, K.G.; Ghorashi, S.A.; Forwood, J.K.; Peters, A.; Raidal, S.R. Evidence of a deep viral host switch event with beak and feather disease virus infection in rainbow bee-eaters (*Merops ornatus*). *Sci. Rep.* **2015**, *5*, 14511. [CrossRef]
44. Kalodimos, N. First account of a nesting population of Monk parakeets (*Myiopsitta monachus*) with nodule-shaped bill lesions in Katehaki, Athens, Greece. *Bird Popul.* **2013**, *12*, 1–6.
45. Berkunsky, I.; Quillfeldt, P.; Brightsmith, D.J.; Abbud, M.C.; Aguilar, J.M.R.E.; Alemán, U.; Aramburú, R.M.; Arce Arias, A.; Balas McNab, R.; Balsby, T.J.S.; et al. Current threats faced by Neotropical parrot populations. *Biol. Conserv.* **2017**, *214*, 278–287. [CrossRef]
46. CITES Trade Database. Available online: https://trade.cites.org/ (accessed on 10 April 2020).
47. Fèvre, E.M.; Bronsvoort, B.M.D.C.; Hamilton, K.A.; Cleaveland, S. Animal movements and the spread of infectious diseases. *Trends Microbiol.* **2006**, *14*, 125–131. [CrossRef] [PubMed]
48. Menchetti, M.; Mori, E. Worldwide impact of alien parrots (Aves Psittaciformes) on native biodiversity and environment: A review. *Ethol. Ecol. Evol.* **2014**, *26*, 172–194. [CrossRef]
49. Sarker, S.; Lloyd, C.; Forwood, J.; Raidal, S.R. Forensic genetic evidence of beak and feather disease virus infection in a powerful owl (*Ninox strenua*). *Emu* **2016**, *116*, 71–74. [CrossRef]
50. Circella, E.; Legretto, M.; Pugliese, N.; Caroli, A.; Bozzo, G.; Accogli, G.; Lavazza, A.; Camarda, A. Psittacine Beak and Feather Disease–like Illness in Gouldian Finches (*Chloebia gouldiae*). *Avian Dis.* **2014**, *58*, 482–487. [CrossRef] [PubMed]
51. Mori, E.; Malfatti, L.; Le Louarn, M.; Brito, D.H.; Ten Cate, B.; Ricci, M.; Menchetti, M. 'Some like it alien': Predation on invasive ring–necked parakeets by the long–eared owl in an urban area. *Anim. Biodiv. Conserv.* **2020**, *43*, 151–158.
52. Hernández-Brito, D.; Blanco, G.; Tella, J.L.; Carrete, M.A. protective nesting association with native species counteracts biotic resistance for the spread of an invasive parakeet from urban to rural habitats. *Front. Zool.* **2020**, *17*, 1–13. [CrossRef]
53. Abellán, P.; Carrete, M.; Anadón, J.D.; Cardador, L.; Tella, J.L. Non-random patterns and temporal trends (1912-2012) in the transport, introduction and establishment of exotic birds in Spain and Portugal. *Divers. Distrib.* **2016**, *22*, 263–273. [CrossRef]
54. Turbé, A.; Strubbe, D.; Mori, E.; Carrete, M.; Chiron, F.; Clergeau, P.; González-Moreno, P.; Le Louarn, M.; Luna, A.; Menchetti, M.; et al. Assessing the assessments: Evaluation of four impact assessment protocols for invasive alien species. *Divers. Distrib.* **2017**, *23*, 297–307. [CrossRef]

Article

Evolution of Beak and Feather Disease Virus across Three Decades of Conservation Intervention for Population Recovery of the Mauritius Parakeet

Deborah J. Fogell [1,2], Simon Tollington [1,3], Vikash Tatayah [4], Sion Henshaw [4], Houshna Naujeer [5], Carl Jones [6], Claire Raisin [7], Andrew Greenwood [8] and Jim J. Groombridge [1,*]

1 Durrell Institute of Conservation and Ecology, School of Anthropology and Conservation, University of Kent, Canterbury CT2 7NZ, UK; djfogell@outlook.com (D.J.F.); simon.tollington@gmail.com (S.T.)
2 Institute of Zoology, Zoological Society of London, Regents Park, London NW1 4RY, UK
3 School of Animal, Rural and Environmental Sciences, Nottingham Trent University, Brackenhurst, Southwell, Nottinghamshire NG25 0QF, UK
4 Mauritian Wildlife Foundation, Grannum Road, Vacoas 73418, Mauritius; vtatayah@mauritian-wildlife.org (V.T.); shenshaw@mauritian-wildlife.org (S.H.)
5 National Parks and Conservation Service, Ministry of Agro-Industry, Government of Mauritius, Reduit 80835, Mauritius; naujeerhb@gmail.com
6 Durrell Wildlife Conservation Trust, Les Augres Manor, Trinity, Jersey, Channel Islands JE3 5BP, UK; carlgjones@btinternet.com
7 North of England Zoological Society, Chester Zoo, Upton-by-Chester, Chester CH2 1EU, UK; c.raisin@chesterzoo.org
8 Wildlife Vets International, Keighley BD21 4NQ, UK; A.Greenwood@izvg.co.uk
* Correspondence: J.Groombridge@kent.ac.uk

Citation: Fogell, D.J.; Tollington, S.; Tatayah, V.; Henshaw, S.; Naujeer, H.; Jones, C.; Raisin, C.; Greenwood, A.; Groombridge, J.J. Evolution of Beak and Feather Disease Virus across Three Decades of Conservation Intervention for Population Recovery of the Mauritius Parakeet. *Diversity* **2021**, *13*, 584. https://doi.org/10.3390/d13110584

Academic Editors: José L. Tella, Guillermo Blanco and Martina Carrete

Received: 11 October 2021
Accepted: 3 November 2021
Published: 16 November 2021

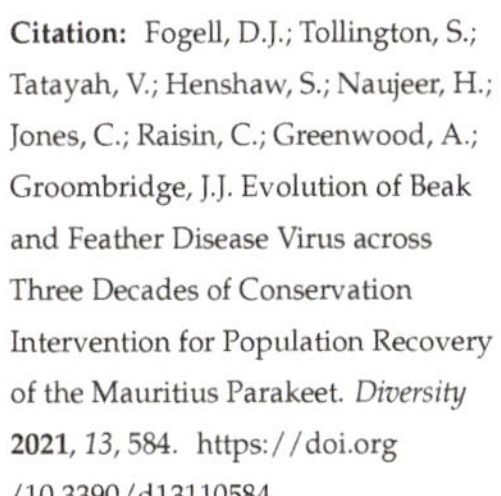

Abstract: Emerging infectious diseases (EIDs) are key contributors to the current global biodiversity crisis. Psittaciformes (parrots) are one of the most vulnerable avian taxa and psittacine beak and feather disease (PBFD) is the most common viral disease in wild parrots. PBFD is caused by the beak and feather disease virus (BFDV), which belongs to the Circoviridae family and comprises a circular, single-stranded DNA genome. BFDV is considered to have spread rapidly across the world and, in 2005, an outbreak of PBFD was documented in the recovering population of the Mauritius parakeet (*Alexandrinus eques*). The Mauritius parakeet was once the world's rarest parrot and has been successfully recovered through 30 years of intensive conservation management. Molecular surveillance for the prevalence of BFDV was carried out across a 24-year sample archive spanning the period from 1993 to 2017, and DNA sequencing of positive individuals provided an opportunity to assess patterns of phylogenetic and haplotype diversity. Phylogenetic analyses show variation in the extent of viral diversification within the replicase protein (Rep). Timeseries of BFDV prevalence and number of haplotypes reveal that two subsequent waves of infection occurred in 2010/2011 and 2013/2014 following the initial outbreak in 2005. Continued disease surveillance to determine the frequency and intensity of subsequent waves of infection may benefit future translocation/reintroduction planning. The continued growth of the Mauritius parakeet population despite the presence of BFDV bodes well for its long-term persistence.

Keywords: BFDV; emerging infectious disease; haplotypes; parrot; PBFD; viral diversification

1. Introduction

Emerging infectious diseases (EIDs) are key contributors to the current global biodiversity crisis [1,2]. Although population biologists recognize infectious pathogens as an integral mechanism for evolutionary change within natural populations [3], the emergence of novel pathogens may increase the risk of extinction for vulnerable species and populations [4]. Viruses are responsible for over 40% of all recently surveyed wildlife EIDs [5,6], and consequently have been highlighted as an important threat to the conservation of global biodiversity. The threats from viruses are in part due to their ability to adapt rapidly to novel hosts [7,8], enabling them to become infectious across a wide host range [7].

Psittaciformes (parrots) are one of the most vulnerable avian taxa, with approximately 30% of all extant species listed as Vulnerable or Threatened by the International Union for Conservation of Nature and more than 75% of species in population decline [9]. One major threat to parrots is the emergence and global spread of Psittacine Beak and Feather Disease (PBFD), the most common viral disease in wild Psittaciformes [10]. PBFD was first described in the 1970s [11] and is thought to have post-Gondwanan origins due to the paucity of ancestral non-Australian clades and infrequent observations across other regions where parrot endemism is high, such as Africa and South America [12]. It is caused by the Beak and Feather Disease Virus (BFDV), which belongs to the Circoviridae family and comprises a circular, single-stranded DNA genome of approximately 2000 nucleotides [13]. Both its small size and structure make BFDV a relatively simple pathogen for studying molecular variation in the context of disease ecology and drivers of spread [14]. The genome consists of a highly conserved replicase (Rep) [15,16] and a capsid (Cap) protein responsible for viral encapsidation and host–cell penetration [16,17]. BFDV is transmissible horizontally, through contact with contaminated feather dust, surfaces, or objects [18], and vertically, from a female to her offspring [19].

BFDV is thought to have spread rapidly across the world owing to its high environmental persistence and ability to shift between closely related host species [20,21]. All Psittaciformes are considered to be susceptible to infection [14] and to date, BFDV or PBFD has been recorded in a total of 78 species (18 New World and 60 Old World) and five subspecies globally [22]. Small, isolated host populations such as parrot species endemic to islands are considered to be particularly vulnerable to EIDs, as their populations often have low genetic diversity [23,24] and have frequently evolved in the presence of an impoverished pathogen community [23,25]. Island species are also increasingly at risk due to human-facilitated biological invasions and the alteration of an often already limited habitat [26], with the number of bird species introduced to oceanic islands being roughly equal to the number of species extirpated from them [27]. Consequently, island-endemic parrot populations infected with BFDV can provide near-ideal study systems for documenting how this virus evolves in its psittacine host.

The Mauritius parakeet (*Alexandrinus eques*) was once the world's rarest parrot [16,28], but by 2017 had recovered to approximately 136 known breeding pairs [29]. PBFD was first recorded in the Mauritius parakeet in the early 1990s [30,31] and low viral prevalence was detected in blood samples taken on an ad hoc basis from 1993 to 2004. However, in the 2005/2006 breeding season, an outbreak of PBFD swept through the population of Mauritius parakeets, coinciding with a viral mutation located in Rep [16]. Since that outbreak, blood samples have been taken from all annually produced offspring. Extracted DNA from each sample has been screened using PCR [16,32] to detect the presence of BFDV and positive samples have been sequenced to distinguish the different viral haplotypes (genetic variants). This process has provided a unique opportunity to characterise the temporal evolution of BFDV in the Mauritius parakeet host population spanning the last three decades before, during, and after the outbreak [16,33].

Here, we assess some of the patterns evident in the BFDV viral haplotypes present in the endemic Mauritius parakeet on Mauritius through phylogenetic and haplotype network analyses. We (i) examine patterns of viral diversification that have occurred in isolation on Mauritius since 1993, (ii) compare the rate of BFDV mutation on Mauritius to that found in other global regions, and (iii) interpret the patterns of BFDV prevalence and viral diversity in a context of multiple outbreak events following the initial outbreak in 2005. Finally, we consider how this EID should be viewed in the context of the future conservation management of the Mauritius parakeet. Our findings provide valuable insights into the evolutionary dynamics of BFDV in a recovering host population of this once critically endangered species, the last remaining endemic parrot of the Mascarene islands [34].

2. Materials and Methods

2.1. Mauritius Parakeet Sampling, DNA Extraction, PCR, and Sequencing

Blood samples were taken by the Mauritius parakeet field team from all accessible 45-day-old nestlings produced each breeding season (September to May) since 2005 and opportunistically from post-fledged birds since 1993 as part of ongoing species management. For this study, a total of 1321 samples were screened for BFDV across all breeding seasons from 2009/2010 to 2016/2017 (comprising 639 breeding attempts where at least one fledgling was produced). The resulting DNA sequences of viral haplotypes were added to an existing viral prevalence dataset for the Mauritius parakeet [16], resulting in a dataset spanning 24 years. Additionally, 70 further Mauritius parakeet blood samples were screened from three cohorts of fledglings that were translocated from the Black River Gorges National Park to Vallée de Ferney during the 2014/2015, 2015/2016, and 2016/2017 breeding seasons as part of a conservation translocation programme to establish a Mauritius parakeet population on the east coast of Mauritius. Prior to screening for BFDV, an ammonium acetate DNA extraction method was used to extract both host and viral DNA [35]. In brief, approximately 50 to 100 μL of whole blood was digested in 250 μL of DIGSOL lysis buffer (20 mM EDTA, 50 mM Tris, 120 mM NaCl, 1% SDS, pH 8.0) with 10 μL of 10 mg/mL proteinase K. Extractions were quantified using a Qubit dsDNA Assay Kit and standardized to approximately 25 ng/μL prior to screening for BFDV using PCR.

Virus-specific PCR primers were then used to determine the presence of viral DNA within that of the host. Screening was carried out through a PCR assay that amplified a 717 bp region of Rep [36]. Reactions comprised 1 μL of extracted DNA template, 5 μL MyTaq^{TM} HS Red Mix (Bioline), 0.2 μL each of the forward and reverse primers at 10 pmol/μL and were made up to 10 μL with double-distilled water. PCR annealing temperature was set to 60 °C for 30 cycles and products were visualized on a 1.5% agarose gel. Both a known BFDV-positive Mauritius parakeet sample and a negative control were included in each PCR batch. All positive PCR products were sequenced using forward and reverse sequencing reactions (Macrogen Europe, Amsterdam, The Netherlands). All sequences obtained from Mauritius parakeet hosts between 2009 and 2017 have been deposited in GenBank (KT753406–KT753526, MZ673091–MZ673140).

2.2. BFDV Phylogeny and Haplotype Network

Geneious 8.1.7 [37] DNA editing software was used to align and edit forward and reverse sequence reads and to produce a consensus sequence for each positive sample. Rep was chosen for analysis because a previous study identified a selective mutation in this gene as being the most likely cause for the initial outbreak of PBFD observed on Mauritius in 2005 [16]. For phylogenetic reconstruction, the programme jModelTest 2.1.7 [38] was used to infer the best-fit nucleotide substitution model. A transition model with gamma-distributed rate variation and a proportion of invariable sites (GTR + I + G) was favoured. We constructed a maximum likelihood (ML) phylogenetic tree using RAxML version 8 [39], which applies a gamma substitution model and a rapid bootstrapping (RBS) heuristic procedure [40]. We collapsed branches with <50% bootstrap support using TreeGraph 2 [41] and edited and annotated the final tree in FigTree version 1.4.4 [42].

Network 10.2.0.0 [43] was used to construct a median-joining nucleotide haplotype network for Rep sequences to analyse patterns in clustering and diversity both temporally and spatially. We used DNAsp 6.12.03 [44] to examine whether the Mauritius BFDV population had experienced demographic changes (significant population expansion) over the assessed period. Departures from mutation–drift equilibrium were tested using Fu's F_S statistic [45], where a negative value would be indicative of diversification and a positive value would be reflective of a recent population bottleneck.

3. Results

The blood samples that amplified a PCR product for BFDV yielded edited sequences of 462 bp of the Rep gene. Since the first observation of BFDV in Mauritius parakeets in the 1993/1994 breeding season, our data indicate that this section of the Rep gene has diverged into 63 observed haplotypes. Of these, 49 haplotypes were detected in single host individuals, whereas seven haplotypes occurred in more than five individuals and persisted over multiple breeding seasons. The maximum likelihood phylogeny given in Figure 1 shows an 'outbreak cluster' of haplotypes that comprises haplotypes sampled during the 2005/2006 outbreak year as well as a mixture of phylogenetically similar haplotypes sampled from across subsequent years, particularly from 2008/2009 and 2010/2011. In contrast, several other large clusters of haplotypes appear to comprise haplotypes that are sampled from just one or two years; for example, 'cluster A' comprises haplotypes almost entirely from 2006/2007 and 2014/2015 years, 'cluster B' comprises predominantly haplotypes from 2013/2014, and 'cluster C' comprises haplotypes entirely from 2015/2016 and 2016/2017. Those sequences obtained from the 2015/2016 and 2016/2017 breeding seasons are the only ones to not have any dispersal throughout the rest of the phylogeny.

Figure 2 shows the changes in BFDV prevalence since the initial outbreak in 2005 against a backdrop of increasing numbers of host–breeding pairs as the host population has continued to recover from its initial low population size as a consequence of intensive conservation management. The changes in BFDV prevalence indicate that the parakeet host population has experienced at least two subsequent 'waves' of BFDV infection; in 2010/2011 (39.4% BFDV prevalence) and in 2013/2014 (41.3% BFDV prevalence) that were equal to or larger than the initial 2005 outbreak, interspersed with periods of low infection. The haplotype network shown in Figure 3 indicates at least three clusters of the most frequently occurring BFDV haplotypes (Figure 3a), with each one of the three most dominant haplotypes being those corresponding to the initial outbreak and the second and third waves of BFDV infection (Figure 3b). The starburst pattern present within the haplotype network is indicative of significant demographic expansion and diversification within the host population since 1993, and is supported by the Fu's F test statistic [45] (-33.30, $p < 0.001$, $k = 4.46$, $h = 0.85 \pm 0.00$, $\pi = 0.01 \pm 0.01$).

There has been a large variation in the number of haplotypes present across breeding seasons, with the largest number seen in the 2010/2011 season ($n = 20$) and the fewest seen in 2006/2007 ($n = 1$) (Figure 3b). We found no geographical/spatial separation of haplotypes and sequences from host individuals in all subpopulations, including captive Mauritius parakeets in the Gerald Durrell Endemic Wildlife Sanctuary (GDEWS) and the newly established subpopulation at Vallée de Ferney, were dispersed throughout the network. A pattern of haplotype dominance was observed across the seasons where a single haplotype comprising the majority of sequences from the initial outbreak in 2005 persisted until 2010/2011. A single base-pair change separates this haplotype from the subsequent dominant haplotype that has persisted from 2011/2012 to the most recently assessed 2016/2017 breeding season (Figure 3a). Although there was a third haplotype group that occurred from 2004/2005, this was only detected at a lower frequency in the Mauritius parakeet population until 2010/2011 and has not been detected since.

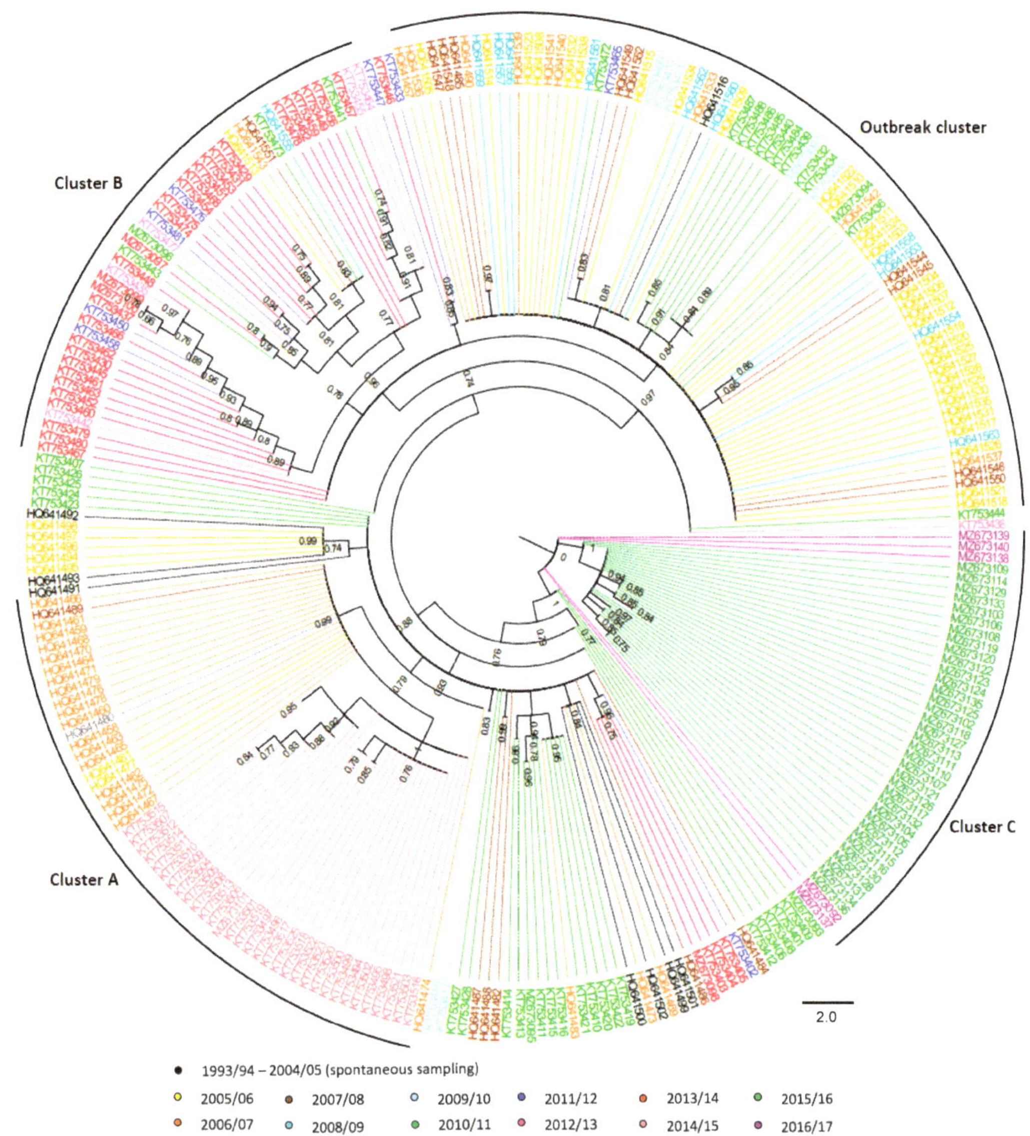

Figure 1. Maximum likelihood phylogenetic tree denoting relationships between BFDV Rep sequences in Mauritius, where branches with <50% branch support have been collapsed. Branches are coloured based on the year of sampling as denoted in the key.

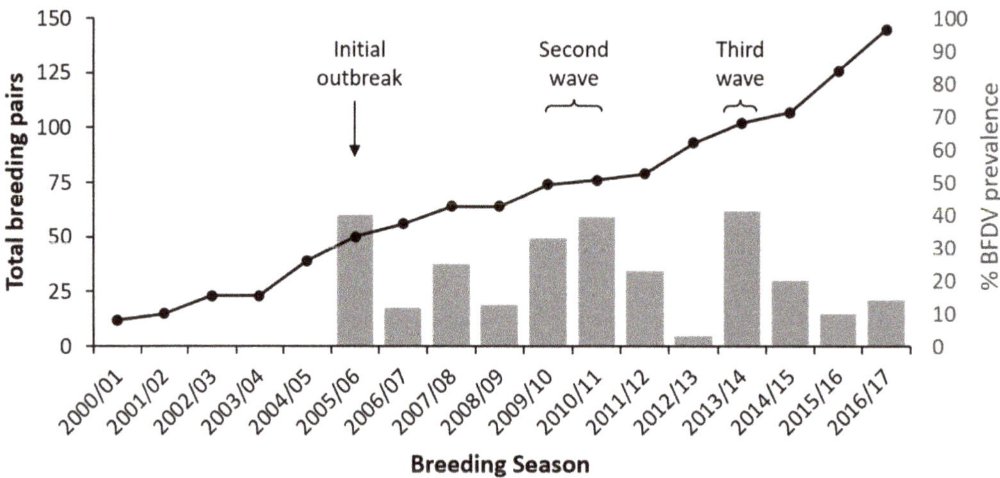

Figure 2. The total number of Mauritius parakeet breeding pairs recorded between 2000 and 2017 (black line) plotted with the percentage BFDV prevalence detected in offspring produced in each breeding season since systematic sampling began in 2005 (grey bars).

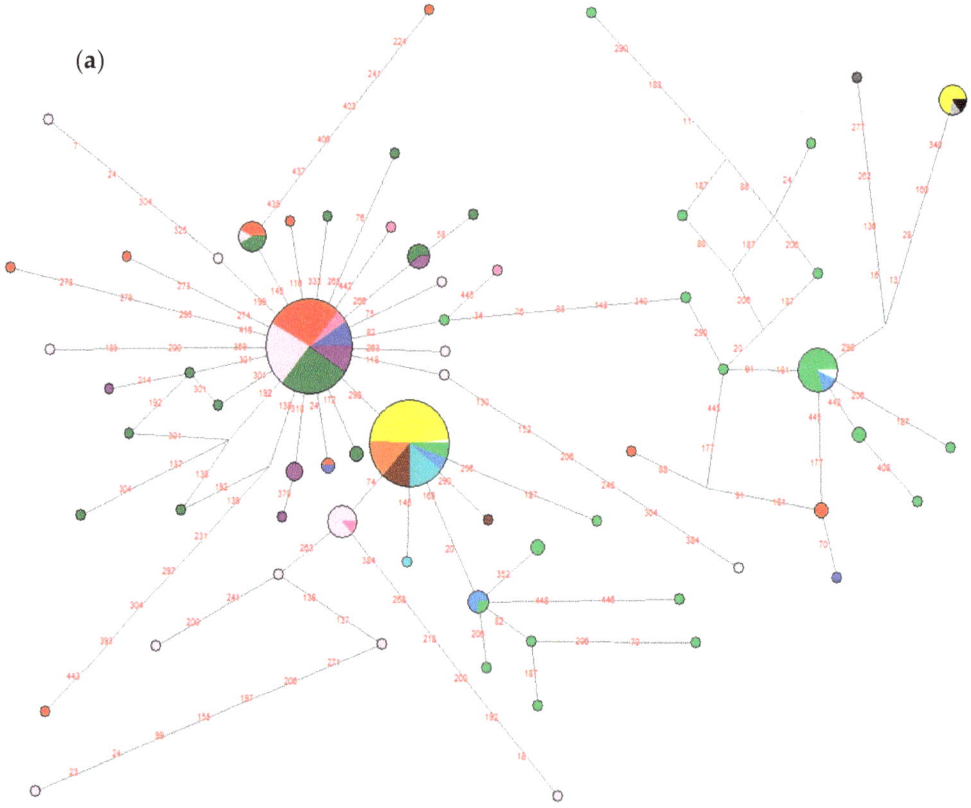

(a)

Figure 3. *Cont.*

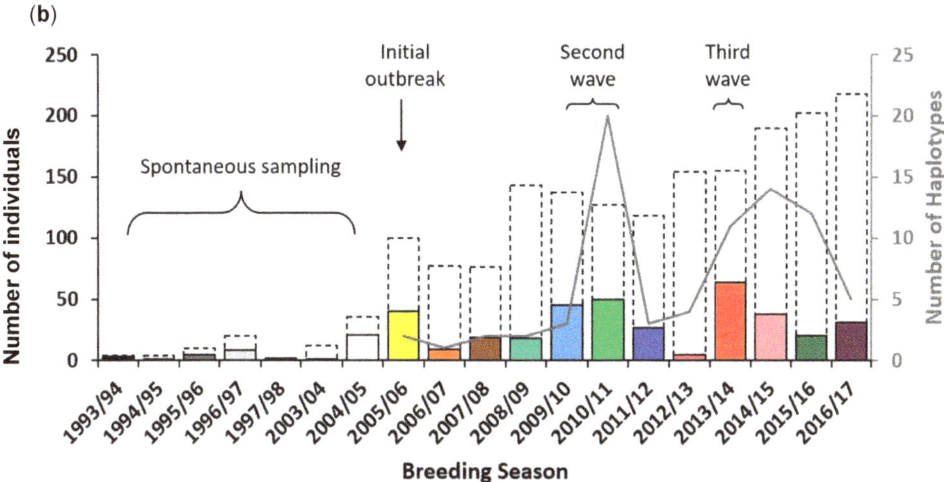

Figure 3. (**a**, **top**) Haplotype network displaying the diversity and evolution of BFDV Rep haplotypes in Mauritius parakeet hosts. The size of each circle is proportional to the number of individuals sharing that viral haplotype and the length of the lines between circles are proportional to the number of base pair changes between each haplotype. (**b**, **bottom**) Number of individuals screened (dashed bars), viral prevalence (solid bars), and number of haplotypes (grey line) detected from 1993 to 2017. Color codes in 3a refer to the sampling year shown in 3b. The screening dataset from 2005 to 2017 comprises a systematic sampling of fledglings.

4. Discussion

Pathogen persistence in large populations is generally regulated by host population size and density, whereas pathogen establishment among small populations is much more likely to be influenced by stochastic factors [3]. Host–parasite relationships are often disrupted in threatened species, which can result in the local elimination of endemic diseases owing to reduced size and increased fragmentation of host species beyond a threshold required to maintain viral transmission [46]. As a result of a very successful conservation initiative, the Mauritius parakeet population has grown and has become less fragmented and, as a consequence, it is probable that pre-bottleneck host–parasite dynamics have been restored to some extent. Furthermore, there has been a rapid increase in the highly invasive rose-ringed parakeet (*Alexandrinus krameri*) population on the island [16], which potentially acts as a reservoir host for BFDV infection and transmission.

The strain of BFDV present within the Mauritius parakeet population has rapidly diversified over the last three decades, with two subsequent waves of high prevalence since the initial outbreak in 2005. The ability of a pathogen to establish in a host population has a direct relationship with its transmission efficiency and an inverse relationship with its virulence [3,46]. Although BFDV has been found to be widely infectious [47] and PBFD is frequently fatal in immature birds up to three years of age [19], infected adults commonly recover from severe clinical presentation of the disease, which usually lasts only for a number of months [48]. These attributes of BFDV appear to have allowed it to become highly prevalent in the Mauritius parakeet population whereas host numbers have continued to increase [33]. A prevalence of BFDV among Mauritius parakeets nestlings of 41.3% in 2013/2014 is one of the highest among wild parrot populations to our knowledge [22].

4.1. Multiple Waves of BFDV Infection Following Initial Outbreak

Our screening for BFDV across three decades has identified two additional 'waves' of BFDV infection in the Mauritius parakeet population. Such peaks and troughs of infection are a common signature of pathogen populations and may be an indication of the host–pathogen dynamics between the host's immune system and the ability of the virus to mutate. The timing of the outbreak in 2005 and subsequent two waves of infection in 2010/2011 and 2013/2014 suggests a periodicity of between 3 and 5 years. This corresponds to the estimated generation time for Mauritius parakeets of 4 years, as determined through observations of the species in the wild [49], where generation time was defined as the average age of reproduction and is also the average time from reproduction in one generation to reproduction in the next generation [50]. As there has been substantial genetic homogenisation of the Mauritius parakeet population due to conservation management for recovery [51], this may have influenced the host–pathogen coevolution cycle [52], which could be an avenue for future research with the potential opportunities available through whole-genome sequencing. Future screening for BFDV, and identification of further waves of infection, will determine whether the regularity at which they occur stabilises and whether host population size has an influence on their frequency and intensity. If the periodicity of waves of infection remains relatively stable, then this knowledge may provide valuable insight for future conservation management of the Mauritius parakeet population. Although it is still important to note that the high prevalence of infection does not necessitate high pathogenicity [47,53], these considerations could assist in timing planned future translocation/reintroduction initiatives to occur between waves of infection.

4.2. BFDV in the Context of Ongoing Conservation Management

During the intensive management and recovery of this host population, interventions such as brood manipulation, captive breeding, and reintroduction were undertaken to help rapidly increase the number of parakeets [51,54]. The Mauritian Wildlife Foundation's parakeet field team has attempted to reduce or eliminate any potential human-mediated transmission of BFDV with a rigorous biosecurity and hygiene protocol since 2005 [32]. However, despite these protocols, the recent translocation of parakeets to Le Vallée de Ferney on Mauritius to initiate a new subpopulation also included the transfer of BFDV to the east coast. The unavoidable regular movement of vehicles, equipment, and field staff between localities for ongoing species management is the most likely reason for this movement of viral populations, which is evident in the lack of within-subpopulation haplotype clustering, despite their geographical separation. In addition to this, some translocated individuals are known to have relocated back to the Black River Gorges National Park (S Henshaw, Pers. Obs.), indicating that these non-breeding sub-adults may now also facilitate the transmission of BFDV between subpopulations.

Although our haplotype network generated for the strain of BFDV present on Mauritius provides a window into the viral population dynamics and diversification, it is unlikely to represent all variants present on the island given that it is based on a fragment of the Rep gene rather than the full BFDV genome. The ability for multiple BFDV infections to persist within a single host, along with its high rate of mutation, allows for the rapid evolution of novel BFDV variants through recombination [54]. Indeed, we have detected some instances of multiple infections in this study system, although from limited screening. Currently, there only appears to be a single strain of BFDV present in Mauritius, and this is shared between both the native and introduced populations [47]. However, if novel BFDV variants are introduced to the Mauritius parakeet population through an accidental leak from captive pet parrot species (as is plausibly the case with the introduction of rose-ringed parakeets to the island [16]), such an event may alter the virus' pathogenicity and subsequently increase the threat imposed by infection [55,56].

Despite the continued presence (and at times high prevalence) of BFDV in the Mauritius parakeet, the population size of this once critically endangered species continues to grow (Figure 2). The species was down-listed by the IUCN to Endangered in 2007 and to Vulnerable in 2019. This positive population trajectory in the face of an EID stands as an encouraging example that runs counter to expectations for bottlenecked island-endemic species which are often suspected to be immunologically naïve and genetically impoverished, and consequently are expected to respond poorly to the challenges of EIDs.

Author Contributions: Conceptualization and methodology, D.J.F., S.T., A.G. and J.J.G.; validation, S.H., C.J., V.T., A.G. and H.N.; formal analysis, D.J.F. and S.T.; sample collection, S.H., S.T., C.R., C.J., V.T., A.G. and H.N.; data curation, D.J.F., S.T., C.R. and J.J.G.; writing—original draft preparation, D.J.F., S.T. and J.J.G.; writing—review and editing, All authors; visualization, D.J.F.; funding acquisition, D.J.F., S.T. and J.J.G. All authors have read and agreed to the published version of the manuscript.

Funding: This work was supported by funds contributed to D.J.F. by the National Environmental Research Council (DTP grant number NE/L002582/1), African Bird Club, the Genetics Society, and the Durrell Institute of Conservation and Ecology. The research was also facilitated by a British Ecological Society Small Research Grant (5163-6205) awarded to S.T.

Institutional Review Board Statement: This research received ethical approval from the Research Ethics Committee of the School of Anthropology and Conservation at the University of Kent on 28th October 2015 (the UK does not have a national registration system for university research ethics committees). Sampling was undertaken in collaboration with and with permission from local wildlife authorities, conservation nongovernmental, and research organizations.

Data Availability Statement: This statement confirms that, should the manuscript be accepted, then data supporting the results will be archived in the Kent Academic Repository.

Acknowledgments: We thank the other MWF Mauritius parakeet project volunteers and staff who contributed to sample collection and nest site management, and NPCS for permission to work within the national park.

Conflicts of Interest: The authors declare no conflict of interest. The funders had no role in the design of the study; in the collection, analyses, or interpretation of data; in the writing of the manuscript, or in the decision to publish the results.

References

1. Brooks, D.R.; Ferrao, A.L. The historical biogeography of co-evolution: Emerging infectious diseases are evolutionary accidents waiting to happen. *J. Biogeogr.* **2005**, *32*, 1291–1299. [CrossRef]
2. Yap, T.A.; Koo, M.S.; Ambrose, R.F.; Wake, D.B.; Vredenburg, V.T. Averting a North American biodiversity crisis. *Science* **2015**, *349*, 481–482. [CrossRef]
3. Lyles, A.M.; Dobson, A.P. Infectious disease and intensive management: Population dynamics, threatened hosts, and their parasites. *J. Zoo Wildl. Med.* **1993**, *24*, 315–326.
4. Lips, K.R.; Brem, F.; Brenes, R.; Reeve, J.D.; Alford, R.A.; Voyles, J.; Carey, C.; Livo, L.; Pessier, A.P.; Collins, J.P. Emerging infectious disease and the loss of biodiversity in a neotropical amphibian community. *Proc. Natl. Acad. Sci. USA* **2006**, *103*, 3165–3170. [CrossRef]
5. Dobson, A.; Foufopoulos, J. Emerging infectious pathogens of wildlife. *Philos. Trans. R. Soc. Lond. B. Biol. Sci.* **2001**, *356*, 1001–1012. [CrossRef] [PubMed]
6. Tompkins, D.M.; Carver, S.; Jones, M.E.; Krkošek, M.; Skerratt, L.F. Emerging infectious diseases of wildlife: A critical perspective. *Trends Parasitol.* **2015**, *31*, 149–159. [CrossRef] [PubMed]
7. Altizer, S.; Harvell, D.; Friedle, E. Rapid evolutionary dynamics and disease threats to biodiversity. *Trends Ecol. Evol.* **2003**, *18*, 589–596. [CrossRef]
8. Jones, K.E.; Patel, N.G.; Levy, M.A.; Storeygard, A.; Balk, D.; Gittleman, J.L.; Daszak, P. Global trends in emerging infectious diseases. *Nature* **2008**, *451*, 990–993. [CrossRef]
9. IUCN. The IUCN Red List of Threatened Species. Version 2020-3. Available online: http://www.iucnredlist.org/ (accessed on 1 August 2021).
10. Khalesi, B.; Bonne, N.; Stewart, M.; Sharp, M.; Raidal, S. A comparison of haemagglutination, haemagglutination inhibition and PCR for the detection of Psittacine beak and feather disease virus infection and a comparison of isolates obtained from loriids. *J. Gen. Virol.* **2005**, *86*, 3039–3046. [CrossRef] [PubMed]
11. Pass, D.A.; Perry, R.A. The pathology of Psittacine beak and feather disease. *Aust. Vet. J.* **1984**, *61*, 69–74. [CrossRef]
12. Raidal, S.R.; Sarker, S.; Peters, A. Review of Psittacine beak and feather disease and its effect on Australian endangered species. *Aust. Vet. J.* **2015**, *93*, 466–470. [CrossRef]
13. Ritchie, B.W.; Niagro, F.D.; Lukert, P.D.; Steffens, W.L., III; Latimer, K.S. Characterization of a new virus from cockatoos with psittacine beak and feather disease. *Virology* **1989**, *171*, 83–88. [CrossRef]
14. Sarker, S.; Ghorashi, S.A.; Forwood, J.K.; Bent, S.J.; Peters, A.; Raidal, S.R. Phylogeny of beak and feather disease virus in cockatoos demonstrates host generalism and multiple-variant infections within Psittaciformes. *Virology* **2014**, *460*, 72–82. [CrossRef]
15. Kondiah, K.; Albertyn, J.; Bragg, R.R. Genetic diversity of the rep gene of Beak and feather disease virus in South Africa. *Arch. Virol.* **2006**, *151*, 2539–2545. [CrossRef] [PubMed]

16. Kundu, S.; Faulkes, C.G.; Greenwood, A.G.; Jones, C.G.; Kaiser, P.; Lyne, O.D.; Black, S.A.; Chowrimootoo, A.; Groombridge, J.J. Tracking viral evolution during a disease outbreak: The rapid and complete selective sweep of a *Circovirus* in the endangered Echo parakeet. *J. Virol.* **2012**, *86*, 5221–5229. [CrossRef]
17. Heath, L.; Martin, D.P.; Warburton, L.; Perrin, M.; Horsfield, W.; Kingsley, C.; Rybicki, E.P.; Williamson, A.-L. Evidence of unique genotypes of Beak and feather disease virus in Southern Africa. *J. Virol.* **2004**, *78*, 9277–9284. [CrossRef] [PubMed]
18. Ritchie, P.A.; Anderson, I.L.; Lambert, D.M. Evidence for specificity of Psittacine beak and feather disease viruses among avian hosts. *Virology* **2003**, *306*, 109–115. [CrossRef]
19. Ritchie, B.W.; Niagro, F.D.; Lukert, P.D.; Latimer, K.S.; Steffens III, W.L.; Pritchard, N. A review of Psittacine beak and feather disease: Characteristics of the PBFD virus. *J. Assoc. Avian Vet.* **1989**, *3*, 143–150. [CrossRef]
20. Peters, A.; Patterson, E.I.; Baker, B.G.B.; Holdsworth, M.; Sarker, S.; Ghorashi, S.A.; Raidal, S.R. Evidence of Psittacine beak and feather disease virus spillover into wild critically endangered Orange-bellied parrots (*Neophema chrysogaster*). *J. Wildl. Dis.* **2014**, *50*, 288–296. [CrossRef]
21. Sarker, S.; Patterson, E.I.; Peters, A.; Baker, G.B.; Forwood, J.K.; Ghorashi, S.A.; Holdsworth, M.; Baker, R.; Murray, N.; Raidal, S.R. Mutability dynamics of an emergent single stranded DNA virus in a naïve host. *PLoS ONE* **2014**, *9*, e8537. [CrossRef]
22. Fogell, D.J.; Martin, R.O.; Groombridge, J.J. Beak and feather disease virus in wild and captive parrots: An analysis of geographic and taxonomic distribution and methodological trends. *Arch. Virol.* **2016**, *161*, 2059–2074. [CrossRef] [PubMed]
23. Wikelski, M.; Foufopoulos, J.; Vargas, H.; Snell, H. Galápagos Birds and Diseases: Invasive Pathogens as Threats for Island Species. *Ecol. Soc.* **2004**, *9*, 5. [CrossRef]
24. Trinkel, M.; Cooper, D.; Packer, C.; Slotow, R. Inbreeding depression increases susceptibility to bovine tuberculosis in lions: An experimental test using an inbred-outbred contrast through translocation. *J. Wildl. Dis.* **2011**, *47*, 494–500. [CrossRef] [PubMed]
25. Spurgin, L.G.; Illera, J.C.; Padilla, D.P.; Richardson, D.S. Biogeographical patterns and co-occurrence of pathogenic infection across island populations of Berthelot's pipit (*Anthus berthelotii*). *Oecologia* **2012**, *168*, 691–701. [CrossRef]
26. Russell, J.C.; Kueffer, C. Island Biodiversity in the Anthropocene. *Annu. Rev. Environ. Resour.* **2019**, *44*, 31–60. [CrossRef]
27. Sax, D.F.; Gaines, S.D.; Brown, J.H. Species invasions exceed extinctions on islands worldwide: A comparative study of plants and birds. *Am. Nat.* **2002**, *160*, 766–783. [CrossRef] [PubMed]
28. Jones, C.G. The larger land-birds of Mauritius. In *Studies of Mascarene Island Birds*; Diamond, A.W., Ed.; Cambridge University Press: Cambridge, UK, 1987; pp. 208–300.
29. Henshaw, S.; Blackwell, L.; Yeung Shi Chung, C.; Bielsa, M.; Bertille, H. *Mauritian Wildlife Foundation: Echo Parakeet (Psittacula eques) Management Report 2017/2018*; Mauritian Wildlife Foundation: Vacoas, Mauritius, 2018.
30. Lovegrove, T.G.; Nieuwland, A.B.; Green, S. *Interim Report on the Echo Parakeet Conservation Project*; Mauritian Wildlife Foundation: Vacoas, Mauritius, 1995.
31. Greenwood, A.G. Veterinary support for in situ avian conservation programmes. *Bird Conserv. Int.* **1996**, *6*, 285–292. [CrossRef]
32. Fogell, D.J.; Groombridge, J.J.; Tollington, S.; Canessa, S.; Henshaw, S.; Zuel, N.; Jones, C.G.; Greenwood, A.; Ewen, J.G. Hygiene and biosecurity protocols reduce infection prevalence but do not improve fledging success in an endangered parrot. *Sci. Rep.* **2019**, *9*, 1–10. [CrossRef] [PubMed]
33. Tollington, S.; Greenwood, A.; Jones, C.G.; Hoeck, P.; Chowrimootoo, A.; Smith, D.; Richards, H.; Tatayah, V.; Groombridge, J.J. Detailed monitoring of a small but recovering population reveals sublethal effects of disease and unexpected interactions with supplemental feeding. *J. Anim. Ecol.* **2015**, *84*, 969–977. [CrossRef]
34. Hume, J.P. *Reappraisal of the Parrots (Aves: Psittacidae) from the Mascarene Islands, with Comments on Their Ecology, Morphology, and Affinities*; Magnolia Press: Auckland, New Zealand, 2007; ISBN 1175-5326.
35. Bruford, M.W.; Hanotte, O.; Brookfield, J.F.Y.; Burke, T. Single-locus and multilocus DNA fingerprinting. In *Molecular Genetic Analysis of Populations: A Practical Approach*; Hoelzel, A.R., Ed.; IRL Press: Oxford, UK, 1998; pp. 287–336.
36. Ypelaar, I.; Bassami, M.R.; Wilcox, G.E.; Raidal, S.R. A universal polymerase chain reaction for the detection of Psittacine beak and feather disease virus. *Vet. Microbiol.* **1999**, *68*, 141–148. [CrossRef]
37. Kearse, M.; Moir, R.; Wilson, A.; Stones-Havas, S.; Cheung, M.; Sturrock, S.; Buxton, S.; Cooper, A.; Markowitz, S.; Duran, C.; et al. Geneious Basic: An integrated and extendable desktop software platform for the organization and analysis of sequence data. *Bioinformatics* **2012**, *28*, 1647–1649. [CrossRef] [PubMed]
38. Posada, D. jModelTest: Phylogenetic model averaging. *Mol. Biol. Evol.* **2008**, *25*, 1253–1256. [CrossRef]
39. Stamatakis, A. RAxML version 8: A tool for phylogenetic analysis and post-analysis of large phylogenies. *Bioinformatics* **2014**, *30*, 1312–1313. [CrossRef] [PubMed]
40. Stamatakis, A.; Hoover, P.; Rougemont, J. A rapid bootstrap algorithm for the RAxML web servers. *Syst. Biol.* **2008**, *57*, 758–771. [CrossRef] [PubMed]
41. Stöver, B.C.; Müller, K.F. TreeGraph 2: Combining and visualizing evidence from different phylogenetic analyses. *BMC Bioinform.* **2010**, *11*, 7. [CrossRef]
42. Rambaut, A. FigTree v1.4.2. Available online: http://tree.bio.ed.ac.uk/software/figtree/ (accessed on 17 July 2015).
43. Fluxus Technology Ltd Network 10.2.0.0. Available online: http://www.fluxus-engineering.com/netwinfo.htm (accessed on 1 August 2021).
44. Rozas, J.; Ferrer-Mata, A.; Sánchez-DelBarrio, J.C.; Guirao-Rico, S.; Librado, P.; Ramos-Onsins, S.; Sánchez-Gracia, A. DNA Sequence Polymorphism v 6.12.03. 2018. Available online: http://www.ub.edu/dnasp/ (accessed on 23 May 2021).

45. Fu, Y.X.; Li, W.H. Statistical tests of neutrality of mutations. *Genetics* **1993**, *133*, 693–709. [CrossRef] [PubMed]
46. Easterday, W.R. The first step in the success or failure of emerging pathogens. *Proc. Natl. Acad. Sci. USA* **2020**, *117*, 29271–29273. [CrossRef]
47. Fogell, D.J.; Martin, R.O.; Bunbury, N.; Lawson, B.; Sells, J.; McKeand, A.M.; Tatayah, V.; Trung, C.T.; Groombridge, J.J. Trade and conservation implications of new Beak and feather disease virus detection in native and introduced parrots. *Conserv. Biol.* **2018**, *32*, 1325–1335. [CrossRef]
48. Todd, D. Circoviruses: Immunosuppressive threats to avian species: A review. *Avian Pathol.* **2000**, *29*, 373–394. [CrossRef]
49. Raisin, C. Conservation Genetics of the Mauritius Parakeet. Ph.D. Thesis, University of Kent, Kent, UK, 2010.
50. Lacy, R.C.; Ballou, J.D.; Pollak, J.P. PMx: Software package for demographic and genetic analysis and management of pedigreed populations. *Methods Ecol. Evol.* **2012**, *3*, 433–437. [CrossRef]
51. Raisin, C.; Frantz, A.C.; Kundu, S.; Greenwood, A.G.; Jones, C.G.; Zuel, N.; Groombridge, J.J. Genetic consequences of intensive conservation management for the Mauritius parakeet. *Conserv. Genet.* **2012**, *13*, 707–715. [CrossRef]
52. Duxbury, E.M.L.; Day, J.P.; Vespasiani, D.M.; Thüringer, Y.; Tolosana, I.; Smith, S.C.L.; Tagliaferri, L.; Kamacioglu, A.; Lindsley, I.; Love, L.; et al. Host-pathogen coevolution increases genetic variation in susceptibility to infection. *Elife* **2019**, *8*, e46440. [CrossRef] [PubMed]
53. McCallum, H.; Dobson, A. Disease, habitat fragmentation and conservation. *Hungarian Q.* **2008**, *49*, 2041–2049. [CrossRef]
54. Tatayah, R.V.V.; Malham, J.; Haverson, P.; Van de Wetering, J. Design and provision of nest boxes for echo parakeets Psittacula eques in Black River Gorges National Park, Mauritius. *Conserv. Evid.* **2007**, *4*, 16–19.
55. Julian, L.; Piasecki, T.; Chrzastek, K.; Walters, M.; Muhire, B.; Harkins, G.W.; Martin, D.P.; Varsani, A. Extensive recombination detected among Beak and feather disease virus isolates from breeding facilities in Poland. *J. Gen. Virol.* **2013**, *94*, 1086–1095. [CrossRef]
56. Jackson, B.; Varsani, A.; Holyoake, C.; Jakob-Hoff, R.; Robertson, I.; McInnes, K.; Empson, R.; Gray, R.; Nakagawa, K.; Warren, K. Emerging infectious disease or evidence of endemicity? A multi-season study of Beak and feather disease virus in wild Red-crowned parakeets (*Cyanoramphus novaezelandiae*). *Arch. Virol.* **2015**, *160*, 2283–2292. [CrossRef] [PubMed]

 MDPI

Article

Satellite Telemetry of Blue-Throated Macaws in Barba Azul Nature Reserve (Beni, Bolivia) Reveals Likely Breeding Areas

Lisa C. Davenport [1,2,*], Tjalle Boorsma [3], Lucas Carrara [4], Paulo de Tarso Zuquim Antas [5], Luciene Faria [4], Donald J. Brightsmith [6], Sebastian K. Herzog [3,7], Rodrigo W. Soria-Auza [3,7,*] and A. Bennett Hennessey [3]

1 Florida Natural History Museum, Department of Biology, University of Florida Gainesville, Gainesville, FL 32611, USA
2 College of Marine and Environmental Sciences, James Cook University, Cairns, QLD 4878, Australia
3 Asociación Civil Armonía, Zona Palmasola, Santa Cruz, Bolivia; tboorsma@armonia-bo.org (T.B.); skherzog@armonia-bo.org (S.K.H.); abhennessey@armonia-bo.org (A.B.H.)
4 Aves Gerais, Belo Horizonte 30431-236, MG, Brazil; lucas.avesgerais@gmail.com (L.C.); luciene.avesgerais@gmail.com (L.F.)
5 Fundação Pró-Natureza (Funatura), Brasilia 70743-520, DF, Brazil; ptzantas@gmail.com
6 Schubot Avian Health Center, Department of Veterinary Pathobiology, Texas A & M University, College Station, TX 77843, USA; Brightsmith1@tamu.edu
7 Museo de Historia Natural Alcide d'Orbigny, Cochabamba, Bolivia
* Correspondence: ldavenport@parkswatch.org (L.C.D.); wilbersa@armonia-bo.org (R.W.S.-A.)

Citation: Davenport, L.C.; Boorsma, T.; Carrara, L.; Antas, P.d.T.Z.; Faria, L.; Brightsmith, D.J.; Herzog, S.K.; Soria-Auza, R.W.; Hennessey, A.B. Satellite Telemetry of Blue-Throated Macaws in Barba Azul Nature Reserve (Beni, Bolivia) Reveals Likely Breeding Areas. *Diversity* **2021**, *13*, 564. https://doi.org/10.3390/d13110564

Academic Editors: José L. Tella, Guillermo Blanco and Martina Carrete

Received: 21 September 2021
Accepted: 29 October 2021
Published: 5 November 2021

Abstract: The Blue-throated Macaw (*Ara glaucogularis)* is a Critically Endangered species endemic to the Llanos de Moxos ecosystem of Beni, Bolivia. To aid conservation of the northwestern population that utilizes the Barba Azul Nature Reserve during the non-breeding season, we set out to learn the sites where these birds breed using satellite telemetry. We describe preliminary tests conducted on captive birds (at Loro Parque Foundation, Tenerife, Spain) that resulted in choosing Geotrak Parrot Collars, a metal, battery-operated unit that provides data through the Argos satellite system. In September 2019, we tagged three birds in Barba Azul with Geotrak collars, and received migration data for two birds, until battery depletion in November and December 2019. Our two migrant birds were tracked leaving Barba Azul on the same date (27 September), but departed in divergent directions (approximately 90 degrees in separation). They settled in two sites approximately 50–100 km from Barba Azul. Some details of the work are restricted out of conservation concern as the species still faces poaching pressures. Knowing their likely breeding grounds, reserve managers conducted site visits to where the birds were tracked, resulting in the discovery of breeding birds, although no birds still carrying a transmitter were seen then. A single individual still carrying its collar was spotted 13 August 2021 at Barba Azul. The work suggests that the Blue-throated Macaws of Barba Azul use breeding sites that are scattered across the Llanos de Moxos region, although within the recognized boundaries of the northwestern subpopulation. We conclude that the use of satellite collars is a feasible option for research with the species and could provide further conservation insights.

Keywords: Psittacidae; *Ara glaucogularis*; migration; daily behavioral patterns; Llanos de Moxos

1. Introduction

The technology used to track animal movements with miniaturized, animal-borne devices is advancing rapidly and opening up many new avenues of animal research [1–5]. In avian studies, tracking technology is typically used to tag birds caught on their breeding grounds, and to track them remotely to learn about annual movements, migrations routes, reproductive parameters, foraging behaviors, and stopover ecology. Examples hail from around the world and for a wide range of taxa including Magellanic Penguins [6], raptors in the Americas [7,8], Black Skimmers migrating over the Andes [9], waterfowl in Africa [10], upland sandpipers in Western Hemisphere [11], and nomadic Banded Stilts of Australia [12]. These and other studies provide valuable information for better understanding birds' lives and conservation needs, as well as providing scientific insight into important questions about birds' sensory abilities, and their behavior in changing environments [3].

Far less common than tracking animals *from* their breeding grounds is tracking them *to* their breeding locations. Some examples exist tracking marine vertebrates (e.g., Atlantic tuna [13] and green turtles [14]). For birds, Takekawa et al. [15] tracked East Asian waterbirds to breeding grounds to study potential routes of transmission of avian influenza outbreaks, and shorebirds have been tracked from flyway locations ([16]. Many parrots have relatively well-known breeding areas, although for some macaw species, including Lear's Macaw (*Anodorhynchus leari*) and Great Green Macaws (*Ara ambiguus*), breeding areas were unknown until late in the 20th century [17–19]. For long-lived birds such as parrots, one of the most critical determinants of a species' demography is breeding success; however, conservationists often need information on rare parrots' reproductive seasons, behaviors, and success. Such is the case for the northern population of the Critically Endangered Blue-throated Macaw (*Ara glaucogularis*—henceforth "BTM") in Beni, Bolivia, for which knowledge of breeding ground locations, and some basic reproductive strategies and parameters are largely unknown. Such information can be aided by telemetry studies, yet these are difficult to implement for most parrot species.

Harvest for the pet trade in the 1960s and 1970s helped to push the BTM to the brink of extinction [20,21], and the location of its remnant wild population was unknown until its re-discovery in 1992 [20]. The species is endemic to the Llanos de Moxos ecosystem of the Beni, a 12,000,000-ha expanse of treeless grasslands, wetlands, cerrado-like savanna, palm savanna, and palm forest islands that is flooded approximately 6 months per year from October to May [22–25]. BTMs require forest islands that harbor one of its preferred feeding and roosting trees, the Motacú palm (*Attalea princeps*) [22]. In the breeding season, it requires large tree cavities for nesting, demonstrating high nest site fidelity to both artificial and natural cavities [26]. Between 312 and 455 BTMs are thought to remain in the wild [22]. These are divided among three potentially isolated subpopulations in the northwestern, northeastern, and southern Beni [22]. In 2008, a private 11,000-ha reserve, the Barba Azul Nature Reserve (henceforth "Barba Azul") was established in the northwestern part of the range to protect the groups that use the area during the nonbreeding season (estimated at approximately 160–200 individuals by [27]).

While satellite telemetry is a potentially useful technology for enhancing conservation goals in Barba Azul, the task of tracking wild parrots across large areas and over many months is a formidable challenge. Few Platform Transmitter Terminal (PTT) units available today are able to resist the great intelligence, strong beaks, and flexible necks and tongues of most large parrots, and many species can easily remove attached transmitters in minutes [28]. However, species and individuals vary, and field researchers have therefore had some successes with a few larger species that have carried telemetry units for months and allowed useful data on landscape-level movements [28–31]. Since no previous telemetry studies exist on the BTM, and it is smaller than macaws previously studied, we felt it vital to test potential tag models in a controlled environment prior to working with wild birds. Moreover, to ensure the Barba Azul Reserve's continued viability as a center of conservation and ecotourism, we needed to trap and study wild birds while causing the least possible disruption to the known roosting, feeding, and tourist–viewing sites.

The overriding goals of this study were to: (1) test available technologies on captive birds to determine the safest and most appropriate techniques, and (2) to safely tag and track wild BTMs living at Barba Azul to determine the location of their breeding sites and understand more about local movement patterns, distances traveled, and resources used.

2. Materials and Methods

2.1. Data Accessibility

Because of ongoing poaching of BTMs, especially on their nests and outside of protected areas, we chose not to share all the tracking data obtained in the study. We do provide access to tracking data within and near to the limits of the Barba Azul Nature

Reserve, where the birds are protected from poaching [32]. We also redacted some of the specifics of our capture methods of wild birds. Researchers with valid requests for additional specifics about methods and tracking data can write to the corresponding author and/or Asociación Armonía (Rodrigo W. Soria-Auza, email: wilbersa@armonia-bo.org) for further details.

2.2. Study Site: Barba Azul Nature Reserve

The study was conducted in the seasonally inundated savannahs and forest islands of the Llanos de Moxos in the northern part of the Beni Department of Bolivia. Mean annual precipitation in the region is approximately 1800 mm with most rain falling between October and May [33]. Nearly the entirety of the Llanos de Moxos habitat in northern Beni (~12,000,000 ha) is subject to cattle ranching and associated land management practices, including areas within declared protected areas. Within the entire Llanos de Moxos, there is one national protected area (Estación Biológica del Beni, of 135,200 ha), municipal reserves (1,222,400 ha total), and private protected areas (106,000 ha total). These municipal reserves were declared on top of private land claims and some are overlapping. Also superimposed on these areas are 6,920,200 ha of RAMSAR sites that confer international recognition to the areas as important waterbird habitats, but do not currently translate into any regulation on the ground [34].

Barba Azul Nature Reserve is an agglomeration of multiple private parcels now under a single management unit. The 11,000-ha reserve was created by Asociación Armonía in 2008 with support from American Bird Conservancy, the International Conservation Fund of Canada, IUCN Netherlands, Rainforest Trust, US Fish and Wildlife Service, and the World Land Trust. The study site is located on the Rio Omi in the extremely flat floodplain between the Mamoré and the Beni Rivers, which flow North through Bolivia to the main trunk of the Amazon (Figure 1). The reserve protects wetlands along the Rio Omi, small forest islands of anthropic origin, naturally fragmented gallery forests, and cerrado-like savanna.

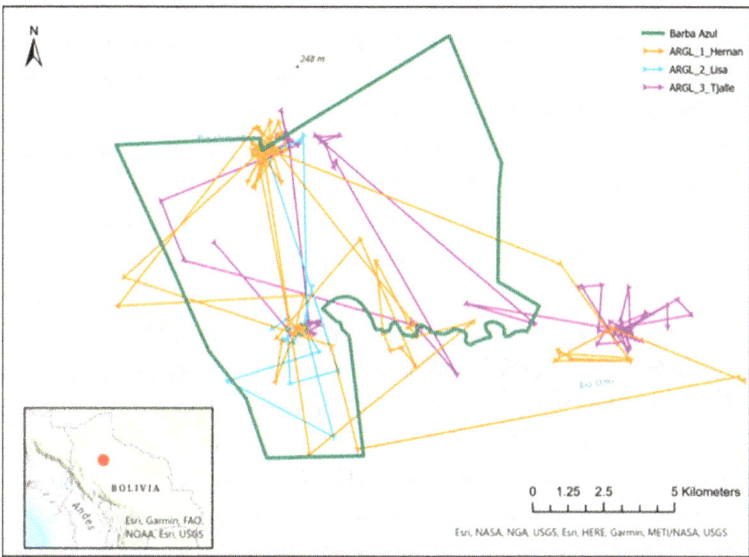

Figure 1. Map of Barba Azul Nature Reserve (inside green boundary) on the Rio Omi, with tracks of 3 Blue-throated macaws in and adjacent to reserve. Argos LC Classes A-3 with DAF filtering. Nicknames of birds used within Movebank and "Animal Tracker" app are appended to codes. Created in ARCGIS Pro 2.7.

There are 6000 ha dedicated to strict conservation use and the other 5000 ha are managed for low-impact cattle ranching, because Bolivia's Función Economico Social (FES) policy ties land titles to a requirement for remunerative agricultural production. A description of the Llanos de Moxos wetland/savanna habitat mosaic, specific characteristics of the reserve's habitats and its history of anthropogenic use, as well as the detrimental local effects of cattle ranching on the Blue-Throated Macaws' main feeding tree, *Atalaea principeps*, have been described elsewhere [22,23,35–37].

Groups of up to 155 BTMs have been counted at Barba Azul and annual counts have showed slow increases in their numbers [22,27]. The birds are almost exclusively present in the non-breeding season, in contrast to the southern population, where they appear to be resident year-round. The departure of BTMs from Barba Azul just before the breeding season suggests a regular migratory movement between this dry season congregation site and unknown breeding grounds. The impulse to migrate from Barba Azul, we suspect, is due to the local rarity of *Mauritia* palms, a known nesting tree, so long as adequate feeding resources are also available in the vicinity [27].

The flocks are present in predictable numbers and locations around the reserve every dry season, and we suspect that these are the same birds that return each year [27]. The BTM is known to use nest boxes in the southern population [22], but nest boxes in Barba Azul have attracted only one pair of Blue-throated Macaws in 2019 and have yet to result in successful reproduction [27].

Wet season surveys conducted by Asociación Armonía to find breeding BTMs in the northwestern population in 2017 and 2020 have located 15 active nests so far [27]. Whether these are the same sites used by the birds using the Barba Azul is still uncertain. The remoteness of the region, the difficulty of access during wet-season flooding, and private ownership concerns are all factors that have hindered locating the breeding grounds of Barba Azul birds.

2.3. Attachment Tests

After approaching multiple institutions unsuccessfully, we were finally allowed to conduct fit and durability tests of satellite transmitters on 5 captive BTMs at the Loro Parque Foundation (Tenerife, Spain). All birds were males housed together in a group. We tested 3 models of dummy collars (made by Ecotone (Gdynia, Poland): n = 1 unit; Telenax (Queretaro, Mexico): n = 3 units and Geotrak (Raleigh, NC USA): n = 1 unit) and 2 types of dummy backpacks (Ecotone n = 1 and Microwave Telemetry, Inc. (Columbia, MD USA): n = 1). Outer shells of the Ecotone and Microwave Telemetry, Inc. units are made of hardened plastic, as were collars used on these units. Telenax and Geotrak collars were made of metal, with rubber coverings. MTI, Geotrak, and 2 Telenax units had external antennae, while the Ecotone units and 1 Telenax unit did not. Backpack harnesses were custom fit with 1/3" width Teflon ribbon using a customizable protocol used with Orinoco geese [38]. One Ecotone backpack unit was painted blue to see if camouflage would reduce the wearer's attacks on the unit. Super glue was added as reinforcement on all knots and attachment points. All harnessing was conducted by Loro Parque Foundation caretakers and LCD. We video-taped the birds' reactions to wearing some of the units (Supplemental Video S1) and allowed Loro Parque Foundation care-takers to determine timing for all removals.

2.4. Capture of Wild BTMs

We assessed the options for trapping over three different dry seasons (2017–2019). Given the importance of this area for large numbers of this Critically Endangered species, our goal was to minimize disruption of important roosting, feeding, and tourism sites. As a result, we aimed to capture a small number of birds at a time in isolated locations. The methods used were approved under University of Florida Gainesville IACUC study permit #201709973. Permits for working in Barba Azul were obtained through the Museum of Natural History Alcide d'Orbigny. The specific method used for final captures is not reported here, out of conservation concern. Ultimately, we trapped four birds at a forest patch used by small groups of birds, over about 1 week, just before the start of the breeding season in August–September 2019. Although all birds were trapped at the same site, the movement data did not suggest that any of the collared birds were traveling in pairs or as parents and offspring.

2.5. *Tagging*

After tests on captive birds proved the suitability of Geotrak Argos units (see Results), we attached active Geotrak satellite parrot PTT collars to 3 of the 4 birds captured between 31 August and 9 September 2019 (see Supplementary Materials Table S1). Each bird was calmed by putting a breathable cotton hood over the head with a drawstring opening that allowed it to breathe unimpeded; legs and beak were immobilized using black electrical tape. All birds were measured and weighed prior to attaching a tag (Supplemental Table S1). All 3 tagged birds' body mass exceeded the minimum required for the transmitter weight to meet the standard limit in avian telemetry studies of <5% body weight [39]. Collars were produced with adjustable neck bands, so were fitted to each bird, with a small bolt permanently inserted and glued at the appropriate neckband hole. We noted start and end times of all procedures, periodically assessed the bird's condition, took photos and video of the procedure, photographed the feather and face patterns, and recorded their behavior and flight ability upon release (Supplemental Video S2). All movement data from within Barba Azul and auxiliary data are archived in the Movebank Data Repository [32].

2.6. *Programming and Location Accuracy*

The Geotrak parrot collar is a battery-operated unit, with no option to have a solar panel or external charger to recharge the internal battery. We therefore had to choose whether to prioritize an extended period of use or higher quality data, with more frequent location fixes. Not knowing how long the units would survive on the birds (especially fearing they might quickly succeed at detaching them), we chose to prioritize collecting higher quality data. The units were therefore programmed by the manufacturer so that about 3 months of data would be collected. A duty cycle of up to 5 h on, followed by 2 days off was programmed, dependent on satellite availability. We assumed from previous observations at Barba Azul that migration to the breeding grounds would occur in October, so that the units should have given us approximately half their data from Barba Azul and half from the breeding grounds. The units provide Doppler based locations calculated by the Argos system, which vary between 500 m and 5 km accuracy depending on Location Class (LC), with LC B being the lowest accuracy and LC 3 the highest accuracy [40,41].

2.7. *Analyses*

All data received from the ARGOS satellite system were filtered with the Kalman filter and uploaded in a live feed to Movebank.org for analysis and archival purposes. We inspected data to determine when tags were likely no longer on a moving animal; these were determined due to (a) failure to travel to night roosts, and (b) abrupt loss of reporting regularity and signal quality (low LC classes only). We performed three types of analyses on the data: (1) For all birds, we calculated "Home Range" estimates for the full dataset, (2) we calculated home range estimates for subsets of data from the dry season and wet season, and (3) we plotted time from capture against accumulated distance traveled by the 2 birds that migrated out of Barba Azul. Distances traveled within dry vs. wet season areas and on migration were calculated by choosing one "Best of Day" point that was the highest LC point provided on that day (either an LC 3 or 2), and applying the Haversine Great Circle distance between daily points. We report on maximum distances between daily points at dry vs. wet sites, and accumulated distances traveled throughout the study for the two birds that undertook an outbound migration.

Home Range Estimation: For the purposes of home range analysis, we retained Argos Location classes 3 through A, and then applied a Douglas Filter (available through

Movebank.org filter options) that removed locations with unrealistic distance and turning angles between 3 points, with a minimum turning angle of 35 degrees and 50 km/h maximum flight speed [42]. For each bird, we calculated 95% Autocorrelated Kernel Density Estimates (AKDE) to represent overall area (in hectares) used by each bird, both for the full dataset, and the subsets during wet season (breeding period) and dry season while at Barba Azul. From the Barba Azul and wet season data, we also present a "core home range" represented by the 70% AKDE at both dry and wet season locations. All AKDE analyses were performed using the R program "ctmm" [43–45]. Within the "ctmm" program, we used the "ctmm.select" procedure to choose the utilization distribution that best fit the data based on AIC comparisons. For both ARGL1 and ARGL3, the selection procedure based on data in and around Barba Azul and for the complete dataset resulted in selecting a model with correlated positions but uncorrelated velocities, or "OU" (for Ornstein–Uhlenbeck) anisotropic utilization distributions [46,47]. For the wet season data, the selection procedure resulted in selecting an "OU" anisotropic model for ARGL3, but an "IID" anisotropic utilization distribution that assumes no correlated locations or velocities [47]. In the Llanos de Moxos habitats, all these range estimates include large areas of what we consider "flyover" space (open savannas, which are not used by macaws), between heavily-used forest patches. While these calculations may be crude estimates of space of importance to the birds, from the perspective of reserve design and management, the analysis is helpful to calculate a rough areal extent required for conservation of the species in situ. The choice of these methods also allows comparisons with similar datasets on other wild macaw species that used comparable methods.

2.8. Wet Season Nest Surveys

After tracking BTMs that left Barba Azul, nest site searches were undertaken around the Department of Beni by Asociación Armonía staff in the following wet season to assess if BTMs remained in the areas indicated by GPS locations. In February and March of 2020 (after all transmissions had stopped), T.B. and other Asociación Armonía staff undertook horseback surveys through large areas of the northern Beni to look for nesting sites there and further afield [27,48]. At the same time, landowners in the region were contacted to discuss the ongoing tracking program and conservation efforts of Armonía.

3. Results

3.1. PTT Dummy Testing

Observations (and photos plus video recordings) documented how captive BTMs at Loro Parque Foundation treated different models of transmitters and attachment methods, and, most importantly, how they managed to render the units inoperable. Both the Ecotone unit placed as a collar and the one placed as a backpack were destroyed in minutes by piercing the casing and breaking off large chunks of the plastic housing (Figure 2b). The MTI backpack model fared somewhat better, but Teflon was quickly removed at the attachment points so that the unit then sat high near the neck rather than centered on the back, and then the antennae and casing were both opened (Supplemental Video S1). With the metal collars made by Telenax and Geotrak, the material and rounded shape of the units did inhibit their immediate destruction, especially of the casings. However, only one of the four metal collars, the Geotrak unit, was allowed to remain on a Loro Parque BTM for more than eight days by Loro Parque caretakers. Antennae on two Telenax were quickly chewed, exposing wires, so those units were removed immediately. The remaining Telenax collar was found after eight days to have been causing cuts on the bird's neck after the wearer chewed off all the protective rubber on the neckband, exposing the end of a metal seam behind the nut (Figure 3d). Somewhat surprisingly, the long straight antenna that extends beyond the Geotrak unit, up behind the head, was not a focus of attention, thus it remained intact. According to M. Weinzetll, curator at Loro Parque, the birds focused attention, especially early on, on breaking the metal collar on the Geotrak unit, but they did not succeed during the test period. Some, but not all, of the protected rubber shielding was removed at the sides. The outer case was deformed, but never pierced (Figure 3e,f). The single Geotrak collar unit tested remained on its captive BTM for 27 days, at which point it was also removed by caretakers. At the time of removal, caretakers reported the bird was tolerating the unit well, and the unit did not seem to limit flying, feeding, or self-grooming. They further determined that the unit caused two small feather cysts, but no significant injury [49]. A comparison of the case of an untested unit (upper) and the tested unit (lower) are shown in Figure 3f. We deemed the Geotrak model to be the only suitable unit for use on wild birds.

Figure 2. Testing dummy units at Loro Parque (Tenerife, Spain): (**a**) attachment of a Telenex collar; (**b**) captive BTM opening the Ecotone unit set as a collar. See also Supplemental Video S1 for video of a testing episode with the MTI backpack unit.

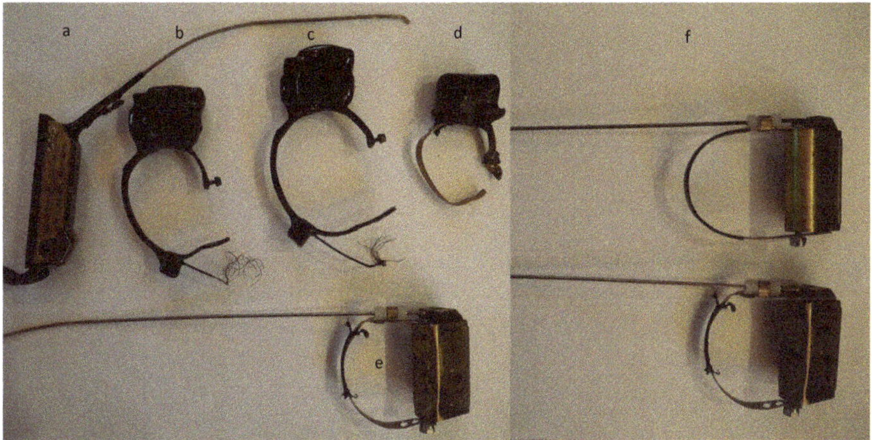

Figure 3. Units retrieved from tests on captive Loro Parque BTMs: (**a**) MTI set as backpack; (**b–d**) Telenax collars; (**e**) Geotrak parrot collar; (**f**) comparison of an untested Geotrak collar (upper) and the collar tested for 27 days on a captive BTM (lower).

3.2. Satellite PTT Performance

We received a total of 655 Argos locations (LC Class B or better, excluding duplicate timestamps) over the course of the study. On average, duty cycles returned locations for 3.1 (±1.6 SD) hours, then cycled off for 2.2 (±0.6) days. Data quality was high, with 54% of the points in the 2 highest Argos Locations Classes (LC 2 and 3). After truncating and filtering data as described above, we obtained 266 locations spanning 117.1 days from ARGL1 (31 August 2019 to 26 December 2019) and 218 locations over 79.5 days from ARGL3 (9 September to 28 November 2019). Subsets of the data used in home range analyses included: (1) 63 points for ARGL1, and 23 points for ARGL3 within the vicinity of Barba Azul; and, (2) 115 points for ARGL1, and 131 points for ARGL3 on purported breeding (wet season) grounds. We only received 24 usable locations over 6.6 days from ARGL2 (1 September–8 September 2019). This collar continued transmitting from a stationary point within the reserve from 10 September until 20 November. Though the area was extensively searched on foot several times (by T.B.), neither the collar, feathers, nor a carcass were found, suggesting that the collar was removed by the bird or its family members. Results from ARGL2 were of such limited duration that they are only considered with respect to the bird's movements relative to the other tracked birds in the earliest days of the study.

3.3. Range Size, Movements, and Observations around Barba Azul

The tracking data collected while the three BTMs were in and near to Barba Azul showed them moving independently while using similar areas of Barba Azul known to be popular with flocks, including the three largest forest islands and two known nighttime roosts (Figure 1). An animation of the data based on interpolated timestamps demonstrated that, while the three tagged macaws used similar areas and often roosted together at night, they mostly foraged separately, suggesting that they belong in distinct family groups (Supplemental Video S3). In addition, birds used one previously unknown daytime feeding area outside the reserve near the Rio Omi and one previously unknown nighttime roost. While in the Barba Azul area, ARGL1 moved on average 4.7 km between consecutive days; ARGL2 3.4 km, and ARGL3 2.7 km. For the two birds with adequate data to calculate AKDE home range size, the overall area of use was similar at 22,593 ha for ARGL1 (see Table 1 for 95% confidence limits)) vs. 27,066 ha for ARGL3, while core area of use (70% AKDE) was 8415 ha for ARGL1 and 12,551 ha for ARGL3. (Also see Supplemental Figure S1 for a diagram of 95% Home Range limits in the Barba Azul region).

Table 1. Home Range Estimates (with 95% CI in parentheses) of BTMs in total area of use, dry season (Barba Azul), and wet season locations by AKDE methods in "ctmm" program.

	Total 95% AKDE (ha)	Wet 95% AKDE (ha)	Dry (Barba Azul) 95% AKDE (ha)	Core Wet 70% AKDE (ha)	Core Dry 70% AKDE (ha)
ARGL1	325,851 (221,430–450,124)	4587 (3551–5773)	22,593 (14,723–32,122)	1292 (1000–1620)	8415 (5484–11,963)
ARGL3	118,970 (79,393–166,441)	2068 (1729–2436)	27,066 (5484–11,963)	506 (423–597)	12,551 (6078–21,332)

During the nearly week-long tagging period at Barba Azul in 2019, observers (L.C.D., L.C., L.F.) twice spotted BTMs carrying transmitters, both in the immediate vicinity of where they were tagged, and also while birds were flying together in the evening over the "bajio" drinking areas. Transmitters appeared to be well-fitting, not limiting the birds' flying, and with antennae intact. Though individual recognition was impossible in these instances, these sightings, as well as the tracking data collected, indicated to us that the experience of being trapped and tagged probably did not seriously alter the birds' use of the areas around the tagging site, nor change their associations with conspecifics.

On 13 August 2021, a BTM was spotted and photographed (by park ranger Miguel Martinez-Diaz) still wearing a collar nearly two years after it was first deployed (Figure 4).

Figure 4. Pair of BTMs photographed in Barba Azul 13 August 2021, with one still carrying a Geotrak collar (right). Rted with permission by Miguel Martinez-Diaz.

3.4. Migration, and Breeding Season Ranges and Movements

Two of the tagged birds (ARGL1 and ARGL3) migrated away from Barba Azul on the same date (27 September 2019), but diverged in their paths by about 90°, and did not interact again during the life of their PTT's. On inspection, ARGL1 was considered to be migrating, i.e., not yet settled, for 24.5 days and ARGL3 for 13.4 days. Distances moved on migration were, on average, 7.5 km/day for ARGL1 and 4.9 km/day for ARGL3. Maximum distances measured between consecutive points (2 day spread) while on migration were 59 km/day for ARGL1 and 18 km/day for ARGL3. Accumulated daily distances moved to the end of transmissions are displayed in Figure 5.

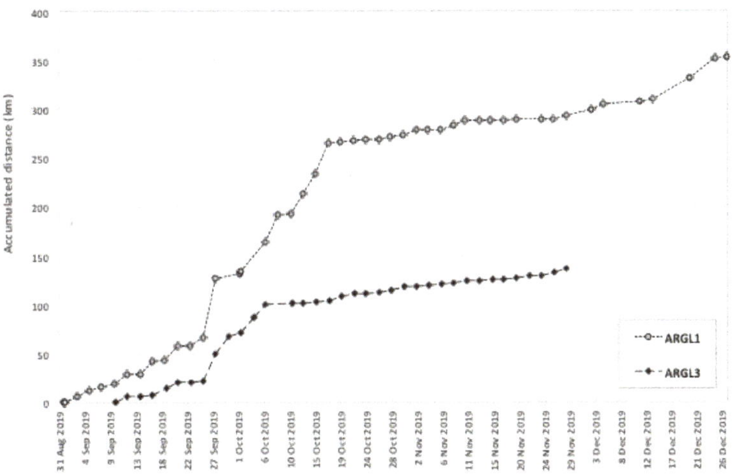

Figure 5. Accumulated daily distances traveled by two migrant BTMs tagged at Barba Azul.

The two migrant birds eventually settled in distinct areas that were both ~50–100 km from Barba Azul, still within the Llanos de Moxos region, and still within the boundaries of the northwestern subpopulation's distribution. ARGL3 used the same general region through to the end of transmissions. ARGL1, however, used two different spots in the same period, these being approximately 30 km apart. While at wet season locations, average daily movements (away from few regular nighttime roosts) were smaller than while at Barba

Azul, at 3.7 km average for ARGL1 and 1.2 km for ARGL3. Similarly, wet season range sizes were smaller than during the dry season. ARGL1 used a 95% AKDE of 4587 ha and a core 70% AKDE of 1292 ha. ARGL3 used a 95% AKDE of 2068 ha and a core 70% AKDE area of 506 ha (Table 1 and Supplemental Figure S2).

3.5. Wet Season Ground–Truthing Surveys

Results, observations, and information on community outreach obtained during the nest surveys undertaken by Armonía are continually reported in regular Armonía newsletters and elsewhere [22,27,48]. Despite gaining access to both final locations where we tracked ARGL1 and ARGL3, we found no birds still wearing tags there. Habitat at the two sites both had a greater availability of *Mauritia flexuosa* palms, which are not present at Barba Azul. The team did succeed in finding 15 BTM nests across three breeding areas, and observed between 93 and 104 BTMs in a survey area of approximately 30,700 ha. In a previously discovered roosting site, we counted 56 BTMs roosting in a small *Motacú*-dominated forest island. Based on the location data of the 15 known nests, the average distance from a nest to the second nearest known nest is 4.48 km. The total distance between the southernmost nest and the northernmost nest is 63.25 km. *Motacú* forests are found on average at a distance of 1.35 km from a nest [27].

4. Discussion

Prior to this study, nesting sites of the vitally important sub-population of Blue-throated macaws from Barba Azul were unknown. In general, tracking birds *to* their breeding areas is a far less common practice than tracking *from* their breeding sites, in part because it offers fewer opportunities to trap in sites that are visited regularly. For this reason, and also because the species is Critically Endangered and the smallest macaw species tracked to date, the work required multiple phases of testing to implement tracking with wild birds in the safest means possible. Ultimately, the methods used allowed us to capture and track two wild BTMs to two breeding areas where nests were later found. The migration data also inspired the initiation of conversations with the relevant landowners about best practices in ranching for macaw conservation [37,48]. We believe that follow-up efforts can now be more focused and effective, and lead to improved chances of our ultimate goal of better protecting the birds of Barba Azul throughout the year, monitoring their reproductive success, and improving the chances for the species' long-term conservation at Barba Azul and more widely in the northern Llanos de Moxos.

4.1. Tracking Results

The two tagged birds that continued to transmit through migration (ARGL1 and ARGL3) demonstrated the durability and effectiveness of the Geotrak units for tracking wild Blue-throated Macaws for many months. The 2021 rediscovery of a wild bird still carrying a now-defunct Geotrak collar also demonstrates that both long and short-term deployments should be feasible with the species. Although the date of outbound migration (27 September) was earlier than expected from previous Barba Azul census data [22,50], it was identical for our two migrant birds. Prior to migrating, birds only occasionally left the immediate area protected by Barba Azul, confirming the critical location and size of the reserve. However, our data are based on a small window of time of only a few weeks, thus it is likely that a longer tracking period of these and other birds within Barba Azul would reveal a wider area of use. Similarly, it is highly likely that a longer tracking period for ARGL1 and ARGL2 could have revealed more distant areas of use beyond the reserve. Yet, relative to the size of protected areas in the region, overall range sizes of 95% AKDE using all locations (from both dry and wet season data and while on migration) were quite large (at 325,851 ha for ARGL1 and 118,970ha for ARGL3 (Table 1)). Still, our home range estimates should clearly be considered minimum estimates for both seasons, and conservation needs assessed with this caveat. A future effort focused on tracking birds for a complete season of residence inside Barba Azul is a priority for future work.

Both birds tracked on migration eventually settled into two distinct areas ~50–100 km from Barba Azul, although after leaving in divergent directions. This result, combined with the subsequent nest discoveries, suggests that BTM breeding sites are probably widely scattered across the Llanos de Moxos landscape. Their scattered distribution could make conservation efforts more difficult, but is important for understanding the ecoregion-level needs for saving the species from extinction. The choice to program for frequent location fixes proved very useful to learn the specific timing, direction, and length of migration periods; moreover, it helped indicate specific areas of the Beni where the birds visited prior to settling down. Future work may require longer, less detailed programming.

These data have and will continue to guide additional overflights and ground-truthing searches for other nesting sites in nearby locations. Adding a banding program of young at nest sites discovered here could also help definitively link reproductive outputs to new recruits at Barba Azul.

4.2. Distances Traveled

The distance an individual macaw can travel has different implications. Besides indicating a birds' physical capacity to cover a given distance, it is also important from a conservation viewpoint. Can a given reserve protect the whole annual cycle for large and critically threatened birds such as Blue-throated Macaws? Does distance traveled tell us anything about the likelihood that Beni's different subpopulations of BTMs are really isolated? The measurements of wild individual macaw movements range are scanty, but comparison between what we learned in this study and a few other known examples are useful.

We found that the BTMs' daily locations averaged about 1 to 8 km apart across all seasons, with the largest distances recorded during migration periods where birds moved up to 59 km between days. The magnitude of these movements is hard to compare with other species as previous studies have used observations or calculated movements as distances among multiple points per day. For example, the last wild Spix's Macaw, *Cyanopsitta spixi*, reportedly moved up to 60 km from its night roost to its feeding place and back during the dry season [51]. Tracked Red-fronted Macaws (*Ara rubrogenys*) in Bolivia moved at least 9 to 28 km per day, and Blue-and-yellow Macaws (*Ara ararauna*) and Scarlet Macaws (*Ara macao*) in southeastern Peru moved on average at least 6 to 13 km per day [30,52]. In addition, one individual *A. ararauna* moved back and forth through the same 160 km path in one week, totaling a 320-km displacement [41].

The BTMs studied here migrated distances of approximately 50 to 100 km from their wintering ranges to their presumed breeding areas (Figure 5). These movements are similar to the majority of seasonal migrations documented in macaws to date. Blue-and-yellow Macaws and Scarlet Macaws moved an average of 80 to 112 km from their nests post breeding in southeastern Peru and adjacent areas of Bolivia [30] A single Red-fronted Macaw in Bolivia moved through areas up to 77 km from its wintering range before returning to the wintering range the next season [52]. In the Brazilian Pantanal, an ecosystem similar to the Llanos del Moxos, UHF studies documented seasonal movements of Hyacinth Macaws (*Anodorhynchus hyacinthinus*) of 36 to 50 km [29,53,54]. Migratory Great Green Macaws (*Ara ambiguous*) in Costa Rica have been found 35–40 km from their nests during the non-breeding season [55].

Similarly, comparisons of 95% AKDE home range estimates are considerably smaller for BTMs both in the dry and in the wet season compared to home range estimates from southeastern Peru [30]. This means that whereas *A. macao* in Tambopata averaged home ranges of 11,900 ha in the breeding season and *A. ararauna* 16,700 [30], our two BTMs ranged only over approximately 4500 and 2000 ha as estimated from their wet season data using the AKDE procedure (Supplemental Table S2). In the non-breeding season, the differences were even more notable, with our 2 tagged BTMs ranging over 23,000 and 27,000 ha respectively, and *A. macao* averaging 206,500 and *A. ararauna* 191,000 [30].

Differences in distances moved may be due to body size and physiology, but may also reflect the density and seasonality of food available in Barba Azul and the Llanos de Moxos compared to other sites. Spix's macaws inhabit a semi-arid region where food abundances should fluctuate widely and they may need to move long distances to dispersed food patches. In Barba Azul, some *Motacú* patches are nearly monodominant, possibly due to past human influences, and can produce fruit nearly all year [25,35], so that the birds there do need not to move far for feeding. In contrast, in Tambopata, Peru, food abundance shows large seasonal fluctuations [56] and feeding trees are often scattered in a more diverse forest matrix. The flight range of BTMs that left Barba Azul suggests that the species is able to fly similar distances as other macaws. However, at least for BTMs using Barba Azul, there seems to be scant evidence from observations, or from our study, for partial migration, use of intermediate stopover sites, or more nomadic behavior. The energetic reserves needed for their longest movements, even while present in high densities, were obtained while in Barba Azul, reinforcing the seasonal importance of the reserve for this BTM population.

4.3. Challenges Working with the Psittacidae

Although tracking BTMs required multiple field seasons of testing equipment and capture methods, ultimately, we did manage to determine methods that allowed us to capture and track three wild BTMs, including to two likely breeding sites. The work shows the potential to use satellite telemetry for biological and conservation questions with this species and within the Psittacidae. While not all our methods or results are reported here, out of an abundance of conservation concern, we hope that sharing our experiences with testing, using satellite telemetry units on macaws, and analyzing habitat features can help other researchers faced with similar challenges, and especially those with conservation questions best answered through the use of satellite telemetry.

Despite our success, several practical and logistical features of studying any parrot species makes the use of satellite telemetry on this family a complicated task. One of the greatest difficulties is the fact that the ideal (and necessarily robust) transmitter design appears to be quite species-specific. Although with some other parrot species, researchers have had good results with backpack models or hardened plastic materials set as collars, our birds destroyed all such models in minutes, piercing antennae and casings through any seams or corners they could find. The successful units, parrot collars by Geotrack, are made by hand by the company's founder, using hard-won skills known best to himself [57] making them quite unique (and also expensive), and therefore hard to deploy on a larger scale. Manufacturers are understandably loathe to invest much time researching and improving any "weakness" in a transmitter model to be used on parrots when a model will see very limited use in small studies. This reality gives researchers limited, mostly expensive options. Moreover, we found that many institutions with captive parrots were unwilling to allow testing with their captive animals; we had to find an institution outside the Americas with which to collaborate on the critical step of first testing out dummy transmitter designs. We also still lack a truly comprehensive understanding of the potential effects of attaching units to parrots in different configurations, as nearly all studies on either captive or wild animals, including our own, involve extremely small sample sizes. Ideally, we would also quickly remove any defunct unit from a wild bird, but we are still seeking a means to do so without a large re-trapping effort. We would welcome suggestions for reliable drop-off mechanisms that could be added to collar designs without adding excessive weight. Overall, our experience with this species has taught us that, despite the great advances in the technology of satellite telemetry in recent years, and with the many projects still needing to be done to study parrots of great conservation concern, the Psittacidae family is still relatively understudied and requiring considerably more innovation for their efficient and effective study. Therefore, as a final recommendation for the future, we urge fellow researchers and tinkerers to consider expending more effort with this particularly underserved group of birds.

Supplementary Materials: The following are available online at https://www.mdpi.com/article/10.3390/d13110564/s1, Figure S1: Map of Barba Azul Nature Reserve, with tracks of 3 BTM;s and 95% Home Range Estimates for 2 BTM's (ARGL1 and ARGL3) with sufficient data for home range analyses., Figure S2: Map of 95% Home Range Estimates for 2 BTMs (ARGL1_Hernan on left and ARGL3_Tjalle on right) on wet season (breeding season) sites. Table S1: Argos Id's and morphometric measures of captured wild Blue-throated Macaws from Barba Azul., Video S1: Video of captive BTM tagged with an MTI dummy transmitter at Loro Parque, Tenerife, Spain. Video S2: Video of release of a wild BTM with a Geotrak transmitter. Video S3: Animation of 3 BTM movements around Barba Azul, using the R program "MoveVis".

Author Contributions: Conceptualization, L.C.D., T.B., L.C., P.d.T.Z.A., D.J.B., S.K.H., R.W.S.-A., and A.B.H.; methodology, L.C.D., T.B., L.C., P.d.T.Z.A., L.F., and D.J.B.; formal analysis, L.C.D. and L.C.; investigation, L.C.D., T.B., L.C., P.d.T.Z.A., and L.F.; data curation, L.C.D., L.C., and L.F.; writing—original draft preparation, L.C.D., T.B., L.C., P.d.T.Z.A., and D.J.B.; writing—review and editing, L.C.D., T.B., L.C., P.d.T.Z.A., D.J.B., S.K.H., R.W.S.-A., and A.B.H.; project administration, T.B., S.K.H., R.W.S.-A., and A.B.H. All authors have read and agreed to the published version of the manuscript.

Funding: This research was funded by the Mohammed bin Zayed Foundation, project number 180517998, the American Bird Conservancy, and anonymous private donors.

Institutional Review Board Statement: The study was conducted according to the guidelines of the Declaration of Helsinki, and approved by the Animal Care and Use Committee of the University of Florida, Gainesville (protocol #201709973). It was also approved by the Ministerio de Medio Ambiente y Agua (Bolivia): MMAYA/VMABCCGDF/DGAP/MEG No 0228/2017.

Data Availability Statement: Public data is redacted after migration due to poaching pressure and concern for private landowners. Requests for data sharing can be directed to Asociación Armonía as noted in the text.

Acknowledgments: The authors wish to thank Loro Parque Foundation (esp. David Waugh and Marcia Weinzetll) for assisting us with testing units on their captive Blue-throated macaws. In addition, we thank Keith Le Sage of Geotrak, Sarah Davidson of Movebank, and Torbjorn Haugaasen of the Norwegian University of Life Sciences for advice and aid to the project. For field assistance, we thank Carlos, Yuri, Hernan, and John Terborgh. Luciano Naka helped in reading and editing an early draft of the manuscript. We also want to thank the Museum of Natural History Alcide d'Orbigny for helping us to obtain government permit to conduct this research.

Conflicts of Interest: The authors declare no conflict of interest. The funders had no role in the design of the study; in the collection, analyses, or interpretation of data; in the writing of the manuscript, or in the decision to publish the results.

References

1. Bridge, E.S.; Thorup, K.; Bowlin, M.S.; Chilson, P.B.; Diehl, R.H.; Fléron, R.W.; Hartl, P.; Kays, R.; Kelly, J.F.; Robinson, W.D.; et al. Technology on the move: Recent and forthcoming innovations for tracking migratory birds. *BioScience* **2011**, *61*, 689–698. [CrossRef]
2. Crossin, G.T.; Cooke, S.J.; Goldbogen, J.A.; Phillips, R.A. Tracking fitness in marine vertebrates: Current knowledge and opportunities for future research. *Mar. Ecol. Prog. Ser.* **2014**, *496*, 1–17. [CrossRef]
3. Guilford, T.; Åkesson, S.; Gagliardo, A.; Holland, R.A.; Mouritsen, H.; Muheim, R.; Wiltschko, R.; Wiltschko, W.; Bingman, V.P. Migratory navigation in birds: New opportunities in an era of fast-developing tracking technology. *J. Exp. Biol.* **2011**, *214*, 3705–3712. [CrossRef]
4. Robinson, W.D.; Bowlin, M.S.; Bisson, I.; Shamoun-Baranes, J.; Thorup, K.; Diehl, R.H.; Kunz, T.H.; Mabey, S.; Winkler, D.W. Integrating concepts and technologies to advance the study of bird migration. *Front. Ecol. Environ.* **2010**, *8*, 354–361. [CrossRef]
5. Wikelski, M.; Kays, R.W.; Kasdin, N.J.; Thorup, K.; Smith, J.A.; Swenson, G.W., Jr. Going wild: What a global small-animal tracking system could do for experimental biologists. *J. Exp. Biol.* **2007**, *210*, 181–186. [CrossRef] [PubMed]
6. Skewgar, E.; Boersma, P.D.; Simeone, A. Winter migration of Magellanic Penguins (*Spheniscus magellanicus*) along the southeastern Pacific. *Waterbirds* **2014**, *37*, 203–209. [CrossRef]
7. Martell, M.S.; Henny, C.J.; Nye, P.E.; Solensky, M.J. Fall migration routes, timing, and wintering sites of North American ospreys as determined by satellite telemetry. *Condor* **2001**, *103*, 715–724. [CrossRef]
8. McIntyre, C.L. Quantifying Sources of Mortality and Wintering Ranges of Golden Eagles from Interior Alaska Using Banding and Satellite Tracking. *J. Raptor Res.* **2012**, *46*, 129–134. [CrossRef]

9. Davenport, L.C.; Goodenough, K.; Haugaasen, T. Birds of two oceans? Trans-Andean and divergent migration of Black Skimmers (*Rynchops niger*) from the Peruvian Amazon. *PLoS ONE* **2016**, *11*, e0144994. [CrossRef] [PubMed]
10. Cappelle, J.; Iverson, S.A.; Takekawa, J.Y.; Newman, S.H.; Dodman, T.; Gaidet, N. Implementing telemetry on new species in remote areas: Recommendations from a large-scale satellite tracking study of African waterfowl. *Ostrich* **2011**, *82*, 17–26. [CrossRef]
11. Hill, J.M.; Sandercock, B.K.; Renfrew, R.B. Migration Patterns of Upland Sandpipers in the Western Hemisphere. *Front. Ecol. Evol.* **2019**, *7*, 426. [CrossRef]
12. Pedler, R.D.; Ribot, R.F.H.; Bennett, A.T.D. Extreme nomadism in desert waterbirds: Flights of the banded stilt. *Biol. Lett.* **2014**, *10*, 20140547. [CrossRef]
13. Lutcavage, M.E.; Brill, R.W.; Skomal, G.B.; Chase, B.C.; Howey, P.W. Results of pop-up satellite tagging of spawning size class fish in the Gulf of Maine: Do North Atlantic Bluefin tuna spawn in the mid-Atlantic? *Can. J. Fish. Aquat. Sci.* **1999**, *56*, 173–177. [CrossRef]
14. Pilcher, N.E.; Rodriguez-Zarate, C.J.; Antonopoulou, M.A.; Mateos-Molina, D.; Sekhar-Das, H.; Abdullah-Bugla, I. Combining laparoscopy and satellite tracking: Successful round-trip tracking of female green turtles from feeding areas to nesting grounds and back. *Glob. Ecol. Cons.* **2020**, *23*, e01169. [CrossRef]
15. Takekawa, J.Y.; Newman, S.H.; Xiao, X.; Prosser, D.J.; Spragens, K.A.; Palm, E.C.; Yan, B.; Li, T.; Lei, F.; Zhao, D.; et al. Migration of waterfowl in the East Asian flyway and spatial relationship to HPAI H5N1 outbreaks. *Avian Dis.* **2010**, *54* (Suppl. 1), 466–476. [CrossRef] [PubMed]
16. Chan, Y.C.; Tibbitts, T.L.; Lok, T.; Hassell, C.J.; Peng, H.B.; Ma, Z.; Zhang, Z.; Piersma, T. Filling knowledge gaps in a threatened shorebird flyway through satellite tracking. *J. Appl. Ecol.* **2019**, *56*, 2305–2315. [CrossRef]
17. Sick, H.; Teixeira, D.M.; Gonzaga, L.P. A nossa descoberta da pátria da arara *Anodorhynchus leari*. *Anais da Acadamia Bras Ciências* **1979**, *51*, 575–576.
18. Styles, F.G.; Skutch, A.F. *A Guide to the Birds of Costa Rica*; Comstock Publishing Associates: Ithaca, NY, USA, 1989.
19. Forshaw, J.M. *Parrots of the World*, 3rd ed.; Landsdowne Editions: Melbourne, Australia, 1989.
20. Jordan, O.C.; Munn, C.A. First observations of the Blue-throated Macaw in Bolivia. *Wilson Bull. Wilson Ornithol. Soc.* **1993**, *105*, 694–695.
21. Ortiz-von Halle, B. *Bird's-Eye View: Lessons from 50 Years of Bird Trade Regulation & Conservation in Amazon Countries*; TRAFFIC: Cambridge, UK, 2018.
22. Herzog, S.K.; Maillard, O.Z.; Boorsma, T.; Sanchez-Avila, G.; Garcia-Soliz, V.H.; Paca-Condori, A.; Vailez de Abajo, M.; Soria-Auza, W. First systematic sampling approach to estimate the global population size of the Critically Endangered Blue-throated Macaw (*Ara glaucogularis*). *Bird Conserv. Int.* **2021**, *31*, 293–311. [CrossRef]
23. Hordijk, I.; Meijer, F.; Nissen, E.; Boorsma, T.; Poorter, L. Cattle affect regeneration of the palm species *Attalea princeps* in a Bolivian forest-savannah mosaic. *Biotropica* **2019**, *51*, 28–38. [CrossRef]
24. Lombardo, U.; Veit, H. The origin of oriented lakes: Evidence from the Bolivian Amazon. *Geomorphology* **2014**, *204*, 502–509. [CrossRef]
25. Lombardo, U.; Iriarte, J.; Hilbert, L.; Ruiz-Pérez, J.; Capriles, J.M.; Veit, H. Early Holocene crop cultivation and landscape modification in Amazonia. *Nature* **2020**, *581*, 190–193. [CrossRef]
26. Berkunsky, I.; Daniele, G.; Kacoliris, F.P.; Díaz-Luque, J.A.; Silva Frias, C.P.; Aramburu, R.M.; Gilardi, J.D. Reproductive parameters in the critically endangered Blue-throated Macaw: Limits to the recovery of a parrot under intensive management. *PLoS ONE* **2014**, *9*, e99941. [CrossRef] [PubMed]
27. Boorsma, T.; Asociación Armonia, Santa Cruz, Bolivia. 2020; Unpublished data.
28. Kennedy, E.M.; Kemp, J.R.; Mosen, C.C.; Perry, G.L.W.; Dennis, T.E. GPS telemetry for parrots: A case study with the Kea (*Nestor notabilis*). *Auk Ornithol. Adv.* **2015**, *132*, 389–396. [CrossRef]
29. Antas, P.T.Z.; Carrara, L.A.; Yabe, R.S.; Ubaid, F.K.; Oliveira-Junior., S.B.; Vasques, E.R.; Ferreira, L.P. *A Arara-Azul na Reserva Particular do Patrimônio Natural SESC Pantanal*; The Hyacinth Macaw in the SESC Pantanal Natural Heritage Private Reserve; Conhecendo o Pantanal 6; SESC Serviço Social do Comércio, Departamento Nacional: Rio de Janeiro, Brazil, 2010; Available online: http://www.sescpantanal.com.br/arquivos/cadastro-itens/layout-6/arquivos/file-635877033528123910.pdf (accessed on 4 November 2021). (In Portuguese)
30. Brightsmith, D.J.; Boyd, J.; Hobson, E.A.; Randal, C.J. Satellite telemetry reveals complex migratory movement patterns of two large macaw species in the western Amazon basin. *Avian Conserv. Ecol.* **2021**, *16*, 14. [CrossRef]
31. Rycken, S.; Warren, K.S.; Yeap, L.; Jackson, B.; Riley, K.; Page, M.; Dawson, R.; Smith, K.; Mawson, P.R.; Shephard, J.M. Assessing Integration of Black Cockatoos Using Behavioral Change Point Analysis. *J. Wildl. Manag.* **2019**, *83*, 334–342. [CrossRef]
32. Davenport, L.C.; Boorsma, T.; Carrara, L.A.; Antas, P.; Faria, L.; Brightsmith, D.J.; Herzog, S.K.; Soria-Auza, R.W.; Hennessey, A.B. Data from: Satellite telemetry of Blue-throated macaws in Barba Azul Nature Reserve (Beni, Bolivia) reveals likely breeding areas. *Movebank Data Repos.* **2021**. [CrossRef]
33. Haase, R.; Beck, G. Structure and composition of savanna vegetation in northern Bolivia: A preliminary report. *Brittonia* **1989**, *41*, 80–100. [CrossRef]
34. Wittmann, F.; Householder, E.; de Oliveira, A.; Lopes, A.; Junk, W.J.; Piedade, M.T. Implementation of the Ramsar Convention on South American wetlands: An update. *Res. Rep. Biodivers. Stud.* **2015**, *4*, 47–58. [CrossRef]

35. Capriles, J.M.; Lombardo, U.; Maley, B.; Zuna, C.; Veit, H.; Kennett, D.J. Persistent Early to Middle Holocene tropical foraging in southwestern Amazonia. *Sci. Adv.* **2019**, *5*, eaav5449. [CrossRef]
36. Mayle, F.E.; Langstroth, R.P.; Fisher, R.A.; Meir, P. Long-term forest-savannah dynamics in the Bolivian Amazon: Implications for conservation. *Philos. Trans. R. Soc. B Biol. Sci.* **2007**, *362*, 291–307. [CrossRef] [PubMed]
37. Callaú, L.N.M.; Boorsma, T. *Guía Práctica Para Ganadería de Armonización: La Ganadería Sostenible Para el Beni*; Asociación Civil Armonía: Santa Cruz de la Sierra, Bolivia, 2019; Available online: http://armoniabolivia.org/wp-content/uploads/2020/01/Gu%C3%ADa-Ganader%C3%ADa-Sostenible-baja.pdf (accessed on 8 August 2020).
38. Davenport, L.C.; Nole, I.; Carlos, N. East with the Night: Longitudinal migration of the Orinoco Goose (*Neochen jubata*) between Manú National Park, Peru and the Llanos de Moxos, Bolivia. *PLoS ONE* **2012**, *7*, e46886. [CrossRef]
39. Barron, D.G.; Brawn, J.D.; Weatherhead, P.J. Meta analysis of transmitter effects on avian behaviour and ecology. *Methods Ecol. Evol.* **2010**, *1*, 180–187. [CrossRef]
40. CLS (Collecte Localisation Satellites). *Argos User's Manual*; Collecte Localisation Satellites: Ramonville Saint-Agne, France, 2011; 69p, Available online: http://www.argos-system.org/manual/ (accessed on 14 February 2021).
41. Boyd, J.D.; Brightsmith, D.J. Error properties of Argos satellite telemetry locations using Least Squares and Kalman Filtering. *PLoS ONE* **2013**, *8*, e63051. [CrossRef]
42. Douglas, D.C.; Weinzierl, R.C.; Davidson, S.; Kays, R.; Wikelski, M.; Bohrer, G. Moderating Argos location errors in animal tracking data. *Methods Ecol. Evol.* **2012**, *3*, 999–1007. [CrossRef]
43. Fleming, C.H.; Calabrese, J.M.; Mueller, T.; Olson, K.A.; Leimgruber, P.; Fagan, W.F. Non-Markovian maximum likelihood estimation of autocorrelated movement processes. *Methods Ecol. Evol.* **2014**, *5*, 462–472. [CrossRef]
44. Calabrese, J.M.; Fleming, C.H.; Gurarie, E. ctmm: An r package for analyzing animal relocation data as a continuous-time stochastic process. *Methods Ecol. Evol.* **2016**, *7*, 1124–1132. [CrossRef]
45. R Core Team. *R: A Language and Environment for Statistical Computing*; R Foundation for Statistical Computing: Vienna, Austria, 2020; Available online: https://www.R-project.org/ (accessed on 9 February 2021).
46. Uhlenbeck, G.E.; Ornstein, L.S. On the theory of the Brownian motion. *Phys. Rev.* **1930**, *36*, 823. [CrossRef]
47. Silva, I.; Fleming, C.H.; Noonan, M.J.; Alston, J.; Folta, C.; Fagan, W.F.; Calabrese, J.M. Autocorrelation-informed home range estimation: A review and practical guide. *EcoEvoRxiv* **2021**. preprint. Available online: https://ecoevorxiv.org/23wq7/ (accessed on 9 February 2021).
48. Asociación Civil Armonía. *Discoveries from 2020 Nest Search Expedition Guide Next Steps for Blue-Throated Macaw Research and Conservation 2020*; Asociación Civil Armonía: Santa Cruz de la Sierra, Bolivia, 2020; Available online: http://armoniabolivia.org/2020/04/28/discoveries-guide-next-steps-for-research-and-conservation (accessed on 9 February 2021).
49. Weinzetll, M.; (Loro Parque Foundation, Tenerife, Spain). Personal communication, 2018.
50. Asociación Civil Armonía. *Barba Azul Nature Reserve*; May 2020 Update Report; Asociación Civil Armonía: Santa Cruz de la Sierra, Bolivia, 2020; Available online: http://armoniabolivia.org/wp-content/uploads/2020/09/Barba-Azul-Nature-Reserve-May-2020-Update-Report.pdf (accessed on 9 February 2021).
51. Da-Ré, M.A.; (Centro de Economia Verde, Fundação Certi Florionopolis, Florianópolis, Brazil). Personal communication, 2020.
52. Meyer, C. Spatial Ecology and Conservation of the Endemic and Endangered Red-Fronted MACAW (*Ara Rubrogenys*) in the Bolivian Andes. Diploma Thesis, Georg-August University Göttingen, Göttingen, Germany, 2010. Available online: http://armoniabolivia.org/wp-content/uploads/2016/07/01_2010-Meyer-DiplomArbeit-GoettingenUniversity.pdf (accessed on 9 February 2021).
53. Guedes, N.M.R.; Seixas, G.H.F. Métodos para estudos de reprodução de Psitacídeos. In *Ecologia e Conservação de Psitacídeos no Brasil*; Melopsittacus Publicações Científicas: Belo Horizonte, Brasil, 2020; pp. 123–140.
54. Antas, P.; Fundação Pró-Natureza (Funatura), Brasilia, Brazil. 2020; Unpublished data.
55. Powell, G.; Wright, P.; Aleman, U.; Guindon, C.; Palminteri, S.; Bjork, R. *Research Findings and Conservation Recommendations for the Great Green Macaw (Ara Ambigua) in Costa Rica*; Centro Cientifico Tropica: San Jose, Costa Rica, 1999.
56. Brightsmith, D.J.; Hobson, E.A.; Martinez, G. Food availability and breeding season as predictors of geophagy in Amazonian parrots. *IBIS* **2018**, *160*, 112–129. [CrossRef]
57. Le Sage, K.; (Geotrak, Raleigh, NC, USA). Personal communication, 2018.

MDPI

Article

Burrowing Parrots *Cyanoliseus patagonus* as Long-Distance Seed Dispersers of Keystone Algarrobos, Genus *Prosopis*, in the Monte Desert

Guillermo Blanco [1,*], Pedro Romero-Vidal [2], Martina Carrete [2], Daniel Chamorro [3], Carolina Bravo [4], Fernando Hiraldo [5] and José L. Tella [5]

[1] Department of Evolutionary Ecology, Museo Nacional de Ciencias Naturales CSIC, 28006 Madrid, Spain
[2] Department of Physical, Chemical and Natural Systems, Universidad Pablo de Olavide, Carretera de Utrera, km 1, 41013 Sevilla, Spain; pedroromerovidal123@gmail.com (P.R.-V.); mcarrete@upo.es (M.C.)
[3] Departamento de Ciencias Ambientales, Universidad de Castilla-La Mancha, Av. Carlos III s/n, 45071 Toledo, Spain; daniel.chamorro.cobo@gmail.com
[4] Centre d'Etudes Biologiques de Chizé, UMR 7372, CNRS and La Rochelle Université, F-79360 Beauvoir-sur-Niort, France; carolina.bravo.parraga@gmail.com
[5] Department of Conservation Biology, Estación Biológica de Doñana CSIC, 41092 Sevilla, Spain; hiraldo@ebd.csic.es (F.H.); tella@ebd.csic.es (J.L.T.)
* Correspondence: g.blanco@csic.es

Abstract: Understanding of ecosystem structure and functioning requires detailed knowledge about plant–animal interactions, especially when keystone species are involved. The recent consideration of parrots as legitimate seed dispersers has widened the range of mechanisms influencing the life cycle of many plant species. We examined the interactions between the burrowing parrot *Cyanoliseus patagonus* and two dominant algarrobo trees (*Prosopis alba* and *Prosopis nigra*) in the Monte Desert, Argentina. We recorded the abundance and foraging behaviour of parrots; quantified the handling, consumption, wasting, and dispersal of ripe and unripe pods; and tested the viability of soft and hard ripe seeds wasted and transported by parrots. We found a high abundance of burrowing parrots. They predated on soft seeds from unripe pods while exclusively feeding upon pulp wrapping hard seeds from ripe pods. Frequent pod wasting beneath the plant or transport at a distance invariably implied the dispersal of multiple seeds in each event. Moreover, soft seeds retained viability after desiccation outside the mother plant, suggesting effective seed dispersal after partial pod predation due to a predator satiation effect. In about half of the foraging flocks, at least one parrot departed in flight with pods in its beak, with 10–34% of the flock components moving pods at distances averaging 238 m (*P. alba*) and 418 m (*P. nigra*). A snapshot sampling of faeces from livestock and wild mammals suggested a low frequency of seed dispersal by endozoochory and secondary dispersal by ants and dung beetles. The nomadic movements and long flights of burrowing parrots between breeding and foraging sites can lead to the dispersal of huge amounts of seeds across large areas that are sequentially exploited. Further research should evaluate the role of the burrowing parrot as a functionally unique species in the structure of the Monte Desert woods and the genetic structure of algarrobo species.

Keywords: algarrobo; drylands; High Monte; parrots; seed dispersal; soft seed viability; stomatochory

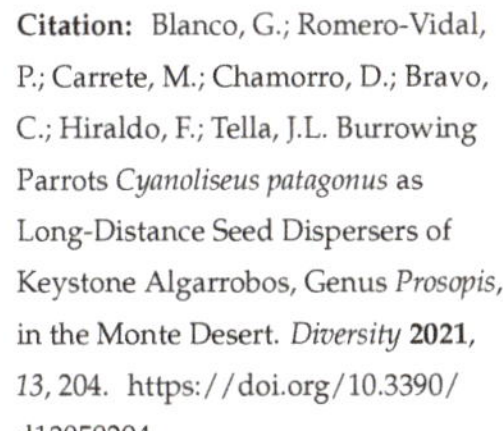

Citation: Blanco, G.; Romero-Vidal, P.; Carrete, M.; Chamorro, D.; Bravo, C.; Hiraldo, F.; Tella, J.L. Burrowing Parrots *Cyanoliseus patagonus* as Long-Distance Seed Dispersers of Keystone Algarrobos, Genus *Prosopis*, in the Monte Desert. *Diversity* **2021**, *13*, 204. https://doi.org/10.3390/d13050204

Academic Editor: Luc Legal

Received: 24 April 2021
Accepted: 10 May 2021
Published: 12 May 2021

1. Introduction

Dry forests are among the most threatened ecosystems worldwide due to direct habitat destruction and fragmentation by fire, overexploitation for fuelwood, and agriculture [1,2]. Livestock ranching has exerted a predominant impact for decades on the loss and trampling of vegetation, alteration of nutrient cycles, and competition with wild plant consumers [3–5]. These impacts are increasingly disrupting many ecological interactions of vital importance for the composition, structure, and functioning of dry ecosystems [5–8], even before they are known and properly understood [9]. Livestock often alter seed dispersal of the plants they feed on, favouring some species over others and disturbing native plant–animal interactions with a function in seed dispersal [5,8,10]. These threats, together with hunting for bush meat, are reducing populations of wild species that interact with the remaining vegetation in multiple ways [1,2,5,8]. Among them, there are several vertebrate groups that have been long neglected or understudied for their seed dispersal function in drylands, including parrots, rodents, carnivore and ungulate mammals, and reptiles [5,7,11–16]. This has contributed to sustaining the old controversy on the role of extinct megafauna as legitimate dispersers of many large-fruited plants [17], which has led to the proposal of domestic ungulates as contemporary substitutes conducting this ecological function [18]. The megafauna syndrome hypothesis gained momentum in the last few decades, but recent research has increasingly challenged its foundations by incorporating overlooked dispersers into seed dispersal webs and by considering ectozoochory in addition to endozoochory (e.g., [8,14,16,19–22]).

Legume trees (Fabaceae) constitute the dominant woody layer of many drylands in America [1,2] and have been exploited by human populations for millennia [23]. In particular, algarrobos (genus *Prosopis*, Mimosoideae, Leguminosae) are numerically-, biomass-, and functionally-dominant species intensively exploited for fruits by wild animals, livestock, and humans [24,25]. Because of the presence of a sweet pulp covering the seeds inside relatively large pods, dispersal has been mostly attributed to endozoochorous syndromes involving extant native and non-native mammals [26,27] and extinct megafauna in the past [17]. External short-distance dispersal by ants (myrmecochory) and potentially dung beetles [28,29] and stomatochory and seed hoarding by rodents [29,30] have also been highlighted. However, molecular reconstructions suggest long-distance dispersal by birds as the most likely hypothesis to explain the phylogeography and population genetics of the genus *Prosopis* and other legumes in America [31,32]. This conflicts with the assumption that avian species exploiting algarrobos and other legume pods are exclusive seed predators [28,33], or short-distance dispersers, in the case of flightless greater rhea (*Rhea americana*) [34,35]. On the contrary, recent research has highlighted parrots as seed dispersers of legumes, including several *Prosopis* species [13,36]. This role may be especially evident and relevant in ecological terms in dry ecosystems with avian frugivore richness lower than expected from overall bird diversity [Kissling et al., 2009] and in the absence of other avian long-distance dispersers, as documented in the dry tropical forest of the inter-Andean valleys [13,37].

Algarrobo seeds show physical dormancy, an adaptation to germinate when the environmental conditions are optimal after the erosion of the external hard coat by factors like temperature, sunlight, and soil abrasion [27,38,39]. This process has also been assumed to be a seed adaptation to resist the chemical and abrasive action of the digestive tract of mammals without losing seed viability while influencing germinability, depending on both legume and consumer species [27,40,41]. Seeds of legumes and other plants may retain viability and can germinate even when fruits end ripening—and seed desiccation—separated from the mother plant (generally termed after-ripening) during the final fruiting stage [42–46]. This implies that the transport of unripe fruits with viable seeds completely formed but not entirely dried and hardened (hereafter, soft seeds) could be considered effective for seed dispersal. This remains a largely overlooked mechanism, with implications in dispersal ecology [47]. To our knowledge, this possibility has been scarcely evaluated for its potential influence in seed dispersal by animals handling unripe fruits [48]. In addition, most experiments on viability and germinability after mammal gut passage have been conducted in laboratory and greenhouse conditions, thus often obviating the critical stage between seed handling by consumers and dispersal and germination in natural conditions.

The recent consideration of parrots as key seed dispersers by multiple and complementary mechanisms, including the wasting and transport of ripe and unripe fruits, have widened the range of mechanisms influencing the life cycle of many plant species [49,50], including legumes [13,36]. The combination of a relatively high abundance of parrots compared to other frugivores in many ecosystems, their generalist trophic habits, and their extensive daily and seasonal movements make them pervasive dispersers of most of their food plants [49]. This includes stomatochory (i.e., seed dispersal transported externally with beaks and dropped after fruit consumption) [14,16,36,49,51], endozoochory (i.e., dispersal of viable seeds after gut passage) [52–55], and epizoochory (i.e., dispersal of seeds adhered to the body surface) [56]. These ecological roles, together with the understudied pollination and predation of invertebrate plant pests, can have a pervasive influence on ecosystem structure and functioning [13,20,37,51,57]. Therefore, the previously denied attention is essential to understand the ecological function of parrots in understudied systems in general and in Neotropical drylands poor in other avian frugivores in particular.

In this study, we examined the ecological interaction between the burrowing parrot (*Cyanoliseus patagonus*) and dominant algarrobo trees at the end of the fruiting period in the Monte Desert ecoregion. We recorded the abundance and foraging behaviour of parrots and quantified the handling, consumption, wasting, and dispersal of ripe and unripe pods. The viability of seeds from unripe and ripe pods wasted and transported by parrots was examined to determine whether their respective dispersal could be considered antagonistic or complementary mechanisms on plant demography. Finally, we conducted a snapshot sampling of faeces of wild mammals and livestock to determine the presence of algarrobo seeds dispersed by endozoochory.

2. Material and Methods

2.1. Study Area

The study was conducted in northwestern Argentina (provinces of Salta, Tucuman, and Catamarca), which is included in the northern range of the Monte Phytogeographic Province [58], comprising two ecoregions (High Monte and Low Monte) (https://ecoregions2017.appspot.com, accessed on 24 April 2021) (Figure 1). In the High Monte (600–3500 m.a.s.l.), the climate is semi-arid to arid, with dry hot and cold rainy well-marked seasons [58,59]. Rainfall shows inter-annual variability but rarely exceeds 200 mm, concentrated in summer. The surveyed area, mostly within the region of Valles Calchaquíes, is occupied by native dry forests that have been converted to savanna- and steppe-like landscapes with variable levels of forest fragmentation for agriculture and soil erosion due to fire and free-ranging livestock [60,61]. More dense arboreal patches and riverine forests remain in valley bottoms and across the Calchaquí and Santa María rivers [61]. The flora is characterised by halophytic species of shrubby steppe and arboreal and shrub layers dominated by legumes (Fabaceae), especially algarrobos [58,62].

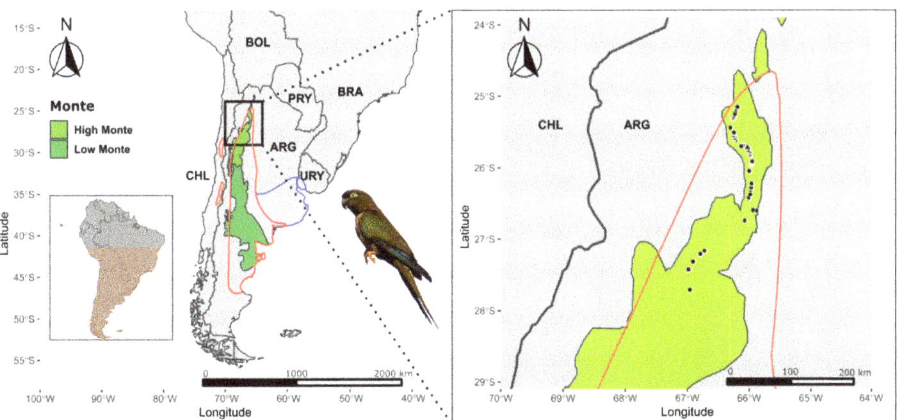

Figure 1. Map showing the distribution range of the burrowing parrot, including breeding (red line) and non-breeding (blue line) areas in southern South America (Argentina and Chile). The Monte Phytogeographic Province [58] comprising the High Monte and Low Monte Ecoregions (https://ecoregions2017.appspot.com, accessed on 24 April 2021) is shown. The study area is indicated by the black box, and locations where burrowing parrots were recorded are shown by black dots (right panel).

2.2. Study Species

Prosopis is a primitive genus of shrubs and trees with currently disjointed natural distribution in Africa, Asia, and America. *Prosopis alba* and *Prosopis nigra* are medium-sized trees (4–16 m height) with widely overlapping distributions. Both species are present in the subtropical plains of north and central Argentina, Paraguay, and Uruguay and patches of semi-arid zones of Peru and Bolivia [24,25]. Genetically, they are close and poorly defined species [63]. Within the Algarobia section, to which both species belong, hybrids and introgression are relatively frequent [31,63–65]. The intraspecific morphological variations are usually high between localities [66]. Fruits are indehiscent, cylindrical, thick, and more or less compressed pods (length × width, 12–25 × 1.1–2 cm and 10–16 × 0.7–0.9 cm for *P. alba* and *P. nigra*, respectively). These may contain multiple seeds (generally 12–30 and 8–27 for *P. alba* and *P. nigra*, respectively) and are wrapped in a sweet spongy mesocarp [24,25,66,67]. The mesocarp (hereafter, pulp) of both species is rich in sugars and relatively low in fat and protein. In the seeds, the proportion of proteins is high and much higher than that of fats [67,68]. Mature seeds of *Prosopis* have a hard and impermeable seed coat that prevents seed imbibition and germination (i.e., physical dormancy). The seeds of different *Prosopis* species are predated when they are still developing in the pods by birds and invertebrates, especially bruchid beetles (Chrysomelidae: Bruchinae), which may consume a large proportion of the seed crop [41,69]. Fallen fruits are preyed on and/or dispersed by endozoochory by greater rhea, wild mammals, reptiles, and livestock [25,34,35,41,70]. Rodents and invertebrates such as ants are seed predators and dispersers by stomatochory, followed by underground seed hoarding [28,33,71].

The burrowing parrot inhabits the Monte Biogeographic Province, located in Argentina and Chile, southern South America (Figure 1). It includes two subspecies with an extensive distribution range across northwestern–central (*C. p. andinus*) and central–southern (*C. p. patagonus*) Argentina, respectively, and a presumable subspecies (*C. p. conlara*) hybrid between them [72]. The southern subspecies also inhabits the Patagonian steppe and the Espinal Biogeographic Provinces, including several ecoregions [73]. These subspecies are categorised as Least Concern, although their populations show variable degrees of threat, depending on the region, due to persecution and habitat loss and degradation [74]. A fourth subspecies (*C. p. bloxami*) inhabits a comparatively small distribution area in central Chile [72,75] and shows a small and declining population categorised as Endangered [74]. Burrowing parrots are colonial cliff nesters, nomadic and partially migratory, and gregarious foragers in variably-sized mobile flocks [76,77]. Depending on the subspecies, they exploit the fruits and seeds of different varieties of plants [72,75], whose crops are tracked through very long-distance flights from the breeding colonies and communal roosts [76]. They have also been recorded feeding on flowers, buds, and bark of native and exotic species [72,75] as well as on seeds of cultivated plants without causing remarkable impacts on crops [77]. No information is available on the role of this parrot as a mutualist or antagonist of their food plants except for observations of seed dispersal by stomatochory [36], the dispersal of the Atacama Desert shrub (*Balsamocarpon brevifolium*, Fabaceae) [32], and the lack of seeds dispersed by endozoochory over a limited snapshot sampling of faeces in southern Patagonia [53].

2.3. Survey of Parrot Abundance and Foraging Observations

During January 2020, coinciding with the fruiting period of algarrobos, we conducted surveys of burrowing parrots of the *andinus* subspecies in the northern Monte Biogeographical Province (hereafter Monte desert). We drove at low speeds (20–40 km/h) along low-transited paved roads and unpaved tracks following the methodology used in previous studies [13,57,78], totalling 529 km that were surveyed only once. The surveys were carried out by three persons, the driver and two observers, during the breeding season of the parrots. During the surveys, we counted the parrots observed and recorded their behaviour. We generally detected burrowing parrots visually, but when the survey track crossed dense woods generally associated with riparian forests with low visibility, we also detected parrots aurally. In these cases, we assigned the mean flock size of parrots detected visually to the aural records [78]. We also measured the perpendicular distance from the road to parrot flocks, using a laser rangefinder for estimating the density of this species through distance sampling modelling (see [79] for more methodological details and results). The obtained relative abundance (number of individuals/km) strongly correlated with density (number of individuals/km^2) estimates obtained through distance-sampling modelling in this and many other parrot species [78,79].

When foraging flocks were detected, we stopped the survey to observe with a telescope and binoculars their feeding behaviour including the food handling, the consumed part of each plant, the ripening state (ripe or unripe) of fruit pulp, and the maturation of seeds (soft or hard) of each exploited plant, identified to species or genera levels with field guides. We also recorded whether parrots dropped each food type beneath the canopy of food plants and confirmed what parrots were eating and wasting by searching for food remains on the ground beneath foraging sites, following the methodology previously detailed [14,57]. This information was also recorded for foraging flocks observed outside roadside surveys conducted to determine parrot abundance.

2.4. Handling, Resource Consumption, and Wasting of Algarrobo Fruits

After each observation of foraging on algarrobos, and when possible, depending on the location of food plants, we searched for pods handled and wasted by parrots on the ground beneath the foraging trees and the sites where the fruits were dispersed at a distance. At accessible sites, we collected a sample of these pods, which were clearly identified by the typical beak marks on them; no other parrot species inhabits the study area. Over a sample of the pods found to be handled by parrots (n = 363), we recorded whether they were entirely or partially consumed, the ripening state, and the part consumed by parrots. For unripe pods, which lacked pulp, we recorded the number of predated and intact seeds. In the case of ripe pods, we counted the number of seeds around which the pulp was consumed (i.e., consumed pulp) and the number of intact seeds remaining within or absent from each handled pod. We also estimated the proportion of the resources (pulp or seed) consumed over the total available in each handled pod.

2.5. Seed Dispersal by Burrowing Parrots

When parrots were located foraging, we recorded whether they flew and transported the fruits in their beaks from the fruiting trees to distant sites, thus dispersing the seeds by stomatochory (e.g., [14,16,57]). When the parrots ended feeding and left the area, we measured the exact dispersal distance (in metres) from the tree where the fruit was collected to where the fruit was released. This included other algarrobos, other tree and shrub species, and electric poles and wires, below which we confirmed, when possible, the presence of discarded entire fruits or fruit remains with seeds. When the foraging parrots moved far away with fruits in flight, we followed them using binoculars with a laser rangefinder (Leica Geovid 10 × 42) to measure the exact distance to which the parrots moved the fruits, either by losing and discarding the fruits in flight or from distant perches. In the particular situations when flying parrots carrying fruits from the food plant were lost from sight or when they were already carrying fruits in their beaks at first sighting, we followed the specifications described in previous studies to conservatively estimate minimum dispersal distances [14,16,51,57].

We used MCMCglmm [80] to test differences in dispersal distances (log-transformed; cengaussian distribution) between the two algarrobo species (fixed effect). The model was run for 100,000 iterations, preceded by a burn-in of 10,000 iterations. Estimates of parameters were stored every 25th iteration to reduce autocorrelation. We tested the statistical support of the fixed effect by evaluating whether its posterior distribution (95% credible interval) overlapped zero.

For a sample of flocks feeding on algarrobos, the rate of fruit movement leading to seed dispersal was estimated by recording the number of parrots that departed simultaneously in flight with pods in their beaks over the total number of parrots that formed each flock [16,51,57]. Given the large size of several of the flocks, which often fed on several nearby trees, the observations were coordinated between three observers, each of whom focussed on different fractions of the flocks to better count parrots moving pods.

2.6. Seed Viability

Samples of unripe and ripe pods of each algarrobo species wasted by parrots were collected and dried with a forced-air heater to simulate the drying conditions in the hot austral summer in the Monte Desert. Once the pods were completely dry, they were stored at room temperature until arrival at the laboratory, where the seeds from each pod were extracted and counted. We determined whether each seed was colonised by bruchid beetles by recording larvae, pupae, or the characteristic adult exit hole in the endocarp [69]. Non-parasitised seeds were tested for viability using the tetrazolium test [81]. After cutting with a scalpel, seeds were incubated in a 1% solution of 2,3,5-triphenyl tetrazolium chloride for 48 h. The tetrazolium reacted with respiring radicles to produce a red stain when the seed was viable, while non-stained white radicles indicated non-viable seeds [81]. This test was conducted to assess whether seeds from unripe pods wasted and dispersed by parrots retained viability compared to seeds from ripe pods, rather than to precisely determine germination potential and rate, which would require germination experiments [55].

2.7. Algarrobo Seeds Excreted by Livestock and Wild Mammals

In the area surveyed for parrot presence and abundance, we conducted a random search for faeces of free-ranging livestock (cows and equids) and wild large-bodied mammals that could act as potential dispersers of algarrobo seeds [41]. A random sample of fresh faeces was shredded in situ to determine the presence of seeds. We opportunistically recorded the presence of invertebrates dispersing seeds wasted by parrots as well as those present in mammal faeces.

3. Results

3.1. Parrot Abundance and Foraging

During the surveys, we recorded 98 flocks of burrowing parrots totalling 1559 observed individuals, which represented a relative abundance of 2.95 parrots per km surveyed. Relative abundance reached 3.61 parrots/km when adding the estimated number of parrots that were only heard. The estimated density obtained through distance sampling modelling was 18.17 individuals/km^2. All observations were recorded in the High Monte ecoregion (Figure 1).

We located 40 flocks foraging on six plant species (Table 1). The flocks for which we were able to determine the number of individuals were composed of 1 to 160 parrots (mean $\pm$ SD = 25.2 $\pm$ 8.1, n = 27). Most foraging records corresponded not only to the consumption of seeds and pulp of algarrobo pods and other legume species (Fabaceae) but also to flowers of a hemiparasite mistletoe (*Ligaria cuneifolia*, Lorenthaceae) and fleshy bark of branches of a columnar cactus (*Trichonocereus atacamensis*, Cactaceae) (Table 1). All foraging observations confirmed food wasting of each type of plant part consumed, including partially consumed and entire unripe and ripe fruits or their components (seeds and pulp) as well as flowers and branch bark (Table 1).

3.2. Pod Handling and Consumption

Overall, we collected 363 algarrobo pods handled and partially or entirely consumed, wasted, or dispersed by burrowing parrots (Figure 2A,B), corresponding to *P. alba* (n = 113 unripe pods from four trees and n = 164 ripe pods from three trees) and *P. nigra* (n = 33 unripe pods from nine trees and n = 53 ripe pods from nine trees); some of the sampled unripe and ripe pods corresponded to the same trees. When feeding on unripe pods, parrots focussed on the consumption of seeds (soft) in all cases while discarding other fruit parts of both species (Figure 2C,E). When feeding on ripe pods, they only consumed the pulp but never the seeds (hard) or other parts in both species (Figure 2D,E).

Table 1. Abundance, food plant species, feeding behaviour, and seed dispersal of burrowing parrots in the Monte Desert, Argentina.

			Plant Part Consumed [a]			Fruit Part Consumed [c]		Seed Dispersal (Pod)		
Family Plant Species	***n* Flocks (%)**	***n* Parrots (%)**	**Fruit (%)**	**Flower (%)**	**Other [b] (%)**	**Seed (%)**	**Pulp (%)**	**Unripe (%) [d]**	**Ripe (%) [d]**	**Total (%) [d]**
Fabaceae										
Geoffroea spinosa	1 (3.7)	5 (0.7)	1(2.6)				1 (4.0)		1 (100)	1 (100)
Parkinsonia praecox	6 (22.2)	30 (4.4)	6 (15.8)			6 (24.0)		0 (0.0)	0 (0.0)	0 (0.0)
Prosopis alba	10 (37.0)	363 (53.4)	14 (36.8)			5 (20.0)	13 (52.0)	1 (20.0)	6 (46.2)	7 (38.9)
Prosopis nigra	7 (25.9)	173 (25.4)	16 (42.1)			14 (56.0)	11 (44.0)	6 (42.9)	3 (27.3)	9 (36.0)
Prosopis sp.	1 (3.7)	22 (3.2)	1 (2.6)							
Loranthaceae										
Ligaria cuneifolia	1 (3.7)	75 (11.0)		1 (100)						
Cactaceae										
Trichonocereus atacamensis	1 (3.7)	12 (1.8)			1 (100)					
Total	27	680	38	1	1	25	25	7	10	17

[a] Refers to flocks, including those for which the number of parrots was not determined accurately. Flocks often fed on several nearby trees, including *Prosopis* trees of both species. [b] Fleshy bark of branches. [c] Consumed seeds correspond to unripe legume pods, while pulp was exploited from ripe pods. Both unripe and ripe pods were simultaneously available in several *Prosopis* trees. Therefore, a single flock could be computed twice when parrots consumed both unripe and ripe pods. [d] Percentage calculated of the total number of foraging observations on seeds (unripe pods) from *P. alba* ($n = 5$) and *P. nigra* ($n = 14$) and pulp (ripe pods) from *P. alba* ($n = 13$) and *P. nigra* ($n = 11$).

Figure 2. (**A**) Ripe and unripe pods of *Prospis alba* (**A**) and *Prosopis nigra* (**B**) accumulated beneath fruiting trees after feeding bouts by burrowing parrots. Details of wasted and partially consumed unripe (**C**) and ripe (**D**) pods of *P. alba* and of *P. nigra* (**E**). Pictures: G. Blanco.

Among the unripe pods handled, most were partially consumed, while a lesser proportion was wasted entirely or partially consumed, and these figures differed between algarrobo species (Fisher's exact test, $p < 0.0001$; Figure 3A). Most ripe pods of both species were also partially consumed, and their handling also differed between species (Fisher's exact test, $p < 0.001$; Figure 3B).

Given the differences in the handling and consumption of pods depending on ripening state and species and due to species-specific differences in pod size-related availability of food for parrots (seeds and pulp) and the number of seeds (see *Study Species* section), we analysed partially and completely consumed or wasted entire pods separately for each species. The mean number of seeds predated per pod (unripe pods), wasted inside pods (unripe and ripe pods), or missing (ripe pods) are shown in Table 2 for each species. Overall, parrots consumed a similar number of seeds (6–7) from unripe pods of both species (Mann–Whitney U test, $z = 1.93$, $p = 0.054$, $n_{alba} = 76$, $n_{nigra} = 145$, Table 2). The number of seeds remaining inside wasted pods after handling (pods partially or completely consumed) was higher in *P. alba* (about 11–14) than *P. nigra* (about 4–8), both for unripe and ripe pods ($z = 7.17$, $n_{alba} = 76$, $n_{nigra} = 145$ and $z = 4.60$, $n_{alba} = 32$, $n_{nigra} = 38$, respectively, both $p < 0.0001$, Table 2). The number of seeds missing from ripe pods (about 2) was much lower than those remaining in them after handling by parrots (Table 2) but was similar between species ($z = 1.72$, $p = 0.08$, $n_{alba} = 32$, $n_{nigra} = 38$, Table 2). Regardless of the fate of seeds, their total numbers were similar between pods partially or completely consumed and those wasted entirely in within-species comparisons (see Table 2).

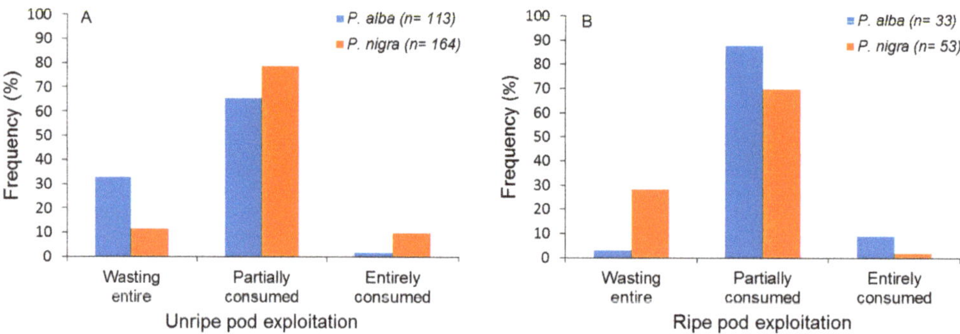

Figure 3. Frequency of each type of handling and consumption of unripe (**A**) and ripe (**B**) pods of two algarrobo species (*Prosopis alba* and *Prosopis nigra*) by burrowing parrots.

Table 2. Number (mean ± SD) and range (in parentheses) of seeds predated, wasted, and missing per pod according to the ripening state and consumption (partial and total) and wasting (entire) of pods of two algarrobo species handled by burrowing parrots. Missing seeds refer to hard seeds from ripe pods that did not remain inside the pod once discarded by parrots (i.e., assumed to be wasted outside the pod or dispersed by epizoochory). Consumed pulp corresponds to the number of seeds around which the pulp was consumed. n_p, n_s indicate sample sizes for pods and seeds, respectively.

	Pods Partially or Completely Consumed						Pods Wasted Entirely	
	Predated Seeds	**Wasted Seeds Inside Pods**	**Missing Seeds**	**Total**	**Consumed Pulp**	n_p, n_s	**Wasted Seeds**	n_p, n_s
Unripe pods								
Prosopis alba	7.3 ± 4.7 (1–22)	11.1 ± 7.7 (0–30)	-	18.5 ± 5.6 (8–31)	-	76, 1405	20.2 ± 7.1 (5–32)	37, 748
Prosopis nigra	5.9 ± 3.5 (1–15)	4.1 ± 3.4 (0–17)	-	10.0 ± 3.4 (2–20)	-	145, 1446	9.8 ± 4.5 (3–17)	19, 187
Ripe pods								
Prosopis alba	-	14.4 ± 5.6 (4–24)	1.7 ± 3.7 (0–15)	16.2 ± 4.3 (9–28)	8.7 ± 4.8 (1–19)	32, 517	10	1, 10
Prosopis nigra	-	8.2 ± 4.1 (1–16)	2.2 ± 2.8 (0–9)	10.3 ± 3.3 (4–16)	4.4 ± 3.0 (1–14)	38, 393	11.1 ± 4.8 (3–18)	15, 167

Considering pods partially or completely consumed, the exploitation of seeds reached an average of 43.7% (SD = 28.5%, range = 3.2–100%) of the available ones in each unripe *P. alba* pod ($n = 558$ seeds from 76 pods) and 60.1% (SD = 28.28%, range = 5.6–100%) in *P. nigra* ($n = 856$ seeds from 145 pods). Regarding ripe pods, parrots consumed 54.5% (SD = 28.6%, range = 5.3–100%) of the available pulp in *P. alba* pods ($n = 277$ seeds from 32 pods) and 44.4% (SD = 26.6%, range = 6.3–100%) in *P. nigra* pods ($n = 169$ seeds from 38 pods).

3.3. *Rate and Distance of Seed Dispersal*

In addition to seed dispersal beneath food plants due to food wasting (Figure 2A–C and Figure 4A), we recorded primary pod dispersal by stomatochory at a distance from the mother plant for both algarrobo species (Figure 4B,C) and for spherical fruits with a single seed each in the case of *Geoffroea spinosa* (Fabaceae) (Table 1). A proportion of the flocks feeding on algarrobos moved pods away, thus dispersing soft and hard seeds of both species (Table 1).

Figure 4. Burrowing parrots feeding on ripe pods of *Prospis nigra* (**A**)and dispersing ripe pods of *Prospis alba* (**B**) and ripe pods of *Prosopis nigra* (**C**). Yellow arrows indicate pods transported in the beak (stomatochory), and red arrows indicate seeds dispersed adhered to throat feathers (epizoochory). Pictures: J.L. Tella.

The proportion of parrots departing with pods or without pods in their beaks was estimated for 20 flocks composed of 36.1 ± 44.5 parrots (range = 2–160) feeding on *P. alba* (n = 12 trees or groups of trees), *P. nigra* (n = 11), and *G. spinosa* (n = 1); note that some particular flocks were feeding on nearby algarrobo trees of both species. In about half of these flocks (55.0%, n = 20), at least one parrot departed with pods in its beak. The mean proportion of dispersing parrot flocks moving pods was 13.1% (SD = 6.2%, range = 6.7–23.1) for *P. alba* (n = 7 flocks), 33.9% (SD = 44.8%, range = 4.5–100) for *P. nigra* (n = 4 flocks), and 20.0% for *G. spinosa* (n = 1 flock of five parrots). In the remaining flocks, parrots departed without fruits in their beaks.

Minimum dispersal distances did not differ between *P. alba* (54.8 ± 93.8 m, median = 16, range = 10–530, n = 65) and *P. nigra* (54.0 ± 62.6 m, median = 30, range = 3–180, n = 71) (Mann–Whitney U test, z = 1.38, p = 0.17). Exact dispersal distances also did not differ between *P. alba* (73.4 ± 28.4 m, median = 76, range = 23–160, n = 15) and *P. nigra* (79.2 ± 66.4 m, median = 92, range = 5–150, n = 6) (z = 0.03, p = 0.97) (Figure 5A). Estimated dispersal distances did not differ between the two algarrobo species (posterior mean: 0.26, 95% credible interval: −0.10–0.68), although they were larger for *P. nigra* (417.8 m, 95% credible interval: 187.50–972.75) compared to *P. alba* (237.7 m, 95% credible interval: 138.0–485.3) (Figure 5B). The single dispersal distance recorded for *G. spinosa* was 93 m.

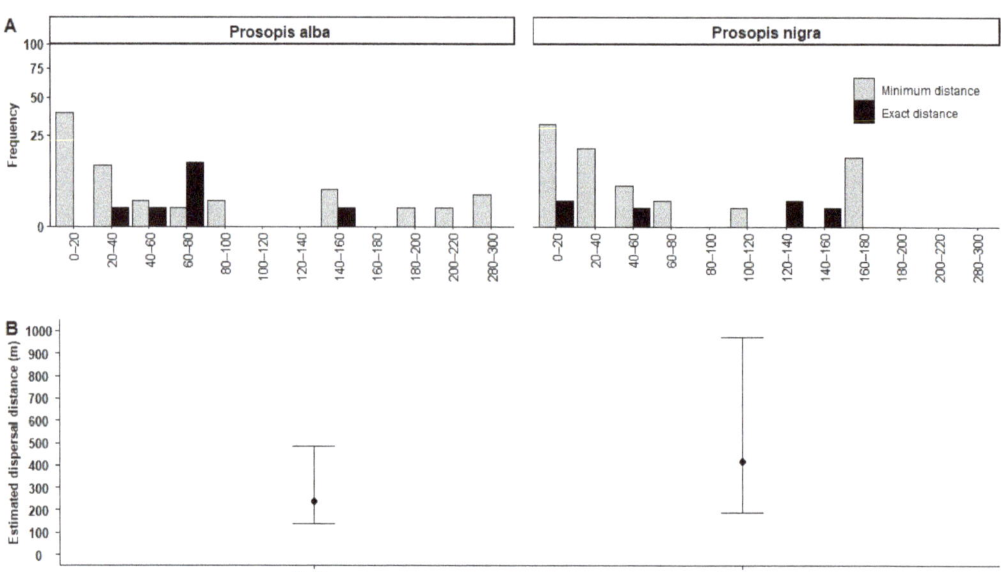

Figure 5. Dispersal distances (in m) for algarrobo pods dispersed by burrowing parrots. (**A**) Minimum (grey bars) and exact (black bars) distances recorded for each species shown in intervals of 20 m. Intervals without data are absent from the graph. (**B**) Estimates of the mean dispersal distance for *P. alba* and *P. nigra*, obtained from the MCMCglmm function. The highest posterior density (HPD) interval at 95% fixed probability is shown for both species.

During foraging observations on the ripe pods of both algarrobo species, we ascertained the presence of seeds adhering to the beak and the face and throat feathers of parrots. After examining pictures of flying individuals dispersing pods by stomatochory, we confirmed seed dispersal by epizoochory as a consequence of feeding on the sticky pulp of ripe pods (Figure 4C). However, we were unable to estimate the frequency of this type of dispersal or the number of seeds dispersed.

3.4. Viability of Seeds Wasted and Dispersed by Parrots

Most seeds from unripe and ripe wasted and dispersed pods of both algarrobo species were predated by bruchid beetles (Table 3). Among non-parasitised seeds, about half were viable and the other half unviable according to the tetrazolium test (Table 3). There was no significant difference in the proportion of viable seeds between unripe and ripe pods of each species wasted by parrots (generalised linear model, binomial error distribution, logit link function; algarrobo species, $\chi^2 = 0.36$, $p = 0.55$; ripening state, $\chi^2 = 0.08$, $p = 0.78$; interaction, $\chi^2 = 0.30$, $p = 0.58$, Table 3).

Table 3. Frequency (mean% ± SD) of algarrobo seeds predated by bruchid beetles per pod handled by burrowing parrots, depending on their ripening state. Non-parasitised seeds were tested for viability using the tetrazolium test.

		Seeds		
Species	**Sample Size**	**Parasitised**	**Non-Parasitised**	
Pod State	**Pods/Seeds**		**Unviable**	**Viable**
Prosopis alba				
Unripe	20/320	92.8 ± 7.3	4.2 ± 4.1	3.0 ± 5.0
Ripe	31/603	89.1 ± 8.9	6.4 ± 7.3	4.5 ± 4.6
Prosopis nigra				
Unripe	5/53	78.7 ± 20.3	12.4 ± 15.7	8.9 ± 8.3
Ripe	8/82	86.5 ± 10.0	6.2 ± 5.2	7.2 ± 6.4

3.5. Seeds in Mammal Faeces and Secondary Dispersal

A small proportion of livestock faeces contained algarrobo seeds, including faeces from cattle (*Bos taurus*; 2.4%, n = 85) and equids (*Equus* spp; 5.0%, n = 101). Among wild animals, we found no seeds in the faeces of the guanaco (*Lama guanicoe*; 0.0%, n = 86), while half of the small number of faeces of the South American grey fox (*Lycalopex griseus*) contained seeds (50.0%, n = 6).

Seeds from algarrobo pods wasted by parrots were observed being secondarily dispersed by columns of unidentified ants, which we did not quantify. Moreover, ants were observed dispersing algarrobo seeds from equid faeces (n = 1 of 5 faeces with seeds, 20.0%), while coprophagus beetles (Scarabaeidae) were observed dispersing seeds within dung balls from cattle (n = 1 of 2 faeces with seeds, 50.0%) and fox faeces (n = 1 of 3 faeces with seeds, 33.3%).

4. Discussion

Understanding the structure and functioning of increasingly human-altered drylands requires detailed knowledge about plant–animal interactions. Special attention has been given to the nature and strength of the interactions involving keystone and unique species in food and mutualistic webs before they become extinct or functionally threatened [9]. Here, we highlight the unrecognised role played by a bird species, the burrowing parrot, as the main disperser of the dominant trees of the Monte Desert. Our results show a very high abundance of burrowing parrots in the northern Monte Desert compared to other Neotropical parrot species [13,57,78]. In fact, the relative abundance and density of burrowing parrots in the Monte Desert are higher than those of the same species in other ecoregions [82] and are among the five highest obtained from 575 abundance surveys of 137 parrot species across the world [79]. No other large- or medium-sized frugivorous birds with the potential to act as seed dispersers were recorded during these surveys. In fact, the Monte Desert is poor in this type of species [83], and only the greater rhea has been suggested as a disperser of legume seeds [34,35]. Other bird species exploiting legume pods have been assumed to be exclusive pre-dispersal seed predators, including the burrowing parrot, the other parrot species partially inhabiting this biome at southern latitudes (i.e., the monk parakeet *Myiopssita monachus*), and some passerine birds [28,69]. Contrary to previous assertions not supported by sound, specific research on bird–legume interactions, we found evidence that burrowing parrots were frequent primary, long-distance dispersers of algarrobo seeds.

Burrowing parrots only predated on soft seeds from unripe pods, while they exclusively fed upon the pulp wrapping hard seeds from ripe pods of both algarrobo species. This food selectivity depending on the fruit ripening state and seed hardening has been cited for other parrot–legume interactions [13,84–86] and could be related to the nutrients provided in each case [87,88]. Unripe pods provide soft seeds rich in proteins and water, while ripe pods provide abundant sweet pulp rich in carbohydrates [67,68]. In addition to these changes in nutritional composition, dry and hard seeds may be less attractive to parrots when abundant pulp is available in ripe pods. Alternatively, burrowing parrots can have difficulties cracking or digesting these hard seeds, even though the strong beaks of parrots enhance the consumption of very hard items [16,49]. The wide phenological range of algarrobo fruiting allows parrots to exploit these nutrients in different or the same trees in each feeding bout. These resources can be selected or exploited depending on the abundance in each tree, spatial area, or period and based on the contrasting physiological needs of individuals for reproduction (nestling provisioning) or maintenance [87,88]. Whatever the case, our results indicate a close dependence of burrowing parrots on fruits of the dominant trees of the northern Monte Desert. Other plant species were exploited much less frequently due to their lower abundance and biomass and likely due to different phenological stages and resources provided, a determination that requires assessment through year-round studies.

The typical wasteful feeding of parrots implies that most handled unripe and ripe pods were partially consumed or discarded entirely. This behaviour has been documented for all parrot species examined and for those feeding on legumes of many species [50]. Wasting implies that burrowing parrots discarded beneath the mother plant or at a distance after dispersal all hard seeds from the ripe pods handled and partially exploited for pulp as well as a proportion of soft seeds from unripe pods. The relatively large size of pods including many seeds (ripe and unripe pods) and abundant pulp (ripe pods) could also contribute to their partial consumption and wasting. This is supported by the similar quantity of food extracted from pods of both species, despite their different size and number of seeds. Moreover, wasting entails a higher number of soft and hard seeds remaining inside after handling the larger *P. alba* pods as compared to *P. nigra* pods. Thus, each partially consumed pod wasted beneath the tree or moved away implies the dispersal of about half of the seeds available for parrots, with average ranges between 4 and 14 seeds per pod depending on species and ripening state and a maximum of up to 30 seeds per pod. These figures were slightly higher for pods wasted entirely. Therefore, the wasting beneath the plant or transport of pods at a distance invariably implies the dispersal of multiple seeds in each event.

Stomatochory has been previously documented for many parrot–plant interactions [14,16,20,36,49,51], including *P. alba, P. kuntzei*, and other legumes moved by four parrot species at minimal distances of up to 400 m in the Bolivian inter-Andean dry forest [13]. The much-longer-than-wider shape of legume pods allows parrots to move these large, multi-seeded fruits in flight, thus relaxing the physical limitations imposed by spherical fruits for transport in the beak, depending on parrot species-specific size [49]. We estimated that in about half of the flocks, at least one parrot departed in flight from the fruiting trees with pods in its beak, with 10–34% of the flock individuals moving pods. These figures were similar or higher than those previously estimated in other parrot–plant interactions (ranging from 4.8 to 29.6%) [51,57], except for the highest seed dispersal rate of bush-layer *Attalea* palms [16]. The exact dispersal distances recorded ranged from 5 to 160 m, while the mean dispersal distances estimated when controlling for the right-censored distribution of our data estimated 238 m for *P. alba* and 418 m for *P. nigra*. These distances were well above 100 m, a distance threshold often used to define long-distance seed dispersal (see [19] and references there). Therefore, similarly to other parrot species [14,16,49], burrowing parrots may act as short- and long-distance seed vectors by stomatochory. In addition, we found that two hard seeds were missing, on average, from ripe pods of both algarrobo species after handling by parrots. This suggests that a small number of hard seeds were wasted outside pods or were primarily dispersed by epizoochory documented for the first time in this species. This dispersal mechanism has been little explored in parrots, but a recent review has shown that it is more widespread than previously thought [56]. Through epizoochory, parrots could presumably disperse seeds at a much longer distance than by stomatochory, as burrowing parrots often conduct non-stop flights of tens of kilometres [76]. Both long-distance dispersal mechanisms by parrots can contribute to explaining the phylogeography and population genetics of the genus *Prosopis* in South America, which was previously hypothesised to be due to bird seed dispersal, although the avian disperser species were unknown [31]. In fact, many parrot species disperse legumes in other Neotropical dry biomes [13,36,50]. In support of this hypothesis, the radiation centre and distribution range of the section Algarobia [31,89] largely overlap with that of the nomadic and partially migratory burrowing parrot [72], coinciding in time with the spreading of arid areas in the Americas after the Andean uplift [72,89–91]. Further research is warranted, including the assessment of epizoochory and potential endozoochory, to determine the dispersal role of burrowing parrots in the phylogeography and population genetics of their food plants.

As in other studies [41,69], a high proportion of algarrobo seeds was predated by bruchid beetles. Among non-predated seeds, no significant difference was found in the proportion of viable seeds from unripe and ripe pods after being handled by parrots. The transport of unripe pods could thus contribute to the dispersal of viable seeds. This agrees with previous studies showing that seeds from the unripe fruits of several plant species may retain viability and germinate after external dispersal [45,48]. This mechanism may add spatiotemporal heterogeneity to the process of seed dispersal as a potential complementary plant adaptation to increase the likelihood of seedling establishment, depending on the feeding behaviour and food selectivity of dispersers. In fact, the predation of a proportion of soft seeds from unripe pods was invariably accompanied by the wasting or dispersal of a similar number of seeds. In agreement with other parrot–plant interactions, multiple-seeded algarrobo pods associated with burrowing parrot stomatochory could be an adaptive interaction with eco-evolutionary consequences for both partners [13,16,37,51,57]. Specifically, large pods providing different food resources may satiate parrots predating soft seeds, thus also promoting their role as dispersers [51,57]. In this form, plants can ensure the movement of a proportion of viable seeds regardless of fruit maturation state [48]. Determining whether this dual parrot–legume interaction evolved in the context of antagonism–mutualism continuums [16,37] and whether it has implications in the evolution of fruit traits would require future studies. Further research is also needed to determine whether seed hardening and desiccation outside the mother plant may influence the dormancy and germinability of soft seeds.

Besides acting as frequent primary dispersers of algarrobo seeds, burrowing parrots wasting pods provided food for secondary dispersers like wild and domestic vertebrates and invertebrates (mainly ants). Even when livestock and wild ungulates can feed on accessible fruits directly from trees, our snapshot sampling of faeces suggests a low frequency of seed dispersal by livestock and guanacos but a higher frequency in a very small sample of grey fox scats. When algarrobo woods consist mainly of an arboreal layer, these dispersers could be limited to the consumption of pods fallen passively when ripe at the very end of the fruiting period. Therefore, wasteful feeding of parrots may increase the food availability for secondary seed dispersers, thus also widening the dispersal period and the range of potential dispersers [49,50]. Seed scarification and physical dormancy-breaks due to consumer gut passage can enhance germination speed after excretion in controlled but artificial greenhouse conditions. However, rather than being an adaptive advantage, seed excretion within large faeces (especially from large livestock) can be an ecological trap wherein seeds can be "forced" to quick germination, rarely developing to viable seedling establishment [29,40,92,93]. This can be enhanced by dormancy breaking coupled with faecal moisture favouring immediate germination at unsuitable times (e.g., before the rainy season) and conditions including high seed density, increasing seedling competition, nitrogen concentration, pathogen loads, and predator attraction, thus overall constituting an unfavourable microhabitat for seedling establishment [29,93]. The higher and long-lasting moisture from large faeces of livestock, especially cattle, as compared to the much smaller and drier faeces of native vertebrates, could drive differences in germinability and seedling establishment [41,92,93]. These factors could also exert an influence on the action of secondary dispersers like scatter-hoarding rodents and ants [29,94] and dung beetles recorded dispersing seeds from livestock and fox faeces during our limited snapshot sampling. On the contrary, the external dispersal by parrots does not imply any direct influence on seed physical dormancy besides a potential de-inhibition effect [95] by defleshing a small proportion of hard seeds. Seeds wasted by parrots are frequently deposited where they could be secondarily dispersed, while seeds moved away can be dispersed at microhabitats and seed banks where the scarification and subsequent germination takes place under natural environmental conditions [96].

5. Conclusions

The high abundance of the burrowing parrot in the Monte Desert suggests that a high number of multiple-seeded ripe and unripe pods are moved daily during their respective periods of availability. The high dispersal rates recorded associated with the typical daily routines of foraging flocks from breeding colonies and communal roosts can clearly contribute to this function. These routines consist of 1–4 daily flights of more than 60 km with multiple foraging stops across their roundtrip flights between the breeding colony and foraging areas and nomadic and partially migratory movements over extended home ranges of thousands of square kilometres [73,76,77]. Wasting and dispersal events occurred in most foraging bouts, both beneath and away from each mother plant. Therefore, the overall strength of seed dispersal can be amplified through successive foraging events across different extensive areas exploited daily, seasonally, and inter-annually [73,76,77]. This additive-like process, encompassing multiple seed dispersal events daily by a proportion of individuals, can represent the movement of huge amounts of seeds across large areas exploited sequentially. Further research is needed to quantify the number of seeds and accumulated dispersal distances attributed to the burrowing parrot and to evaluate its role as a functionally unique species (*sensu* [9]) in the structure and functioning of algarrobo woods in the Monte Desert.

Author Contributions: Conceptualization, G.B., P.R.-V., M.C., F.H. and J.L.T.; methodology, G.B., P.R.-V., M.C., D.C., C.B., F.H. and J.L.T.; software, P.R.-V. and M.C.; validation, G.B., P.R.-V., M.C., F.H. and J.L.T.; formal analysis, G.B., P.R.-V., M.C. and J.L.T. investigation, G.B., P.R.-V., M.C., D.C., C.B., F.H. and J.L.T.; resources, J.L.T.; data curation, G.B., P.R.-V., M.C., D.C., C.B., F.H. and J.L.T.; writing—original draft preparation, G.B.; writing—review and editing, G.B., P.R.-V., M.C., D.C., C.B., F.H. and J.L.T.; visualization, G.B., P.R.-V., M.C., F.H. and J.L.T.; supervision, G.B., M.C., F.H. and J.L.T.; project administration, M.C., J.L.T.; funding acquisition, J.L.T. All authors have read and agreed to the published version of the manuscript.

Funding: This research was funded by the Loro Parque Fundación (grant reference: PP-146-2018-1).

Institutional Review Board Statement: Not applicable. This work did not require approval by our respective Ethical Committees because it did not involve invasive methods or experimental work with live animals.

Informed Consent Statement: Not applicable.

Data Availability Statement: Most data are provided in the text in the form of percentages and sample sizes. Additional detailed data sets can be requested from the corresponding author.

Acknowledgments: We thank two anonymous reviewers, and I. Afán (LAST-EBD) for helping in the design of road surveys using satellite maps. Logistic and technical support were provided by Doñana ICTS-RBD.

Conflicts of Interest: The authors declare no conflict of interest. The funders had no role in the design of the study; in the collection, analyses, or interpretation of data; in the writing of the manuscript; or in the decision to publish the results.

References

1. Miles, L.; Newton, A.C.; DeFries, R.S.; Ravilious, C.; May, I.; Blyth, S.; Gordon, J.E. A global overview of the conservation status of tropical dry forests. *J. Biogeogr.* **2006**, *33*, 491–505. [CrossRef]
2. Sanchez-Azofeifa, A.; Powers, J.S.; Fernandes, G.W.; Quesada, M. (Eds.) *Tropical Dry Forests in the Americas: Ecology, Conservation, and Management*; CRC Press: Boca Raton, FL, USA, 2013.
3. Guevara, J.C.; Grünwaldt, E.G.; Estevez, O.R.; Bisigato, A.J.; Blanco, L.J.; Biurrun, F.N.; Passera, C.B. Range and livestock production in the Monte Desert, Argentina. *J. Arid Environ.* **2009**, *73*, 228–237. [CrossRef]
4. Kunst, C.; Ledesma, R.; Bravo, S.; Albanesi, A.; Anriquez, A.; van Meer, H.; Godoy, J. Disrupting woody steady states in the Chaco region (Argentina): Responses to combined disturbance treatments. *Ecol. Eng.* **2012**, *42*, 42–53. [CrossRef]
5. Mazzini, F.; Relva, M.A.; Malizia, L.R. Impacts of domestic cattle on forest and woody ecosystems in southern South America. *Plant Ecol.* **2018**, *219*, 913–925. [CrossRef]
6. Villagra, P.E.; Defossé, G.E.; Del Valle, H.F.; Tabeni, S.; Rostagno, M.; Cesca, E.; Abraham, E. Land use and disturbance effects on the dynamics of natural ecosystems of the Monte Desert: Implications for their management. *J. Arid Environ.* **2009**, *73*, 202–211. [CrossRef]
7. Chillo, V.; Ojeda, R.A. Mammal functional diversity loss under human-induced disturbances in arid lands. *J. Arid Environ.* **2012**, *87*, 95–102. [CrossRef]
8. Periago, M.E.; Chillo, V.; Ojeda, R.A. Loss of mammalian species from the South American Gran Chaco: Empty savanna syndrome? *Mamm. Rev.* **2015**, *45*, 41–53. [CrossRef]
9. O'Gorman, E.J.; Yearsley, J.M.; Crowe, T.P.; Emmerson, M.C.; Jacob, U.; Petchey, O.L. Loss of functionally unique species may gradually undermine ecosystems. *Proc. R. Soc. B Biol. Sci.* **2011**, *278*, 1886–1893. [CrossRef]
10. Borman, M.M. Forest stand dynamics and livestock grazing in historical context. *Conserv. Biol.* **2005**, *19*, 1658–1662. [CrossRef]
11. Varela, O.; Bucher, E. The lizard *Teius teyou* (Squamata: Teiidae) as a legitimate seed disperser in the dry Chaco Forest of Argentina. *Stud. Neotrop. Fauna Environ.* **2002**, *37*, 115–117. [CrossRef]
12. Varela, O.; Bucher, E.H. Seed dispersal by *Chelonoidis chilensis* in the Chaco dry woodland of Argentina. *J. Herpetol.* **2002**, *36*, 137–140. [CrossRef]
13. Blanco, G.; Hiraldo, F.; Rojas, A.; Dénes, F.V.; Tella, J.L. Parrots as key multilinkers in ecosystem structure and functioning. *Ecol. Evol.* **2015**, *5*, 4141–4160. [CrossRef]
14. Blanco, G.; Tella, J.L.; Díaz-Luque, J.A.; Hiraldo, F. Multiple external seed dispersers challenge the megafaunal syndrome anachronism and the surrogate ecological function of livestock. *Front. Ecol. Evol.* **2019**, *7*, 328. [CrossRef]
15. Maldonado, D.E.; Loayza, A.P.; Garcia, E.; Pacheco, L.F. Qualitative aspects of the effectiveness of Culpeo foxes (*Lycalopex culpaeus*) as dispersers of *Prosopis alba* (Fabaceae) in a Bolivian dry valley. *Acta Oecol.* **2018**, *87*, 29–33. [CrossRef]
16. Tella, J.L.; Hiraldo, F.; Pacífico, E.; Díaz-Luque, J.A.; Dénes, F.V.; Fontoura, F.M.; Guedes, N.; Blanco, G. Conserving the diversity of ecological interactions: The role of two threatened macaw species as legitimate dispersers of "megafaunal" fruits. *Diversity* **2020**, *12*, 45. [CrossRef]
17. Janzen, D.H.; Martin, P.S. Neotropical anachronisms: The fruits the gomphotheres ate. *Science* **1982**, *215*, 19–27. [CrossRef] [PubMed]
18. Janzen, D.H. Differential seed survival and passage rates in cows and horses, surrogate Pleistocene dispersal agents. *Oikos* **1982**, *38*, 150–156. [CrossRef]
19. Jansen, P.A.; Hirsch, B.T.; Emsens, W.J.; Zamora-Gutierrez, V.; Wikelski, M.; Kays, R. Thieving rodents as substitute dispersers of megafaunal seeds. *Proc. Natl. Acad. Sci. USA* **2012**, *109*, 12610–12615. [CrossRef] [PubMed]
20. Baños-Villalba, A.; Blanco, G.; Díaz-Luque, J.A.; Dénes, F.V.; Hiraldo, F.; Tella, J.L. Seed dispersal by macaws shapes the landscape of an Amazonian ecosystem. *Sci. Rep.* **2017**, *7*, 7373. [CrossRef] [PubMed]
21. Rebein, M.; Davis, C.N.; Abad, H.; Stone, T.; del Sol, J.; Skinner, N.; Moran, M.D. Seed dispersal of *Diospyros virginiana* in the past and the present: Evidence for a generalist evolutionary strategy. *Ecol. Evol.* **2017**, *7*, 4035–4043. [CrossRef]
22. McConkey, K.R.; Nathalang, A.; Brockelman, W.Y.; Saralamba, C.; Santon, J.; Matmoon, U.; Somnuk, R.; Srinoppawan, K. Different megafauna vary in their seed dispersal effectiveness of the megafaunal fruit *Platymitra macrocarpa* (Annonaceae). *PLoS ONE* **2018**, *13*, e0198960. [CrossRef]
23. McRostie, V.B.; Gayo, E.M.; Santoro, C.M.; De Pol-Holz, R.; Latorre, C. The pre-Columbian introduction and dispersal of Algarrobo (*Prosopis*, Section Algarobia) in the Atacama Desert of northern Chile. *PLoS ONE* **2017**, *12*, e0181759. [CrossRef]
24. Burkart, A. A monograph of the genus *Prosopis* (Leguminosae subfam. Mimosoidae). *J. Arnold Arbor.* **1976**, *57*, 219–525.
25. Galera, F.M. *Las Especies del Genero Prosopis (Algarrobos) en América Latina con Especial Énfasis en Aquellas de Interés Económico*; FAO/SECYT: Córdoba, Argentina, 2000; Available online: http://www.fao.org/docrep/006/ad314s/ad314s00.HTM (accessed on 6 April 2021).
26. Peinetti, R.; Pereyra, M.; Kin, A.; Sosa, A. Effects of cattle ingestion on viability and germination rate of caldén (*Prosopis caldenia*) seeds. *J. Range Manag.* **1993**, *46*, 483–486. [CrossRef]
27. Villagra, P.E.; Vilela, A.; Giordano, C.; Alvarez, J.A. Ecophysiology of *Prosopis* species from the arid lands of Argentina: What do we know about adaptation to stressful environments? In *Desert Plants*; Ramawat, K.G., Ed.; Springer: Berlin/Heidelberg, Germany, 2010; pp. 321–340.
28. Milesi, F.A.; Lopez De Casenave, J. Unexpected relationships and valuable mistakes: Non-myrmecochorous *Prosopis* dispersed by messy leafcutting ants in harvesting their seeds. *Austral Ecol.* **2004**, *29*, 558–567. [CrossRef]

29. Velez, S.; Chacoff, N.P.; Campos, C.M. Seed predation and removal from faeces in a dry ecosystem. *Basic Appl. Ecol.* **2016**, *17*, 145–154. [CrossRef]
30. Giannoni, S.M.; Campos, V.; Andino, N.; Ramos-Castilla, M.; Orofino, A.; Borghi, C.; de Los Rios, C.; Campos, C.M. Hoarding patterns of sigmodontine rodent species in the Central Monte desert (Argentina). *Austral Ecol.* **2013**, *38*, 485–492. [CrossRef]
31. Bessega, C.; Vilardi, J.C.; Saidman, B.O. Genetic relationships among American species of the genus *Prosopis* (Mimosoideae, Leguminosae) inferred from ITS sequences: Evidence for long-distance dispersal. *J. Biogeogr.* **2006**, *33*, 1905–1915. [CrossRef]
32. Stoll, A.; Harpke, D.; Schütte, C.; Jimenez, L.; Letelier, L.; Blattner, F.R.; Quandt, D. Landscape genetics of the endangered Atacama Desert shrub *Balsamocarpon brevifolium* in the context of habitat fragmentation. *Glob. Planet. Chang.* **2020**, *184*, 103059. [CrossRef]
33. Villagra, P.E.; Marone, L.; Cony, M.A. Mechanisms affecting the fate of *Prosopis flexuosa* (Fabaceae, Mimosoideae) seeds during early secondary dispersal in the Monte Desert, Argentina. *Austral Ecol.* **2002**, *27*, 416–421. [CrossRef]
34. Pratolongo, P.; Quintana, R.; Malvárez, I.; Cagnoni, M. Comparative analysis of variables associated with germination and seedling establishment *for Prosopis nigra* (Griseb.) Hieron and *Acacia caven* (Mol.) Mol. *For. Ecol. Manag.* **2003**, *179*, 15–25. [CrossRef]
35. Renison, D.; Valladares, G.; Martella, M.B. The effect of passage through the gut of the Greater Rhea (*Rhea americana*) on germination of tree seeds: Implications for forest restoration. *Emu Austral Ornithol.* **2010**, *110*, 125–131. [CrossRef]
36. Tella, J.L.; Baños-Villalba, A.; Hernández-Brito, D.; Rojas, A.; Pacífico, E.; Díaz-Luque, J.A.; Carrete, M.; Blanco, G.; Hiraldo, F. Parrots as overlooked seed dispersers. *Front. Ecol. Environ.* **2015**, *13*, 338–339. [CrossRef]
37. Montesinos-Navarro, A.; Hiraldo, F.; Tella, J.L.; Blanco, G. Network structure embracing mutualism-antagonism continuums increases community robustness. *Nat. Ecol. Evol.* **2017**, *1*, 1661. [CrossRef]
38. Catalán, L.; Balzarini, M. Improved laboratory germination condition for several arboreal *Prosopis* species: *P. chilensis, P. flexuosa, P. nigra, P. alba, P. caldenia* and *P. affinis*. *Seed Sci. Technol.* **1992**, *29*, 293–298.
39. Vilela, A.E.; Ravetta, D.A. The effect of seed scarification and soil-media on germination, growth, storage, and survival of seedlings of five species of Prosopis, L.(Mimosaceae). *J. Arid Environ.* **2001**, *8*, 171–184. [CrossRef]
40. van Klinken, R.D.; Goulier, J.B. Habitat-specific seed dormancy-release mechanisms in four legume species. *Seed Sci. Res.* **2013**, *23*, 181–188. [CrossRef]
41. Campos, C.M.; Velez, S. Almacenadores y frugívoros oportunistas: El papel de los mamíferos en la dispersión del algarrobo (*Prosopis flexuosa* DC) en el desierto del Monte, Argentina. *Rev. Ecosist.* **2015**, *24*, 28–34. [CrossRef]
42. Samarah, N.H. Effect of drying methods on germination and dormancy of common vetch (*Vicia sativa* L.) seed harvested at different maturity stages. *Seed Sci. Technol.* **2005**, *33*, 733–740. [CrossRef]
43. Gresta, F.; Avola, G.; Anastasi, U.; Miano, V. Effect of maturation stage, storage time and temperature on seed germination of *Medicago* species. *Seed Sci. Technol.* **2007**, *35*, 698–708. [CrossRef]
44. Angelovici, R.; Galili, G.; Fernie, A.R.; Fait, A. Seed desiccation: A bridge between maturation and germination. *Trends Plant Sci.* **2010**, *15*, 211–218. [CrossRef]
45. Klinger, Y.P.; Eckstein, R.L.; Horlemann, D.; Otte, A.; Ludewig, K. Germination of the invasive legume *Lupinus polyphyllus* depends on cutting date and seed morphology. *NeoBiota* **2020**, *60*, 79. [CrossRef]
46. Baskin, C.C.; Baskin, J.M. *Seeds: Ecology, Biogeography, and Evolution of Dormancy and Germination*; Academic Press: San Diego, CA, USA, 2014.
47. Fenner, M.K.; Fenner, M.; Thompson, K. *The Ecology of Seeds*; Cambridge University Press: Cambridge, UK, 2005.
48. Barnett, A.A.; Boyle, S.A.; Pinto, L.P.; Lourenço, W.C.; Almeida, T.; Silva, W.S.; Spironello, W.R. Primary seed dispersal by three Neotropical seed-predating primates (*Cacajao melanocephalus ouakary, Chiropotes chiropotes* and *Chiropotes albinasus*). *J. Trop. Ecol.* **2012**, *28*, 543–555. [CrossRef]
49. Blanco, G.; Hiraldo, G.; Tella, J.L. Ecological functions of parrots: An integrative perspective from plant life cycle to ecosystem functioning. *Emu Austral Ornithol.* **2018**, *118*, 36–49. [CrossRef]
50. Sebastián-González, E.; Hiraldo, F.; Blanco, G.; Hernández-Brito, D.; Romero-Vidal, P.; Carrete, M.; Gómez-Llanos, E.; Pacífico, E.; Díaz-Luque, J.A.; Denés, F.V.; et al. The extent, frequency and ecological functions of food wasting by parrots. *Sci. Rep.* **2019**, *9*, 15280. [CrossRef]
51. Tella, J.L.; Blanco, G.; Dénes, F.V.; Hiraldo, F. Overlooked parrot seed dispersal in Australia and South America: Insights on the evolution of dispersal syndromes and seed size in Araucaria trees. *Front. Ecol. Evol.* **2019**, *7*, 82. [CrossRef]
52. Young, L.M.; Kelly, D.; Nelson, X.J. Alpine flora may depend on declining frugivorous parrot for seed dispersal. *Biol. Conserv.* **2012**, *147*, 133–142. [CrossRef]
53. Blanco, G.; Bravo, C.; Pacífico, E.; Chamorro, D.; Speziale, K.; Lambertucci, S.; Hiraldo, F.; Tella, J.L. Internal seed dispersal by parrots: An overview of a neglected mutualism. *PeerJ* **2017**, *4*, e1688. [CrossRef]
54. Blanco, G.; Bravo, C.; Chamorro, D.; Lovas-Kiss, A.; Hiraldo, F.; Tella, J.L. Herb endozoochory by cockatoos: Is "foliage the fruit"? *Austral Ecol.* **2020**, *45*, 122–126. [CrossRef]
55. Bravo, C.; Chamorro, D.; Hiraldo, F.; Speziale, K.; Lambertucci, S.A.; Tella, J.L.; Blanco, G. Physiological dormancy broken by endozoochory: Austral parakeets (*Enicognathus ferrugineus*) as legitimate dispersers of Calafate (*Berberis microphylla*) in the Patagonian Andes. *J. Plant Ecol.* **2020**, *13*, 538–544. [CrossRef]
56. Hernández-Brito, D.; Romero-Vidal, P.; Hiraldo, F.; Blanco, G.; Díaz-Luque, J.A.; Barbosa, J.; Symes, C.; White, T.W.; Pacifico, E.G.; Sebastián-González, E.; et al. Epizoochory in parrots as an overlooked yet widespread plant-animal mutualism. *Plants* **2021**, *10*, 760. [CrossRef]

57. Tella, J.L.; Dénes, F.V.; Zulian, V.; Prestes, N.P.; Martínez, J.; Blanco, G.; Hiraldo, F. Endangered plant-parrot mutualisms: Seed tolerance to predation makes parrots pervasive dispersers of the Parana pine. *Sci. Rep.* **2016**, *6*, 31709. [CrossRef] [PubMed]
58. Cabrera, A.L. Regiones fitogeográficas argentinas. In *Enciclopedia Argentina de Agricultura y Jardinería*; Editorial Acme: Buenos Aires, Argentina, 1976; Volume 2, pp. 1–85.
59. Labraga, J.C.; Villalba, R. Climate in the Monte Desert: Past trends, present conditions, and future projections. *J. Arid Environ.* **2009**, *73*, 154–163. [CrossRef]
60. Salfity, J.A. Geología regional del Valle Calchaquí, Argentina. *Anal. Acad. Nac. Cien. Exact. Fis. Nat.* **2004**, *56*, 133–150.
61. Varela Ituarte, M.A.; Perea, M.C.; Neder, L.D.V.; Rios, R.D.V. Unidades del Paisaje del valle de Santa María, Cafayate (Salta, Argentina). *Rev. Asoc. Argent. Ecol. Paisajes* **2018**, *8*, 1–16.
62. Godoy-Bürki, A.C.; Biganzoli, F.; Sajama, J.M.; Ortega-Baes, P.; Aagesen, L. Tropical high Andean drylands: Species diversity and its environmental determinants in the Central Andes. *Biodivers. Conserv.* **2017**, *26*, 1257–1273. [CrossRef]
63. Fontana, M.L.; Pérez, V.R.; Luna, C.V. Características evolutivas en *Prosopis* spp., citogenética, genética e hibridaciones. *Rodriguésia* **2018**, *69*, 409–421. [CrossRef]
64. Hunziker, J.H.; Poggio, L.; Naranjo, C.A.; Palacios, R.A.; Andrada, A.B. Cytogenetic of some species and natural hybrids in *Prosopis* (Leguminosae). *Can. J. Gen. Cytol.* **1975**, *17*, 253–262. [CrossRef]
65. Ramírez, L.; de la Vega, A.; Razkin, N.; Luna, V.; Harris, P.J. Analysis of the relationships between species of the genus *Prosopis* revealed by the use of molecular markers. *Agronomie* **1999**, *19*, 31–43. [CrossRef]
66. Giamminola, E.M.; Viana, M.L. Caracterización morfológica de frutos y semillas de dos accesiones de *Prosopis nigra* (Griseb.) Hieron. Y *Cesalpinia paraguarensis* (D.Parodi) Burkart., conservadas en el Banco de Germoplasma de Especies Nativas de la Universidad Nacional de Salta, Argentina. *Lhawet Nuestro Entorno* **2013**, *2*, 23–29.
67. Prokopiuk, D.; Cruz, G.; Grados, N.; Garro, O. Estudio comparativo entre frutos de *Prosopis alba* y *Prosopis pallida*. *Multequina* **2000**, *9*, 35–45.
68. Sciammaro, L.; Ferrero, C.; Puppo, C. Agregado de valor al fruto de *Prosopis alba*. Estudio de la composición química y nutricional para su aplicación en bocaditos dulces saludables. *Rev. Agronom. La Plata* **2015**, *114*, 115–123.
69. Velez, S.; Chacoff, N.P.; Campos, C.M. Pre-dispersal seed loss in two *Prosopis* species (Fabacea: Mimosoidea) from the Monte Desert, Argentina. *Ecol. Austral* **2018**, *28*, 361–373. [CrossRef]
70. Campos, C.M.; Ojeda, R.A. Dispersal and germination of *Prosopis flexuosa* (Fabaceae) seeds by desert mammals in Argentina. *J. Arid Environ.* **1997**, *35*, 707–714. [CrossRef]
71. Campos, C.M.; Giannoni, S.M.; Taraborelli, P.; Borghi, C.E. Removal of mesquite seeds by small rodents in the Monte desert, Argentina. *J. Arid Environ.* **2007**, *69*, 228–236. [CrossRef]
72. Masello, J.F.; Quillfeldt, P.; Munimanda, G.K.; Klauke, N.; Segelbacher, G.; Schaefer, H.M.; Moodley, Y. The high Andes, gene flow and a stable hybrid zone shape the genetic structure of a wide-ranging South American parrot. *Front. Zool.* **2011**, *8*, 1–17. [CrossRef] [PubMed]
73. Masello, J.F.; Quillfeldt, P. ¿Cómo reproducirse exitosamente en un ambiente cambiante? Biología reproductiva del Loro Barranquero (*Cyanoliseus patagonus*) en el noreste de la Patagonia. *Hornero* **2012**, *27*, 73–88.
74. BirdLife International. *Cyanoliseus patagonus*. The IUCN Red List of Threatened Species. Available online: 2018:e.T22685779A132255876. (accessed on 6 April 2021).
75. Vargas-Rodríguez, R.; Squeo, F.A. *Historia Natural del Loro Tricahue en el Norte de Chile*; Universidad de La Serena Press: La Serena, Chile, 2014.
76. Masello, J.F.; Pagnossin, M.L.; Sommer, C.; Quillfeldt, P. Population size, provisioning frequency, flock size and foraging range at the largest known colony of Psittaciformes: The Burrowing Parrots of the north-eastern Patagonian coastal cliffs. *Emu Austral Ornithol.* **2006**, *106*, 69–79. [CrossRef]
77. Sánchez, R.; Ballari, S.A.; Bucher, E.H.; Masello, J.F. Foraging by burrowing parrots has little impact on agricultural crops in northeastern Patagonia, Argentina. *Int. J. Pest Manag.* **2016**, *62*, 326–335. [CrossRef]
78. Romero-Vidal, P.; Hiraldo, F.; Rosseto, F.; Blanco, G.; Carrete, M.; Tella, J.L. Opportunistic or Non-Random Wildlife Crime? Attractiveness rather than Abundance in the Wild Leads to Selective Parrot Poaching. *Diversity* **2020**, *12*, 314. [CrossRef]
79. Tella, J.L.; Romero-Vidal, P.; Dènes, F.V.; Hiraldo, F.; Toledo-González, B.; Rossetto, F.; Blanco, G.; Hernández-Brito, D.; Pacífico, E.; Díaz, J.A.; et al. Roadside car surveys: Methodological constraints and solutions for estimating parrot abundances across the world. *Diversity* **2021**. (in review).
80. Hadfield, J.D. MCMC methods for multi-response generalized linear mixed models: The MCMCglmm R package. *J. Stat. Softw.* **2010**, *33*, 1–22. [CrossRef]
81. Moore, R.P. *Handbook on Tetrazolium Testing*; The International Seed Testing Association: Zurich, Switzerand, 1985.
82. Grilli, P.G.; Soave, G.E.; Arellano, M.L.; Masello, J.F. Abundancia relativa del Loro Barranquero (*Cyanoliseus patagonus*) en la provincia de Buenos Aires y zonas limítrofes de La Pampa y Río Negro, Argentina. *Hornero* **2012**, *27*, 63–71.
83. Kissling, W.D.; Bohning-Gaese, K.; Jetz, W. The global distribution of frugivory in birds. *Glob. Ecol. Biogeogr.* **2009**, *18*, 150–162. [CrossRef]
84. Renton, K. Lilac-crowned parrot diet and food resource availability: Resource tracking by a parrot seed predator. *Condor* **2001**, *103*, 62–69. [CrossRef]

85. Selman, R.G.; Hunter, M.L.; Perrin, M.R. The feeding ecology of Ruppell's parrot *Poicephalus rueppellis* in Namibia. *Ostrich* **2002**, *73*, 127–134. [CrossRef]
86. Boyes, R.S.; Perrin, M.R. The feeding ecology of Meyer's parrot, Poicephalus meyeri, in the Okavango Delta, Botswana. *Ostrich* **2009**, *80*, 153–164. [CrossRef]
87. White, T.C. The significance of unripe seeds and animal tissues in the protein nutrition of herbivores. *Biol. Rev.* **2011**, *86*, 217–224. [CrossRef] [PubMed]
88. Nuevo, Á. *Manual Técnico de Dietología de Papagayos;* Castel Negrino: Aicurzio, Italy, 2015.
89. Catalano, S.A.; Vilardi, J.C.; Tosto, D.; Saidman, B.O. Molecular phylogeny and diversification history of *Prosopis* (Fabaceae: Mimosoideae). *Biol. J. Linn. Soc.* **2008**, *93*, 621–640. [CrossRef]
90. Masello, J.F.; Montano, V.; Quillfeldt, P.; Nuhlíčková, S.; Wikelski, M.; Moodley, Y. The interplay of spatial and climatic landscapes in the genetic distribution of a South American parrot. *J. Biogeogr.* **2015**, *42*, 1077–1090. [CrossRef]
91. Luebert, F.; Weigend, M. Phylogenetic insights into Andean plant diversification. *Front. Ecol. Evol.* **2014**, *2*, 27. [CrossRef]
92. Campos, C.M.; Peco, B.; Campos, E.; Malo, J.E.; Giannoni, S.M.; Suárez, F. Endozoochory by native and exotic herbivores in dry areas: Consequences for germination and survival of *Prosopis* seeds. *Seed Sci. Res.* **2008**, *18*, 91–100. [CrossRef]
93. Campos, C.M.; Campos, V.E.; Mongeaud, A.; Borghi, C.E.; De los Rios, C.; Giannoni, S.M. Relationships between *Prosopis flexuosa* (Fabaceae) and cattle in the Monte desert: Seeds, seedlings and saplings on cattle-use site classes. *Rev. Chil. Hist. Nat.* **2011**, *84*, 289–299. [CrossRef]
94. Vander Wall, S.B.; Longland, W.S. Diplochory: Are two seed dispersers better than one? *Trends Ecol. Evol.* **2004**, *19*, 155–161. [CrossRef] [PubMed]
95. Robertson, A.W.; Trass, A.; Ladley, J.J.; Kelly, D. Assessing the benefits of frugivory for seed germination: The importance of the deinhibition effect. *Funct. Ecol.* **2006**, *20*, 58–66. [CrossRef]
96. Marone, L.; Horno, M.; González del Solar, R. Postdispersal fate of seeds in the Monte desert of Argentina: Patterns of germination in successive wet and dry years. *J. Ecol.* **2000**, *88*, 940–949. [CrossRef]

MDPI

Review

Advancing Genetic Methods in the Study of Parrot Biology and Conservation

George Olah [1,2,*], Brian Tilston Smith [3], Leo Joseph [4], Samuel C. Banks [5] and Robert Heinsohn [1]

1 Fenner School of Environment and Society, The Australian National University, Canberra, ACT 2601, Australia; robert.heinsohn@anu.edu.au
2 Wildlife Messengers, Richmond, VA 23230, USA
3 Department of Ornithology, American Museum of Natural History, New York, NY 10024, USA; bsmith1@amnh.org
4 Australian National Wildlife Collection, National Research Collections Australia, CSIRO, Canberra, ACT 2601, Australia; leo.joseph@csiro.au
5 Research Institute for the Environment and Livelihoods, College of Engineering, IT and Environment, Charles Darwin University, Darwin, NT 0810, Australia; sam.banks@cdu.edu.au
* Correspondence: george.olah@anu.edu.au

Abstract: Parrots (Psittaciformes) are a well-studied, diverse group of birds distributed mainly in tropical and subtropical regions. Today, one-third of their species face extinction, mainly due to anthropogenic threats. Emerging tools in genetics have made major contributions to understanding basic and applied aspects of parrot biology in the wild and in captivity. In this review, we show how genetic methods have transformed the study of parrots by summarising important milestones in the advances of genetics and their implementations in research on parrots. We describe how genetics helped to further knowledge in specific research fields with a wide array of examples from the literature that address the conservation significance of (1) deeper phylogeny and historical biogeography; (2) species- and genus-level systematics and taxonomy; (3) conservation genetics and genomics; (4) behavioural ecology; (5) molecular ecology and landscape genetics; and (6) museomics and historical DNA. Finally, we highlight knowledge gaps to inform future genomic research on parrots. Our review shows that the application of genetic techniques to the study of parrot biology has far-reaching implications for addressing diverse research aims in a highly threatened and charismatic clade of birds.

Keywords: Psittaciformes; conservation genetics; ecology; evolution; genomics; museomics

Citation: Olah, G.; Smith, B.T.; Joseph, L.; Banks, S.C.; Heinsohn, R. Advancing Genetic Methods in the Study of Parrot Biology and Conservation. *Diversity* **2021**, *13*, 521. https://doi.org/10.3390/d13110521

Academic Editors: José L. Tella, Guillermo Blanco and Martina Carrete

Received: 23 July 2021
Accepted: 9 October 2021
Published: 23 October 2021

1. Introduction

The order of parrots (Psittaciformes) contains a diverse group of species distributed mainly in tropical and subtropical regions [1,2]. Around one third of the nearly 400 parrot species are threatened, and they are declining faster than other comparable groups of birds, making them one of the bird orders of greatest concern [3]. The most important threats affecting parrots are anthropogenic and include agricultural expansion, the wildlife trade, logging, climate change, and invasive alien species [3,4]. However, the relative importance of these threats differs geographically. In the Neotropics, agriculture is the greatest threat followed by the illegal pet trade and logging [5]. In the Afrotropics, the illicit wildlife trade has the biggest impact followed by agriculture and logging [3,6,7], and in the regions of Oceania and Indomalaya, logging and invasive species are the most critical threats to the survival of the endemic parrot species [8]. Some species have been introduced to regions outside their natural ranges, including cities worldwide [9], where they may be perceived as pests [10].

The discipline of genetics (using it in this review for all methods that include molecular analysis of DNA) has made a major contribution to understanding the natural world. With the advancement of new DNA sequencing technologies in the past two decades, genetic research has been revolutionised and now has a wide range of applications to the field of biology and beyond. Genetics has contributed to the study of parrots in the wild and in captivity by helping to construct precise phylogenies [11,12], tracking the history of their early diversification [13], contributing important information at the population and individual levels to help conservation efforts [14–16], and revealing insights into their ecology and health [17,18]. Molecular genetic approaches have even been also used to further our understanding of long extinct parrot species [19–21]. Here, we review what has been learned through the use of different genetic methods applied to parrot studies in past decades and in the current era of genomics. The aim of this review is to provide a comprehensive overview of this field and highlight knowledge gaps to inform future genomic research on parrots.

2. Short History of Advances in Genetic Studies of Parrots

The word "genetic" was used for the first time in 1819 by Hungarian nobleman Imre Festetics who formulated a number of rules of heredity [22], laying the groundwork for the discovery of Mendelian genetics in the mid-19th century [23]. However, the molecular background of these ground-breaking theories was unknown until the determination of the structure of the DNA molecule in 1953 [24], leading to deciphering of the genetic code and the central tenets of molecular biology [25]. The invention of the polymerase chain reaction (**PCR**; see Glossary) in 1983 enabled the amplification of DNA and revolutionised genetic research. Even though many bird studies made use of DNA fingerprinting from the late 1980s, molecular studies of wild parrots started more slowly.

The first scientific publications on parrot genetics used karyotypes and allozymes to study the chromosomal and protein evolution of parrots at the taxonomic levels of species, family, and order [26–29]. These were followed by molecular sexing with gel electrophoresis [30]. Then came the advent of mitochondrial DNA (mtDNA) analysis initially studied by enrichment and cloning, and later by sequencing of individual mitochondrial and nuclear genes (Sanger sequencing) eased by PCR technology. The initial research focus on parrots with these methods was on phylogeny and systematics [12,31,32] and then work increased on species-level taxonomy and phylogeographic scales [33–36].

Studies of detailed population structure and individual-based behaviour in wild populations of parrots began with the advent of DNA fingerprinting by **minisatellites**. Minisatellites (complex tandem repeat regions of DNA) were used, for instance, on the Burrowing Parrot *Cyanoliseus patagonus* [17], some macaw species [37–39], and in the Palm Cockatoo *Probosciger aterrimus* [40]. Later, the discovery of **microsatellite** genetic markers (simple sequence repeat) transformed the application of genetics to many biological research projects including parrot studies. The length of these markers can be measured precisely by capillary electrophoresis, providing a great advantage over the original fingerprinting methods of minisatellites visualised by gel electrophoresis. Various studies have identified and published species-specific microsatellites for parrots [41–47]. This advance resulted in important tools for a wide variety of genetic research via cross-species amplification to other parrot species. Microsatellites were mainly used for fine-scale studies of individuals including family relationships. For example, Klauke et al. [48,49] used these markers to report a cooperative breeding system, not widely known in parrots (e.g., [50]), and estimated fine-scale population structure in the recently discovered El Oro Parakeet *Pyrrhura orcesi*.

Early, or first-generation, sequencing technologies (e.g., Sanger sequencing) made it possible to read the genetic code of specific DNA sequences. Later, molecular genetic technology advanced and phased into the second- or **next-generation sequencing (NGS)** or genomics era. The massively parallel high-throughput feature (i.e., sequencing multiple fragments and individuals at once) of these new sequencing platforms pushed down the price of sequencing and sped up the process of whole **genome** sequencing. The first complete mitochondrial genome (mitogenome) of a parrot was published in 2004 for the Kākāpō *Strigops habroptila* [51], followed by many more (e.g., [52,53]). The first draft of a full parrot genome, the Budgerigar *Melopsittacus undulatus*, was uploaded to the National Center for Biotechnology Information (NCBI) database in 2011 (www.ncbi.nlm.nih.gov/genome/10765; accessed on 16 October 2021). This was followed by the Puerto Rican Amazon *Amazona vittata* in 2012 [54], and the Scarlet Macaw *Ara macao* in 2013 [55]. Whole genome sequencing aided the discovery of new microsatellite markers, for example the Orange-bellied Parrot *Neophema chrysogaster* [56] and the Scarlet Macaw [44]. At the time of this publication, there are whole genomes available from second-generation technologies for 36 parrot species and complete mitochondrial genomes for 69 parrot species in the NCBI genome database (www.ncbi.nlm.nih.gov/genome; accessed on 19 July 2021). This is approximately 10% and 20% of parrot species for nuclear and mitochondrial genomes, respectively. Currently, the best available parrot genomes assembled at the chromosome level belong to the Budgerigar (61x coverage, scaffold N50 size of 104 Mb, annotated with 16,458 protein-coding genes; GenBank assembly accession: GCA_012275295.1), Kākāpō (76x, scaffold N50 = 83 Mb, 16,053 protein-coding genes; GCF_004027225.2), Blue-fronted Amazon *Amazona aestiva* (60x, scaffold N50 = 89 Mb; GCA_017639355.1), and Monk Parakeet *Myiopsitta monachus* (67x, scaffold N50 = 76 Mb; GCA_017639245.1). However, international consortia of scientists continue to sequence the genomes of many more species. The Genome 10K project aims to sequence the genomes of representatives from all genera of vertebrates [57]. The B10K project [58,59] and OpenWings Project (openwings.org) aim to sequence all extant bird species and understand their evolutionary histories and relationships.

NGS opened new research pathways to genome-wide association studies aimed at understanding the underlying genetic variants determining traits [60]. Microsatellite, mitochondrial, and multi-locus studies have transitioned into analyses of many more (sometimes thousands) polymorphic sites of **single nucleotide polymorphisms (SNPs)**, which are found throughout the coding- and non-coding parts of the genome, giving them a further advantage over microsatellites. SNPs can be generated in several ways. In restriction-site-associated DNA sequencing (RAD-seq), one or two restriction enzymes cut the genome at enzyme-specific restriction sites, and these fragments are then barcoded, filtered, and sequenced [61,62]. In sequence capture methods, oligonucleotide probes (baits) are designed to hybridise with specific regions of interest. These are then captured, barcoded, and enriched before sequencing [63]. For this technique, the sequences of interest need to be known (e.g., from a complete genome of the same or related species), or the baits can be generated by other techniques such as double-digest RAD-seq [64]. The sequencer generated data can then be analysed in different bioinformatic pipelines [65]. This more comprehensive sampling of the genome has enabled more detailed examination of signatures of selection and local adaptation on the genome [66]. Sequencing RNA shows which genes are being expressed (transcriptomics) and can have an important role in reintroductions by predicting the potential for local adaptation and tolerance capacity in the source population [67].

In the past decade, further advancements in genome sequencing technology have pushed the boundaries of data collection. For example, nanopore technology has enabled portable sequencers as small as a USB drive [68,69]. Of great help to parrot biologists interested in conducting genetic research in the field, these sequencers were shown to work even in harsh environments [70]. Parrots are notoriously hard to capture in the wild, so making use of non-invasive sampling methods (e.g., feathers, eggshells, faeces, or even residual saliva) with the new technologies will provide further advances [71–73]. Metagenomic and metabarcoding applications, where all DNA materials are extracted from environmental samples, allow bioinformatic pipelines to be used to find and match sequenced DNA of species with reference to online databases (e.g., NCBI GenBank, European Variation Archive). This way, the presence or absence of species can be detected in the environment, and abundance estimates might be derived in some cases [74,75], although this technique needs further development. One of the major limitations of current parrot genomes is that they were produced with short-read sequencing. These short-reads make genome assembly more challenging by causing genomes to be more fragmented (with smaller scaffold sizes) and incomplete, and limit the accuracy of some downstream uses of genomic data (e.g., studying structural variants). The advent of long-read or third-generation sequencing technologies can produce reads greater than 10 Kb in length, which allow for the assembly of chromosomal-level genomes [76]. These advances still present challenges including high-cost, specialised bioinformatic expertise, and access to high-quality genetic samples (i.e., DNA samples with high molecular weight), but these limitations are likely to be overcome in the near future.

3. Research Fields

3.1. Deeper Phylogeny and Historical Biogeography

An obvious contribution of genetics to parrot research is the construction of an accurate molecular phylogeny of the group. The first compendium of DNA-based molecular systematics for all birds was published in 1990 and was based on DNA–DNA hybridization of the whole genome [77]. Early sequencing studies used only a handful of genes mainly from mtDNA to study the phylogenetic relationship among some species [78]. Later studies included DNA sequences of both mitochondrial and nuclear origin to gain better resolution within certain taxa, like the genera *Amazona* [79], *Forpus* [36], or the broad-tailed (platycercine) parrots [80–82] and cockatoos (Cacatuidae) [83]. Using more genes and eventually whole genome resequencing [84], phylogenomics helped to resolve previously conflicting relationships on the phylogenetic tree. A surprising higher level result showed that parrots are the sister group of passerines (Passeriformes) and that falcons (Falconidae) are the sister group of both [85]. This was later robustly confirmed by other studies [86–88]. One of the most complete recent phylogenies for parrots was published in 2018 using a 30-gene supermatrix (12 mitochondrial and 18 nuclear genes) and included 307 species [11]. This study highlighted that phylogeographic or population genetic studies were only available for about a third of the extant parrot species [11].

Phylogenetic data indicate that parrots originated from the southern supercontinent Gondwana [32,89–91], while the fossil record has been interpreted to indicate a northern origin [92]. Similarly, the time of the origin of parrots is under debate, where molecular dating is used in addition to the fossil record and biogeographic distributions [93]. Cretaceous origin, before the Cretaceous-Paleogene extinction event 66 million years ago (Mya), was proposed by an early study based on multilocus phylogeny and a splitting of New Zealand from Gondwana calibration [12]. Other studies based on three nuclear genes coupled with divergence dates from non-parrot bird fossil evidence also suggested dispersal from Australasia and Antarctica, but later in the Paleogene (66–23 Mya) period [13]; initial vicariance events (i.e., continental breakups) were followed by local radiations and crown group diversification around 58 Mya [94]. Taking into consideration the split between falcons and parrots/passerines (57–62 Mya), and between parrots and passerines (51.8–66.5 Mya), we note that current data suggest that parrot crown-group diversification probably happened in the early Oligocene, around 28–34 Mya [95].

There is consensus that the Strigopoidea superfamily (containing the New Zealand Kākā *Nestor meridionalis*, Kea *Nestor notabilis*, and Kākāpō) is sister to all other parrots, i.e., the clade containing Psittacoidea and Cacatuoidea [96]. Rheindt et al. [97] argued that within Strigopoidea the *Strigops* and *Nestor* lineages diverged probably ca. 28–29 Mya. This would have coincided with the potential Oligocene submergence of Zealandia when much of its landmass may have been fragmented into smaller islands, providing a setting for allopatric diversification [98]. Since their origination in the Neotropics, the Arini tribe diversified by early adaptive radiation, the rate of which has remained constant [99]. Constant diversification was also shown at a shallower phylogenetic scale in the Neotropical parrotlet genus *Forpus* over the past 5 Myr [36] but the pattern was dependent on how species were delimited. Also in the Neotropics, most of the speciation events in the genus *Aratinga* (*sensu lato*) occurred during the Pliocene (5.3–2.5 Mya) and Pleistocene (2.5–0.01 Mya), possibly related to climatic oscillations [100]. In what now comprises the genera *Pionopsitta* and *Pyrilia*, however, diversification was attributed more to geotectonic events and river dynamics between 8.7 and 0.6 Mya than to glacial cycles [101,102]. A study on *Pionus* spp. (a genus occurring both in the Andes and the lowland Amazonian rainforest) showed that the elevation of mountains explained their disjunct diversification, while subsequent speciation within the mountains was linked to climatic oscillations and their effects on habitat change [103]. This was also confirmed with the other parrot species, implying a dynamic climatic history for South American biomes since the Pliocene [104]. With the increased availability of genetic datasets of parrots and other taxa with which they co-occur, it will be possible to directly test these proposed speciation hypotheses. Without genetics, it would have been impossible to reconstruct the historical biogeography of parrots. However, there are still many questions left about the exact routes and time of their early diversification, and the incongruency regarding the fossil record. With expanding detailed genomic data of parrot species, these questions might be better answered soon.

3.2. Species- and Genus-Level Systematics and Taxonomy

The species is the widely accepted default unit used for evaluating conservation status (e.g., in the IUCN Red List), hence defining species and resolving taxonomic uncertainties by genetic techniques is important for conservation [105]. Active speciation of parrots on islands is most readily evident in Australasia, as shown by the *Eclectus roratus* and *Trichoglossus haematodus* complexes [106]. In such cases of dynamic evolution, wider sampling and genetic data of finer resolution are often needed to resolve phylogenetic relationships [107]. The extinction of island-endemic parrot species and replacement by invasive alien species led to loss of phylogenetic diversity, but understanding these frameworks can aid conservation strategies to restore island ecosystem function [108].

In some parrots, the traditional taxonomy based on plumage might need some revision, as shown with a genetic study on amazon parrots in the Neotropics [109]. **Cryptic species** of parrots were suggested by genetic studies for various taxa, including the mealy amazons *Amazona* spp. (*A. farinosa, A. guatemalae*) in the Neotropics [110] and the ground parrots *Pezoporus* spp. (*P. wallicus, P. flaviventris*) in Australia [111]. The need to recognize subspecies within the Mulga Parrot *Psephotellus varius*, generally considered monotypic, was also evident from phylogeographic structure either side of a well-known biogeographic barrier in southern Australia in their mitogenomic diversity and genome-wide nuclear markers [112]. Notably in contrast, recognition of *Amazona gomezgarzai* by Silva et al. [113] has been roundly debunked by Escalante et al. [114].

Defining **management units (MUs)** within species also holds important merits for conservation [115], however a refinement to the original definition, which was framed in terms of allele frequency differentiation, would be to define MUs with reference to the management issue in question, such as identifying demographically independent units for population monitoring, or genetically differentiated units for mixed-source introductions. For example, a genetic study revealed cryptic diversity within the Bahama Amazon *Amazona leucocephala bahamensis* between populations living on two remote islands [116]. A study on the Blue-fronted Amazon suggested treating its two subspecies as separate MUs [33], and a recent study argued for MU consideration for the Atlantic Forest population of the Southern Mealy Amazon *Amazona farinosa farinosa* [117]. Similarly, another study on Military Macaws *Ara militaris* in Mexico proposed two MUs in the country based on genetic data [118]. In Africa, a study warned that a population of Grey Parrot *Psittacus erithacus* living on Príncipe Island, São Tome and Príncipe, should be treated as an independent MU from the continental African populations, given their evolutionary dynamics and heavy local poaching pressure [119]. The Cape Parrot *Poicephalus robustus* was previously considered to comprise three subspecies until a study using multilocus DNA analyses concluded that *P. r. robustus* diverged from *P. r. suahelicus* and *P. r. fuscicollis* around 2.4 Mya [120]. Accordingly, it is now usually treated as a monotypic species *P. robustus* and has been uplisted to Vulnerable by the IUCN Red List, while the other two subspecies now form the Brown-necked Parrot *P. fuscicollis* complex of Least Concern (e.g., [121–123]).

Evolutionarily significant units (ESUs) are independently evolving units of genetic variation [115]. These units were proposed for the two subspecies of the Orange-fronted Parakeet *Eupsittula canicularis* in Mexico [124]. A comprehensive genetic analysis (using genome-wide SNPs and mitochondrial data) of the Red-tailed Black-Cockatoo *Calyptorhynchus banksii* identified five ESUs over their large distribution, and advised taxonomic reassessments including recognition of a new subspecies [125]. Distinctions between ESUs and MUs were made during a genetic assessment of Major Mitchell's Cockatoo *Lophochroa leadbeateri* [126]. An analysis employing mtDNA and microsatellite data failed to detect genetic evidence for the two subspecies of Kākā in New Zealand, instead it is hypothesised that phenotypic diversity was due to an adaptive latitudinal size cline consistent with Bergmann's rule [127], an important consideration for possible translocation attempts. In contrast, another study using similar genetic evidence argued that the current genetic clusters of Kea should not be considered as independent conservation units because the structure evolved through very recent postglacial recolonisation processes [128]. In these and similar cases, appropriate taxonomic rank is debatable, but conservation and management units can be assigned where appropriate. Again, as shown with the example studies, these units of conservation can only be revealed with the help of genetic studies, which also have an ever-growing role in defining taxonomic units.

3.3. Conservation Genetics and Genomics

Conservation genetics is an interdisciplinary science dealing with the genetic factors affecting extinction risk of species and how to minimise these risks [129]. It is transitioning into using genomic techniques [66]. In the previous section, we discussed the importance of phylogeny to conservation. Here, we provide an overview of other major areas where the transition to genomics has contributed to the conservation of parrots.

Preventing the loss of **genetic diversity** is an essential aim of any conservation project. Genetic monitoring can provide important tools to quantify this diversity before, during, or after management efforts on threatened parrot species or populations [130,131]. In small remaining populations of species, diversity can be lost due to **genetic drift**, which can override natural selection [132]. Intensive management restored the Echo Parakeet *Psittacula echo* population from 20 remaining individuals in 1987. Genetic research showed that re-distribution of genetic material among its populations has reduced the likelihood of losing private alleles that could otherwise be lost due to the random effect of genetic drift in small, isolated populations [16]. On the island of Tasmania and its own offshore islands, a study of the migratory Swift Parrot *Lathamus discolor* could not detect genetic differentiation among breeding populations in consecutive years and across multiple islands [133]. Genetic estimations were used to calculate the **effective population size** of their single, panmictic population, and after combining it with demographic data, the study calculated a potential contemporary population size as low as 300 individuals [134].

Contemporary population fragmentation due to anthropogenic factors can lead to reduction in **gene flow** among the fragments resulting in genetic structure detectable via genetic testing. It is important to detect early signs of genetic fragmentation as it could lead to loss of genetic diversity and eventually to inbreeding. However, these effects take time, depending on habitat corridors, migration rates, and the mobility, dispersal, and lifespan of the species. For instance, at least a 35-year-long lag was shown between deforestation in the Brazilian Cerrado biome and changes in the genetic structure of Goias Parakeet *Pyrrhura pfrimeri* populations [135], corresponding to about five generations of the species. Genetic structure was also found in the Scarlet Macaw in the highly fragmented landscape of Costa Rica [136]. Historical population structure can also have important implications for present day conservation efforts. A broader genetic analysis of the Scarlet Macaw for instance suggested a distinct conservation unit for its Central American subspecies *A. m. cyanoptera* [137]. A population genetic study on the Palm Cockatoo on Cape York Peninsula, Australia found genetic differences among the studied populations, probably due to a mountain barrier [138]. Incorporating this population genetic data, especially the connectivity between populations, into a **population viability analysis (PVA)** model predicted that dispersal between populations is not enough to buffer decline given their extremely low breeding success. The study concluded that Palm Cockatoos in Australia should be uplisted from Vulnerable to Endangered [139].

Genetic studies can have an important role in *ex situ* conservation management of threatened species to avoid **inbreeding** and to maintain maximum genetic diversity among captive individuals. Genetic testing can accurately identify relatedness among birds, which can be useful for the mixing of breeding pairs as demonstrated by *Amazona* parrots [140]. However, in socially monogamous species like parrots, natural mate choice can result in higher reproductive success than forced choice based solely on genetics [141], as shown for Cockatiels *Nymphicus hollandicus* [142], where pairs with higher behavioural compatibility were better parents [143]. The effect of **inbreeding depression** was first explicitly studied in parrots with respect to clutch size of captive budgerigars in the 1980s [144]. It has been used to guide the Puerto Rican Amazon recovery program through genetic fingerprinting since the 1990s [14]. Low levels of inbreeding were detected for the Red-tailed Amazon *Amazona brasiliensis,* indicating that more direct threats, like habitat destruction and illegal wildlife trade, should be the focus of conservation efforts [145]. Genetics has also helped to identify the pedigrees of the remaining Kākāpō population *in situ* [15] and inform conservation strategies [146]. It can also detect signs of genetic adaptation to captivity, which can have negative effects on reintroduction success. A recent genetic study on wild and captive populations of Blue-throated Macaws *Ara glaucogularis* and Thick-billed Parrots *Rhynchopsitta pachyrhyncha* highlighted the need for both *in situ* and *ex situ* conservation strategies [147].

Establishing captive populations of endangered species is often used by conservation management programmes. However, rapid genetic adaptation to captivity (within a few generations), low founder diversity, and potential inbreeding are of concern for future recovery goals, but these have been rarely studied in parrots. A captive population of Orange-bellied Parrots was founded in 1985 and later supplemented with wild individuals. A recent study found low diversity in their toll-like receptors (TLR), partially responsible for the innate immune response and so the first line of defence against pathogens, highlighting that they might be unable to adapt to novel disease outbreaks [148]. For instance, a spillover of beak and feather disease virus (BFDV) to the remaining wild population almost wiped out the entire species [149]. The psittacine beak and feather disease (PBFD) was first reported on Red-rumped Parrots *Psephotus haematonotus* in 1907 near Adelaide, Australia [150]. BFDV was isolated and characterised much later from cockatoos [151]. PCR tests were developed for the detection of BFDV [152,153], helping to identify cases in psittacines. A recent study provides an excellent overview of the ecology of PBFD in parrots and highlights the importance of mitigating its effects on threatened parrot species [18]. BFDV is also an ongoing threat to many other Australian parrot species [154]. Another study, using SNP data of the wild and captive populations of the Orange-bellied Parrot, showed that their genetic diversity could be retained in the captive population [155], possibly improving their health for future reintroductions. Retaining diversity at the major histocompatibility complex (MHC) is also important, as it is responsible for the adaptive immune response in birds and other vertebrates [156]. However, the MHC has been studied in only a handful of parrot species, including the Budgerigar [157], the Green-rumped Parrotlet *Forpus passerinus* [158], and the Red-crowned Parakeet *Cyanoramphus novaezelandiae* [159].

Outbreeding depression occurs when distinct species hybridise or isolated populations of the same species are mixed and the results are adverse [160]. One proposed underlying mechanism is that species have coadapted gene complexes nearby on the same chromosomes and that recombination during hybridization disrupts their adaptive functions [161]. Alternatively, outbreeding depression is likely to be rare and its effects restricted to the first few generations of crossing among evolutionarily diverged lineages [162]. Around 8% of parrot species have been recorded to hybridise in the wild [163] and almost half of all parrot species have been reported to hybridise in captivity [164]. Genetic screening of the last remaining population of the Critically Endangered Forbes' Parakeet *Cyanoramphus forbesi* helped to determine the magnitude of hybridisation with the Chatham Island Red-crowned Parakeet *C. novaezelandiae chathamensis* and to identify cryptic hybrids [165]. A complex hybrid zone was studied involving the phenotypically distinct non-sister species Pale-headed Rosella *Platycercus adscitus* and Eastern Rosella *P. eximius*, and showed a lack of post-zygotic barriers to gene flow between these species [166]. The last remaining male individual of the Spix's Macaw *Cyanopsitta spixii* was breeding with a Blue-winged Macaw *Primolius maracana* and genetic sequencing showed that the resultant embryo was indeed a hybrid of the two species, but it never hatched [167].

Molecular genetic techniques can be applied in **wildlife forensic** investigations. Molecular genotyping helped Australian authorities to match DNA extracted from eggshells found in the wild to a nestling of Red-tailed Black-Cockatoo at a nearby property [168]. During the investigation, forensic scientists concluded that the nestling was hatched from the eggshell recovered from a tree hollow and this led to a criminal conviction. In another case, eggs were seized from an alleged trafficker arriving in Australia. Comparing the extracted mtDNA to the genetic database of the NCBI, researchers identified several threatened parrot and cockatoo species, and the smuggler was prosecuted [169]. Poachers were also arrested in Brazil intending to fly to Europe, one in 2003 with avian eggs later identified by molecular genetic techniques as of parrots and owls [170], and another in 2018 with eggs identified as of Short-tailed Parrot *Graydidascalus brachyurus* [171]. Ewart et al. [126] developed a forensic test with 20 nuclear SNPs for the Major Mitchell's Cockatoo and demonstrated its application for subspecies identification. A similar toolkit combining various forensic techniques was developed earlier for the Glossy Black-Cockatoo *Calyptorhynchus lathami* [172]. A set of microsatellites were developed in the Cape Parrot with sufficient discriminatory power to distinguish captive versus wild birds via parentage analyses [173], and similar markers proved to be successful in determining the geographic origin of a captive individual of Military Macaw [174]. The control regions of mtDNA of Blue-and-yellow Macaws *Ara ararauna* confiscated from the illegal wildlife trade in Brazil were sequenced and compared to reference sequences of the species, in order to find their provenance and advise on reintroduction planning [175].

3.4. Behavioural Ecology

Genetic techniques have revealed many interesting aspects of behaviour in parrots. Wirthlin et al. [176] looked at the genomic basis of high cognitive abilities, vocal communication, and longevity in parrots by generating an annotated genome for the Blue-fronted Amazon and comparing it to 30 other bird species. They discovered new lifespan-influencing genes, parrot-specific genes critical for brain function, and even indications of convergent evolution of cognition relative to changes in the human genome. Phylogenetic analysis was used to study another cognitive function of parrots, cerebral lateralisation, which is also closely linked to the development of human language [177]. This underpins the well-established behaviour in many parrot species of using the left foot for holding food [178,179], and which may have a fitness benefit deeply rooted in their evolutionary history. Similarly, Benavidez et al. [180] applied phylogenetic analyses to look at diet and range size of Neotropical parrots. They found that diet was independent of phylogenetic history and that range and body size explained diet composition.

Genetic evidence has been often used to reveal an unexpected diversity of breeding systems and individual dispersal patterns in parrots. Using DNA fingerprinting for paternity testing on parrots, Masello et al. [17] found that the Burrowing Parrot is an example of both social and genetic monogamy. This social structure was also shown to be the case for Palm Cockatoos reusing nests in Australia [40] and Blue-and-yellow Macaws in Brazil [181]. When mtDNA and nuclear microsatellite genetic markers were compared for the same species, the observed patterns were best explained by male-biased dispersal and female philopatry [182]. Through application of microsatellite genetic markers, Heinsohn et al. [183] revealed cooperative polyandry and polygynandry in Eclectus Parrots *Eclectus roratus* in northern Australia. Another study showed remarkably similar cooperative polyandry in the Greater Vasa Parrot *Coracopsis vasa* in Madagascar [184].

A 6-year-long study incorporating genetic sampling of nestlings, eggshells, and adults of the Swift Parrot proved that their clutches had high levels (50%) of multiple paternity of the nestlings although the birds remained socially monogamous [185]. Molecular sexing showed that Swift Parrots have adaptive sex allocation with mothers biasing their early hatched nestlings towards males. This is interpreted to allow the males to get extra food and gain greater fitness when they later compete for rare females [186]. The study used population viability analysis to predict a dramatic decline in population size due to an introduced predator to Tasmania. Extra pair paternity was also confirmed in the Echo Parakeet in Mauritius [187]. In the Monk Parakeet, sexual monogamy was shown in their native and invasive sites [188], while a later study found evidence for extra pair paternity in their native range in Argentina and intra-brood parasitism at invasive sites [189]. A recent study on their breeding colonies showed fine-scale genetic structure, high breeding site fidelity, absence of inbreeding, and female-biased natal dispersal by genotyping individuals [190].

In Ecuador, breeding pairs of El Oro Parakeets have been shown to have helpers, whose genetic quality (measured as heterozygosity by microsatellite markers) increased reproductive success of the breeding pairs [49]. The above studies all questioned the widely held notion that parrots are monogamous, and instead showed that parrots have flexible mating systems. In both Eclectus Parrots and Swift Parrots, polyandry is believed to be a result of strong, male-biased adult sex ratios [183,185]. Conversely, it is unknown whether the similarly biased sex ratio in Glossy Black-Cockatoos on Kangaroo Island [191] is associated with polyandry.

A study examined the association between genetic structure and song culture in the Yellow-naped Amazon *Amazona auropalliata*. It found that the factors are not closely associated and that there is high, possibly female-biased gene flow across dialect boundaries [192,193]. There is little evidence that dialects in *Amazona* parrots would isolate populations, which would eventually generate genetic differences among the populations [194]. A recent study on these species showed that their call and genetic divergence did not correspond, which indicated that vocal dialects are not the best surrogates for genetic structure in lifelong local learners like *Amazona* parrots [117]. A study using SNP data of Palm Cockatoos found an association between the nuclear genomic structure of the populations and vocal dialect boundaries, however, these possibly originated from the separation of populations by mountains in the late-Quaternary [138]. In Budgerigars, their life-long vocal learning was found to be associated with the expression levels of specific transcription factors, hence their regulation seems to be essential for vocal mimicry [195,196]. Genetics will further our understanding of the mating system, song culture, and even cognitive abilities of parrots. So far only a handful of genetic studies have focused on these topics and implementing them to other parrot species could reveal important insights into the behavioural ecology of this diverse group.

3.5. Molecular Ecology and Landscape Genetics

Molecular ecology has illuminated the origin of some introduced parrot species, which has been recently reviewed [197]. For instance, Russello et al. [198] sequenced the mtDNA control regions of Monk Parakeet museum specimens from the species' native range and of individuals from their naturalised range in the United States. Their results confirmed that the geographic origins of the U.S. populations overlapped with past trapping records, so the naturalised populations possibly originated from the international pet trade whether from accidental or purposeful releases. A global study of their invasive populations also supported the pet trade hypothesis and observed low genetic diversity, indicating that invasiveness might not be linked to high genetic variation and the role of selection should be further investigated in allowing the birds to adapt to novel urban settings [199]. The success of the Ring-necked Parakeet *Psittacula krameri* as an invasive species to Europe was also studied by determining the genetic origin of the invasive populations [200]. The study showed admixture between individuals from different origins and argued that morphological changes in the introduced parrots might be attributed to their rapid adaptation to European environments over the past 50 years.

Molecular techniques can be used to track individuals in the landscape using their genotype, analogous to telemetry studies. Termed genetic tagging, this technique has been applied to macaws in Peru using shed feathers in the landscape as the source of genetic material [201]. The study revealed how macaws used clay licks and it enabled group size estimates based on genetic capture-mark-recapture analysis [72]. Such non-invasive genetic sampling provides an important tool for studying wild parrot populations, negating the need to capture the birds [202].

Landscape ecology is an interdisciplinary science focusing on the ecological understanding of spatial heterogeneity. Incorporating genetic studies into landscape ecology can reveal the complexity of genetic structure compared to the simpler approach using comparisons of populations selected *a priori*. A landscape genetic study, applying the theory of electrical circuits and resistance surfaces, on Scarlet Macaw populations in Peru showed that outlying ridges of the Andes mountains can limit gene flow between populations [203]. Similar findings were made on the same species in Costa Rica [136]. In the Ecuadorian Andes, limited dispersal was found in the El Oro Parakeet in a fine-scale landscape genetic study [48]. The genetic divergence between populations was again attributed to geographic barriers. The authors argued that climate change might explain upslope movement of this already endangered species eventually leading to isolation of populations. Another study also used a landscape genetics approach to look at climatic and geographic effects on the genetic structure of the Burrowing Parrot in the Southern Andes [204], and revealed that climate (precipitation and temperature) indeed drove changes in their genetic structure.

In Australia, dispersal of Palm Cockatoos is inhibited by narrow corridors of rainforest habitat, the two major populations being poorly connected due to a mountain barrier [205]. In contrast, no geographical or ecological barriers were found for the Red-fronted Macaw *Ara rubrogenys* across inter-Andean valleys in Bolivia. This suggests that social factors might reinforce their philopatry-related genetic structure, as cliffs with nest sites are not continuously distributed across the landscape [206]. Landscape genetics was also used to study the historical and current distributions of the Crimson Rosella *Platycercus elegans* complex, showing that population expansion followed by secondary contact and hybridization might be responsible for their present genetic structure [207]. A recent study looked at functional genomic differences between the alpine Kea and the forest adapted Kākā in New Zealand, and showed that these adaptations are not driving the ecological differentiation between the two species [208].

Understanding the drivers of genetic structure of parrots in the natural environment can be important for understanding the impacts of anthropogenic and natural dispersal barriers and help guide decisions about important corridors for maintaining population connectivity and gene flow. Genetics at the landscape level also helps us to understand the environmental correlates of population boundaries, assign MUs, and inform better decisions on connectivity plans.

3.6. Museomics and Historical DNA

Museum collections are becoming increasingly important in genomic studies as they are repositories of genetic material from the past [209,210]. Using historical DNA (hDNA) of birds can be challenging but nevertheless offers important insights into their evolution, ecology, and conservation [211]. Museum samples have been used to study the subspecies- [212] and population-level structures [138] of the Palm Cockatoo in New Guinea and Australia, of the *Pezoporus* ground parrots [111], and of Red-tailed Black-Cockatoos and Major Mitchell's Cockatoos in Australia [125,126,213]. Jackson et al. [108] extracted mtDNA from toepad samples of three extinct *Psittacula* parrots (*P. exsul*, *P. eques*,

P. wardi). They resolved the species' taxonomic placement and quantified how their replacement on Indian Ocean islands by the invasive Ring-necked Parakeet led to the loss of endemic phylogenetic diversity. Conversely, another study of Indian Ocean parrots [214] involving the extinct Mascarene Parrot *Mascarinus mascarin* was misled by technical errors, which led to generation of a false hypothesis about its taxonomic placement [215].

Several hDNA studies have recently used genome-scale data to look at whether now extinct or endangered species were declining prior to the Anthropocene. The first mtDNA sequences from the extinct Carolina Parakeet *Conuropsis carolinensis* museum specimens were obtained in 2012 [216], and their analysis found robust support for placing the species in a clade of long-tailed parrots, including the genus *Aratinga*. Gelabert et al. [19] generated the whole genome of this species and found no evidence of a dramatic demographic decline in the past or of excess homozygosity, reinforcing anthropogenic causes of the species' extinction. Another study looking at the extinction of eastern North American birds found lower genetic diversity in Carolina Parakeets and a lower effective population size, but a similar demographic history compared to species that persisted; this study also suggested their disappearance was due to anthropogenic factors [21]. In addition, the western subspecies *C. c. ludovicianus* went extinct about 30 years earlier than the eastern *C. c. carolinensis* possibly driven by different pressures [217].

The first whole mitochondrial genome of an extinct parrot species was published by Anmarkrud and Lifjeld [218] for the Paradise Parrot *Psephotellus pulcherrimus*, a species of central eastern Australia that went extinct in about 1928; the sequenced museum specimen was collected in 1881. A whole genome resequencing study used another museum specimen collected during the period when the species started to decline (in the second half of the 19th century). It argued that the species had relatively high effective population size and had not declined before the major expansion of pastoral settlements in its range. That expansion led to destruction of the parrots' nesting habitat and subsequent trapping for the avicultural trade, so excluding causes of extinction related to genetics [84]. The mitogenome of the extinct Cuban Macaw *Ara tricolor* was published in 2018 and showed that the species was closely related to the extant Military Macaw and the Great Green Macaw *Ara ambiguus*, possibly diverging from them around 4 Mya [20].

Museum and contemporary specimens were used to study the underlying processes leading to the collapse of the historically widespread and abundant Kākāpō in New Zealand. A study analysing mtDNA, microsatellites, and models of their demographic history concluded that a population bottleneck linked to the European colonisation ruled out earlier Polynesian settlement as a cause of the species' decline [219]. Another study sequenced full mitogenomes of the species and confirmed the previous study's conclusions, and found no evidence for fixation of deleterious mutations [220]. However, it argued that despite high pre-decline genetic diversity, a rapid decline combined with the species' lek mating system and its life-history traits contributed to a rapid loss of genetic diversity. By sequencing historical and modern genomes of the Kākāpō, a recent study showed that the remining island population has a reduced number of harmful mutations compared to the extinct mainland individuals, providing key insights into their recovery [146].

Parrots have been appreciated and traded since historic times [221]. For instance, Scarlet Macaw bones were recovered from archaeological sites in northern Mexico and the southern United States, over a thousand kilometres outside their endemic range [222]. Low genetic diversity found after sequencing the mitochondrial genomes of the macaw remains pointed towards a macaw breeding colony translocated by humans possibly from Mexico or Guatemala [223]. The first study relying solely on ancient parrot feathers, recovered from a pre-Hispanic religious site in the Atacama Desert in Peru, has successfully obtained and sequenced hDNA and identified various parrot species native to the Amazonian region of the country [224]. Captive rearing of macaws and amazon parrots was also shown to have occurred in the Atacama Desert in Chile around the years 1100–1450, at least 500 km outside their present-day native range [225].

Capitalising on less destructive sampling methods, trace DNA, and technological advances in museomics, genome-wide markers can now be generated from old museum specimens. A study generated thousands of SNP markers from museum (up to 123 years old) and contemporary specimens by a RAD approach and highlighted higher error rates and missing data in SNPs from the museum samples of Red-tailed Black-Cockatoos [213]. Another study used a hybridisation RAD (hyRAD) technique where probes generated from fresh samples were used to hybridise to fragmented museum hDNA (up to 140 years old), and similarly indicated lower diversity of SNPs in older samples of a songbird [226]. Hence, studies using low-quality museum samples to generate phylogenomic data must be careful and follow best practices for assembling, processing, and analysing such data to avoid misinterpretations [107].

4. Conclusions

Our overview has shown wide application of molecular genetic- and genomic techniques for studies of parrots in their global distribution. There is increasing interest by field biologists studying parrots in incorporating genetics as part of their research agenda. Given the high proportion of threatened species in the group, and the extraordinarily high level of interest in parrots among humans (including the wildlife trade and captive breeding), one or more centralised parrot genetics laboratories, perhaps on different continents, might be advantageous for future collaborative research. This could also consolidate expertise and boost efficacy in sample collection, DNA extraction, sequencing, and genomic analysis. It would be important to include genetics as a component of studies on parrot species with high conservation concern, as this could help to find populations with low genetic diversity and the most appropriate source populations to "rescue" them.

Recent breakthroughs in technology and consolidation of approaches will allow genetic techniques to be used more extensively in wildlife forensic investigations. The lack of validated DNA reference sequences is hindering our ability to accurately assign species identity. A focus on establishing DNA reference databases for the most traded wildlife species will assist in forensic casework. Building a baseline reference genomic database of wild parrot populations could help to determine the provenance of confiscated birds, aid rewilding and translocation projects, and resolve questions about captive or wild origins. As part of the licensing agreement to maintain some protected species in captivity, DNA samples could be taken with the explicit intention of using them to verify parentage and identity in the future [227,228]. Genetics has also been effective in disease testing. Studying the interactions between the TLR, MHC, and resistance to diseases would be important for both captive and wild parrot populations.

Choosing the correct markers for genetic analyses is very important as different conclusions might be reached without a genome-wide investigation. For instance, using RAD-seq data, Shipham et al. [229] confirmed a sister relationship between the Pale-headed Rosella and Northern Rosella *Platycercus venustus*, which was previously all but overlooked based purely on mtDNA sequences in which there had been a mtDNA capture event between non-sister taxa. However, the switch from solely sequencing mtDNA regions to relatively cheap and easy SNP genotyping methods has limited the capacity for comparative studies among species as different marker panels are used, optimised for each species. Absolute metrics of genetic structure and diversity are therefore not readily comparable, so approaches that produce DNA sequences may be preferable. For example, ultraconserved elements (UCEs) that target portions of the genome that remain similar across divergent clades but contain variable sequences in the flanking regions are a common approach used in avian phylogenomic studies [230]. UCEs have been used in studies on the phylogenomics of lorikeets [107] and historical demography of the Carolina Parakeet [21]. There is increasing interest in applying comparative genomic techniques to conservation studies [231,232]. These are limited with current data types but perhaps the increasing use of whole genome sequencing will make independent datasets more comparable among individuals, populations, and species. This would open up interesting opportunities for questions from behavioural, conservation, and evolutionary perspectives. The field of genetics has always been at the forefront of data sharing through repositories such as GenBank, so the opportunities for comparative analyses and insights as data comparability increases are enormous. However, sequence data alone are not enough to understand genome evolution and function, and entirely new approaches, like chromosomics [233] with superior bioinformatics like pangenome models [234], are needed in the future.

In conclusion, genetics has aided parrot research substantially in the past and will continue to do so as exciting new applications emerge in the advancing genomic era. We certainly encourage parrot researchers to consider implementing genetics as part of their research agenda, given the wide array of questions genetics can help to answer as demonstrated in this review. We realise that these research projects often do not have the capacity, expertise, or funds to do genetic research. However, many commercial laboratories now provide sequencing services at ever-dropping costs, so researchers might consider using these services to generate data from their samples. For genomic data interpretation, we propose a consortium of scientists sharing their experience in conservation genomics, analysis pipelines, and mentorship of students in genetic research on parrots. This consortium could work as a specialist group within the well-established Parrot Researchers Group.

Author Contributions: Conceptualization, G.O., B.T.S., L.J. and R.H.; Investigation, G.O.; Data curation, L.J. and B.T.S.; Writing—original draft preparation, G.O., B.T.S., L.J. and R.H.; Writing—review and editing, G.O., B.T.S., L.J., S.C.B. and R.H.; Supervision, R.H.; Project administration, G.O.; Funding acquisition, G.O. and R.H. All authors have read and agreed to the published version of the manuscript.

Funding: B.T.S. was supported by awards from the National Science Foundation US (DEB-1655736; DBI-2029955). MDPI waived the publication fee of this publication.

Acknowledgments: We thank Jose L. Tella for inviting us to contribute this review paper to the special issue on parrots. We are grateful to Juan Masello for building a comprehensive online library on parrot research, which is available via membership of the Parrot Researchers Group. We thank Rod Peakall and four anonymous reviewers, whose comments largely helped to improve this manuscript.

Conflicts of Interest: The authors declare no conflict of interest.

Glossary

Cryptic species	Morphologically often indistinguishable but genetically distinct species, following the evolutionary species concept.
Effective population size (N_e)	The size of the ideal, panmictic population that would experience the same loss of genetic variation, through genetic drift, as the observed population.
Gene flow	The exchange of genetic information between randomly mating populations through migration, measured in allele frequencies.
Genetic diversity	The extent of genetic variation in a population, species, or across species, measured in heterozygosity, allelic diversity, or heritability.
Genetic drift	Random changes in the genetic composition of a small population between generations. It results in loss of genetic diversity, random changes in allele frequencies, and diversification among populations.
Genome	The complete genetic material of an organism, including nuclear and mitochondrial DNA.
Inbreeding	The accumulation of deleterious mutations due to breeding among close relatives.
Inbreeding depression	Reduction in reproduction, survival, or related characters due to inbreeding
Management units (MUs)	Populations with significant divergence of allele frequencies at nuclear or mitochondrial loci, regardless of the phylogenetic distinctiveness of the alleles.
Microsatellite	A locus with a short tandem repeat DNA sequence, typically showing variable number of repeats across individuals. Consequently, they are highly informative genetic markers.

Minisatellite	Typically, between 6–100 bp section of DNA, repeated many times in a long string with no gaps between the repeats. These were the first type of DNA markers used in human identification and later in wildlife genetics.
Next generation sequencing (NGS)	Includes technologies that use short-read, massively parallel, high-throughput sequencing of the genetic material (e.g., Illumina, Ion Torrent).
Outbreeding depression	Reduction in reproductive fitness due to crossing of two populations, subspecies, or species.
Polymerase chain reaction (PCR)	A method to replicate copies (amplify) of specific segments of DNA, with thermostable Taq polymerase enzyme in a thermocycler.
Population viability analysis (PVA)	A model to predict the extinction risk of a population by using information about population size and structure, birth and death rates, risks and severity of catastrophes, levels of inbreeding depression, rate of habitat loss, etc. PVA can be used as a management tool to examine different management op-tions to recover threatened species.
Single nucleotide polymorphism (SNP)	A nucleotide site (base pair) in a DNA sequence that is polymorphic in a population and can be used as a marker to assess genetic variation within and among populations.
Wildlife forensics	Application of science to the law, including detection of illegal wildlife trade with DNA-based methods.

References

1. Kosman, E.; Burgio, K.R.; Presley, S.J.; Willig, M.R.; Scheiner, S.M. Conservation prioritization based on trait-based metrics illustrated with global parrot distributions. *Divers. Distrib.* **2019**, *25*, 1156–1165. [CrossRef]
2. Barbosa, J.M.; Hiraldo, F.; Romero, M.Á.; Tella, J.L. When does agriculture enter into conflict with wildlife? A global assessment of parrot–agriculture conflicts and their conservation effects. *Divers. Distrib.* **2021**, *27*, 4–17. [CrossRef]
3. Olah, G.; Butchart, S.H.M.; Symes, A.; Guzmán, I.M.; Cunningham, R.; Brightsmith, D.J.; Heinsohn, R. Ecological and socio-economic factors affecting extinction risk in parrots. *Biodivers. Conserv.* **2016**, *25*, 205–223. [CrossRef]
4. Vergara-Tabares, D.L.; Cordier, J.M.; Landi, M.A.; Olah, G.; Nori, J. Global trends of habitat destruction and consequences for parrot conservation. *Glob. Change Biol.* **2020**, *26*, 4251–4262. [CrossRef]
5. Berkunsky, I.; Quillfeldt, P.; Brightsmith, D.J.; Abbud, M.C.; Aguilar, J.M.R.E.; Alemán-Zelaya, U.; Aramburú, R.M.; Arce Arias, A.; Balas McNab, R.; Balsby, T.J.S.; et al. Current threats faced by Neotropical parrot populations. *Biol. Conserv.* **2017**, *214*, 278–287. [CrossRef]
6. Martin, R.O.; Senni, C.; D'Cruze, N.C. Trade in wild-sourced African grey parrots: Insights via social media. *Glob. Ecol. Conserv.* **2018**, *15*, e00429. [CrossRef]
7. Martin, R.O.; Perrin, M.R.; Boyes, R.S.; Abebe, Y.D.; Annorbah, N.D.; Asamoah, A.; Bizimana, D.; Bobo, K.S.; Bunbury, N.; Brouwer, J.; et al. Research and conservation of the larger parrots of Africa and Madagascar: A review of knowledge gaps and opportunities. *Ostrich* **2014**, *85*, 205–233. [CrossRef]
8. Olah, G.; Theuerkauf, J.; Legault, A.; Gula, R.; Stein, J.; Butchart, S.; O'Brien, M.; Heinsohn, R. Parrots of Oceania—A comparative study of extinction risk. *Emu-Austral Ornithol.* **2018**, *118*, 94–112. [CrossRef]
9. Calzada Preston, C.E.; Pruett-Jones, S. The number and distribution of introduced and naturalized parrots. *Diversity* **2021**, *13*, 412. [CrossRef]
10. Cardador, L.; Abellán, P.; Anadón, J.D.; Carrete, M.; Tella, J.L. The World Parrot Trade. In *Naturalized Parrots of the World: Distribution, Ecology, and Impacts of the World's Most Colorful Colonizers*; Princeton University Press: Princeton, NJ, USA, 2021; ISBN 0-691-20441-1.
11. Provost, K.L.; Joseph, L.; Smith, B.T. Resolving a phylogenetic hypothesis for parrots: Implications from systematics to conservation. *Emu-Austral Ornithol.* **2018**, *118*, 7–21. [CrossRef]
12. Wright, T.F.; Schirtzinger, E.E.; Matsumoto, T.; Eberhard, J.R.; Graves, G.R.; Sanchez, J.J.; Capelli, S.; Müller, H.; Scharpegge, J.; Chambers, G.K.; et al. A multilocus molecular phylogeny of the parrots (Psittaciformes): Support for a Gondwanan origin during the Cretaceous. *Mol. Biol. Evol.* **2008**, *25*, 2141–2156. [CrossRef]
13. Schweizer, M.; Seehausen, O.; Güntert, M.; Hertwig, S.T. The evolutionary diversification of parrots supports a taxon pulse model with multiple trans-oceanic dispersal events and local radiations. *Mol. Phylogenet. Evol.* **2010**, *54*, 984–994. [CrossRef]
14. Brock, M.K.; White, B.N. Application of DNA fingerprinting to the recovery program of the endangered Puerto Rican Parrot. *Proc. Natl. Acad. Sci. USA* **1992**, *89*, 11121–11125. [CrossRef]
15. Miller, H.C.; Lambert, D.M.; Millar, C.D.; Robertson, B.C.; Minot, E.O. Minisatellite DNA profiling detects lineages and parentage in the endangered Kakapo (*Strigops habroptilus*) despite low microsatellite DNA variation. *Conserv. Genet.* **2003**, *4*, 265–274. [CrossRef]
16. Raisin, C.; Frantz, A.C.; Kundu, S.; Greenwood, A.G.; Jones, C.G.; Zuel, N.; Groombridge, J.J. Genetic consequences of intensive conservation management for the Mauritius Parakeet. *Conserv. Genet.* **2012**, *13*, 707–715. [CrossRef]

17. Masello, J.F.; Sramkova, A.; Quillfeldt, P.; Epplen, J.T.; Lubjuhn, T. Genetic monogamy in Burrowing Parrots *Cyanoliseus patagonus*? *J. Avian Biol.* **2002**, *33*, 99–103. [CrossRef]
18. Raidal, S.R.; Peters, A. Psittacine Beak and Feather Disease: Ecology and implications for conservation. *Emu-Austral Ornithol.* **2018**, *118*, 80–93. [CrossRef]
19. Gelabert, P.; Sandoval-Velasco, M.; Serres, A.; De Manuel, M.; Renom, P.; Margaryan, A.; Stiller, J.; De-Dios, T.; Fang, Q.; Feng, S.; et al. Evolutionary history, genomic adaptation to toxic diet, and extinction of the Carolina Parakeet. *Curr. Biol.* **2020**, *30*, 108–114.e5. [CrossRef]
20. Johansson, U.S.; Ericson, P.G.P.; Blom, M.P.K.; Irestedt, M. The phylogenetic position of the extinct Cuban Macaw *Ara tricolor* based on complete mitochondrial genome sequences. *IBIS* **2018**, *160*, 666–672. [CrossRef]
21. Smith, B.T.; Gehara, M.; Harvey, M.G. The demography of extinction in eastern North American birds. *Proc. R. Soc. B Biol. Sci.* **2021**, *288*, 20201945. [CrossRef]
22. Poczai, P.; Bell, N.; Hyvönen, J. Imre Festetics and the Sheep Breeders' Society of Moravia: Mendel's forgotten "research network". *PLoS Biol.* **2014**, *12*, e1001772. [CrossRef]
23. Weiling, F. Historical study: Johann Gregor Mendel 1822-1884. *Am. J. Med. Genet.* **1991**, *40*, 1–25. [CrossRef]
24. Watson, J.D.; Crick, F.H.C. Molecular structure of nucleic acids: A structure for Deoxyribose Nucleic Acid. *Nature* **1953**, *171*, 737–738. [CrossRef]
25. Crick, F. Central dogma of molecular biology. *Nature* **1970**, *227*, 561–563. [CrossRef]
26. Adams, M.; Baverstock, P.; Saunders, D.; Schodde, R.; Smith, G. Biochemical systematics of the Australian cockatoos (Psittaciformes: Cacatuinae). *Aust. J. Zool.* **1984**, *32*, 363. [CrossRef]
27. Christidis, L.; Schodde, R.; Shaw, D.D.; Mayness, S.F. Relationships among the Australo-Papuan parrots, lorikeets, and cockatoos (Aves: Psittaciformes): Protein evidence. *Condor* **1991**, *93*, 302–317. [CrossRef]
28. Christidis, L.; Shaw, D.D.; Schodde, R. Chromosomal evolution in parrots, lorikeets and cockatoos (Aves: Psittaciformes). *Hereditas* **1991**, *114*, 47–56. [CrossRef]
29. Joseph, L.; Hope, R. Aspects of genetic relationships and variation in parrots of the Crimson Rosella *Platycercus elegans* complex (Aves: Psittacidae). *Trans. R. Soc. S. Aust.* **1984**, *108*, 77–84.
30. Miyaki, C.Y.; Wajntal, A. Sex identification of South American parrots (Psittacidae, Aves) using the human minisatellite probe 33.15. *The Auk* **1997**, *114*, 516–520. [CrossRef]
31. De Kloet, R.S.; De Kloet, S.R. The evolution of the spindlin gene in birds: Sequence analysis of an intron of the spindlin W and Z gene reveals four major divisions of the Psittaciformes. *Mol. Phylogenet. Evol.* **2005**, *36*, 706–721. [CrossRef] [PubMed]
32. Tavares, E.S.; Baker, A.J.; Pereira, S.L.; Miyaki, C.Y. Phylogenetic relationships and historical biogeography of Neotropical parrots (Psittaciformes: Psittacidae: Arini) inferred from mitochondrial and nuclear DNA sequences. *Syst. Biol.* **2006**, *55*, 454–470. [CrossRef]
33. Caparroz, R.; Seixas, G.H.F.; Berkunsky, I.; Collevatti, R.G. The role of demography and climatic events in shaping the phylogeography of *Amazona aestiva* (Psittaciformes, Aves) and definition of management units for conservation. *Divers. Distrib.* **2009**, *15*, 459–468. [CrossRef]
34. Dolman, G.; Joseph, L. Evolutionary history of birds across southern Australia: Structure, history and taxonomic implications of mitochondrial DNA diversity in an ecologically diverse suite of species. *Emu-Austral Ornithol.* **2015**, *115*, 35–48. [CrossRef]
35. Rocha, A.V.; Rivera, L.O.; Martinez, J.; Prestes, N.P.; Caparroz, R. Biogeography of speciation of two sister species of Neotropical *Amazona* (Aves, Psittaciformes) based on mitochondrial sequence data. *PLoS ONE* **2014**, *9*, e108096. [CrossRef] [PubMed]
36. Smith, B.T.; Ribas, C.C.; Whitney, B.M.; Hernández-Baños, B.E.; Klicka, J. Identifying biases at different spatial and temporal scales of diversification: A case study in the Neotropical parrotlet genus *Forpus*. *Mol. Ecol.* **2013**, *22*, 483–494. [CrossRef]
37. Caparroz, R.; Guedes, N.M.R.; Bianchi, C.A.; Wajntal, A. Analisis of the genetic variability and breeding behaviour of wild populations of two macaw species (Psittaciformes, Aves) by DNA fingerprinting. *Ararajuba* **2001**, *9*, 43–49.
38. Caparroz, R.; Miyaki, C.Y.; Bampi, M.I.; Wajntal, A. Analysis of the genetic variability in a sample of the remaining group of Spix's Macaw (*Cyanopsitta spixii*, Psittaciformes: Aves) by DNA fingerprinting. *Biol. Conserv.* **2001**, *99*, 307–311. [CrossRef]
39. Nader, W.; Werner, D.; Wink, M. Genetic diversity of Scarlet Macaws *Ara macao* in reintroduction studies for threatened populations in Costa Rica. *Biol. Conserv.* **1999**, *87*, 269–272. [CrossRef]
40. Murphy, S.; Legge, S.; Heinsohn, R. The breeding biology of Palm Cockatoos (*Probosciger aterrimus*): A case of a slow life history. *J. Zool.* **2003**, *261*, 327–339. [CrossRef]
41. Afanador, Y.; Velez-Valentín, J.; Valentín de la Rosa, R.; Martínez-Cruzado, J.-C.; VonHoldt, B.; Oleksyk, T.K. Isolation and characterization of microsatellite loci in the critically endangered Puerto Rican Parrot (*Amazona vittata*). *Conserv. Genet. Resour.* **2014**, *6*, 885–889. [CrossRef]
42. Caparroz, R.; Miyaki, C.Y.; Baker, A.J. Characterization of microsatellite loci in the Blue-and-gold Macaw, *Ara ararauna* (Psittaciformes: Aves). *Mol. Ecol. Notes* **2003**, *3*, 441–443. [CrossRef]
43. Chan, C.; Ballantyne, K.N.; Lambert, D.M.; Chambers, G.K. Characterization of variable microsatellite loci in Forbes' Parakeet (*Cyanoramphus forbesi*) and their use in other parrots. *Conserv. Genet.* **2005**, *6*, 651–654. [CrossRef]
44. Olah, G.; Heinsohn, R.G.; Espinoza, J.R.; Brightsmith, D.J.; Peakall, R. An evaluation of primers for microsatellite markers in Scarlet Macaw (*Ara macao*) and their performance in a Peruvian wild population. *Conserv. Genet. Resour.* **2015**, *7*, 157–159. [CrossRef]

45. Russello, M.; Calcagnotto, D.; DeSalle, R.; Amato, G. Characterization of microsatellite loci in the endangered St. Vincent Parrot, *Amazona guildingii*. *Mol. Ecol. Notes* **2002**, *1*, 162–164. [CrossRef]
46. Taylor, T.D.; Parkin, D.T. Characterization of 12 microsatellite primer pairs for the African Grey Parrot, *Psittacus erithacus* and their conservation across the Psittaciformes. *Mol. Ecol. Notes* **2006**, *7*, 163–167. [CrossRef]
47. Russello, M.A.; Saranathan, V.; Buhrman-Deever, S.; Eberhard, J.; Caccone, A. Characterization of polymorphic microsatellite loci for the invasive Monk Parakeet (*Myiopsitta monachus*). *Mol. Ecol. Notes* **2007**, *7*, 990–992. [CrossRef]
48. Klauke, N.; Schaefer, H.M.; Bauer, M.; Segelbacher, G. Limited dispersal and significant fine-scale genetic structure in a tropical montane parrot species. *PLoS ONE* **2016**, *11*, e0169165. [CrossRef] [PubMed]
49. Klauke, N.; Segelbacher, G.; Schaefer, H.M. Reproductive success depends on the quality of helpers in the endangered, cooperative El Oro parakeet (*Pyrrhura orcesi*). *Mol. Ecol.* **2013**, *22*, 2011–2027. [CrossRef]
50. Theuerkauf, J.; Rouys, S.; Mériot, J.M.; Gula, R.; Kuehn, R. Cooperative breeding, mate guarding, and nest sharing in two parrot species of New Caledonia. *J. Ornithol.* **2009**, *150*, 791–797. [CrossRef]
51. Harrison, G.L.A. Four new avian mitochondrial genomes help get to basic evolutionary questions in the Late Cretaceous. *Mol. Biol. Evol.* **2004**, *21*, 974–983. [CrossRef]
52. Urantowka, A.D. Complete mitochondrial genome of Blue-headed Macaw (*Primolius couloni*): Its comparison with mitogenome of Blue-throated Macaw (*Ara glaucogularis*). *Mitochondrial DNA* **2014**, *27*, 1–2. [CrossRef] [PubMed]
53. Urantowka, A.D.; Kroczak, A.M.; Strzała, T. Complete mitochondrial genome of endangered Socorro Conure (*Aratinga brevipes*): Taxonomic position of the species and its relationship with Green Conure. *Mitochondrial DNA* **2014**, *25*, 365–367. [CrossRef]
54. Oleksyk, T.K.; Pombert, J.-F.; Siu, D.; Mazo-Vargas, A.; Ramos, B.; Guiblet, W.; Afanador, Y.; Ruiz-Rodriguez, C.T.; Nickerson, M.L.; Logue, D.M.; et al. A locally funded Puerto Rican Parrot (*Amazona vittata*) genome sequencing project increases avian data and advances young researcher education. *GigaScience* **2012**, *1*, 14. [CrossRef] [PubMed]
55. Seabury, C.M.; Dowd, S.E.; Seabury, P.M.; Raudsepp, T.; Brightsmith, D.J.; Liboriussen, P.; Halley, Y.; Fisher, C.A.; Owens, E.; Viswanathan, G.; et al. A multi-platform draft de novo genome assembly and comparative analysis for the Scarlet Macaw (*Ara macao*). *PLoS ONE* **2013**, *8*, e62415. [CrossRef]
56. Miller, A.D.; Good, R.T.; Coleman, R.A.; Lancaster, M.L.; Weeks, A.R. Microsatellite loci and the complete mitochondrial DNA sequence characterized through next generation sequencing and de novo genome assembly for the critically endangered Orange-bellied Parrot, *Neophema chrysogaster*. *Mol. Biol. Rep.* **2013**, *40*, 35–42. [CrossRef]
57. Koepfli, K.-P.; Paten, B.; O'Brien, S.J. The Genome 10K Project: A way forward. *Annu. Rev. Anim. Biosci.* **2015**, *3*, 57–111. [CrossRef]
58. Feng, S.; Stiller, J.; Deng, Y.; Armstrong, J.; Fang, Q.; Reeve, A.H.; Xie, D.; Chen, G.; Guo, C.; Faircloth, B.C.; et al. Dense sampling of bird diversity increases power of comparative genomics. *Nature* **2020**, *587*, 252–257. [CrossRef]
59. Zhang, G. Bird sequencing project takes off. *Nature* **2015**, *522*, 34. [CrossRef]
60. Brandies, P.; Peel, E.; Hogg, C.J.; Belov, K. The value of reference genomes in the conservation of threatened species. *Genes* **2019**, *10*, 846. [CrossRef]
61. Andrews, K.R.; Good, J.M.; Miller, M.R.; Luikart, G.; Hohenlohe, P.A. Harnessing the power of RADseq for ecological and evolutionary genomics. *Nat. Rev. Genet.* **2016**, *17*, 81–92. [CrossRef] [PubMed]
62. Elshire, R.J.; Glaubitz, J.C.; Sun, Q.; Poland, J.A.; Kawamoto, K.; Buckler, E.S.; Mitchell, S.E. A robust, simple genotyping-by-sequencing (GBS) approach for high diversity species. *PLoS ONE* **2011**, *6*, e19379. [CrossRef]
63. Davey, J.W.; Hohenlohe, P.A.; Etter, P.D.; Boone, J.Q.; Catchen, J.M.; Blaxter, M.L. Genome-wide genetic marker discovery and genotyping using next-generation sequencing. *Nat. Rev. Genet.* **2011**, *12*, 499–510. [CrossRef]
64. Suchan, T.; Pitteloud, C.; Gerasimova, N.S.; Kostikova, A.; Schmid, S.; Arrigo, N.; Pajkovic, M.; Ronikier, M.; Alvarez, N. Hybridization capture using RAD probes (hyRAD), a new tool for performing genomic analyses on collection specimens. *PLoS ONE* **2016**, *11*, e0151651. [CrossRef]
65. Andrews, K.R.; Luikart, G. Recent novel approaches for population genomics data analysis. *Mol. Ecol.* **2014**, *23*, 1661–1667. [CrossRef]
66. Allendorf, F.W.; Hohenlohe, P.A.; Luikart, G. Genomics and the future of conservation genetics. *Nat. Rev. Genet.* **2010**, *11*, 697–709. [CrossRef]
67. He, X.; Johansson, M.L.; Heath, D.D. Role of genomics and transcriptomics in selection of reintroduction source populations. *Conserv. Biol.* **2016**, *30*, 1010–1018. [CrossRef] [PubMed]
68. Jain, M.; Olsen, H.E.; Paten, B.; Akeson, M. The Oxford Nanopore MinION: Delivery of nanopore sequencing to the genomics community. *Genome Biol.* **2016**, *17*, 239. [CrossRef]
69. Liang, F.; Zhang, P. Nanopore DNA sequencing: Are we there yet? *Sci. Bull.* **2015**, *60*, 296–303. [CrossRef]
70. Pomerantz, A.; Peñafiel, N.; Arteaga, A.; Bustamante, L.; Pichardo, F.; Coloma, L.A.; Barrio-Amorós, C.L.; Salazar-Valenzuela, D.; Prost, S. Real-time DNA barcoding in a rainforest using Nanopore sequencing: Opportunities for rapid biodiversity assessments and local capacity building. *GigaScience* **2018**, *7*, 1–14. [CrossRef]
71. Monge, O.; Dumas, D.; Baus, I. Environmental DNA from avian residual saliva in fruits and its potential uses in population genetics. *Conserv. Genet. Resour.* **2020**, *12*, 131–139. [CrossRef]
72. Olah, G.; Heinsohn, R.G.; Brightsmith, D.J.; Peakall, R. The application of non-invasive genetic tagging reveals new insights into the clay lick use by macaws in the Peruvian Amazon. *Conserv. Genet.* **2017**, *18*, 1037–1046. [CrossRef]

73. Spitzer, R.; Norman, A.J.; Königsson, H.; Schiffthaler, B.; Spong, G. De novo discovery of SNPs for genotyping endangered Sun Parakeets (*Aratinga solstitialis*) in Guyana. *Conserv. Genet. Resour.* **2020**, *12*, 631–641. [CrossRef]
74. Andersen, K.; Bird, K.L.; Rasmussen, M.; Haile, J.; Breuning-Madsen, H.; Kjaer, K.H.; Orlando, L.; Gilbert, M.T.P.; Willerslev, E. Meta-barcoding of 'dirt' DNA from soil reflects vertebrate biodiversity. *Mol. Ecol.* **2012**, *21*, 1966–1979. [CrossRef]
75. Ji, Y.; Ashton, L.; Pedley, S.M.; Edwards, D.P.; Tang, Y.; Nakamura, A.; Kitching, R.; Dolman, P.M.; Woodcock, P.; Edwards, F.A.; et al. Reliable, verifiable and efficient monitoring of biodiversity via metabarcoding. *Ecol. Lett.* **2013**, *16*, 1245–1257. [CrossRef] [PubMed]
76. Amarasinghe, S.L.; Su, S.; Dong, X.; Zappia, L.; Ritchie, M.E.; Gouil, Q. Opportunities and challenges in long-read sequencing data analysis. *Genome Biol.* **2020**, *21*, 30. [CrossRef]
77. Sibley, C.G.; Ahlquist, J.E. *Phylogeny and Classification of Birds: A Study in Molecular Evolution*; Yale University Press: New Haven, CT, USA, 1990; ISBN 0-300-04085-7.
78. Miyaki, C.Y.; Matioli, S.R.; Burke, T.; Wajntal, A. Parrot evolution and paleogeographical events: Mitochondrial DNA evidence. *Mol. Biol. Evol.* **1998**, *15*, 544–551. [CrossRef]
79. Russello, M.A.; Amato, G. A molecular phylogeny of *Amazona*: Implications for Neotropical parrot biogeography, taxonomy, and conservation. *Mol. Phylogenet. Evol.* **2004**, *30*, 421–437. [CrossRef]
80. Joseph, L.; Toon, A.; Schirtzinger, E.E.; Wright, T.F. Molecular systematics of two enigmatic genera *Psittacella* and *Pezoporus* illuminate the ecological radiation of Australo-Papuan parrots (Aves: Psittaciformes). *Mol. Phylogenet. Evol.* **2011**, *59*, 675–684. [CrossRef] [PubMed]
81. Schweizer, M.; Güntert, M.; Hertwig, S.T. Out of the Bassian province: Historical biogeography of the Australasian *Platycercine* parrots (Aves, Psittaciformes). *Zool. Scr.* **2013**, *42*, 13–27. [CrossRef]
82. Shipham, A.; Schmidt, D.J.; Joseph, L.; Hughes, J.M. Phylogenetic analysis of the Australian rosella parrots (*Platycercus*) reveals discordance among molecules and plumage. *Mol. Phylogenet. Evol.* **2015**, *91*, 150–159. [CrossRef]
83. White, N.E.; Phillips, M.J.; Gilbert, M.T.P.; Alfaro-Núñez, A.; Willerslev, E.; Mawson, P.R.; Spencer, P.B.S.; Bunce, M. The evolutionary history of cockatoos (Aves: Psittaciformes: Cacatuidae). *Mol. Phylogenet. Evol.* **2011**, *59*, 615–622. [CrossRef]
84. Irestedt, M.; Ericson, P.G.P.; Johansson, U.S.; Oliver, P.; Joseph, L.; Blom, M.P.K. No signs of genetic erosion in a 19th century genome of the extinct Paradise Parrot (*Psephotellus pulcherrimus*). *Diversity* **2019**, *11*, 58. [CrossRef]
85. Hackett, S.J.; Kimball, R.T.; Reddy, S.; Bowie, R.C.K.; Braun, E.L.; Braun, M.J.; Chojnowski, J.L.; Cox, W.A.; Han, K.-L.; Harshman, J.; et al. A phylogenomic study of birds reveals their evolutionary history. *Science* **2008**, *320*, 1763–1768. [CrossRef]
86. Jarvis, E.D.; Mirarab, S.; Aberer, A.J.; Li, B.; Houde, P.; Li, C.; Ho, S.Y.W.; Faircloth, B.C.; Nabholz, B.; Howard, J.T.; et al. Whole-genome analyses resolve early branches in the tree of life of modern birds. *Science* **2014**, *346*, 1320–1331. [CrossRef]
87. McCormack, J.E.; Harvey, M.G.; Faircloth, B.C.; Crawford, N.G.; Glenn, T.C.; Brumfield, R.T. A phylogeny of birds based on over 1500 loci collected by target enrichment and high-throughput sequencing. *PLoS ONE* **2013**, *8*, e54848. [CrossRef]
88. Suh, A.; Paus, M.; Kiefmann, M.; Churakov, G.; Franke, F.A.; Brosius, J.; Kriegs, J.O.; Schmitz, J. Mesozoic retroposons reveal parrots as the closest living relatives of passerine birds. *Nat. Commun.* **2011**, *2*, 443. [CrossRef]
89. Astuti, D.; Azuma, N.; Suzuki, H.; Higashi, S. Phylogenetic relationships within parrots (Psittacidae) inferred from mitochondrial cytochrome-b gene sequences. *Zoolog. Sci.* **2006**, *23*, 191–198. [CrossRef]
90. Claramunt, S.; Cracraft, J. A new time tree reveals Earth history's imprint on the evolution of modern birds. *Sci. Adv.* **2015**, *1*, e1501005. [CrossRef]
91. Cracraft, J. Avian evolution, Gondwana biogeography and the Cretaceous–Tertiary mass extinction event. *Proc. R. Soc. Lond. B Biol. Sci.* **2001**, *268*, 459–469. [CrossRef]
92. Waterhouse, D.M. Parrots in a nutshell: The fossil record of Psittaciformes (Aves). *Hist. Biol.* **2006**, *18*, 227–238. [CrossRef]
93. Toft, C.A.; Wright, T.F. *Parrots of the Wild: A Natural History of the World's Most Captivating Birds*; University of California Press: Berkeley, CA, USA, 2015; ISBN 978-0-520-96264-4.
94. Schweizer, M.; Seehausen, O.; Hertwig, S.T. Macroevolutionary patterns in the diversification of parrots: Effects of climate change, geological events and key innovations. *J. Biogeogr.* **2011**, *38*, 2176–2194. [CrossRef]
95. Kimball, R.T.; Oliveros, C.H.; Wang, N.; White, N.D.; Barker, F.K.; Field, D.J.; Ksepka, D.T.; Chesser, R.T.; Moyle, R.G.; Braun, M.J.; et al. A phylogenomic supertree of birds. *Diversity* **2019**, *11*, 109. [CrossRef]
96. Joseph, L.; Toon, A.; Schirtzinger, E.E.; Wright, T.F.; Schodde, R. A revised nomenclature and classification for family-group taxa of parrots (Psittaciformes). *Zootaxa* **2012**, *3205*, 26. [CrossRef]
97. Rheindt, F.E.; Christidis, L.; Kuhn, S.; De Kloet, S.; Norman, J.A.; Fidler, A. The timing of diversification within the most divergent parrot clade. *J. Avian Biol.* **2014**, *45*, 140–148. [CrossRef]
98. Worthy, T.H.; De Pietri, V.L.; Scofield, R.P. Recent advances in avian palaeobiology in New Zealand with implications for understanding New Zealand's geological, climatic and evolutionary histories. *N. Z. J. Zool.* **2017**, *44*, 177–211. [CrossRef]
99. Schweizer, M.; Hertwig, S.T.; Seehausen, O. Diversity versus disparity and the role of ecological opportunity in a continental bird radiation. *J. Biogeogr.* **2014**, *41*, 1301–1312. [CrossRef]
100. Ribas, C.C.; Miyaki, C.Y. Molecular systematics in Aratinga parakeets: Species limits and historical biogeography in the 'solstitialis' group, and the systematic position of *Nandayus nenday*. *Mol. Phylogenet. Evol.* **2004**, *30*, 663–675. [CrossRef]
101. Ribas, C.C.; Gaban-Lima, R.; Miyaki, C.Y.; Cracraft, J. Historical biogeography and diversification within the Neotropical parrot genus *Pionopsitta* (Aves: Psittacidae). *J. Biogeogr.* **2005**, *32*, 1409–1427. [CrossRef]

102. Eberhard, J.R.; Bermingham, E. Phylogeny and comparative biogeography of *Pionopsitta* parrots and *Pteroglossus* toucans. *Mol. Phylogenet. Evol.* **2005**, *36*, 288–304. [CrossRef]
103. Ribas, C.C.; Moyle, R.G.; Miyaki, C.Y.; Cracraft, J. The assembly of montane biotas: Linking Andean tectonics and climatic oscillations to independent regimes of diversification in *Pionus* parrots. *Proc. R. Soc. B Biol. Sci.* **2007**, *274*, 2399–2408. [CrossRef]
104. Ribas, C.C.; Miyaki, C.Y.; Cracraft, J. Phylogenetic relationships, diversification and biogeography in Neotropical *Brotogeris* parakeets. *J. Biogeogr.* **2009**, *36*, 1712–1729. [CrossRef]
105. Boon, W.M.; Kearvell, J.C.; Daugherty, C.H.; Chambers, G.K. Molecular systematics of New Zealand *Cyanoramphus* parakeets: Conservation of Orange-fronted and Forbes' Parakeets. *Bird Conserv. Int.* **2000**, *10*, 211–239. [CrossRef]
106. Braun, M.P.; Reinschmidt, M.; Datzmann, T.; Waugh, D.; Zamora, R.; Häbich, A.; Neves, L.; Gerlach, H.; Arndt, T.; Mettke-Hofmann, C.; et al. Influences of oceanic islands and the Pleistocene on the biogeography and evolution of two groups of Australasian parrots (Aves: Psittaciformes: *Eclectus roratus, Trichoglossus haematodus* complex). Rapid evolution and implications for taxonomy and conservation. *Eur. J. Ecol.* **2016**, *3*, 47–66. [CrossRef]
107. Smith, B.T.; Mauck, W.M.; Benz, B.W.; Andersen, M.J. Uneven missing data skew phylogenomic relationships within the lories and lorikeets. *Genome Biol. Evol.* **2020**, *12*, 1131–1147. [CrossRef]
108. Jackson, H.; Jones, C.G.; Agapow, P.-M.; Tatayah, V.; Groombridge, J.J. Micro-evolutionary diversification among Indian Ocean parrots: Temporal and spatial changes in phylogenetic diversity as a consequence of extinction and invasion. *IBIS* **2015**, *157*, 496–510. [CrossRef]
109. Ribas, C.C.; Tavares, E.S.; Yoshihara, C.; Miyaki, C.Y. Phylogeny and biogeography of Yellow-headed and Blue-fronted Parrots (*Amazona ochrocephala* and *Amazona aestiva*) with special reference to the South American taxa. *IBIS* **2007**, *149*, 564–574. [CrossRef]
110. Wenner, T.J.; Russello, M.A.; Wright, T.F. Cryptic species in a Neotropical parrot: Genetic variation within the *Amazona farinosa* species complex and its conservation implications. *Conserv. Genet.* **2012**, *13*, 1427–1432. [CrossRef]
111. Murphy, S.A.; Joseph, L.; Burbidge, A.H.; Austin, J. A cryptic and critically endangered species revealed by mitochondrial DNA analyses: The Western Ground Parrot. *Conserv. Genet.* **2011**, *12*, 595–600. [CrossRef]
112. McElroy, K.; Beattie, K.; Symonds, M.R.E.; Joseph, L. Mitogenomic and nuclear diversity in the Mulga Parrot of the Australian arid zone: Cryptic subspecies and tests for selection. *Emu-Austral Ornithol.* **2018**, *118*, 22–35. [CrossRef]
113. Silva, T.; Guzmán, A.; Urantówka, A.D.; Mackiewicz, P. A new parrot taxon from the Yucatán Peninsula, Mexico: Its position within genus *Amazona* based on morphology and molecular phylogeny. *PeerJ* **2017**, *5*, e3475. [CrossRef]
114. Escalante, P.; Arteaga-Rojas, A.E.; Gutiérrez-Sánchez-Rüed, M. A new species of Mexican parrot? Reasonable doubt on the status of *Amazona gomezgarzai* (Psittaciformes: Psittacidae). *Zootaxa* **2018**, *4420*, 139. [CrossRef]
115. Moritz, C. Defining 'Evolutionarily Significant Units' for conservation. *Trends Ecol. Evol.* **1994**, *9*, 373–375. [CrossRef]
116. Russello, M.A.; Stahala, C.; Lalonde, D.; Schmidt, K.L.; Amato, G. Cryptic diversity and conservation units in the Bahama Parrot. *Conserv. Genet.* **2010**, *11*, 1809–1821. [CrossRef]
117. Hellmich, D.L.; Saidenberg, A.B.S.; Wright, T.F. Genetic, but not behavioral, evidence supports the distinctiveness of the Mealy Amazon parrot in the Brazilian Atlantic Forest. *Diversity* **2021**, *13*, 273. [CrossRef]
118. Rivera-Ortíz, F.A.; Solórzano, S.; Arizmendi, M. del C.; Dávila-Aranda, P.; Oyama, K. Genetic diversity and structure of the Military Macaw (*Ara militaris*) in Mexico. *Trop. Conserv. Sci.* **2017**, *10*, 194008291668434. [CrossRef]
119. Melo, M.; O'Ryan, C. Genetic differentiation between Príncipe Island and mainland populations of the Grey Parrot (*Psittacus erithacus*), and implications for conservation. *Mol. Ecol.* **2007**, *16*, 1673–1685. [CrossRef]
120. Coetzer, W.G.; Downs, C.T.; Perrin, M.R.; Willows-Munro, S. Molecular systematics of the Cape Parrot (*Poicephalus robustus*): Implications for taxonomy and conservation. *PLoS ONE* **2015**, *10*, e0133376. [CrossRef]
121. Clements, J.F.; Schulenberg, T.S.; Iliff, M.J.; Billerman, S.M.; Fredericks, T.A.; Sullivan, B.L.; Wood, C.L. *The eBird/Clements Checklist of Birds of the World*, 2019th ed.; IBIS Publishing: Vista, CA, USA, 2019.
122. Gill, F.; Donsker, D.; Rasmussen, P. (Eds.) *IOC World Bird List (v 11.1)*; IOC: Lausanne, Switzerland, 2021. [CrossRef]
123. Perrin, M. *Parrots of Africa, Madagascar and the Mascarene Islands: Biology, Ecology and Conservation*; Wits University Press: Johannesburg, South Africa, 2012.
124. Padilla-Jacobo, G.; Monterrubio-Rico, T.C.; Camacho, H.C.; Zavala-Páramo, M.G. Use of phylogenetic analysis to identify evolutionarily significant units for the Orange-fronted Parakeet (*Eupsittula canicularis*) in Mexico. *Ornitol. Neotrop.* **2016**, *26*, 325–335.
125. Ewart, K.M.; Lo, N.; Ogden, R.; Joseph, L.; Ho, S.Y.W.; Frankham, G.J.; Eldridge, M.D.B.; Schodde, R.; Johnson, R.N. Phylogeography of the iconic Australian Red-tailed Black-Cockatoo (*Calyptorhynchus banksii*) and implications for its conservation. *Heredity* **2020**, *125*, 85–100. [CrossRef]
126. Ewart, K.M.; Johnson, R.N.; Joseph, L.; Ogden, R.; Frankham, G.J.; Lo, N. Phylogeography of the iconic Australian Pink Cockatoo, *Lophochroa leadbeateri*. *Biol. J. Linn. Soc.* **2021**, *132*, 704–723. [CrossRef]
127. Dussex, N.; Sainsbury, J.; Moorhouse, R.; Jamieson, I.G.; Robertson, B.C. Evidence for Bergmann's rule and not allopatric subspeciation in the threatened Kaka (*Nestor meridionalis*). *J. Hered.* **2015**, *106*, 679–691. [CrossRef]
128. Dussex, N.; Wegmann, D.; Robertson, B.C. Postglacial expansion and not human influence best explains the population structure in the endangered Kea (*Nestor notabilis*). *Mol. Ecol.* **2014**, *23*, 2193–2209. [CrossRef]
129. Frankham, R.; Briscoe, D.A.; Ballou, J.D. *Introduction to Conservation Genetics*, 2nd ed.; Cambridge University Press: Cambridge, UK, 2010; ISBN 978-0-521-70271-3.

130. Chan, K.; Glover, D.R.; Ramage, C.M.; Harrison, D.K. Low genetic diversity in the Ground Parrot (*Pezoporus wallicus*) revealed by randomly amplified DNA fingerprinting. *Ann. Zool. Fenn.* **2008**, *45*, 211–216. [CrossRef]
131. Faria, P.J.; Guedes, N.M.R.; Yamashita, C.; Martuscelli, P.; Miyaki, C.Y. Genetic variation and population structure of the endangered Hyacinth Macaw (*Anodorhynchus hyacinthinus*): Implications for conservation. *Biodivers. Conserv.* **2008**, *17*, 765–779. [CrossRef]
132. Taylor, T.D.; Parkin, D.Y. Preliminary insights into the level of genetic variation retained in the endangered Echo Parakeet (*Psittacula eques*) towards assisting its conservation management. *Afr. Zool.* **2010**, *45*, 189–194. [CrossRef]
133. Stojanovic, D.; Olah, G.; Webb, M.; Peakall, R.; Heinsohn, R. Genetic evidence confirms severe extinction risk for critically endangered swift parrots: Implications for conservation management. *Anim. Conserv.* **2018**, *21*, 313–323. [CrossRef]
134. Olah, G.; Stojanovic, D.; Webb, M.H.; Waples, R.S.; Heinsohn, R. Comparison of three techniques for genetic estimation of effective population size in a critically endangered parrot. *Anim. Conserv.* **2021**, *24*, 491–498. [CrossRef]
135. Miller, M.P.; Bianchi, C.A.; Mullins, T.D.; Haig, S.M. Associations between forest fragmentation patterns and genetic structure in Pfrimer's Parakeet (*Pyrrhura pfrimeri*), an endangered endemic to central Brazil's dry forests. *Conserv. Genet.* **2013**, *14*, 333–343. [CrossRef]
136. Monge, O.; Schmidt, K.; Vaughan, C.; Gutiérrez-Espeleta, G. Genetic patterns and conservation of the Scarlet Macaw (*Ara macao*) in Costa Rica. *Conserv. Genet.* **2016**, *17*, 745–750. [CrossRef]
137. Schmidt, K.L.; Aardema, M.L.; Amato, G. Genetic analysis reveals strong phylogeographical divergences within the Scarlet Macaw *Ara macao*. *IBIS* **2020**, *162*, 735–748. [CrossRef]
138. Keighley, M.V.; Heinsohn, R.; Langmore, N.E.; Murphy, S.A.; Peñalba, J.V. Genomic population structure aligns with vocal dialects in Palm Cockatoos (*Probosciger aterrimus*); evidence for refugial late-Quaternary distribution? *Emu-Austral Ornithol.* **2019**, *119*, 24–37. [CrossRef]
139. Keighley, M.V.; Haslett, S.; Zdenek, C.N.; Heinsohn, R. Slow breeding rates and low population connectivity indicate Australian Palm Cockatoos are in severe decline. *Biol. Conserv.* **2021**, *253*, 108865. [CrossRef]
140. Ringler, E. The use of cross-species testing of microsatellite markers and sibship analysis in ex situ population management. *Conserv. Genet. Resour.* **2012**, *4*, 815–819. [CrossRef]
141. Ihle, M.; Kempenaers, B.; Forstmeier, W. Fitness benefits of mate choice for compatibility in a socially monogamous species. *PLoS Biol.* **2015**, *13*, e1002248. [CrossRef]
142. Fox, R.A.; Millam, J.R. Personality traits of pair members predict pair compatibility and reproductive success in a socially monogamous parrot breeding in captivity. *Zoo Biol.* **2014**, *33*, 166–172. [CrossRef]
143. Spoon, T.R.; Millam, J.R.; Owings, D.H. The importance of mate behavioural compatibility in parenting and reproductive success by Cockatiels, *Nymphicus hollandicus*. *Anim. Behav.* **2006**, *71*, 315–326. [CrossRef]
144. Daniell, A.; Murray, N.D. Effects of inbreeding in the Budgerigar *Melopsittacus undulatus* (Aves: Psittacidae). *Zoo Biol.* **1986**, *5*, 233–238. [CrossRef]
145. Caparroz, R.; Martuscelli, P.; Scherer-Neto, P.; Miyaki, C.Y.; Wajntal, A. Genetic variability in the Red-tailed Amazon (*Amazona brasiliensis*, Psittaciformes) assessed by DNA fingerprinting. *Rev. Bras. Ornitol.* **2006**, *14*, 15–19.
146. Dussex, N.; van der Valk, T.; Morales, H.E.; Wheat, C.W.; Díez-del-Molino, D.; von Seth, J.; Foster, Y.; Kutschera, V.E.; Guschanski, K.; Rhie, A.; et al. Population genomics of the critically endangered Kākāpō. *Cell Genom.* **2021**, *1*, 100002. [CrossRef]
147. Campos, C.I.; Martinez, M.A.; Acosta, D.; Diaz-Luque, J.A.; Berkunsky, I.; Lamberski, N.L.; Cruz-Nieto, J.; Russello, M.A.; Wright, T.F. Genetic diversity and population structure of two endangered neotropical parrots inform in situ and ex situ conservation strategies. *Diversity* **2021**, *13*, 386. [CrossRef]
148. Morrison, C.E.; Hogg, C.J.; Gales, R.; Johnson, R.N.; Grueber, C.E. Low innate immune-gene diversity in the critically endangered Orange-bellied Parrot (*Neophema chrysogaster*). *Emu-Austral Ornithol.* **2020**, *120*, 56–64. [CrossRef]
149. Peters, A.; Patterson, E.I.; Baker, B.G.B.; Holdsworth, M.; Sarker, S.; Ghorashi, S.A.; Raidal, S.R. Evidence of Psittacine Beak and Feather Disease Virus spillover into wild critically endangered Orange-bellied Parrots (*Neophema chrysogaster*). *J. Wildl. Dis.* **2014**, *50*, 288–296. [CrossRef] [PubMed]
150. Ashby, E. Parrakeets moutling. *Emu-Austral Ornithol.* **1907**, *6*, 193–194. [CrossRef]
151. Ritchie, B.W.; Niagro, F.D.; Lukert, P.D.; Steffens, W.L.; Latimer, K.S. Characterization of a new virus from cockatoos with psittacine beak and feather disease. *Virology* **1989**, *171*, 83–88. [CrossRef]
152. Ogawa, H.; Yamaguchi, T.; Fukushi, H. Duplex shuttle PCR for differential diagnosis of Budgerigar Fledgling Disease and Psittacine Beak and Feather Disease. *Microbiol. Immunol.* **2005**, *49*, 227–237. [CrossRef] [PubMed]
153. Ypelaar, I.; Bassami, M.R.; Wilcox, G.E.; Raidal, S.R. A universal polymerase chain reaction for the detection of psittacine beak and feather disease virus. *Vet. Microbiol.* **1999**, *68*, 141–148. [CrossRef]
154. Australian Government; Department of Agriculture, Water and the Environment. *Psittacine Beak and Feather Disease and Other Identified Threats to Australian Threatened Parrots*; Department of the Environment: Canberra, Australia, 2015.
155. Morrison, C.E.; Johnson, R.N.; Grueber, C.E.; Hogg, C.J. Genetic impacts of conservation management actions in a critically endangered parrot species. *Conserv. Genet.* **2020**, *21*, 869–877. [CrossRef]
156. Kaspers, B.; Schat, K.A. *Avian Immunology*; Elsevier Science: Amsterdam, The Netherlands, 2012; ISBN 978-0-12-397272-9.
157. Edwards, S.V.; Hess, C.M.; Gasper, J.; Garrigan, D. Toward an evolutionary genomics of the avian MHC. *Immunol. Rev.* **1999**, *167*, 119–132. [CrossRef]

158. Hughes, C.R.; Miles, S.; Walbroehl, J.M. Support for the minimal essential MHC hypothesis: A parrot with a single, highly polymorphic MHC class II B gene. *Immunogenetics* **2008**, *60*, 219–231. [CrossRef]
159. Knafler, G.J.; Jamieson, I.G. Primers for the amplification of major histocompatibility complex class I and II loci in the recovering Red-crowned Parakeet. *Conserv. Genet. Resour.* **2014**, *6*, 37–39. [CrossRef]
160. Frankham, R.; Ballou, J.D.; Eldridge, M.D.B.; Lacy, R.C.; Ralls, K.; Dudash, M.R.; Fenster, C.B. Predicting the probability of outbreeding depression. *Conserv. Biol.* **2011**, *25*, 465–475. [CrossRef]
161. Huff, D.D.; Miller, L.M.; Chizinski, C.J.; Vondracek, B. Mixed-source reintroductions lead to outbreeding depression in second-generation descendents of a native North American fish. *Mol. Ecol.* **2011**, *20*, 4246–4258. [CrossRef] [PubMed]
162. Ralls, K.; Sunnucks, P.; Lacy, R.C.; Frankham, R. Genetic rescue: A critique of the evidence supports maximizing genetic diversity rather than minimizing the introduction of putatively harmful genetic variation. *Biol. Conserv.* **2020**, *251*, 108784. [CrossRef]
163. Grant, P.R.; Grant, B.R. Hybridization of bird species. *Science* **1992**, *256*, 193–197. [CrossRef] [PubMed]
164. McCarthy, E.M. *Handbook of Avian Hybrids of the World*; Oxford University Press: Oxford, UK, 2006.
165. Chan, C.-H.; Ballantyne, K.N.; Aikman, H.; Fastier, D.; Daugherty, C.H.; Chambers, G.K. Genetic analysis of interspecific hybridisation in the world's only Forbes' Parakeet (*Cyanoramphus forbesi*) natural population. *Conserv. Genet.* **2006**, *7*, 493–506. [CrossRef]
166. Shipham, A.; Joseph, L.; Schmidt, D.J.; Drew, A.; Mason, I.; Hughes, J.M. Dissection by genomic and plumage variation of a geographically complex hybrid zone between two Australian non-sister parrot species, *Platycercus adscitus* and *Platycercus eximius*. *Heredity* **2019**, *122*, 402–416. [CrossRef] [PubMed]
167. Miyaki, C.Y.; Faria, P.J.; Griffiths, R.; Araújo, J.C.C.; Barros, Y.M. The last wild Spix's Macaw and an Illiger's Macaw produced a hybrid. *Conserv. Genet.* **2001**, *2*, 53–55. [CrossRef]
168. White, N.E.; Dawson, R.; Coghlan, M.L.; Tridico, S.R.; Mawson, P.R.; Haile, J.; Bunce, M. Application of STR markers in wildlife forensic casework involving Australian black-cockatoos (*Calyptorhynchus* spp.). *Forensic Sci. Int. Genet.* **2012**, *6*, 664–670. [CrossRef]
169. Coghlan, M.L.; White, N.E.; Parkinson, L.; Haile, J.; Spencer, P.B.S.; Bunce, M. Egg forensics: An appraisal of DNA sequencing to assist in species identification of illegally smuggled eggs. *Forensic Sci. Int. Genet.* **2012**, *6*, 268–273. [CrossRef]
170. Gonçalves, P.F.M.; Oliveira-Marques, A.R.; Matsumoto, T.E.; Miyaki, C.Y. DNA barcoding identifies illegal parrot trade. *J. Hered.* **2015**, *106*, 560–564. [CrossRef] [PubMed]
171. Formentão, L.; Saraiva, A.S.; Marrero, A.R. DNA barcoding exposes the need to control the illegal trade of eggs of non-threatened parrots in Brazil. *Conserv. Genet. Resour.* **2021**, *13*, 275–281. [CrossRef]
172. Lee, J. A Forensic Toolkit for the Examination of Wildlife Crime Using the Glossy Black-Cockatoo as a Model Species. Ph.D. Thesis, University of Canberra, Canberra, Australia, 2013.
173. Coetzer, W.G.; Downs, C.T.; Perrin, M.R.; Willows-Munro, S. Testing of microsatellite multiplexes for individual identification of Cape Parrots (*Poicephalus robustus*): Paternity testing and monitoring trade. *PeerJ* **2017**, *5*, e2900. [CrossRef]
174. Rivera-Ortíz, F.A.; Juan-Espinosa, J.; Solórzano, S.; Contreras-González, A.M.; Arizmendi, M.D.C. Genetic assignment tests to identify the probable geographic origin of a captive specimen of Military Macaw (*Ara militaris*) in Mexico: Implications for conservation. *Diversity* **2021**, *13*, 245. [CrossRef]
175. Fernandes, G.A.; Caparroz, R. DNA sequence analysis to guide the release of Blue-and-yellow Macaws (*Ara ararauna*, Psittaciformes, Aves) from the illegal trade back into the wild. *Mol. Biol. Rep.* **2013**, *40*, 2757–2762. [CrossRef] [PubMed]
176. Wirthlin, M.; Lima, N.C.B.; Guedes, R.L.M.; Soares, A.E.R.; Almeida, L.G.P.; Cavaleiro, N.P.; de Morais, G.L.; Chaves, A.V.; Howard, J.T.; de Melo Teixeira, M.; et al. Parrot genomes and the evolution of heightened longevity and cognition. *Curr. Biol.* **2018**, *28*, 4001–4008.e7. [CrossRef] [PubMed]
177. Brown, C.; Magat, M. The evolution of lateralized foot use in parrots: A phylogenetic approach. *Behav. Ecol.* **2011**, *22*, 1201–1208. [CrossRef]
178. Friedmann, H.; Davis, M. "Left-handedness" in parrots. *Auk* **1938**, *55*, 478–480. [CrossRef]
179. Harris, L.J. Footedness in parrots: Three centuries of research, theory, and mere surmise. *Can. J. Psychol. Can. Psychol.* **1989**, *43*, 369–396. [CrossRef]
180. Benavidez, A.; Palacio, F.X.; Rivera, L.O.; Echevarria, A.L.; Politi, N. Diet of Neotropical parrots is independent of phylogeny but correlates with body size and geographical range. *IBIS* **2018**, *160*, 742–754. [CrossRef]
181. Caparroz, R.; Miyaki, C.Y.; Baker, A.J. Genetic evaluation of the mating system in the Blue-and-yellow Macaw (*Ara ararauna*, Aves, Psittacidae) by DNA fingerprinting. *Genet. Mol. Biol.* **2010**, *34*, 161–164. [CrossRef]
182. Caparroz, R.; Miyaki, C.Y.; Baker, A.J. Contrasting phylogeographic patterns in mitochondrial DNA and microsatellites: Evidence of female philopatry and male-biased gene flow among regional populations of the Blue-and-yellow Macaw (Psittaciformes: *Ara ararauna*) in Brazil. *Auk* **2009**, *126*, 359–370. [CrossRef]
183. Heinsohn, R.; Ebert, D.; Legge, S.; Peakall, R. Genetic evidence for cooperative polyandry in reverse dichromatic Eclectus Parrots. *Anim. Behav.* **2007**, *74*, 1047–1054. [CrossRef]
184. Ekstrom, J.M.M.; Burke, T.; Randrianaina, L.; Birkhead, T.R. Unusual sex roles in a highly promiscuous parrot: The Greater Vasa Parrot *Caracopsis vasa*. *IBIS* **2007**, *149*, 313–320. [CrossRef]
185. Heinsohn, R.; Olah, G.; Webb, M.; Peakall, R.; Stojanovic, D. Sex ratio bias and shared paternity reduce individual fitness and population viability in a critically endangered parrot. *J. Anim. Ecol.* **2019**, *88*, 502–510. [CrossRef] [PubMed]

186. Heinsohn, R.; Au, J.; Kokko, H.; Webb, M.H.; Deans, R.M.; Crates, R.; Stojanovic, D. Can an introduced predator select for adaptive sex allocation? *Proc. R. Soc. B Biol. Sci.* **2021**, *288*, 20210093. [CrossRef] [PubMed]
187. Taylor, T.D.; Parkin, D.T. Preliminary evidence suggests extra-pair mating in the endangered echo parakeet, *Psittacula eques*. *Afr. Zool.* **2009**, *44*, 71–74. [CrossRef]
188. Da Silva, A.G.; Eberhard, J.R.; Wright, T.F.; Avery, M.L.; Russello, M.A. Genetic evidence for high propagule pressure and long-distance dispersal in Monk Parakeet (*Myiopsitta monachus*) invasive populations. *Mol. Ecol.* **2010**, *19*, 3336–3350. [CrossRef]
189. Martínez, J.J.; De Aranzamendi, M.C.; Masello, J.F.; Bucher, E.H. Genetic evidence of extra-pair paternity and intraspecific brood parasitism in the monk parakeet. *Front. Zool.* **2013**, *10*, 68. [CrossRef]
190. Dawson Pell, F.S.E.; Senar, J.C.; Franks, D.W.; Hatchwell, B.J. Fine-scale genetic structure reflects limited and coordinated dispersal in the colonial Monk Parakeet, *Myiopsitta monachus*. *Mol. Ecol.* **2021**, *30*, 1531–1544. [CrossRef]
191. Lee, J.; Pedler, L.; Sarre, S.D.; Robertson, J.; Joseph, L. Male sex-ratio bias in the endangered South Australian Glossy Black-Cockatoo *Calyptorhynchus lathami halmaturinus*. *Emu-Austral Ornithol.* **2015**, *115*, 356–359. [CrossRef]
192. Wright, T.F.; Rodriguez, A.M.; Fleischer, R.C. Vocal dialects, sex-biased dispersal, and microsatellite population structure in the parrot *Amazona auropalliata*. *Mol. Ecol.* **2005**, *14*, 1197–1205. [CrossRef]
193. Wright, T.F.; Wilkinson, G.S. Population genetic structure and vocal dialects in an *Amazon* parrot. *Proc. R. Soc. Lond. B Biol. Sci.* **2001**, *268*, 609–616. [CrossRef] [PubMed]
194. Wright, T.F.; Dahlin, C.R. Vocal dialects in parrots: Patterns and processes of cultural evolution. *Emu-Austral Ornithol.* **2018**, *118*, 50–66. [CrossRef] [PubMed]
195. Hara, E.; Perez, J.M.; Whitney, O.; Chen, Q.; White, S.A.; Wright, T.F. Neural FoxP2 and FoxP1 expression in the Budgerigar, an avian species with adult vocal learning. *Behav. Brain Res.* **2015**, *283*, 22–29. [CrossRef]
196. Whitney, O.; Voyles, T.; Hara, E.; Chen, Q.; White, S.A.; Wright, T.F. Differential FoxP2 and FoxP1 expression in a vocal learning nucleus of the developing Budgerigar. *Dev. Neurobiol.* **2015**, *75*, 778–790. [CrossRef] [PubMed]
197. Russello, M.A.; Smith-Vidaurre, G.; Wright, T.F. Genetics of invasive parrot populations. In *Naturalized Parrots of the World: Distribution, Ecology, and Impacts of the World's Most Colorful Colonizers*; Princeton University Press: Princeton, NJ, USA, 2021; ISBN 0-691-20441-1.
198. Russello, M.A.; Avery, M.L.; Wright, T.F. Genetic evidence links invasive Monk Parakeet populations in the United States to the international pet trade. *BMC Evol. Biol.* **2008**, *8*, 217. [CrossRef]
199. Edelaar, P.; Roques, S.; Hobson, E.A.; Gonçalves Da Silva, A.; Avery, M.L.; Russello, M.A.; Senar, J.C.; Wright, T.F.; Carrete, M.; Tella, J.L. Shared genetic diversity across the global invasive range of the Monk Parakeet suggests a common restricted geographic origin and the possibility of convergent selection. *Mol. Ecol.* **2015**, *24*, 2164–2176. [CrossRef]
200. Le Gros, A.; Samadi, S.; Zuccon, D.; Cornette, R.; Braun, M.P.; Senar, J.C.; Clergeau, P. Rapid morphological changes, admixture and invasive success in populations of Ring-necked Parakeets (*Psittacula krameri*) established in Europe. *Biol. Invasions* **2016**, *18*, 1581–1598. [CrossRef]
201. Olah, G.; Heinsohn, R.G.; Brightsmith, D.J.; Espinoza, J.R.; Peakall, R. Validation of non-invasive genetic tagging in two large macaw species (*Ara macao* and *A. chloropterus*) of the Peruvian Amazon. *Conserv. Genet. Resour.* **2016**, *8*, 499–509. [CrossRef]
202. Presti, F.T.; Meyer, J.; Antas, P.T.Z.; Guedes, N.M.R.; Miyaki, C.Y. Non-invasive genetic sampling for molecular sexing and microsatellite genotyping of Hyacinth Macaw (*Anodorhynchus hyacinthinus*). *Genet. Mol. Biol.* **2013**, *36*, 129–133. [CrossRef]
203. Olah, G.; Smith, A.L.; Asner, G.P.; Brightsmith, D.J.; Heinsohn, R.G.; Peakall, R. Exploring dispersal barriers using landscape genetic resistance modelling in Scarlet Macaws of the Peruvian Amazon. *Landsc. Ecol.* **2017**, *32*, 445–456. [CrossRef]
204. Masello, J.F.; Montano, V.; Quillfeldt, P.; Nuhlíčková, S.; Wikelski, M.; Moodley, Y. The interplay of spatial and climatic landscapes in the genetic distribution of a South American parrot. *J. Biogeogr.* **2015**, *42*, 1077–1090. [CrossRef]
205. Keighley, M.V.; Langmore, N.E.; Peñalba, J.V.; Heinsohn, R. Modelling dispersal in a large parrot: A comparison of landscape resistance models with population genetics and vocal dialect patterns. *Landsc. Ecol.* **2020**, *35*, 129–144. [CrossRef]
206. Blanco, G.; Morinha, F.; Roques, S.; Hiraldo, F.; Rojas, A.; Tella, J.L. Fine-scale genetic structure in the critically endangered Red-fronted Macaw in the absence of geographic and ecological barriers. *Sci. Rep.* **2021**, *11*, 556. [CrossRef] [PubMed]
207. Joseph, L.; Dolman, G.; Donnellan, S.; Saint, K.M.; Berg, M.L.; Bennett, A.T.D. Where and when does a ring start and end? Testing the ring-species hypothesis in a species complex of Australian parrots. *Proc. R. Soc. B Biol. Sci.* **2008**, *275*, 2431–2440. [CrossRef] [PubMed]
208. Martini, D.; Dussex, N.; Robertson, B.C.; Gemmell, N.J.; Knapp, M. Evolution of the "world's only alpine parrot": Genomic adaptation or phenotypic plasticity, behaviour and ecology? *Mol. Ecol.* **2021**. [CrossRef]
209. Bi, K.; Linderoth, T.; Vanderpool, D.; Good, J.M.; Nielsen, R.; Moritz, C. Unlocking the vault: Next-generation museum population genomics. *Mol. Ecol.* **2013**, *22*, 6018–6032. [CrossRef]
210. Rowe, K.C.; Singhal, S.; Macmanes, M.D.; Ayroles, J.F.; Morelli, T.L.; Rubidge, E.M.; Bi, K.E.; Moritz, C.C. Museum genomics: Low-cost and high-accuracy genetic data from historical specimens. *Mol. Ecol. Resour.* **2011**, *11*, 1082–1092. [CrossRef]
211. Billerman, S.M.; Walsh, J. Historical DNA as a tool to address key questions in avian biology and evolution: A review of methods, challenges, applications, and future directions. *Mol. Ecol. Resour.* **2019**, *19*, 1115–1130. [CrossRef]
212. Murphy, S.A.; Double, M.C.; Legge, S.M. The phylogeography of Palm Cockatoos, *Probosciger aterrimus*, in the dynamic Australo-Papuan region. *J. Biogeogr.* **2007**, *34*, 1534–1545. [CrossRef]

213. Ewart, K.M.; Johnson, R.N.; Ogden, R.; Joseph, L.; Frankham, G.J.; Lo, N. Museum specimens provide reliable SNP data for population genomic analysis of a widely distributed but threatened cockatoo species. *Mol. Ecol. Resour.* **2019**, *19*, 1578–1592. [CrossRef] [PubMed]
214. Kundu, S.; Jones, C.G.; Prys-Jones, R.P.; Groombridge, J.J. The evolution of the Indian Ocean parrots (Psittaciformes): Extinction, adaptive radiation and eustacy. *Mol. Phylogenet. Evol.* **2012**, *62*, 296–305. [CrossRef] [PubMed]
215. Podsiadlowski, L.; Gamauf, A.; Töpfer, T. Revising the phylogenetic position of the extinct Mascarene Parrot *Mascarinus mascarin* (Linnaeus 1771) (Aves: Psittaciformes: Psittacidae). *Mol. Phylogenet. Evol.* **2017**, *107*, 499–502. [CrossRef] [PubMed]
216. Kirchman, J.J.; Schirtzinger, E.E.; Wright, T.F. Phylogenetic relationships of the extinct Carolina Parakeet (*Conuropsis carolinensis*) inferred from DNA sequence data. *Auk* **2012**, *129*, 197–204. [CrossRef]
217. Burgio, K.R.; Carlson, C.J.; Bond, A.L.; Rubega, M.A.; Tingley, M.W. The two extinctions of the Carolina Parakeet *Conuropsis carolinensis*. *Bird Conserv. Int.* **2021**. [CrossRef]
218. Anmarkrud, J.A.; Lifjeld, J.T. Complete mitochondrial genomes of eleven extinct or possibly extinct bird species. *Mol. Ecol. Resour.* **2017**, *17*, 334–341. [CrossRef]
219. Bergner, L.M.; Dussex, N.; Jamieson, I.G.; Robertson, B.C. European colonization, not Polynesian arrival, impacted population size and genetic diversity in the critically endangered New Zealand Kākāpō. *J. Hered.* **2016**, *107*, 593–602. [CrossRef]
220. Dussex, N.; Von Seth, J.; Robertson, B.; Dalén, L. Full mitogenomes in the critically endangered Kākāpō reveal major post-glacial and anthropogenic effects on neutral genetic diversity. *Genes* **2018**, *9*, 220. [CrossRef]
221. Boehrer, B.T. *Parrot Culture: Our 2500-Year-Long Fascination with the World's Most Talkative Bird*; University of Pennsylvania Press: Philadelphia, PA, USA, 2004; ISBN 0-8122-3793-5.
222. Watson, A.S.; Plog, S.; Culleton, B.J.; Gilman, P.A.; Leblanc, S.A.; Whiteley, P.M.; Claramunt, S.; Kennett, D.J. Early procurement of Scarlet Macaws and the emergence of social complexity in Chaco Canyon, NM. *Proc. Natl. Acad. Sci. USA* **2015**, *112*, 8238–8243. [CrossRef]
223. George, R.J.; Plog, S.; Watson, A.S.; Schmidt, K.L.; Culleton, B.J.; Harper, T.K.; Gilman, P.A.; LeBlanc, S.A.; Amato, G.; Whiteley, P.; et al. Archaeogenomic evidence from the southwestern US points to a pre-Hispanic Scarlet Macaw breeding colony. *Proc. Natl. Acad. Sci. USA* **2018**, *115*, 8740–8745. [CrossRef]
224. Bover, P.; Llamas, B.; Heiniger, H.; Olah, G.; Segura, R.; Shimada, I. Ancient DNA Analysis of Feathers from Funerary Bundles at the Pre-Hispanic Religious Center of Pachacamac (Perú). In Proceedings of the 9th International Symposium on Biomolecular Archaeology, ISBA9, Toulouse, France, 1–4 June 2021.
225. Capriles, J.M.; Santoro, C.M.; George, R.J.; Flores Bedregal, E.; Kennett, D.J.; Kistler, L.; Rothhammer, F. Pre-Columbian transregional captive rearing of Amazonian parrots in the Atacama Desert. *Proc. Natl. Acad. Sci. USA* **2021**, *118*, e2020020118. [CrossRef] [PubMed]
226. Crates, R.; Olah, G.; Adamski, M.; Aitken, N.; Banks, S.; Ingwersen, D.; Ranjard, L.; Rayner, L.; Stojanovic, D.; Suchan, T.; et al. Genomic impact of severe population decline in a nomadic songbird. *PLoS ONE* **2019**, *14*, e0223953. [CrossRef]
227. Groth, D.M.; Coates, D.; Wetherall, J.D.; Mell, D.; Hall, G.P. A Study of the genetic diversity and relatedness in the parrot, Naretha Blue Bonnet *Northiella haematogaster narethae*. Conservation biology in Australia and Oceania at the University of Queensland, Brisbane, Australia, 30 September–4 October 1991; p. 85.
228. Mell, D.; Wetherall, J. To catch a thief. *Landscope* **1992**, *7*, 28–32.
229. Shipham, A.; Schmidt, D.J.; Joseph, L.; Hughes, J.M. A genomic approach reinforces a hypothesis of mitochondrial capture in Eastern Australian Rosellas. *Auk* **2017**, *134*, 181–192. [CrossRef]
230. Faircloth, B.C.; Mccormack, J.E.; Crawford, N.G.; Harvey, M.G.; Brumfield, R.T.; Glenn, T.C. Ultraconserved elements anchor thousands of genetic markers spanning multiple evolutionary timescales. *Syst. Biol.* **2012**, *61*, 717–726. [CrossRef]
231. Benestan, L.M.; Ferchaud, A.-L.; Hohenlohe, P.A.; Garner, B.A.; Naylor, G.J.P.; Baums, I.B.; Schwartz, M.K.; Kelley, J.L.; Luikart, G. Conservation genomics of natural and managed populations: Building a conceptual and practical framework. *Mol. Ecol.* **2016**, *25*, 2967–2977. [CrossRef] [PubMed]
232. Garner, B.A.; Hand, B.K.; Amish, S.J.; Bernatchez, L.; Foster, J.T.; Miller, K.M.; Morin, P.A.; Narum, S.R.; O'Brien, S.J.; Roffler, G.; et al. Genomics in conservation: Case studies and bridging the gap between data and application. *Trends Ecol. Evol.* **2016**, *31*, 81–83. [CrossRef] [PubMed]
233. Deakin, J.E.; Potter, S.; O'Neill, R.; Ruiz-Herrera, A.; Cioffi, M.B.; Eldridge, M.D.B.; Fukui, K.; Marshall Graves, J.A.; Griffin, D.; Grutzner, F.; et al. Chromosomics: Bridging the gap between genomes and chromosomes. *Genes* **2019**, *10*, 627. [CrossRef]
234. Eizenga, J.M.; Novak, A.M.; Sibbesen, J.A.; Heumos, S.; Ghaffaari, A.; Hickey, G.; Chang, X.; Seaman, J.D.; Rounthwaite, R.; Ebler, J.; et al. Pangenome graphs. *Annu. Rev. Genom. Hum. Genet.* **2020**, *21*, 139–162. [CrossRef] [PubMed]

MDPI

Article

Genetic, but Not Behavioral, Evidence Supports the Distinctiveness of the Mealy Amazon Parrot in the Brazilian Atlantic Forest

Dominique L. Hellmich [1,*], Andre B. S. Saidenberg [2] and Timothy F. Wright [1]

1 Department of Biology, New Mexico State University, Las Cruces, NM 88003, USA; wright@nmsu.edu
2 Department of Veterinary Pathology, University of Sao Paulo, Sao Paulo 05508-060, SP, Brazil; andresaidenberg@usp.br
* Correspondence: dlh263@nmsu.edu

Citation: Hellmich, D.L.; Saidenberg, A.B.S.; Wright, T.F. Genetic, but Not Behavioral, Evidence Supports the Distinctiveness of the Mealy Amazon Parrot in the Brazilian Atlantic Forest. *Diversity* **2021**, *13*, 273. https://doi.org/10.3390/d13060273

Academic Editors: Luc Legal, José L. Tella, Guillermo Blanco and Martina Carrete

Received: 11 May 2021
Accepted: 12 June 2021
Published: 17 June 2021

Abstract: The presence of unidentified cryptic species within a species complex can obscure demographic trends of vulnerable species, impacting potential species conservation and management decisions. Previous work identified a taxonomic split between Central and South American populations of the mealy amazon (*Amazona farinosa*) that subsequently resulted in the elevation of these two populations to full species status (*Amazona guatemalae* and *A. farinosa*, respectively). In that study, however, a third, geographically disjunct population from the Brazilian Atlantic Forest was insufficiently sampled, limiting the ability of researchers to fully evaluate its genetic distinctiveness. Given that significant levels of biodiversity and endemism are found in this region, we aimed to use genetic and behavioral data to determine if the Atlantic Forest population of *A. f. farinosa* represents a third cryptic species within the complex. We sequenced 6 genes (4 mitochondrial and 2 nuclear introns) from the Atlantic Forest population of *A. f. farinosa* to measure the genetic relationships between this population and all other recognized species and subspecies of the mealy amazon. In addition, we use spectrographic cross-correlation and an analysis of 29 acoustic parameters to determine whether the taxa diverge in their learned contact call structure and if the degree of vocal differentiation correlates to genetic structure. We found that the Atlantic Forest population of *A. f. farinosa* was genetically distinct from that of the greater Amazon basin, but the degree of differentiation was less than that separating the Central and South American taxa. Acoustic analysis revealed substantial variation in contact call structure within each clade. This variation created substantial overlap in acoustic space between the clades. In all, the degree of call divergence between clades did not correspond to the degree of genetic divergence between the same clades. The results suggest that in taxa with substantial geographic variation in learned calls, such as the mealy amazon, vocalizations may not be a useful tool in the identification of cryptic species that are lifelong vocal learners. While these results do not support the elevation of the Brazilian Atlantic Forest population of the mealy amazon to full species status, given current trends of habitat loss in the Atlantic Forest as well as the imperiled status of large parrot species globally, we argue that this population nonetheless warrants special conservation and management consideration as a pool of unique genetic diversity within the southern mealy amazon species.

Keywords: cryptic species; vocal variation; parrot; genetic differentiation; open-ended learning; *Amazona farinosa*

1. Introduction

A primary challenge for conservationists and wildlife managers is understanding the demographic trends of vulnerable species. This task is complicated if the basic taxonomic relationships between species are not well understood. This issue commonly arises within species complexes that contain unidentified cryptic species: morphologically indistinguishable but genetically distinct species that are mistakenly classified under a single species name [1,2]. The presence of unidentified cryptic species can lead to artificially low assessments of overall biodiversity and inflated estimations of the health or distribution of species of special conservation concern [3–5]. Historically, species delineations were made primarily using morphological features [2], but the current widespread availability of genetic sequencing technologies has given researchers new insight into the level of genetic diversity that exists within many of these morphologically defined species. Ideally, a combination of morphological, ecological, behavioral, and genetic data are used to describe potential cryptic species, as this integrative approach reduces subjective biases and allows for more informed conservation decisions [1].

The quantification of behavioral variation within a species complex may be a particularly powerful component of cryptic species identification, as these data can be collected remotely and at significantly lower cost than most traditional trapping methods [6,7]. Specifically, variation in acoustic signals was found to correlate with genetic divergence in a variety of taxa, including bats [8,9], primates [10], insects [11], anurans [12], and birds [13–16]. Learned or culturally derived vocalizations in particular might facilitate speciation, as these signals can rapidly accumulate mutations through the transmission process both within and between generations, leading to increases in assortative mating [6,17]. However, evidence for the impact of learning on the speciation process is variable and contradictory [18–20]. This association may be particularly complicated in species with lifelong vocal plasticity, such as humans [21], some parrots [22], and bats [23]. Individuals with open-ended vocal learning are able to modify their vocalizations into adulthood and may do so to serve a variety of social functions (e.g., pair formation and maintenance, social group integration) [24]. This flexibility poses the question: If individual vocalizations are not wedded to the underlying population genetic structure, can we use the vocalizations of these open-ended vocal learners to identify cryptic species within a species complex?

Parrots (Order Psittaciformes) are an interesting group with which to study the relationship between the acoustic and genetic variation found within a cryptic species complex. Nearly 43% of known species are considered threatened with extinction [25], primarily due to habitat loss and poaching for the pet trade [26,27], making them a primary conservation concern. In addition, parrots are open-ended vocal learners with many members of the group displaying patterns of geographic variation (i.e., vocal dialects) in their socially learned contact calls [24,28]; these dialects may or may not correlate to patterns of genetic differentiation within a species [29–32]. Finally, there are numerous examples of species with cryptic genetic diversity within the parrot order [33–35], including the focus of our study, the mealy amazon (*Amazona farinosa*) [36].

The mealy amazon is a large-bodied (540–700 g) parrot in the Genus Amazona that ranges widely through Central and South America. It was long classified as a single species, *Amazona farinosa*, with five recognized subspecies [37]. Wenner and colleagues used DNA sequence data from several regions of mitochondrial and nuclear genomes to examine the geographic patterns of genetic variation within the species [36]. Based on the evidence of deep splits between the Central and South American subspecies the nominate mealy amazon species was formally reclassified into the northern mealy amazon (*A. guatemalae*) containing the *guatemalae* and *virenticeps* subspecies, and the southern mealy amazon (*A. farinosa*) containing the *farinosa*, *inornata*, and *chapmani* subspecies [38]. Subsequently, in 2014 the IUCN Red List up-listed both newly split species to Near Threatened from Least Concern [25]. Then, in 2018, the government of the Brazilian state of São Paulo listed local populations of the Atlantic Forest *A. f. farinosa*, the largest surviving populations in the region, as critically endangered [39]. With populations of both species continuing to decline, there is an urgent need to fully understand the taxonomic divisions within this species complex to inform future conservation decisions. Here, we look to rectify a sampling gap from the original [36] study, namely the lack of substantial genetic data from the disjunct Brazilian Atlantic Forest population of the southern mealy amazon subspecies *A. f. farinosa*.

The Atlantic Forest is the second largest rainforest system in South America, stretching across a broad strip of Brazil's southeastern Atlantic coastline, and is a hotspot of endemic biodiversity in the Neotropics [40]. Currently, the Atlantic Forest is separated from the greater Amazon basin by the dry Caatinga region to the northeast, the upland, wooded savanna of the Cerrado region to the north/northwest, and from the Andean forests to the west by the lowland, arid Chaco region of central South America [41]. Historically, this forest has undergone several periods of connection and separation from these other major forest regions, leading to the isolation of some species over both short (e.g., 10,000–20,000 years ago) and long (e.g., >3 million years ago) time scales [41,42]. This dynamic evolutionary history has created a region rich with endemism and cryptic species complexes [43–47] that is threatened today by high levels of habitat loss and fragmentation [48]. For this reason, the Atlantic Forest is an area of upmost concern for biodiversity conservation [48,49].

We aimed to address two main questions. First, is the Atlantic Forest population of *A. f. farinosa* a cryptic species within the mealy amazon species complex? Given the possible timeframes when the Atlantic Forest population of *A. f. farinosa* could have been separated from the larger Amazon basin population of *A. f. farinosa*, we hypothesize that it is as genetically dissimilar from the other southern mealy amazon subspecies as the northern species is from the southern species. Second, can we use the vocalizations of an open-ended vocal learner to identify potential cryptic species? If so, we hypothesize that the degree of divergence in contact call structure between and within the mealy amazon species, subspecies, and the Atlantic Forest population of *A. f. farinosa* will correlate with the genetic distance between these groups. To measure genetic differentiation, we expanded on the original dataset from [36] by sequencing six gene regions from newly sampled individuals of the Atlantic Forest population of *A. f. farinosa* to create both a phylogeny and a haplotype network that includes this group. The similarity of contact calls within and between groups, as well as their relationship to the genetic distance between groups, was then assessed using 29 acoustic parameter measurements and spectrographic cross-correlation values. We predicted that our genetic and behavioral (i.e., contact call) measures will show related degrees of differentiation, with the Atlantic Forest population of *A. f. farinosa* falling out as divergent from the Amazon basin population of *A. f. farinosa* and the southern mealy amazon species as a whole. These results will inform the study and preservation of this geographically distinct population of the mealy amazon.

2. Materials and Methods

2.1. Study Species and Data Summary

The mealy amazon species complex includes two recognized species and five subspecies. The northern mealy amazon *A. guatemalae* (including the *virenticeps* and *guatemalae* subspecies) ranges from southern Mexico to central Panama, while the southern mealy amazon *A. farinosa* is distributed in three geographically distinct locations: (1) the *inornata* subspecies extends from southern Panama through western Ecuador and eastern Colombia, (2) the *chapmani* and *farinosa* subspecies range from eastern Ecuador, eastern Peru, and northern Bolivia through the greater Amazon Basin, and (3) in an isolated population of the *farinosa* subspecies along Brazil's Atlantic coast [38] (Figure 1). Seven samples from the Atlantic Forest population of *A. f. farinosa* were unique to this study as all other clades within the species complex were sampled adequately by [36]; see Table S1 for details on the origins of the genetic samples.

Recordings of contact calls from individuals of all subspecies and the Atlantic Forest population of *A. f. farinosa* were obtained from the Macaulay library (Cornell Laboratory of Ornithology, Ithaca, NY, U.S.) and Xeno-canto (Xeno-canto Foundation for Nature Sounds, Netherlands). We used 148 high-quality sound files from 1954 to 2019 in our analysis; sound file metadata can be found in Table S2.

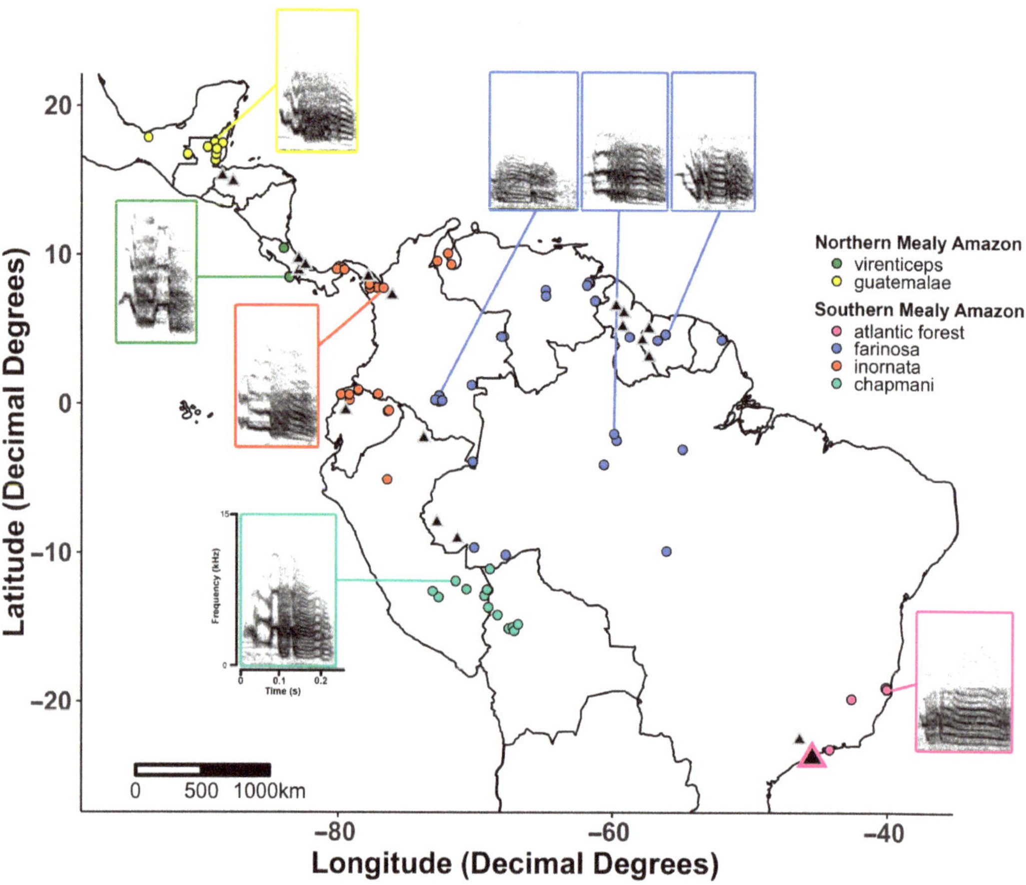

Figure 1. Map of vocal sampling locations. Spectrograms of representative calls from each clade are shown at their corresponding recording location. Spectrograms were created using the 'spec_param' and 'specreator' functions in warbler to optimize the visualization parameters (overlap = 90, window length = 475, color levels = (−50, 0, 5)). Genetic samples from the [36] study are indicated by grey-outlined triangles on the map. The location of the new genetic samples used in our study is indicated by the purple-outlined triangle.

2.2. Phylogenetic Analysis and Haplotype Network

Blood samples were collected, according to SISBIO permit #71978-1 protocols following Brazilian environmental legislation, on Whatman FTA elute micro cards from seven individuals being held in captivity at the ASM Cambaquara rescue center on the island of Ilhabela, São Paulo, Brazil, after seizure as illegal pets or after being found injured in the wild. These individuals represent a single Atlantic Forest *A. f. farinosa* population (for complete sample metadata see Table S1). For each sample, genomic DNA was isolated and then PCR amplification was performed for the four mitochondrial and two nuclear intron gene fragments previously used in [36]: 12S rDNA (12S), 16S rDNA (16S), Cytochrome oxidase subunit I (COI), cytochrome *b* (CytB), tropomyosin alpha-subunit intron 5 (TROP), and transforming growth factor β-2 intron I (TGFB2). An additional mitochondrial gene fragment for NADH dehydrogenase 2 (ND2) was amplified for both the [36] samples and the new *A. f. farinosa* Atlantic Forest samples. A complete description of the number of individuals sampled from each population, the genes sequences per individual, GenBank accession numbers, and primers can be found in Tables S1, S3 and S4. All PCR products were directly sequenced at the University of Texas at El Paso's Border Biomedical Research

Center genomic analysis core facility with the same primers used for PCR. Bidirectional reads were screened for quality, assembled to an *Amazona* reference sequence (Table S3), and trimmed to the appropriate length to enable gene alignment using Sequencher v5.4.6 (Gene Codes Corporation, Ann Arbor, MI, U.S.). The individual gene alignments were concatenated into a supermatrix using the Bio.Nexus module from the Biopython project v1.76 [50]. Three outgroups used in the [36] analysis—*A. kawalli*, *A. auropalliata*, and *A. amazonica*—were included to root our phylogenetic tree and to provide a measure of relative genetic distance with which to compare recognized *Amazona* species to the clades within the mealy amazon species complex.

A maximum-likelihood phylogenetic analysis with rapid bootstrapping was performed on the concatenated gene alignment using RAxML-HPC v8.2.12 [51] and a GTR-CAT model on XSEDE through the Cipres Science Gateway v3.3 [52]. The maximum likelihood, majority rule, best tree results and bootstrap values were visualized using FigTree v1.4.4 [53].

Two median joining haplotype networks [54] were constructed and visualized using PopART v1.7 [55]. The first network includes all of the available mitochondrial genes (12S, 16S, CytB, ND2) because we were able to obtain sequence data for both recognized species, at least four of the five subspecies (all minus *virenticeps*), and Atlantic Forest population of *A. f. farinosa*. The second network was created using only CytB data, as this most closely resembles the data set used by [36].

2.3. Genetic Differentiation

To quantify inter- and intraspecific genetic variation we calculated the between group mean genetic distance (i.e., the average number of nucleotide substitutions between clades) using the concatenated gene alignment and a p-distance model in MEGA X [56,57].

2.4. Acoustic Analysis

High-quality (i.e., low background noise, no signal overlap) contact calls were manually selected from the sound files using Raven Pro v1.5.0 (Cornell Laboratory of Ornithology, Ithaca, NY, U.S.). The accurate documentation of caller ID was lacking in most sound files with multiple calling individuals; therefore, only one call from one individual was randomly selected per call file. In call files where the researcher was able to identify more than one individual from counter-calling (i.e., the repeated, predictable response of a second individual to the calls of a first), one call was randomly selected from up to two individuals. In all, only one randomly selected high-quality (i.e., determined by each database to be "A" quality out of an A–F rating scale) call per individual was used in all downstream analyses. After call selection, the warbleR package v1.1.19 [58] in R v3.6.2 (The R Foundation) was used to perform additional quality control and data processing. Individual spectrograms were created with the 'specreator' function using a window length of 510 samples, a 90% overlap between windows, and a frequency limit of 0 kHz to 9kHz to reassess call quality. Poorly selected calls (i.e., calls where the temporal coordinates being read by a function did not match the actual start and end times of the calls) were reselected using the 'seltailor' function.

Differences in contact call structure between all populations were assessed both quantitatively and qualitatively. First, 26 acoustic parameter measurements—including various frequency and duration characteristics, entropy, skew, and kurtosis—were calculated for each call using the warbleR 'specan' function with a 90% overlap between windows and a bandpass filter from 0 kHz to 7 kHz (Table S5). The extracted data were then used in a principal component analysis performed with the 'prcomp' function from the base stats R package. The first two principal components were plotted to visualize the clustering patterns of calls by population. Next, the 'xcorr' function of the warbleR package was used to create a spectrographic cross-correlation matrix of all the calls. This matrix was transformed into distance measurements so that a multidimensional scaling analysis using the stats base package in R could be used to visualize the clustering patterns of calls by clade.

A Mantel test was performed using the vegan package v2.5-6 in R to determine whether the degree of genetic differentiation (measured as the average number of sequence differences) between clades correlates with vocal differentiation (measured as the average cross-correlation dissimilarity values).

3. Results

3.1. Phylogenetic Tree and Haplotype Network

A total of 3444 bp from 6 genes (393 bp of 12S, 525 bp of 16S, 868 bp of CytB, 510 bp of ND2, 625 bp of TGFB2, 523 bp of TROP) were included in the maximum-likelihood phylogenetic analysis, though not all genes were able to be sequenced for every individual; see Table S3 for a complete list genes per individual. Notably, the forward reads for COI failed to sequence for all of the Atlantic Forest *A. f. farinosa* samples, so this gene was removed from the analysis. With the addition of the ND2 mtDNA sequence and the exclusion of the COI mtDNA sequence, the topology of our maximum likelihood majority rule consensus tree is nearly identical to the phylogeny reported by the original [36] study for taxa included in both studies (Figure 2). The Atlantic Forest population of *A. f. farinosa* is recovered as a monophyletic group sister to the assemblage that includes the three recognized subspecies of *A. farinosa* (*A. f. inornata*, *A. f. chapmani*, and *A. f. farinosa*). This group does not, however, show the same level of genetic differentiation as the two currently recognized species of mealy amazon (*A. guatemalae* and *A. farinosa*; see insert of Figure 2).

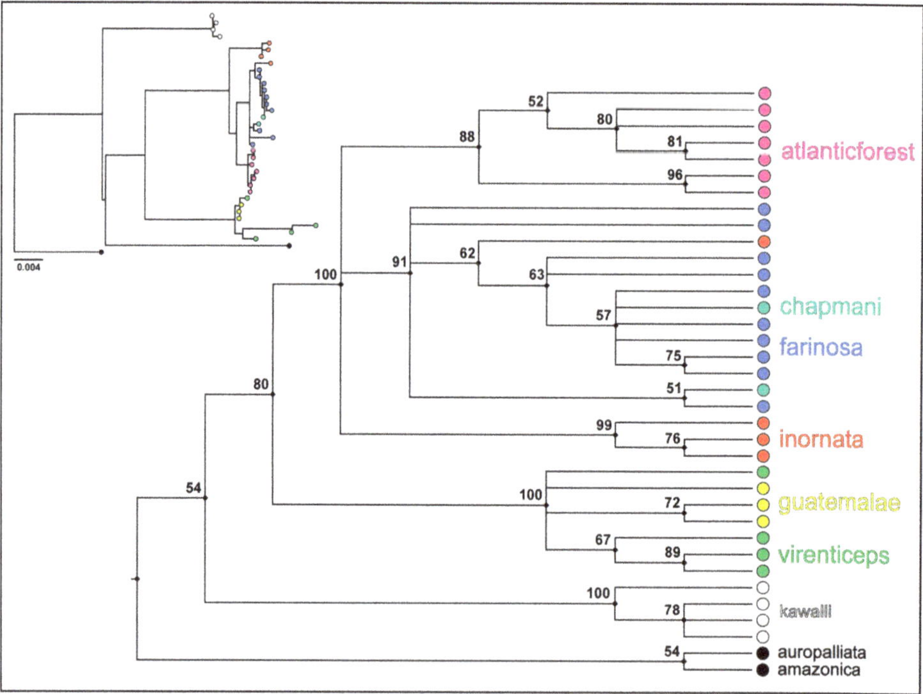

Figure 2. Maximum likelihood majority rule consensus tree (cladogram) and median joining haplotype networks. Maximum likelihood majority rule consensus tree based on an analysis of the combined nuclear and mtDNA from all northern and southern mealy amazon clades. Numbers to the left of each node are bootstrap consensus values. The insert to the left of the consensus tree is a phylogram of the best tree recovered from the same maximum likelihood search and illustrates the deep split in the evolutionary timeline of the northern and southern species, as well as the relatively shallow split between the Atlantic Forest clade and the other southern mealy amazon clades. The relative branch lengths of the phylogram represent sequence divergence.

The median joining haplotype network created using the CytB data returned a total of 16 haplotypes, with 28 sequence differences between the northern and southern species and 5 sequence differences between the Atlantic Forest populations of *A. f. farinosa* and their closest Amazon basin *A. f. farinosa* relative (Figure 3). In addition, the Atlantic Forest populations form a distinct cluster within the greater southern mealy amazon cluster. The haplotype network created with the reduced dataset of the four mitochondrial genes shows a similar pattern. Thirteen haplotypes were recovered, with 67 sequence differences separating the northern and southern mealy amazon species haplotypes and 10 sequence differences separating the Atlantic Forest populations of *A. f. farinosa* from their closest relative from the Amazon basin population of *A. f. farinosa* (Figure 3). This network lacks a representative of the *A. g. virenticeps* northern mealy amazon subspecies (Table S3) due to an insufficient amount of sequence data for the analysis software. However, we assume any *virenticeps* individuals would be closely related to the other northern mealy amazon subspecies *A. g. guatamalae* based on the topology of the CytB network. Together, these results consistently show that the Atlantic Forest population is a distinct group when compared to the other *A. farinosa* subspecies, albeit one that lacks the genetic distinctiveness of previously recognized full species.

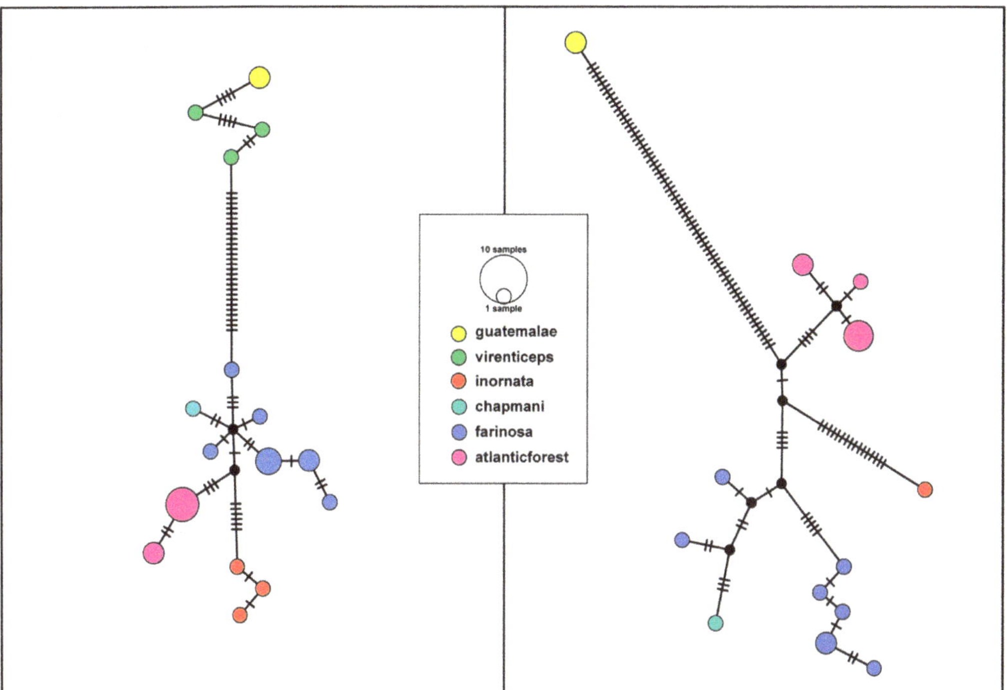

Figure 3. (**Left panel**) Median joining haplotype network based on CytB data. Size of each circle corresponds to the number of individuals sharing that haplotype and color to each clade. Ticks on each branch represent the number of sequence differences between each haplotype. (**Right panel**) Median joining haplotype network based on 4 mitochondrial genes (CytB, ND2, 12S, 16S). Size of each circle corresponds to the number of individuals sharing that haplotype and color to each clade. Ticks on each branch represent the number of sequence differences between each haplotype.

3.2. Genetic Differentiation

Between-group mean genetic distances based on data from all genes included in this study indicate 2.9% average sequence divergence between northern and southern mealy amazon subspecies, compared to 0.4% average sequence divergence between southern mealy amazon subspecies (including the Atlantic Forest population of *A. f. farinosa*) and 0.3% average sequence divergence between northern mealy amazon subspecies.

3.3. Call Similarity between Subspecies

A total of 150 calls (110 from the Macaulay Library and 40 from Xeno-canto) were used in our analysis of call similarity between the 5 recognized mealy amazon subspecies and the Atlantic Forest population of *A. f. farinosa*. Sampling locations and spectrograms of representative calls from each subspecies are shown in Figure 1. An initial visual inspection of the call spectrograms suggested there is some variation in the structure of contact calls between subspecies (Figure 1), however, plots of the first two components from a principal component analysis of acoustic parameter measurements and a multidimensional scaling analysis of the spectrographic cross-correlation matrix did not reveal distinct clusters of calls among the subspecies. Instead, we saw significant overlap of each clades' calls in acoustic space (Figure 4) due to a large amount of variation in calls within each subspecies. This pattern suggests that (a) substantial structural differences exist across each subspecies' range (e.g., the three example spectrograms of the *A. f. farinosa* subspecies of southern mealy amazon in Figure 1), and (b) variation within a subspecies is as great as variation among subspecies (Figure 4).

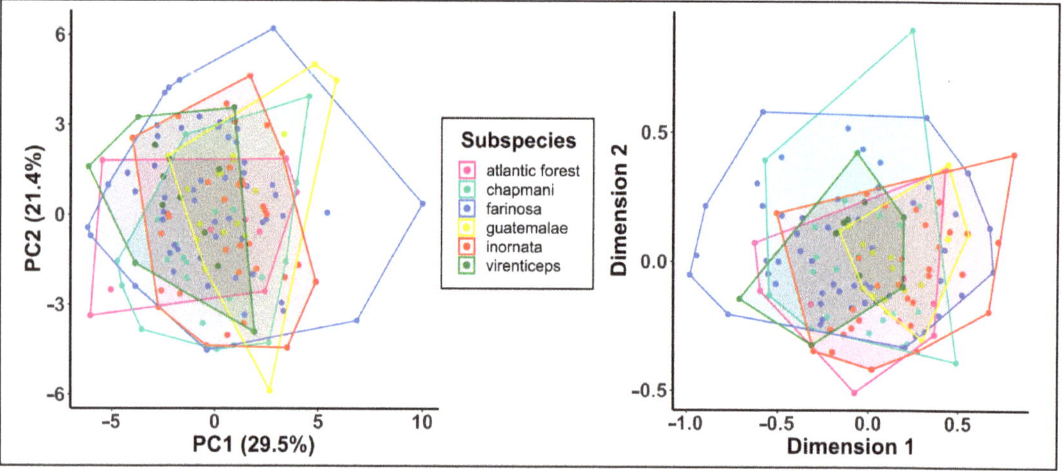

Figure 4. Acoustic variation in call data. Plots of acoustic variation in contact calls based on principle components analysis of 27 call measures (**left**) and a multidimensional scaling of spectrogram cross-correlation values (**right**). The points represent individual calls, and the polygons represent the total area of occupied by each clade's set of calls in acoustic space.

3.4. Assessment of Genetic Distance and Vocal Divergence

A qualitative assessment of the relationship between group mean genetic distance and average cross-correlation vocal similarity among clades of *A. farinosa* and *A. guatemalae* (Figure 5) showed no clear pattern of association. This apparent lack of a correlation between vocal and genetic differentiation was supported by the Mantel test, which did not detect any statistical association between the two matrices (Mantel test: r = −0.012, matrix size = 6 clades, $p = 0.451$).

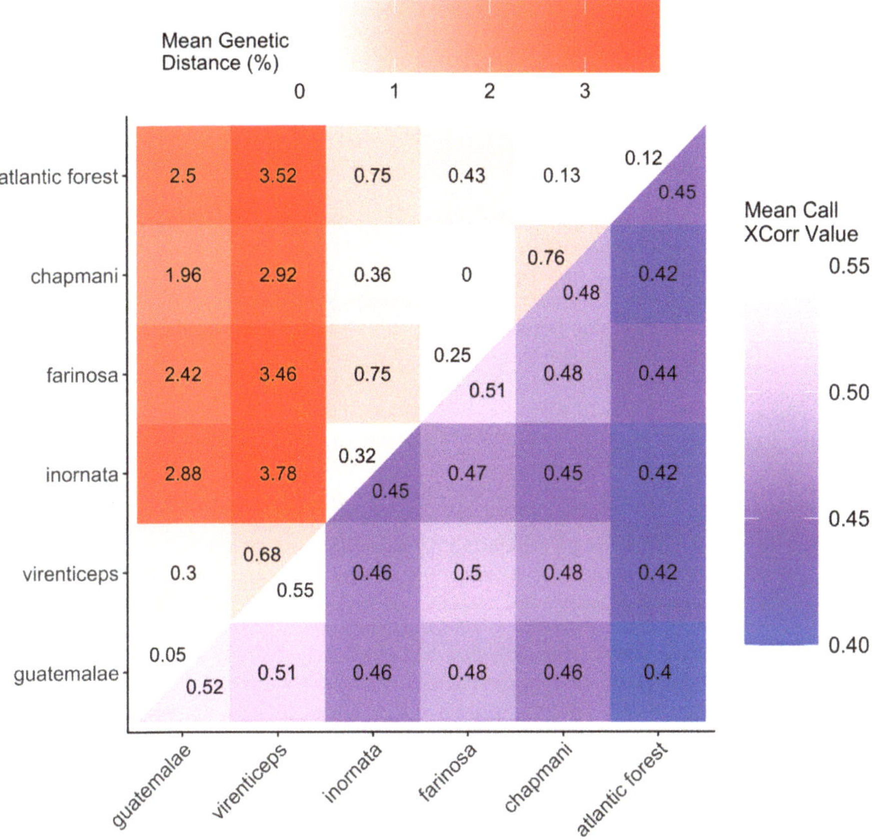

Figure 5. Matrix comparing net between group genetic distance with mean call cross-correlation values. For both sets of values, darker colors indicate greater dissimilarity between clades and lighter colors indicate greater similarity between clades. 'Mean genetic distance' is the numeric average of all pairwise distances (i.e., number of nucleotide substitutions) between clades, computed with the Kimura 2-parameter distance estimation model.

4. Discussion

We expanded upon a genetic dataset representing all recognized subspecies of the mealy amazon species complex to examine their relationships to the disjunct Atlantic Forest population of *A. f. farinosa*. In addition, we examined contact call similarity within and between these groups to determine if variation in this behavioral trait was correlated with underlying genetic structure. We found that while the Atlantic Forest population is recovered as a distinct clade, the degree of genetic differentiation between this population and the Amazon basin population of *A. f. farinosa* is substantially less than the degree of separation between the recognized northern and southern mealy amazon species. We also found that variation in a learned behavioral trait did not correlate to underlying genetic population structure. We discuss these results and their conservation implications in more detail below.

4.1. Genetic Relationships of the Mealy Amazon Clades

Our phylogeny and haplotype networks suggest that while the Brazilian Atlantic Forest population of *A. f. farinosa* is genetically and geographically distinct from other populations and subspecies of *A. farinosa*, it is not genetically distant enough to be considered a third cryptic species within the mealy amazon species complex. Specifically, the between-group genetic distances for the three recognized southern mealy amazon subspecies and Atlantic Forest population had a range of 0.13–0.75% compared to genetic distances with a range of 1.9–3.8% between the northern and southern mealy amazon subspecies. Based on the magnitude of difference between the recently reclassified northern and southern mealy amazon species, we conclude that the Atlantic Forest population is best considered a genetically distinct population warranting formal recognition as a subspecies within the *A. farinosa* species.

Other studies aiming to identify potential cryptic species in parrots have employed comparable thresholds at which a genetically distinct clade should be considered a new subspecies or full cryptic species. In studies of three populations of the Cuban parrot subspecies *Amazona leucocephala bahamensis*, researchers concluded that sequence differences of 1.7–2.2% in the CR1 region warranted a subspecies designation for the Abaco phylogenetic species [33]. In the mulga parrot *Psephotellus varius*, an average sequence difference of 1.9% in the majority of the mitogenome (~84%) between the eastern and western populations was also justification for an elevation of the populations to separate subspecies [34]. The same conclusion was reached for northern and southern subspecies of the scarlet macaw (*Ara macao*) showing a 1.8% average sequence difference in a combined dataset of 12S, 16S, COI, and CytB [59], and for three recognized species within the *Amazona ochrocephala* complex with only 2% sequence differences in a combined dataset of ATP synthase 6 and 8, COI, ND2, and CytB despite their varied morphology [60]. In contrast, the authors of a study finding 4.4–5.1% sequence divergence in CytB between eastern and western populations of the ground parrot (*Pezoporus wallicus*) concluded that the western population should be reclassified as a new, cryptic species [35]. While the results of these studies are not directly comparable due to the various genes and analysis models used, the relative sequence differences serve as useful benchmarks when we consider the threshold at which a genetically distinct population is assigned a new taxonomic status. We should note that the range of genetic distances between the Atlantic Forest population of *A. f. farinosa* and all other *A. farinosa* subspecies for the CytB gene alone (0.76–0.99%) is equivalent to the interspecific ranges previously calculated for CytB in 88 avian genera [61]. However, given our more extensive data set and the established level of sequence divergence between the northern and southern mealy amazon species, we do not believe the Atlantic Forest population merits classification as a full species.

While it is likely our results represent the true phylogenetic relationship between the Atlantic Forest and Amazon basin populations of *A. f. farinosa*, there is a possibility our data suffer from a lack of geographic sampling diversity from the Atlantic Forest population (i.e., all individuals sampled originated from the same location). Wider geographic sampling of the mealy amazon across the Atlantic Forest may reveal additional genetic structure between populations within the Atlantic Forest (e.g., between the north-eastern region of the Atlantic Forest, often referred to as the Pernambuco Center of Endemism, and the southern or central regions of the Atlantic Forest) as well as between *A. f. farinosa* populations in the Amazon basin and Atlantic Forest. Patterns of intraspecific genetic differentiation within the Atlantic Forest are seen in other avian [62–64] and non-avian taxa [65] within the Atlantic Forest, and may be driven by the historic expansion of species from the Amazon basin to the central Atlantic Forest through gallery forests of the Cerrado region [42,66,67] or the relative stability and connectivity of the central and western regions of the Atlantic Forest in relation to historical climate patterns [68,69].

4.2. Utility of Behavioral Variation for Detecting Cryptic Species

The second goal of our study was to evaluate the potential utility of vocalization data in the identification of cryptic species. Given that behavioral data can often be collected with relative ease and at a very low cost, we aimed to determine whether variation in call structure correlated with genetic variation among groups. Such a relationship has been observed for a number of songbird species such as the variable antshrike (*Thamnophilus caertulescens*), red crossbill (*Loxia curvirostra*), and greenish warbler (*Phylloscopus trochiloides*) [13–16]. However, parrots and other open-ended vocal learners are able to alter their calls into adulthood, a phenomenon that may potentially lead to rapid call divergence or convergence irrespective of the underlying genetic structure of a population. For example, individually distinct call signatures may promote the recognition of social group members in fission–fusion societies and lead to a wide distribution of calls across the available acoustic space of a species or population [70]. Alternatively, because convergence on local call types (i.e., vocal dialects) signals group membership, the increased flexibility provided by open-ended learning may facilitate social integration throughout an individual's lifetime, especially after dispersal to unrelated populations [24], leading to an overall reduction in the total number of call types in a species or population.

Our results do not indicate a high degree of differentiation in contact call structure amongst groups. Cluster plots of each subspecies calls indicate substantial overlap of call structure and call features in acoustic space, and no correlation of call similarity with genetic structure (e.g., average cross-correlation similarity value for the Atlantic Forest population of *A. f. farinosa* and closely related *A. f. chapmani* is nearly identical, 0.419, to the similarity value of the Atlantic Forest population of *A. f. farinosa* and *A. g. virenticeps*, 0.418, from the northern species). Similarly, in yellow-naped amazons (*Amazona auropalliata*), a species well known for its regionally distinct vocal dialects, call diversity is unlinked to the underlying population genetic structure, indicating a strong preference of individuals to conform to local call types after dispersal or movement across dialect boundaries [71]. In contrast, the crimson rosella (*Platycercus elegans*) is one parrot species that shows clinal variation in their vocalizations and microsatellite genetic data [72]. Songbirds such as the white-crowned sparrow (*Zonotrichia leucophrys*) and suboscine passerines such as the variable antshrike (*Thamnophilus caerulescens*), which use innate vocalizations, also show greater correlation between acoustic and genetic variation [15,73,74], though this is not a consistent trend amongst all species (e.g., [65]).

Potential methodological limitations to our current study include inconsistencies in the quality and quantity of the available data from citizen science databases (e.g., lack of standardized recording protocols, variation in the type of recording equipment used, the unequal spatial distribution of sampling across each population's range). In addition, the call data were collected over a substantial timeframe (1950s to 2010s). Both factors may mask patterns of variation at the regional versus the subspecies level if call structures have changed over the decades or if information was lost in poor-quality recordings. These issues might be remedied by more aggressive pre-analysis filtering of the available data; however, attempting to apply stricter quality standards to our study led to a severe reduction in sample size that would have inhibited our current analyses. Overall, we suggest that vocalizations may be less useful as tools in the identification of cryptic species when those species are open-ended vocal learners, though further study is warranted.

4.3. Conservation Implications

Typically, conservation action is focused at the species level, creating a critical need for understanding the true relationship of disjunct and vulnerable populations to the species as a whole. In the case of the Atlantic Forest population of *A. f. farinosa*, our results do not support an elevation in species status but do indicate this population represents a unique pool of genetic diversity within the mealy amazon species complex. Currently, it is estimated that 7–16% of historic Atlantic Forest cover remains, and rates of habitat degradation or destruction have increased significantly in the past three decades [48,49], leading to restrictions in the ranges of local mealy amazon populations and increased conflict with humans (A. Saidenberg, *unpublished data*). Shrinking refuges could further endanger a population that, like most large parrot species, is already facing substantial threats from poaching and harvest for the pet trade [26,27]. For these reasons, we advocate for the special consideration of the Atlantic Forest population of mealy amazons in conservation and management decisions for the species, including the formal recognition of this population as a subspecies of the southern mealy amazon. Finally, we strongly encourage the continued study of parrot species complexes to help identify additional taxonomic groups warranting classification and protection.

Supplementary Materials: The following are available online at https://www.mdpi.com/article/10.3390/d13060273/s1, Table S1: Metadata and gene regions sequenced of the samples used in this study. Table S2: Metadata for the sound files used in this study. Table S3: Primer sequences used to amplify and sequence the gene regions used in this study. Table S4: Genbank accession numbers for new Atlantic forest *A. f. farinosa* gene sequences used in this study. Table S5: Acoustic parameters generated by the 'specan' function and used in our principal component analysis.

Author Contributions: Conceptualization, T.F.W. and A.B.S.S.; methodology, T.F.W. and D.L.H.; formal analysis, D.L.H.; resources, T.F.W. and A.B.S.S.; data curation, D.L.H.; writing—original draft preparation, D.L.H.; writing—review and editing, D.L.H., T.F.W., and A.B.S.S.; funding acquisition, T.F.W. All authors have read and agreed to the published version of the manuscript.

Funding: This research was funded by the World Parrot Trust.

Institutional Review Board Statement: Not applicable.

Informed Consent Statement: Not applicable.

Data Availability Statement: The data presented in this study are available in the article and supplementary materials file.

Acknowledgments: The authors would like to thank C. Campos for lab training, A. Betancourt and N. Joffe for sequencing and troubleshooting, and S. Davino and P. Melero from ASM Cambaquara for their help and logistical support.

Conflicts of Interest: The authors declare no conflict of interest. The funders had no role in the design of the study; in the collection, analyses, or interpretation of data; in the writing of the manuscript, or in the decision to publish the results.

References

1. Struck, T.H.; Feder, J.L.; Bendiksby, M.; Birkeland, S.; Cerca, J.; Gusarov, V.I.; Kistenich, S.; Larsson, K.H.; Liow, L.H.; Nowak, M.D.; et al. Finding evolutionary processes hidden in cryptic species. *Trends Ecol. Evol.* **2018**, *33*, 153–163. [CrossRef]
2. Bickford, D.; Lohman, D.J.; Sodhi, N.S.; Ng, P.K.L.; Meier, R.; Winker, K.; Ingram, K.K.; Das, I. Cryptic species as a window on diversity and conservation. *Trends Ecol. Evol.* **2007**, *22*, 148–155. [CrossRef] [PubMed]
3. Angulo, A.; Icochea, J. Cryptic species complexes, widespread species and conservation: Lessons from Amazonian frogs of the *Leptodactylus marmoratus* group (Anura: Leptodactylidae). *Syst. Biodivers.* **2010**, *8*, 357–370. [CrossRef]
4. Davidson-Watts, I.; Walls, S.; Jones, G. Differential habitat selection by *Pipistrellus pipistrellus* and *Pipistrellus pygmaeus* identifies distinct conservation needs for cryptic species of echolocating bats. *Biol. Conserv.* **2006**, *133*, 118–127. [CrossRef]
5. Delić, T.; Trontelj, P.; Rendoš, M.; Fišer, C. The importance of naming cryptic species and the conservation of endemic subterranean amphipods. *Sci. Rep.* **2017**, *7*, 1–12. [CrossRef]
6. Wilkins, M.R.; Seddon, N.; Safran, R.J. Evolutionary divergence in acoustic signals: Causes and consequences. *Trends Ecol. Evol.* **2013**, *28*, 156–166. [CrossRef] [PubMed]
7. Jones, G. Acoustic signals and speciation: The roles of natural and sexual selection in the evolution of cryptic species. *Adv. Study Behav.* **1997**, *26*, 317–354. [CrossRef]
8. Thabah, A.; Rossiter, S.J.; Kingston, T.; Zhang, S.; Parsons, S.; Mya, K.M.; Akbar, Z.; Jones, G. Genetic divergence and echolocation call frequency in cryptic species of *Hipposideros larvatus* s.l. (Chiroptera: Hipposideridae) from the Indo-Malayan region. *Biol. J. Linn. Soc.* **2006**, *88*, 119–130. [CrossRef]
9. Kingston, T.; Lara, M.C.; Jones, G.; Akbar, Z.; Kunz, T.H.; Schneider, C.J. Acoustic divergence in two cryptic *Hipposideros* species: A role for social selection? *Proc. R. Soc. B Biol. Sci.* **2001**, *268*, 1381–1386. [CrossRef]
10. Braune, P.; Schmidt, S.; Zimmermann, E. Acoustic divergence in the communication of cryptic species of nocturnal primates (*Microcebus* ssp.). *BMC Biol.* **2008**, *6*, 9–11. [CrossRef]
11. Henry, C.S.; Brooks, S.J.; Duelli, P.; Johnson, J.B. Discovering the true *Chrysoperla carnea* (Insecta: Neuroptera: Chrysopidae) using song analysis, morphology, and ecology. *Ann. Entomol. Soc. Am.* **2006**, *95*, 172–191. [CrossRef]
12. Funk, C.W.; Caminer, M.; Ron, S.R. High levels of cryptic species diversity uncovered in Amazonian frogs. *Proc. R. Soc. B Biol. Sci.* **2012**, *279*, 1806–1814. [CrossRef] [PubMed]
13. Irwin, D.E.; Alström, P.; Olsson, U.; Benowitz-Fredericks, Z.M. Cryptic species in the genus *Phylloscopus* (Old World leaf warblers). *Ibis (Lond. 1859)* **2001**, *143*, 233–247. [CrossRef]
14. Groth, J.G. Resolution of cryptic species in Appalachian red crossbills. *Condor* **1988**, *90*, 745–760. [CrossRef]

15. Isler, M.L.; Isler, P.R.; Brumfield, R.T. Clinal variation in vocalizations of an antbird (Thamnophilidae) and implications for defining species limits. *Auk* **2005**, *122*, 433–444. [CrossRef]
16. Irwin, D.E.; Thimgan, M.P.; Irwin, J.H. Call divergence is correlated with geographic and genetic distance in greenish warblers (*Phylloscopus trochiloides*): A strong role for stochasticity in signal evolution? *J. Evol. Biol.* **2008**, *21*, 435–448. [CrossRef]
17. Lachlan, R.F.; Servedio, M.R. Song learning accelerates allopatric speciation. *Evolution (N. Y.)* **2004**, *58*, 2049–2063. [CrossRef]
18. Olofsson, H.; Frame, A.M.; Servedio, M.R. Can reinforcement occur with a learned trait? *Evolution (N. Y.)* **2011**, *65*, 1992–2003. [CrossRef]
19. Verzijden, M.N.; ten Cate, C.; Servedio, M.R.; Kozak, G.M.; Boughman, J.W.; Svensson, E.I. The impact of learning on sexual selection and speciation. *Trends Ecol. Evol.* **2012**, *27*, 511–519. [CrossRef]
20. Mason, N.A.; Burns, K.J.; Tobias, J.A.; Claramunt, S.; Seddon, N.; Derryberry, E.P. Song evolution, speciation, and vocal learning in passerine birds. *Evolution (N. Y.)* **2017**, *71*, 786–796. [CrossRef]
21. Doupe, A.J.; Kuhl, P.K. Birdsong and human speech: Common themes and mechanisms. *Annu. Rev. Neurosci.* **1999**, *22*, 567–631. [CrossRef]
22. Farabaugh, S.M.; Linzenbold, A.; Dooling, R.J. Vocal plasticity in budgerigars (*Melopsittacus undulatus*): Evidence for social factors in the learning of contact calls. *J. Comp. Psychol.* **1994**, *108*, 81–92. [CrossRef] [PubMed]
23. Genzel, D.; Desai, J.; Paras, E.; Yartsev, M.M. Long-term and persistent vocal plasticity in adult bats. *Nat. Commun.* **2019**, *10*, 1–12. [CrossRef]
24. Sewall, K.B.; Young, A.M.; Wright, T.F. Social calls provide novel insights into the evolution of vocal learning. *Anim. Behav.* **2016**, *120*, 163–172. [CrossRef]
25. IUCN The IUCN Red List of Threatened Species. Available online: https://www.iucnredlist.org (accessed on 15 May 2020).
26. Berkunsky, I.; Quillfeldt, P.; Brightsmith, D.J.; Abbud, M.C.; Aguilar, J.M.R.E.; Alemán-Zelaya, U.; Aramburú, R.M.; Arce Arias, A.; Balas McNab, R.; Balsby, T.J.S.; et al. Current threats faced by Neotropical parrot populations. *Biol. Conserv.* **2017**, *214*, 278–287. [CrossRef]
27. Wright, T.F.; Toft, C.A.; Enkerlin-Hoeflich, E.; Gonzalez-Elizondo, J.; Albornoz, M.; Rodríguez-Ferraro, A.; Rojas-Suárez, F.; Sanz, V.; Trujillo, A.; Beissinger, S.R.; et al. Nest poaching in Neotropical parrots. *Conserv. Biol.* **2001**, *15*, 710–720. [CrossRef]
28. Wright, T.F.; Dahlin, C.R. Vocal dialects in parrots: Patterns and processes of cultural evolution. *Emu Austral Ornithol.* **2018**, *118*, 50–66. [CrossRef] [PubMed]
29. Slabbekoorn, H.; Smith, T.B. Bird song, ecology and speciation. *Philos. Trans. R. Soc. B Biol. Sci.* **2002**, *357*, 493–503. [CrossRef] [PubMed]
30. Rendell, L.; Mesnick, S.L.; Dalebout, M.L.; Burtenshaw, J.; Whitehead, H. Can genetic differences explain vocal dialect variation in sperm whales, *Physeter macrocephalus*? *Behav. Genet.* **2012**, *42*, 332–343. [CrossRef]
31. Keighley, M.V.; Heinsohn, R.; Langmore, N.E.; Murphy, S.A.; Peñalba, J.V. Genomic population structure aligns with vocal dialects in Palm Cockatoos (*Probosciger aterrimus*); evidence for refugial late-Quaternary distribution? *Emu* **2019**, *119*, 24–37. [CrossRef]
32. Nevo, E.; Heth, G.; Beiles, A.; Frankenberg, E. Geographic dialects in blind mole rats: Role of vocal communication in active speciation. *Proc. Natl. Acad. Sci. USA* **1987**, *84*, 3312–3315. [CrossRef]
33. Russello, M.A.; Stahala, C.; Lalonde, D.; Schmidt, K.L.; Amato, G. Cryptic diversity and conservation units in the Bahama parrot. *Conserv. Genet.* **2010**, *11*, 1809–1821. [CrossRef]
34. McElroy, K.; Beattie, K.; Symonds, M.R.E.; Joseph, L. Mitogenomic and nuclear diversity in the Mulga Parrot of the Australian arid zone: Cryptic subspecies and tests for selection. *Emu Austral Ornithol.* **2018**, *118*, 22–35. [CrossRef]
35. Murphy, S.A.; Joseph, L.; Burbidge, A.H.; Austin, J. A cryptic and critically endangered species revealed by mitochondrial DNA analyses: The western Ground Parrot. *Conserv. Genet.* **2011**, *12*, 595–600. [CrossRef]
36. Wenner, T.J.; Russello, M.A.; Wright, T.F. Cryptic species in a Neotropical parrot: Genetic variation within the *Amazona farinosa* species complex and its conservation implications. *Conserv. Genet.* **2012**, *13*, 1427–1432. [CrossRef]
37. Juniper, T.; Parr, M. *Parrots: A Guide to Parrots of the World*, 1st ed.; Yale University Press: New Haven, CT, USA, 1998; ISBN 978-0300074536.
38. Collar, N.; del Hoyo, J.; Bonan, A.; Kirwan, G.M.; Boesman, P.F.D. *Mealy Parrot (Amazona farinosa)*; Version 1; Billerman, S.M., Keeney, B.K., Rodewald, P.G., Schulenberg, T.S., Eds.; Cornell Lab of Ornithology: Ithaca, NY, USA, 2020.
39. Government of São Paulo. Decree on the Declaration of the Species of Wild Fauna in the State of São Paulo Regionally Extinct, Those Threatened with Extinction, Those Almost Threatened and Those with Insufficient Data for Evaluation. Available online: https://www.al.sp.gov.br/repositorio/legislacao/decreto/2018/decreto-63853-27.11.2018.html (accessed on 27 November 2018).
40. Ribeiro, M.C.; Martensen, A.C.; Metzger, J.P.; Tabarelli, M.; Scarano, F.; Fortin, M.-J. *The Brazilian Atlantic Forest: A Shrinking Biodiversity Hotspot. Biodiversity Hotspots*, 1st ed.; Springer: Berlin, Germany, 2011.
41. Silva, J.M.C.; Casteleti, C.H.M. Status of the biodiversity of the Atlantic Forest of Brazil. In *The Atlantic Forest of South America: Biodiversity Status, Threats, and Outlook*; Galindo-Leal, C., de Gusmão Câmara, I., Eds.; Island Press: Washington, DC, USA, 2003; pp. 43–59.
42. Batalha-Filho, H.; Fjeldså, J.; Fabre, P.H.; Miyaki, C.Y. Connections between the Atlantic and the Amazonian forest avifaunas represent distinct historical events. *J. Ornithol.* **2013**, *154*, 41–50. [CrossRef]

43. Fusinatto, L.A.; Alexandrino, J.; Haddad, C.F.B.; Brunes, T.O.; Rocha, C.F.D.; Sequeira, F. Cryptic genetic diversity is paramount in small-bodied amphibians of the genus *Euparkerella* (Anura: Craugastoridae) endemic to the Brazilian Atlantic forest. *PLoS ONE* **2013**, *8*, 1–12. [CrossRef] [PubMed]
44. Costa, W.J.E.M.; Amorim, P.F. Delimitation of cryptic species of *Notholebias*, a genus of seasonal miniature killifishes threatened with extinction from the Atlantic Forest of outh-eastern Brazil (Cyprinodontiformes: Rivulidae). *Ichthyol. Explor. Freshwaters* **2013**, *24*, 63–72.
45. Mata, H.; Fontana, C.S.; Maurício, G.N.; Bornschein, M.R.; de Vasconcelos, M.F.; Bonatto, S.L. Molecular phylogeny and biogeography of the eastern Tapaculos (Aves: Rhinocryptidae: Scytalopus, Eleoscytalopus): Cryptic diversification in Brazilian Atlantic Forest. *Mol. Phylogenet. Evol.* **2009**, *53*, 450–462. [CrossRef]
46. Forlani, M.C.; Tonini, J.F.R.; Cruz, C.A.G.; Zaher, H.; de Sá, R.O. Molecular and morphological data reveal three new cryptic species of *Chiasmocleis* (Mehely 1904) (Anura, Microhylidae) endemic to the Atlantic Forest, Brazil. *PeerJ* **2017**, *2017*, 1–43. [CrossRef]
47. Nemésio, A.; Engel, M.S. Three new cryptic species of *Euglossa* from Brazil (Hymenoptera, apidae). *Zookeys* **2012**, *222*, 47–68. [CrossRef]
48. Ribeiro, M.C.; Metzger, J.P.; Martensen, A.C.; Ponzoni, F.J.; Hirota, M.M. The Brazilian Atlantic Forest: How much is left, and how is the remaining forest distributed? Implications for conservation. *Biol. Conserv.* **2009**, *142*, 1141–1153. [CrossRef]
49. Tabarelli, M.; Pinto, L.P.; Silva, J.M.C.; Hirota, M.; Bedê, L. Challenges and opportunities for biodiversity conservation in the Brazilian Atlantic forest. *Conserv. Biol.* **2005**, *19*, 695–700. [CrossRef]
50. Cock, P.J.A.; Antao, T.; Chang, J.T.; Chapman, B.A.; Cox, C.J.; Dalke, A.; Friedberg, I.; Hamelryck, T.; Kauff, F.; Wilczynski, B.; et al. Biopython: Freely available Python tools for computational molecular biology and bioinformatics. *Bioinformatics* **2009**, *25*, 1422–1423. [CrossRef]
51. Stamatakis, A. RAxML version 8: A tool for phylogenetic analysis and post-analysis of large phylogenies. *Bioinformatics* **2014**, *30*, 1312–1313. [CrossRef] [PubMed]
52. Miller, M.A.; Pfeiffer, W.; Schwartz, T. Creating the CIPRES Science Gateway for inference of large phylogenetic trees. *2010 Gatew. Comput. Environ. Work. GCE 2010* **2010**. [CrossRef]
53. *FigTree*; Version 1.4.4; A Graphical Viewer of Phylogenetic Trees; Andrew Rambaut: Edinburgh, UK, 2007.
54. Bandelt, H.-J.; Forster, P.; Röhl, A. Median-joining networks for inferring intraspecific phylogenies. *Mol. Biol. Evol.* **1999**, *16*, 37–48. [CrossRef]
55. Leigh, J.W.; Bryant, D. POPART: Full feature software for haplotype network construction. *Methods Ecol. Evol.* **2015**, *6*, 1110–1116. [CrossRef]
56. Kumar, S.; Stecher, G.; Li, M.; Knyaz, C.; Tamura, K. MEGA X: Molecular evolutionary genetics analysis across computing platforms. *Mol. Biol. Evol.* **2018**, *35*, 1547–1549. [CrossRef]
57. Stecher, G.; Tamura, K.; Kumar, S. Molecular Evolutionary Genetics Analysis (MEGA) for macOS. *Mol. Biol. Evol.* **2020**, *37*, 1237–1239. [CrossRef]
58. Araya-Salas, M.; Smith-Vidaurre, G. warbleR: An R package to streamline analysis of animal acoustic signals. *Methods Ecol. Evol.* **2016**, 184–191. [CrossRef]
59. Schmidt, K.L.; Aardema, M.L.; Amato, G. Genetic analysis reveals strong phylogeographical divergences within the Scarlet Macaw *Ara macao*. *Ibis* **2020**, *162*, 735–748. [CrossRef]
60. Eberhard, J.R.; Bermingham, E. Phylogeny and biogeography of the *Amazona ochrocephala* (Aves: Psittacidae) complex. *Auk* **2004**, *121*, 318–332. [CrossRef]
61. Johns, G.C.; Avise, J.C. A comparative summary of genetic distances in the vertebrates from the mitochondrial cytochrome b gene. *Mol. Biol. Evol.* **1998**, *15*, 372–382. [CrossRef] [PubMed]
62. Bocalini, F.; Bolívar-Leguizamón, S.D.; Silveira, L.F.; Bravo, G.A. Comparative phylogeographic and demographic analyses reveal a congruent pattern of sister relationship between bird populations of the northern and south-central Atlantic Forest. *Mol. Phylogenet. Evol.* **2020**, *154*, 106973. [CrossRef]
63. Capurucho, J.M.G.; Ashley, M.V.; Ribas, C.C.; Bates, J.M. Connecting Amazonian, Cerrado, and Atlantic forest histories: Paraphyly, old divergences, and modern population dynamics in tyrant-manakins (Neopelma/Tyranneutes, Aves: Pipridae). *Mol. Phylogenet. Evol.* **2018**, *127*, 696–705. [CrossRef]
64. Cabanne, G.S.; d'Horta, F.M.; Sari, E.H.R.; Santos, F.R.; Miyaki, C.Y. Nuclear and mitochondrial phylogeography of the Atlantic forest endemic *Xiphorhynchus fuscus* (Aves: Dendrocolaptidae): Biogeography and systematics implications. *Mol. Phylogenet. Evol.* **2008**, *49*, 760–773. [CrossRef]
65. Costa, L.P. The historical bridge between the Amazon and the Atlantic Forest of Brazil: A study of molecular phylogeography with small mammals. *J. Biogeogr.* **2003**, *30*, 71–86. [CrossRef]
66. Trujillo-Arias, N.; Calderón, L.; Santos, F.R.; Miyaki, C.Y.; Aleixo, A.; Witt, C.C.; Tubaro, P.L.; Cabanne, G.S. Forest corridors between the central Andes and the southern Atlantic Forest enabled dispersal and peripatric diversification without niche divergence in a passerine. *Mol. Phylogenet. Evol.* **2018**, *128*, 221–232. [CrossRef]
67. Lavinia, P.D.; Barreira, A.S.; Campagna, L.; Tubaro, P.L.; Lijtmaer, D.A. Contrasting evolutionary histories in Neotropical birds: Divergence across an environmental barrier in South America. *Mol. Ecol.* **2019**, *28*, 1730–1747. [CrossRef]

68. Carnaval, A.C.; Moritz, C. Historical climate modelling predicts patterns of current biodiversity in the Brazilian Atlantic forest. *J. Biogeogr.* **2008**, *35*, 1187–1201. [CrossRef]
69. Cheng, H.; Sinha, A.; Cruz, F.W.; Wang, X.; Edwards, R.L.; D'Horta, F.M.; Ribas, C.C.; Vuille, M.; Stott, L.D.; Auler, A.S. Climate change patterns in Amazonia and biodiversity. *Nat. Commun.* **2013**, *4*. [CrossRef]
70. Smith-Vidaurre, G.; Araya-Salas, M.; Wright, T.F. Individual signatures outweigh social group identity in contact calls of a communally nesting parrot. *Behav. Ecol.* **2021**, *31*, 448–458. [CrossRef]
71. Wright, T.F.; Wilkinson, G.S. Population genetic structure and vocal dialects in an amazon parrot. *Proc. R. Soc. B Biol. Sci.* **2001**, *268*, 609–616. [CrossRef] [PubMed]
72. Ribot, R.F.H.; Buchanan, K.L.; Endler, J.A.; Joseph, L.; Bennett, A.T.D.; Berg, M.L. Learned vocal variation is associated with abrupt cryptic genetic change in a parrot species complex. *PLoS ONE* **2012**, *7*, 1–9. [CrossRef] [PubMed]
73. MacDougall-Shackleton, E.A.; MacDougall-Shackleton, S.A. Cultural and genetic evolution in mountain white-crowned sparrows: Song dialects are associated with population structure. *Evolution (N. Y.)* **2001**, *55*, 2568–2575. [CrossRef]
74. Lipshutz, S.E.; Overcast, I.A.; Hickerson, M.J.; Brumfield, R.T.; Derryberry, E.P. Behavioural response to song and genetic divergence in two subspecies of white-crowned sparrows (*Zonotrichia leucophrys*). *Mol. Ecol.* **2017**, *26*, 3011–3027. [CrossRef] [PubMed]

MDPI

Article

Genetic Diversity and Population Structure of Two Endangered Neotropical Parrots Inform *In Situ* and *Ex Situ* Conservation Strategies

Carlos I. Campos [1], Melinda A. Martinez [1], Daniel Acosta [1], Jose A. Diaz-Luque [2], Igor Berkunsky [3], Nadine L. Lamberski [4], Javier Cruz-Nieto [5], Michael A. Russello [6] and Timothy F. Wright [1,*]

1 Department of Biology, New Mexico State University, Las Cruces, NM 88003, USA; campos73@nmsu.edu (C.I.C.); melinda_martinez@fws.gov (M.A.M.); daniel19.acosta86@gmail.com (D.A.)
2 Fundación CLB (FPCILB), Estación Argentina, Calle Fermín Rivero 3460, Santa Cruz de la Sierra, Bolivia; jose.diaz@endangeredconservation.org
3 Instituto Multidisciplinario sobre Ecosistemas y Desarrollo Sustenable, CONICET-Universidad Nacional del Centro de la Provincia de Buenos Aires, Tandil 7000, Argentina; igorberkunsky@gmail.com
4 San Diego Zoo Wildlife Alliance, San Diego, CA 92027, USA; NLamberski@sdzwa.org
5 Organización Vida Silvestre A.C. (OVIS), San Pedro Garza Garciá 66260, Mexico; jcruzpictus@gmail.com
6 Department of Biology, University of British Columbia, Kelowna, BC V1V 1V7, Canada; michael.russello@ubc.ca
* Correspondence: wright@nmsu.edu

Citation: Campos, C.I.; Martinez, M.A.; Acosta, D.; Diaz-Luque, J.A.; Berkunsky, I.; Lamberski, N.L.; Cruz-Nieto, J.; Russello, M.A.; Wright, T.F. Genetic Diversity and Population Structure of Two Endangered Neotropical Parrots Inform *In Situ* and *Ex Situ* Conservation Strategies. *Diversity* **2021**, *13*, 386. https://doi.org/10.3390/d13080386

Academic Editors: Luc Legal, José L. Tella, Guillermo Blanco and Martina Carrete

Received: 18 June 2021
Accepted: 11 August 2021
Published: 17 August 2021

Publisher's Note: MDPI stays neutral with regard to jurisdictional claims in published maps and institutional affiliations.

Abstract: A key aspect in the conservation of endangered populations is understanding patterns of genetic variation and structure, which can provide managers with critical information to support evidence-based status assessments and management strategies. This is especially important for species with small wild and larger captive populations, as found in many endangered parrots. We used genotypic data to assess genetic variation and structure in wild and captive populations of two endangered parrots, the blue-throated macaw, *Ara glaucogularis*, of Bolivia, and the thick-billed parrot, *Rhynchopsitta pachyrhyncha*, of Mexico. In the blue-throated macaw, we found evidence of weak genetic differentiation between wild northern and southern subpopulations, and between wild and captive populations. In the thick-billed parrot we found no signal of differentiation between the Madera and Tutuaca breeding colonies or between wild and captive populations. Similar levels of genetic diversity were detected in the wild and captive populations of both species, with private alleles detected in captivity in both, and in the wild in the thick-billed parrot. We found genetic signatures of a bottleneck in the northern blue-throated macaw subpopulation, but no such signal was identified in any other subpopulation of either species. Our results suggest both species could potentially benefit from reintroduction of genetic variation found in captivity, and emphasize the need for genetic management of captive populations.

Keywords: genetic diversity; demographic history; population structure; captive breeding; blue-throated macaw; thick-billed parrot

1. Introduction

One important step in designing effective conservation plans for endangered species is the quantification of genetic diversity and population structure. Underlying genetic factors can have important ramifications on the susceptibility of wild populations to extinction [1,2], success or failure of captive breeding programs [3], and reintroductions of captive individuals to wild populations [4]. Many conservation plans focus on reintroducing captive individuals back into their native range [5]. Reintroductions can often benefit wild populations by increasing the amount of breeding individuals and raising the effective population size [6]. However, the result of reintroductions is not always favorable; such interventions may be detrimental and cause a reduction in the fitness of the population through increased inbreeding or introduction of deleterious alleles due to artificial selection to the captive environment [6–8]. To better understand whether a reintroduction event will lead to a desired result, the underlying genetics of both [9] wild and captive populations should be quantified.

Nearly half of all Neotropical parrots (Order Psittaciformes) are threatened with some level of extinction [10–12] due in large part to habitat loss and capture for the pet trade [13–15]. Neotropical parrots represent a special case for conservation because they are often found in large numbers in captivity due to an intensive pet trade that removes individuals from the wild and places them in zoos and private homes. As a result of this trade, captive populations may be as large as wild populations, or even larger, with the most extreme cases being ones such as the Spix's macaw, *Cyanopsitta spixii*, which is extinct in the wild [16,17]. Properly managed captive populations aim to serve as a reservoir of genetic diversity or preserve unique genetic variants no longer detected in wild population(s) through selective breeding [18]. However, captive breeding programs may face unique challenges such as a reduction in genetic diversity due to small founder size [19], or increased inbreeding from incomplete pedigree data [20]. This situation makes it valuable to reconstruct patterns of genetic variation in both captive and wild populations.

One strategy that has often been debated for parrot conservation is the reintroduction of captive individuals into the wild [21]. One rationale for doing so is to reintroduce valuable gene variants back into declining populations [22]. However, this does not take into account some potential risks in doing so, including disease introduction [23] and behavioral issues affecting successful integration of released individuals [24]. One example of a successful reintroduction is that of the orange-bellied parrot, *Neophema chrysogaster*, in 2010 [25]. In this case reintroductions were deemed to be the best option as captive populations displayed novel genetic diversity not found in wild populations [26] and extinction in the wild was thought to be imminent. Alternatively, some reintroductions fail as was the case for releases of the thick-billed parrot, *Rhynchopsitta pachyrhyncha*, from 1986–1992 due to factors including poor condition of released birds that lead to to increased rates of predation and disease [27]. A metanalysis of parrot reintroductions showed that high predation rates, longer periods of supplementary food supply, and selection of high-quality release sites are some of the most important factors affecting reintroduction success [28].

Two parrot species that have disproportionally large captive populations relative to their remaining wild populations are the critically endangered blue-throated macaw, *Ara glaucogularis*, and the endangered thick-billed parrot. The blue-throated macaw is endemic to the tropical savannahs of the Beni region of Bolivia and is one of the most endangered species of macaws still found in the wild [10]. Population estimates over the last decade had placed an upper limit of 250 individuals in the wild divided into two subpopulations: a northern subpopulation to the east of the Mamoré river near Trinidad in the Mamoré province and a southern subpopulation near Loreto in the Marbán province [29]. With the addition of a recently discovered third subpopulation to the west of the Mamoré River in Yacuma and José Ballivián provinces, estimates for the global wild population have risen to 312–455 individuals [30]. The captive population is estimated to consist of approximately 1000 individuals throughout the private sector and zoos in North America, as recorded through the Association of Zoos and Aquariums (AZA)-recognized captive "studbook" (i.e., the official record of the pedigree and demographic history of all animals managed among AZA member institutions and their partners) [31]. Recent studies on the persistence of the wild blue-throated macaw population have suggested that the population is stable, with survival and breeding success of adults being the most influential aspect affecting population growth [32,33]. Current conservation strategies are aimed at increasing the long-term persistence of the wild population and include protection from poaching and predation, installation and monitoring of nest boxes, and the creation of the private Barba Azul Nature Reserve [29]. Future reintroductions are a goal of private breeding programs in Bolivia, the United States, Canada, and the United Kingdom.

The thick-billed parrot is found in the high-elevation forests of the Sierra Madre Occidental mountain range throughout the northern Mexican states of Sonora, Durango, and Chihuahua [34]. Population numbers for this species vary due to infrequent surveys and difficulties in accessing remote sites [35], but current estimates suggest between 2000–2800 mature individuals remain [34]. There is also a well-established captive population consisting of some 100 birds held at zoos that are managed as part of an AZA-recognized studbook, as well as an undetermined number of individuals in private hands [36]. Threats faced by the thick-billed parrot include habitat loss due to logging operations throughout its range as well as historical pet trade capture and shooting [37–39]. A recent study discovered evidence of increased rates of predation by bobcats and recommended increased antipredation methods at nesting sites [40]. Studies have suggested that reduction of logging and favorable ecological conditions in the thick-billed parrot habitat range are key to its recovery [41].

Both the blue-throated macaw and thick-billed parrot face similar challenges of low or declining populations in the wild and sizeable populations in captivity. Direct comparisons between the two species can be valuable to elucidate general patterns that other endangered species may be facing. Small population sizes may influence the long-term persistence of each species through increased inbreeding and susceptibility to demographic and environmental stochasticity [42]. Nonetheless, little is known about the underlying level of genetic diversity and population structure in the wild and captivity as well as the genetic consequences of recent population declines in both species. Quantifying these population genetic parameters can help to inform the best plan of action for each species moving forward.

Our project aims to improve understanding of blue-throated macaw and thick-billed parrot population genetics for aiding the conservation of these species. To achieve these goals, we genotyped individuals from wild and captive populations using panels of polymorphic microsatellite loci. We used these multi-locus genotypes to first test for the presence of genetic structure in wild and captive populations of each species. We then quantified the genetic diversity of all populations (wild and captive) and tested for signatures of a population bottleneck in the wild population of each species. We subsequently used these results to assess the suitability of reintroductions from the captive to wild populations from a standpoint of genetic diversity and health.

2. Materials and Methods

2.1. Sample Collection

All procedures were approved by New Mexico State University (NMSU) Institutional Animal Care and Use Committee (protocols: thick-billed parrot 2007-07 and 2008-028; blue-throated macaw 2015-033 and 2018-025). Samples were exported under CITES permits (blue-throated macaw: Bolivian export 001128 and USA import US15671C/9; thick-billed parrot: Mexico export MX43843 and USA import 06US118407/9).

Blood samples were collected from 60 wild blue-throated macaws from the northern and southern subpopulations in Bolivia from 2007–2017 (Figure 1, Supplementary Materials Table S1) and 46 wild thick-billed parrots from three nesting colonies in Mexico from 2007–2009 (Figure 1, Supplementary Materials Table S1). Individuals sampled in the northern part of the blue-throated macaw range came from the northeastern subpopulation and not the newly discovered northwestern population; here we refer to these samples simply as the northern subpopulation. Blood samples were taken from 60 captive blue-throated macaws held in zoos in the United States, Canada, and Bolivia and 73 captive thick-billed parrots held in zoos in the United States (Supplementary Materials Table S2). All Bolivian blood samples were mixed with lysis buffer and stored at room temperature short-term and at −80 °C long-term. All other blood samples were stored on FTA paper (Whatman^{TM}) at room temperature.

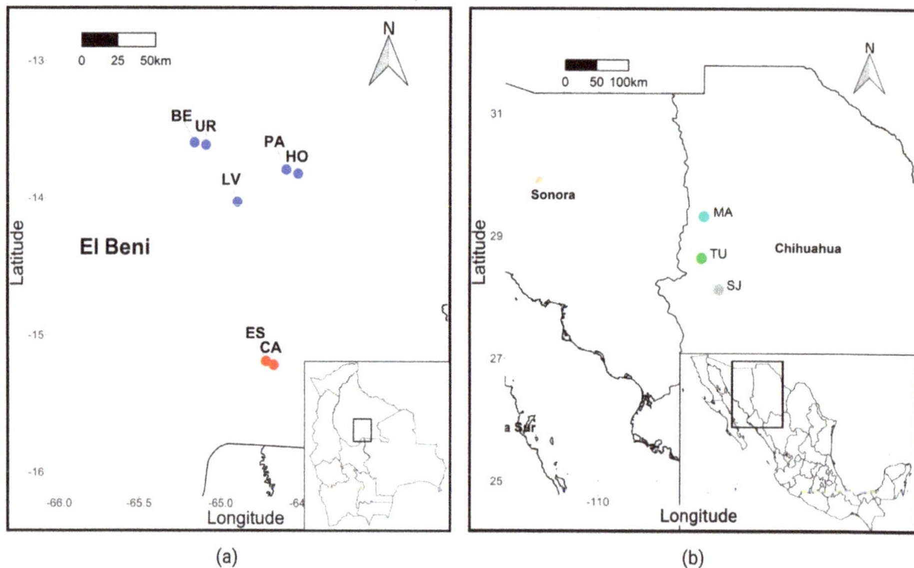

Figure 1. (**a**) Wild blue-throated macaw Bolivian sample sites. Northern subpopulation = blue; southern subpopulation = red. Sampling sites (individuals sampled): ES = Esperancita (25), BE = Bethel (9), HO = Holanda (8), LV = La Verde (6), PA = Palma Sola (5), CA = Cantina (5); UR = Urkupina (2). (**b**) Wild thick-billed parrot sampling sites. MA = Madera (37); TU = Tutuaca (8); SJ = San Juanito (1).

2.2. DNA Extraction

We used two methods of DNA extraction depending on the sample medium. DNA was extracted from blood stored in lysis buffer using the Qiagen DNeasy Blood and Tissue Kit (Qiagen, Valencia, Santa Clarita, CA, USA) following the standard protocol for extraction from nucleated erythrocytes (https://www.qiagen.com (accessed on 20 September 2017)). DNA extraction from blood stored on FTA paper was performed following a standard Whatman™ elution protocol [43].

2.3. Microsatellite Genotyping

Microsatellite primers developed for the scarlet macaw, *Ara macao* [44], were tested for their cross-amplification and variability in a subset of captive blue-throated macaws. A panel of 12 of these primers were chosen for use in this study based on consistent amplification and presence of multiple alleles. Separately, we tested microsatellites for the thick-billed parrot from an array of primers developed for several parrot species including the blue-and-yellow macaw, *Ara ararauna* [45], monk parakeet, *Myiopsitta monachus* [46], burrowing parrot, *Cyanoliseus patagonus* [47], and previously unpublished primers for the thick-billed parrot [48,49] (Supplementary Materials Table S3). We selected 11 of these primers for use in this study.

Each forward primer was labelled with a fluorescent dye and PCR was carried out in a 15 µL reaction with the following concentrations: 2.5 mM MgCl, 10× Gold PCR Buffer, 0.3 U AmpliTaq Gold™ (ThermoFisher Scientific), 8 mM dNTP's, 1–3 µL DNA template (1:10 dilution of DNA extract), and water to dilute. The reaction was carried out with an initial denaturation of 95 °C for 25 s followed by 30 touchdown cycles with annealing temperature starting at 60 °C and decreasing 0.5 °C per cycle, the reaction ended with five final annealing cycles at 45 °C and an extension at 72 °C for 10 min. PCR products were purified using the QIAquick PCR purification kit (Qiagen, Valencia, Santa Clarita, CA, USA), following the multi-well PCR purification protocol. Purified PCR products from the blue-throated macaw were sent to the UTEP Genomic Analysis Core Facility for fragment analysis on an ABI 3500 Genetic Analyzer and allele scoring was performed in Geneious R8.1.7 (http://www.geneious.com accessed on 24 January 2018). Purified PCR fragments from the thick-billed parrot were analyzed on an ABI 3100 Genetic Analyzer at NMSU and allele scoring was performed in GeneMapper v3.5 (ABI). All PCR fragments were analyzed twice to confirm results obtained from scoring.

For all loci, MICRO-CHECKER v2.2.3 [50] was used to check for the presence of null alleles, large allele dropout, and scoring errors. Some wild samples were obtained from different individuals at the same nest. To eliminate bias from the presence of first-order relatives in the analysis we used data from field observations to identify potential parent–offspring and full sibling relationships in each species wild sample pool. The associated studbook for each species' captive population was used to identify first-order relatives in the captive population via the construction of a captive pedigree using the R package kinship2 [51]. We then used ML-RELATE [52] as a secondary measure to confirm that none of the remaining individuals in any population were first-order relatives. We randomly removed one of each pair of first-order relatives to create a reduced dataset for each species. Global deviations from Hardy–Weinberg equilibrium and linkage disequilibrium between all locus pairs (Markov chain parameters: 1000 randomizations, 100 batches, and 1000 iterations per batch) were tested in Genepop 4.7 [53].

2.4. Population Structure and Genetic Diversity

We used the Bayesian clustering approach implemented in STRUCTURE v2.3 [54] to infer putative population structure in our wild datasets. Using the admixture model and the LOCPRIOR option, we tested a range of population numbers (K) from 1 to 10 using 10 independent Markov Chain Monte Carlo (MCMC) runs of 500,000 repetitions and a burn-in period of 500,000. Results were evaluated using STRUCTURE HARVESTER [55]. To infer the optimal K value, we employed the ΔK method [56] and the plotting of the log probability of the data [57] to assess where $\ln \Pr(X \mid K)$ plateaued (see STRUCTURE manual). CLUMPP [58] and DISTRUCT [59] were used to create plots of the STRUCTURE results. Self-assignment rates of individuals to the population from which they were sampled were determined as implemented in GenAlEx v6.5 [60]. We also used this analysis to assign captive individuals to the referenced wild subpopulations. We investigated evidence for first-generation migrants between wild subpopulations using the approach of [61] as implemented in GENECLASS2 [62]. Principal Coordinates Analysis (PCoA) plots were generated for each species from genetic distances as implemented in GenAlEx v6.5. Pairwise Fst values among populations were calculated in GenAlEx v6.5 and p-values were calculated in Hierfstat [63]. We used GenAlEx v6.5 to calculate the number of alleles, number of private alleles, observed heterozygosity, and expected heterozygosity for each locus.

2.5. Demographic History

The presence of a recent population bottleneck in the wild subpopulations was determined using the heterozygote excess test [64] with significance determined using the Sign test and the Wilcoxon signed-rank test (1000 iterations) implemented in BOTTLENECK v1.2.02 [65]. We used the assumptions of a two-phase and stepwise microsatellite mutation model. For the two-phase model, estimates in birds suggest 60 to 80% of mutations occur via a stepwise change [66,67]. Therefore, we used the assumption of either 60 or 80% stepwise mutation for the analysis using this model.

3. Results

3.1. Microsatellite Validation

We genotyped a total of 120 blue-throated macaws at 12 microsatellite loci for this study. Four individuals did not amplify at any locus and were excluded from further analysis. For the remaining individuals, amplification success rates per locus varied from 70 to 97% (Supplementary Materials Table S4). Using field observations, the captive pedigree (Supplementary Materials Figure S1), and ML-Relate, 31 wild and 37 captive individuals were judged to be first-order relatives of other birds in the study (Supplementary Materials Table S5a–c) and were removed to create the reduced dataset used in all remaining analyses. MicroChecker detected no evidence of null alleles in this reduced dataset. Loci SCMA12 and SCMA34 were removed due to levels of missing data exceeding 24%. There was no evidence of deviation from Hardy–Weinberg or linkage equilibria for the remaining loci. After the removal of first-order relatives, the dataset included 48 blue-throated macaws (19 wild and 29 captive) genotyped at 10 microsatellite loci with 3.95% missing data.

We genotyped a total of 119 thick-billed parrots at 11 microsatellite loci. Four samples were deemed to be duplicates of another individual already in the study and two other samples were removed due to unrecorded sample origin. Only one individual was sampled from the San Juanito subpopulation; because we could not perform any population-level comparative analyses with a single individual we excluded this sample from further analysis. For the remaining individuals, amplification success rate per locus varied from 93 to 99% (Supplementary Materials Table S4). Using field observations, the captive pedigree (Supplementary Materials Figure S1), and ML-Relate, 31 wild and 39 captive thick-billed parrots were deemed to be first-order relatives with other birds in the study (Supplementary Materials Table S6a–c) and were removed to create the reduced dataset which was used for all remaining analyses. In this reduced dataset there was evidence of null alleles at locus MmGT090, which was subsequently removed from further analyses. There was no deviation from Hardy–Weinberg or linkage equilibria for the remaining loci. As a result, our dataset contained 42 individuals (15 wild and 27 captive) at 10 microsatellite loci with 2.85% missing data.

3.2. Population Structure

We found evidence for weak, but significant, population differentiation between the northern and southern subpopulations of the wild blue-throated macaw as evidenced by a pairwise *Fst* value of 0.048 (p-value < 0.016; Table 1) and significant evidence of genotypic differentiation (chi2 = infinity, p-value = highly sign.). Likewise, 94% of individuals self-assigned to the subpopulation in which they were sampled (Supplementary Materials Table S7), and no first-generation migrants (i.e., individuals not born in their current subpopulation) were detected. However, Bayesian clustering analysis did not find evidence for $K > 1$ based on ΔK or the plot of the log probability of the data (Supplementary Materials Table S8).

Table 1. Pairwise *Fst* values for the blue-throated macaw and thick-billed parrot. p-values denoted by: * = p-value < 0.05; ** = p-value < 0.016. NA denotes comparisons that cannot be made between a population and one of the subpopulations that compose it.

Blue-Throated Macaw				Thick-Billed Parrot			
Population	Captive	North Wild	South Wild	Population	Captive	Madera	Tutuaca
Captive	-			Captive	-		
North Wild	0.026	-		Madera	0.017	-	
South Wild	0.047 **	0.048 **	-	Tutuaca	0.045	0.053	-
Whole Wild	0.025 *	NA	NA	Whole Wild	0.014	NA	NA

All individuals in the blue-throated macaw captive population were assigned to either the northern ($n = 22$) or southern ($n = 7$) wild subpopulation. The pairwise *Fst* value was significant between the captive population and the southern wild subpopulation but not between the captive population and the northern wild subpopulation. Approximately 16.3% of variation in the blue-throated macaw data was explained by the first two coordinates of the PCoA; the resulting ordination plot suggested that the captive population contained genetic diversity not found in either wild subpopulation (Figure 2). Furthermore, we detected 16 private alleles in the captive population that were not present in either wild population.

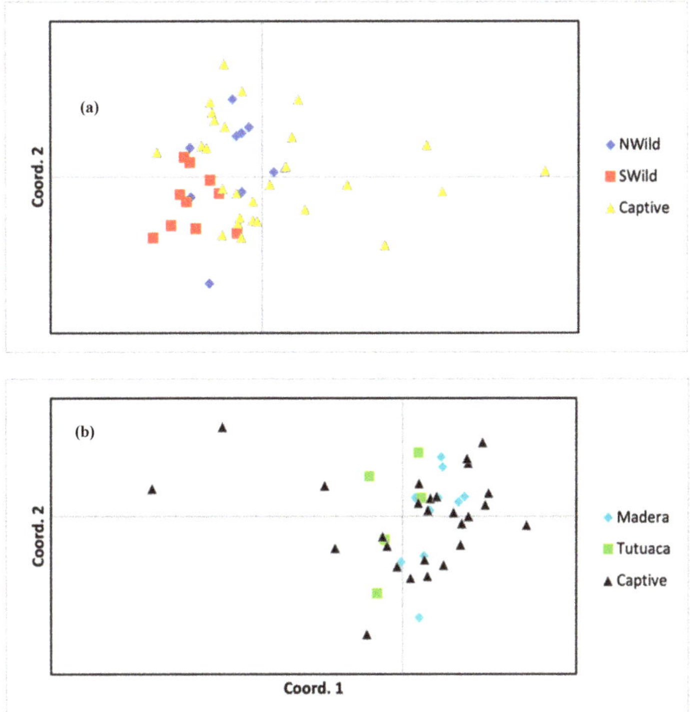

Figure 2. Principal coordinates analysis (PCoA) between genetic distance of captive, northern wild, and southern wild populations in the (**a**) blue-throated macaw and whole captive and wild populations in the (**b**) thick-billed parrot. Plots based on the reduced datasets with first order relatives removed (see Methods).

We found no evidence of population differentiation between the Madera and Tutuaca subpopulations of the thick-billed parrot (Table 1). Similarly, Bayesian clustering analysis did not find evidence for $K > 1$ based on ΔK or the plot of the log probability of the data (Supplementary Materials Table S8). Self-assignment to the subpopulation in which an individual was sampled was low (74%; Table S7), however, no first-generation migrants were detected.

All individuals in the thick-billed parrot captive population were assigned to either the Madera ($n = 21$) or Tutuaca ($n = 6$) subpopulation. Pairwise *Fst* values between the captive and wild populations were not significant (Table 1). Approximately 21.6% of variation in the thick-billed parrot data was explained by the first two coordinates of the PCoA; the resulting ordination plot of these two dimensions revealed little evidence of differentiation between the wild and captive populations (Figure 2). We did detect eight private alleles in the captive population and four in the entire wild population.

3.3. Genetic Diversity of Captive and Wild Populations

Mean observed heterozygosity for wild blue-throated macaws was 0.578 and 0.660 for the northern and southern subpopulations, respectively, while allelic richness ranged from 4.1–4.4 (5.5 combined). Both wild subpopulations exhibited negative inbreeding coefficients, although the northern population was not significantly different from 0 (Table 2). Mean observed heterozygosity for the captive population was similar to the wild (0.618), however, allelic richness was higher (7.0) and the inbreeding coefficient was slightly positive (0.025), but not significantly different from 0.

Table 2. Estimates of population genetic parameters for the blue-throated macaw and thick-billed parrot including number of individuals (*N*), number of alleles (*Na*), observed heterozygosity (*Ho*), expected heterozygosity (*He*), and inbreeding coefficient (*F*).

Blue-Throated Macaw Locus	Captive					North Wild					South Wild					Whole Wild				
	N	*Na*	*Ho*	*He*	*F*	*N*	*Na*	*Ho*	*He*	*F*	*N*	*Na*	*Ho*	*He*	*F*	*N*	*Na*	*Ho*	*He*	*F*
SCMA09 [25]	29	2	0.034	0.034	−0.018	9	1	0	0	NA	10	2	0.200	0.180	−0.111	19	2	0.105	0.100	−0.056
SCMA46 [25]	27	9	0.852	0.846	−0.007	9	7	0.889	0.821	−0.083	10	5	0.700	0.725	−0.034	19	7	0.789	0.812	0.027
SCMA19 [25]	23	9	0.609	0.635	0.026	9	5	0.667	0.728	0.085	10	2	0.500	0.455	−0.099	19	5	0.579	0.614	0.056
SCMA31 [25]	25	8	0.800	0.814	0.017	9	7	0.778	0.815	0.045	10	6	0.800	0.740	−0.081	19	8	0.789	0.812	0.027
SCMA41 [25]	28	10	0.750	0.839	0.106	9	7	0.778	0.765	−0.016	10	8	1.000	0.820	−0.220	19	10	0.895	0.820	−0.091
SCMA27 [25]	29	6	0.759	0.707	−0.072	9	4	0.444	0.648	0.314	10	2	0.700	0.455	−0.538	19	4	0.579	0.611	0.052
SCMA11 [25]	28	2	0.179	0.219	0.184	9	2	0.111	0.105	−0.059	10	2	0.400	0.320	−0.250	19	2	0.263	0.229	−0.152
SCMA22 [25]	28	5	0.679	0.668	−0.015	9	2	0.444	0.444	0.000	10	2	0.500	0.455	−0.099	19	2	0.474	0.450	−0.052
SCMA26 [25]	29	10	0.724	0.825	0.122	9	5	0.889	0.735	−0.210	10	8	0.900	0.805	−0.118	19	9	0.895	0.817	−0.095
SCMA02 [25]	25	7	0.800	0.731	−0.094	9	5	0.778	0.716	0.086	10	4	0.900	0.655	−0.374	19	6	0.842	0.720	−0.169
Mean (SE)	27	7	0.618 (0.089)	0.631 (0.088)	0.025 (0.028)	9	4.5	0.578 (0.100)	0.578 (0.094)	−0.001 (0.046)	10	4.1	0.660 (0.081)	0.561 (0.069)	−0.186 (0.053)	19	5.5	0.621 (0.087)	0.598 (0.082)	−0.045 (0.026)

Thick-billed Parrot Locus	Captive					Madera					Tutuaca					Whole Wild				
	N	*Na*	*Ho*	*He*	*F*	*N*	*Na*	*Ho*	*He*	*F*	*N*	*Na*	*Ho*	*He*	*F*	*N*	*Na*	*Ho*	*He*	*F*
UnaCT74 [26]	25	3	0.120	0.114	−0.049	10	2	0.100	0.095	−0.053	5	3	0.200	0.460	0.565	15	3	0.133	0.240	0.444
UnaCT21 [26]	26	3	0.500	0.536	0.066	10	3	0.700	0.505	−0.386	5	2	0.400	0.320	−0.250	15	3	0.600	0.464	−0.292
UnaCT55 [26]	27	8	0.815	0.767	−0.063	10	7	0.700	0.700	0.000	5	5	0.800	0.680	−0.176	15	7	0.733	0.702	−0.044
UnaCT43 [26]	25	2	0.120	0.114	−0.064	10	2	0.100	0.095	−0.053	5	1	0.000	0.000	NA	15	2	0.067	0.064	−0.034
TBP2−39 [29]	27	10	0.778	0.684	−0.136	10	7	0.900	0.785	−0.146	5	6	1.000	0.800	−0.250	15	10	0.933	0.856	−0.091
TBP2−61 [29]	25	4	0.640	0.578	−0.107	10	4	0.900	0.625	−0.440	5	3	0.600	0.460	−0.304	15	5	0.800	0.616	−0.300
Rhpac149 [30]	26	5	0.654	0.604	−0.082	10	5	0.600	0.700	0.143	5	4	0.800	0.700	−0.143	15	5	0.667	0.722	0.077
Rhpac074 [30]	25	8	0.800	0.766	−0.044	10	5	0.900	0.690	−0.304	5	5	0.800	0.680	−0.176	15	7	0.867	0.744	−0.164
MmGT057 [27]	26	3	0.500	0.452	−0.106	10	2	0.500	0.495	−0.010	5	2	0.400	0.480	0.167	15	2	0.467	0.491	0.050
CyanP05 [28]	26	7	0.808	0.794	−0.017	10	5	0.700	0.765	0.085	5	5	0.600	0.680	0.118	15	5	0.667	0.771	0.135
Mean (SE)	26	5.3	0.573 (0.084)	0.541 (0.079)	−0.060 (0.018)	10	4.2	0.610 (0.095)	0.546 (0.081)	−0.116 (0.063)	5	3.6	0.560 (0.098)	0.526 (0.075)	−0.050 (0.089)	15	4.9	0.593 (0.092)	0.567 (0.080)	0.022 (0.069)

Mean observed heterozygosity for wild thick-billed parrots was 0.610 and 0.560 for the Madera and Tutuaca subpopulations, respectively, while allelic richness ranged from 3.6–4.2 (4.9 combined). Both wild subpopulations exhibited negative inbreeding coefficients, although that of the Tutuaca subpopulation was not significantly different from 0 (Table 2). Mean observed heterozygosity (0.573), allelic richness (5.3), and inbreeding coefficient (−0.060) for the captive population were both similar to values recovered for the wild subpopulations (Table 2).

3.4. Demographic History

To detect evidence of a recent population bottleneck, we implemented the sign test and Wilcoxon signed-rank test in the program BOTTLENECK. Both wild subpopulations of the blue-throated macaw were tested independently. Although BOTTLENECK recommends population sizes of 10 or greater individuals, we chose to test our northern wild (n = 9) subpopulation as well. The thick-billed parrot subpopulations were combined into a single population as no significant evidence of population structure between them was detected. We found evidence of a recent population bottleneck in the northern subpopulation of the blue-throated macaw under the 60% stepwise assumption of two-phase model (Wilcoxon 1-tail: p = 0.009; Wilcoxon 2-tail: p = 0.018) and the 80% stepwise assumption of the two-phase model (Wilcoxon 1-tail: p = 0.016, Wilcoxon 2-tail: 0.032). None of the other tests revealed evidence of a bottleneck in either species (Table 3).

Table 3. Heterozygote excess test results indicating p-values for the blue-throated macaw and thick-billed parrot * = p-value < 0.05.

	Blue-Throated Macaw: (Left) Southern and (Right) Northern Population					
	Stepwise Mutation Model	Two-Phase Model		Stepwise Mutation Model	Two-Phase Model	
		60% Stepwise	80% Stepwise		60% Stepwise	80% Stepwise
Sign test	0.158	0.153	0.176	0.535	0.074	0.092
Wilcoxon signed rank test (1-tail)	0.285	0.082	0.125	0.161	0.009*	0.016 *
Wilcoxon signed rank test (2-tail)	0.570	0.164	0.250	0.322	0.018*	0.032 *
	Thick-billed parrot: whole wild population					
	Stepwise Mutation Model	Two-Phase Model				
		60% Stepwise	80% Stepwise			
Sign test	0.193	0.449	0.428			
Wilcoxon signed rank test (1-tail)	0.838	0.422	0.577			
Wilcoxon signed rank test (2-tail)	0.375	0.845	0.921			

4. Discussion

Here, we investigated the underlying genetic diversity and structure of the critically endangered blue-throated macaw and endangered thick-billed parrot. Similar levels of genetic diversity were detected between the wild and captive populations of both species, with only allelic richness in the wild blue-throated macaw population being measurably less (5.5) than found in captivity (7.0). Private alleles were detected in captivity for both species. We found evidence of weak population differentiation both between the two wild subpopulations of the blue-throated macaw, and among the blue-throated macaw wild and captive populations. We found no evidence of population differentiation between any populations of the thick-billed parrot. We recovered weak evidence of a recent population bottleneck in the northern wild subpopulation of the blue-throated macaw with four of eight total tests returning significant evidence of a bottleneck and no evidence of a recent population bottleneck in the thick-billed parrot. Below, we discuss these results in more depth and their implications for ongoing conservation efforts.

4.1. Population Structure

We found that the wild subpopulations of blue-throated macaws exhibited a low amount of genetic distinctiveness as evidenced by significant *Fst* values, high self-assignment to the location from which individuals were sampled, and from private allele tests. However, this level of differentiation was not high enough to be detected by Bayesian clustering analysis. Thick-billed parrot subpopulations showed no significant evidence of differentiation between any population pair. One of the problems that many parrot species face is habitat fragmentation [68]. As reviewed in [69], habitat fragmentation caused by humans has led to population differentiation in several species of parrots while undisturbed habitat tends to promote gene flow between populations. The presence of population structure in the blue-throated macaw and not the thick-billed parrot may be indicative of a stronger effect of human-caused habitat fragmentation in the former species, as well as its smaller estimated population size. Alternatively, population structure may arise from social factors promoting high philopatry, as suggested for the cooperatively breeding El Oro parakeet, *Pyrrhura orcesi*, of Ecuador [70] and the colonially breeding red-fronted macaw, *Ara rubrogenys*, of Bolivia [71]. The lack of population differentiation in the thick-billed parrot may also be explained by the migratory nature of the species. The thick-billed parrot undergoes seasonal migration from breeding sites in northern Chihuahua to overwintering sites in central Mexico. This migration event has been thought to be a contributor to the lack of geographic variation in vocalizations in thick-billed parrots [37] and could also promote gene flow among populations. An additional consideration in our findings of genetic structure is the influence of the extensive time period over which samples were collected. This is especially important in the blue-throated macaw, where sampling efforts occurred over the span of ten years. It is possible that the genetic structure of initially sampled populations in 2007 could be different from that of the most recent efforts in 2017 and this may have influenced our findings.

One limitation in our ability to detect population differentiation may be low sample size. We sampled over 100 individuals from each species, however, each dataset was reduced to less than 50 individuals due to relatedness. Further sampling efforts for the blue-throated macaw could lead to the addition of individuals from the recently discovered northwestern population in the Yacuma and José Ballivián provinces [30], which may uncover novel population structure not currently observed in our sample. Likewise, thick-billed parrot sampling could be expanded to include new individuals from captivity and other breeding colonies in the wild as our most recent sampling of thick-billed parrots was in 2008.

The detection of private alleles and the lack of PCoA clustering suggests that the captive populations contain genetic variation not sampled in the wild populations of each species. Similarly, private alleles in both captive and wild populations have been observed in the critically endangered orange-bellied parrot [26]. Novel genetic diversity in each population can prove critical when extinction appears imminent in the wild. If absolutely necessary, taking in new individuals as fledglings from declining wild populations can bolster genetic diversity in captivity which, in turn, may help the success of captive releases [25]. Conversely, reintroductions of captive-bred individuals into the wild could potentially reintroduce genetic variants back to locations from which they originally derived, as seen in the Mauritius parakeet, *Psittacula eques*, where genetic panmixia was restored with relatively small-scale reintroductions [72,73].

4.2. Genetic Diversity

Low levels of genetic diversity have long been associated with factors such as increased inbreeding [74], and fixation of deleterious mutations [75] in small populations. We found overall levels of genetic diversity within the populations of each species [blue-throated macaw (mean Ho range: 0.578–0.660) and thick-billed parrot (mean Ho range 0.560–0.610) that were comparable to other members of the order Psittaciformes with heightened conservation status, including the critically endangered swift parrot, *Lathamus discolor* (population size = ~2400; mean Ho = 0.679; [76,77] and subspecies of the near-threatened Cuban amazon, *Amazona leucocephala* (population size = ~23,000; mean Ho range 0.64–0.77; [9,78], but higher than the critically endangered kakapo, *Strigops habroptilus* (population size = ~116, mean Ho = 0.489; [79,80]. While heterozygosity values may not be directly comparable across studies, all of the cited works used microsatellite loci which provide a general sense of relative levels of genetic variation across a range of threatened and endangered parrot species.

The similarity in genetic diversity of the blue-throated macaw and thick-billed parrot raises interesting comparisons of these species' natural and human-impacted history. Historically, human-mediated range decline and poaching have been the two major threats faced by both species. The blue-throated macaw relies on palms as a food source and for breeding [81], and suffered a drastic population decline during the 1970s and 1980s due to poaching for the pet trade. The thick-billed parrot also has a specific association with high-elevation pine forest habitat [39], but the extent of historical poaching is thought to have been lower. In contrast, the critically endangered kakapo historically faced different threats, including high rates of predation from cats [82] and introduced rats, *Rattus rattus* [83].

Our results show that neither the blue-throated macaw nor the thick-billed parrot are experiencing significant levels of inbreeding with a population average inbreeding coefficient range of −0.186 to 0.025 and −0.116 to −0.022, respectively, and with the highest values recorded in the southern population of the blue-throated macaw. It is important to note that we removed 68 blue-throated macaws and 70 thick-billed parrots from our study as they were either known to be, or statistically defined as, first-order relatives of other individuals in the sampling pool. Therefore, true levels of inbreeding in the entire population may be higher than our estimates and there may be substantial family structure within each population. High inbreeding levels have been shown to increase deleterious effects such as reduced clutch size, as seen in the kakapo [84], and hatching failure, as observed in the Puerto Rican parrot, *Amazona vittata* [85].

While microsatellites can provide estimates of genetic diversity and population structure, they are selectively neutral and do not provide information about functional genetic variation. On the other hand, while the number of individuals sampled in our study was small, it was likely representative given that it encompassed a sizeable proportion of the known populations for each parrot species in question.

4.3. Demographic History

We found evidence of a significant population bottleneck in the Northern wild subpopulation of the blue-throated macaw. However, we should note that this result was detected in a sample of nine individuals, which is less than that recommended ($n \geq 10$) for the heterozygote excess test implemented in BOTTLENECK. Although preliminary, our finding is still notable, as bottlenecks could potentially have a negative impact on population fitness through reduced genetic diversity and increased inbreeding [86,87]. Severe inbreeding in the wild could also be a concern if the release of unrelated captive individuals leads to outbreeding depression [88] which has been found to increase disease susceptibility in song sparrows, *Melospiza melodia* [89]. Importantly, no evidence of increased hatching failure has been detected in the blue-throated macaw [10], suggesting that while the population size is small, this species may not yet be suffering from detectable deleterious effects of inbreeding.

4.4. Conservation Implications

Our estimates of genetic diversity and population structure in wild and captive populations of the blue-throated macaw and thick-billed parrot should help inform conservation efforts in both species (but see [90]). While protection of key habitats and resources have been the primary focus of conservation efforts to date for both species, reintroductions and translocations have also been considered as approaches for bolstering wild populations [86]. Our genetic findings suggest that introduction of variants currently found in the captive populations of each species could be beneficial should such reintroductions be deemed necessary by population managers. For the blue-throated macaw, managers should be mindful of the limited structure we detected in the wild population and consider the origin of captive individuals if planning any reintroductions so as to find the best fit population for release of any given individual. In contrast, we found no differentiation in the wild population of the thick-billed parrot, suggesting that any reintroductions can be designed to maximize the enhancement of genetic diversity across the entire wild population.

Although we evaluate the suitability of captive populations for each species for use in reintroductions, these significant interventions should only be considered after comprehensive feasibility and risk assessments have been conducted and balanced by consideration of costs and benefits relative to other conservation actions [91]. In the meantime, the *ex situ* management programs of the blue-throated macaw and thick-billed parrot should strive to maintain genetic diversity and minimize kinship in the captive populations so as to maximize their current and future value for the conservation of these two iconic parrot species.

Supplementary Materials: The following are available online at https://www.mdpi.com/article/10.3390/d13080386/s1, Table S1: Wild sampling information for the blue-throated macaw and thick-billed parrot, Table S2: Captive sampling information for the blue-throated macaw and thick-billed parrot, Table S3: Previously unpublished primer sequences developed for the thick-billed parrot, Table S4: Primer success rate in PCR for blue-throated macaw and thick-billed parrot, Table S5a–c: Relatedness values of individuals from the blue-throated macaw from MLRelate, Table S6a–c: Relatedness values of individuals from the thick-billed parrot from MLRelate Table S7: Proportions of self-assignment of wild and captive individuals to the wild populations, Table S8: Evanno Table output from STRUCTURE Harvester based on 10 iterations of STRUCTURE output and number of clusters (K) ranging from 1 to 10. Figure S1: Captive pedigrees for the blue-throated macaw and thick-billed parrot.

Author Contributions: Project conception and presentation: T.F.W., M.A.R.; Bolivian sample collection and metadata management: I.B., J.A.D.-L.; Mexican sample collection and metadata management: N.L.L., J.C.-N.; blue-throated macaw and thick-billed parrot data collection and analysis: C.I.C., M.A.M., D.A.; manuscript preparation: C.I.C.; initial revisions: T.F.W., M.A.R., C.I.C.; final manuscript revisions: all. All authors have read and agreed to the published version of the manuscript.

Funding: Funding for this project has been provided by Parrot Action Grants from the World Parrot Trust to T.F.W. and M.A.R.

Institutional Review Board Statement: The use of animals in this study was approved by the Institutional Animal Care and Use Committee of New Mexico State University (protocols 2007-07, 2008-028, 2015-033 and 2018-025). Samples were exported under authority from CITES (blue-throated macaw: Bolivian export 001128 and USA import US15671C/9; thick-billed parrot: Mexico export MX43843 and USA import 06US118407/9).

Data Availability Statement: The data presented in this study are openly available in Dryad at https://doi.org/10.5061/dryad.jq2bvq897.

Acknowledgments: D.A. and C.I.C. were supported by the NMSU-HHMI Research Scholars program (Howard Hughes Medical Institute) and M.A.M. was supported by NMSU's NRCT Career Tracks program (USDA). Samples from Bolivia were collected under the approval of the General Direction of Biodiversity and Protected Areas for a scientific project authorized to the Centro Biodiversidad Genética from University Mayor San Simón. We would like to thank the staff of the Border Biomedical Research Center (BBRC) Genomics Analysis Core Facility for services and facilities provided, Valerie Brewer for assistance with PCR, Grace Smith-Vidaurre for help constructing sample maps, and Kari Schmidt and George Amato of the American Museum of Natural History for sharing two unpublished primer pairs for the thick-billed parrot. Thank you to studbook managers Gen Anderson (BTM) and Susan Healy (TBP). Thank you to all individuals involved in sample collection, especially Amy Chalbot, Gareth Morgan, Marcelo Ruiz, and Mario Daniel Zambrana. Thank you to all institutions who provided samples of captive individuals including: African Lion Safari, Akron Zoo, Albuquerque Zoo, Dallas World Aquarium, Ft. Worth Zoo, Living Desert Zoo, Moody Gardens, Natural Encounters Zoo, Phoenix Zoo, Pinola Conservancy, Riverbanks Zoo, Sacramento Zoo, San Diego Wild Animal Park, San Diego Zoo, Santa Cruz Zoo, Sedgewick Zoo, St. Augustine Alligator Farm, and Tulsa Zoo.

Conflicts of Interest: The authors declare no conflict of interest.

References

1. Frankham, R. Genetics and extinction. *Biol. Conserv.* **2005**, *126*, 131–140. [CrossRef]
2. Palstra, F.; Ruzzante, D. Genetic estimates of contemporary effective population size: What can they tell us about the importance of genetic stochasticity for wild population persistence? *Mol. Ecol.* **2008**, *17*, 3428–3447. [CrossRef]
3. Witzenberger, K.A.; Hochkirch, A. Ex situ conservation genetics: A review of molecular studies on the genetic consequences of captive breeding programmes for endangered animal species. *Biodiv. Conserv.* **2011**, *20*, 1843–1861. [CrossRef]
4. Haig, S.M.; Ballou, J.D.; Derrickson, S.R. Management options for preserving genetic diversity: Reintroduction of guam rails to the wild. *Conserv. Biol.* **1990**, *4*, 290–300. [CrossRef]
5. Seddon, P.; Armstrong, D.; Maloney, R. Developing the science of reintroduction biology. *Conserv. Biol.* **2007**, *21*, 303–312. [CrossRef]
6. Hedrick, P.W.; Garcia-Dorado, A. Understanding inbreeding depression, purging, and genetic rescue. *Trends Ecol. Evol.* **2016**, *31*, 940–952. [CrossRef]
7. Briskie, J.V.; Mackintosh, M. Hatching failure increases with severity of population bottlenecks in birds. *Proc. Natl. Acad. Sci. USA* **2004**, *101*, 558–561. [CrossRef]
8. Frankham, R.; Hendry, H.; Margan, S.; Briscoe, D. Does equalization of family sizes reduce genetic adaption to captivity? *Anim. Conserv.* **2006**, *3*, 357–363. [CrossRef]
9. Milián-García, Y.; Jensen, E.L.; Madsen, J.; Álvarez Alonso, S.; Serrano Rodríguez, A.; Espinosa López, G.; Russello, M.A. Founded: Genetic reconstruction of lineage diversity and kinship informs ex situ conservation of cuban amazon parrots (Amazona leucocephala). *J. Hered.* **2015**, *106*, 573–579. [CrossRef] [PubMed]
10. Berkunsky, I.; Daniele, G.; Kacoliris, F.P.; Díaz-Luque, J.A.; Silva Frias, C.P.; Aramburu, R.M.; Gilardi, J.D. Reproductive parameters in the critically endangered blue-throated macaw: Limits to the recovery of a parrot under intensive management. *PLoS ONE* **2014**, *9*, e99941. [CrossRef] [PubMed]
11. Clarke, R.; By, R. Poaching, habitat loss and the decline of neotropical parrots: A comparative spatial analysis. *J. Exp. Criminol.* **2013**, *9*, 333–353. [CrossRef]
12. Collar, N.J. Globally threatened parrots: Criteria, characteristics and cures. *Int. Zoo Yearb.* **2000**, *37*, 21–35. [CrossRef]
13. The IUCN Red List of Threatened Species. Available online: https://www.iucnredlist.org (accessed on 18 July 2021).
14. Berkunsky, I.; Quillfeldt, P.; Brightsmith, D.J.; Abbud, M.C.; Aguilar, J.M.R.E.; Alemán-Zelaya, U.; Aramburú, R.M.; Arce-Arias, A.; Balas-McNab, R.; Balsby, T.J.S.; et al. Current threats faced by Neotropical parrot populations. *Biol. Conserv.* **2017**, *214*, 278–287. [CrossRef]
15. Wright, T.; Toft, C.; Enkerlin-Hoeflich, E.; Gonzalez-Elizondo, J.; Albornoz, M.; Rodríguez-Ferraro, A.; Rojas-Suárez, F.; Sanz D'Angelo, V.; Trujillo, A.; Beissinger, S.; et al. Nest poaching in neotropical parrots. *Conserv. Biol.* **2001**, *15*, 710–720. [CrossRef]
16. Watson, R. *International Studbook: Annual Report and Recommendations for 2011, Spix's Macaw (Cyanopsitta spixii)*; Al Wabra Wildlife Preservation: Al Shahaniya, Qatar, 2011.
17. BirdLife International. *Cyanopsitta spixii*. Available online: http://datazone.birdlife.org/species/factsheet/spixs-macaw-cyanopsitta-spixii (accessed on 18 July 2021).
18. Ballou, J.D.; Lacy, R.C. Identifying genetically important individuals for management of genetic diversity in captive populations. In *Population Management for Survival and Recovery*; Ballou, J.D., Gilpin, M., Foose, T.J., Eds.; Columbia University Press: New York, NY, USA, 1995; pp. 76–111.
19. Jamieson, I.G. Founder effects, inbreeding, and loss of genetic diversity in four avian reintroduction programs. *Conserv. Biol.* **2011**, *25*, 115–123. [CrossRef] [PubMed]

20. Lutaaya, B.E.; Misztal, I.; Bertrand, J.K.; Mabry, J.W. Inbreeding in populations with incomplete pedigrees. *J. Anim. Breed. Genet.* **1999**, *116*, 475–480. [CrossRef]
21. Collar, N.J. Parrot reintroduction: Towards a synthesis of best practice. In Proceedings of the 6th International Parrot Convention, Loro Parque, Puerto de la Cruz, Tenerife, Spain, 27–30 September 2006; pp. 82–107.
22. Ebenhard, T. Conservation breeding as a tool for saving animal species from extinction. *Trends Ecol. Evol.* **1995**, *10*, 438–443. [CrossRef]
23. Viggers, K.L.; Lindenmayer, D.; Spratt, D. The importance of disease in reintroduction programmes. *Wildl. Res.* **1993**, *20*, 687–698. [CrossRef]
24. McPhee, M. Generations in captivity increases behavioral variance: Considerations for captive breeding and reintroduction programs. *Biol. Conserv.* **2004**, *115*, 71–77. [CrossRef]
25. Morrison, C.E.; Johnson, R.N.; Grueber, C.E.; Hogg, C.J. Genetic impacts of conservation management actions in a critically endangered parrot species. *Conserv. Genet.* **2020**, *21*, 869–877. [CrossRef]
26. Coleman, R.; Weeks, A. *Population Genetics of the Endangered Orange-Bellied Parrot (Neophema chrysogaster): Project Update*; University of Melbourne: Melbourne, Australia, 2013.
27. Snyder, N.F.R.; Koenig, S.E.; Koschmann, J.; Snyder, H.A.; Johnson, T.B. Thick-billed parrot releases in Arizona. *Condor* **1994**, *96*, 845–862. [CrossRef]
28. White, T.H.; Collar, N.J.; Moorhouse, R.J.; Sanz, V.; Stolen, E.D.; Brightsmith, D.J. Psittacine reintroductions: Common denominators of success. *Biol. Conserv.* **2012**, *148*, 106–115. [CrossRef]
29. BirdLife International. Ara Glaucogularis. Available online: https://dx.doi.org/10.2305/IUCN.UK.2018-2.RLTS.T22685542A130868462.en (accessed on 1 June 2021).
30. Herzog, S.K.; Maillard, Z.O.; Boorsma, T.; Sánchez-Ávila, G.; García-Solíz, V.H.; Paca-Condori, A.C.; De Abajo, M.V.; Soria-Auza, R.W. First systematic sampling approach to estimating the global population size of the critically endangered blue-throated macaw Ara glaucogularis. *Bird Conserv. Int.* **2021**, *31*, 293–311. [CrossRef]
31. Anderson, G. *North American Regional Studbook for the Blue-Throated Macaw (Ara glaucogularis)*; St. Augustine Alligator Farm Zoological Park: St. Augustine, FL, USA, 2017.
32. Maestri, M.L.; Ferrati, R.; Berkunsky, I. Evaluating management strategies in the conservation of the critically endangered Blue-throated Macaw (Ara glaucogularis). *Ecol. Model.* **2017**, *361*, 74–79. [CrossRef]
33. Strem, R.; Bouzat, J. Population viability analysis of the blue-throated macaw (*Ara glaucogularis*) using individual-based and cohort-based PVA programs. *Open Conserv. Biol. J.* **2012**, *6*, 12–24. [CrossRef]
34. BirdLife International. Rhynchopsitta Pachyrhyncha (Amended Version of 2016 Assessment). Available online: https://dx.doi.org/10.2305/IUCN.UK.2017-1.RLTS.T22685766A110475642.en (accessed on 1 June 2021).
35. U.S. Fish and Wildlife Service. *Thick-Billed Parrot (Rhynchopsitta pachyrhyncha) Recovery Plan Addendum*; U.S. Fish and Wildlife Service, Southwest Region: Albuquerque, NM, USA, 2013.
36. Putnam, A. *Thick-Billed Parrot (Rhynchopsitta pachyrhyncha) AZA Species Survival Plan*; San Diego Zoo Global: San Diego, CA, USA, 2014.
37. Guerra, J.E.; Cruz-Nieto, J.; Ortiz-Maciel, S.G.; Wright, T.F. Limited geographic variation in the vocalizations of the endangered thick-billed parrot: Implications for conservation strategies. *Condor* **2008**, *110*, 639–647. [CrossRef] [PubMed]
38. Monterrubio-Rico, T.C.; Enkerlin-Hoeflich, E. Present use and characteristics of thick-billed parrot nest sites in northwestern Mexico. *J. Field Ornithol.* **2004**, *75*, 96–103. [CrossRef]
39. Snyder, N.F.; Enkerlin-Hoeflich, E.C.; Cruz-Nieto, M.A.; Valdes-Peña, R.A.; Ortiz-Maciel, S.G.; Cruz-Nieto, J. Thick-billed parrot (rhynchopsitta pachyrhyncha). In *Birds of the World*, 1st ed.; Billerman, S.M., Ed.; Cornell Lab of Ornithology: Ithaca, NY, USA, 2020.
40. Sheppard, J.K.; Rojas, J.I.G.; Cruz, J.; GonzÁLez, L.F.T.; Nieto, M.Á.C.; Lezama, S.D.J.; Juarez, E.A.; Lamberski, N. Predation of nesting thick-billed parrots Rhychopsitta pachyrhyncha by bobcats in northwestern Mexico. *Bird Conserv. Int.* **2020**, 1–9. [CrossRef]
41. Monterrubio-Rico, T.C.; Charre-Medellin, J.F.; Sáenz-Romero, C. Current and future habitat availability for thick-billed and maroon-fronted parrots in northern Mexican forests. *J. Field Ornithol.* **2015**, *86*, 1–16. [CrossRef]
42. Frankham, R. Inbreeding and extinction: A threshold effect. *Conserv. Biol.* **1995**, *9*, 792–799. [CrossRef]
43. GE Healthcare Life Sciences. Whatman™ FTA Elute™. Available online: https://at.vwr.com/assetsvc/asset/de_AT/id/17887357/contents (accessed on 20 September 2017).
44. Olah, G.; Heinsohn, R.G.; Espinoza, J.R.; Brightsmith, D.J.; Peakall, R. An evaluation of primers for microsatellite markers in scarlet macaw (Ara macao) and their performance in a Peruvian wild population. *Conserv. Genet. Resour.* **2015**, *7*, 157–159. [CrossRef]
45. Caparroz, R.; Miyaki, C.Y.; Baker, A.J. Characterization of microsatellite loci in the blue-and-gold macaw, Ara ararauna (Psittaciformes: Aves). *Mol. Ecol. Notes* **2003**, *3*, 441–443. [CrossRef]
46. Russello, M.A.; Saranathan, V.; Buhrman-Deever, S.; Eberhard, J.; Caccone, A. Characterization of polymorphic microsatellite loci for the invasive monk parakeet (Myiopsitta monachus). *Mol. Ecol. Notes* **2007**, *7*, 990–992. [CrossRef]
47. Klauke, N.; Masello, J.F.; Quillfeldt, P.; Segelbacher, G. Isolation of tetranucleotide microsatellite loci in the burrowing parrot (*Cyanoliseus patagonus*). *J. Ornithol.* **2009**, *150*, 921–924. [CrossRef]
48. Schmidt, K.; Amato, G. Thick-Billed Parrot Primers. 2009; unpublished.

49. Acosta, D. *Development and Characterization of Microsatellite Markers for the Endangered Thick-Billed Parrot (Rhynchopsitta pachyrhyncha)*; New Mexico State University: Las Cruces, NM, USA, 2009.
50. Van Oosterhout, C.; Hutchinson, W.; Wills, D.; Shipley, P. MICRO-CHECKER: Software for identifying and correcting genotyping errors in microsatellite data. *Mol. Ecol. Notes* **2004**, *4*, 535–538. [CrossRef]
51. Sinnwell, J.P.; Therneau, T.M.; Schaid, D.J. The kinship2 R package for pedigree data. *Hum. Hered.* **2014**, *78*, 91–93. [CrossRef]
52. Kalinowski, S.; Wagner, A.; Taper, M. ML-RELATE: A computer program for maximum likelihood estimation of relatedness and relationship. *Mol. Ecol. Notes* **2006**, *6*, 576–579. [CrossRef]
53. Raymond, M.; Rousset, F. GENEPOP (version-1.2)—Population-genetics software for exact tests and ecumenicsim. *J. Hered.* **1995**, *86*, 248–249. [CrossRef]
54. Falush, D.; Stephens, M.; Pritchard, J.K. Inference of population structure using multilocus genotype data: Linked loci and correlated allele frequencies. *Genetics* **2003**, *164*, 1567–1587. [CrossRef]
55. Earl, D.; Vonholdt, B. STRUCTURE HARVESTER: A website and program for visualizing STRUCTURE output and implementing the Evanno method. *Conserv. Genet. Resour.* **2012**, *4*, 359–361. [CrossRef]
56. Evanno, G.; Regnaut, S.; Goudet, J. Detecting the number of clusters of individuals using the software structure: A simulation study. *Mol. Ecol.* **2005**, *14*, 2611–2620. [CrossRef] [PubMed]
57. Pritchard, J.; Mj, S.; Donnelly, P.J. Inference of population structure using multilocus genotype data. *Genetics* **2000**, *155*, 945–959. [CrossRef] [PubMed]
58. Jakobsson, M.; Rosenberg, N. CLUMPP: A cluster matching and permutation program for dealing with label switching and multimodality in analysis of population structure. *Bioinformatics* **2007**, *23*, 1801–1806. [CrossRef] [PubMed]
59. Rosenberg, N. DISTRUCT: A program for the graphical display of population structure. *Mol. Ecol. Notes* **2004**, *4*, 137–138. [CrossRef]
60. Peakall, R.; Smouse, P. GenAlEx 6.5: Genetic analysis in excel. Population genetic software for teaching and research-an update. *Bioinformatics* **2012**, *28*, 2537–2539. [CrossRef] [PubMed]
61. Paetkau, D.; Slade, R.; Burden, M.; Estoup, A. Genetic assignment methods for the direct, real-time estimation of migration rate: A simulation-based exploration of accuracy and power. *Mol. Ecol.* **2004**, *13*, 55–65. [CrossRef]
62. Piry, S.; Alapetite, A.; Cornuet, J.; Paetkau, D.; Baudouin, L.; Estoup, A. GENECLASS2: A software for genetic assignment and first-generation migrant detection. *J. Hered.* **2004**, *95*, 536–539. [CrossRef]
63. Goudet, J. hierfstat, a package for R to compute and test hierarchical F-statistics. *Mol. Ecol. Notes* **2005**, *5*, 184–186. [CrossRef]
64. Cornuet, J.M.; Luikart, G. Description and power analysis of two tests for detecting recent population bottlenecks from allele frequency data. *Genetics* **1996**, *144*, 2001–2014. [CrossRef]
65. Piry, S.; Luikart, G.; Cornuet, J.M. BOTTLENECK: A computer program for detecting recent reductions in the effective population size using allele frequency data. *J. Hered.* **1999**, *90*, 502–503. [CrossRef]
66. Beck, N.; Double, M.; Cockburn, A. Microsatellite evolution at two hypervariable loci revealed by extensive avian pedigrees. *Mol. Biol. Evol.* **2003**, *20*, 54–61. [CrossRef]
67. Miller, M.; Haig, S.; Mullins, T.; Popper, K.; Green, M. Evidence for population bottlenecks and subtle genetic structure in the yellow rail. *Condor* **2012**, *114*, 100–112. [CrossRef]
68. Vergara-Tabares, D.L.; Cordier, J.M.; Landi, M.A.; Olah, G.; Nori, J. Global trends of habitat destruction and consequences for parrot conservation. *Glob. Chang. Biol.* **2020**, *26*, 4251–4262. [CrossRef]
69. Russello, M.A.; Smith-Vidaurre, G.; Wright, T.F. Genetics of invasive parrot populations. In *Naturalized Parrots of the World: Distribution, Ecology, and Impacts of the World's Most Colorful Colonizers*; Pruitt-Jones, S., Ed.; Princeton University Press: Princeton, NJ, USA, 2021.
70. Klauke, N.; Schaefer, H.M.; Bauer, M.; Segelbacher, G. Limited dispersal and significant fine - scale genetic structure in a tropical montane parrot species. *PLoS ONE* **2016**, *11*, e0169165. [CrossRef]
71. Blanco, G.; Morinha, F.; Roques, S.; Hiraldo, F.; Rojas, A.; Tella, J.L. Fine-scale genetic structure in the critically endangered red-fronted macaw in the absence of geographic and ecological barriers. *Sci. Rep.* **2021**, *11*, 556. [CrossRef]
72. Raisin, C.; Frantz, A.C.; Kundu, S.; Greenwood, A.G.; Jones, C.G.; Zuel, N.; Groombridge, J.J. Genetic consequences of intensive conservation management for the Mauritius parakeet. *Conserv. Genet.* **2012**, *13*, 707–715. [CrossRef]
73. Tollington, S.; Jones, C.G.; Greenwood, A.; Tatayah, V.; Raisin, C.; Burke, T.; Dawson, D.A.; Groombridge, J.J. Long-term, fine-scale temporal patterns of genetic diversity in the restored Mauritius parakeet reveal genetic impacts of management and associated demographic effects on reintroduction programmes. *Biol. Conserv.* **2013**, *161*, 28–38. [CrossRef]
74. Lacy, R.; Petric, A.; Warneke, M. Inbreeding and outbreeding depression in captive populations of wild animals. In *The Natural History of Inbreeding and Outbreeding: Theoretical and Empirical Perspectives*; Thornhill, N.W., Ed.; University of Chicago Press: Chicago, IL, USA, 1993.
75. Lande, R. Anthropogenic, ecological and genetic factors in extinction and conservation. *Popul Ecol.* **1998**, *40*, 259–269. [CrossRef]
76. Stojanovic, D.; Olah, G.; Webb, M.; Peakall, R.; Heinsohn, R. Genetic evidence confirms severe extinction risk for critically endangered swift parrots: Implications for conservation management. *Anim. Conserv.* **2018**, *21*, 313–323. [CrossRef]
77. BirdLife International. Lathamus Discolor. Available online: https://dx.doi.org/10.2305/IUCN.UK.2018-2.RLTS.T22685219A130886700.en (accessed on 1 June 2021).

78. BirdLife International. Amazona Leucocephala. Available online: https://dx.doi.org/10.2305/IUCN.UK.2020-3.RLTS.T22686201A179212864.en (accessed on 1 June 2021).
79. Bergner, L.M.; Jamieson, I.G.; Robertson, B.C. Combining genetic data to identify relatedness among founders in a genetically depauperate parrot, the Kakapo (Strigops habroptilus). *Conserv. Genet.* **2014**, *15*, 1013–1020. [CrossRef]
80. BirdLife International. Strigops Habroptila. Available online: https://dx.doi.org/10.2305/IUCN.UK.2018-2.RLTS.T22685245A129751169.en (accessed on 1 June 2021).
81. Yamashita, C.; Barros, Y.M. The Blue-throated Macaw (*Ara glaucogularis*), characterization of its distinctive habitat in the savannas of Beni, Bolivia. *Ararajuba* **1997**, *5*, 141–150.
82. Clout, M.N.; Merton, D.V. Saving the kakapo: The conservation of the world's most peculiar parrot. *Bird Conserv. Int.* **1998**, *8*, 281–296. [CrossRef]
83. Collar, N.; Sharpe, C.J.; Boesman, P.F.D. Kakapo (Strigops habroptila). In *Birds of the World*, 1st ed.; del Hoyo, J., Elliott, A., Sargatal, J., Christie, D.A., de Juana, E., Eds.; Cornell Lab of Ornithology: Ithaca, NY, USA, 2020.
84. White, K.L.; Eason, D.K.; Jamieson, I.G.; Robertson, B.C. Evidence of inbreeding depression in the critically endangered parrot, the kakapo. *Anim. Conserv.* **2015**, *18*, 341–347. [CrossRef]
85. Beissinger, S.R.; Wunderle, J.M.; Meyers, J.M.; Sæther, B.E.; Engen, S. Anatomy of a bottleneck: Diagnosing factors limiting population growth in the Puerto Rican parrot. *Ecol. Monogr.* **2008**, *78*, 185–203. [CrossRef]
86. Frankham, R.; Ballou, J.D.; Briscoe, D.A. *Introduction to Conservation Genetics*; Cambridge University Press: Cambridge, UK, 2002.
87. Shields, W.M. The natural and unnatural history of inbreeding and outbreeding. In *The Natural History of Inbreeding and Outbreeding*; Thornhill, N.W., Ed.; University of Chicago Press: Chicago, IL, USA, 1993; pp. 143–169.
88. Edmands, S. Between a rock and a hard place: Evaluating the relative risks of inbreeding and outbreeding for conservation and management. *Mol. Ecol.* **2007**, *16*, 463–475. [CrossRef]
89. Marr, A.B.; Keller, L.F.; Arcese, P. Heterosis and outbreeding depression in descendants of natural immigrants to an inbred population of song sparrows (Melospiza melodia). *Evolution* **2002**, *56*, 131–142. [CrossRef]
90. Herzog, S.; Bürger, J.; Troncoso, A.; Vargas-Rodriguez, R.; Boorsma, T.; Soria-Auza, R.W. Deterministic population growth models and conservation translocation as a management strategy for the critically endangered blue-throated macaw (*Ara glaucogularis*): A critique of Maestri et al. *Ecol. Model.* **2018**, *388*, 145–148. [CrossRef]
91. IUCN/SSP. *Guidelines for Reintroductions and Other Conservation Translocations*; Version 1.0; IUCN Species Survival Commission: Gland, Switzerland, 2013; p. 57.

MDPI

Article

Genetic Assignment Tests to Identify the Probable Geographic Origin of a Captive Specimen of Military Macaw (*Ara militaris*) in Mexico: Implications for Conservation

Francisco A. Rivera-Ortíz [1,*], Jessica Juan-Espinosa [1], Sofía Solórzano [1], Ana M. Contreras-González [2] and María del C. Arizmendi [2]

[1] Laboratorio de Ecología Molecular y Evolución, Unidad de Biotecnología y Prototipos FES Iztacala, Universidad Nacional Autónoma de México, Avenida de los Barrios 1, Los Reyes Iztacala, Tlalnepantla de Baz 54090, Mexico; puki_pink@hotmail.com (J.J.-E.); solorzanols@unam.mx (S.S.)

[2] Laboratorio de Ecología, Unidad de Biotecnología y Prototipos, Facultad de Estudios Superiores Iztacala, Universidad Nacional Autónoma de México, Avenida de los Barrios No. 1, Los Reyes Ixtacala, Tlalnepantla de Baz 54090, Mexico; acontrerasgonzalez@iztacala.unam.mx (A.M.C.-G.); coro@unam.mx (M.d.C.A.)

* Correspondence: francisco.rivera@iztacala.unam.mx

Abstract: The Military Macaw (*Ara militaris*) faces a number of serious conservation threats. The use of genetic markers and assignment tests may help to identify the geographic origin of captive individuals and improve conservation and management programs. The purpose of this study was to identify the possible geographic origin of a captive individual using genetic markers. We used a reference database of genotypes of 86 individuals previously shown to belong to two different genetic groups to determine the genetic assignment of the captive individual of unknown origin (captive specimen) and five individuals of known geographic origin (as positive controls). We evaluated the accuracy of three assignment/exclusion criteria to determine the success of correct assignment of the individual of unknown origin and the five positive control individuals. WICHLOCI estimated that eight loci were required to achieve an assignment success of 83%. The correct geographic origin of positive controls was identified with 83% confidence. All of the analyses assigned the captive individual to the genetic group from the Sierra Madre Oriental. Bayesian assignment tests, tests for genetic distance and allele frequency tests assigned the unknown individual to the locations from the Sierra Madre Oriental with a probability of 71.2–82.4%. We show that the use of genetic markers provides a promising tool for determining the origin of pets and individuals seized from the illegal animal trade to better inform decisions on reintroduction and improve conservation programs.

Keywords: conservation genetics; genetic assignment tests; probable geographic origin; Military Macaw

Citation: Rivera-Ortíz, F.A.; Juan-Espinosa, J.; Solórzano, S.; Contreras-González, A.M.; Arizmendi, M.d.C. Genetic Assignment Tests to Identify the Probable Geographic Origin of a Captive Specimen of Military Macaw (*Ara militaris*) in Mexico: Implications for Conservation. *Diversity* **2021**, *13*, 245. https://doi.org/10.3390/d13060245

Academic Editor: Michael Wink

Received: 7 May 2021
Accepted: 29 May 2021
Published: 2 June 2021

1. Introduction

The order Psittaciformes contains some of the most charismatic and recognizable bird species in the world [1]. However, of the order's approximately 352 species, 26% face some degree of extinction risk [2]. For example, out of the 22 Psittacidae species recorded in Mexico [1,3], 20 are at risk according to Mexican law [4], and at the international level, the Red List of the International Union for the Conservation of Nature (IUCN) places eight of those species in some risk category [5]. The Military Macaw (*Ara militaris*) is one of the endangered psittacid species in Mexico, and faces two main threats: (1) habitat transformation (loss, fragmentation and degradation) [6,7], and (2) illegal collection for the national and international illegal pet trade [7–15]. Indeed, illegal trafficking has led to the extirpation of populations from conserved areas [11,12,16].

In Mexico, the illegal wildlife trade has threatened 19 out of the 22 Psittacidae species [7]. The capture of any wild Psittacidae species was outlawed in Mexico in 2003, and the current General Wildlife Law (LGVS) prohibits the extraction of psittacid species, only granting permits for conservation or scientific research purposes [17]. One of the objectives of the LGVS was to guide management efforts, including the recovery, reproduction, research, release, and/or relocation of individuals [18]. One of the problems faced by reintroduction and recovery efforts is that in most cases, the geographic origin of animals recovered from the illegal pet trade is unknown. Information on the geographic origin of rehabilitated individuals is crucial in order to avoid mixing individuals from genetically distinct populations, which can lead to genetic problems (e.g., local maladaptation and outbreeding depression) [19–22].

Molecular tools make it possible to answer questions concerning evolutionary history, define taxonomic uncertainties, and identify release locations using molecular markers (e.g., microsatellites) and statistical approaches [23–26]. However, these techniques are not often used for the identification of release locations for rehabilitated birds illegally taken from the wild [21,27]. The use of molecular tools to establish the origin of individuals for conservation purposes is increasing in reintroduction plans and for identifying illegal trade sites, as demonstrated by studies of several species, such as the Hyacinth Macaw (*Anodorhynchus hyacinthinus*) [28], the Blue-and-Yellow Macaw (*Ara ararauna*) [21] and the European Pond Turtle (*Emys orbicularis*) [29].

The purpose of this study was to determine the probable geographic origin of a captive Military Macaw of unknown origin using different molecular statistical analyses and test the accuracy of these techniques using individuals of known origin, in order to generate a protocol that can be used for reintroduction programs, for management and conservation.

The Military Macaw is one of the most charismatic species in the New World. Its distribution is fragmented, ranging from northern Mexico to northwestern Argentina [1,6,30]. In Mexico, the Military Macaw is distributed in apparently isolated colonies in two separate areas. One includes the Sierra Madre Occidental and the Sierra Madre del Sur (from southern Sonora to Chiapas); the other is in the Sierra Madre Oriental, where the macaws are reported in Tamaulipas, San Luis Potosi, Guanajuato, and Querétaro [31,32]. The geographic distribution of the Military Macaw in Mexico declined by 43% over 16 years (2000–2016) [10,13,14]. It is endangered under Mexican law [4,13], vulnerable on the IUCN Red List [5], and listed in Appendix I of the Convention on International Trade in Species [33].

A study conducted by Rivera-Ortíz et al. [34] on the genetics of the Military Macaw used microsatellites from samples collected in seven Mexican locations and found strong genetic structuring, showing two groups presenting geographic concordance. Group one corresponded to the locations found in the Sierra Madre Occidental and the Sierra Madre del Sur (Pacific slope), and Group two corresponded to the locations found in the Sierra Madre Oriental (Gulf of Mexico slope). Given these results, the authors proposed the protection of the two genetic groups found in the three physiographic regions as independent conservation units. In this study, we used both classification (correspondence analysis) and genetic assignment methods to evaluate whether the Military Macaw individual of unknown origin belonged to any of those previously identified genetic groups, and if possible, assign it to a particular location.

2. Materials and Methods

2.1. DNA Extraction and Genotyping

We extracted DNA from five feathers collected from a captive individual housed at the AFP OCEAN Foundation A.C. The rachises of the collected feathers were cleaned with 75% molecular grade ethyl [34]. This same procedure was performed with positive control samples from five individuals of known geographic origin.

Total genomic DNA was extracted using the standard digestion protocol with Proteinase-K/sodium dodecyl sulfate (SDS), followed by chloroform and alcohol purification as described by Leeton and Christidis [35]. Eight loci were amplified from nuclear microsatellites using primers designed for other parrot species (five for *Ara ararauna* and three for *Amazona guildinguii*) [36–38]. These polymorphic microsatellites were those previously used by Rivera-Ortíz et al. [34] for Military Macaw individuals from seven different locations.

These eight loci were amplified by polymerase chain reaction (PCR) according to the parameters used by Rivera-Ortíz et al. [34]. Electrophoresis was carried out using an ABI PRISM 3100 Avant sequencer (Applied Biosystems) with Gene Scan LIZ 500 to determine fragment size. Fragments and their final size were analyzed using GENE MAPPER 4.0 software (Applied Biosystems). Since this program automatically determines allele size, we visually checked the electropherograms of microsatellites from the eight loci to corroborate their size and number. PCR sequences were repeated for samples with unclear electropherograms to resolve uncertainties [39].

2.2. Data Analysis

Genetic assignment analyses were conducted using the genotypes of 86 individuals from seven locations grouped into the two genetic groups reported by Rivera-Ortíz et al., [34]. Those genetic groups considered the candidate places of origin of the captive specimen of unknown origin and the five specimens of known origin (one individual from Sinaloa, two from Nayarit, one from Oaxaca and one from Tamaulipas).

Theoretical studies have examined how the number of loci and alleles relate to the success of assignment [40,41]. WICHLOCI 1.0 software [41] chooses the best combination of loci to assign a captive individual by analyzing the data extracted from candidate locations. Combinations of the eight microsatellites were used to test the minimum loci needed for a successful assignment [41,42].

2.3. Factorial Correspondence Analysis (FCA)

First, we did a factorial correspondence analysis (FCA) using GENETIX 4.05.4 software [43]. This analysis is a multivariate interdependence statistical method that is well adapted to describe associations between variables [44] and provides a graphical display of the genetic relationships between the individuals of interest and those of the reference populations in a multidimensional space based on allelic data. FCA was performed using three data combinations: (i) positive controls + unknown individual + individuals from the Sierra Madre Occidental/Sierra Madre del Sur populations + individuals from the Sierra Madre Oriental populations and (ii) positive controls + unknown individual + individuals from the Sierra Madre Oriental. These combinations were created to determine whether any differences existed between the unknown individual and the locations or candidate genetic groups.

2.4. Genetic Assignment Analysis

A Bayesian approach was used to assign the unknown individual and positive controls to genetic groups or populations, implemented in STRUCTURE 2.3.1 software [45,46]. This approach was designed to infer the number of genetic groups or populations of individuals (K) according to their genotypes and estimate the proportional membership of each individual's genotype to one or more of the inferred genetic clusters.

We used the results of the population structure analysis of Military Macaws from Rivera-Ortíz et al. [34], which identified two genetic groups: (1) Sierra Madre Occidental/Sierra Madre del Sur populations and (2) Sierra Madre Oriental populations. Within these groups, we explored the possibility of hierarchical structuring as recommended by Jombart [47]. We repeated these analyses until no additional structure was found within clusters, i.e., until the optimal K value was 1. The burn-in length for each repetition consisted of 500,000 steps, followed by 10,000,000 iterations, under the admixture assumption. No clear substructure was detected in the two genetic groups reported by Rivera-Ortíz et al. [34] (Supplementary Materials, Figure S1). A similar analysis, the discriminant analysis of principal components (DAPC) also showed that the studied seven candidate locations grouped into two inferred clusters (K = 2), according to Bayesian information criterion

(BIC) [47] (Supplementary Materials, Figure S2). Therefore, we assumed that there was no hierarchical structure and we continued the genetic assignment analysis under these conditions.

To assign the unknown individual and positive controls of known origin back to their genetic groups, we used the USEPOPINFO function within an admixture framework. We performed this analysis with all of the controls and the two genetic groups.

We used 10 repetitions in a range of K = 1 to K = 10. The burn-in length for each repetition consisted of 500,000 steps, followed by 10,000,000 iterations, under the admixture assumption in order to determine the maximum value of the posteriori probability (lnP (D)), to detect the true K [45]. CLUMPP 1.1.2 software [48] was used to eliminate label switching, using the greedy algorithm with 1,000 random input orders. These values were visualized using bar plots prepared with DISTRUCT software [49], which showed how the test individuals were assigned relative to the grouping of the reference set of individuals, and to determine the probability of their assignment to one of the two genetic groups identified by Rivera-Ortíz et al. [34].

Genetic relationships between the unknown individual and positive controls and the genetic groups/locations were also examined by applying discriminant analysis of principal components (DAPC) [47] using the "adegenet 2.1.3" package [26] in R 4.0.5 software [50], with the number of principle components set to 35 following alpha-score indication. DAPC is a multivariate, model-free approach designed to generate clusters based on prior population information [26]. DAPC allowed us to analyze the population structure by assigning the unknown individual and positive controls to the genetic groups or locations.

We used three genetic assignment/exclusion approaches implemented in GENECLASS 2.0 software [51]. The analyses were carried out for the two genetic groups and for each of the locations that contain them. The first approach used allele frequencies [52]; the unknown individual and positive controls were assigned to the genetic groups and candidate locations where each of their genotypic frequencies was expected to be the highest. We calculated the probability of the genotype of the controls and then applied the simulation algorithm proposed by Paetkau et al. [53] with a Monte Carlo (MC) resampling of 10,000 steps and an exclusion threshold of $p < 0.05$. In the second approach, we used a partially Bayesian test based on Rannala and Mountain [54], which estimates the population's allele frequencies and individual assignment's statistical significance (unknown individual and positive controls). We used the simulation algorithm proposed by Paetkau et al. [53] with an MC resampling of 10,000 steps and an exclusion threshold of $p < 0.05$. In the third approach, we calculated the genetic distance between the unknown individual and candidate groups and locations. The same was done for the positive controls [55].

We used three measurements of genetic distance: (1) Nei's minimum genetic distance [56], (2) Nei's DA genetic distance [57], and (3) the simulation algorithm proposed by Paetkau et al. [53] with an MC resampling of 10,000 steps and an exclusion threshold of $p < 0.05$. A self-assignment of all of the individuals from the reference populations was also performed, with an exclusion threshold of $p < 0.05$, with an MC resampling of 10,000 steps.

3. Results

The summary statistics per locus and per-locations locus, positive controls and unknown captive individual are given in the Supplementary Materials (Tables S1 and S2).

The assignment scores estimated by WICHLOCI 1.0 software indicated that the contribution to the genetic assignment of each of the eight loci varied between 9.0% (locus UnaCT21) and 15.55% (locus UnaCT21). Using eight loci produced the highest score, assigning the captive specimen to the candidate populations with an accuracy of 83% (Table 1). The eight loci were used in all subsequent assignment/exclusion tests to improve the assignment success for the unknown captive individual and positive controls.

Table 1. Ranking carried out in WICHLOCI 1.0 for the eight loci. The loci are in order of highest to the lowest score obtained.

Locus	Score	Score (%)	A (%)
UnaCT21	139.474	15.556	
UnaCT32	118.573	13.225	
UnaCT74	116.432	12.986	
UnaCT43	114.071	12.723	
UnaGT55	110.096	12.279	83%
AgGT17	109.722	12.237	
AgGT19	98.694	11.007	
AgGT32	89.51	9.983	

A = Correct assignment with the eight loci combined.

3.1. FCA

Using the combination of the positive controls + unknown individual + individuals from the Sierra Madre Occidental/Sierra Madre del Sur locations + individuals from the Sierra Madre Oriental locations, the FCA produced a data cloud showing the position of each of the 86 individuals from the candidate genetic groups in a two-dimensional space (taken from the study by Rivera-Ortíz et al. [34]) including the unknown individual and the positive controls. Two distinct genetic groups were differentiated in the data cloud: (1) individuals from the candidate locations from the Sierra Madre Occidental/Sierra Madre del Sur and (2) individuals from the candidate locations from the Sierra Madre Oriental (Figure 1). We observed that the positive controls of the individuals of Sinaloa, Nayarit, and Oaxaca were located in the point cloud of the candidate locations from the Sierra Madre Occidental/Sierra Madre del Sur. In contrast, the unknown individual and the positive control from Tamaulipas were located in the point cloud of the candidate locations from the Sierra Madre Oriental (Figure 1).

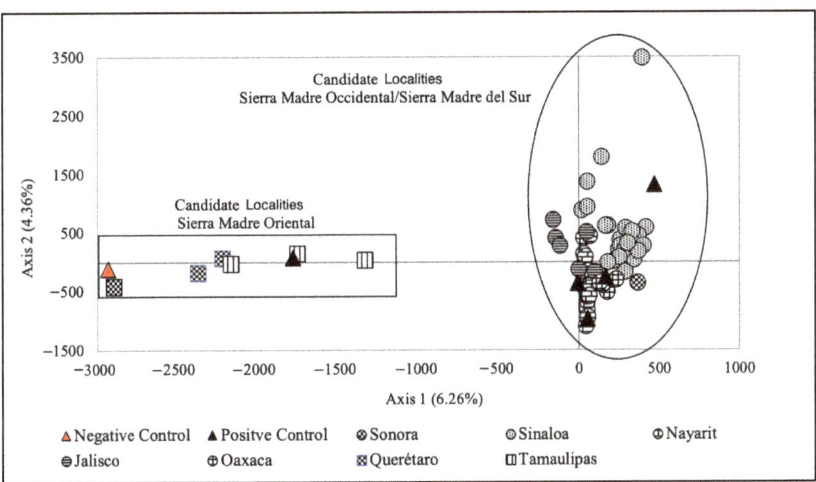

Figure 1. Results of the test to analyze the behavior of a bidimensional FCA of all candidate locations of the Military Macaw with respect to the unknown individual and positive control.

The combination of the positive controls + unknown individual + individuals from the Sierra Madre Oriental showed that the unknown individual was placed within the data cloud corresponding to the Sierra Madre Oriental candidate locations and was closest to the individuals from the Querétaro location. The positive control of the individual from Tamaulipas was also placed in this cloud but was closer to the individuals from the location from Tamaulipas. In contrast, positive control individuals from Sinaloa, Nayarit, and Oaxaca formed a cloud of points that was quite distant from the individuals from the reference locations in the Sierra Madre Oriental (Figure 2).

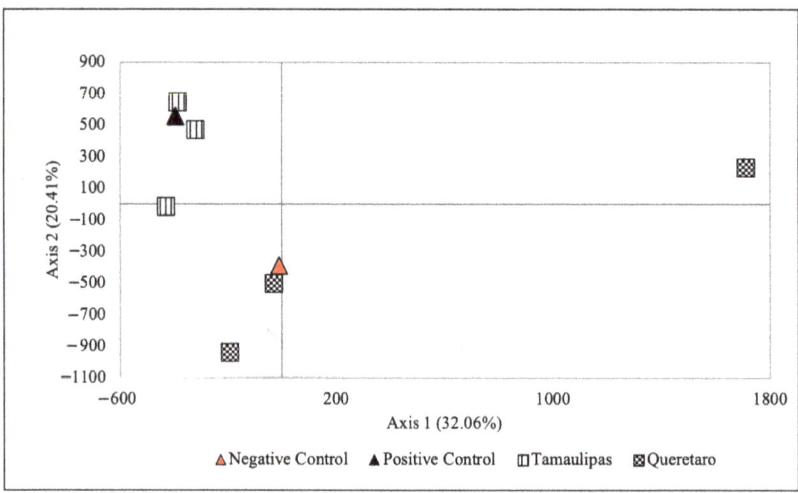

Figure 2. Results of the test to analyze the behavior of a bidimensional FCA of candidate locations of the Military Macaw of the Sierra Madre Oriental with respect to the unknown individual and positive control.

3.2. Genetic Assignment

STRUCTURE showed the two genetic groups (K = 2) reported previously by Rivera-Ortíz et al. [34]. The results of this analysis (Figure 3) indicates that the positive control individuals were correctly assigned to the locations corresponding to their known geographic origins. The unknown individual was assigned to the genetic group of the Sierra Madre Oriental with a genetic allocation ratio of 97.1% to 99.6% (Table 2).

Table 2. Variation in the genetic allocation percentages of the unknown individual of Military Macaw, with the USEPOPINFO function of STRUCTURE for 10 runs with K = 2.

Run	Genetic Group 1	Genetic Group 2
	Assignment Percentages	
1	2.9	97.1
2	0.4	99.6
3	97.2	92.8
4	97.2	92.8
5	2.7	98.3
6	1.6	98.4
7	0.8	99.2
8	97.3	90.7
9	0.5	99.5
10	1.8	98.2

Genetic group 1 = Sierra Madre Occidental/Sierra Madre del Sur. Genetic group 2 = Sierra Madre Oriental.

DAPC supported STRUCTURE, identifying two genetic groups (K = 2) (Figure 4), according to Bayesian information criterion (BIC). The DAPC plot also reflects the assignment probabilities of the positive control individuals as well as the individual of unknown origin to the Sierra Madre Oriental (Figure 4).

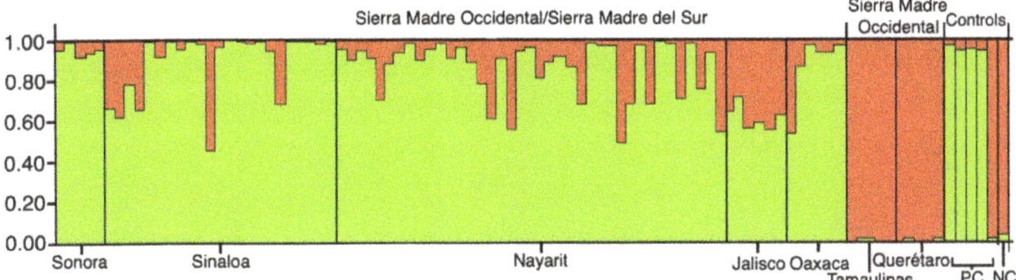

Figure 3. Proportional membership of unknown individual (NC) and positive controls (PC) and the relationship with the two genetic groups of Military Macaw (86 individuals). Each bar represents the genotype of each individual; the colors (green and red) represent the likelihood of membership of each of the two genetic clusters identified in the STRUCTURE analysis. The vertical black lines show divisions between sampling locations.

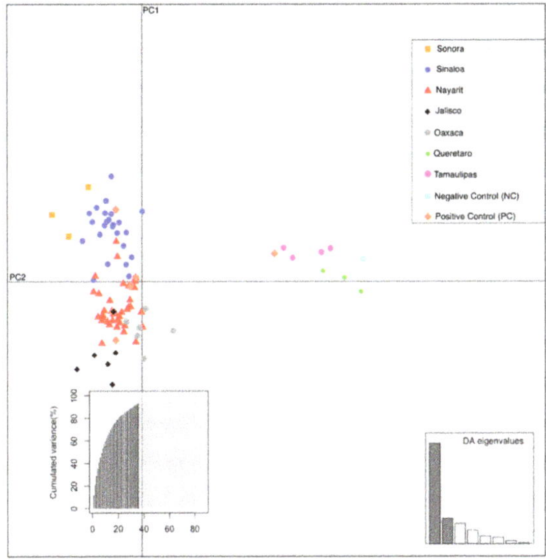

Figure 4. Scatter plots of the DAPC of the microsatellite data for Military Macaw candidate locations, positive controls and unknown individual from Mexico. The axes represent the first two linear discriminants (LD). Each dot represents an individual. Eigenvalues of the analysis are displayed in the inset.

In the four approaches implemented by GENECLASS, positive controls from Sinaloa, Nayarit and Oaxaca were assigned to the Sierra Madre Occidental/Sierra Madre del Sur genetic group, with values ranging from 70% to 82.4%. The positive control from Tamaulipas and the unknown individual were assigned to the genetic group from the Sierra Madre Oriental with values of 71.2% to 82.4% (Table 3).

The unknown captive individual was assigned to the candidate locations from Tamaulipas and Querétaro (Table 4). The analysis of Nei's genetic distances and the minimum Nei distance proposed by Cournuet et al. [55–57] showed the probabilities of assignment of the unknown individual to the locations from Querétaro and Tamaulipas (Table 4). The partially Bayesian approach taken by Rannala and Mountain [54] and the allele frequency approach described by Paetkau et al. [53] both assigned the unknown individual to the location from Querétaro, with probabilities of 32.7% and 30.3%, respectively (Table 4). All positive control individuals were assigned correctly to each of their locations of origin; the four approaches showed assignment probabilities between 29.8% and 55.0% for individuals from Sinaloa, Nayarit, Oaxaca, and Tamaulipas (Table 5).

Table 3. Assignment probabilities of the positive control samples and the captive individual of unknown origin, according to four types of analysis.

Genetic Groups	Controls	Analysis Type			
		Frequencies	**Bayesian**	**Nei's Genetic Distance (1983)**	**Minimal Nei Distance (1973)**
Sierra Madre Occidental/Sierra Madre del Sur	Sinaloa	0.756 *	0.705 *	0.771 *	0.816 *
	Nayarit (Ind. 1)	0.745 *	0.723 *	0.700 *	0.801 *
	Nayarit (Ind. 2)	0.742 *	0.737 *	0.715 *	0.810 *
	Oaxaca	0.778 *	0.756 *	0.789 *	0.836 *
Sierra Madre Oriental	Tamaulipas	0.764 *	0.758 *	0.764 *	0.799 *
	Unknown individual	0.712 *	0.748 *	0.79 *	0.824 *

* The probability of exclusion calculated with the Monte Carlo method of Paetkau et al. (2004) is $p < 0.05$.

Table 4. Assignment probabilities for the captive individual of unknown origin to each candidate location, according to four types of analysis.

Candidate Populations	Analysis Type			
	Frequencies	**Bayesian**	**Genetic Distance of Nei (1983)**	**Minimal Distance of Nei (1973)**
Sonora	0.00	0.00	0.00	0.00
Sinaloa	0.00	0.00	0.00	0.00
Nayarit (Ind. 1)	0.00	0.00	0.00	0.00
Nayarit (Ind. 2)	0.00	0.00	0.00	0.00
Jalisco	0.00	0.00	0.00	0.00
Oaxaca	0.00	0.00	0.00	0.00
Querétaro	0.303 *	0.327 *	0.373 *	0.461 *
Tamaulipas	0.00	0.00	0.224 *	0.418 *

* The probability of exclusion calculated with the Monte Carlo method of Paetkau et al. (2004) is $p < 0.05$.

Table 5. Assignment probabilities for the positive control samples to each candidate location, according to four types of analysis.

Candidate Populations	Analysis Type			
	Positive Controls (• = TAM, + = SIN, ♢ = NAY and ► = Oax)			
	Frequencies	**Bayesian**	**Genetic Distance of Nei (1983)**	**Minimal Distance of Nei (1973)**
Sonora	0.00	0.00	0.00	0.00
Sinaloa	0.306 +*	0.298 +*	0.471 +*	0.501 +*
Nayarit (Ind. 1)	0.349 ♢*	0.376 ♢*	0.401 ♢*	0.550 ♢*
Nayarit (Ind. 2)	0.449 ♢*	0.376 ♢*	0.401 ♢*	0.505 ♢*
Jalisco	0.00	0.00	0.00	0.00
Oaxaca	0.408 ►*	0.378 ►*	0.439 ►*	0.451 ►*
Querétaro	0.00	0.00	0.00	0.00
Tamaulipas	0.35.7 •*	0.310 •*	0.309 •*	0.457 •*

* The probability of exclusion calculated with the Monte Carlo method of Paetkau et al. (2004) is $p < 0.05$.

Self-assignment tests of the candidate locations panel correctly assigned between 65.1% and 90.6% of the individuals to their population of origin (Table 6).

Table 6. Allocation criteria implemented in GENECLASS, self-assignment test of the 86 wild individuals of the seven candidate locations.

The Criterion of the Algorithm	Number of Individuals	Percentage (%)
Frequencies	67	77.9
Bayesian	74	86.0
Nei's Genetic distance (1983)	78	90.6
Minimal Nei distance (1973)	56	65.1

4. Discussion

Identification of an individual's geographic origin by means of genetic analysis depends on the ability to assign it to a particular location, which in turn depends on the level of genetic structuring among reference locations [58]. Here, we tested the ability of molecular genetic assignment to identify the likely location of origin of one individual of the Military Macaw of unknown origin and five individuals of known origin, in order to evaluate the method's utility in future conservation efforts for the species. In this study, the results showed that the methods tested were useful in identifying the geographic areas from which individuals likely originated, for both the unknown individual and the five positive controls.

The results of the FCA, STRUCTURE and DAPC tests grouped the unknown individual with the Sierra Madre Oriental genetic group with high confidence. However, it was impossible to assign it to a specific geographic location because there is no differentiation between individuals from different reference locations in this genetic group, indicating gene flow. These FCA, STRUCTURE and DAPC results are reliable because the reference sample of individuals used in the study and provided by Rivera-Ortíz et al. [34] presents a marked genetic structure and differentiation across the distribution range of the Military Macaw in Mexico, showing a pattern that was also found by Eberhard et al. [59] with mitochondrial markers. These previously documented patterns of genetic structure are important in the context of the present study because structure and differentiation among the reference locations must be high if there is to be reasonable success in geographic allocation using grouping methods (with 80–100% correct allocation) [26,60,61].

The allocation/exclusion analyses carried out using GENCLASS suggest that the likely origin of the unknown individual is the Sierra Madre Oriental metapopulation, as determined by the grouping analyses. Three of the four criteria used for the allocation/exclusion analyses show some probability that the unknown individual belongs to the Querétaro location, although with relatively low certainty (30–46% probability). These low probability values should be interpreted with caution, since they may be affected by small sample sizes in some of the reference locations. Some authors suggest that a sample of 30–50 individuals per reference location is necessary to allow accurate estimates [26,55,58]. Unfortunately, obtaining large sample sizes in studies of endangered species is extremely difficult due to small population sizes, restricted areas, and difficulty accessing their distributional areas [13,62,63], as in the case of the Military Macaw.

The different methods used to identify the probable location of origin of the individual Military Macaw of unknown origin proved to be effective and complementary, as demonstrated in this study. When carrying out this type of analysis, we recommend graphically showing the genetic similarity of the individuals as a first step that reveals if the samples of unknown origin are grouped in the reference localities [64]. Then, consider a Bayesian approach to determine the probability that the individuals of unknown origin originate from a population, considering all reference localities together [26]. Finally, use the tests to exclude or identify individuals of unknown origin in the reference localities, to determine the probability that individuals of unknown origin are rejected or belong to the reference localities [65].

Our study shows that given the degree of population genetic structure in Military Macaw locations in Mexico, it is possible to use microsatellite data to identify the probable location of origin of an individual of unknown provenance. This, in turn, makes it possible to make a more informed selection of locations at which the individual could be released. The captive specimen was geographically assigned to the Sierra Madre Oriental, and according to our results, is a candidate for release in that zone. It is essential that the programs for reintroducing and releasing Military Macaw individuals into the wild make proper use of this kind of molecular tools [42,66,67], given that for an endangered species, such as the Military Macaw, the strong genetic structuring of wild locations may reflect local adaptations that would be lost if they were to be managed as a single group [34].

To improve the accuracy of assigning individuals of unknown origin to their correct populations, it is crucial to continue genetic studies of wild locations and increase the number of molecular markers used in genetic analyses. Relatively low numbers of microsatellites were used in this study, but microsatellites have provided sufficient power for geographic assignment of a variety of wild species due to their high level of polymorphism and genetic structure of the populations [68–70]. The use of other markers such as mitochondrial DNA would be very informative and complementary since it might allow us to distinguish lineages that correspond to particular geographic areas [21]. Identification of single nucleotide polymorphisms (SNPs) from genomic data also have a significant advantage for geographic assignment, since information from hundreds or thousands of SNPs could potentially provide improved resolution of patterns of genetic structure, and thus, the more precise assignment of an individual's geographic origin [61].

Our study demonstrates that in combination with the reference samples analyzed by Rivera-Ortiz et al. [34], currently available molecular markers and statistical assignment and exclusion software can help identify the geographic origin of captive individuals or specimens confiscated from illegal trade [50]. No studies have been conducted to analyze the number of Military Macaw individuals trapped each year, but Cantú et al. [16] estimated that 65,000 to 78,000 psittacid individuals are poached for illegal trade and suffer a mortality rate of 77%. Only about 2% of poached individuals are seized by the Mexican Federal Environmental Protection Agency (PROFEPA) [16], but given how many are poached, this small percentage still represents several hundred individuals. In this sense, identifying the geographic origin of captive individuals or specimens confiscated from illegal trade helps biodiversity managers to detect locations with intense poaching, and thus, focus efforts and resources on these sites to prevent poaching. It will also support and guide restoration or demographic translocation programs if they are deemed necessary to increase genetic variability [23,28].

A crucial component of this study was the availability of the set of reference samples of known geographic origin [34]. We recommend the establishment of large DNA reference collections and large public databases containing allele frequencies from many populations, and the use of museum collections, which can play an essential role since DNA can be extracted from museum skins. Any genetic analysis that attempts to identify geographic origin of an individual/sample depends on having good data on georeferenced genetic variation. These databases would be extremely valuable in efforts to conserve endangered species [26], by helping to detect and reduce illegal trade and informing conservation management plans.

Supplementary Materials: The following are available online at https://www.mdpi.com/article/10.3390/d13060245/s1, Table S1: Locus name and sequences of the eight microsatellite loci used for assignment analysis, Table S2: Descriptive statistics over all loci for each reference locations, positive controls, and unknown individual Military Macaw, Figure S1: Ln(DK) values plotted from 1 to 10 for the genetic group of the (a) Sierra Madre Occidental / Sierra Madre del Sur and (b) Sierra Madre Oriental, Figure S2: Changes in mean Bayesian information criterion (BIC) values in successive K-means clustering.

Author Contributions: Conceptualization, F.A.R.-O.; methodology, F.A.R.-O. and J.J.-E.; software, F.A.R.-O. and J.J.-E.; validation, F.A.R.-O. and S.S.; formal analysis, F.A.R.-O., J.J.-E. and S.S.; investigation, F.A.R.-O., A.M.C.-G. and M.d.C.A.; resources, F.A.R.-O., A.M.C.-G. and S.S.; writing—original draft preparation, F.A.R.-O., J.J.-E. and S.S.; writing—review and editing, F.A.R.-O. and M.d.C.A. All authors have read and agreed to the published version of the manuscript.

Funding: The author(s) received the following financial support for the research, authorship, and/or publication of this article: this study was funded by a CONACYT (National Council of Science and Technology) scholarship for "Complementary Support for Research Groups", granted to F.A.R.-O.

Institutional Review Board Statement: Not applicable.

Data Availability Statement: Not applicable.

Acknowledgments: We thank the Organization for the Conservation, Study and Analysis of Nature A.C. for providing the sample of the individual in captivity. Also, thanks to the CONACYT (National Council of Science and Technology) scholarship for "Complementary Support for Research Groups". Special thanks to Patricia Davila and Ken Oyama for their support. We thank Karl Phillips and second reviewer for his constructive suggestions to improve an earlier version of the manuscript, and Lynna Kiere for the review of the English language style.

Conflicts of Interest: The authors declare no conflict of interest.

References

1. Juniper, T.; Parr, M. *Parrots: A Guide to the Parrots of the World*; Yale University Press: New Haven, CT, USA, 1998.
2. Collar, N.J. Globally threatened parrots; criteria, characteristics and cures. *Int. Zoo. Yearbk.* **2000**, *37*, 21–35. [CrossRef]
3. Berlanga, H.; Gómez de Silva, H.; Vargas-Canales, V.M.; Rodríguez-Contreras, V.; Sánchez-González, L.A.; Ortega-Álvarez, R.; Claderón-Parra, R. *Aves de México: Lista Actualizada de Especies y Nombres Communes*; CONABIO Press: Mexico City, Mexico, 2015.
4. Norma Oficial Mexicana 059 SEMARNAT. Protección ambiental—Especies Nativas de México de Flora y Fauna Silvestres—Categorías de Riesgo y Especificaciones Para su Inclusión, Exclusión o Cambio. Lista de Especies en Riesgo. México. Available online: https://www.gob.mx/profepa/documentos/norma-oficial-mexicana-nom-059-semarnat-2010 (accessed on 13 December 2020).
5. IUCN. Red List of Threatened Species. Available online: https://www.iucnredlist.org/ (accessed on 24 April 2020).
6. Collar, N.J. Family Psittacidae (parrots). In *Handbook of the Birds of the World*; del Hoyo, J., Elliott, A., Sargatal, J., Eds.; Bird Life Internationa: Barcelona, Spain, 1997; pp. 280–477.
7. Macías, C.C.; Iñigo-Elías, E.E.; Enkerlin-Hoeflich, E.C. *Proyecto de Recuperación de Especies Prioritarias: Proyecto Nacional para la Conservación, Manejo y Aprovechamiento Sustentable de los Psitácidos de México*; SEMARNAT Press: Mexico City, Mexico, 2000.
8. Iñigo-Elias, E.; Ramos, M. The psittacine trade in Mexico. In *Neotropical Wildlife Use and Conservation*; Robinson, J.G., Redford, K.H., Eds.; University of Chicago: Chicago, IL, USA, 1991; pp. 380–392.
9. Collar, N.J.; Juniper, A.T. Dimensions and causes of the parrot conservation crisis. In *New World Parrots in Crisis: Solutions from Conservation Biology*; Beissinger, S.R., Snyder, N.F.R., Eds.; Smithsonian Institution: Washington, DC, USA, 1992; pp. 1–24.
10. Ríos-Muñoz, C.A.; Navarro-Sigüenza, A.G. Efectos del cambio de uso de suelo en la disponibilidad hipotética de hábitat para los psitácidos de México. *Ornitol. Neotrop.* **2009**, *20*, 491–509.
11. Monterrubio-Rico, T.C.; de Labra-Hernández, M.A.; Ortega-Rodríguez, J.M.; Cancino-Murillo, R.; Villaseñor-Gómez, F. Distribución actual y potencial de la guacamaya verde en Michoacán, México. *Rev. Mex. Biodivers.* **2011**, *82*, 1311–1319. [CrossRef]
12. Marín-Togo, M.C.; Monterrubio-Rico, T.C.; Renton, K.; Rubio-Rocha, Y.; Macías-Caballero, C.; Ortega-Rodríguez, J.M.; Cancino-Murillo, R. Reduced current distribution of Psittacidae on the Mexican Pacific coast: Potential impacts of habitat loss and capture for trade. *Biodivers. Conserv.* **2012**, *21*, 451–473. [CrossRef]
13. Rivera- Ortíz, F.A.; Oyama, K.; Ríos-Muñoz, C.A.; Solórzano, S.; Navarro-Sigüenza, A.G.; Arizmendi, M.C. Habitat characterization and modeling of the potential distribution of the Military Macaw (*Ara militaris*) in Mexico. *Rev. Mex. Biodivers.* **2013**, *84*, 1200–1215. [CrossRef]
14. Monterrubio-Rico, T.C.; Charre-Medellín, J.F.; Pacheco-Figueroa, C.; Arriaga-Weiss, S.; Valdez-Leal, J.D.; Cancino-Murillo, R.; Escalona-Segura, G.; Bonilla-Ruiz, C.; Rubio-Rocha, Y. Distribución potencial histórica y contemporánea de la familia Psittacidae en México. *Rev. Mex. Biodivers.* **2016**, *87*, 1103–1117. [CrossRef]
15. Monterrubio-Rico, T.C.; Álvarez-Jara, M.; Téllez-García, L. Hábitat de anidación de *Amazona oratrix* (Psittaciformes: Psittacidae) en el Pacífico Central, México. *Rev. Mex. Biodivers.* **2014**, *62*, 1053–1072. [CrossRef]
16. Cantú, J.C.; Sánchez, M.E.; Grosselet, M.; Silva, J. *Tráfico Ilegal de Pericos en México: Una Evaluación Detallada*; Defenders of Wildlife/Teyeliz, Press: Washington, DC, USA, 2007; pp. 3–74.
17. SEMARNAT. Ley General del Equilibrio Ecológico y la Protección al Ambiente. México. Available online: https://biblioteca.semarnat.gob.mx/janium/Documentos/Ciga/agenda/DOFsr/148.pdf (accessed on 24 April 2020).
18. Moya, H.; Peters, E.; Zamorano, P. La importancia de un enfoque regional para la conservación del hábitat natural en la frontera norte de México. In *Temas Sobre la Conservación de Vertebrados Silvestres en México*; Sánchez, O., Zamorano, P., Peters, E., Moya, H., Eds.; SEMARNAT: Mexico City, Mexico, 2011; pp. 1–67.

19. Lynch, M. The genetic interpretation of inbreeding depression and outbreeding depression. *Evolution* **1991**, *45*, 622–629. [CrossRef] [PubMed]
20. Edmands, S. Between a rock and a hard place: Evaluating the relatives risks of inbreeding and outbreeding for conservation and management. *Mol. Ecol.* **2007**, *16*, 463–475. [CrossRef]
21. Fernandes, G.A.; Caparroz, R. DNA sequence analysis to guide the release of blue-and-yellow macaws (*Ara ararauna*, Psittaciformes, Aves) from the illegal trade back into the wild. *Mol. Biol. Rep.* **2013**, *40*, 2757–2762. [CrossRef]
22. Pertoldi, C.; Bijlsma, R.; Loeschke, V. Conservation genetics in a globally changing environment: Present problems, paradoxes and future challenges. *Biodivers. Conserv.* **2007**, *16*, 4147–4163. [CrossRef]
23. Frankham, R.; Ballou, J.D.; Briscoe, D.A. *Introduction to Conservation Genetics*; Cambridge University Press: Cambridge, UK, 2010; pp. 161–181.
24. Allendorf, F.W.; Luikart, G.H.; Aitken, S.N. *Conservation and the Genetics of Populations*; Wiley-Blackwell Press: Oxford, UK, 2012; pp. 52–73.
25. Qiu-Hong, W.; Hua, W.; Tsutomu, F.; Sheng-Guo, F. Which genetic marker for which conservation issue? *Electrophoresis* **2004**, *25*, 2165–2176. [CrossRef]
26. Manel, S.; Berthier, P.; Luikart, G. Detecting Wildlife Poaching: Identifying the Origin of Individuals with Bayesian Assignment Test and Multilocus Genotypes. *Conserv. Biol.* **2002**, *16*, 650–659. [CrossRef]
27. Burns, C.E.; Ciofi, C.; Beheregaray, L.B.; Fritts, T.H.; Gibbs, J.P.; Márquez, C.; Milinkovitch, M.C.; Powell, J.R.; Caccone, A. The origin of captive Galápagos tortoises based on DNA analysis: Implications for the management of natural populations. *Anim. Conserv.* **2003**, *6*, 329–337. [CrossRef]
28. Presti, F.T.; Guedes, N.M.R.; Antas, P.T.Z.; Miyaki, C.Y. Population genetic structure in Hyacinth Macaws (*Anodorhynchus hyacinthinus*) and identification of the probable origin of confiscated individuals. *J. Heredity* **2015**, *106*, 491–502. [CrossRef] [PubMed]
29. Velo-Antón, G.; Godinho, R.; Ayres, C.; Ferrand, N.; Cordero-Rivera, A. Assignment tests applied to relocate individuals of unknown origin in a threatened species, the European pond turtle (*Emys orbicularis*). *Amphibia-Reptilia* **2007**, *28*, 475–484. [CrossRef]
30. Iñigo, E.E. Las Guacamayas verde y escarlata en México. *Biodiversitas* **1999**, *25*, 7–11.
31. Howell, S.N.G.; Webb, S. *A Guide to the Birds of Mexico and Northern Central America*; Oxford University Press: Oxford, UK, 1995; p. 337.
32. Arizmendi, M.C.; Marquez-Valdelamar, L. *Áreas de Importancia para la Conservación de las Aves en México*; CIPAMEX AC Press: Mexico City, Mexico, 2000; pp. 7, 23, 57, 59, 76, 79, 127.
33. CITES. Appendices I, II and II to the Convention on International Trade in Endangered Species of wild Fauna and Flora. USA. Available online: https://www.cites.org/ (accessed on 13 December 2020).
34. Rivera-Ortíz, F.A.; Solórzano, S.; Arizmendi, M.C.; Dávila-Aranda, P.; Oyama, K. Genetic diversity and structure of the Military Macaw (*Ara militaris*) in Mexico: Implications for conservation. *Trop. Conserv. Sci.* **2017**, *10*, 1–12. [CrossRef]
35. Leeton, P.; Christidis, L. Feathers from Museum Bird Skins—A good source of DNA for Phylogenetic Studies. *Condor* **1993**, *95*, 465–466. [CrossRef]
36. Caparroz, R.; Miyaki, C.Y.; Baker, A.J. Characterization of microsatellite loci in Blue-and-Gold Macaw, *Ara ararauna* (Psittaciforme: Aves). *Mol. Ecol. Notes* **2003**, *3*, 441–443. [CrossRef]
37. Russello, M.A.; Calcagnotto, D.; DeSalle, R.; Amato, G. Characterization of microsatellite loci in the endangered St. Vicent Parrot, *Amazona guildingii*. *Mol. Ecol. Notes* **2001**, *1*, 162–164. [CrossRef]
38. Russello, M.A.; Lin, K.; Amato, G.; Caccone, A. Additional microsatellite loci for the endangered St. Vicent Parrot, *Amazona guildingii*. *Conserv. Genet.* **2005**, *6*, 643–645. [CrossRef]
39. Morin, P.A.; Chambers, K.E.; Boesch, C.; Vigilant, L. Quantitative polymerase chain reaction analysis of DNA from noninvasive samples for accurate microsatellite genotyping of wild chimpanzees (*Pantroglodytes verus*). *Mol. Ecol.* **2001**, *10*, 1835–1844. [CrossRef]
40. Garrido-Garduño, T.; Vázquez-Domínguez, E. Métodos de análisis genéticos, espaciales y de conectividad en genética del paisaje. *Rev. Mex. Biodivers.* **2013**, *84*, 1031–1054. [CrossRef]
41. Banks, M.; Eichert, W.; Olsen, J.B. Which genetic loci have greater population assignment power? *Bioinf. Appl. Note* **2003**, *19*, 1436–1438. [CrossRef] [PubMed]
42. Yue, G.H.; Xia, J.H.; Liu, P.; Liu, F.; Sun, F.; Lin, G. Tracing asian aeabass individuals to single fish farms using microsatellites. *PLoS ONE* **2012**, *7*, e52721. [CrossRef]
43. Belkhir, K.; Borsa, P.; Chikhi, L.; Raufaste, N.; Bonhomme, F. GENETIX 4.05, Logiciel Sous Windows TM Pour la Génetique des Populations. Francia. Available online: https://kimura.univ-montp2.fr/genetix/ (accessed on 20 August 2020).
44. Hair, J.F.; Anderson, R.E.; Tatham, R.L.; Black, W.C. *Análisis Multivariante*; Pearson Prentice Hall Press: Madrid, Spain, 1999; pp. 92–98.
45. Pritchard, J.K.; Stephens, M.; Donnelly, P. Inference of population structure using multilocus genotype data. *Genetics* **2000**, *155*, 945–959. [CrossRef] [PubMed]
46. Falush, D.; Stephens, M.; Pritchard, J. Inference of population structure using multilocus genotypes data: Linked loci and correlated allele frequencies. *Genetics* **2003**, *164*, 1567–1587. [CrossRef] [PubMed]

47. Jombart, T. Adegenet: A R package for the multivariate analysis of genetic markers. *Bioinformatics* **2008**, *24*, 1403–1405. [CrossRef]
48. Jakobsson, M.; & Rosenberg, N.A. CLUMPP: A cluster matching and permutation program for dealing with label switching and multimodality in analysis of population structure. *Bioinformatics* **2007**, *23*, 1801–1806. [CrossRef]
49. Rosenberg, N.A. DISTRUCT: A program for the graphical display of population structure. *Mol. Ecol. Notes* **2004**, *4*, 137–138. [CrossRef]
50. *R Core Team R: A Language and Environment for Statistical Computing*; R Foundation for Statistical Computing: Vienna, Austria; Available online: https://www.R-project.org/ (accessed on 3 April 2021).
51. Piry, S.; Alapetite, A.; Cornuet, J.M.; Paetkau, D.; Baudouin, L.; Estoup, A. GENECLASS2: A Software for Genetic Assignment and First-Generation Migrant Detection. *J. Heredity* **2004**, *95*, 536–539. [CrossRef] [PubMed]
52. Paetkau, D.; Calvert, W.; Stirling, I.; Strobeck, C. Microsatellite analysis of population structure in Canadian polar bears. *Mol. Ecol.* **1995**, *4*, 347–354. [CrossRef] [PubMed]
53. Paetkau, D.; Slade, R.; Burden, M.; Estoup, A. Genetic assignment methods for the direct, real-time estimation of migration rate: A simulation-based exploration o accuracy and power. *Mol. Ecol.* **2004**, *13*, 55–65. [CrossRef] [PubMed]
54. Rannala, B.; Mountain, J. Detecting immigration by using multilocus genotypes. *Proc. Natl. Acad. Sci. USA* **1997**, *94*, 9197–9201. [CrossRef] [PubMed]
55. Cornuet, J.M.; Piry, S.; Luikart, G.; Estoup, A.; Solignac, M. New methods employing multilocus genotypes to select or exclude populations as origins of individuals. *Genetics* **1999**, *153*, 1989–2000. [CrossRef]
56. Nei, M. The theory and estimation of genetic distance. In *Genetic Structure of Population*; Morton, N.E., Ed.; University Press: Honolulu, HI, USA, 1973; pp. 45–54.
57. Nei, M.; Tajima, F.; Tateno, Y. Accuracy of estimated phylogenetic trees from molecular data. *J. Mol. Evol.* **1983**, *9*, 153–170. [CrossRef]
58. Ogden, R.; Dawnay, N.; McEwing, R. Wildlife DNA forensics—Bridging the gap between conservation genetics and law enforcement. *Endanger. Species Res.* **2009**, *9*, 179–195. [CrossRef]
59. Eberhard, J.R.; Iñigo-elias, E.; Enkerlin-Hoeflich, E.; Cun, E.P. Phylogeography of the Military Macaw (*Ara militaris*) and the Great Green Macaw (*A. Ambiguus*) Based on MTDNA Sequence Data. *Wilson J. Ornithol.* **2015**, *127*, 661–669. [CrossRef]
60. Hauser, L.; Seamons, T.R.; Dauer, M.; Naish, K.A.; Quinn, T.P. An empirical verification of population assignment methods by marking and parentage data: Hatchery and wild steelhead (*Oncorhynchus mikiss*) in Forks Creek, Washington, USA. *Mol. Ecol.* **2006**, *15*, 3157–3173. [CrossRef]
61. Ogden, R.; Linacre, A. Wildlife forensic science: A review of genetic geographic origin assignment. *Forensic Sci. Int. Genet.* **2015**, *18*, 152–159. [CrossRef] [PubMed]
62. Palomera-García, C.; Santana, E.; Amparán-Salido, R. Patrones de distribución de la avifauna en tres estados del occidente de México. *An. Inst. Biol.* **1994**, *5*, 137–175.
63. Gaucín, R.N. Biología de la Conservación de la Guacamaya Verde (*Ara militaris*) en el Sótano del Barro, Querétaro. Master's Thesis, Universidad Autónoma de Querétaro, Santiago de Querétaro, Mexico, March 2000.
64. Honnen, A.C.; Petersen, B.; Kabler, L.; Elmeros, M.; Ross, A.; Sommer, R.S.; Zachos, F.E. Genetic structure of Eurasian otter (Lutra lutra, Carnivora: Mustelidae) populations from the western Baltic sea region and its implications for the recolonization of north-western Germany. *J. Zool. Syst. Evol. Res.* **2011**, *49*, 169–175. [CrossRef]
65. García-Navas, V.; Ferrer, E.S.; Sanz, J.J.; Ortego, J. The role of immigration and local adaptation on fine-scale genotypic and phenotypic population divergence in a less mobile passerine. *J. Evol. Biol.* **2014**, *27*, 1590–1603. [CrossRef] [PubMed]
66. Mateus, J.C.; Russo-Almeida, P.A. Traceability of 9 Portuguese cattle breed with PDO products in the market using microsatellites. *Food Control* **2015**, *47*, 487–492. [CrossRef]
67. Tonteri, A.; Je, A.; Zubchenko, A.V.; Lumme, J.; Primmer, C.R. Microsatellites reveal clear genetic boundaries among Atlantic salmon (*Salmo salar*) populations from the Barents and White seas, northwest Russia. *Can. J. Fish. Aquat. Sci.* **2009**, *66*, 717–735. [CrossRef]
68. Wasser, K.S.; Clark, W.J.; Drori, O.; Kisamo, E.S.; Mailand, C.; Mutayoba, B.; Stephens, M. Combating the illegal trade in African Elephant Ivory whit DNA forensics. *Conserv. Biol.* **2008**, *22*, 1065–1071. [CrossRef]
69. Millions, D.G.; Swanson, B.J. An application of Manel's model: Detecting Bobcat Poaching in Michigan. *Wildl. Soc. Bull.* **2006**, *34*, 150–155. [CrossRef]
70. Gonclaves-Nazareno, A.; Sedrez dos Reis, M. Where did they come from? Genetic diversity and forensic investigation of the threatened palm species *Butia eriospatha*. *Conserv. Gent.* **2014**, *15*, 441–452. [CrossRef]

MDPI

Review

A Literature Synthesis of Actions to Tackle Illegal Parrot Trade

Ada Sánchez-Mercado [1,2,3,*], José R. Ferrer-Paris [2], Jon Paul Rodríguez [1,4,5] and José L. Tella [6]

[1] Provita, Calle La Joya, Edificio Unidad Técnica del Este, Chacao, Caracas 1060, Venezuela; jprodriguez@provitaonline.org
[2] School of Biological, Earth and Environmental Sciences, University of New South Wales, Kensington, NSW 2052, Australia; j.ferrer@unsw.edu.au
[3] Ciencias Ambientales, Universidad Espíritu Santo, Samborondón 092301, Ecuador
[4] IUCN Species Survival Commission, Caracas, Venezuela
[5] Centro de Ecología, Instituto Venezolano de Investigaciones Científicas (IVIC), Apartado 20632, Caracas 1020-A, Venezuela
[6] Department of Conservation Biology, Doñana Biological Station CSIC, 41092 Sevilla, Spain; tella@ebd.csic.es
* Correspondence: ay.sanchez.mercado@gmail.com; Tel.: +61-48-120-3171

Abstract: The order Psittaciformes is one of the most prevalent groups in the illegal wildlife trade. Efforts to understand this threat have focused on describing the elements of the trade itself: actors, extraction rates, and routes. However, the development of policy-oriented interventions also requires an understanding of how research aims and actions are distributed across the trade chain, regions, and species. We used an action-based approach to review documents published on illegal Psittaciformes trade at a global scale to analyze patterns in research aims and actions. Research increased exponentially in recent decades, recording 165 species from 46 genera, with an over representation of American and Australasian genera. Most of the research provided basic knowledge for the intermediary side of the trade chain. Aims such as the identification of network actors, zoonosis control, and aiding physical detection had numerous but scarcely cited documents (low growth rate), while behavior change had the highest growth rate. The Americas had the highest diversity of research aims, contributing with basic knowledge, implementation, and monitoring across the whole trade chain. Better understanding of the supply side dynamics in local markets, actor typology, and actor interactions are needed. Protecting areas, livelihood incentives, and legal substitutes are actions under-explored in parrots, while behavior change is emerging.

Keywords: illegal wildlife trade; conservation actions; literature review; poaching; wildlife markets

Citation: Sánchez-Mercado, A.; Ferrer-Paris, J.R.; Rodríguez, J.P.; L. Tella, J. A Literature Synthesis of Actions to Tackle Illegal Parrot Trade. *Diversity* **2021**, *13*, 191. https://doi.org/10.3390/d13050191

Academic Editor: Luc Legal

Received: 22 March 2021
Accepted: 27 April 2021
Published: 29 April 2021

1. Introduction

Parrots (order Psittaciformes including parakeets, macaws, cockatoos, and allies) are among the groups of vertebrates with the largest proportion of species involved in the wildlife trade [1]. Parrots are mostly traded to supply the demand for pets and cage birds, and since 1982, the entire order (with the exception of four relatively common species) has been listed in the Appendices of the Convention on International Trade in Endangered Species of Wild Fauna and Flora (CITES) in an attempt to make this trade sustainable and avoid illegal trade [2]. However, illegal trade may run in parallel with CITES-regulated international trade [3], and illegal domestic trade remains substantial in some countries, representing an important threat to parrot populations [4]. Aside from conservation impacts on the harvested species, and despite CITES regulations and international bans, both the legal and illegal trade have contributed to the establishment of alien and invasive populations of parrots worldwide [5,6]. In some instances, these non-native populations may cause ecological, economic, and even human health problems [7] including the potential transmission of zoonotic diseases associated with illegally traded specimens [8,9].

Efforts to summarize heterogeneous and disperse information on the illegal parrot trade including literature reviews and CITES database analyses have focused mostly on documenting the number of individuals and species as well as trade mechanisms and routes involved [10–14]. However, the development of coordinated and effective policies to tackle the illegal parrot trade requires not only understanding the temporal and geographic patterns of the problem itself, but also their proposed solutions. Actions aimed at regulating different levels of the illegal trade chain cover the reduction of harvesting by patrolling to controlling trade by enforcement as well as efforts to reduce the demand [15]. The extent to which these solutions are implemented greatly depends on the financial, capacity building, and legal contexts within source and recipient countries [16]. Recent multifaceted, interdisciplinary approaches have simultaneously reduced extraction and demand within source countries [17]. However, it is not clear whether these policy-oriented initiatives are common or the exception in the practice of tackling illegal parrot trade. Tallying the frequency of actions across the trade chain including an evaluation of the base-line information available related to each action, taking into account regional and temporal contexts, is critical for the development of evidence-based, policy-oriented interventions [18].

In this study, we used an action-based approach to review published research on the illegal parrot trade at the global scale to analyze the distribution of conservation aims and action types among regions and species. We aimed to generate a 'road map' for future research and implementation of anti-trafficking efforts by: (1) understanding how different actions have been conducted in different geographic, temporal, and taxonomic contexts, and (2) identifying existing knowledge gaps and highlighting areas where further research is needed. Furthermore, we discuss how well integrated and consistent actions have been taken at different points in the trade chain across regions and species in order to better inform regional policies.

2. Materials and Methods

2.1. Literature Search Strategy

We conducted a specific and a general literature search on the database Web of Science (WoS). For the specific search, we used terms in English and Spanish: 'illegal wildlife trade', 'extraction', or 'poaching' combined with terms related with the focal taxonomic group ('Psittaciformes', 'Psittacidae', or 'parrot*s') in the themes section. We limited the search until March 2020. This search resulted in a 'WoS dataset1′ with 166 documents. The general search included only search terms in English related with the focal taxonomic group (Psittaci *, parrot *, macaw *, parakeet *, amazon *, cockatoo *). This resulted in a bigger dataset (12.095 documents, 'WoS dataset2′).

We also searched in the web pages of international non-governmental organizations related with the topic (TRAFFIC, WWF, WCS) and in the Mendeley database, in order to include gray literature not represented in WoS (e.g., reports, books, and thesis). This search resulted in the 'gray dataset' with 88 documents.

We combined the three datasets and removed duplicate documents, resulting in a final dataset with 11,948 documents published between 1990 and 2020. We then applied three types of filters. In the first filter, we did an automatic screen of the title, abstract, and authors' keywords looking for eight topic specific words (exotic, extract*, illegal, trade, pet, illicit, market, poach*). We then performed a manual interactive check of the actual keyword phrases to discard false positives or non-informative keyword combinations, and to manually add overlooked publications for some countries or taxa of special interest. After this step, 11,375 documents were discarded as unlikely to have information related to the wildlife trade. In the second filter, we reviewed the title and abstracts, and if necessary, also the full text of the 573 remaining documents, and classified them into three main categories: included in the review (163 documents with original data about illegal parrot trade), not available (four without abstract or for which no document was found), and rejected (406). Rejected documents included those evidently off topic of either parrot or illegal trade (359), opinion articles or overviews (17), or those mentioning illegal trade only circumstantially as a threat to the species (30).

2.2. Document Classification

For the 163 documents included in the review, we reviewed the full text and extracted the information about the countries where the studies were conducted and aggregated them into five main regions following ISO classification: Africa (Eastern, Northern, Southern and Western Africa), the Americas (North America, Latin America, and the Caribbean), Asia, Europe, and Oceania [19]. We also extracted the parrot species reported using the species list of BirdLife International [20] to unify the species scientific names across documents.

We classified each document according to three variables: (1) level of the trade chain addressed (supply, transactional or demand); (2) research contribution level (basic knowledge, implementation, or monitoring); and (3) aims and types of conservation actions implemented (Table 1). Categories within variables were not exclusive, so documents with multifaceted approaches were included in more than one category (Table S1 in Supplementary Materials).

Table 1. Illegal wildlife trade mitigation measures scheme used to classify published research about the illegal parrot trade.

Side	Actions Aims	Action Types	Action Examples
Supply side	Reduce harvesting	Area based	Protected areas, private areas
		Species based	Bans, extractions quotes
		Enforcement	Patrols, surveillance, fences, seizing, prosecutions, extraction bans
		Incentives	Sustainable use, alternative livelihood
		Legal substitutes	Captive breeding, ranching
		Modelling	Population Viability Analysis, CPUE models
Transactional	Aid physical detections	Forensic analysis	Forensic analyses
		Molecular methods	Genetic markers
		Citizen science	Identification and reporting applications
		Locator device	Radio tracking, nano locators
		Certification schemes	Captive breeding certification
	Identify network actors	Trade structure	Social network analysis, actors description, actors identification
		Market dynamic	Open market surveys, internet markets, trade routes, parallel trade, dark web, local market dynamic, import/export dynamic
		Extraction dynamic	Extraction scope, extraction amount, extraction dynamic, extraction methods
		Demand dynamic	Demand scope, demand amount
	Legislation	International	International convention (CITES), international bans
		Domestic	Nation acts, updated legislation
		Consortia collaboration	Consortia and collaborations, stakeholders collaboration
	Zoonosis control	Detection of infectious diseases	Prevalence on wildlife or humans
		Monitoring outbreak	Transmission dynamics, epidemiology
		Effect	Mortality rates, survival rates
	Invasion control	Risk assessment	Driven factors
		Status evaluation	Status of exotic population

Table 1. *Cont.*

Side	Actions Aims	Action Types	Action Examples
Demand	Behavior change	Limits on purchase and possession	Keeping bans
		Social marketing campaigns	Attitudes or perceptions of pet owners, behavior models
		Education	Education campaigns
		Awareness-raising campaigns	Pride campaigns, awareness-raising campaigns

For the level of the trade chain addressed, we classified as 'supply' those documents addressing how poached individuals enter the trade chain including poaching dynamics and motivations. We classified as 'transactional' the documents describing how the product is processed as well as how trade is operated, facilitated, or moderated, involving different intermediaries such as transporters, smugglers, traders, enforcement agents, etc. We also included in this category documents describing trade chain structure and dynamics. Documents describing how and why parrots are purchased were classified as 'demand' [21,22].

We defined three broad categories to describe the research contributions. We classified as 'basic knowledge' those studies focusing on understanding patterns and processes including magnitude and scope of trade, and development of monitoring tools. We classified as 'implementation' those documents describing which and how specific actions (see below) were implemented. We classified as 'monitoring' those evaluating whether the actions implemented helped to tackle illegal trade, usually implying before–after and treatment–control comparisons [21,22].

We adopted the illegal wildlife trade mitigation measures scheme proposed by 't Sas-Rolfes et al. (2019) to define the aims and types of actions that could be implemented to tackle the parrot illegal trade. This scheme classifies aims into the following categories (Table 1): (1) to reduce illegal harvesting (including actions like protecting areas, extraction bans, sustainable use, alternative livelihood approaches, etc.); (2) to aid in the physical detection of illegal products (e.g., forensic, genetic tools, locator devices); (3) to identify wider networks of actors and address the enabling environment for illegal wildlife trade (e.g., local and international market dynamic, extraction scope, extraction amount, extraction dynamic, etc.); (4) to regulate trade with high-level measures and national legislation (e.g., CITES, national acts) as well as the establishment of conservation initiatives, consortia, and specialist groups (e.g., Parrot Researchers Group); (5) to evaluate or control impact on biodiversity (e.g., zoonosis, invasive species); and (6) to reduce demand by behavior change either with coercive measures (e.g., imposing limits on purchase and possession) or encourage behavior change using awareness, education, or social marketing campaigns (Table 1).

We used a PostgreSQL database and customized PHP and R clients to manage all steps of filtering, data curation, and annotation. Source codes are available in a public repository.

2.3. Data Analysis

We evaluated temporal patterns in illegal parrot trade publications by aggregating the number of published documents by year. Additionally, we calculated changes in the mean of document citations by action aim and by year [23]. We excluded the last two years to reduce the number of zeros in the sample. We fitted a Poisson mixed model, where the expected response is given by:

$$\log(E(y \mid u)) = \alpha + (\beta + b)year + u\ (year \mid action) \quad (1)$$

where $(E(y \mid u))$ is the expected response conditional on u; α is the fixed intercept, and β is the fixed slope; u and b are the random intercepts and slopes (respectively) that are normally distributed with mean zero; *year* is the publication year, and *action* is the study aim as described in Table 1. In such a model, aims with positive random intercepts can be interpreted as reaching higher than average cited articles in the period. Similarly, aims with positive random slopes have higher-than-average growth rate (i.e., larger change in cited publications during the same period) [23].

To visualize taxonomic patterns, we aggregated the number of documents by genera, aim, and region and represented these relationships with a bar plot. We used the taxonomic list of BirdLife [24] to aggregate the species reported in their respective genera. We also used the IUCN conservation status categories reported by BirdLife to describe the distribution of conservation status of the species by region.

To visualize geographical patterns in the aims reported, we followed a double approach. We first created an incidence matrix by region where columns were the three variables assessed (trade chain level, research contribution, and aims and action type) and rows were the combinations of levels for each variable. We used Sankey diagrams to represent the distribution of combinations in our multivariate dataset. In Sankey diagrams, variables are assigned to vertical axes that are parallel. Levels for each variable are represented by blocks with its size proportional to the frequency of observations. Flow lines join co-occurring categories in adjacent levels, and flow widths are proportional to their frequency. Some combinations were not represented in our dataset (i.e., flow = 0).

Finally, we created a country scientific collaboration network with author affiliation countries as nodes and the number of co-authorships among countries as links. All affiliations of a given author were considered. Node attributes included the number of publications and research contribution level. To visualize the relationship between study location and authorship at the country level, we overlapped the proportion of research developed in a given country and the proportion of author affiliations for the same country and represented them in a map.

Analyses were performed using the packages *alluvial*, *lme4*, and *igraph* of R [25,26].

3. Results

3.1. Temporal Patterns

The number of publications related to the illegal parrot trade showed a sharp increase after 2000, with a mean publication rate of 1 ± 1.66 publication/year between 1990–2000, which increased sharply after 2001 (7.39 ± 0.38; Figure 2a).

Temporal patterns in aims suggest that documents about behavior change had the highest number of citations (rate of change in cited publications; Figure 1b) even though it only accounts for four published documents. Identification of network actors (143 documents), harvesting reduction (18), and aid physical detection (36) had low growth rates, with numerous but low cited documents. Actions aimed to control invasion (four documents) and zoonosis diseases (51 documents) had the lowest rate of change in the cited publications, even though research about zoonosis control was the second aim with the highest number of publications (Figure 1b).

3.2. Taxonomic Patterns

We found 165 from 46 genera reported in the illegal parrot trade literature. The top 10 reported species were Psittacus erithacus, Amazona aestiva, Ara ararauna, Ara macao, Anodorhynchus hyacinthinus, Myiopsitta monachus, Aratinga solstitialis, Amazona ochrocephala, Amazona finschi, Amazona auropalliata, and Amazona farinosa.

Research focused on an average of 4.2 ± 0.6 species per document, although 58% of published documents reported only one species (median = 1, range 0–44; Figure 2a). Traded species recorded in the published literature were evenly distributed across conservation status categories in all regions, but in the Americas, where Less Concern and Near Threatened species reached larger percentages (Figure 2b).

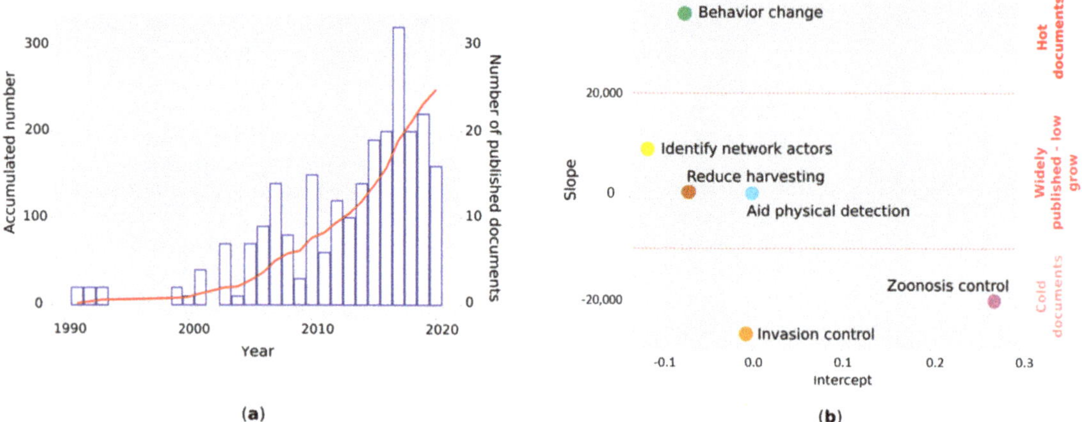

Figure 1. Temporal pattern in the published illegal parrot trade literature. (**a**) Published production across the years. The number of published documents by year (blue bars) and the accumulated number (red line) are shown. (**b**) Temporal pattern in action aims reported in the published literature. Hot, medium, and cold documents represent coarse groupings defined for example purposes only, and should not be considered as statistically robust.

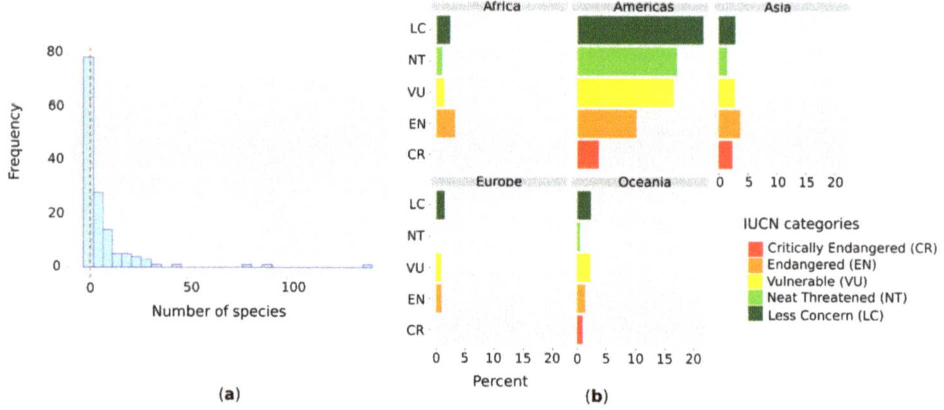

Figure 2. (**a**) Number species reported in published documents on illegal parrot trade. Median value shown as a dotted red line. (**b**) Percentage of species reported as illegally traded for each IUCN conservation status category by region.

Almost half of the genera reported (46%) are under-represented in the illegal parrot trade literature with none or only one document published (Figure 3). In general, identifying network actors was the most frequent aim reported across all species (Figure 3), but *Amazona* and *Ara* were the genera with higher diversity in aims, with research on aiding physical detection, identifying network actors, harvesting reduction, and zoonosis control. *Amazona* was the only genus for which research aimed to reduce demand through behavior change has been reported (Figure 3). The second pair of well-studied genera were *Brotogeris* and *Cacatua*, with research on identifying network actors, aiding physical detection, and reducing harvesting.

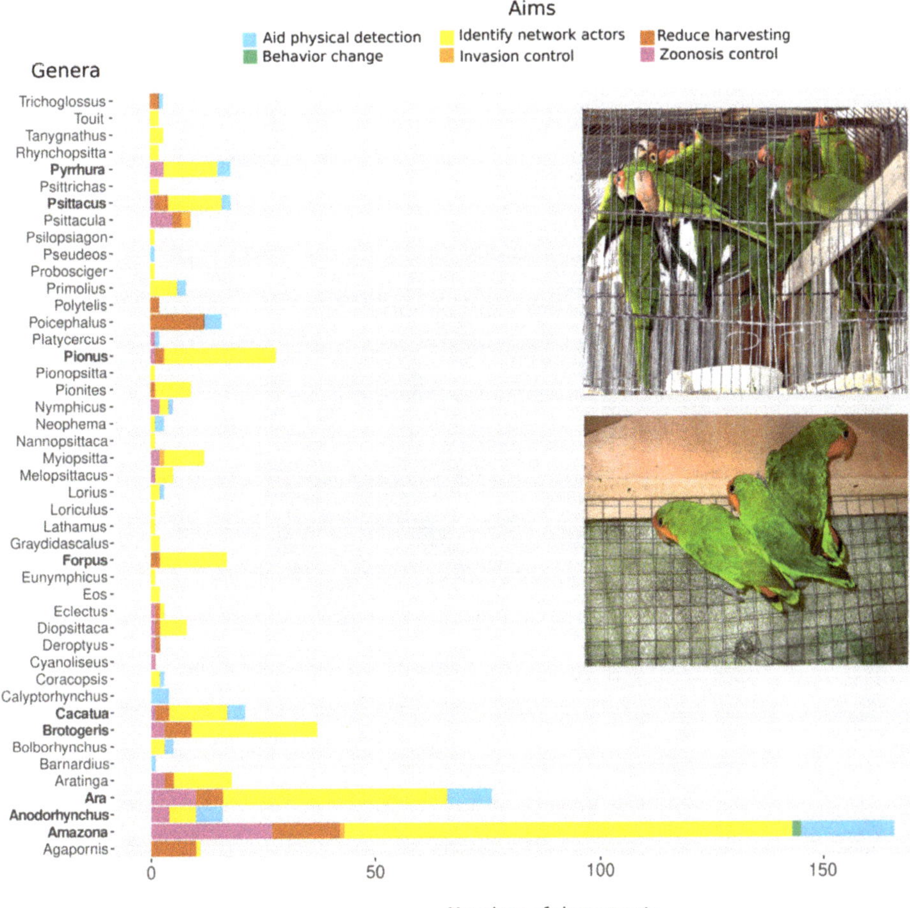

Figure 3. Taxonomic pattern in conservation aims in the published illegal parrot trade literature. The number of documents by genus and aim are shown. The top 10 most studied genera (with more than 10 documents published) are in bold. Genera are in alphabetical order from bottom to top. The former genus *Aratinga*, as reported in the literature, currently comprises four different genera (*Aratinga*, *Eupsittula*, *Psittacara*, and *Thectocercus*). Insert: red-masked parakeets (*Psittacara erythrogenys*, top) and red-faced lovebirds (*Agapornis pullarius*, bottom) involved in the domestic and international illegal trade in Peru and Senegal, respectively (Pictures: José L. Tella).

3.3. Geographic Patterns

The Americas was the region with the highest number of documents regarding the illegal parrot trade: 129 documents from 22 countries, with Brazil, Bolivia, and Peru holding the most frequent study locations. Asia was the second best represented region with 52 documents from 18 countries, with Indonesia, India, Japan, and Singapore as the most frequent study locations. We recorded 34 documents from 14 African countries, mainly from South Africa, Guinea, Mali, and Congo. We only recorded six and five documents for four European and Oceania countries, respectively. The Netherlands and Australia were the most frequently reported study locations in those regions (Figure 4).

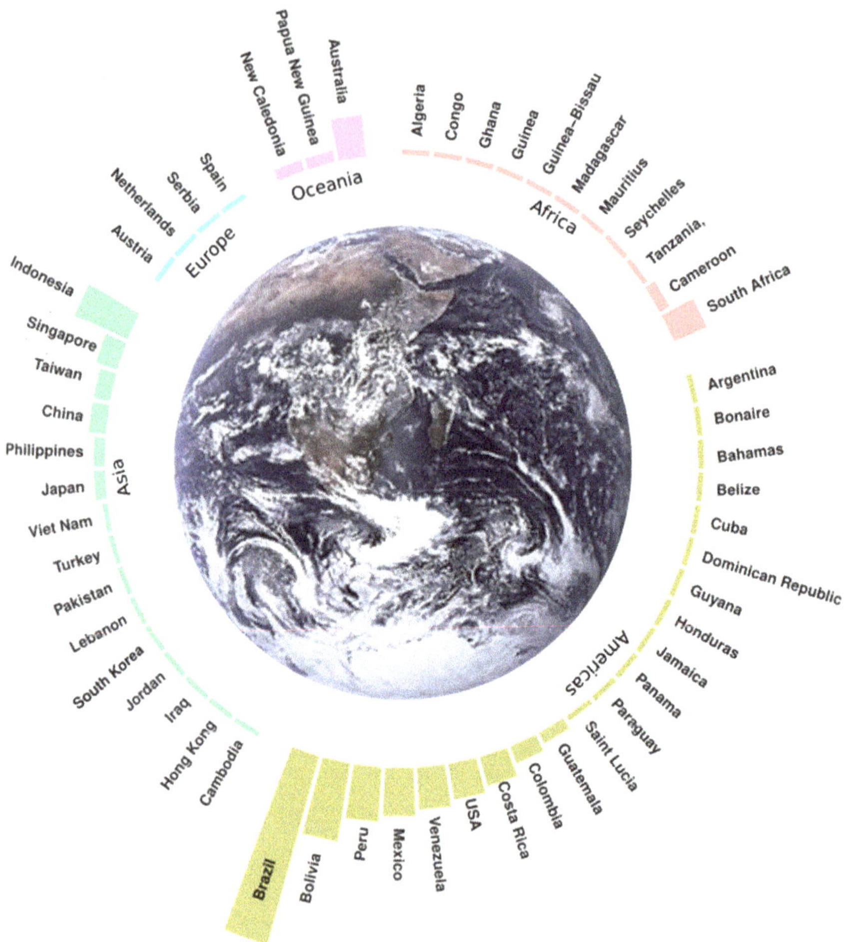

Figure 4. Geographical representation of published literature on the illegal parrot trade. Countries where studies were located are grouped by regions and ordered by number of documents recorded.

In general, 90% of research focused on the transactional side of the trade chain, while the supply and demand side research only represented 8% and 2% of the published research, respectively. Most of the research (86%) provided basic knowledge, while 6–7% contributed with monitoring and implementation. About half of the research (55%) focused on identifying network actors, followed by zoonosis control (20%), aid physical detection (14%), and harvesting reduction (7%). Both invasion control and behavior change represented 4% of the published research. We only detected one document aimed to evaluate the local legislation to tackle the illegal parrot trade. There were important regional variations of this general pattern (Figure 5). At the contribution level, basic knowledge was the only research contribution detected in Europe (Figure 5d), while in the Americas, Asia, Africa, and Oceania regions, we also detected examples of implementation and monitoring (Figure 5a–c,e). At the trade chain level, the Americas was the only region with research on all sides of the trade chain (Figure 5b), while the supply side research was also present in Asia and Oceania (Figure 5c,e). Finally, at the aims level, the Americas was the only region with a research focus on behavior change (Figure 5b), while works about harvesting reduction were lacking in Africa (Figure 5a) and Europe (Figure 5d). Research about invasion control was detected in Africa, Asia, and Oceania (Figure 5a).

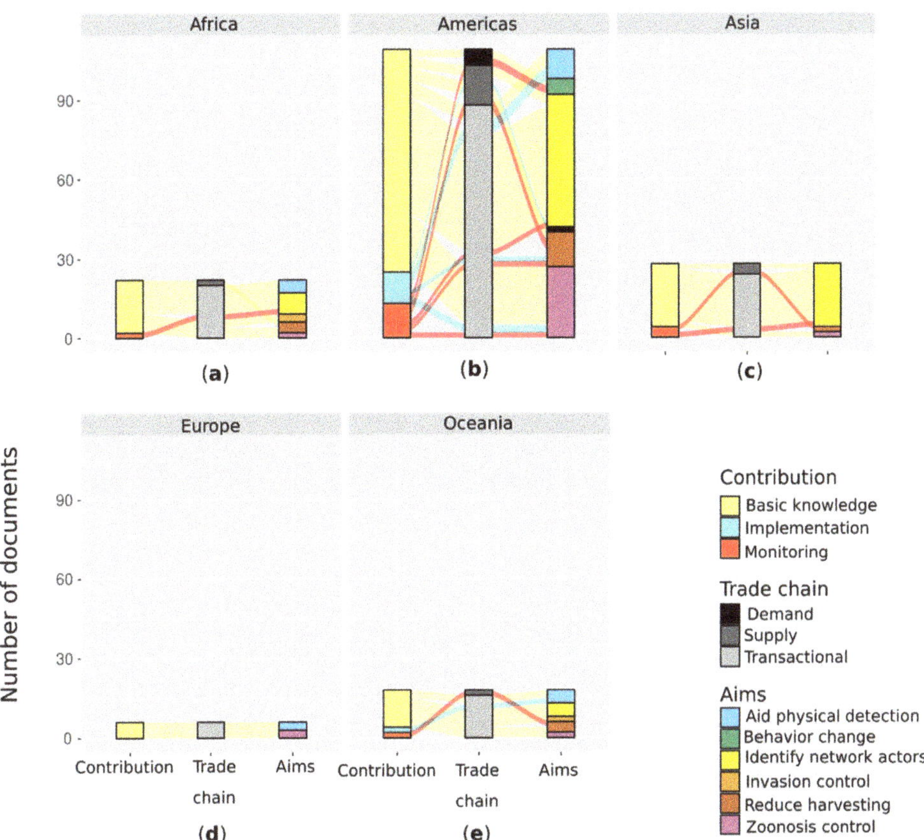

Figure 5. Trends in the illegal parrot trade literature in (**a**) Africa, (**b**) Americas, (**c**) Asia, (**d**) Europe, and (**e**) Oceania, showing the combination of research contribution, trade chain focus, and action aims. Each column represents the variables analyzed about research contribution, trade chain, and actions. Column length is proportional to the number of documents classified under each variable category. Flows across columns are proportional to the frequencies of variable combinations. Color flow traces the research contribution level of basic knowledge (beige), action implementation (cyan), and monitoring (red).

Across regions, the Americas had the highest diversity in research, with five aims (Figure 5b) and 11 action types (Figure 6b). The most prevalent aim was the identification of network actors, mainly through basic knowledge (Figure 5b) on market, extraction, and demand dynamics (Figure 6b). Zoonosis control and aid physical detection were the second most prevalent aims, the former contributing with basic knowledge and implementation to detect infectious diseases, and the later in genetic methods (Figures 5b and 6b). Reducing harvesting was the third most frequent action aim with contributions in knowledge, implementation, and monitoring (Figure 5b) of species-based and enforcement measures (Figure 6b). Behavior change was far less prevalent, but with documented examples of monitoring (Figures 5b and 6b).

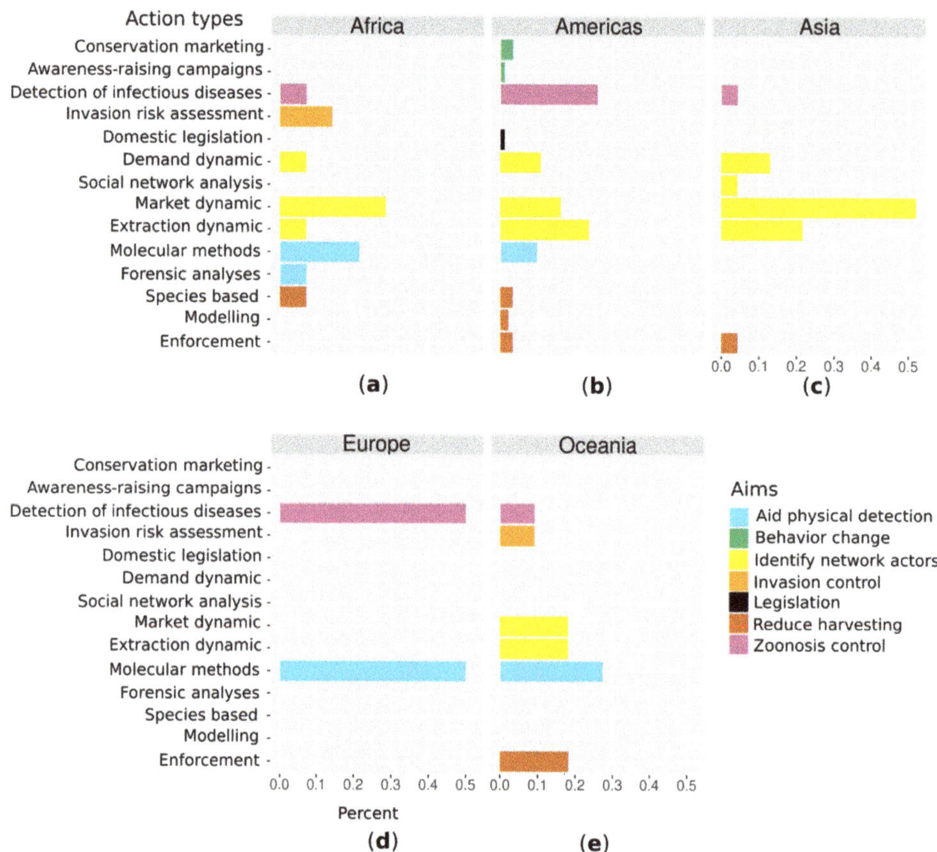

Figure 6. Action types used to tackle illegal parrot trade in in (**a**) Africa, (**b**) Americas, (**c**) Asia, (**d**) Europe, and (**e**) Oceania. Percentages of documents reporting each action type are shown. Actions are grouped by aims.

Asia was the second region in research diversity with four aims (Figure 5c) and nine action types (Figure 6c). Again, the identification of network actors was the main aim, mostly contributing with basic knowledge, but also with the monitoring of markets, extraction, and demand market dynamics, and to a lesser extent, with network analysis (Figure 6c). Reducing harvesting was the second aim recorded, notably contributing with monitoring (Figure 6c) at the supply level in enforcement and species-based measures (Figure 6c). Zoonosis control was also an aim in the anti-trafficking efforts recorded for Asia, with basic knowledge and implementation efforts (Figure 6c) for the detection of infectious diseases (Figure 6c).

Africa was in the third position of research diversity with four aims (Figure 5a) and eight action types (Figure 6a), notably, contributing with monitoring in market dynamic and with basic knowledge for invasion risk assessment. Research in Oceania was characterized by providing basic knowledge in four aims and monitoring experience for reducing harvesting (Figures 5e and 6e).

3.4. Research Collaboration

Global authorship in the illegal parrot trade seems to be highly collaborative, with most of the research authored by researchers affiliated to institutions in the same country. The countries with the best balance between number of studies and authorship from the same country were Brazil, Australia, and China (Figure 7).

Figure 7. Geographic distribution of the study locations and authorship on the illegal parrot trade. Red circles indicate country-level illegal parrot trade research and blue circles indicate country-level author affiliations, with purple circles where both overlap. Circle sizes are proportional to the maximum value in each dataset (logarithm). Wider blue rings indicate disproportionately higher number of researchers than research specific to that country (e.g., the United Kingdom and Canada), whereas wider orange rings (e.g., Bolivia, Peru) indicate the opposite. Purple circles with no external rings indicate a proportionally similar number of studies and authors from a given country (e.g., Brazil, Australia, and China).

Authors affiliated with institutions in the UK, USA, and Spain were more prevalent, but their contribution focused on other countries, generating three predominant collaboration nodes (Figure 8). The first was the American group dominated by authors affiliated with institutions in the USA, collaborating mainly with authors in South America. The UK group, dominated by authors from the UK, collaborated with African, Asian, and European institutions. The connection between the American and the UK groups was low (Figure 8). The third was the collaboration node formed by Spain–Argentina–Colombia–Paraguay (Figure 8). We additionally detected two isolated nodes, one formed by Mexico–Cuba–Ecuador and another by Netherlands–Italy–Malaysia–Singapore (Figure 8).

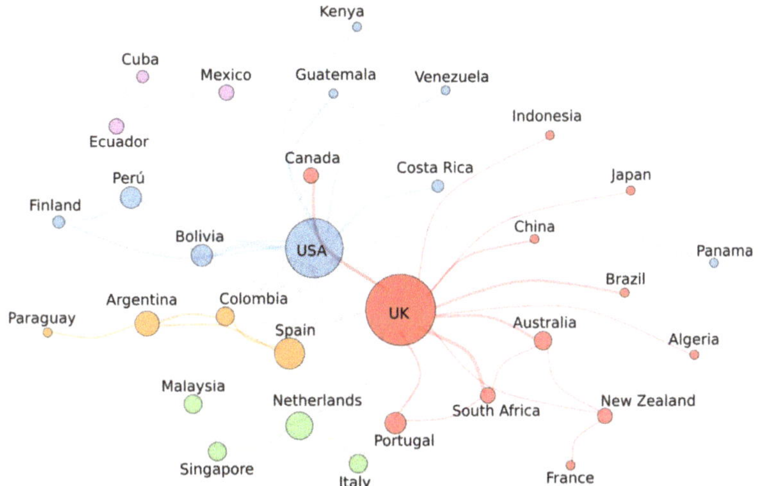

Figure 8. International collaboration network recorded in the illegal parrot trade literature. Circle size is proportional to the number of authors with a given country affiliation. Only the 40 most frequent country affiliations are shown. Collaboration nodes are represented by different colors.

4. Discussion

Globally, birds are the group with more species facing the illegal wildlife trade among all vertebrates (45% of their species) and estimates of future trade suggest the addition of 230–1475 bird species [1]. Aside from this alarming prevalence, the fact that the wildlife trade has caused a 62% decline in species [27] calls for a strategic plan to combat this threat with policies that are proactive rather than reactive [1]. Developing such a strategic plan requires tracing the actions implemented, understanding how well integrated and consistent these actions are regarding to local market dynamics, and evaluating their effectiveness. Our review takes a step forward to build this plan for parrots, one of the most traded bird orders, by providing the first literature synthesis of the illegal parrot trade using an action-based approach. This approach not only describes the current geographic, temporal, and taxonomic pattern of the conservation aims and actions taken, but also allows us to visualize how articulated the actions and market patterns are. Furthermore, our action-based approach allows us to identify strategies used to tackle illegal trade in other taxa (e.g., rhinos, elephants, other bird groups) that have been little or not recorded in the illegal parrot trade literature. Although we did not evaluate the effectiveness of the implemented actions, this baseline will support the future development of quantitative meta-analyses estimating action-driven recovery to inform the much needed implementation and monitoring interventions to reduce the impact of illegal trade on parrot populations.

4.1. Relevant Topics: Extraction Dynamics

Illegal parrot trade research has largely focused on identifying actor networks: this aim represented half of the published literature and was the most prevalent aim across regions and genera (Figures 5 and 6). The two most relevant topics were the scope of the traded product (extraction dynamic, Figure 6) and the scale of market operations in terms of source-destination countries and trade routes (market dynamic, Figure 6).

Beyond how much and which species are traded, there is an active discussion of whether the scope of the traded product (live wild-caught parrots) is opportunistic, with more abundant and available species facing higher extractions, or is selective, thus focusing on particular species [28]. Disentangling these hypotheses requires testing whether species are poached proportionally or not to their abundance in the wild, and both the opportunistic [29] and selective poaching [30] of parrots were supported when using rough proxies of their abundance in the wild. However, Romero-Vidal et al. [27] recently demonstrated, by simultaneously measuring the relative abundance of parrot species in the wild and as poached pets, that those species preferred as pets (due to their coloration, size, and ability to imitate human speech) were selectively poached. The over-exploitation of selected species, rather than the opportunistic harvesting of the commonest species, increases the concerns on the impact of poaching and the illegal trade and the challenges of conservation actions aimed to halt it [31].

Thus far, actions used to reduce the harvesting of wild-caught parrots has been more diverse in the Americas, where examples of species-based actions like quota systems [32], and local enforcement measures like seizure [33] and nest protection [34–37] have been implemented and monitored (Figure 6). The widespread use of enforcement measures in the Americas agrees with the perception among researchers and practitioners in the region that enforcement is the most efficient measure to combat the illegal bird trade [38]. Examples of prosecution [39] and nest protection in Oceania, Asia [40,41], and Africa [42] have been less frequent.

Alternative actions used in other illegally traded species such as protecting areas [43,44], livelihood incentives, and using legal substitutes [45,46] are scarcely recorded in the illegal parrot trade literature. The impact of extraction for trade in vertebrates in general is significantly lower in protected areas than in unprotected ones, meaning that successful conservation of many traded species is intertwined with improved integrity of protected areas and the maintenance of true wilderness [27]. That the role of area protection in preventing the illegal parrot trade has been little evaluated is of particular concern given that the distribution of several threatened parrot species facing trade occurs into protected areas [24]. The impacts of protection against the nest poaching of parrots have been evaluated in different countries and continents [10,40]. However, as authors have used a wide definition of protection, covering nest-site protection to national bans, tribal laws banning exploitation and reserve designation, it is difficult to disentangle the contribution of the different protection actions. Nonetheless, the fact that the numbers of Lear's macaws (*Anodorhynchus leari*) annually seized by the authorities have significantly decreased after protecting their main nesting areas suggests a positive effect of area protection, at least for an extremely range-restricted species [47].

Another 'missing' action in the illegal parrot trade literature is the use of ecotourism incentives for local communities aimed to reduce poaching. Interestingly, examples in other taxa of successful ecotourism incentives mainly depend on protected areas [48,49]. For parrots, ecotourism initiatives have been used to increase general public awareness toward parrot conservation problems and as a source of funding to support research [50], but not as a way to generate direct payments to reduce illegal hunting and trade [49].

Although the role of captive breeding operations in providing legal substitutes to cover parrot demand for the pet market has been mentioned [31,51,52], we did not detect in-depth analyses of the real capacity and scope of the current captive breeding facilities to cover the current parrot demand, or an evaluation of the legal and illegal trade relationship (but see [32]). An in-depth and quantitative analysis of the scope, size, and extent of captive breeding and their role in the legal and illegal trade of parrot species across regions could help to understand the opportunities and limitations of market-driven conservation approaches [45,47].

4.2. Relevant Topics: Market Dynamics

The second topic largely discussed in the illegal parrot trade literature was market dynamics including the actors involved and the scale at which market operations occur (Figure 6). In general, research has focused on describing the elements comprising the market itself: actors involved, extraction rates, routes, and market value. Less attention has been paid to understanding their dynamics, or how changes in socio-economic contexts or conservation interventions affect them. An exception exists, however, in Africa [53], where long-term monitoring of transactions of the Grey and Timneh parrots (*Psittacus erithacus* and *P. timneh*) have been developed.

Nevertheless, the accumulated knowledge of illegal parrot trade markets allows us to draw a bigger picture of the different dynamics occurring across regions. In the Americas, for example, the current illegal parrot trade is largely driven by local markets with small-scale activity [54]. The trade network seems to be composed of widespread but not organized intermediaries, working independently [29,54]. Moreover, in Colombia and some areas of Ecuador and Venezuela, most parrots are poached locally to satisfy the demand of household pets without entering markets [28,31,55,56], a fact that could be extended across the Americas. Further research is thus needed to estimate the actual volumes of poached parrots, which may be much higher than those estimated when only surveying illicit markets [57]. Moreover, the increasing professionalization of criminal groups in wildlife trafficking [58] in the region may be creating new markets and routes [59].

African and Asian markets are less documented than the American ones [11], but insights from the most traded African species, *P. erithacus* and *P. timneh*, show complex markets with shifting geographical patterns of imports, exports, and re-exports of wild-sourced and captive-bred birds across time [3]. In contrast to American markets, the role of criminal actors exploiting the legal trade in parrots to traffic threatened and protected species in international markets is more evident in African and Asian contexts [60]. Trade of wild-caught parrots at local African markets seems to be extremely low and largely opportunistic [11,40,61].

Oceanian markets, dominated by Australian research, provide a very interesting and contrasting scenario: local and international illegal trade of native Australian parrots is insignificant, and 89% of demand for Australian parrot species is supplied by overseas captive breeding populations [51,62,63]. Effective national trade bans and successful captive breeding programs have been proposed as the main explanation for this achievement [62].

At any case, the legal or illegal nature of parrot market dynamics could be affected by how the trade in parrots is perceived in different countries, which may differ substantially across regions [38]. For example, in South America, enforcement staff perceive that wildlife trade is a minor offense, and frequently release minor offenders without issuing any further notification or providing basic information about the incident (e.g., species used, number of specimens, locality, date, etc.) to administrative officers [64]. Similarly, difficulties associated with law enforcement, monitoring, and discerning between legal and illegal trade have been identified in other regions as critical issues in wildlife trade [38]. Legal wildlife trade remains largely unexplored despite its scale, with 34% of the trade described with broad code descriptions and without detailed taxonomic information, despite encompassing thousands of species [65]. Clearer documentation of the quantity and identity of imports, together with more funding, personnel, and training in species identification, would improve the staff's ability to detect irregularities [65].

4.3. *Actions across Regions: Facts, Gaps, and Opportunities*

Regional differences in the market dynamics of the illegal parrot trade highlight the need for regional tailored actions. For example, actions focused on reducing extraction (e.g., nesting site surveillance, seizures, prosecution) and reducing demand of wild-caught parrots through behavior change campaigns could be best suited to tackle the prevalent local markets in the Americas.

The behavior change approach to reduce the local demand of threatened parrots is an emerging topic (Figure 1b), with the Americas the only region on which this topic has been developed (Figure 5b). Few but highly cited studies provide baseline knowledge about people's attitudes and motivations to keep parrots as pets [56,66–68], and examples of implementation and monitoring of social marketing campaigns to reduce demand and poaching of the threatened *Amazona barbadensis* in Bonaire [17,69]. Given the cultural nature of parrot ownership [68,69], there is an increasing need for more in-depth and culturally sensitive research to inform and develop interventions targeted at changing consumer preferences and purchasing behaviors [70–72]. While identifying the attitudes and motivations of consumers is a relevant first step, further efforts should include the use of behavior models such as the theory of planned behavior [73]. Behavior models allow for the identification and prioritization of the underlying factors influencing the behavior to be changed (e.g., attitudes, social norms, perceived control; [74]), and for this information to be used to develop effective interventions targeted at the key actors identified [55].

For African, Asian, and Oceanian markets where the risk of laundering illegally caught parrots into the legal trade is higher, reliable and effective methods to identify species and their origin could help to distinguish between legal and illegal trade, and whether the specimen comes from a threatened population [75–77]. Genetic methods to accurately identify species, kinship, and geographic origin of illegally traded parrots have been developed for several *Amazona*, *Anodorhynchus*, *Cacatua*, and *Ara* species in the Americas and Europe [78–82], and for *Poicephalus* [83,84] and *Psittacus* [85,86] in Africa. Encouragingly, beyond basic knowledge generation, these genetics tools have been tested in Australia [87–89], Brazil [78], and Colombia [90] (Figures 5 and 6). However, this implementation experience, and worryingly, even baseline knowledge seems to be absent in Asia (Figure 3, Figure 5, and Figure 6), where countries such as Singapore are well known important trans-shipment hubs where wild-caught parrots are laundered as captive bred to fuel the pet trade market [91].

Development of tools for identifying the geographic origin of a specimen in a forensic context remains in its early stages for most species [77,92], and parrots are not the exception. Low availability of parrot genomes, and the lack of reference databases, especially for rare species or species with distribution ranges located in remote areas [76,77], help explain why genetic and forensic methods to aid physical detection have been developed for only a few species (Figure 4), and why the popularity of this topic has decreased across time (Figure 1b). However, recent developments in genomics tools and stable isotope analyses for African grey parrots [85] could provide innovative solutions to cross the bridge across research–implementation, allowing a wider implementation of forensic tools to tackle the illegal parrot trade [59].

Nonetheless, the illegal parrot trade could benefit from diversification in the methods used to aid in the physical detection of traded parrots. Passive integrated transponder devices (PIT tags) and closed bands [93] have been used by CITES to verify that an animal is captive bred, as opposed to wild caught, as a mechanism for monitoring illegal harvest of animals in international trade [94]. Additionally, multidisciplinary approaches using machine learning and citizen science have been proposed to monitor the illegal trade in social media [95].

Beyond species and trade markets, generating a comprehensive picture of the illegal parrot trade requires linking this information across actors in the trade chain and evaluating the economic and social factors shaping the actors' decisions [96–98]. That is, it requires an understanding of the network structure, which is poorly known for the illegal parrot trade across all regions (Figures 5 and 6). We detected only one study aimed at evaluating changes in the network structure in Indonesia, where parrot keeping has shifted from an older person's hobby to increasingly involving younger people [99]. Besides general descriptions about poaching methods and smuggling routes [52,100], there is not a nuanced description of actor typology [21] involved in the parrot trade, their roles, interactions, levels of economic reliance, and knowledge. Social network analysis has already been used to identify key countries that play crucial roles in the illegal trade network of African parrots [3,98]. A wider application of this approach could help to improve our understanding of the interaction among actors and products, which in turn could help to identify opportunities for conservation intervention tailored to the specific actor group [101–104]. Recent research in the Red Siskin (*Spinus cucullatus*), a globally Endangered finch threatened by illegal trade, combined tools from social network analysis, interviews, social media monitoring, and the literature to describe the trade network for this species [101], which could be applied to the illegal parrot trade.

Finally, the link between the illegal parrot trade and transmission of zoonotic diseases has been largely explored (20% of the documents in the illegal parrot trade literature) in the Americas (mainly Brazil) and in Europe. Basic knowledge about the prevalence of Newcastle disease, *Chlamydiophila psittaci*, avian influenza virus, new beak and feather disease, and Psittacine Herpesvirus have been reported for the Americas [105–107] and Europe [108,109]. Additionally, outbreaks affecting wildlife have been reported for the Americas [110–112] and Oceania [113,114], while those affecting humans have been only recorded in the Americas [9,115]. In Africa, only studies about the prevalence of beak and feather disease were detected [116]. In contrast, the relationship between illegal trade and the establishment of non-native populations has been less studied. While there is clear evidence for the role of the international legal trade on the establishment of several parrot species out of their native ranges [116,117], we only detected one study that clearly related illegal trade events with the establishment of *Psittacula krameri* populations in South Africa [117]. Nonetheless, non-native populations in the Americas are probably related to the domestic illegal trade, but have been scarcely reported [6], a fact that merits further research.

4.4. Biases and Pitfalls

Our literature review had geographic, linguistic, and temporal sampling biases, which could have affected the results in two main ways: (1) causing an underestimate of the magnitude of research, and (2) detecting a smaller diversity of aims and actions than what actually exists. The fact that our search strategy used only terms in English and Spanish likely under detected the published literature in Asian languages. The Asian documents recorded were published in collaboration with the UK and Netherlands-based institutions (Figure 8), but the high prevalence of local researchers (red halo in Figure 7) suggests that part of the research could be under detected because it is published in local languages. Overcoming the under detection of documents published in local languages is important because they could be those making the greatest impact on policy change and the implementation of conservation actions at local contexts. Clearly, greater monitoring effort, using a wider battery of languages including French, Chinese, Bahasa Indonesia, and Bahasa Malaya, would be necessary to better understand trade in Africa and Asia, which appears to be influencing demand for wildlife in the Americas, creating new markets and routes [59], and emerging as an important transit point for the illegal trade of wild-caught Grey parrots [53].

Detectability of the aims and action types was also likely reduced by the incompleteness of sources. Additional implementation and monitoring research results are likely hidden in the gray literature (i.e., reports and theses), which was under-represented in WoS. Although we were able to include reports from international NGOs working on the topics, gray literature represented only 5% of the analyzed documents. However, this high emphasis in the generation of basic knowledge but lower effort in implementation or monitoring, agrees with description of the knowledge–implementation gap observed in other conservation topics [18,118,119].

Besides detectability biases related to the limitations of our searching strategy, we also identified intrinsic geographic and taxonomic biases related to the dynamics of illegal parrot trade research. Although our review is representative of parrot species occurring globally (37% of species included in CITES; [2]), the over representation of a handful of them (Figure 3), mostly genera with American (*Amazona*, *Anodorhynchus*, *Ara*, *Aratinga*) and Australasian (*Cacatua*) distribution, suggests a taxonomic bias. As expected, for many rare, range restricted species, there are few studies, and even fewer implementation and monitoring examples, while the most conspicuous species with large distributions might be over-represented in the analysis likely because they are easier to detect. This pattern may represent a combination of: (1) a higher diversity of American parrots compared to other regions (233 spp in the Americas *versus* 128 spp in Asia and 129 in Oceania; [120]), (2) higher scientific capacity in the Americas both in terms of number of countries with research in the topic (39%) and number of documents published (65%; Figure 4), and (3) preferences toward highly attractive species for both consumers [30] and researchers [121]. In any case, the threat status seems to vary across regions, with the Americas showing an over representation of less threatened species, while the others have focused on more threatened ones (Figure 2b). In any case, the focus on American attractive parrots observed in the illegal parrot trade research agrees with those observed in mammals, for whom the scientific capacity of the countries where a species occurs is a strong driver of conservation research bias [122].

Filling the gap in information about the illegal trade for rare and endemic parrot species is clearly an important issue in order to obtain a comprehensive dataset. Supporting research in countries with low scientific capacity and high biodiversity in close collaboration between practitioners and academics could be an important first step [123].

5. Conclusions

The illegal parrot trade research has been largely collaborative and interdisciplinary, incorporating concepts and methods from criminology, veterinary, human sciences, and genetics, but most of those tools have focused on describing the trade process itself. Description of the component of trade is, however, only the first step to understand the illegal parrot trade and identify timely and effective actions. Our review shows that the illegal parrot trade research has compiled enough information to build a sketch of trade patterns. However, there are increasing calls to adopt multifaceted approaches that go beyond description, can integrate the information available, and build a comprehensive picture of trade networks including addressing the drivers of illegal trade by acknowledging market conditions, consumer preferences, and the socioeconomic needs of communities at the local level [96,124].

Additional efforts are required to improve the actor typology and how they interact as well as how products and money fluxes into the network vary responding to socio-economic and conservation contexts. The predominant local market dynamics highlight that more effort is needed to improve our knowledge at the supply side of the trade chain including measuring the current volume of poached parrots instead of traded ones.

This review represents a baseline compilation of information about the aims and actions for tackling the illegal parrot trade at a global scale, allowing for the identification of alternative actions in other illegally traded species that have not yet been properly explored in the parrot trade literature. Protecting areas, livelihood incentives, and legal substitutes have proven to be effective in reducing poaching and harvesting in other species and are worthy of exploring in parrots. In addition, the use of tools and concepts from the social sciences are emerging as a promising approach to better understand the actors' motivations across the trade chain and design culturally sensitive, behavior-based interventions. Nonetheless, a more comprehensive evaluation of the effectiveness of the implemented actions will require measuring their effect-size on relevant illegal wildlife trade indicators [125].

Supplementary Materials: The following are available online at https://www.mdpi.com/article/10.3390/d13050191/s1, Table S1: Bibliographic information of the documents manually reviewed as provided in bibtex format. Classification of the document included in the review follows the action-based approach described in this manuscript.

Author Contributions: Conceptualization, A.S.-M., J.R.F.-P., J.L.T., and J.P.R.; Data curation, A.S.-M.; Formal analysis, A.S.-M.; Investigation, A.S.-M. and J.R.F.-P.; Methodology, A.S.-M.; Writing–original draft, A.S.-M.; Writing–review & editing, A.S.-M., J.R.F.-P., J.L.T., and J.P.R. All authors have read and agreed to the published version of the manuscript.

Funding: This research was funded by the Whitley-Segre Conservation Fund, Kilverstone Wildlife Charitable Trust, and World Land Trust.

Institutional Review Board Statement: Not applicable.

Informed Consent Statement: Not applicable.

Data Availability Statement: Not applicable.

Conflicts of Interest: The authors declare no conflict of interest. The funders had no role in the design of the study; in the collection, analyses, or interpretation of data; in the writing of the manuscript, or in the decision to publish the results.

References

1. Scheffers, B.R.; Oliveira, B.F.; Lamb, L.; Edwards, D.P. Global wildlife trade across the tree of life. *Science* **2019**, *366*, 71–76. [CrossRef] [PubMed]
2. CITES Trade Database, UNEP World Conservation Monitoring Centre, Cambridge, UK. Available online: http://www.unep-wcmc.org/citestrade/trade.cfm (accessed on 22 September 2020).
3. Martin, R.O. Grey areas: Temporal and geographical dynamics of international trade of grey and timneh parrots (*Psittacus erithacus* and *P. timneh*) under CITES. *Emu* **2017**, *118*, 113–125. [CrossRef]
4. Berkunsky, I.; Quillfeldt, P.; Brightsmith, D.; Abbud, M.; Aguilar, J.; Alemán-Zelaya, U.; Aramburú, R.; Arce Arias, A.; Balas McNab, R.; Balsby, T.; et al. Current threats faced by Neotropical parrot populations. *Biol. Conserv.* **2017**, *214*, 278–287. [CrossRef]
5. Cardador, L.; Lattuada, M.; Strubbe, D.; Tella, J.L.; Reino, L.; Figueira, R.; Carrete, M. Regional bans on wild-bird trade modify invasion risks at a global scale. *Conserv. Lett.* **2017**, *10*, 717–725. [CrossRef]
6. Mori, E.; Grandi, G.; Menchetti, M.; Tella, J.L.; Jackson, H.A.; Reino, L.; van Kleunen, A.; Figueira, R.; Ancillotto, L. Worldwide distribution of non–native Amazon parrots and temporal trends of their global trade. *Anim. Biodivers. Conserv.* **2017**, *40*, 49–62. [CrossRef]
7. Menchetti, M.; Mori, E. Worldwide impact of alien parrots (Aves Psittaciformes) on native biodiversity and environment: A review. *Ethol. Ecol. Evol.* **2014**, *26*, 172–194. [CrossRef]
8. Matias, C.A.R.; Pereira, I.A.; dos Santos de Araújo, M.; Santos, A.F.M.; Lopes, R.P.; Christakis, S.; dos Prazeres Rodrigues, D.; Siciliano, S. Characteristics of Salmonella spp. Isolated from wild birds confiscated in illegal trade markets, Rio de Janeiro, Brazil. *BioMed Res. Int.* **2016**, *2016*, 3416864. [CrossRef] [PubMed]

9. Lopes, E.S.; Maciel, W.C.; Medeiros, P.H.Q.S.; Bona, M.D.; Bindá, A.H.; Lima, S.V.G.; Gaio, F.C.; Teixeira, R.S.C. Molecular diagnosis of diarrheagenic *Escherichia coli* isolated from Psittaciformes of illegal wildlife trade. *Pesqui. Vet. Bras.* **2018**, *38*, 762–766. [CrossRef]
10. Wright, T.F.; Toft, C.A.; Enkerlin-Hoeflich, E.; Gonzalez-Elizondo, J.; Albornoz, M.; Rodríguez-Ferraro, A.; Rojas-Suárez, F.; Sanz, V.; Trujillo, A.; Beissinger, S.R.; et al. Nest poaching in Neotropical parrots. *Conserv. Biol.* **2001**, *15*, 710–720. [CrossRef]
11. Martin, R.O.; Perrin, M.R.; Boyes, R.S.; Abebe, Y.D.; Annorbah, N.D.; Asamoah, A.; Bizimana, D.; Bobo, K.S.; Bunbury, N.; Brouwer, J.; et al. Research and conservation of the larger parrots of Africa and Madagascar: A review of knowledge gaps and opportunities. *Ostrich* **2014**, *85*, 205–233. [CrossRef]
12. Roldán-Clarà, B.; López-Medellín, X.; Espejel, I.; Arellano, E. Literature review of the use of birds as pets in Latin-America, with a detailed perspective on Mexico. *Ethnobiol. Conserv.* **2014**, *3*, 1–18. [CrossRef]
13. Pires, S.F. The illegal parrot trade: A literature review. *Glob. Crime* **2012**, *13*, 176–190. [CrossRef]
14. Weston, M.K.; Memon, M.A. The illegal parrot trade in Latin America and its consequences to parrot nutrition, health and conservation. *Bird Popul.* **2009**, *9*, 76–83.
15. 't Sas-Rolfes, M.; Challender, D.W.S.; Hinsley, A.; Veríssimo, D.; Milner-Gulland, E.J. Illegal wildlife trade: Patterns, processes, and governance. *Annu. Rev. Environ. Resour.* **2019**, *44*, 1–28. [CrossRef]
16. Broad, S.; Mulliken, T.; Roe, D. The nature and extent of legal and illegal trade in wildlife. In *The Trade in Wildlife: Regulation for Conservation*; Oldfield, S., Ed.; Earthscan Publications, Ltd.: London, UK, 2003; pp. 3–22.
17. Salazar, G.; Mills, M.; Verissimo, D. Qualitative impact evaluation of a social marketing campaign for conservation. *Conserv. Biol.* **2018**, *33*, 634–644. [CrossRef]
18. Toomey, A.H.; Knight, A.T.; Barlow, J. Navigating the space between research and implementation in conservation. *Conserv. Lett.* **2017**, *10*, 619–625. [CrossRef]
19. International Organization for Standardization. ISO 3166 Country Codes. Available online: https://www.iso.org/iso-3166-country-codes.html (accessed on 2 January 2021).
20. BirdLife International Data Zone. Available online: http://datazone.birdlife.org/species/search (accessed on 3 May 2020).
21. Phelps, J.; Biggs, D.; Webb, E.L. Tools and terms for understanding illegal wildlife trade. *Front. Ecol. Environ.* **2016**, *14*, 479–489. [CrossRef]
22. Cooney, R.; Kasterine, A.; MacMillan, D.; Milledge, S.; Nossal, K.; Roe, D.; Sas-Rolfes, M.t. *The Trade in Wildlife: A Framework to Improve Biodiversity and Livelihood Outcomes*; International Trade Centre: Geneva, Switzerland, 2015.
23. Westgate, M.J.; Barton, P.S.; Pierson, J.C.; Lindenmayer, D.B. Text analysis tools for identification of emerging topics and research gaps in conservation science. *Conserv. Biol.* **2015**, *29*, 1606–1614. [CrossRef]
24. BirdLife International IUCN Red List for birds. Available online: http://www.birdlife.org (accessed on 10 October 2019).
25. Csardi, G.; Nepusz, T. The Igraph Software Package for Complex Network Research. *Inter J. Complex Syst.* **2006**, *1695*, 1–9.
26. R Development Core Team. *R: A Language and Environment for Statistical Computing*; V. 3.6.2.; R Foundation for Statistical Computing: Viena, Austria, 2019.
27. Morton, O.; Scheffers, B.R.; Haugaasen, T.; Edwards, D.P. Impacts of wildlife trade on terrestrial biodiversity. *Nat. Ecol. Evol.* **2021**, *5*, 540–548. [CrossRef]
28. Biddle, R.; Solis-Ponce, I.; Jones, M.; Pilgrim, M.; Marsden, S. Parrot ownership and capture in coastal Ecuador: Developing a trapping pressure index. *Diversity* **2021**, *13*, 15. [CrossRef]
29. Pires, S.; Clarke, R.V. Are parrots craved? An analysis of parrot poaching in Mexico. *J. Res. Crime Delinq.* **2012**, *49*, 122–146. [CrossRef]
30. Tella, J.L.; Hiraldo, F. Illegal and legal parrot trade shows a long-term, cross-cultural preference for the most attractive species increasing their risk of extinction. *PLoS ONE* **2014**, *9*, e107546. [CrossRef]
31. Romero-Vidal, P.; Hiraldo, F.; Rosseto, F.; Blanco, G.; Carrete, M.; Tella, J.L. Opportunistic or Non-Random Wildlife Crime? Attractiveness rather than abundance in the wild leads to selective parrot poaching. *Diversity* **2020**, *12*, 314. [CrossRef]
32. Daut, E.F.; Brightsmith, D.J.; Mendoza, A.P.; Puhakka, L.; Peterson, M.J. Illegal domestic bird trade and the role of export quotas in Peru. *J. Nat. Conserv.* **2015**, *27*, 44–53. [CrossRef]
33. de Oliveira, E.S.; de Freitas Torres, D.; da Nóbrega Alves, R.R. Wild animals seized in a state in Northeast Brazil: Where do they come from and where do they go? *Environ. Dev. Sustain.* **2020**, *22*, 2343–2363. [CrossRef]
34. Hanks, C. *Spatial Patterns in Guyanas's Wild Bird Trade*; University of Texas: Austin, TX, USA, 2005.
35. Vaughan, C.; Nemeth, N.M.; Cary, J.; Temple, S. Response of a Scarlet Macaw *Ara macao* population to conservation practices in Costa Rica. *Bird Conserv. Int.* **2005**, *15*, 119–130. [CrossRef]
36. Briceño-Linares, J.M.; Rodríguez, J.P.; Rodríguez-Clark, K.M.; Rojas-Suárez, F.; Millán, P.A. Adapting to changing poaching intensity of yellow-shouldered parrot (*Amazona barbadensis*) nestlings in Margarita Island, Venezuela. *Biol. Conserv.* **2011**, *144*, 1188–1193. [CrossRef]
37. González, J.A. Harvesting, local trade, and conservation of parrots in the Northeastern Peruvian Amazon. *Biol. Conserv.* **2003**, *114*, 437–446. [CrossRef]
38. Ribeiro, J.; Reino, L.; Schindler, S.; Strubbe, D.; Vall-llosera, M.; Araújo, M.B.; Capinha, C.; Carrete, M.; Mazzoni, S.; Monteiro, M.; et al. Trends in legal and illegal trade of wild birds: A global assessment based on expert knowledge. *Biodivers. Conserv.* **2019**, *28*, 3343–3369. [CrossRef]

39. Alacs, E.; Georges, A. Wildlife across our borders: A review of the illegal trade in Australia. *Aust. J. Forensic Sci.* **2008**, *40*, 147–160. [CrossRef]
40. Pain, D.J.; Martins, T.L.F.; Boussekey, M.; Diaz, S.H.; Downs, C.T.; Ekstrom, J.M.M.; Garnett, S.; Gilardi, J.D.; McNiven, D.; Primot, P.; et al. Impact of protection on nest take and nesting success of parrots in Africa, Asia and Australasia. *Anim. Conserv.* **2006**, *9*, 322–330. [CrossRef]
41. Barré, N.; Theuerkauf, J.; Verfaille, L.; Primot, P.; Saoumoé, M. Exponential population increase in the endangered Ouvéa Parakeet (*Eunymphicus uvaeensis*) after community-based protection from nest poaching. *J. Ornithol.* **2010**, *151*, 695–701. [CrossRef]
42. Baker, N.E. *The Live Bird Trade in Tanzania*; IUCN Species Survival Commission: Daar Es Salaam, Tanzania, 1991; ISBN 2831703654.
43. McCallum, J.W.; Rowcliffe, J.M.; Cuthill, I.C. Conservation on international boundaries: The impact of security barriers on selected terrestrial mammals in four protected areas in Arizona, USA. *PLoS ONE* **2014**, *9*, e93679. [CrossRef]
44. Watson, F.; Becker, M.S.; McRobb, R.; Kanyembo, B. Spatial patterns of wire-snare poaching: Implications for community conservation in buffer zones around National Parks. *Biol. Conserv.* **2013**, *168*, 1–9. [CrossRef]
45. Jepson, P.; Ladle, R.J. Bird-keeping in Indonesia: Conservation impacts and the potential for substitution-based conservation responses. *Oryx* **2005**, *39*, 442–448. [CrossRef]
46. Jepson, P.; Ladle, R.J.; Sujatnika. Assessing market-based conservation governance approaches: A socio-economic profile of Indonesian markets for wild birds. *Oryx* **2011**, *45*, 482–491. [CrossRef]
47. Barbosa, A.E.A.; Tella, J.L. How much does it cost to save a species from extinction? Costs and rewards of conserving the Lear's macaw. *R. Soc. Open Sci.* **2019**, *6*, 190190. [CrossRef] [PubMed]
48. Aseres, S.A.; Sira, R.K. Ecotourism development in Ethiopia: Costs and benefits for protected area conservation. *J. Ecotourism* **2020**, 1–26. [CrossRef]
49. Eshoo, P.F.; Johnson, A.; Duangdala, S.; Hansel, T. Design, monitoring and evaluation of a direct payments approach for an ecotourism strategy to reduce illegal hunting and trade of wildlife in Lao PDR. *PLoS ONE* **2018**, *13*, e0186133. [CrossRef]
50. Brightsmith, D.J.; Stronza, A.; Holle, K. Ecotourism, conservation biology, and volunteer tourism: A mutually beneficial triumvirate. *Biol. Conserv.* **2008**, *141*, 2832–2842. [CrossRef]
51. Vall-llosera, M.; Cassey, P. 'Do you come from a land down under?' Characteristics of the international trade in Australian endemic parrots. *Biol. Conserv.* **2017**, *207*, 38–46. [CrossRef]
52. Ortiz-von Halle, B. *Bird's-Eye View: Lessons from 50 Years of Birds Trade Regulation & Conservation in Amazon Countries*; TRAFFIC: Cambridge, UK, 2018; ISBN 9781911646044.
53. Martin, R.O.; Senni, C.; D'Cruze, N.C. Trade in wild-sourced African grey parrots: Insights via social media. *Glob. Ecol. Conserv.* **2018**, *15*, e00429. [CrossRef]
54. Reuter, P.; O'Regan, D. Smuggling wildlife in the Americas: Scale, methods, and links to other organised crimes. *Glob. Crime* **2017**, *18*, 77–99. [CrossRef]
55. Sánchez-Mercado, A.; Blanco, O.; Sucre-Smith, B.; Briceño-Linares, J.M.; Peláez, C.; Rodríguez, J.P. When good attitudes are not enough: Understanding intentions to keep Yellow Shouldered Amazons as pets in Margarita Island, Venezuela. *Oryx* **2021**, in press.
56. Sánchez-Mercado, A.; Blanco, O.; Sucre-Smith, B.; Briceño-Linares, J.M.; Peláez, C.; Rodríguez, J.P. Using peoples' perceptions to improve conservation programs: The Yellow-shouldered Amazon in Venezuela. *Diversity* **2020**, *12*, 342. [CrossRef]
57. Gastañaga, M.; MacLeod, R.; Hennessey, B.; Núñez, J.U.; Puse, E.; Arrascue, A.; Hoyos, J.; Chambi, W.M.; Vasquez, J.; Engblom, G. A study of the parrot trade in Peru and the potential importance of internal trade for threatened species. *Bird Conserv. Int.* **2011**, *21*, 76–85. [CrossRef]
58. Gluszek, S.; Ariano-Sánchez, D.; Cremona, P.; Goyenechea, A.; Luque Vergara, D.A.; McLoughlin, L.; Morales, A.; Reuter Cortes, A.; Rodríguez Fonseca, J.; Radachowsky, J.; et al. Emerging trends of the illegal wildlife trade in Mesoamerica. *Oryx* **2020**, 1–9. [CrossRef]
59. Esmail, N.; Wintle, B.C.; Sas-Rolfes, M.; Athanas, A.; Beale, C.M.; Bending, Z.; Dai, R.; Fabinyi, M.; Gluszek, S.; Haenlein, C.; et al. Emerging illegal wildlife trade issues: A global horizon scan. *Conserv. Lett.* **2020**, *13*, e12715. [CrossRef]
60. Martin, R.O.; Senni, C.; D'cruze, N.; Bruschi, N. Tricks of the trade—Legal trade used to conceal Endangered African grey parrots on commercial flights. *Oryx* **2019**, *53*, 213. [CrossRef]
61. Annorbah, N.N.D.; Collar, N.J.; Marsden, S.J. Trade and habitat change virtually eliminate the Grey Parrot *Psittacus erithacus* from Ghana. *IBIS* **2016**, *158*, 82–91. [CrossRef]
62. Low, B.W. The global trade in native Australian parrots through Singapore between 2005 and 2011: A summary of trends and dynamics. *Emu* **2014**, *114*, 277–282. [CrossRef]
63. Wyatt, T. A comparative analysis of wildlife trafficking in Australia, New Zealand and the United Kingdom. *J. Traffick. Organ. Crime Secur.* **2016**, *2*, 62–81. [CrossRef]
64. Ojasti, J. *Utilización de la Fauna Silvestre en América Latina. Situación y Perspectiva Para un Manejos Sostenible*; Organización de Las Naciones Unidas Para la Agricultura y la Alimentación: Roma, Italia, 1993.
65. Andersson, A.A.; Tilley, H.B.; Lau, W.; Dudgeon, D.; Bonebrake, T.C.; Dingle, C. CITES and beyond: Illuminating 20 years of global, legal wildlife trade. *Glob. Ecol. Conserv.* **2021**, *26*, e01455. [CrossRef]
66. White, T.H., Jr.; Jhonson Camacho, A.; Bloom, T.; Lancho Diéguez, P.; Sellares, R. Human perceptions regarding endangered species conservation: A case study of Saona Island, Dominican Republic. *Rev. Latinoam. Conserv.* **2011**, *2*, 18–29.

67. Anderson, P.K. A bird in the house: An anthropological perspective on companion parrots. *Soc. Anim.* **2008**, *11*, 393–418. [CrossRef]
68. Collard, R.C. Putting animals back together, taking commodities apart. *Ann. Assoc. Am. Geogr.* **2014**, *104*, 151–165. [CrossRef]
69. Jenks, B.; Vaughan, P.W.; Butler, P.J. The evolution of rare pride: Using evaluation to drive adaptive management in a biodiversity conservation organization. *Eval. Program Plann.* **2010**, *33*, 186–190. [CrossRef]
70. Anderson, P.K. Social dimensions of the human—Avian bond: Parrots and their persons. *Anthrozoos* **2015**, *27*, 371–387. [CrossRef]
71. Shuttlewood, C.Z.; Greenwell, P.J.; Montrose, V.T. Pet ownership, attitude toward pets, and support for wildlife management strategies. *Hum. Dimens. Wildl.* **2016**, *21*, 180–188. [CrossRef]
72. Biggs, D.; Cooney, R.; Roe, D.; Dublin, H.; Allan, J.; Challender, D.; Skinner, D. *Engaging Local Communities in Tackling Illegal Wildlife Trade: Can a 'Theory of Change' Help?* TRAFFIC, IUCN: Gland, Switzerland, 2015; ISBN 9781784312367.
73. Zain, S. *Behaviour Change We Can Believe in: Towards a Global Demand Reduction Strategy for Tigers*; TRAFFIC: Gland, Switzerland, 2012.
74. Miller, Z.D. A Theory of Planned Behavior approach to developing belief-based communication: Day hikers and bear spray in Yellowstone National Park. *Hum. Dimens. Wildl.* **2019**, *24*, 1–15. [CrossRef]
75. Summerell, A.E.; Frankham, G.J.; Gunn, P.; Johnson, R.N. DNA based method for determining source country of the short beaked echidna (*Tachyglossus aculeatus*) in the illegal wildlife trade. *Forensic Sci. Int.* **2019**, *295*, 46–53. [CrossRef] [PubMed]
76. Hogg, C.J.; Dennison, S.; Frankham, G.J.; Hinds, M.; Johnson, R.N. Stopping the spin cycle: Genetics and bio-banking as a tool for addressing the laundering of illegally caught wildlife as 'captive-bred. ' *Conserv. Genet. Resour.* **2018**, *10*, 237–246. [CrossRef]
77. Bourret, V.; Albert, V.; April, J.; Côté, G.; Morissette, O. Past, present and future contributions of evolutionary biology to wildlife forensics, management and conservation. *Evol. Appl.* **2020**, *13*, 1420–1434. [CrossRef] [PubMed]
78. Pacífico, E.C.; Sánchez-Montes, G.; Miyaki, C.Y.; Tella, J.L. Isolation and characterization of 15 new microsatellite markers for the globally endangered Lear's macaw *Anodorhynchus leari*. *Mol. Biol. Rep.* **2020**, *47*, 8279–8285. [CrossRef] [PubMed]
79. Jan, C.; Fumagalli, L. Polymorphic DNA microsatellite markers for forensic individual identification and parentage analyses of seven threatened species of parrots (family Psittacidae). *PeerJ* **2016**, *2016*. [CrossRef]
80. Lima-Rezende, C.A.; Fernandes, G.A.; da Silva, H.E.; Dobkowski-Marinho, S.; Santos, V.F.; Rodrigues, F.P.; Caparroz, R. In silico identification and characterization of novel microsatellite loci for the Blue-and-yellow Macaw *Ara ararauna* (Linnaeus, 1758) (Psittaciformes, Psittacidae). *Genet. Mol. Biol.* **2019**, *42*, 68–73. [CrossRef]
81. Fernandes, G.A.; Dobkowski-Marinho, S.; Santos, V.F.; Lima-Rezende, C.A.; da Silva, H.E.; Rodrigues, F.P.; Caparroz, R. Development and characterization of novel microsatellite loci for the Blue-fronted Amazon (*Amazona aestiva*, Psittaciformes, Aves) and cross-species amplification for other two threatened Amazona species. *Mol. Biol. Rep.* **2019**, *46*, 1377–1382. [CrossRef]
82. Taylor, T.D.; Parkin, D.T. Characterisation of 13 microsatellite loci for the Moluccan Cockatoo, Cacatua moluccensis, and Cuban Amazon, Amazona leucocephala, and their conservation and utility in other parrot species (Psittaciformes). *Conserv. Genet.* **2007**, *8*, 991–994. [CrossRef]
83. Pillay, K.; Dawson, D.A.; Horsburgh, G.J.; Perrin, M.R.; Burke, T.; Taylor, T.D. Twenty-two polymorphic microsatellite loci aimed at detecting illegal trade in the Cape Parrot, *Poicephalus robustus* (Psittacidae, AVES). *Mol. Ecol. Resour.* **2010**, *10*, 142–149. [CrossRef]
84. Coetzer, W.G.; Downs, C.T.; Perrin, M.R.; Willows-Munro, S. Testing of microsatellite multiplexes for individual identification of Cape Parrots (*Poicephalus robustus*): Paternity testing and monitoring trade. *PeerJ* **2017**. [CrossRef]
85. Alexander, J.; Downs, C.T.; Butler, M.; Woodborne, S.; Symes, C.T. Stable isotope analyses as a forensic tool to monitor illegally traded African grey parrots. *Anim. Conserv.* **2019**, *22*, 134–143. [CrossRef]
86. Willows-Munro, S.; Kleinhans, C. Testing microsatellite loci for individual identification of captive African grey parrots (*Psittacus erithacus*): A molecular tool for parentage analysis that will aid in monitoring legal trade. *Conserv. Genet. Resour.* **2020**, *12*, 489–495. [CrossRef]
87. Miller, A.D.; Good, R.T.; Coleman, R.A.; Lancaster, M.L.; Weeks, A.R. Microsatellite loci and the complete mitochondrial DNA sequence characterized through next generation sequencing and de novo genome assembly for the critically endangered Orange-bellied Parrot, *Neophema chrysogaster*. *Mol. Biol. Rep.* **2013**, *40*, 35–42. [CrossRef] [PubMed]
88. Coghlan, M.L.; White, N.E.; Parkinson, L.; Haile, J.; Spencer, P.B.S.; Bunce, M. Egg forensics: An appraisal of DNA sequencing to assist in species identification of illegally smuggled eggs. *Forensic Sci. Int. Genet.* **2012**, *6*, 268–273. [CrossRef] [PubMed]
89. White, N.E.; Dawson, R.; Coghlan, M.L.; Tridico, S.R.; Mawson, P.R.; Haile, J.; Bunce, M. Application of STR markers in wildlife forensic casework involving Australian black-cockatoos (*Calyptorhynchus* spp.). *Forensic Sci. Int. Genet.* **2012**, *6*, 664–670. [CrossRef]
90. Mendoza, Á.M.; Torres, M.F.; Paz, A.; Trujillo-Arias, N.; López-Alvarez, D.; Sierra, S.; Forero, F.; Gonzalez, M.A. Cryptic diversity revealed by DNA barcoding in Colombian illegally traded bird species. *Mol. Ecol. Resour.* **2016**, *16*, 862–873. [CrossRef]
91. Poole, C.M.; Shepherd, C.R. Shades of grey: The legal trade in CITES-listed birds in Singapore, notably the globally threatened African Grey Parrot *Psittacus erithacus*. *Oryx* **2017**, *51*, 411–417. [CrossRef]
92. Alacs, E.A.; Georges, A.; FitzSimmons, N.N.; Robertson, J. DNA detective: A review of molecular approaches to wildlife forensics. *Forensic Sci. Med. Pathol.* **2010**, *6*, 180–194. [CrossRef]
93. Cardador, L.; Tella, J.L.; Anadón, J.D.; Abellán, P.; Carrete, M. The European trade ban on wild birds reduced invasion risks. *Conserv. Lett.* **2019**, *12*, e12631. [CrossRef]
94. Gibbons, W.J.; Andrews, K.M. PIT tagging: Simple technology at its best. *Bioscience* **2004**, *54*, 447–454. [CrossRef]

95. Di Minin, E.; Fink, C.; Hiippala, T.; Tenkanen, H. A framework for investigating illegal wildlife trade on social media with machine learning. *Conserv. Biol.* **2019**, *33*, 210–213. [CrossRef]
96. Challender, D.W.S.; MacMillan, D.C. Poaching is more than an enforcement problem. *Conserv. Lett.* **2013**, *7*, 1–11. [CrossRef]
97. Challender, D.W.S.; Harrop, S.R.; Macmillan, D.C. Towards informed and multi-faceted wildlife trade interventions. *Glob. Ecol. Conserv.* **2015**, *3*, 129–148. [CrossRef]
98. Patel, N.G.; Rorres, C.; Joly, D.O.; Brownstein, J.S.; Boston, R.; Levy, M.Z.; Smith, G. Quantitative methods of identifying the key nodes in the illegal wildlife trade network. *Proc. Natl. Acad. Sci. USA.* **2015**, *112*, 7948–7953. [CrossRef]
99. Budiani, I.; Raharningrum, F. *Illegal Online Trade in Indonesian Parrots*; The Global Initiative against Transnational Organized Crime: Geneva, Switzerland, 2018.
100. Cantú Guzmán, J.C.; Sánchez Saldaña, M.E.; Grosselet, M.; Silve Gamez, J. *The Illegal Parrot Trade in Mexico. A Comprehensive Assessment*; Defenders of Wildlife: Mexico City, Mexico, 2007.
101. Sánchez-Mercado, A.; Cardozo-Urdaneta, A.; Moran, L.; Ovalle, L.; Arvelo, M.Á.; Morales-Campos, J.; Coyle, B.; Braun, M.J.; Rodríguez-Clark, K.M. Social network analysis reveals specialized trade in an Endangered songbird. *Anim. Conserv.* **2019**, *23*, 132–144. [CrossRef]
102. Kahler, J.S. Alarm call: Innovative study highlights the need for robust conservation crime science to effectively impede specialized songbird trafficking. *Anim. Conserv.* **2020**, *23*, 149–150. [CrossRef]
103. Farine, D. Mapping illegal wildlife trade networks provides new opportunities for conservation actions. *Anim. Conserv.* **2020**, *23*, 145–146. [CrossRef]
104. Tsang, S.M. A social network approach to detect parallel wildlife trafficking: A novel tool, with potential limitations in a closed, specialist system. *Anim. Conserv.* **2020**, *23*, 147–148. [CrossRef]
105. De Freitas Raso, T.; Godoy, S.; Milanelo, L.; De Souza, C.; Matuschima, E.; Araújo, J.; Pinto, A. An Outbreak of Chlamydiosis in Captive Blue-Fronted Amazon Parrots (*Amazona aestiva*) in Brazil. *J. Zoo Wildl. Med.* **2004**, *35*, 94–96. [CrossRef]
106. Deem, S.L.; Noss, A.J.; Cuéllar, R.L.; Karesh, W.B. Health evaluation of free-rangeing and captive Blue-fronted Amazon (*Amazona aestiva*) in the Gran Chaco, Bolivia. *J. Zoo Wildl. Med.* **2005**, *36*, 598–605. [CrossRef]
107. Bruning-Fann, C.; Kaneene, J.; Heamon, J. Investigation of an outbreak of velogenic viscerotropic Newcastle disease in pet birds in Michigan, Indiana, Illinois, and Texas. *J. Am. Vet. Med. Assoc.* **1992**, *201*, 1709–1714.
108. Lee, D.-H.; Killian, M.L.; Torchetti, M.K.; Brown, I.; Lewis, N.; Berhane, Y.; Swayne, D.E. Intercontinental spread of Asian-origin H7 avian influenza viruses by captive bird trade in 1990′s. *Infect. Genet. Evol.* **2019**, *73*, 146–150. [CrossRef]
109. Morinha, F.; Carrete, M.; Tella, J.L.; Blanco, G. High prevalence of novel beak and feather disease virus in sympatric invasive parakeets introduced to Spain from Asia and South America. *Diversity* **2020**, *12*, 192. [CrossRef]
110. Ribas, J.M.; Sipinski, E.A.B.; Serafini Pereira, P.; Lindmayer Ferreira, V.; de Freitas Raso, T.; Pinto, A.A. *Chlamydophila psittaci* assessment in threatened Red-tailed Amazon (*Amazona brasiliensis*) parrots in Paraná, Brazil. *Ornithologia* **2014**, *6*, 144–147.
111. Hawkins, M.G.; Crossley, B.M.; Osofsky, A.; Webby, R.J.; Lee, C.-W.; Suarez, D.L.; Hietala, S.K. Avian influenza A virus subtype H5N2 in a red-lored Amazon parrot. *J. Am. Vet. Med. Assoc.* **2006**, *228*, 236–241. [CrossRef]
112. Vilela, D.A.R.; Marin, S.Y.; Resende, M.; Coelho, H.L.G.; Resende, J.S.; Ferreira-Junior, F.C.; Ortiz, M.C.; Araujo, A.V.; Raso, T.F.; Martins, N.R.S. Phylogenetic analyses of *Chlamydia psittaci* ompA gene sequences from captive blue-fronted Amazon parrots (*Amazona aestiva*) with hepatic disease in Brazil. *Rev. Sci. Tech.* **2019**, *38*, 711–719. [CrossRef] [PubMed]
113. Doneley, R.J.T. Acute Beak and Feather Disease in juvenile African grey parrots—An uncommon presentation of a common disease. *Aust. Vet. J.* **2003**, *81*, 206–207. [CrossRef] [PubMed]
114. Araújo, A.V.; Andery, D.A.; Ferreira, F.C., Jr.; Ortiz, M.C.; Marques, M.V.R.; Marin, S.Y.; Vilela, D.A.R.; Resende, J.S.; Resende, M.; Donatti, R.V.; et al. Molecular diagnosis of beak and feather disease in native Brazilian Psittacines. *Braz. J. Poult. Sci.* **2015**, *17*, 451–458. [CrossRef]
115. Davies, Y.M.; Cunha, M.P.V.; Oliveira, M.G.X.; Oliveira, M.C.V.; Philadelpho, N.; Romero, D.C.; Milanelo, L.; Guimarães, M.B.; Ferreira, A.J.P.; Moreno, A.M.; et al. Virulence and antimicrobial resistance of *Klebsiella pneumoniae* isolated from Passerine and Psittacine birds. *Avian Pathol.* **2016**, *45*, 194–201. [CrossRef] [PubMed]
116. Fogell, D.J.; Martin, R.O.; Bunbury, N.; Lawson, B.; Sells, J.; McKeand, A.M.; Tatayah, V.; Trung, C.T.; Groombridge, J.J. Trade and conservation implications of new beak and feather disease virus detection in native and introduced parrots. *Conserv. Biol.* **2018**, *32*, 1325–1335. [CrossRef] [PubMed]
117. Symes, C.T. Founder populations and the current status of exotic parrots in South Africa. *Ostrich* **2014**, *85*, 235–244. [CrossRef]
118. Esler, K.J.; Prozesky, H.; Sharma, G.P.; McGeoch, M. How wide is the "knowing-doing" gap in invasion biology? *Biol. Invasions* **2010**, *12*, 4065–4075. [CrossRef]
119. Fabian, Y.; Bollmann, K.; Brang, P.; Heiri, C.; Olschewski, R.; Rigling, A.; Stofer, S.; Holderegger, R. How to close the science-practice gap in nature conservation? Information sources used by practitioners. *Biol. Conserv.* **2019**, *235*, 93–101. [CrossRef]
120. IUCN the International Union for Conservation of Nature's Red List of Threatened Species. Available online: https://www.iucnredlist.org/ (accessed on 20 September 2020).
121. Frynta, D.; Lišková, S.; Bültmann, S.; Burda, H. Being attractive brings advantages: The case of parrot species in captivity. *PLoS ONE* **2010**, *5*, e12568. [CrossRef]
122. Santos, J.; Correia, R.; Malhado, A.; Campos-Silva, J.; Teles, D.; Jepson, P.; Ladle, R. Drivers of taxonomic bias in conservation research: A global analysis of terrestrial mammals. *Anim. Conserv.* **2020**, *23*, 679–688. [CrossRef]

123. Mair, L.; Mill, A.C.; Robertson, P.A.; Rushton, S.P.; Shirley, M.D.F.; Rodriguez, J.P.; McGowan, P.J.K. The contribution of scientific research to conservation planning. *Biol. Conserv.* **2018**, *223*, 82–96. [CrossRef]
124. Roe, D.; Dickman, A.; Kock, R.; Milner-Gulland, E.J.; Rihoy, E.; Sas-Rolfes, M. Beyond banning wildlife trade: COVID-19, conservation and development. *World Dev.* **2020**, *136*, 105121. [CrossRef]
125. Sutherland, W.J.; Dicks, L.V.; Ockendon, N.; Smith, R.K. *What Works in Conservation*; Open Book Publisher: Cambridge, UK, 2015; Volume 1, ISBN 2059-4240.

MDPI

Article

Parrot Ownership and Capture in Coastal Ecuador: Developing a Trapping Pressure Index

Rebecca Biddle [1,*], Ivette Solis-Ponce [1], Martin Jones [2], Mark Pilgrim [1] and Stuart Marsden [2]

1 The North of England Zoological Society, Upton, Chester CH2 1LH, UK; ivisolisponce@gmail.com (I.S.-P.); m.pilgrim@chesterzoo.org (M.P.)
2 Department of Natural Sciences, Manchester Metropolitan University, Manchester M15 6BH, UK; m.jones@mmu.ac.uk (M.J.); S.Marsden@mmu.ac.uk (S.M.)
* Correspondence: b.biddle@chesterzoo.org; Tel.: +44-124-438-9729 or +44-739-551-4945

Abstract: We located rural communities with pet parrots and used these locations to predict the probability of illegal parrot ownership across coastal Ecuador, using variables related to demand for pets, parrot availability, and trapping accessibility. In 12 pet keeping communities, we carried out in-depth interviews with 106 people, to quantify ownership, trapping, and interviewees' attitudes towards these behaviours. We combined these data to calculate a trapping pressure index for four key roosting, feeding and nesting sites for the Critically Endangered Lilacine or Ecuadorian Amazon Parrot *Amazona lilacina*. We found that 66% of all communities had pet parrots and 31% had pet Lilacines. Our predictive models showed that pet parrot ownership occurs throughout coastal Ecuador, but ownership of Lilacines by rural communities, is more likely to occur within the natural distribution of the species. The number of people per community who had owned Lilacines in the last three years varied from 0–50%, as did the number of people who had trapped them—from 0–26%. We interviewed 10 people who had captured the species in the last three years who reported motives of either to sell or keep birds as pets. Attitudes towards pet keeping and trapping differed among the 12 communities: 20–52% believed it was acceptable to keep pet parrots, and for 32–74%, it was acceptable to catch parrots to sell. This being said, most people believed that wild parrots were important for nature and that local people had a responsibility to protect them. We conclude that trapping pressure is greatest in the southern part of the Lilacine's range, and urgent conservation measures such as nest and roost protection, and local community engagement are needed.

Keywords: *Amazona lilacina*; poaching; conservation threats; mangrove; dry forest; local knowledge; attitudes; Lilacine Amazon

Citation: Biddle, R.; Solis-Ponce, I.; Jones, M.; Pilgrim, M.; Marsden, S. Parrot Ownership and Capture in Coastal Ecuador: Developing a Trapping Pressure Index. *Diversity* **2021**, *13*, 15. https://doi.org/10.3390/d13010015

Received: 25 November 2020
Accepted: 21 December 2020
Published: 5 January 2021

1. Introduction

Parrots (Psittaciformes) are among the most endangered and rapidly declining bird groups, with 28% of their species classified as threatened [1]. Globally, over a third of parrot species are caught to fulfil the demand of the international wildlife trade [1–3]. In the Neotropics, over half of the studied parrot populations are in decline [4], and one reason for this is the high demand for the pet trade [5]. Neotropical species are particularly favoured as pets [2,6], and it is suggested that trapping is a stronger threat to their conservation than habitat loss [7]. Amazon parrots and macaws are preferred due to their attractiveness and ability to mimic the human voice [8]; this is illustrated in Costa Rica, where nearly 20% of households have a pet parrot and half of these are *Amazona* species [9]. Consequently, the rate at which Amazon parrots and macaws are trapped is much higher than expected considering their availability in the wild [10].

Trapping risk is highest where parrots are abundant in the wild, where demand is high and where parrots are relatively easy to catch and sell [11]; therefore, trapping pressure may differ across a species' range and also between species. Additional factors found to drive hunting and trapping include overlap with human population [11,12] and proximity to infrastructure or towns [13]. Attitudes and subjective norms are also factors that influence decision making [14], and are therefore likely to affect the level of pet keeping and capture in different areas. In Ecuador, wild bird keeping is illegal [15], and whilst ownership appears to be declining in major cities [16] demand is still high in rural areas, where over half of coastal communities still keep pet parrots [17]. The most frequently reported confiscated bird species in the country are those with wild distributions exclusive to this coastal region [16,18]. Moreover, this region is one of the most densely populated and impoverished [19] parts of Ecuador, the habitats here have been drastically reduced [20] and are greatly underrepresented in the country's national protected areas system [21].

The Critically Endangered Lilacine or Ecuadorian Amazon *Amazona lilacina*, a species recently split from the *A. autumnalis* group, is found exclusively within the coastal region of Ecuador [22]. CITES reported thousands of individuals of this species being trapped and exported in the early 1980s [23] and although frequency of trapping is likely to have reduced significantly in recent years, there are still multiple reports of capture and pet-keeping within rural communities [17]. An average of 392 wild-caught parrots, including 30 *A. autumnalis*, were confiscated annually in Ecuador between 2003 and 2016 [16]. Although some of these may be older birds, and they may be either *A. lilacina* or *A. a. salvini*, this figure suggests that some level of trapping is still occurring to fulfil the demand for pets. The goal of this study was to understand the risk of trapping in rural communities and formulate a strategy for conservation support. Specific objectives were:

1. Locate communities with pet parrots by conducting surveys across coastal Ecuador, and use these locations to predict the distribution of pet parrots, and the likelihood of local parrot trapping, using variables related to parrot availability, opportunity and demand;
2. Within communities that keep pet parrots, interview local people to quantify the level of parrot ownership, trapping and the attitudes towards these behaviours;
3. Develop a trapping pressure index based on model predictions, locally reported incidence and attitudes towards parrot capture and ownership.

2. Materials and Methods

2.1. Surveys to Locate Communities with Pet Parrots

In order to locate rural communities with pet parrots, we conducted surveys between January and July 2017. The study area encompassed the extent of occurrence of the Lilacine Amazon *Amazona lilacina* and communities close (<10 km) to forest patches, where wild parrots may occur were selected. Participants were asked to confirm if they knew of pet parrots in their community, and if possible to identify the species. Prior verbal consent was obtained from each participant and full ethical approval of survey content and methods was gained from The North of England Zoological Ethical Review Committee. We aimed to survey at least four households per community; however, some communities were made up of just a few houses, so this was not always possible. We recorded the geographic coordinates of communities with all pets, pet parrots and pet Lilacines, and calculated how many communities each species was recorded in. We used IUCN Red List range maps provided by BirdLife International [24] in order to determine if species were native to the study area. Range maps are frequently updated so we report the year of update for each range map in the results. ArcGIS (version 10.8.1) [25] was used, clipping the distribution shape files, to calculate the size of each species range within our study area.

It is illegal to keep native bird species as pets in Ecuador [15], yet in our experience, people speak openly about their parrots and are proud to show them off. However, it was important that participants did not feel threatened or that we were collecting information to inform the authorities. Therefore, surveys were conducted by a local Ecuadorian researcher, in Spanish, with only the researcher and interviewee present, and it was made clear that all information given was anonymous, and only to be used for scientific research.

Although we refer to *A. lilacina* as the Lilacine or Ecuadorian Amazon Parrot, neither of these common names have Spanish translations that are used in Ecuador. Most local communities refer to "loro frentirrojo" (Red-lored Parrot), which in English describes the *A. autumnalis* group and includes *A. a. salvini* in northern Ecuador. To avoid confusion, we use *A. lilacina* in our communication with local communities and use photographs to confirm identity, but refer to the Lilacine Amazon in this manuscript.

2.2. Distribution Models to Predict Parrot Ownership

From our surveys we created two groups of geographic coordinates to represent (1) communities with pet parrots, and (2) communities with pet Lilacines. The MaxEnt package in R (version 4.0.3) [26,27] was used to build distribution models based on these coordinates combined with random background points within 30 km buffers of community locations, to predict the distribution of pet parrots, and the distribution of pet Lilacines. Variables were extracted to match each corresponding location and were chosen due to their influence over parrot ownership and trapping [11]: opportunity (presence of parrots and their desirability); demand (presence of people and the infrastructure for trade); and accessibility (into the forest).

For each location, we calculated a "species value" to represent parrot trapping opportunity. For the pet model, this was calculated based on the presence of wild parrot species at that location using species range maps [24], combined with the frequency of the species being reported as a pet; 0.1 was allocated for each species present in that area, and an additional 0.1 was added if that species was reported in a single community, 0.2 if in two communities, etc. This value was used just for comparative purposes within the study and we gave equal weighting to wild species presence and popularity in captivity, as we had no evidence that either was more important than the other. For the pet Lilacine model, this value was replaced with the predicted occupancy area from our distribution models created using observations of the wild population [28]. For both models, we also used the estimated human population [29], the Euclidean distance to the nearest town and nearest road calculated in ArcGIS using OpenStreetMap [30] data, and the mean annual Normalised Difference Vegetation Index (NDVI) from the monthly MODIS product over 2010–2015 as a proxy of vegetation cover.

For each group of points, spatial autocorrelation was controlled for by limiting them to one per 1 km using the R package spThin [31]. Predictors were checked for pairwise correlation across random points within the study area, using pair plot for collinearity [32]. Model evaluation was performed with five-fold cross validation and the mean AUC +/− SD are presented to demonstrate the predictive ability. An AUC of 0.7 means there is a 70% chance that the fitted model will be able to correctly distinguish between presence and absence [33]. All data were included in the final models. We present the permutation importance (%) of variables, with a high value indicating that the final model depends heavily on that variable [27].

2.3. Interviews to Quantify Parrot Ownership, Trapping and Attitudes

We selected 12 communities where pet parrots were present to conduct interviews with community members about their experiences and attitudes towards parrot ownership and capture. These communities ranged in size from 50 to 300 people. The reason these sites were chosen was because our focus was on understanding risk to Lilacine Amazons, so the selected communities fell within the species extent of occurrence and were <15 km away from key roosting, nesting and feeding grounds [17]. These 12 communities were grouped into four clusters (Figure 1). We interviewed at least six participants from different households in each community. Participants were outdoor workers (i.e., agriculturalists, fishers and crab fishers) selected for their familiarity with parrots in their local area. The same methods regarding informed consent and data anonymity as described in Section 2.1 were followed. Due to low literacy levels amongst participants, all questions were read out aloud and the answer provided was recorded by the researcher. Age and gender of each participant was recorded. The interview consisted of eight questions and seven attitude statements arranged on a five point symmetric Likert scale (Table 1). The Likert package [34] in R (version 4.0.3) [26] was used to visualise attitude statements. Responses were grouped into positive, neutral or negative and a non-parametric test (Kruskal–Wallis) was used to determine significant differences in responses between the four community clusters.

Table 1. Interviews about parrot ownership and capture asking eight questions and seven attitude statements, which were read out aloud by the researcher in Spanish. Interviews were anonymous and participants could decline to answer any questions.

	Interview Questions						
1.	How often do you see *Amazona lilacina*?	daily/weekly/monthly/yearly/never					
2.	In your opinion, have *A. lilacina* numbers changed in the last three years?	increased/decreased/stayed the same/not sure					
3.	Have you ever owned a pet parrot?	yes/no/prefer not to say					
4.	Have you ever owned a pet *A. lilacina*?	yes/no/not sure/prefer not to say					
	- If yes, how did you get it?	caught it/bought it/given it/prefer not to say					
	- If yes, where did you get it?						
	- If yes, how long ago did you get it?	last year/two years/three years/> three years/prefer not to say					
	- If yes, how many *A. lilacina* have you owned in the last three years?						
5.	How many other people in the village have a pet *A. lilacina*?						
6.	Have you ever taken *A. lilacina* from a nest or caught one from the wild?	yes/no/not sure/prefer not to say					
	- If yes, how many in the last three years?						
	- If yes, for what purpose did you catch it?						
7.	Have you ever sold *A. lilacina*?	yes/no/not sure/prefer not to say					
8.	To your knowledge, do other people in your village take *A. lilacina* from nests/the wild?	yes/no/not sure/prefer not to say					
	Attitude Statements						
Response categories were: strongly disagree (1), disagree (2), neutral (3), agree (4), strongly agree (5), I don't know (NA).		1	2	3	4	5	NA
I think that it is OK to keep a parrot as a pet.							
Catching parrots from the wild can make them extinct in my local area.							
Wild parrots are important for nature.							
I have a responsibility to protect the environment.							
I am comfortable with outsiders catching parrots in my local area.							
It is OK to catch wild parrots to sell to the pet trade.							
It is OK if parrots disappeared from the wild.							

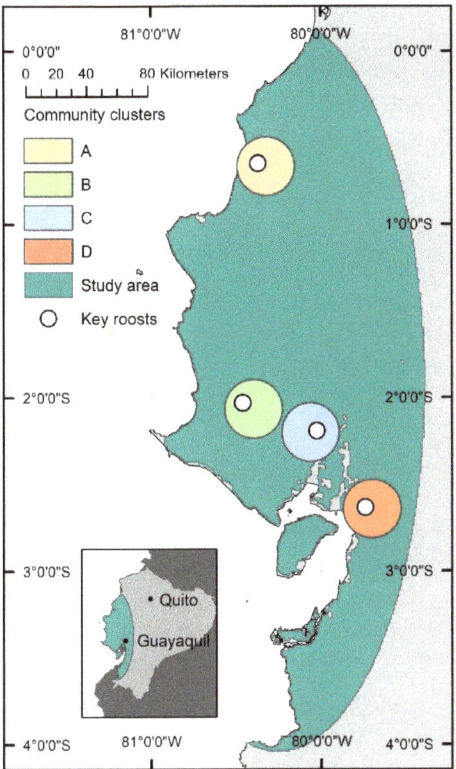

Figure 1. Interviews about parrot ownership and capture were conducted in 12 communities, grouped into four clusters (A, B, C, and D) near key Lilacine Amazon roost sites. Each cluster contains three communities <10 km apart.

2.4. Trapping Pressure Index

To prioritise areas for conservation support, we calculated a trapping pressure value for each of the four community clusters, to represent the level of risk to the wild Lilacine Amazon population from capture and local desire for pet keeping. This risk value was calculated by combining the following six factors: (1) the mean model value for pet parrot keeping (which represents a probability that pets occur at that location); (2) the mean model value for pet Lilacine keeping; (3) the percentage of people who have owned a pet Lilacine in the last three years; (4) the percentage who have trapped Lilacines in the last three years; (5) the percentage of people who think it is OK to keep a parrot as a pet; and (6) the percentage who think it is OK to catch wild parrots to sell.

3. Results

3.1. Locations and Species of Pet Parrots

Surveys were carried out in 65 communities (mean = 6 interviewed people per community; range 3–20). In 43 (66%), pet parrots were confirmed, and in 20 (31%), pet Lilacines were confirmed. Of the 19 wild parrot species, nine were reported in at least one community, with the most frequently reported being Lilacine Amazons (Figure 2) and Grey-cheeked Parakeets *Brotogeris pyrrhoptera*. The mean range size within the study area of parrots found as pets was 27,370 km^2, compared to 8677 km^2 for those not kept as pets (Table 2).

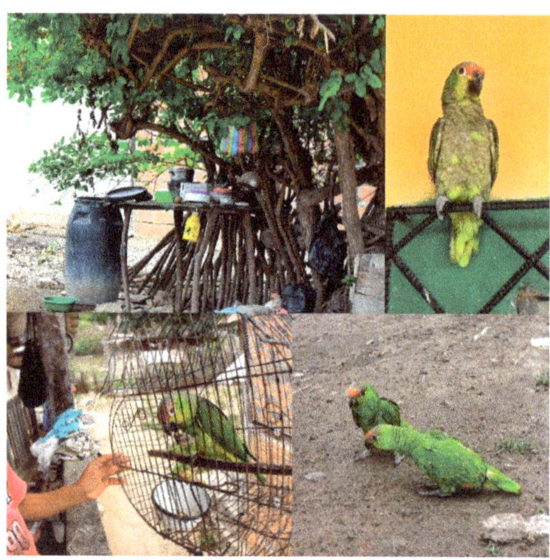

Figure 2. Examples of pet Lilacine Amazons in four rural communities in coastal Ecuador. Parrots were kept in a variety of situations; indoors or outdoors, caged or with clipped wings. In some cases pet parrots that were housed in gardens were not initially considered to be captive by the owner, but for the purposes of this study any parrot living in the locality of people was classed as a pet.

Table 2. The number of communities in which each of the 19 species was reported as a pet and the range size within the study area according to BirdLife International [24].

Parrot Species (Year of Update)	Range within Study Area SA (km^2)	Number of Communities Reporting the Species (Out of 65)
Lilacine or Ecuadorian Amazon *Amazona lilacina* (2018)	38,860	20
Grey-cheeked Parakeet *Brotogeris pyrrhoptera* (2014)	8645	20
Red-masked Parakeet *Psittacara erythrogenys* (2007)	54,327	17
Pacific Parrotlet *Forpus coelestis* (2017)	55,300	8
Red-lored Amazon *Amazona autumnalis* (2017)	5583	6
Blue-headed Parrot *Pionus menstruus* (2013)	27,943	2
Southern Mealy Amazon *Amazona farinosa* (2013)	8612	1
Bronze-winged Parrot *Pionus chalcopterus* (2014)	46,508	1
White-winged Parakeet *Brotogeris versicolurus* (2018)	549	1
Chestnut-fronted Macaw *Ara severus* (2014)	49,329	0
Blue-fronted Parrotlet *Touit dilectissimus* (2014)	13,470	0
White-capped Parrot *Pionus seniloides* (2012)	1482	0
Rose-faced Parrot *Pyrilia pulchra* (2002)	12,828	0
Great Green Macaw *Ara ambiguus* (2014)	3899	0
Red-faced Parrot *Hapalopsittaca pyrrhops* (2000)	49	0
Cordilleran Parakeet *Psittacara frontatus* (2014)	1347	0
Barred Parakeet *Bolborhynchus lineola* (2014)	2183	0
Red-billed Parrot *Pionus sordidus* (2014)	1565	0
El Oro Parakeet *Pyrrhura orcesi* (1999)	615	0
Kept by communities but non-native		
Orange-winged Amazon *Amazona amazonica*	NA	1
Yellow-crowned Amazon *Amazona ochrocephala*	NA	5

3.2. Predicted Distribution of Pet Parrots

The locations of the 43 communities with pet parrots and the 20 communities with pet Lilacines were reduced to 42 and 19, respectively, after limiting each group of locations to one per 1 km. A total of 3803 background points were randomly allocated. The mean AUC of resulting models was 0.69 ± 0.06 (sd) for pet parrots and 0.62 ± 0.20 (sd) for pet Lilacines. The most important variables predicting the presence of pet parrots were distance to nearest road (permutation importance, PI = 40%) and distance to nearest town (PI = 28%); the key factors for the presence of pet Lilacines were the mean annual NDVI (PI = 33%) and species value, representing the native distribution (PI = 27%) (Table 3). Predictions show that pet parrots are likely to be widespread throughout the study area, whereas pet Lilacines seem to be more likely within the species range. Both models show a high probability of occurrence of pets to the west of Guayaquil and out towards the coast (Figure 3).

Table 3. Permutation importance values for variables used to create models predicting the distribution of pet parrots and pet Lilacines in coastal Ecuador.

Variable	Permutation Importance (%)	
	Pet Parrot Model	**Pet Lilacine Model**
Mean annual NDVI	11	33
Distance to road	40	1
Human population density	18	23
Distance to town	28	16
Species value	3	27

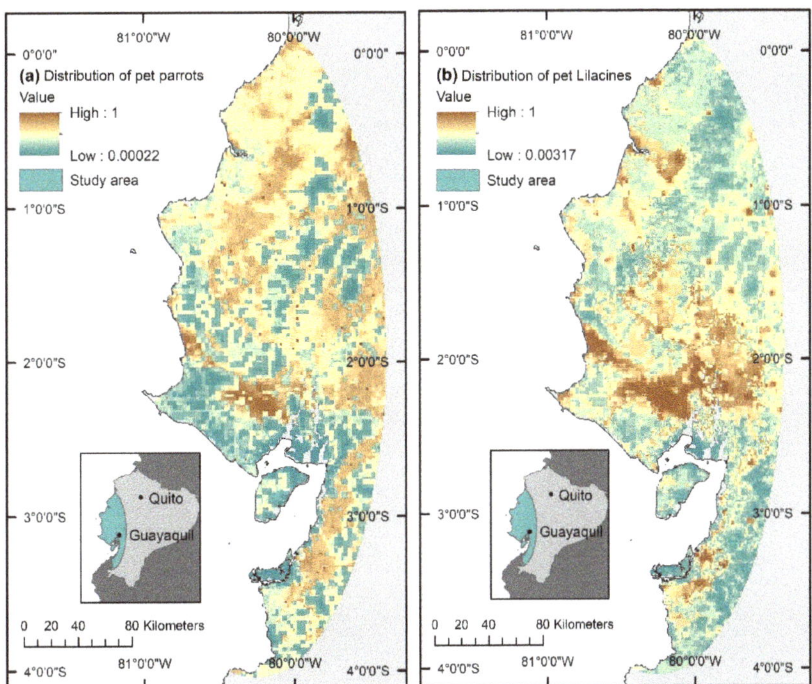

Figure 3. Model predictions showing the distribution of pet parrots (**a**) and pet Lilacines (**b**).

3.3. Incidence of Parrot Ownership and Trapping

Within 12 selected communities where pet parrots occur, 106 (96 men/10 women) participants (min 6, max 13, mean 8.8 per community) took part in interviews. All participants worked outdoors as farmers (57), fishers (25), crab fishers (18), bee keepers (3) or wildlife guides (3). Participants were familiar with *A. lilacina*, the majority seeing them daily (68%), weekly (19%), or monthly (8%), with the remaining 5% just a few times per year. Of all participants, 66% (70) had owned a pet parrot either previously or currently, and 36% (38) a pet Lilacine. The majority (74%, 28) of Lilacine pets had been caught by the owner themselves, with the remainder received as gifts (16%, 6), bought (2%, 6), or found (2%, 6). In the last three years, 15 people have owned a total of 24 Lilacines. In total, 34 people (32%) confirmed that they had previously captured Lilacines, the majority (76%, 26) to keep as a pet themselves, the others to sell (9%, 26) or for undisclosed reasons (15%, 5). Pet ownership and trapping varied between community clusters, with the highest rates of historic and current ownership and trapping of Lilacine Amazons occurring in the crab fishing communities (D) in the southern part of the range (Table 4).

Table 4. The number, age and occupations of people interviewed from each community cluster and the number who reported owning parrots or catching parrots, either previously or in the last three years.

Community Cluster	n	Mean Age (Years)	Occupation: Farmer (F), Fisher (Fi), Crab Fisher (CF), Other (O)				In Life Time:			In the Last Three Years:	
			F	Fi	CF	O	Owned Parrot	Owned Lilacine	Caught Lilacine	Owned Lilacine	Caught Lilacine
A	31	53	8	19	1	3	23	11	9	0	0
B	23	48	23	0	0	0	11	8	8	4	3
C	29	53	26	0	0	3	18	4	4	1	1
D	23	46	0	6	17	0	18	15	13	10	6
Total	106	50	57	25	18	6	70	38	34	15	10

In the last three years, 10 interviewees reported that they had caught Lilacines, with at least 16 birds among them, to either keep the bird as a pet (7), to sell it (1), or for an undisclosed reason (2). All had either no or primary level schooling, and were men 23–72 years old. They reported seeing wild Lilacines daily (9) or weekly (1), and all but one believed the wild population was stable or increasing. In cases where the capture location was given, this always corresponded to the person's occupation, i.e., farmers reported catching parrots in the forest, fishers and crab fishers reported trapping parrots in mangroves (Table 5). Seven of the 10 people who had caught Lilacines in the last three years reported that multiple other people within their community also catch Lilacines, and all 10 knew of multiple pet Lilacines in their community (mean 5.2 Lilacines).

Table 5. The age, gender, schooling, and occupation of all interviewees who reported catching Lilacines in the last three years. We report the trapping location, reason for capture and how many were caught.

Community Cluster	Age (Years)	Gender	Level of Schooling	Occupation	Location of Capture	Reason for Capture	Number of Lilacines Caught in Last Three Years
B	41	Male	Primary	Farmer	Dry forest	Pet	1
	23	Male	Primary	Farmer	Dry forest	Pet	1
	72	Male	Primary	Farmer	Dry forest	Pet	1
C	68	Male	None	Farmer	Undisclosed	Undisclosed	1
D	32	Male	Primary	Crab fisher	Mangrove	Pet	2
	54	Male	Primary	Crab fisher	Mangrove	Pet	1
	40	Male	Primary	Crab fisher	Mangrove	Pet	1
	47	Male	Primary	Crab fisher	Undisclosed	Undisclosed	Unknown
	51	Male	Primary	Fisher	Mangrove	Pet	1
	67	Male	None	Fisher	Mangrove	Sell	7

3.4. *Attitudes towards Parrot Ownership and Trapping*

Across all communities, responses to attitude statements show a strong feeling that wild parrots are important for nature and participants indicated that local people have a responsibility to protect the environment. This is mirrored by a strong feeling of discomfort with outsiders coming to catch parrots and with parrots disappearing from their area. On the contrary, 46% of all participants believe it is OK to catch wild parrots to sell and 32% that it is acceptable to keep a pet parrot. Furthermore, 17% of people did not believe that catching wild parrots could make them become extinct in the local area (Figure 4). There were no significant differences between communities in the distribution of positive, neutral and negative responses to all attitude statements apart from one: "I think it is OK to keep a parrot as a pet". For this, there was a significant difference between mean responses of the community groups (H = 6.613, $p = 0.022$), with 52% of community cluster D believing this is acceptable, and just 20% of cluster A believing so.

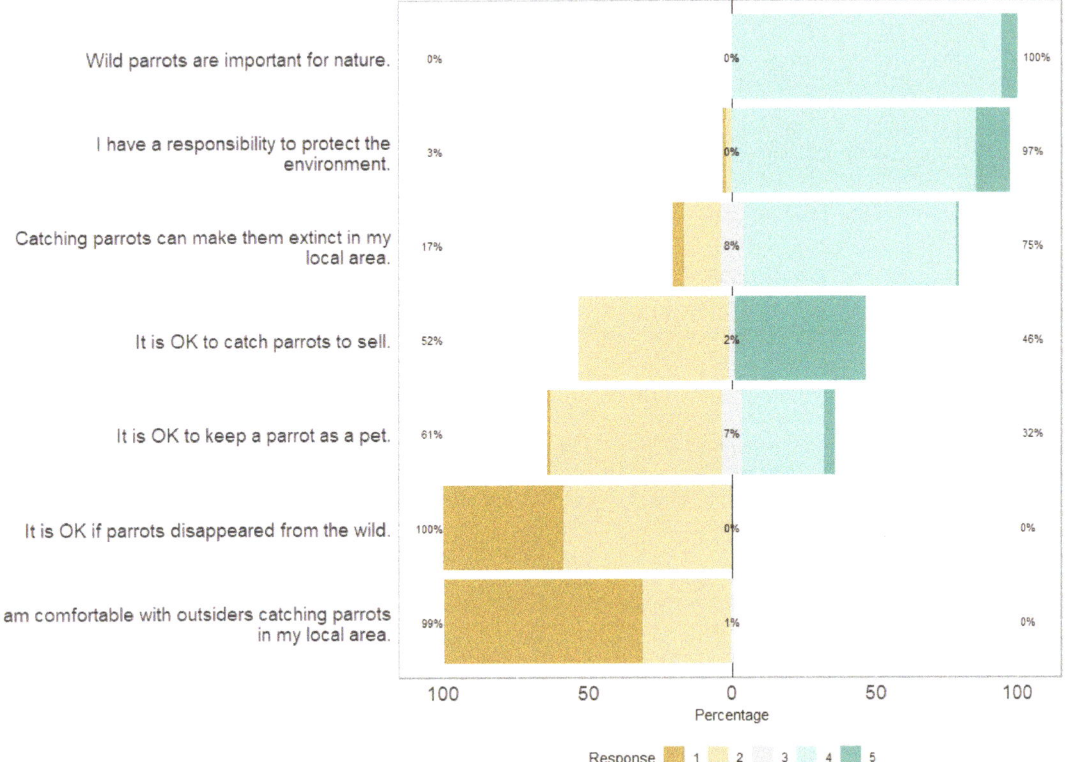

Figure 4. Responses to attitude statements are reported on a five point Likert scale (1 = strongly disagree, 2 = disagree, 3 = neutral, 4 = agree, 5 = strongly agree).

3.5. *Trapping Pressure Index*

When combining our results into a trapping pressure index, we can see variation between the four community clusters, with higher values suggesting a higher risk to the wild Lilacine Amazon population in that area (Table 6). Wild Lilacines occurring around community cluster D are at greatest risk, due to the high level of reported parrot ownership and capture, and a strong local attitude that this is acceptable. Those occurring around community cluster C are also at high risk, as model predictions here show a high probability of pet Lilacine occurrence (0.78), which suggests a high probability of trapping as pet owners mostly report catching their pet themselves. The wild population occurring around community cluster A appears to be at the lowest risk from trapping, as there were no reports here of current Lilacine ownership or capture; however, this is the smallest remaining subpopulation of the species within its range, which could also explain the low prevalence of pets. When considered against participants' responses to their perceived status of the wild population locally, we see more negative responses from the southern community clusters, with the most frequent response in cluster C being 'decreasing' (76%), in cluster D 'stable' (39%), whilst 'increasing' in cluster B (83%) and A (42%).

Table 6. Trapping pressure index for each community cluster, calculated by adding together factors of predicted or reported level of pet ownership and trapping, and attitudes towards pet keeping, capture and trade. Predicted probabilities were converted into percentages for this calculation.

Trapping Pressure Factor	Community Cluster			
	A	B	C	D
Mean probability of predicted parrot ownership (0–1)	0.47	0.35	0.70	0.36
Mean probability of predicted Lilacine ownership (0–1)	0.51	0.59	0.78	0.19
Percentage of participants who owned pet Lilacines in the last three years	0	17	3	43
Percentage of participants who caught Lilacines in last three years	0	13	3	26
Percentage of participants believing it is OK to keep a pet parrot	20	23	37	52
Percentage of participants believing it is OK to catch wild parrots to sell	39	32	41	74
Overall trapping pressure index (rank)	157 (4)	179 (3)	232 (2)	250 (1)

4. Discussion

This study found that 66% of rural coastal communities in Ecuador have pet parrots and 31% have pet Lilacine Amazons *A. lilacina*. Within these communities, 66% of people had owed a pet parrot during their lifetime, and 14% currently owned Lilacines. This is similar to Costa Rica, where 18% of households owned a pet parrot in 2001 [9]. Our current ownership questions focused on just one species, so we expect the level of current ownership of all parrot species to be much higher and similar to Colombia where 58% of all people had pet parrots [10]. Current ownership and reports of Lilacine trapping in the last three years varied between communities, with 0% to 50% and 0% to 26%, respectively. Ten participants confirmed that they had taken Lilacines from the wild in the last three years to keep birds as pets, or to sell them, so we suggest that, similarly to Yellow-shouldered Amazon *Amazona barbadensis* harvesting in Venezuela, there are at least two categories of trappers—"poacher-keepers" and "poacher-sellers" [35], with only the latter having the contacts and logistics to sell birds. People in occupations with an established and frequent trade link—for example, fishers or crab fishers—may have more opportunity to transport trapped birds to other towns to sell. We also found that capture location corresponded to the occupation of the poacher, which may explain the variation between rural communities as occupation depends heavily on location, i.e., in-land or coastal. Our trapping pressure index identified that the southern distribution of the species is likely to be at greatest risk, which agrees with earlier work showing a vast population decline in this area [17] and provides further evidence that this area should be prioritised for conservation support.

Understanding whether taking parrots from the wild is opportunistic or selective is important because selective capture can lead to the extinction of species through over-harvesting [10]. Our results suggest that trapping is selective given the differences in the popularity of species, with some kept in 20 communities and some in none. The two most frequently reported pet parrot species differed greatly in body mass, which, in general, is linked to longevity in captivity [36], suggesting that variation in popularity is not a side effect of survival rates in captivity. Similarly to previous research, we have noted a preference for Amazon parrots, with all three wild occurring species and two non-native species being kept as pets [8]. However, parrot ownership and capture, at least within rural communities, may also be opportunistic. Most parrot owners had caught their bird locally, within areas they visit during a normal days' work, and our predictions showed that pet Lilacines were more likely within the species' wild distribution. Moreover, parrot species kept as pets had a larger average wild range size than those that are not. This all suggests that ownership and capture are driven in part by parrot availability and accessibility [11], but more research including a true measure of wild parrot abundance, and surveys and interviews in larger towns and cities are needed. According to anecdotal reports in some rural communities, orders are placed by outsiders from cities such as Guayaquil or Quito, which fits the typical multi-level chain involving trappers, middlemen and markets described by Pires [37] and needs investigation.

Our interviews revealed that Lilacine Amazons were trapped both in mangroves, where they roost, and dry forests, where they feed and nest [38,39], suggesting that both adults and chicks are being taken from the wild. Anecdotal reports from communities suggest past events of outsiders casting nets over mangrove islands to remove an entire roost of Lilacines at a time. Research has shown that the removal of adults from a population can have more drastic consequences on population size and growth rate, than removal of chicks [40]. In a study of illegal wildlife trade markets in Bolivia, contrary to the idea that most parrots come from nest poaching, 70% of parrots were adults [41]. Our results also suggest that 60% of Lilacines caught in the last three years were from mangroves, so are likely to be adult or juvenile birds. A number of studies have shown that anti-poaching efforts, in the form of additional human presence, can benefit bird populations [42–44] and that recruitment of young people (who may be facilitators in parrot trapping) from the local community to act as nest monitors, can significantly decrease poaching rate [45]. In some cases, nest protection implemented at the correct time of year can have a significant effect [42], but we suggest that year-round protection is needed to safeguard both vulnerable roosting and nesting sites for this species.

The lack of environmental education in Ecuadorian schools is a barrier to reducing parrot ownership and capture [16]. The main purpose of any environmental education strategy is to change people's knowledge and attitudes, and ultimately behaviours [46]. Alone, or in combination with other conservation interventions, environmental education projects can result in a decrease in the persecution of parrots and consequently an increase in population size [43,47]. Most people in our study believed that wild parrots were important for nature and that they themselves had a responsibility to protect parrots. Local people do not want parrots to disappear and are strongly opposed to outsiders coming in to their community to catch them. Contrastingly, up to 74% per community agreed that it was OK to take parrots from the wild to sell, and up to 52% believed that it was OK to keep them as a pet. Furthermore, up to 30% disagreed that catching parrots could make them locally extinct. We found similarity between attitudes and reported behaviours. In areas with more pet Lilacines and reports of parrot trapping, there was also a stronger belief that this was acceptable, compared to areas with fewer pets and trapping. This suggests that changing these attitudes could have an impact on future behaviour, and that the implementation of a targeted behaviour change education project could have conservation benefits to the Lilacine Amazon. We suggest following the practices of the successful PRIDE campaigns [48] which inspire people to take pride in the species and habitats that make their communities so unique, whilst introducing viable alternatives to environmentally destructive practices.

We therefore recommend that a combination of environmental education to change attitudes towards parrot ownership and trapping, and increased protection of wild birds through nest and roost guarding, particularly in the southern part of its range, are conservation priorities for the Lilacine Amazon.

Author Contributions: Conceptualization, R.B., M.J., M.P. and S.M.; Data curation, R.B. and I.S.-P.; Formal analysis, R.B.; Funding acquisition, R.B. and M.P.; Investigation, R.B. and I.S.-P.; Methodology, R.B., M.J. and S.M.; Project administration, R.B.; Resources, R.B. and M.P.; Supervision, M.J. and S.M.;

Visualization, R.B.; Writing—original draft, R.B.; Writing—review and editing, R.B., M.J. and S.M. All authors have read and agreed to the published version of the manuscript.

Funding: This research was funded by the North of England Zoological Society.

Institutional Review Board Statement: The study was approved virtually by The North of England Zoological Society Ethical Review Committee, on Friday 17 November 2017.

Informed Consent Statement: Informed consent was obtained from all subjects involved in the study.

Data Availability Statement: The data presented in this study are available on request from the corresponding author. The data are not publicly available due to anonymity and sensitive content.

Acknowledgments: We would like to extend our thanks to the Amazona Lilacina Education Project volunteers who assisted us with community surveys and interviews. This study would not have been possible without the support of local communities and we thank every person who volunteered their time, valuable knowledge and personal experiences to this research. Thank you to Paolo Escobar for providing one of the pet parrot photographs and to three reviewers for their constructive comments.

Conflicts of Interest: The authors declare no conflict of interest.

References

1. Olah, G.; Butchart, S.H.M.; Symes, A.; Guzmán, I.M.; Cunningham, R.; Brightsmith, D.J.; Heinsohn, R. Ecological and socio-economic factors affecting extinction risk in parrots. *Biodivers. Conserv.* **2016**, *25*, 205–223. [CrossRef]
2. Wright, T.F.; Toft, C.A.; Enkerlin-Hoeflich, E.; Gonzalez-Elizondo, J.; Albornoz, M.; Rodríguez-Ferraro, A.; Rojas-Suárez, F.; Sanz, V.; Trujillo, A.; Beissinger, S.R.; et al. Nest Poaching in Neotropical Parrots. *Conserv. Biol.* **2001**, *15*, 710–720. [CrossRef]
3. Dahlin, C.R.; Blake, C.; Rising, J.; Wright, T.F. Long-term monitoring of Yellow-naped Amazons (*Amazona auropalliata*) in Costa Rica: Breeding biology, duetting, and the negative impact of poaching. *J. Field Ornithol.* **2018**, *89*, 1–10. [CrossRef]
4. Berkunsky, I.; Quillfeldt, P.; Brightsmith, D.; Abbud, M.; Aguilar, J.; Alemán-Zelaya, U.; Aramburú, R.; Arias, A.A.; McNab, R.B.; Balsby, T.; et al. Current threats faced by Neotropical parrot populations. *Biol. Conserv.* **2017**, *214*, 278–287. [CrossRef]
5. Bush, E.R.; Baker, S.E.; Macdonald, D.W. Global Trade in Exotic Pets 2006–2012. *Conserv. Biol.* **2014**, *28*, 663–676. [CrossRef]
6. Ecuador´s Wildlife Trade. Available online: https://www.researchgate.net/profile/Pablo-Sinovas/publication/312600490_Ecuador_s_Wildlife_Trade/links/5885fbd04585150dde4a8016/Ecuador-s-Wildlife-Trade.pdf (accessed on 1 July 2020).
7. Clarke, R.V.; De By, R.A. Poaching, habitat loss and the decline of neotropical parrots: A comparative spatial analysis. *J. Exp. Criminol.* **2013**, *9*, 333–353. [CrossRef]
8. Tella, J.L.; Hiraldo, F. Illegal and Legal Parrot Trade Shows a Long-Term, Cross-Cultural Preference for the Most Attractive Species Increasing Their Risk of Extinction. *PLoS ONE* **2014**, *9*, e107546. [CrossRef] [PubMed]
9. Drews, C. Wild Animals and Other Pets Kept in Costa Rican Households: Incidence, Species and Numbers. *Soc. Anim.* **2001**, *9*, 107–126. [CrossRef]
10. Romero-Vidal, P.; Hiraldo, F.; Rosseto, F.; Blanco, G.; Carrete, M.; Tella, J.L. Opportunistic or Non-Random Wildlife Crime? Attractiveness rather than Abundance in the Wild Leads to Selective Parrot Poaching. *Diversity* **2020**, *12*, 314. [CrossRef]
11. Pires, S.; Clarke, R.V. Are Parrots CRAVED? An Analysis of Parrot Poaching in Mexico. *J. Res. Crime Delinq.* **2011**, *49*, 122–146. [CrossRef]
12. Harrison, R.D.; Sreekar, R.; Brodie, J.F.; Brook, S.; Luskin, M.; O'Kelly, H.; Rao, M.; Scheffers, B.; Velho, N. Impacts of hunting on tropical forests in Southeast Asia. *Conserv. Biol.* **2016**, *30*, 972–981. [CrossRef] [PubMed]
13. Benítez-López, A.; Alkemade, R.; Schipper, A.M.; Ingram, D.J.; A Verweij, P.; Eikelboom, J.; Huijbregts, M.A.J. The impact of hunting on tropical mammal and bird populations. *Science* **2017**, *356*, 180–183. [CrossRef]
14. John, F.A.V.S.; Edwards-Jones, G.; Jones, J.P.G. Conservation and human behaviour: Lessons from social psychology. *Wildl. Res.* **2010**, *37*, 658–667. [CrossRef]
15. Ecuadorian National Assembly. Article 3: Crimes against Wild Fauna. Comprehensive Organic Criminal Code in the Classification of Crimes against Wild Flora and Fauna. 2017. Available online: https://observatoriolegislativo.ec/media/archivos_leyes/Contra_de_la_Flora_y_Fauna_Tr._305545.pdf (accessed on 11 December 2020).
16. Ortiz-von Halle, B. *Bird's-Eye View: Lessons from 50 Years of Bird Trade Regulation & Conservation in Amazon Countries*; TRAFFIC: Cambridge, UK, 2018; p. 198.
17. Biddle, R.; Ponce, I.S.; Cun, P.; Tollington, S.; Jones, M.; Marsden, S.; Devenish, C.; Horstman, E.; Berg, K.; Pilgrim, M. Conservation status of the recently described Ecuadorian Amazon parrot Amazona lilacina. *Bird Conserv. Int.* **2020**, *30*, 586–598. [CrossRef]
18. Athanas, N.; Greenfield, P.J. *Birds of Western Ecuador: A Photographic Guide*; Princeton University Press: Princeton, NJ, USA, 2016.
19. Mideros, M.A. Ecuador: Defining and measuring multidimensional poverty, 2006–2010. *CEPAL Rev.* **2012**, *2012*, 49–67. [CrossRef]
20. Dodson, C.H.; Gentry, A.H. Biological Extinction in Western Ecuador. *Ann. Mo. Bot. Gard.* **1991**, *78*, 273. [CrossRef]
21. Cuesta, F.; Peralvo, M.; Merino-Viteri, A.; Bustamante, M.; Baquero, F.; Freile, J.F.; Muriel, P.; Torres-Carvajal, O. Priority areas for biodiversity conservation in mainland Ecuador. *Neotrop. Biodivers.* **2017**, *3*, 93–106. [CrossRef]

22. BirdLife International. Amazona Lilacina. *IUCN Red List Threat. Species* **2020**, e.T22728296A181432250. Available online: https://dx.doi.org/10.2305/IUCN.UK.2020-3.RLTS.T22728296A181432250.en (accessed on 16 December 2020).
23. IUCN. *Significant Trade in Wildlife: A Review of Selected Species in CITES Appendix II-Volume 3 Birds*; IUCN Conservation Monitoring Centre: Cambridge, UK, 1988; pp. 36–41.
24. BirdLife International and Handbook of the Birds of the World (2020) Bird Species Distribution Maps of the World. Version 2020.1. Available online: http://datazone.birdlife.org/species/requestdis (accessed on 31 July 2020).
25. ArcGIS Desktop, Environmental Systems Research Institute. Available online: https://www.esri.com/en-us/arcgis/products/arcgis-desktop/resources (accessed on 1 November 2020).
26. R Core Team. *R: A Language and Environment for Statistical Computing*; R Core Team: Vienna, Austria, 2020.
27. Phillips, S.J.; Anderson, R.P.; Schapire, R.E. Maximum entropy modeling of species geographic distributions. *Ecol. Model.* **2006**, *190*, 231–259. [CrossRef]
28. Biddle, R.; Solis-Ponce, I.; Jones, M.; Marsden, S.; Pilgrim, M.; Devenish, C. Assessing the value of local community knowledge in species distribution modelling for a threatened Neotropical parrot. *Biodivers. Conserv.* **2020**, in press.
29. Sorichetta, A.; Hornby, G.M.; Stevens, F.R.; Gaughan, A.E.; Linard, C.; Tatem, A.J. High-resolution gridded population datasets for Latin America and the Caribbean in 2010, 2015, and 2020. *Sci. Data* **2015**, *2*, 150045. [CrossRef]
30. OpenStreetMap Foundation, OSMF. OpenStreetMap. Available online: https://www.openstreetmap.org/copyright (accessed on 1 November 2019).
31. Aiello-Lammens, M.E.; Boria, R.A.; Radosavljevic, A.; Vilela, B.; Anderson, R.P. spThin: An R package for spatial thinning of species occurrence records for use in ecological niche models. *Ecography* **2015**, *38*, 541–545. [CrossRef]
32. Zuur, A.F.; Ieno, E.N.; Walker, N.; Saveliev, A.A.; Smith, G.M. *Mixed Effects Models and Extensions in Ecology with R*; Springer: New York, NY, USA, 2009.
33. Wisz, M.S.; Hijmans, R.J.; Elith, J.; Peterson, A.T.; Graham, C.H.; Guisan, A.; NCEAS Predicting Species Distributions Working Group. Effects of sample size on the performance of species distribution models. *Divers. Distrib.* **2008**, *14*, 763–773. [CrossRef]
34. Package 'likert'. Available online: https://cran.r-project.org/web/packages/likert/likert.pdf (accessed on 15 September 2020).
35. Sanchez-Mercado, A.; Blanco, O.; Sucre-Smith, B.; Briceño-Linares, J.M.; Peláez, C.; Rodriguez, J.P.; Sucre-Smith, B. Using Peoples' Perceptions to Improve Conservation Programs: The Yellow-Shouldered Amazon in Venuzuela. *Diversity* **2020**, *12*, 342. [CrossRef]
36. Brouwer, K.; Jones, M.L.; King, C.E.; Schifter, H. Longevity records for Psittaciformes in captivity. *Int. Zoo Yearb.* **2000**, *37*, 299–316. [CrossRef]
37. Pires, S.F. The illegal parrot trade: A literature review. *Glob. Crime* **2012**, *13*, 176–190. [CrossRef]
38. Berg, K.S.; Angel, R.R. Seasonal roosts of Red-lored Amazons in Ecuador provide information about population size and structure. *J. Field Ornithol.* **2006**, *77*, 95–103. [CrossRef]
39. Dupin, M.K.; Dahlin, C.R.; Wright, T.F. Range-Wide Population Assessment of the Endangered Yellow-Naped Amazon (*Amazona auropalliata*). *Diversity* **2020**, *12*, 377. [CrossRef]
40. Valle, S.; Collar, N.J.; Harris, W.E.; Marsden, S. Trapping method and quota observance are pivotal to population stability in a harvested parrot. *Biol. Conserv.* **2018**, *217*, 428–436. [CrossRef]
41. Pires, S.F.; Schneider, J.L.; Herrera, M.; Tella, J.L. Spatial, temporal and age sources of variation in parrot poaching in Bolivia. *Bird Conserv. Int.* **2015**, *26*, 293–306. [CrossRef]
42. A González, J. Harvesting, local trade, and conservation of parrots in the Northeastern Peruvian Amazon. *Biol. Conserv.* **2003**, *114*, 437–446. [CrossRef]
43. Vaughan, C.L.; Nemeth, N.M.; Cary, J.; Temple, S. Response of a Scarlet Macaw Ara macao population to conservation practices in Costa Rica. *Bird Conserv. Int.* **2005**, *15*, 119–130. [CrossRef]
44. Granadeiro, J.P.; Dias, M.P.; Rebelo, R.; Santos, C.D.; Catry, P. Numbers and Population Trends of Cory's Shearwater Calonectris diomedea at Selvagem Grande, Northeast Atlantic. *Waterbirds* **2006**, *29*, 56–60. [CrossRef]
45. Briceño-Linares, J.M.; Rodríguez, J.P.; Rodríguez-Clark, K.M.; Rojas-Suárez, F.; Millán, P.A.; Vittori, E.G.; Carrasco-Muñoz, M. Adapting to changing poaching intensity of yellow-shouldered parrot (*Amazona barbadensis*) nestlings in Margarita Island, Venezuela. *Biol. Conserv.* **2011**, *144*, 1188–1193. [CrossRef]
46. Jacobson, S.K.; McDuff, M.D.; Monroe, M.C. *Conservation Education and Outreach Techniques*; Oxford University Press: Oxford, UK, 2015.
47. Sanz, V.; Grajal, A. Successful Reintroduction of Captive-Raised Yellow-Shouldered Amazon Parrots on Margarita Island, Venezuela. *Conserv. Biol.* **2008**, *12*, 430–441. [CrossRef]
48. The Principles of Pride. Available online: https://www.rare.org/wp-content/uploads/2019/02/Rare-Principles-of-Pride.pdf (accessed on 15 October 2020).

Article

Using Peoples' Perceptions to Improve Conservation Programs: The Yellow-Shouldered Amazon in Venezuela

Ada Sánchez-Mercado [1,2,*], Oriana Blanco [1], Bibiana Sucre-Smith [1], José Manuel Briceño-Linares [1], Carlos Peláez [1] and Jon Paul Rodríguez [1,3,4]

1 Provita, Calle La Joya, Edificio Unidad Técnica del Este, Piso 10, Oficina 30, Chacao, Caracas 1060, Venezuela; oblanco@provitaonline.org (O.B.); bsucre@provitaonline.org (B.S.-S.); jbriceno@provitaonline.org (J.M.B.-L.); capelaez@gmail.com (C.P.); JonPaul.RODRIGUEZ@ssc.iucn.org (J.P.R.)

2 School of Biological, Earth and Environmental Sciences, University of New South Wales, Kensington, NSW 2052, Australia

3 Centro de Ecología, Instituto Venezolano de Investigaciones Científicas, Apdo. 20632, Caracas 1020-A, Venezuela

4 IUCN Species Survival Commission, 28 rue Mauverney, CH-1196 Gland, Switzerland

* Correspondence: ay.sanchez.mercado@gmail.com; Tel.: +61-48-1203171

Received: 31 July 2020; Accepted: 29 August 2020; Published: 5 September 2020

Abstract: The perceptions and attitudes of local communities help understand the social drivers of unsustainable wildlife use and the social acceptability of conservation programs. We evaluated the social context influencing illegal harvesting of the threatened yellow-shouldered Amazon (*Amazona barbadensis*) and the effectiveness of a longstanding conservation program in the Macanao Peninsula, Margarita Island, Venezuela. We interviewed 496 people from three communities and documented their perceptions about (1) status and the impact of threats to parrot populations, (2) acceptability of the conservation program, and (3) social processes influencing unsustainable parrot use. Approval of the program was high, but it failed to engage communities despite their high conservation awareness and positive attitudes towards the species. People identified unsustainable use as the main threat to parrots, but negative perceptions were limited to selling, not harvesting or keeping. Harvesters with different motivations (keepers, sellers) may occur in Macanao, and social acceptability of both actors may differ. Future efforts will require a stakeholder engagement strategy to manage conflicts and incentives to participation. A better understanding of different categories of harvesters, as well as their motives and role in the illegal trade network would provide insights to the design of a behavior change campaign.

Keywords: conservation management; conservation threats; drivers of extinction; illegal wildlife trade; parrot conservation; Psittacidae conservation; threatened species; unsustainable use of wildlife

1. Introduction

Conservation programs often focus on reducing the unsustainable use of wildlife in highly complex social–ecological environments, where local communities are key actors in both the trade chain and the conservation actions implemented [1–4]. Studying the perceptions and attitudes of local communities towards the unsustainable use of wildlife has been key to understanding social and cultural drivers of unsustainable use [5,6], social acceptability of conservation management [7–9], and the design of culturally suitable and more tenable conservation actions [10].

Due to the cultural nature of the illegal parrot trade, local people's perceptions are particularly important to assess performance of conservation programs aimed at tackling this threat. People have been keeping parrots as pets for centuries [11] (Psittaciformes, which include parrots, macaws, parakeets, parrotlets, and cockatoos), and today 28% (111 of 398) of extant species are listed as threatened on the International Union for Conservation of Nature's Red List of Threatened Species [12]. Unsustainable use, including harvest, trade, and keeping, is highly influenced by the species' attractiveness to humans [13,14], but also by cultural and social factors: parrots owners often regard their animals as "family members", perceived and treated as children [15]. This social role may also influence understanding of psittacid conservation challenges and attitudes towards conservation actions.

Here, we evaluate the social context influencing the use of the yellow-shouldered Amazon (*Amazona barbadensis*) and the effectiveness of a longstanding conservation program led by Provita and aimed at restoring their population in the Macanao Peninsula, Margarita Island, Venezuela. The yellow-shouldered Amazon ("parrots" hereafter) is classified as vulnerable internationally [16] and endangered regionally [17], due to the capture of nestlings for the pet trade (both domestic and international) [18] and the destruction of nesting and feeding habitats [19]. The main population (ca. 1600 individuals) inhabits Macanao Peninsula [20]. Provita is a Venezuelan non-governmental organization that has implemented the Yellow-Shouldered Amazon Conservation Program in Macanao over the last 31 years. The program includes school-age environmental education activities, and full-time surveillance of natural and artificial nests in the main breeding site of this parrot population (La Chica). The Ecoguardians, a cooperative of local young people recruited, trained, and hired by Provita, have implement most of these actions in the field [20,21]. However, after 31 years of implementation, illegal harvesting persists, and it is unclear whether this unsustainable use points towards the need to strengthen enforcement strategies [20] or aim for a more holistic approach focused on behavioral change. We specifically explore local perceptions about (1) the status and impact of threats to parrot populations, (2) acceptability of the conservation program in terms of support and responsibilities, and (3) social processes influencing unsustainable parrot use and the performance of conservation actions.

2. Materials and Methods

2.1. Study Area and Socioeconomic Context

Macanao Peninsula is located in the western portion of Margarita Island and is less developed for tourism than the eastern part, resulting in ecosystems that are in relatively good condition (Figure 1) [20]. By 2011, there were approximately 24,419 inhabitants in Macanao (a tenth of Margarita's population). Employment opportunities are scarce, with fishing being the primary economic activity [22].

2.2. Interview Instruments and Survey Process

Between March and September 2017, we interviewed 496 people from three communities across Macanao (Boca del Río, El Horcón, and Robedal; Figure 1b) using a self-reporting questionnaire. All participants were adults (>18 years old), and had lived in Macanao for at least one year. We obtained verbal informed consent from each subject, after explaining the research objectives and assuring subjects that information would be used only for research, and presented the data in aggregate analyses, protecting each participant's identity [23]. The survey protocol was approved by the Laboratory of Political Ecology of the Venezuelan Institute of Scientific Research (February 2017), who acted as the external ethical committee. Households were chosen randomly from community maps, by selecting every fourth house. The sample size represented 21–30% of households in each community.

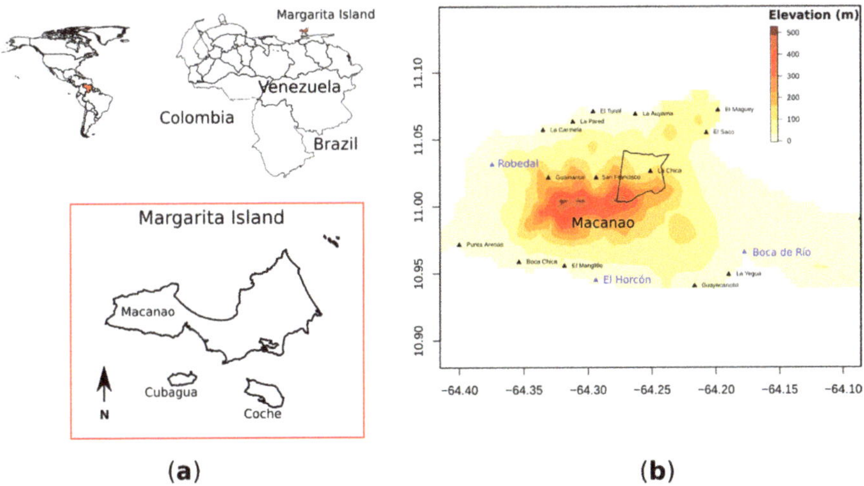

Figure 1. Study area. (**a**) Relative position of Venezuela, Margarita Island, and the Macanao Peninsula. (**b**) Elevation gradient in Macanao. Communities surveyed are highlighted in blue. The main nesting site where Provita implements nest monitoring and surveillance, Hato San Francisco, Quebrada La Chica, is delimited by a black polygon.

We evaluated the general socioeconomic context of participants by asking about their age, gender, level of education, employment status, source of income, and whether this income was enough to cover family monthly expenses. The survey instrument evaluated three distinct aspects related to conservation practice [1]: (1) ecological outcomes, (2) acceptability of conservation management, and (3) social processes influencing the effectiveness of conservation actions (Table 1). To assess perceived ecological outcomes, we evaluated three aspects: awareness of conservation status, perceived threats and their impact on wild populations, and the success of surveillance in preventing fledgling poaching (Table 1). We measured awareness by asking two closed questions: whether the participants keep parrots at home (owners) or not (non-owners), and if they think there are more parrots in captivity than in the wild (yes/no). We evaluated the perceived impact on the wild population by asking two closed questions: "Do you think that the wild parrot population will go extinct in the next 10 years?" and "Do you think that the wild parrot population is stable, declining, or increasing?" To assess people's knowledge about threats faced by the wild parrot population, we asked an open question—"What is the main threat faced by parrots?"—and then reclassified the answers into four categories: "unsustainable use", "deforestation", "drought", and "predators." We asked "Where do you think your parrot comes from?" as a closed question, with the names of the most important nesting sites as options. We used this question as a measure of surveillance effectiveness, as La Chica has been the only nesting site under protection during the last 31 years (Table 1).

To measure the acceptability of conservation management, we evaluated three aspects: support for the conservation program, perceptions of other stakeholders in the process, and perceived responsibilities and roles. We measured support for the conservation program by asking awareness of Provita's work with a closed question "Do you know Provita's work?", and whether it addresses the main threats to the species: "What do you think is the main conservation problem that Provita cares for?" For this latter question, we reclassified the answers into the same four categories we used to assess people's knowledge about threats, so that responses were comparable. Given that the Ecoguardians are a key stakeholder, we inquired about perceptions towards Ecoguardians with an open question, and then reclassified the answers into positive and negative perceptions. To assess perceived responsibilities, we asked which are the institutions responsible for parrot conservation (Provita, communities, or government authorities). To evaluate the role that people have in the illegal parrot trade chain, we used an open question "How did you get your parrot?" We then reclassified the answers into four categories "harvested", "bought", "rescued", and "present/gift" (Table 1).

Table 1. Evaluation of the yellow-shouldered Amazon Conservation Program, based on perceptions in three communities of the Macanao Peninsula, Margarita Island, Venezuela.

Conservation Issue	Aspect Evaluated	Questions
Ecological outcomes of conservation	Awareness about species conservation status	Do you keep a parrot at home? Do you think that there are more parrots in captivity than in the wild?
	Perceived impact on wild population	Do you think that the wild parrot population will go extinct in the next 10 years? Do you think that the wild parrot population is stable, declining, or increasing?
	Perception of species threats Effects of the conservation action (surveillance)	What is the main threat faced by parrots? What is the main location for fledgling extraction?
Acceptability of conservation management	Support for conservation program	Do you know Provita's work? What do you think is the main conservation problem that Provita cares for?
	Perceptions of other stakeholders in the process	What do you think about the work of Ecoguardians?
	Perceived responsibilities and roles	Who is the entity/organization responsible for parrot conservation? How did you get your parrot? (role in the trade chain)
Social processes affecting conservation actions	Social value of the species	What does your parrot mean to you? Who gave you your parrot?
	Attitudes towards stages in the trade chain	Do you agree with this statement? "I will always want to keep a parrot as pet." Fledgling parrot extraction is ... Do you report poachers? Selling fledgling parrots is ...

To understand social processes affecting conservation action, we evaluated two aspects: the social value of parrots and attitudes towards harvesting, selling, and keeping. We used an open question "What does your parrot mean to you?", and then we reclassified the answers into three categories: "pet", "a family member", or "symbol." We also asked whether their parrot was provided by a member of the community, a relative, or an outsider. We used a statement to measure attitudes toward keeping parrots as pets, which was "I will always want to keep a parrot as pet"; we assessed answers on a five-point scale that ranged from 1 (strongly disagree) to 5 (strongly agree). We asked about attitudes towards reporting poaching with a closed question "Do you report poachers?", and in the instances with negative replies we additionally asked "Why not?" and aggregated the answers into four categories: "denounce", "not denounce", "indifferent", and "support." We evaluated attitudes toward extraction and selling using open statements, such as "Fledgling parrot extraction is ... " and "Selling fledgling parrots is ... ", and then we classified the answers into positive or negative attitudes. (Table 1). Finally, we asked how much their parrots were worth in national currency, and converted it into USD using the weekly mean of the currency exchange rate, and how many individuals they currently keep captive.

We summarized the responses (number of records and percentages) for each variable at the community level and for the overall sample.

3. Results

3.1. Characteristics of the Sample

Survey participants were 69% female, with a mean age of 43.7 years old (SD = 15.0). Half of survey participants were unemployed (53%), and 51% of them had a high school diploma, while 26% have had university studies (Table 2).

Family income comes mainly from government social support (34%), salary (24%), or retirement pension (23%). For the majority of participants (79%), their income was not enough to cover monthly basic expenses (Table 2). Among the communities surveyed, El Horcón had the most critical socioeconomic condition, with the highest unemployment rate, more dependence on government help (70%), and a lower education level (only 8% of participants hold a university degree, compared to 35% in Boca del Río).

3.2. Perceptions about Species Conservation Status and Ecological Outcomes

Twenty-two percent of participants keep at least one parrot at home (Figure 2a). The majority (79%) believed that there are more parrots in the wild than in captivity. Although most participants (80%) believed that wild populations are decreasing, 53% thought that it may lead to extinction in 10 years (Figure 2a).

Unsustainable use was identified as the main threat to parrots (69% of participants; Figure 2b), and people believed that harvest occurred mainly at sites other than La Chica (61% of participants; Figure 2b).

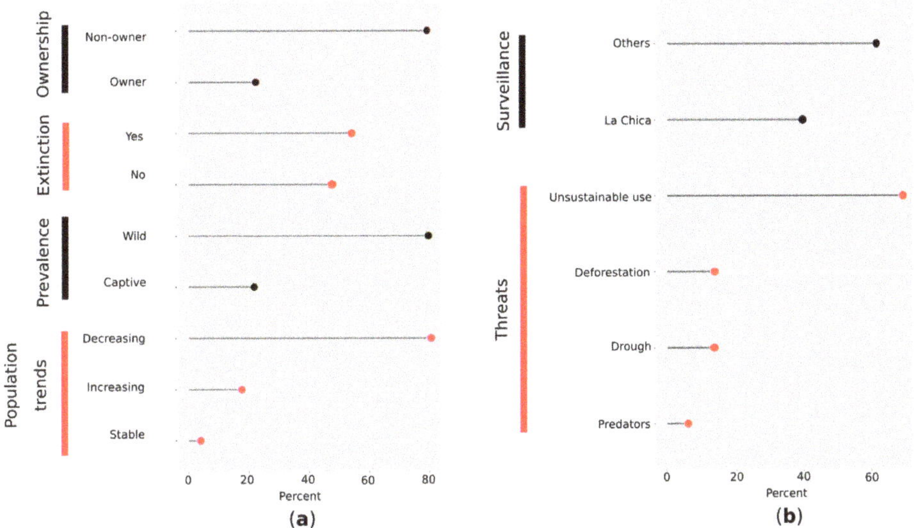

Figure 2. Perceived conservation outcomes from Macanao inhabitants regarding (**a**) population status of the yellow-shouldered Amazon and (**b**) main threats and the impact of nest surveillance.

Table 2. Socio-economic characteristics of the three communities surveyed in Macanao Peninsula, Margarita Island, Venezuela. Percentage of people by gender, employment status, educational level, income level, and source of income are shown for each community (number of answers in each category). Also, mean and standard deviation of age by community is shown.

Community	Gender		Employment Status		Age		Education Level				Income Level					Source of Income			
	F	M	Unemployed	Employed	Mean	SD	None	Primary	High School	University	All	Almost All	The Half	Few	Nothing	Social Program	Own Revenues	Retirement Pension	Salary
Boca del Río	64 (190)	36 (107)	48 (141)	52 (151)	47.7	14.6	1 (3)	14 (42)	50 (147)	35 (102)	0 (0)	2 (5)	8 (23)	28 (84)	62 (184)	26 (66)	19 (48)	26 (65)	29 (73)
El Horcón	84 (41)	16 (8)	68 (32)	32 (15)	43.4	15.8	2 (1)	31 (15)	59 (29)	8 (4)	2 (1)	16 (8)	20 (10)	37 (18)	24 (12)	70 (33)	2 (1)	19 (9)	9 (4)
Robledal	74 (111)	26 (39)	58 (84)	42 (62)	44.0	15.1	5 (7)	31 (45)	50 (73)	15 (22)	5 (8)	16 (23)	16 (24)	31 (45)	32 (46)	36 (40)	27 (79)	18 (94)	19 (98)
All	69 (342)	31 (154)	53	47	43.7	15.0	2	21	51	26	2	7	12	30	49	34	19	23	24

3.3. Acceptability of Conservation Management

Perceptions about Ecoguardians were mixed, with 47% of participants holding negative opinions, mostly because they believed Ecoguardians participate in poaching (Figure 3).

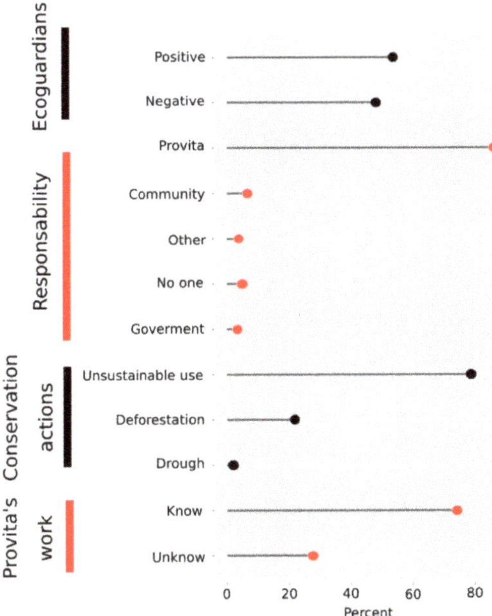

Figure 3. Acceptability of conservation management in three communities of the Macanao Peninsula regarding the yellow-shouldered Amazon Conservation Program.

People had a high level of awareness of Provita's work (73% of participants) and perceived that their conservation actions are coherent with the primary threat, being focused on the reduction of unsustainable harvest (68%; Figure 3). For the majority of participants, Provita is the organization responsible for parrot conservation (85%), while the role or responsibility of the government is low (9%) and the community even lower (6%; Figure 3).

3.4. Social Processes

Affective values—parrots as part of the family—were predominant (82%), and in most cases, parrots were presents from relatives (56%) or provided by people from the community (40%). Most interviewees participated in the trade chain as keepers, whether accepting parrots as presents (39%), directly buying (26%), or rescuing abandoned fledglings (21%). About 15% of interviewees admitted being directly involved in extracting their parrot from the wild (Figure 4a). In a few instances (4%), captive parrots came from people outside of Macanao (Figure 4a). Half of interviewees would not report poachers, because either they are relatives or they feared retaliation (Figure 4a).

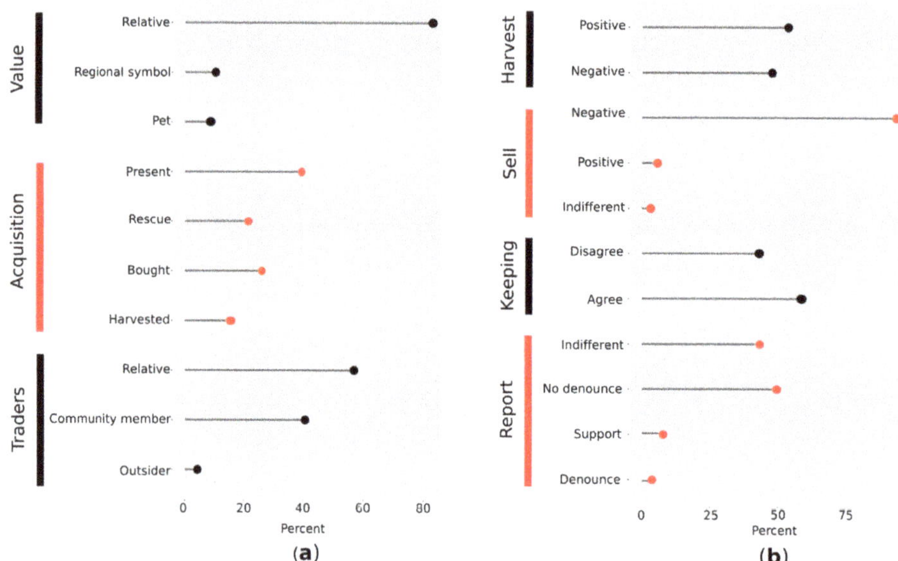

Figure 4. Social processes affecting conservation actions implemented by Provita's yellow-shouldered Amazon Conservation Program in Macanao: (**a**) social values and (**b**) attitudes towards stages in the trade chain.

More than half of participants (53%) had positive attitudes towards poaching, as they mostly considered it as a "child prank", an additional source of income, or a tradition (Figure 4b). Participants with negative attitudes towards poaching (47%) considered this practice as "reproachable" or "improper" (Figure 4b).

Almost all participants had negative attitudes towards selling parrots (93%; Figure 4b). Sixty percent of participants showed positive attitudes towards keeping parrots, but a significant proportion (40%) also disagreed with this behavior (Figure 4b). The mean price was USD 1.70 (value range USD 0.30–7.10).

4. Discussion

Information on the perceptions and attitudes of local communities is important to identify strategies that best suit conservation objectives, alongside the social and cultural context of local communities that use wildlife. The yellow-shouldered Amazon Conservation Program in Macanao has influenced both positive relations between conservation practitioners and local communities, as well as positive perceptions and attitudes towards species conservation. The absence of a historic baseline prevents before–after or control–treatment comparisons, but in general Macanao communities recognized the leading role of Provita, and perceived that their efforts are articulated by species conservation needs.

Social acceptance of the yellow-shouldered Amazon Conservation Program relates to trust and the long-term commitment of Provita. Although the approval of the program is high, it failed to engage and empower local communities in conservation activities. People were not willing to participate in spite of their high conservation awareness and positive attitudes toward the species.

Future efforts in Macanao will require a stakeholder engagement strategy that fosters high quality management decisions and cost-effective implementations [24]. Given the heterogeneity and changing socio-economic conditions in Macanao, this strategy should include identification of stakeholder relationships and understanding of the incentives to promote their participation. A key stakeholder in

Macanao are the Ecoguardians, a cooperative of local young people that implements conservation actions in the field, mostly related to monitoring and surveillance of parrot nests during the breeding season. The scheme has successfully converted ex-harvesters into parrot protectors [20]. As they are of the age and background of current harvesters, this was perceived as an opportunity for Ecoguardians to become "peer multipliers" of positive conservation attitudes among poachers and the general community [20]. However, half of the community believe them to be poachers. Social network analysis could serve to map the relationships between the Ecoguardians and other stakeholders [25], and these insights, combined with social marketing campaigns, could be used to improve the image of Ecoguardians within the communities of Macanao.

Development of an engagement strategy in Macanao will require identification of the most effective incentive to promote stakeholder participation. Economic benefits are the most frequent incentive [24]. The yellow-shouldered Amazon Conservation Program offered communities alternative livelihoods as Ecoguardians. More than 90 local young people have received income in the last 16 years. Expanding the current strategy to include volunteers, using personal and social benefits as incentives, may enhance parrot breeding success. People can participate in surveillance and enforcement, annual censuses, and building nest boxes to foster parrot nesting opportunities [20].

Social Processes in the Yellow-Shouldered Amazon Trade

Evaluation of perceptions also allowed us to interpret how people understand the trade chain and what their role is in it. People from Macanao were aware of the yellow-shouldered Amazon population decline, and they clearly identified unsustainable use as the main cause. However, people seem to relate unsustainable use only to selling, but not to harvesting and keeping.

Selling was the main motivation to harvest fledgling Amazons. Our data partly supports the notion that harvesting for profit is a primary motivation. Although yellow-shouldered Amazons are relatively inexpensive compared to other *Amazona* species in South American markets (e.g., the mealy parrot (*Amazona farinose*) sells for between USD 500–875 in Bolivia) [26], their trade is an opportunity to improve family income through an activity with low risk. However, the profile of poachers does not typically coincide with those of commercial intermediaries [27]. In contrast, in our study poachers kept the parrot as a pet, and professed the same affectionate, non-utilitarian values as keepers. If a sub-group of organized poachers with profit motivation exists, they will require different anti-harvesting strategies than the rest of the poacher community. The fact that people had opposite perceptions about sellers and poachers suggests that different categories (poacher-keepers, poacher-sellers) occur in Macanao, and social acceptability of both actors may differ [28]. For example, "poacher-keepers" could be those harvesting parrots for personal consumption or for relatives and friends. On the other hand, "poacher-sellers" will be able to sell parrots locally, and have the contacts and logistics to sell them in the rest of the country. The former may be more socially accepted than the latter, but they will not report each other because there are relatives involved or fear of retaliation. Future studies should account for the different categories of poachers, their diverse motives, and their role in the illegal trade network, in order to design more effective conservation interventions [9].

As expected, keeping was a widely accepted behavior, but the fact that a significant proportion of interviewees did not agree with parrot keeping suggests that the conservation program has successfully increased conservation concern among people, and there is a good opportunity to change consumptive behavior [29,30]. Local people's knowledge about species conservation status and threats is likely the result of the environmental education program implemented over the last 31 years in 13 schools across Macanao [20,21]. By itself, high levels of knowledge and awareness are not enough for people to engage in actions relevant to species conservation, yet this is the ideal scenario for implementing behavioral change campaigns focused on reducing the unsustainable use of wildlife [31].

Attitudes more consistent with conservation objectives could act as a "seed" in a behavioral change campaign focused on promoting more sustainable uses. For example, citizen science and volunteer programs might help reduce parrot demand while conservation awareness is increased.

"Seed" members may recruit new members into their social networks, who subsequently encourage additional people to participate, and so on [32,33].

The contradiction between improving attitudes and continued high levels of unsustainable parrot harvesting suggests that law enforcement should remain a central activity of the species conservation program. However, the program must not over-rely on enforcement measures, because they fail to address social and cultural factors driving the parrot trade. We suggest that a more bottom-up approach that recognizes the views and motivations of local actors and promotes their engagement in conservation actions could translate to not only lower harvesting and keeping rates, but also in an increase of public support and successful implementation.

5. Conclusions

This study contributes to parrot conservation by incorporating perceptions and attitudes to improve adaptive and evidence-based conservation programs. By improving our understanding of key systemic drivers, we are now able to design demand-focused interventions to better tackle the illegal trade of the yellow-shouldered Amazon in Macanao. The fact that despite high conservation awareness and understanding of threats to parrots, people still fail to link their consumption with the illegal trade chain, suggests that a greater effort is needed to demonstrate behavior's impact in conservation, and importantly, which other behavioral options people can pursue to reduce that impact [34]. Behavioral change campaigns based on social marketing must be evidence-based [35], but also nuanced, with cultural, social, logistic, and funding challenges in order to set realistic objectives and efficient outcomes [36]. In the particular case of the yellow-shouldered Amazon trade in Macanao, such nuanced views will imply (1) improved stakeholder engagement strategies, to both manage conflicts and incentive participation and empowerment; (2) the creation of flexible and creative implementation strategies that take into account the widespread poverty, as well as the prevalence of adult women, who are mostly single mothers; and (3) future research to improve our understanding of different categories of harvesters, as well as their motives and role in the illegal trade network [28].

Author Contributions: Conceptualization, A.S.-M., O.B., B.S.-S., and J.P.R.; data curation, O.B.; formal analysis, A.S.-M.; investigation, O.B., J.M.B.-L., and C.P.; methodology, A.S.-M.; writing–original draft, A.S.-M. and O.B.; writing–review & editing, A.S.-M., B.S.-S., J.M.B.-L., C.P., and J.P.R. All authors have read and agreed to the published version of the manuscript.

Funding: This research was funded by the Whitley-Segré Conservation Fund, the Kilverstone Wildlife Charitable Trust, and the World Land Trust.

Acknowledgments: We are grateful to the anonymous participants in this research for their cooperation, and to Hato San Francisco–Arenera La Chica, Fogones y Banderas, the Macanao Marine Museum, and Fundefir-Bankomunal for their support during field work.

Conflicts of Interest: The authors declare no conflict of interest.

References

1. Bennett, N.J. Using perceptions as evidence to improve conservation and environmental management. *Conserv. Biol.* **2016**, *30*, 582–592. [CrossRef]
2. Bennett, N.J.; Roth, R.; Klain, S.C.; Chan, K.; Christie, P.; Clark, D.A.; Cullman, G.; Curran, D.; Durbin, T.J.; Epstein, G.; et al. Conservation social science: Understanding and integrating human dimensions to improve conservation. *Biol. Conserv.* **2017**, *205*, 93–108. [CrossRef]
3. Schultz, P.W. Conservation means behavior. *Conserv. Biol.* **2011**, *25*, 1080–1083. [CrossRef] [PubMed]
4. Dobson, A.D.M.; De Lange, E.; Keane, A.; Ibbett, H.; Milner-Gulland, E.J. Integrating models of human behaviour between the individual and population levels to inform conservation interventions. *Philos. Trans. R. Soc. B Biol. Sci.* **2019**, *374*. [CrossRef] [PubMed]
5. Ebua, V.B.; Agwafo, T.E.; Fonkwo, S.N. Attitudes and perceptions as threats to wildlife conservation in the Bakossi area, South West Cameroon. *Int. J. Biodivers. Conserv.* **2011**, *3*, 631–636.

6. Gandiwa, E.; Zisadza-Gandiwa, P.; Mango, L.; Jakarasi, J. Law enforcement staff perceptions of illegal hunting and wildlife conservation in Gonarezhou National Park, southeastern Zimbabwe. *Trop. Ecol.* **2014**, *55*, 119–127.
7. Gandiwa, E.; Heitkönig, I.M.A.; Lokhorst, A.M.; Prins, H.H.T.; Leeuwis, C. CAMPFIRE and human-wildlife conflicts in local communities bordering northern Gonarezhou National Park, Zimbabwe. *Ecol. Soc.* **2013**, *18*. [CrossRef]
8. Duncker, L.; Gonçalves, D. Community perceptions and attitudes regarding wildlife crime in South Africa. *CSIR* **2017**, *11*, 191–197.
9. Jenkins, H.M.; Mammides, C.; Keane, A. Exploring differences in stakeholders' perceptions of illegal bird trapping in Cyprus. *J. Ethnobiol. Ethnomed.* **2017**, *13*, 67. [CrossRef]
10. Gebregziabher, D.; Soltani, A. Exclosures in people's minds: Perceptions and attitudes in the Tigray region, Ethiopia. *For. Policy Econ.* **2019**, *101*, 1–14. [CrossRef]
11. Anderson, P.K. A bird in the house: An anthropological perspective on companion parrots. *Soc. Anim.* **2008**, *11*, 393–418. [CrossRef]
12. Berkunsky, I.; Quillfeldt, P.; Brightsmith, D.; Abbud, M.; Aguilar, J.; Alemán-Zelaya, U.; Aramburú, R.; Arce Arias, A.; Balas McNab, R.; Balsby, T.; et al. Current threats faced by Neotropical parrot populations. *Biol. Conserv.* **2017**, *214*, 278–287. [CrossRef]
13. Tella, J.L.; Hiraldo, F. Illegal and legal parrot trade shows a long-term, cross-cultural preference for the most attractive species increasing their risk of extinction. *PLoS ONE* **2014**, *9*, e107546. [CrossRef] [PubMed]
14. Vidal, P.R.; Hiraldo, F.; Rosseto, F.; Blanco, G.; Carrete, M.; Tella, J.L. Opportunistic or Non—Random Wildlife Crime? Attractiveness rather than Abundance in the Wild Leads to Selective Parrot Poaching. *Diversity* **2020**, *12*, 314. [CrossRef]
15. Sollund, R. Expressions of speciesism: The effects of keeping companion animals on animal abuse, animal trafficking and species decline. *Crime Law Soc. Chang.* **2011**, *55*, 437–451. [CrossRef]
16. BirdLife International IUCN Red List for Birds. Available online: http://www.birdlife.org (accessed on 10 October 2019).
17. Rojas-Suárez, F.; Rodríguez, J.P. Cotorra cabeciamarilla, Amazona barbadensis. In *Libro Rojo de la Fauna Venezolana*, 4th ed.; Provita & Fundación Empresas Polar: Caracas, Venezuela, 2015.
18. Sánchez-Mercado, A.; Asmüssen, M.; Rodríguez, J.P.; Moran, L.; Cardozo-Urdaneta, A.; Morales, L.I. Illegal trade of the Psittacidae in Venezuela. *Oryx* **2017**, 1–7. [CrossRef]
19. Rodriguez, J.P.; Fajardo, L.; Herrera, I.; Sánchez, A.; Reyes, A. Conservation and management of the yellow-shouldered parrot (*Amazonas barbadensis*) on the islands of Margarita and La Blanquilla. In *Species Conservation and Management: Case Studies*; Akcakaya, H.R., Burgman, M.A., Kindvall, O., Wood, C., Sjoren-Gulve, P., Hattfield, J., McCarthy, M.A., Eds.; Oxford University Press: Oxford, UK, 2004; pp. 361–370.
20. Briceño-Linares, J.M.; Rodríguez, J.P.; Rodríguez-Clark, K.M.; Rojas-Suárez, F.; Millán, P.A. Adapting to changing poaching intensity of yellow-shouldered parrot (*Amazona barbadensis*) nestlings in Margarita Island, Venezuela. *Biol. Conserv.* **2011**, *144*, 1188–1193. [CrossRef]
21. Rodríguez, J.P.; Rodríguez-Clark, K.M.; Faría-Romero, M.A.; Briceño-Linares, J.M.; Dashiell, S.; Neugarten, R.; Millán, P.A. Education and the threatened parrots of Margarita Island. In Proceedings of the VI International Parrot Convention, Tenerife, Spain, 27–30 September 2006; Loro Parque Fundación: Puerto de La Cruz, Tenerife, Spain, 2006; pp. 159–173.
22. Instituto Nacional de Estadística Censo Poblacional. Available online: http://www.ine.gov.ve/ (accessed on 5 May 2019).
23. Buppert, T.; McKeehan, A. *Guidelines for Applying Free, Prior and Informed Consent: A Manual for Conservation International*; Conservation International: Arlington, VA, USA, 2013; ISBN 0959-8138.
24. Sterling, E.J.; Betley, E.; Sigouin, A.; Gomez, A.; Toomey, A.; Cullman, G.; Malone, C.; Pekor, A.; Arengo, F.; Blair, M.; et al. Assessing the evidence for stakeholder engagement in biodiversity conservation. *Biol. Conserv.* **2017**, *209*, 159–171. [CrossRef]
25. Veríssimo, D.; Campbell, B. Understanding stakeholder conflict between conservation and hunting in Malta. *Biol. Conserv.* **2015**, *191*, 812–818. [CrossRef]
26. Herrera, M.; Hennessey, B. Quantifying the illegal parrot trade in Santa Cruz de la Sierra, Bolivia, with emphasis on threatened species. *Bird Conserv. Int.* **2007**, *17*, 295–300. [CrossRef]

27. Phelps, J.; Biggs, D.; Webb, E.L. Tools and terms for understanding illegal wildlife trade. *Front. Ecol. Environ.* **2016**, *14*, 479–489. [CrossRef]
28. Marshall, H.; Yuda, P.; Collar, N.J.; Lees, A.C.; Marsden, S.J.; Moss, A. Characterising bird-keeping user-groups on Java reveals distinct behaviours, profiles and potential for change. *People Nat.* **2020**, 1–12. [CrossRef]
29. Veríssimo, D.; Bianchessi, A.; Arrivillaga, A.; Cadiz, F.C.; Mancao, R.; Green, K. Does it work for biodiversity? Experiences and challenges in the evaluation of social marketing campaigns. *Soc. Mar. Q.* **2018**, *24*. [CrossRef]
30. Salazar, G.; Mills, M.; Verissimo, D. Qualitative impact evaluation of a social marketing campaign for conservation. *Conserv. Biol.* **2018**, *33*, 634–644. [CrossRef] [PubMed]
31. Chunwang Li, Z.M. Consumer behavior change we believe in: Demanding reduction strategy for endangered wildlife. *J. Biodivers. Endanger. Species* **2015**, *3*, 1–3. [CrossRef]
32. Sánchez-Mercado, A.; Cardozo-Urdaneta, A.; Moran, L.; Ovalle, L.; Arvelo, M.; Morales-Campo, J.; Coyle, B.; Braun, M.J.; Rodriguez-Clark, K.M. Social network analysis reveals specialized trade in an Endangered songbird. *Anim. Conserv.* **2019**, 1–13. [CrossRef]
33. Sánchez-Mercado, A.; Cardozo-Urdaneta, A.; Rodríguez-Clark, K.M.; Moran, L.; Ovalle, L.; Arvelo, M.; Morales-Campo, J.; Coyle, B.; Braun, M.J. Illegal wildlife trade networks: Finding creative opportunities for conservation intervention in challenging circumstances. *Anim. Conserv.* **2020**, *23*, 151–152. [CrossRef]
34. Clayton, S.; Myers, G. *Conservation Psychology: Understanding and Promoting Human Care for Nature*; Wiley-Blackwell: Hoboken, NJ, USA, 2011; ISBN 978-1-444-35641-0.
35. Veríssimo, D.; Wan, A.K.Y. Characterising the efforts to reduce consumer demand for wildlife products. *Conserv. Biol.* **2018**, *33*, 623–633. [CrossRef]
36. Thomas-Walters, L.; Veríssimo, D.; Gadsby, E.; Roberts, D.; Smith, R. Taking a more nuanced look at behavior change for demand reduction in the illegal wildlife trade. *Conserv. Sci. Pract.* **2020**, 1–10. [CrossRef]

Article

Opportunistic or Non-Random Wildlife Crime? Attractiveness Rather Than Abundance in the Wild Leads to Selective Parrot Poaching

Pedro Romero-Vidal [1,*], Fernando Hiraldo [2], Federica Rosseto [3], Guillermo Blanco [4], Martina Carrete [1] and José L. Tella [2]

1 Department of Physical, Chemical and Natural Processes, Universidad Pablo de Olavide, 41013 Sevilla, Spain; mcarrete@upo.es

2 Department of Conservation Biology, Doñana Biological Station CSIC, 41092 Sevilla, Spain; hiraldo@ebd.csic.es (F.H.); tella@ebd.csic.es (J.L.T.)

3 Unidad Mixta de Investigacion en Biodiversidad, Universidad de Oviedo, CSIC, 33600 Mieres, Spain; federica.rossetto1991@gmail.com

4 Department of Evolutionary Ecology, Museo Nacional de Ciencias Naturales, CSIC, 28006 Madrid, Spain; gblanco@mncn.csic.es

* Correspondence: pedroromerovidal123@gmail.com

Received: 21 July 2020; Accepted: 13 August 2020; Published: 14 August 2020

Abstract: Illegal wildlife trade, which mostly focuses on high-demand species, constitutes a major threat to biodiversity. However, whether poaching is an opportunistic crime within high-demand taxa such as parrots (i.e., harvesting proportional to species availability in the wild), or is selectively focused on particular, more desirable species, is still under debate. Answering this question has important conservation implications because selective poaching can lead to the extinction of some species through overharvesting. However, the challenges of estimating species abundances in the wild have hampered studies on this subject. We conducted a large-scale survey in Colombia to simultaneously estimate the relative abundance of wild parrots through roadside surveys (recording 10,811 individuals from 25 species across 2221 km surveyed) and as household, illegally trapped pets in 282 sampled villages (1179 individuals from 21 species). We used for the first time a selectivity index to test selection on poaching. Results demonstrated that poaching is not opportunistic, but positively selects species based on their attractiveness, defined as a function of species size, coloration, and ability to talk, which is also reflected in their local prices. Our methodological approach, which shows how selection increases the conservation impacts of poaching for parrots, can be applied to other taxa also impacted by harvesting for trade or other purposes.

Keywords: CRAVED; conservation criminology; defaunation; harvesting; wildlife trade; parrot abundance; pets; poaching; Savage selectivity index

---

1. Introduction

Defaunation (defined as the global, local or functional extinction of animal populations or species from ecological communities) differs from extinction, as it includes both the disappearance of species as well as their declines in abundance, and has profound ecological consequences, ranging from local to global coextinctions of interacting species to the loss of ecological services critical for humanity [1]. Understanding the causes of defaunation is a growing priority for ecologists, wildlife managers, and conservation biologists, and is important to try to reduce its pace. The drivers of defaunation range from threats operating at global scales, such as climate change, to those that are mostly local, including direct harvest and habitat loss. However, after analyzing the information gathered by the IUCN for more than 8000 threatened or near-threatened species, Maxwell et al. [2] concluded that by far the biggest drivers of biodiversity decline are overexploitation (the harvesting of species from the wild at rates that cannot be compensated for by reproduction or regrowth) and landscape conversion for food production. Moreover, wildlife overexploitation to meet local and global markets was ranked second of five key drivers of harmful ecosystem change by the Intergovernmental Science-Policy Platform on Biodiversity and Ecosystem Services [3].

Wildlife trade is one of the main causes of overexploitation in some taxonomic groups [1], given that some animal products (e.g., ivory and tiger bones) or groups of species (e.g., cage birds) are highly demanded across the world. Considering closely related species, consumers prefer some over others. For instance, buyers prefer multiflowered species among traded orchids [4], while sale prices of traded songbirds are determined by their body size, coloration and song attractiveness [5]. Body size, coloration and the ability to imitate human speech are traits that make parrots highly valued pets [6], thus making them the most traded vertebrate taxa worldwide [7]. The extent to which these consumer preferences determine poaching activities and their impacts are, however, poorly known. Poachers may supply species according to their availability in the wild or could selectively focus on the most demanded among closely related species. This is a key question with important conservation implications, as selective poaching could cause overexploitation and accelerate the defaunation and even extinction of the most demanded species.

The above question was assessed by wildlife criminologists using the CRAVED model. This model proposes that "hot products" sought by thieves are concealable, removable, available, valuable, enjoyable and disposable (CRAVED) [8] and was applied to data available on the parrot trade in Mexico [9], Bolivia and Peru [10]. These authors concluded that parrot poaching was an opportunistic crime where more widely available species were poached in greater numbers than rare and threatened ones, thus lowering concerns for the conservation impacts of poaching on threatened species [8,10]. However, rough proxies of parrot abundances in the wild were used in these studies, including the number of years each species was allowed to be legally trapped [8] or the detectability of species indicated in field guides [10]. Moreover, these authors recognized that a multivariate approach could have let to different conclusions [8]. Indeed, new analyses [11] challenged these conclusions when applying multivariate analyses to the same data [9], showing that amazons and macaws, the most attractive species as reflected by their body size, coloration and ability to imitate human speech, were disproportionally more traded considering the number of years they were legally trapped, thus contributing to their population declines. Nevertheless, these results could be flawed due to the use of the only available proxies for estimating both wild parrot availability and poaching pressure. The number of years each species was allowed to be legally trapped should reflect their abundance in the wild, assuming that scarcer species were allowed to be trapped for fewer years [8], although international markets, local economics, and political pressures could influence this. On the other hand, the use of seizures as a proxy of poaching pressure [11] may have affected the results, given that seizures usually represent <10% of poaching volumes and are often biased towards certain species [12]. In fact, the proportion of amazons and macaws among all parrots seized in Costa Rica (50%) is significantly higher than among those actually poached and kept as household pets (33%), showing that seizures are biased to the most valuable and threatened species (authors' data, in prep.). Thus, a positive selection of amazons and macaws [11] could at least partially result from seizure biases. Given the limitations of these proxies [8,10,11], reliable information on both the abundance of the species in the wild and poaching pressure is needed to properly test whether parrot poaching is selective or opportunistic [11].

To disentangle whether parrot poaching is a selective or opportunistic activity, we designed a large-scale survey in Colombia, where trapping and keeping native animals as pets is a rooted tradition punished by law since 1977 [13]. We simultaneously measured the relative abundance of 28 parrot species in the wild and as household poached pets. We then applied a selectivity index widely used in habitat selection studies, the Savage index, to ascertain whether pet abundances mirrored wild abundances or, conversely, whether some species were found as pets more than expected. Our results show a strong selection for more attractive parrot species to be kept as pets, despite their lower abundances in the wild. Positively selected species, but not those less abundant in the wild, were thus the most expensive. This selection has important consequences for the conservation of parrots and their ecosystem services, as well as for our understanding of overharvesting and defaunation, and the management of illegal trade in general.

2. Materials and Methods

2.1. Study Area

Colombia, with a surface area of 1.1 million km^2 and 45.4 million people [14], is one of the most biodiverse countries on Earth [15]. Differences in elevation and latitude produce large climatic variations across the country, which are responsible for the high diversity of habitats. Colombia can be divided into five continental regions (Andean, Caribbean, Pacific, Orinoco, and Amazon), with remarkable biogeographic, socio-cultural, economic, and demographic differences [16]. Using satellite maps, we designed an a priori road itinerary (4232 km in total) to cover the main biomes of the Andean, Caribbean, and Pacific regions across a wide altitudinal range (4–3520 m.a.s.l., Figure 1).

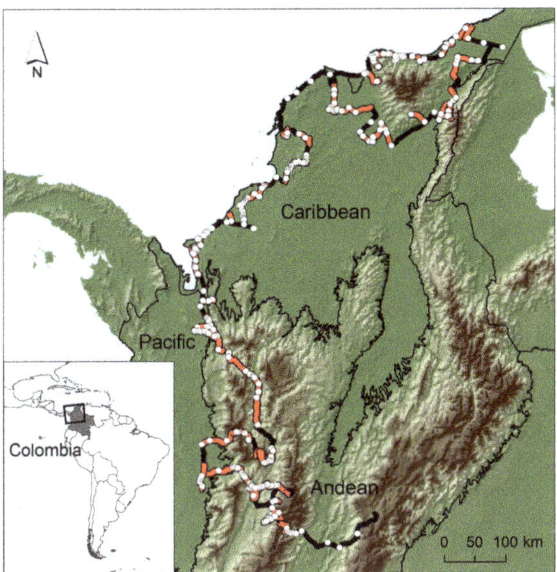

Figure 1. Study area showing the itinerary (black line) crossing the Andean, Pacific and Caribbean regions of Colombia, the roadside parrot surveys (in red), and the localities where poached pets were recorded (white dots).

The itinerary (Figure 1) was designed to cover the spatial distribution of many parrot species (35 species in total, see [17]) and to visit villages where we looked for poached pets (see below), thus maximizing the chances of finding a large variety of poached and wild parrot species at a large geographic scale.

2.2. Wild Parrot Surveys

We estimated the abundance of parrots in the wild through roadside car surveys, a method adequate for parrots as it allows coverage of large areas, thus increasing the probability of detecting individuals of species occurring at very low densities or spatially aggregated [18–24]. Within the designed itinerary, we selected 2221 km of low-transit and unpaved roads to record all parrots detected (Figure 1). The beginning and the end of each habitat patch (categorized as pristine natural, degraded

2.3. Poached Parrot Surveys

As a direct measure of domestic poaching pressure, we recorded the number of all wild and exotic pets, and how many of them were poached native parrots, in 282 villages crossed by our itinerary (Figure 1). We did not conduct systematic surveys using questionnaires [29], as answers to questions related to illegal activities that are prosecuted in the country would be unreliable [23]. Nonetheless, most people did not hide their pets, nor were they afraid to keep them illegally. Therefore, we recorded many visible pets while driving and walking through streets or entering public establishments, such as shops, hotels or gas stations. We combined these direct observations with informal conversations [23] with randomly chosen local people (N = 358), indicating our interest to see and take pictures of their pets. In about half of the cases (55.6%), people told us they had pets at home, and in 62% of cases provided us with information about other people who poached or owned pets. We then confirmed this information by visiting their homes, taking pictures of the pets (Figure 3) and, at the same time, engaging in informal conversation to obtain additional information, such as the price they paid for the parrot and its ability to imitate human speech (see below). We could not obtain prices from all species, as in many cases the owners did not buy the pets but poached themselves. Pet owners confirmed that all native parrots were poached, with no evidence of attempts to breed them in captivity (contrary to a few exotic, small parrot species). All informants and pet owners shared the information with us freely and were kept anonymous.

Figure 3. Some pet parrots observed in public spaces such as streets (**a**–**c**) or within homes (**d**–**f**). Pictures also illustrate species positively ((**a**): scarlet macaw *Ara macao*, (**b**) yellow-crowned amazon *Amazona ochrocephala*, (**c**) blue-and-yellow macaw *Ara ararauna*, (**d**) blue-crowned parakeet *Thectocercus acuticaudatus*) and negatively selected by people as pets ((**e**) brown-throated parakeet *Eupsittula pertinax*, and (**f**) orange-chinned parakeet *Brotogeris jugularis*). Pictures: P. Romero-Vidal and J.L. Tella.

2.4. Parrot Attractiveness

The attractiveness of each parrot species was rated based on its body size (obtained from [30]), coloration, and ability to imitate human speech [8,11]. Parrot coloration was described as the proportion of the body (bright body) and head (bright head) covered by bright colors (i.e., other than the dominant green or brown coloration), scored from 0 to 2 following [8], and the total number of colors (N colors) observed when the bird is perched, using plates in [29]. The ability of each individual pet to imitate human speech was ranked into five categories using the information provided by local pet owners (0: individuals not able to make imitations, 0.5: individuals able to whistle or imitate one or two words, 1: individuals able to imitate several words but poorly pronounced, 1.5: individuals able to imitate several words, with good pronunciation, and 2: individuals able to imitate human speech, using a wide repertoire of words and making up short sentences, singing songs, imitating other domestic animals or sounds such as telephone, TV, radio, etc.). Scores from different individual pets were averaged within species to obtain a rank describing the ability of each species to speak. However, the opinion of local pet owners could be biased by their experience, which is usually limited to their own pets. Therefore, we asked the same question to five people from USA, France, Germany and Spain with >20 years of experience breeding and keeping a large variety of parrot species in captivity. The average scores provided by these experts correlated well with those provided by local pet owners (Spearman rank correlation = 0.85, $p < 0.0001$, Figure 4), thus validating the use of local knowledge for measuring the mimicry ability of different parrot species.

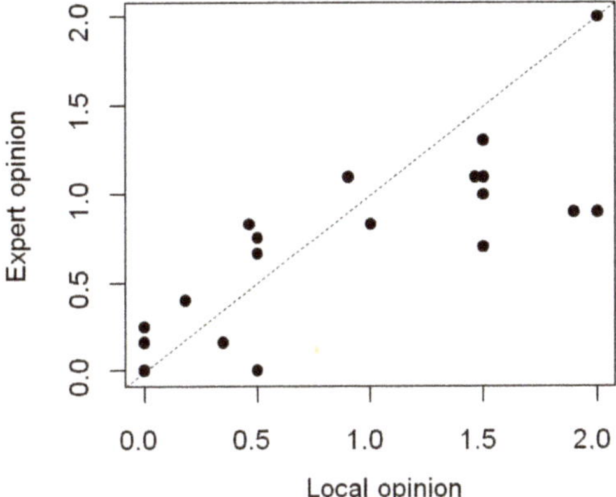

Figure 4. Scatterplot of the averaged opinion on the ability of different parrot species from Colombia to imitate human speech (0 = lowest, 2 = highest) provided by international experts and by local pet owners. Each dot represents a species, and the dashed line represents the theoretic perfect correlation.

2.5. Statistical Analyses

We used the Savage selectivity index [31] to assess whether parrot species are poached proportionally to their abundances in the wild. This index is widely used in resource selection studies (e.g., [32]) and allows us to infer the statistical significance of selection [31]. We used the number of parrots of each species recorded in the wild as units of resource availability and numbers recorded as pets as units of the resource used. The Savage selectivity index was calculated for each species as $W = U_i/p_i$, where U_i is the proportion of a given species (among all poached parrots) recorded as a pet (i.e., used) and p_i is the proportion of that species (among all wild parrots) recorded in the wild (i.e., available). A few species were so scarce that we did not record a single individual in the wild (Table 2) despite their known presence in the study area [17] and finding them locally as pets. In those cases, as the availability of a used resource cannot be zero [31], we conservatively considered that at least one wild individual was recorded to allow calculating the Savage index. The Savage index theoretically varies from zero (full negative selection) to infinite (full positive selection), with one being the expected value by chance (i.e., used in proportion to its availability) [31]. The statistical significance of this index is obtained by comparing the statistic $(wi - 1)^2/se_{wi}{}^2$ with the corresponding critical value of a χ^2 distribution with one degree of freedom [31], the null hypothesis being that species are poached in proportion to their availability in the wild. The standard error of the index (se_{wi}) is calculated as $\sqrt{[(1 - p_i)/(u_+ \times p_i)]}$, where u_+ is the total number of poached parrots recorded. Statistical significance was obtained after applying the Bonferroni correction for multiple tests. We did not calculate the Savage index for four species (Table 2) since they are unable to survive in captivity more than a few days or weeks [6] and thus they are rarely poached. However, results including these species were nearly identical (Spearman correlation, r = 1, $p < 0.0001$, $n = 24$).

Table 2. Parrot species included in the study, their body size (in cm), the scores (0–2) for the brightness of body and head coloration, the total number of colors (Color), their ability to imitate human speech (speech, 0–2, with ranges), their price in US$ (with SD), the number of individuals recorded in the wild (N wild) and as poached pets (N pet), and the Savage selectivity index (W). * Statistically significant W values after Bonferroni correction ($p < 0.002$).

Species	Size	Body	Head	Color	Speech	Price	N Wild	N Pet	W	
Amazona amazonica (Aam)	31	0.2	1	3	1.5 (1–2)	34.85 (-)	93	12	1.17	
Amazona autumnalis (Aau)	34	0.2	1	3	1.5 (1–2)		61	25	3.73	*
Amazona farinosa (Afa)	38	0.2	0.2	1	1.9 (1–2)	38.72 (13.4)	20	18	8.18	*
Amazona mercenarius (Ame)	34	0.2	0	1	1.0 (1–1)		93	0	0.00	*
Amazona ochrocephala (Aoc)	31	0.2	0.8	2	2.0 (1–2)	34.41 (16.0)	136	359	24.00	*
Ara ambiguus (Aab)	85	1.5	0.5	3	1.5 (1–2)		0	3	27.28	*
Ara ararauna (Aar)	85	2	1.8	3	0.9 (0–2)	43.57 (20.5)	80	76	8.64	*
Ara chloropterus (Ach)	90	2	2	3	1.5 (1–2)		0	14	127.28	*
Ara macao (Ama)	85	2	2	3	2.0 (2–2)	145.22 (0.00)	4	74	168.20	*
Ara militaris (Ami)	75	1.5	0.5	3	1.5 (0–2)		10	7	6.36	*
Ara severus (Ase)	46	1	0	2	1.0 (1–1)		54	5	0.84	
Bolborhynchus lineola (Blin)	16	0	0	1	0.0 (0–0)		1	0	0.00	
Brotogeris jugularis (Bju)	18	0	0	1	0.4 (0–2)	6.53 (1.45)	6230	344	0.50	*
Eupsittula pertinax (Epe)	25	0.1	0	1	0.5 (0–2)	5.68 (3.67)	2445	189	0.70	*
Forpus conspicillatus (Fco)	12	0.1	0	1	0.0 (0–0)		83	0	0.00	
Forpus passerinus (Fpa)	12	0.1	0	1	0.0 (0–0)	5.81 (0.00)	35	6	1.56	
Forpus spengeli (Fsp)	12	0.1	0	1	0.0 (0–0)		80	9	1.02	
Hapalopsittaca fuertesi (Hfu)	23	0.5	0.9	3	0.0 (0–0)		1	0	-	
Ognorhynchus icterotis (Oic)	42	0.8	0.8	2	0.5 (0–1)		85	2	0.21	
Pionus chalcopterus (Pch)	29	0.2	0	2	0.0 (0–0)		134	3	0.21	
Pionus menstruus (Pme)	28	0.3	1.5	2	0.2 (0–2)	16.46 (9.22)	497	19	0.35	*
Pionus seniloides (Pse)	30	0.2	0	2	0.0 (0–0)		2	0	0.00	
Pionus sordidus (Pso)	28	0.2	0	1	0.0 (0–0)		17	1	0.52	
Psittacara wagleri (Pwa)	36	0	0.5	2	0.5 (0–1)		628	10	0.14	*
Pyrilia haematotis (Pha)	21	0.1	0	2	0.0 (0–0)		0	1	-	
Pyrilia pyrilia (Ppy)	24	0.1	2	3	0.0 (0–0)		1	0	-	
Thectocercus acuticaudatus (Tac)	37	0	0.1	1	0.5 (0–1)		20	13	5.91	*
Touit batavicus (Tba)	14	0.1	0	3	0.0 (0–0)		1	0	-	

We used principal component analysis (PCA) to obtain a composite variable that describes the attractiveness of each parrot species as a function of its color, body size, and ability to speak. Variables, which were positively correlated (Pearson correlations: 0.50–0.94, all $p < 0.0001$), were scaled before analysis. Bartlett's test of sphericity was computed to establish the validity of the data set. Eigenvalues > 1 were used to assess the number of factors to extract.

We used generalized linear models (GLMs) to test whether the preference of people for certain species (measured through the Savage index; log-transformed; normal error distribution and identity link function) was related to their attractiveness. We then assessed whether preferred or rare species were the most valuable in monetary terms by relating the Savage indexes and abundances of species in the wild (ind./km) to their price (log-transformed; normal error distribution and identity link function). We used the average prices of species provided by pet owners (local currency transformed to US$, Table 2). All statistical analyses were performed in the R v.3.6.1 statistical platform [33].

3. Results

We recorded 10,811 wild individuals from 25 parrot species (Table 1) across the 2221 km of roadside surveys conducted, covering a wide variety of biomes with different degrees of human alteration. Overall abundance reached 4.87 ind./km, although most records (80.31%) corresponded to just two parakeets (orange-chinned parakeet *Brotogeris jugularis* and brown-throated parakeet *Eupsittula pertinax*). The other species were present in low numbers, were extremely rare or even unrecorded in the wild (Table 2). Simultaneously, we recorded a total of 2465 pets from 124 species, kept by 818 owners in 92.9% of the 282 villages surveyed (Figure 1), from which 1179 (47.8%) were pets from 21 native parrot species (Table 2). The rest of the pets were mostly songbirds (Passeriformes, 32.7%), non-native parrots (12.7%), other birds (3.0%), mammals (1.3%), and reptiles (0.4%). Among the 358 local people we met, 58.4% of them kept poached native parrot pets at the time of our survey or recently, and 38.0% knew other people also keeping them.

In absolute numbers, *B. jugularis* and *E. pertinax* made up almost half (45.20%) of all pet parrots. However, these species were actually negatively selected when considering their abundances in the wild (Table 2, Figure 5). On the contrary, most amazons (*Amazona* spp.), large macaws (*Ara* spp.) and *Thectocercus acuticaudatus*, mostly uncommon or extremely rare in the wild, were strongly positively selected as pets (significant W > 1). The other species showed non-significant selection (i.e., were kept as pets in proportion to their availability in the wild; Table 2, Figure 5).

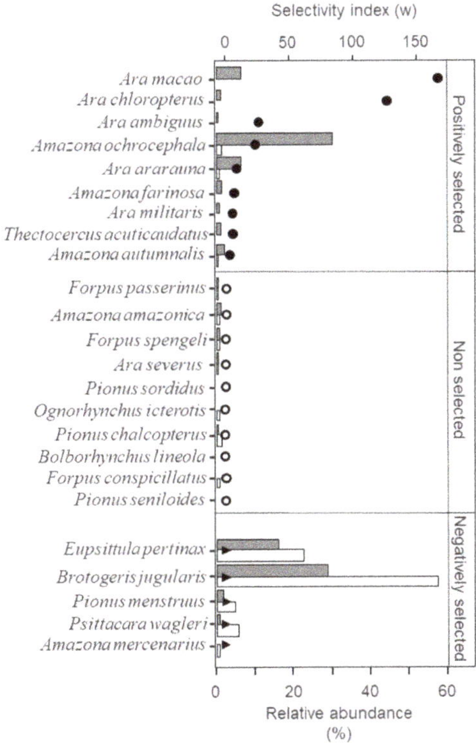

Figure 5. Relative abundance of parrots in Colombia as pets (dark gray bars) and in the wild (white bars), and Savage selectivity index (W; black dots: significant positive selection; black triangles: significant negative selection; white dots: not selected).

The PCA analyses (Bartlett's Test of Sphericity: $\chi^2 = 99.24$, $p < 0.0001$, df = 10) rendered a single dimension with an eigenvalue > 1 (3.40), which positively correlated with body size (0.92), coloration (bright body: 0.91, bright head: 0.79, number of colors: 0.77) and ability to imitate human speech (0.71), explaining 68.08% of the total variance. Thus, PC1 can be interpreted as a descriptor of parrot attractiveness, large, colorful and talkative species being more attractive (positive values) than their counterparts (negative values; Figure 6).

PC1 was positively related to the Savage index (estimate: 0.27, SE: 0.04, t = 6.40, $p < 0.0001$, adjusted-$R^2 = 0.63$), showing that the most attractive species were poached in larger numbers than expected based on their availability in the wild (Figure 7a). The price of the species increased with their attractiveness (estimate: 16.24, SE: 4.06, t = 4.00, $p < 0.0052$, adjusted-$R^2 = 0.65$, Figure 7b) but was unrelated to their abundances in the wild (estimate: 0.85, SE: 0.78, t = −1.09, $p = 0.3139$, Figure 7c), indicating that the most attractive but not the rarest species were more valuable.

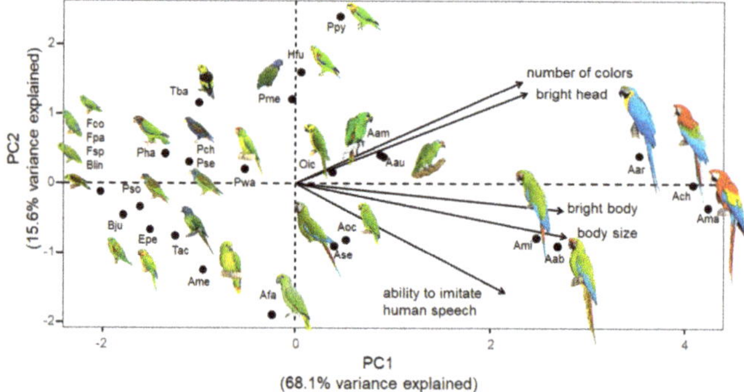

Figure 6. Principal component analysis (PCA) of Colombian parrot traits, namely: body size, coloration (bright body, bright head and number of colors) and ability to imitate human speech. See Tabl 2 for species abbreviations. Drawings of parrots are not scaled. PC2 is plotted to allow better visualization of species across the PC1 axis, which reflects parrot attractiveness, but was not used for further analysis (eigenvalue < 1).

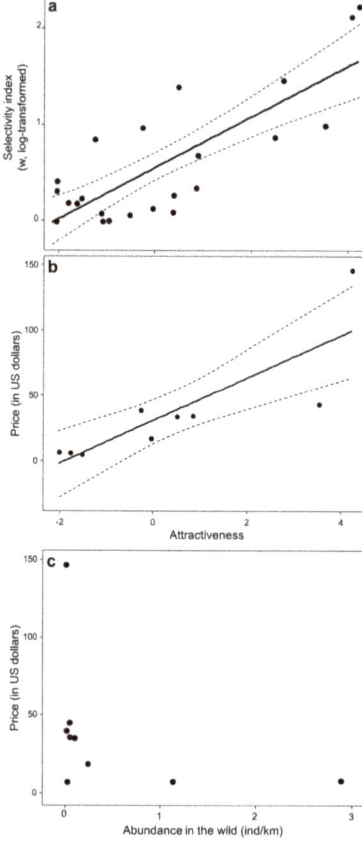

Figure 7. Preferred parrots in Colombia (measured through their Savage selective index) were the most attractive (i.e., large, colorful, and able to imitate human speech) species (**a**), which were also the most expensive (**b**), independently of their abundance in the wild (**c**).

4. Discussion

4.1. Parrot Poaching Is Not an Opportunistic, but a Selective Wildlife Crime

Wildlife trafficking is increasingly recognized as both a specialized area of organized crime and a significant threat to many plant and animal species [2,34,35]. However, due to its intrinsic illegal nature, it is difficult to fully know its actual extent and consequences for wildlife [12,36]. Here, we provide the first reliable and simultaneous large-scale estimation of poaching pressure and abundance in the wild of a community of parrot species, showing that poaching of this taxonomic group is not opportunistic, but largely focused on species with particular traits that make them more attractive to people. Following the CRAVED model approach [8], our data would have suggested that parrot poaching is an opportunistic crime, as the numbers of poached parrots per species positively correlates with numbers recorded in the wild (Kendall's Tau-b = 0.33, $p = 0.018$, N = 27). However, this slight trend, which is markedly influenced by two parakeet species (which jointly made up >80% and >45% of the individuals recorded in the wild and as poached pets, respectively), turns out to be non-significant when they are removed from the analysis (Kendall's Tau-b = 0.23, $p = 0.125$, N = 25). Moreover, as recognized by authors using the CRAVED model [8], conclusions derived from simple univariate analyses could change when simultaneously testing the effects of other variables, such as species attractiveness in multivariate models. This possibility was later confirmed when reanalyzing the same parrot poaching data from Mexico using generalized linear models: attractive species were more poached than expected when controlling for the number of years they were allowed to be legally trapped [11].

To identify selection, it is not only important to assess resource availability, but also be able to calculate its statistical significance. Here, we provide direct estimates of parrot availability and poaching and a key analytical advance, the application of the Savage selectivity index [31] to quantitatively measure poaching selection. To our knowledge, this is the first time that a selectivity index is used to statistically evaluate whether any given species is positively, negatively, or not selected at all. A further advantage of this index is that it can be used as a continuous response variable to ascertain drivers of poaching selection. In this sense, we found that 63% of the variance in parrot poaching selection is explained by the attractiveness of the species, thus confirming that poaching is not a taxonomically random, but a species-specific activity that preferentially focuses on the most attractive species for people [11].

Attractiveness in this taxonomic group has been found related to body size and coloration, which determines for instance which species are kept in zoos [37]. Meanwhile, the ability to imitate human speech can be particularly appealing when parrots are kept as pets at close contact with people [6]. Thus, the combination of these traits can describe species attractiveness and, therefore, predict their selection as pets and their prices [11]. As found in other countries [11,38,39], in Colombia macaws and amazons were much more expensive than other poached parrot species. We also show that higher prices in the domestic pet trade are not related to the rarity of a species in the wild, but strongly related (65% of the variance explained) to its attractiveness. While both rarity and physical attractiveness influence the prices of internationally traded birds [40], our results show that local demand focuses on attractive rather than on rare parrot species.

The quantitative measurement of poaching selection also allows deeper investigation of the unexpected preference of some species and additional cultural drivers of selection. For example, the high positive selection of the scarlet macaw *Ara macao* among Colombian people is surprising (Table 2), while its attractiveness is not much higher than that of similar macaw species (Figure 6). Local knowledge provided us with the answer: people explained to us that this species is sought after because its plumage resembles the Colombian national flag, hence its local name "guacamaya bandera" (flag macaw). Moreover, they also described that the Colombian guerrillas (revolutionary armed forces) persecuted its capture and use as pets because it was considered as unpatriotic; thus, poaching pressure on this species has increased since the guerrillas ceased their warlike activities.

As this species became extremely rare because of overharvesting [17], poachers seemed to switch efforts towards the similarly sized and colored green-winged macaw *Ara chloropterus* (Figure 6) as a substitute species, thus also explaining its outstanding selection (Table 2). Another case that merits attention is the positive selection of the blue-crowned parakeet *T. acuticaudatus* (Table 2), despite its low attractiveness rating (Figure 6). This species is restricted to very dry forests of the Guajira region, where the most preferred species such as macaws and amazons are absent [17], and thus it is the largest and most colorful species available. Other potential covariates of poaching selection could be assessed in further studies, such as the accessibility of nests and life expectancy of parrot species as pets.

4.2. Conservation Implications of Selective Parrot Poaching

The colorful plumage of parrots and their ability to imitate human speech have made them highly popular as pets [6], thus leading to the international trade of at least 259 species of parrots, involving millions of individuals in recent decades [7,41,42]. In the near absence of long-term monitoring programs of wild populations [43] and analyses of sustainable harvesting [44,45], international trade of wild-caught individuals may constitute a threat to many parrot species worldwide [11,46,47]. A concerning example is the African grey parrot *Psittacus erithacus*, considered the best at imitating human speech among all extant parrot species [6]. Overharvesting due to trapping for the international trade has caused large range contractions and decimated the populations, to the point that the species was included in Appendix 1 of CITES in 2017, prohibiting international trade on wild specimens for commercial purposes, and was listed as globally Endangered by IUCN in 2018 [48]. Although international bans have largely reduced the legal trade on parrots [42,49] and the upsurge of captive-breeding [6,49] has reduced the demand of wild-caught traded birds, illegal trade is still active [50], although at much lower volumes, including illegal trade on African grey parrots [51]. Nonetheless, while international trade is a matter of concern, less attention has been paid to the conservation impact of domestic trade on parrots, even though it is known to occur in different regions of the world, such as Madagascar [52], Asia [34], and all across the Neotropical region [9,23,35,38,39,53–57]. Due to its illegal nature, the true scale and impact of parrot poaching are often underestimated [34] and based mainly on counts from pet markets [53,55,58], government seizures, or other information sources difficult to verify [9,35,53].

In the Neotropics, expert knowledge indicated that 68% of the studied parrot populations are threatened due to their capture for the domestic pet trade [59]. However, it is unknown whether and to what extent poaching threatens these species differentially. Based on conclusions obtained through the CRAVED model, parrot poaching would mostly affect common species the most, thus alleviating concerns on its conservation impacts [8,10,60] since harvesting individuals of common species could be even considered as sustainable resource use [55,61,62]. Our results lead to the opposite conclusion: the two most common species are poached half as much as expected based on their availability in the wild, while a few, highly attractive species are poached in larger numbers than expected according to their abundance and are likely to be overharvested. Pet owners indicated that these still abundant parakeets are poached as substitutes of more preferred species, when the former are not available because of their scarcity in the wild and/or their high prices. Therefore, our concern is not the absolute number of individuals poached of a given species but the proportion of the wild population size. In fact, the poaching of as few as 70 individuals per year constitutes a major threat to the Critically Endangered red siskin *Spinus cucullatus* in Venezuela [63]. The trade of some attractive parrot species has been shown to cause negative population trends and affect their conservation status [11]. A proper test of the overharvesting effects of parrot poaching would be to relate the selection on each species to their population trends. While detailed information on population trends is not available for Colombian parrots, they could be estimated through expert knowledge [58]. However, there is evidence that poaching has caused large population declines and range contractions of the yellow-crowned amazon *Amazona ochrocephala*, considered as the species that best imitates human speech in Colombia [17], and of the highly demanded scarlet macaw, while species we identified as less preferred or not selected at all by poachers have not suffered large declines in Colombia [17]. It is worth mentioning that the scarlet macaw was considered the most abundant macaw species in the region in the 1950s, in contrast with its current rarity [17] in this study.

Although adult parrots are also eventually trapped [64], parrot poaching mostly focuses on nestlings [38,64,65] as hand-reared chicks make better pets than birds caught as adults in terms of docility and ability to learn human speech [6]. This has different implications on the population dynamics of the poached species [64], as lifespan generally increases with the size of parrot species [66]. Nest poaching of the largest species such as amazons and large macaws could alleviate concerns about its impact as they have the longest lifespans among parrots (at least 34–63 years in captivity [66]), and thus small reductions in their breeding success due to poaching could have less impact on their population dynamics compared to small, short-lived species. However, we learned from local people that the last remaining nests of the preferred amazon and macaw species are located and poached year after year, often for decades. Indeed, local poachers compete for the same nests to the point that nests are surveyed daily to avoid robbing by others. Therefore, breeding pairs may occupy the same areas for decades, giving the wrong impression of apparent population stability to local people acting as regular or occasional poachers, birdwatchers, and wildlife managers. Ultimately, if current poaching pressure is not halted, the remaining populations will collapse due to the senescence and death of breeding adults in the absence of population recruitment.

4.3. Ecological Implications of Selective Parrot Poaching

Selective parrot poaching severely affects the conservation of the preferred species, as well as the ecosystem services they provide. Parrots have been long considered as plant antagonists, given their undoubted role as seed predators [67]. However, they also provide several ecological functions [68] within an antagonist-mutualism continuum [69]. Particularly, parrots can act as effective seed dispersers through complementary mechanisms, such as stomatochory [21,23,70–73], endozoochory [24,74,75], and epizoochory [76], further facilitating secondary seed dispersal by a variety of other species [77]. Altogether, they may play an important role in the structure of networks, communities, and ecosystems [19,21,69]. Poaching reduces the population size of parrots, thereby quantitatively reducing and threatening their ecological functions. In fact, the selective poaching of the largest species (amazons and macaws) may have the strongest impact, as these species are the main—and sometimes the only—effective long-distance seed dispersers of palms and trees with large-sized fruits, which are biomass-dominant and key species in several ecosystems [21,71–73]. The defaunation of these large-sized parrot species, which could be considered as megafauna attending to a new functional definition [78], further reduces the dispersal of large-fruited plants that previously was only attributed to the decimated large-sized mammals and those extinct in the Pleistocene in South America [72]. The dispersal of some of these tree and palm species has already been disrupted after the large-scale extirpation and population declines of some amazon and macaw species [20,73].

4.4. Suggested Conservation Actions

Keeping parrots as pets in Colombia, as in other Neotropical countries, seems to be ancestrally rooted [17]. This cultural tradition could have been sustainable in the past but not today, given the large human population and economic power increase in recent decades [14]. These two factors have increased the demand for pets while promoting habitat loss, also affecting parrot populations [17]. Therefore, conservation actions are urgently needed to halt parrot defaunation. Based on the conclusions derived from CRAVED model analyses, conservation actions should focus on the most heavily poached species by protecting and preventing poaching in their breeding areas [8,10,60]. In Colombia, this would mostly apply to two parakeet species with a wide distribution [17], thus making the protection of breeding sites unfeasible. Moreover, these species have large, non-threatened populations [17], and thus their conservation should be not a priority. Our results on selective poaching provide a completely different conservation management scenario, as actions must focus on the most preferred, currently overexploited species. The protection of breeding sites to avoid nest poaching [8,10,60] may be efficient in the case of species with restricted breeding ranges [18,79,80], but it is not feasible for Colombian macaws and amazons, with large distribution ranges and low population densities [17] in this study. The attraction of ecotourism and the creation of eco-lodges may increase local incomes and reduce poaching [60], and favor research and the conservation of large parrots and macaws [81]. Colombia has great potential and should promote these conservation-friendly economic activities, but these local activities cannot prevent parrot poaching at a national scale. Paradoxically, in the absence of law enforcement in Colombia, we found tourist establishments displaying captive macaws to attract tourists.

As in other Neotropical countries [82], we learnt from pet owners that parrot poaching in Colombia is generally not an organized crime, but is performed by local people to obtain their own pets or supply pets to neighbors and relatives. A large proportion of the population (c. 60%) is involved in the illegal activity of keeping native parrots as pets, often acting simultaneously as poachers and consumers, and this activity is widespread across the country. Law enforcement and reducing the demand are two strategies to reduce wildlife poaching that must be balanced in terms of cost-effectiveness, especially when conservation resources are limited [83]. Police control should be strengthened to dissuade people from keeping pets at least of overexploited species, while educational campaigns for public awareness on the consequences of poaching should reduce the demand [18]. Alternative sources can also be offered to satisfy the cultural tradition of owning pet parrots. Breeding parrots in captivity is well established [6], and can successfully supply the previous demand for internationally traded wild parrots [49]. Thus, breeding native parrots for local sale [9] could reduce the pressure on wild populations. However, the low-reproductive rates of preferred species (amazons and macaws) in captivity [6] make it difficult to supply enough individuals and at prices low enough to counteract poaching. Moreover, this activity is prone to fraud, as chicks of preferred species could be poached and sold as captive bred (see [84] for traded Asian songbirds). The genetic control of supposedly captive-bred individuals [85] requires great surveillance efforts, the development of genetic markers, and the availability of molecular laboratories [86], which are difficult to implement at a large scale in countries such as Colombia.

An alternative is to supply the pet demand with captive-bred exotic parrot species that are easier to reproduce [6], can be bought at competitive prices, and show low risks of invasion when they are accidentally released to the wild [87,88]. A combination of these actions seems to have been successful at halting parrot poaching in a small Colombian region, within the Andean distribution of the yellow-eared parrot *Ognorhynchus icterotis*, a globally endangered species for which conservation programs and awareness campaigns were implemented over decades [89]. In this area, most people have non-native pet parrots, such as budgerigars *Melopsittacus undulatus*, cockatiels *Nymphicus hollandicus* and lovebirds *Agapornis* spp., often after the seizure of their native pets by the police, and are very aware of the illegal nature of this activity. Wildlife authorities should realize that law enforcement and demand reduction must be urgently extended to the whole country to avoid, at least, the predicted population collapse of overexploited species. Considering cost-effectiveness [83], law enforcement is probably the most effective action at a national scale, since police are widespread across the country and should simply apply current laws without the need for additional economic costs. However, the seizure of all parrot pets is unfeasible due to the economic costs of creating and maintaining wildlife rescue centers [35,65] to hold them. In fact, seized birds are often returned to the wild to reduce costs and to create space for newly confiscated individuals, in the absence of reintroduction programs [35]. Thus, seizures should focus on overexploited species and should be combined with well-designed awareness campaigns [90,91] to reduce demand. On the other hand, captive breeding of exotic parrots to supply the demand should also be promoted, but under strict sanitary control, as exotic parrots can carry pathogens (e.g., [92]) that could spread and negatively impact native populations.

4.5. Further Prospects for Assessing Selective Harvesting

Solving the dichotomy between opportunistic or selective poaching has profound conservation implications, since the overexploitation of preferred species may be causing their decline, pushing their populations toward regional [11,58] and global [84] extinctions. Several lines of evidence show that any form of harvesting (including legal fishing and hunting) is selective toward individuals of a certain sex, size, morphology, or behavior, with long-term population and evolutionary consequences [93–99]. However, to our knowledge, the hypothesis of selective harvest at the community level (i.e., on species with particular characteristics over others) has not yet been properly tested, mainly due to the difficulty of assessing their availability in the wild. The application of a selectivity index allows a quantitative measure and statistical test of harvesting selection in both intra- and interspecific studies. Therefore, it is a powerful tool for assessing selection and investigating the factors driving it, not only in other poached parrot communities and heavily traded birds, such as Asian songbirds [58,84,100,101], but also in other animal and plant species harvested, for example, through deforestation, fisheries, game hunting, or bush-meat exploitation.

Author Contributions: Conceptualization, J.L.T., P.R.-V., G.B., F.H. and M.C.; methodology, P.R.-V., J.L.T. and F.H.; statistical analysis, P.R.-V., F.R. and M.C.; writing—original draft preparation, P.R.-V., M.C. and J.L.T.; writing—review and editing, J.L.T., M.C., P.R.-V., G.B. and F.H.; visualization, P.R.-V., M.C., F.R., J.L.T., G.B. and F.H.; supervision, J.L.T. and M.C.; project administration, M.C.; funding acquisition, J.L.T. and M.C. All authors have read and agreed to the published version of the manuscript.

Funding: This research was funded by Loro Parque Fundación (PP-146-2018-1).

Acknowledgments: E. Arrondo, I. Paredes, L. López-Ricarte and M.C. Díaz helped during fieldwork, and I. Afán (LAST-EBD) elaborated the maps. Logistic and technical support were provided by Doñana ICTS-RBD. Two anonymous reviewers greatly helped to improve the manuscript. This work did not require approval by our Ethical Committee (EBD-CSIC) since it did not involve invasive methods nor experimental work with live animals.

Conflicts of Interest: The authors declare no conflict of interest.

References

1. Young, H.S.; McCauley, D.J.; Galetti, M.; Dirzo, R. Patterns, Causes, and Consequences of Anthropocene Defaunation. *Annu. Rev. Ecol. Evol. Syst.* **2016**, *47*, 333–358. [CrossRef]
2. Maxwell, S.L.; Fuller, R.A.; Brooks, T.M.; Watson, J.E. Biodiversity: The ravages of guns, nets and bulldozers. *Nature* **2016**, *536*, 143–145. [CrossRef]
3. IPBES. Intergovernment Platform on Biodiversity and Ecosystem Services. Summary for Policymakers of the Global Assessment Report. 2019. Available online: https://www.ipbes.net/sites/default/files (accessed on 12 June 2020).
4. Hinsley, A.; Verissimo, D.; Roberts, D.L. Heterogeneity in consumer preferences for orchids in international trade and the potential for the use of market research methods to study demand for wildlife. *Biol. Conserv.* **2015**, *190*, 80–86. [CrossRef]
5. Su, S.; Cassey, P.; Vall-Llosera, M.; Blackburn, T.M. Going cheap: Determinants of bird price in the Taiwanese pet market. *PLoS ONE* **2015**, *10*, e0127482. [CrossRef]
6. Silva, T. *Psittaculture: A Manual for the Care and Breeding of Parrots*; Nová Exota, Czech Republic & AAL Pty. Ltd.: Bellingen, NSW, Australia, 2018; p. 597.
7. Bush, E.R.; Baker, S.E.; Macdonald, D.W. Global trade in exotic pets 2006–2012. *Conserv. Biol.* **2014**, *28*, 663–676. [CrossRef]
8. Pires, S.; Clarke, R.V. Are Parrots CRAVED? An Analysis of Parrot Poaching in Mexico. *J. Res. Crime Delinq.* **2012**, *49*, 122–146. [CrossRef]
9. Cantú, J.C.; Sánchez, M.E.; Grosselet, M.; Silve, J. *The Illegal Parrot Trade in Mexico: A Comprehensive Assessment*; Defenders of Wildlife: Washington, DC, USA, 2007; p. 121.
10. Pires, S.F. A CRAVED analysis of multiple illicit parrot markets in Peru and Bolivia. *Eur. J. Crim. Pol. Res.* **2015**, *21*, 321–336. [CrossRef]

11. Tella, J.L.; Hiraldo, F. Illegal and legal parrot trade shows a long-term, cross-cultural preference for the most attractive species increasing their risk of extinction. *PLoS ONE* **2014**, *9*, e107546. [CrossRef]
12. Esmail, N.; Wintle, B.C.; Sas-Rolfes, M.; Athanas, A.; Beale, C.M.; Bending, Z.; Dai, R.; Fabinyi, M.; Gluszek, S.; Haenlein, C.; et al. Emerging illegal wildlife trade issues: A global horizon scan. *Conserv. Lett.* **2020**, e12715. [CrossRef]
13. Rodríguez, N.J.M.; García, O.R. Comercio de fauna silvestre en Colombia. *Rev. Fac. Nacio. Agron. Medellín* **2008**, *61*, 4618–4645.
14. DANE. Available online: http://www.dane.gov.co (accessed on 12 June 2020).
15. Myers, N.; Mittermeier, R.A.; Mittermeier, C.G.; da Fonseca, G.A.B.; Kent, J. Biodiversity hotspots for conservation priorities. *Nature* **2000**, *403*, 853–858. [CrossRef] [PubMed]
16. Sánchez-Cuervo, A.M.; Aide, T.M.; Clark, M.L.; Etter, A. Land cover change in Colombia: Surprising forest recovery trends between 2001 and 2010. *PLoS ONE* **2012**, *7*, e43943. [CrossRef] [PubMed]
17. Rodríguez-Mahecha, J.V.; Hernández-Camacho, J.I. *Loros de Colombia*; Conservation International: Bogotá D.C., Colombia, 2002.
18. Tella, J.L.; Rojas, A.; Carrete, M.; Hiraldo, F. Simple assessments of age and spatial population structure can aid conservation of poorly known species. *Biol. Conserv.* **2013**, *167*, 425–434. [CrossRef]
19. Blanco, G.; Hiraldo, F.; Rojas, A.; Dénes, F.V.; Tella, J.L. Parrots as key multilinkers in ecosystem structure and functioning. *Ecol. Evol.* **2015**, *5*, 4141–4160. [CrossRef]
20. Tella, J.L.; Dénes, F.V.; Zulian, V.; Prestes, N.P.; Martínez, J.; Blanco, G.; Hiraldo, F. Endangered plant-parrot mutualisms: Seed tolerance to predation makes parrots pervasive dispersers of the Parana pine. *Sci. Rep.* **2016**, *6*, 31709. [CrossRef]
21. Baños-Villalba, A.; Blanco, G.; Díaz-Luque, J.A.; Dénes, F.V.; Hiraldo, F.; Tella, J.L. Seed dispersal by macaws shapes the landscape of an Amazonian ecosystem. *Sci. Rep.* **2017**, *7*, 7373. [CrossRef]
22. Dénes, F.V.; Tella, J.L.; Beissinger, S.R. Revisiting methods for estimating parrot abundance and population size. *Emu* **2018**, *118*, 67–79. [CrossRef]
23. Luna, Á.; Romero-Vidal, P.; Hiraldo, F.; Tella, J.L. Cities may save some threatened species but not their ecological functions. *PeerJ* **2018**, *6*, e4908. [CrossRef]
24. Blanco, G.; Bravo, C.; Chamorro, D.; Lovas-Kiss, A.; Hiraldo, F.; Tella, J.L. Herb endozoochory by cockatoos: Is "foliage the fruit20"? *Aust. Ecol.* **2020**, *45*, 122–126. [CrossRef]
25. Wirminghaus, J.O.; Downs, C.T.; Perrin, M.R.; Symes, C.T. Abundance and activity patterns of the Cape parrot (*Poicephalus robustus*) in two afromontane forests in South Africa. *Afr. Zool.* **2001**, *36*, 71–77. [CrossRef]
26. Salinas-Melgoza, A.; Renton, K. Seasonal variation in activity patterns of juvenile Lilac-crowned parrots in tropical dry forest. *Wilson Bull.* **2006**, *117*, 291–295. [CrossRef]
27. Tella, J.L.; Hernández-Brito, D.; Blanco, G.; Hiraldo, F. Urban sprawl, food subsidies and power lines: An ecological trap for large frugivorous bats in Sri Lanka? *Diversity* **2020**, *12*, 94. [CrossRef]
28. Young, J.C.; Rose, D.C.; Mumby, H.S.; Benitez-Capistros, F.; Derrick, C.J.; Finch, T.; Garcia, C.; Home, C.; Marwaha, E.; Morgans, C.; et al. A methodological guide to using and reporting on interviews in conservation science research. *Methods Ecol. Evol.* **2018**, *9*, 10–19. [CrossRef]
29. Thomas, L.; Buckland, S.T.; Rexstad, E.A.; Laake, J.L.; Strindberg, S.; Hedley, S.L.; Bishop, J.R.; Marques, T.A.; Burnham, K.P. Distance software: Design and analysis of distance sampling surveys for estimating population size. *J. Appl. Ecol.* **2010**, *47*, 5–14. [CrossRef] [PubMed]
30. Forshaw, J.M. *Parrots of the World: Helm Field Guides*; A & C Black Publishers Ltd.: London, UK, 2010.
31. Manly, B.; McDonald, L.; Thomas, D.; McDonald, T. *Resource Selection by Animals: Statistical Design and Analysis for Field Studies*, 2nd ed.; Springer Science & Business Media: Berlin/Heidelberg, Germany, 2007.
32. Rebolo-Ifrán, N.; Tella, J.L.; Carrete, M. Urban conservation hotspots: Predation release allows the grassland-specialist burrowing owl to perform better in the city. *Sci. Rep.* **2017**, *7*, 3527. [CrossRef]
33. R Development Core Team. *R: A Language and Environment for Statistical Computing (R-3.6.1)*; R Foundation for StatisticalComputing: Vienna, Austria, 2019.
34. Sodhi, N.S.; Koh, L.P.; Brook, B.W.; Ng, P.K.L. Southeast Asian biodiversity: An impending disaster. *Trends Ecol. Evol.* **2004**, *19*, 654–660. [CrossRef]

35. Ortiz-von Halle, B. *Bird's-Eye View: Lessons from 50 Years of Bird Trade Regulation*; TRAFFIC: Cambridge, UK, 2018.
36. Tittensor, D.P.; Harfoot, M.; McLardy, C.; Britten, G.L.; Kecse-Nagy, K.; Landry, B.; Outhwaite, W.; Price, B.; Sinovas, P.; Blanc, J.; et al. Evaluating the relationships between the legal and illegal international wildlife trades. *Conserv. Lett.* **2020**, e12724. [CrossRef]
37. Frynta, D.; Lišková, S.; Bültmann, S.; Burda, H. Being attractive brings advantages: The case of parrot species in captivity. *PLoS ONE* **2010**, *5*, e12568. [CrossRef]
38. Wright, T.F.; Toft, C.A.; Enkerlin-Hoeflich, E.; Gonzalez-Elizondo, J.; Albornoz, M.; Rodríguez-Ferraro, A.; Rojas-Suárez, F.; Sanz, V.; Trujillo, A.; Beissinger, S.R.; et al. Nest Poaching in Neotropical Parrots. *Conserv. Biol.* **2001**, *15*, 710–720. [CrossRef]
39. Herrera, M.; Hennessey, B. Quantifying the illegal parrot trade in Santa Cruz de la Sierra, Bolivia, with emphasis on threatened species. *Bird Conserv. Int.* **2007**, *17*, 295–300. [CrossRef]
40. Vall-llosera, M.; Cassey, P. Physical attractiveness, constraints to the trade and handling requirements drive the variation in species availability in the Australian cagebird trade. *Ecol. Econ.* **2017**, *131*, 407–413. [CrossRef]
41. Beissinger, S.R. Trade in live wild birds: Potentials, Principles and Practices of Sustainable Use. In *Conservation of Exploited Species*; Reynolds, J.D., Mace, G.M., Redford, K.H., Robinson, J.G., Eds.; Cambridge University Press: Cambridge, UK, 2001; pp. 182–202.
42. Cardador, L.; Lattuada, M.; Strubbe, D.; Tella, J.L.; Reino, L.; Figueira, R.; Carrete, M. Regional bans on wild-bird trade modify invasion risks at a global scale. *Conserv. Lett.* **2017**, *10*, 717–725. [CrossRef]
43. Marsden, S.J.; Royle, K. Abundance and abundance change in the world's parrots. *Ibis* **2015**, *157*, 219–229. [CrossRef]
44. Beissinger, S.R.; Bucher, E.H. Can parrots be conserved through sustainable harvesting? *BioScience* **1992**, *42*, 164–173. [CrossRef]
45. Valle, S.; Collar, N.J.; Harris, W.E.; Marsden, S.J. Trapping method and quota observance are pivotal to population stability in a harvested parrot. *Biol. Conserv.* **2018**, *217*, 428–436. [CrossRef]
46. Olah, G.; Butchart, S.H.M.; Symes, A.; Guzmán, I.M.; Cunningham, R.; Brightsmith, D.J.; Heinsohn, R. Ecological and socio-economic factors affecting extinction risk in parrots. *Biodivers. Conserv.* **2016**, *25*, 205–223. [CrossRef]
47. The IUCN Red List of Threatened Species. Version 2019-2. Available online: http://www.iucnredlist.org (accessed on 18 July 2019).
48. Martin, R.O. The wild bird trade and African parrots: Past, present and future challenges. *Ostrich* **2018**, *89*, 139–143. [CrossRef]
49. Cardador, L.; Tella, J.L.; Anadón, J.D.; Abellán, P.; Carrete, M. The European trade ban on wild birds reduced invasion risks. *Conserv. Lett.* **2019**, *12*, e12631. [CrossRef]
50. Ribeiro, J.; Reino, L.; Schindler, S.; Strubbe, D.; Vall-llosera, M.; Bastos Araújo, M.; Capinha, C.; Carrete, M.; Mazzoni, S.; Monteiro, M.; et al. Trends in legal and illegal trade of wild birds: A global assessment based on expert knowledge. *Biodiv. Conserv.* **2019**, *28*, 3343–3369. [CrossRef]
51. Martin, R.O.; Senni, C.; D'cruze, N.; Bruschi, N. Tricks of the trade—Legal trade used to conceal Endangered African grey parrots on commercial flights. *Oryx* **2019**, *53*, 213. [CrossRef]
52. Reuter, K.E.; Rodriguez, L.; Hanitriniaina, S.; Schaefer, M.S. Ownership of parrots in Madagascar: Extent and conservation implications. *Oryx* **2019**, *53*, 582–588. [CrossRef]
53. Regueira, R.F.S.; Bernard, E. Wildlife sinks: Quantifying the impact of illegal bird trade in street markets in Brazil. *Biol. Conserv.* **2012**, *149*, 16–22. [CrossRef]
54. Alves, R.R.N.; Lima, J.R.D.F.; Araujo, H.F.P. The live bird trade in Brazil and its conservation implications: An overview. *Bird Conserv. Int.* **2013**, *23*, 53–65. [CrossRef]
55. Daut, E.F.; Brightsmith, D.J.; Mendoza, A.P.; Puhakka, L.; Peterson, M.J. Illegal domestic bird trade and the role of export quotas in Peru. *J. Nat. Conserv.* **2015**, *27*, 44–53. [CrossRef]
56. Biddle, R.; Ponce, I.S.; Cun, P.; Tollington, S.; Jones, M.; Marsden, S.; Devenish, C.; Horstman, E.; Berg, K.; Pilgrim, M. Conservation status of the recently described Ecuadorian Amazon parrot *Amazona lilacina*. *Bird Conserv. Int.* **2020**, in press. [CrossRef]

57. Sánchez-Mercado, A.; Asmüssen, M.; Rodríguez, J.P.; Moran, L.; Cardozo-Urdaneta, A.; Morales, L.I. Illegal trade of the Psittacidae in Venezuela. *Oryx* **2020**, *54*, 77–83. [CrossRef]
58. Harris, J.B.C.; Green, J.M.; Prawiradilaga, D.M.; Giam, X.; Hikmatullah, D.; Putra, C.A.; Wilcove, D.S. Using market data and expert opinion to identify overexploited species in the wild bird trade. *Biol. Conserv.* **2015**, *187*, 51–60. [CrossRef]
59. Berkunsky, I.; Quillfeldt, P.; Brightsmith, D.J.; Abbud, M.C.; Aguilar, J.M.R.E.; Alemán-Zelaya, U.; Aramburú, R.M.; Arce Arias, A.; Balas McNab, R.; Balsby, T.J.S.; et al. Current threats faced by Neotropical parrot populations. *Biol. Conserv.* **2017**, *214*, 278–287. [CrossRef]
60. Pires, S.F.; Moreto, W.D. Preventing Wildlife Crimes: Solutions That Can Overcome the "Tragedy of the Commons". *Eur. J. Crim. Policy Res.* **2011**, *17*, 101–123. [CrossRef]
61. Robinson, J.G.; Bennett, E.L. Having your wildlife and eating it too: An analysis of hunting sustainability across tropical ecosystems. *Anim. Conserv.* **2004**, *7*, 397–408. [CrossRef]
62. Fryxell, J.M.; Packer, C.; McCann, K.; Solberg, E.J.; Sæther, B.E. Resource management cycles and the sustainability of harvested wildlife populations. *Science* **2010**, *328*, 903–906. [CrossRef] [PubMed]
63. Sánchez-Mercado, A.; Cardozo-Urdaneta, A.; Moran, L.; Ovalle, L.; Arvelo, M.Á.; Morales-Campos, J.; Coyle, B.; Braun, M.J.; Rodríguez-Clark, K.M. Social network analysis reveals specialized trade in an Endangered songbird. *Anim. Conserv.* **2020**, *23*, 132–144. [CrossRef]
64. Pires, S.; Herrera, M.; Tella, J.L. Spatial, temporal and age sources of variation in parrot poaching in Bolivia. *Bird Conserv. Int.* **2016**, *26*, 293–306. [CrossRef]
65. Carrascal, J.; Chacón, J.; Ochoa, V. Ingreso de psittacidos al centro de atención de fauna (CAV–CVS), durante los años 2007–2009. *Rev. MVZ Córdoba* **2013**, *18*, 3414–3419. [CrossRef]
66. Young, A.M.; Hobson, E.A.; Lackey, L.B.; Wright, T.F. Survival on the ark: Life-history trends in captive parrots. *Anim. Conserv.* **2012**, *15*, 28–43. [CrossRef]
67. Toft, C.A.; Wright, T.F. *Parrots of the Wild: A Natural History of the World's Most Captivating Birds*; University of California Press: Berkeley, CA, USA, 2015.
68. Blanco, G.; Hiraldo, G.; Tella, J.L. Ecological functions of parrots: An integrative perspective from plant life cycle to ecosystem functioning. *Emu* **2018**, *118*, 36–49. [CrossRef]
69. Montesinos-Navarro, A.; Hiraldo, F.; Tella, J.L.; Blanco, G. Network structure embracing mutualism-antagonism continuums increases community robustness. *Nat. Ecol. Evol.* **2017**, *1*, 1661. [CrossRef]
70. Tella, J.L.; Baños-Villalba, A.; Hernández-Brito, D.; Rojas, A.; Pacífico, E.; Díaz-Luque, J.A.; Carrete, M.; Blanco, G.; Hiraldo, F. Parrots as overlooked seed dispersers. *Front. Ecol. Environ.* **2015**, *13*, 338–339. [CrossRef]
71. Tella, J.L.; Blanco, G.; Dénes, F.V.; Hiraldo, F. Overlooked parrot seed dispersal in Australia and South America: Insights on the evolution of dispersal syndromes and seed size in Araucaria trees. *Front. Ecol. Evol.* **2019**, *7*, 82. [CrossRef]
72. Blanco, G.; Tella, J.L.; Díaz-Luque, J.A.; Hiraldo, F. Multiple external seed dispersers challenge the megafaunal syndrome anachronism and the surrogate ecological function of livestock. *Front. Ecol. Evol.* **2019**, *7*, 328. [CrossRef]
73. Tella, J.L.; Hiraldo, F.; Pacífico, E.; Díaz-Luque, J.A.; Dénes, F.V.; Fontoura, F.M.; Guedes, N.; Blanco, G. Conserving the diversity of ecological interactions: The role of two threatened macaw species as legitimate dispersers of "megafaunal" fruits. *Diversity* **2020**, *12*, 45. [CrossRef]
74. Blanco, G.; Bravo, C.; Pacífico, E.; Chamorro, D.; Speziale, K.; Lambertucci, S.; Hiraldo, F.; Tella, J.L. Internal seed dispersal by parrots: An overview of a neglected mutualism. *PeerJ* **2017**, *4*, e1688. [CrossRef] [PubMed]
75. Bravo, C.; Chamorro, D.; Hiraldo, F.; Speziale, K.; Lambertucci, S.A.; Tella, J.L.; Blanco, G. Physiological dormancy broken by endozoochory: Austral parakeets (*Enicognathus ferrugineus*) as legitimate dispersers of Calafate (*Berberis microphylla*) in the Patagonian Andes. *J. Plant Ecol.* **2020**, in press. [CrossRef]
76. Hernández-Brito, D.; Romero-Vidal, P.; Hiraldo, F.; Blanco, G.; Díaz-Luque, J.A.; Barbosa, J.M.; Symes, C.T.; White, T.K.; Pacífico, E.C.; Sebastián-González, E.; et al. Caught on camera: Epizoochory in parrots as an overlooked yet widespread plant-animal mutualism. *Ecology* **2020**. under review.

77. Sebastián-González, E.; Hiraldo, F.; Blanco, G.; Hernández-Brito, D.; Romero-Vidal, P.; Carrete, M.; Gómez-Llanos, E.; Pacífico, E.; Díaz-Luque, J.A.; Denés, F.V.; et al. The extent, frequency and ecological functions of food wasting by parrots. *Sci. Rep.* **2019**, *9*, 15280. [CrossRef]
78. Moleón, M.; Sánchez-Zapata, J.A.; Donázar, J.A.; Revilla, E.; Martín-López, B.; Gutiérrez-Cánovas, C.; Getz, W.M.; Morales-Reyes, Z.; Campos-Arceiz, A.; Crowder, L.B.; et al. Rethinking megafauna. *Proc. Roy. Soc. Lond. B* **2020**, *287*, 20192643. [CrossRef]
79. Barbosa, A.E.A.; Tella, J.L. How much does it cost to save a species from extinction? Costs and rewards of conserving the Lear's macaw. *Roy. Soc. Open Sci.* **2019**, *6*, 190190. [CrossRef]
80. Briceño-Linares, J.M.; Rodríguez, J.P.; Rodríguez-Clark, K.M.; Rojas-Suárez, F.; Millán, P.A.; Vittori, E.G.; Carrasco-Muñoz, M. Adapting to changing poaching intensity of yellow-shouldered parrot (*Amazona barbadensis*) nestlings in Margarita Island, Venezuela. *Biol. Conserv.* **2011**, *144*, 1188–1193. [CrossRef]
81. Brightsmith, D.J.; Stronza, A.; Holle, K. Ecotourism, conservation biology, and volunteer tourism: A mutually beneficial triumvirate. *Biol. Conserv.* **2008**, *141*, 2832–2842. [CrossRef]
82. Reuter, P.; O'Regan, D. Smuggling wildlife in the Americas: Scale, methods, and links to other organised crimes. *Glob. Crime* **2017**, *18*, 77–99. [CrossRef]
83. Holden, M.H.; Biggs, D.; Brink, H.; Bal, P.; Rhodes, J.; McDonald-Madden, E. Increase anti-poaching law-enforcement or reduce demand for wildlife products? A framework to guide strategic conservation investments. *Conserv. Lett.* **2019**, *12*, e12618. [CrossRef]
84. Nijman, V.; Langgeng, A.; Helene, B.; Imron, M.A.; Nekaris, K.A.I. Wildlife trade, captive breeding and the imminent extinction of a songbird. *Glob. Ecol. Conserv.* **2018**, *15*, e00425. [CrossRef]
85. Coetzer, W.G.; Downs, C.T.; Perrin, M.R.; Willows-Munro, S. Testing of microsatellite multiplexes for individual identification of Cape Parrots (*Poicephalus robustus*): Paternity testing and monitoring trade. *PeerJ* **2017**, *5*, e2900. [CrossRef]
86. Hogg, C.J.; Dennison, S.; Frankham, G.J.; Hinds, M.; Johnson, R.N. Stopping the spin cycle: Genetics and bio-banking as a tool for addressing the laundering of illegally caught wildlife as 'captive-bred'. *Conserv. Gen. Res.* **2018**, *10*, 237–246. [CrossRef]
87. Carrete, M.; Tella, J.L. Rapid loss of antipredatory behaviour in captive-bred birds is linked to current avian invasions. *Sci. Rep.* **2015**, *5*, 18274. [CrossRef] [PubMed]
88. Abellán, P.; Tella, J.L.; Carrete, M.; Cardador, L.; Anadón, J.D. Climatic matching drives spread rate but not establishment success in recent unintentional bird introductions. *Proc. Nat. Acad. Sci. USA* **2017**, *114*, 201704815. [CrossRef]
89. Botero-Delgadillo, E.; Páez, C.A. Estado actual del conocimiento y conservación de los loros amenazados de Colombia. *Conserv. Colomb.* **2011**, *14*, 86–151.
90. Botero-Delgadillo, E.; Páez, C.A. Plan de acción para la conservación de los loros amenazados de Colombia 2010–2020: Avances, logros y perspectivas. *Conserv. Colomb.* **2011**, *14*, 7–16.
91. Thomas-Walters, L.; Veríssimo, D.; Gadsby, E.; Roberts, D.; Smith, R.J. Taking a more nuanced look at behavior change for demand reduction in the illegal wildlife trade. *Conserv. Sci. Pract.* **2020**, *10*, e248. [CrossRef]
92. Morinha, F.; Carrete, M.; Tella, J.L.; Blanco, G. High prevalence of novel beak and feather disease virus in sympatric invasive parakeets introduced to Spain from Asia and South America. *Diversity* **2020**, *12*, 192. [CrossRef]
93. Harris, R.B.; Wall, W.A.; Allendorf, F.W. Genetic consequences of hunting: What do we know and what should we do? *Wildl. Soc. Bull.* **2002**, *30*, 634–643.
94. Coltman, D.W.; O'Donoghue, P.; Jorgenson, J.T.; Hogg, J.T.; Strobeck, C.; Festa-Bianchet, M. Undesirable evolutionary consequences of trophy hunting. *Nature* **2003**, *426*, 655–658. [CrossRef] [PubMed]
95. Allendorf, F.W.; England, P.R.; Luikart, G.; Ritchie, P.A.; Ryman, N. Genetic effects of harvest on wild animal populations. *Trends Ecol. Evol.* **2008**, *23*, 327–337. [CrossRef] [PubMed]
96. Allendorf, F.W.; Hard, J.J. Human-induced evolution caused by unnatural selection through harvest of wild animals. *Proc. Nat. Acad. Sci. USA* **2009**, *106*, 9987–9994. [CrossRef]
97. Mysterud, A. Selective harvesting of large mammals: How often does it result in directional selection? *J. Appl. Ecol.* **2011**, *48*, 827–834. [CrossRef]
98. Segura, L.N.; Perelló, M.; Gress, N.H.; Ontiveros, R. The lack of males due to illegal trapping is causing polygyny in the globally endangered Yellow Cardinal *Gubernatrix cristata*. *Ornithol. Res.* **2019**, *27*, 40–43. [CrossRef]

99. Corlatti, L.; Sanz-Aguilar, A.; Tavecchia, G.; Gugiatti, A.; Pedrotti, L. Unravelling the sex-and age-specific impact of poaching mortality with multievent modeling. *Front. Zool.* **2019**, *16*, 20. [CrossRef]
100. Heinrich, S.; Ross, J.V.; Gray, T.N.; Delean, S.; Marx, N.; Cassey, P. Plight of the commons: 17 years of wildlife trafficking in Cambodia. *Biol. Conserv.* **2020**, *241*, 108379. [CrossRef]
101. Marshall, H.; Collar, N.J.; Lees, A.C.; Moss, A.; Yuda, P.; Marsden, S.J. Spatio-temporal dynamics of consumer demand driving the Asian Songbird Crisis. *Biol. Conserv.* **2020**, *241*, 108237. [CrossRef]

MDPI

Article

Wildlife Trade Influencing Natural Parrot Populations on a Biodiverse Indonesian Island

Dudi Nandika [1,†], Dwi Agustina [1,†], Robert Heinsohn [2,*] and George Olah [2,3]

1 Perkumpulan Konservasi Kakatua Indonesia (KKI), Perum Griya Gandasari Indah Kav. C/03, Gandasari, Cikarang Barat, Bekasi 17530, West Java, Indonesia; dudi.nandika@gmail.com (D.N.); agustina.dwi@gmail.com (D.A.)
2 Fenner School of Environment and Society, The Australian National University, Canberra, ACT 2601, Australia; george.olah@anu.edu.au
3 Wildlife Messengers, Richmond, VA 23230, USA
* Correspondence: robert.heinsohn@anu.edu.au
† Authors contributed equally to the manuscript.

Abstract: Indonesia has been identified as the highest priority country for parrot conservation based on the number of species, endemics, and threats (trapping and smuggling). It is crucial to understand the current population status of parrots in the wild in relation to the illegal wildlife trade but the ecology and population dynamics of most parrot species in this region remain poorly understood. We conducted a parrot survey around an area of high biodiversity in the Manusela National Park, in Seram Island, Indonesia. We used a combination of fixed-radius point counts and fixed-width line transects to count multiple species of parrots. We recorded nearly 530 wild parrots from 10 species in and around Manusela National Park. The dominant parrot species were *Eos bornea*, *Trichoglosus haematodus*, and *Geoffroyus geoffroyi*. We applied the Savage selectivity index to evaluate poaching of parrot species in proportion to their abundance and which species had higher than expected poaching pressure. This study has important implications for the conservation status of endemic parrots (*Cacatua moluccensis, Lorius domicella,* and *Eos semilarvata*) and shows that parrots in the Manusela NP are largely threatened by poaching.

Keywords: parrots; conservation; ecology; wildlife trade; density; endemism; poaching; IUCN Red List; CITES

Citation: Nandika, D.; Agustina, D.; Heinsohn, R.; Olah, G. Wildlife Trade Influencing Natural Parrot Populations on a Biodiverse Indonesian Island. *Diversity* **2021**, *13*, 483. https://doi.org/10.3390/d13100483

Academic Editors: José L. Tella, Guillermo Blanco and Martina Carrete

Received: 9 August 2021
Accepted: 28 September 2021
Published: 30 September 2021

1. Introduction

Parrots (order Psittaciformes) are the most threatened taxon of birds, with one-third of the nearly 400 species classified as threatened under IUCN criteria [1]. In a global analysis, Indonesia was identified as the highest priority country for parrot conservation based on the number of species, endemics, and threats [2]. Further, among CITES-listed species, parrots are by far the most traded and are declining faster than any other comparable groups of birds [2]. In Indonesia, all parrot species (except *Psittinus abbotti*) have been protected by law since 2018 [3], but this legislation is poorly enforced [4]. A higher proportion (almost half) of the parrot species in the Wallacean region of Indonesia are affected by trapping than in neighbouring regions or compared to the world average [5].

Seram Island is in central Wallacea, between New Guinea and Sulawesi. It hosts 11 parrot species, three of which are endemic to the Maluku archipelago: the Salmon-crested Cockatoo *Cacatua moluccensis*, Purple-naped Lory *Lorius domicella*, and the Blue-eared Lory *Eos semilarvata* (Table 1). The island is also home to five subspecies of Wallacean endemic parrots: the Moluccan King-parrot *Alisterus amboinensis amboinensis*, Eclectus Parrot *Eclectus roratus roratus*, Great-billed Parrot *Tanygnathus megalorynchos affinis*, Red-cheeked Parrot *Geoffroyus geoffroyi rhodops*, and the Chattering Lory *Eos bornea rothschildi* (Table 1). Finally, there are also three non-endemic parrot species: the Red-breasted Pygmy-parrot *Micropsitta bruijnii*, Red-flanked Lorikeet *Charmosyna placentis*, and the Coconut Lorikeet *Trichoglossus haematodus* (Table 1).

Table 1. Endemism and conservation status of the parrot fauna in Seram, Indonesia. Presented are endemism, scientific and English names, national protection [3], CITES status (Appendix I or II), and IUCN Red List conservation status (LC = Least Concern, NT = Near Threatened, VU = Vulnerable, EN = Endangered) [1]. * The current study includes recommendations to change these statuses (under Section 4.3).

Endemism	Scientific Name	English Name	National Protection	CITES	IUCN RedList
Species endemic to Seram	*Cacatua moluccensis*	Salmon-crested Cockatoo	protected	I	VU *
	Lorius domicella	Purple-naped Lory	protected	II *	EN *
	Eos semilarvata	Blue-eared Lory	protected	II *	NT *
Subspecies endemic to the Maluku archipelago	*Alisterus amboinensis amboinensis*	Moluccan King-parrot	protected	II	LC
	Eclectus roratus roratus	Eclectus Parrot	protected	II	LC
	Tanygnathus megalorynchos affinis	Great-billed Parrot	protected	II	LC
	Geoffroyus geoffroyi rhodops	Red-cheeked Parrot	protected	II	LC
	Eos bornea rothschildi	Chattering Lory	protected	II	LC
Non endemic	*Micropsitta bruijnii*	Red-breasted Pygmy-parrot	protected	II	LC
	Charmosyna placentis	Red-flanked Lorikeet	protected	II	LC
	Trichoglossus haematodus	Coconut Lorikeet	protected	II	LC

Parrot trade in Indonesia is driven by both demand and opportunity-based factors, and serves both national and international markets [4]. In the illegal wildlife market, there is a high demand for at least two parrot species endemic to the Moluccas, *C. moluccensis* and the *L. domicella*. Local authorities have processed cases of illegal parrot trade originating from the outskirts of Manusela National Park in Seram [6]. Although legislation regulating the wildlife trade in Indonesia is in place, law enforcement, monitoring, and awareness are often lacking. Knowledge of trading routes and source populations is sorely needed to help governments shut down the illegal wildlife trade and to aid the rewilding of the increasing number of confiscated parrots.

The high demand for parrots in the illegal trade threatens parrot populations in the wild, so it is crucial to understand the current population status of parrots in their natural habitat and how the trade affects them directly. We conducted a two-month population census of parrots in 2020 at three locations throughout Seram. The aims of this study are to (1) record current demographic measures of wild parrot populations, including density, species richness, diversity, and evenness, and (2) investigate the effects of poaching on the wild populations. Our results will serve as an important reference for the local authorities in making decisions and management programs regarding law enforcement in the local parrot trade and future rewilding work.

2. Materials and Methods

2.1. Study Area

The study was conducted in the Manusela National Park, on Seram Island, Central Maluku District, Maluku Province, Indonesia (129°9′3″–129°46′14″ E and 2°48′24″–3°18′24″ S). Manusela NP was established in 2014, covering an area of 174,545.59 ha in the central region of the island (Figure 1; map was constructed with QGIS 3.20). The topography of Manusela NP consists of lowland (0–500 m), highland (500–1500 m), mountain (1500–2500 m), and sub-alpine (up to 3000 m) habitats [7]. The extraordinary biodiversity of the Manusela NP includes endemic and unique flora and fauna [8], including a high diversity of orchids, pteridophytes [9], ferns [7], marsupials, bats [9], reptiles [10], and insects [11]. Bowler and Taylor [12] recorded 197 species of birds in the national park, including 124 resident species and 73 migrant bird species.

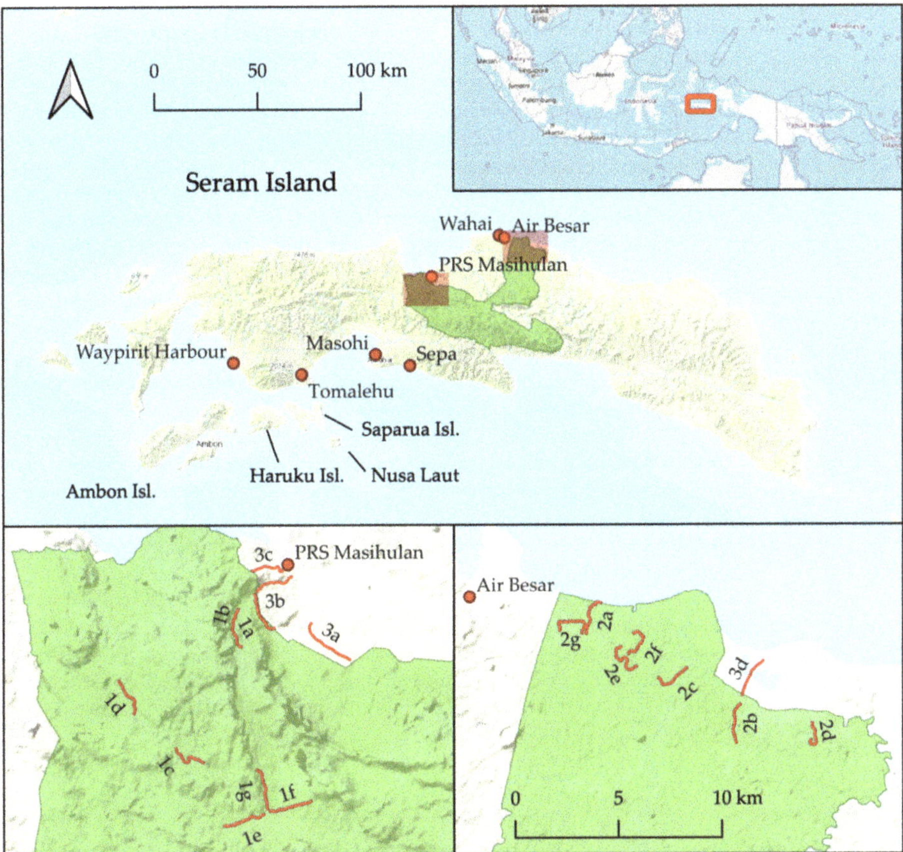

Figure 1. Map of the study area in Seram Island, Indonesia with the Manusela National Park (green). Census transects represent (1) Masihulan Camp sites: (1a) Masihulan camp track, (1b) Pos pantau, (1c) Km 20, (1d) Hatusaka track, (1e) Illie Camp track 2, (1f) -track 3, (1g) -track 1; (2) Sasarata Camp sites: (2a) Mangrove track, (2b) Wae Masin, (2c) Wae Mual, (2d) Pasahari fire track, (2e) Wae Patan, (2f) Wae Faung, (2g) Wae Masinatu; and (3) Manusela NP buffer zone (BZ) sites: (3a) Negeri Masihulan (NM) Galian C, (3b) NM shelter track, (3c) NM cultivation, (3d) Negeri Pasahari Wae Masin. Red dots represent locations mentioned in the main text.

The census was conducted from late March to early May 2020 at three locations including seven observation tracks in Masihulan Camp (1) covering 1.25 km^2; seven observation tracks in Sasarata Camp (2) of 1.25 km^2; and the buffer zone of the Manusela NP (3) including three observation tracks in Negeri Masihulan of 0.54 km^2 and one observation track in Negeri Pasahari of 0.18 km^2 (Figure 1). The survey locations also represented different habitat types, including mountain forests (in Masihulan- and Illie Camp tracks) [13], habitats over 1000 m dominated by mosses and ferns, lowland habitats (alluvial land, mangrove, and savanna in Sasarata Camp and Pasahari), secondary forest, seashore, and cultivated land (in the buffer zone with plants including clove *Syzygium aromaticum* nutmeg plant *Myristica fragrans*).

2.2. Survey Methods

The surveys were conducted using a combination of fixed-radius point counts (FRPC) [14] and fixed-width line transects (FWLT) [15]. We counted multiple species of parrots as targets, using two skilled teams of two parrot specialists to identify location and species, and to minimise observer error rate. The teams included experts with 30 years of experience with visual and auditory identification of parrot species and with excellent local knowledge of nesting and roosting locations. We encountered the birds along the transects with 2 km distance between tracks and 50 m of visibility at each point. We conducted the census on seven transects in Masihulan Camp and Illie Camp, on seven transects in Sasarata Camp and Pasahari, on three transects in the buffer zone of the Manusela NP at Negeri Masihulan, and one transect in Negeri Pasahari (Figure 1). Between each transect there was at least 2 km distance. In total, we conducted 180 FRPC transects and 180 FWLT transects.

Parrot censuses were carried out in the early mornings during a consistent time period from 07:30 to 12:00, but in some locations, we spent more time (until 15:00) because of the difficult path and remoteness. Data were collected by two teams to estimate population size by direct count with the birds. We detected and identified the birds by means of their calls, because parrots are noisy birds and have a specifically distinct loud call [16]. Moreover, no other birds have calls similar to those of parrots in the location. The best time for observation was 30 min after sunrise and closer to the sunset as these times are the peak of activity in birds [17]. In cockatoos, activity begins as soon as the sun rises, and they immediately start looking for food [16]. Competition between species of birds that have the same types of food will lead to separation of resource use and ecological niches. The FRPC data were collected for 15 min on each occasion, identifying all parrot species visually and by their calls. During the FWLT method we walked slowly and steadily, while detecting parrot species.

The data on confiscated parrots were collected by the Maluku office of the Natural Resource Conservation Center of the Ministry of Forestry (Balai Konservasi Sumber Daya Alam (Kota Ambon, Indonesia), BKSDA Maluku hereafter) from 2016 to 2018 [6]. We also included data on birds confiscated by BKSDA Maluku in the Manusela NP and sent to the Pusat Rehabilitasi Satwa Kembali Bebas Avian Rehabilitation and Reintroduction Center Masihulan (PRS Masihulan hereafter) between 2019 and 2020. Currently, these data represent the best available measure of the recent poaching pressure on parrots in Seram Island.

2.3. Data Analyses

Density of birds (D) was calculated based on the total birds observed for each species and divided by the total plot area (A). If the total number of birds observed is S, and the total number of individuals observed in each transect is N, then the index of species richness follows the Menhinick index [18]:

$$R_2 = S/\sqrt{N} \tag{1}$$

If total individuals observed of a species is n and the species proportion is pi, then the species diversity follows the Shannon-Wiener index [19,20]:

$$H' = -\sum \frac{n}{N} ln \frac{n}{n} = -\sum pi\ ln\ pi \tag{2}$$

The Pielou index was used to calculate species evenness [21]:

$$E = H'/lnS \tag{3}$$

Species frequency (*F*) was calculated based on the total number of each species divided by the total number of birds observed in a sample plot [22].

In order to evaluate if parrot species were poached proportionally to their abundance in the wild, we calculated the Savage selectivity index (*W*) [23], following a previously established protocol for the parrot trade [24]:

$$W = u_i/p_i \quad (4)$$

where u_i is the proportion of a given species recorded on the trade (by BKSDA Maluku and PRS Masihulan) and p_i is the proportion of the same species observed in the wild (by the current study). We tested the null hypothesis that parrot species are poached in proportion to their availability in the wild by comparing the statistics of *W* to the corresponding critical value of the chi-squared distribution after Bonferroni correction, with one degree of freedom [24]. We excluded *E. semilarvata* from this analysis, given that we have not observed any individuals in the wild.

3. Results

3.1. Wild Parrot Populations on Seram

During the surveys we recorded 530 individual parrots from 10 species. Of these, 97 individuals were observed at Masihulan Camp, 294 at Sasarata Camp, and 139 in the buffer zone of the Manusela NP (105 in Masihulan and 34 in Sasarata buffer zones; Figure A1). The dominant parrot species observed were *E. bornea*, *T. haematodus*, and *G. geoffroyi*. The encounter rates for each species at each transect were: *E. bornea* 25.7%, 29% and 24.5%; *T. haematodus* 23%, 26.4%, and 18.7%; and *G. geoffroyi* 14.4%, 29.6%, and 17.1% respectively.

During our study, we detected *E. bornea* with the highest density (*D*) of 51.8 individuals/km^2 (ranged between 19.2 and–104.8 individuals/km^2; in Table 2, we also recorded high densities for *G. geoffroyi* (39.9 individuals/km^2), *T. haematodus* (37.1 individuals/km^2), and *C. moluccensis* (10.8 individuals/km^2). The lowest density was for *L. domicella* (2.3 individuals/km^2).

Table 2. Parrot densities (*D* = individuals/km^2) and species evenness (*E*) at different transects conducted in the Manusela National Park, Seram, Indonesia, including Masihulan Camp (1), Sasarata Camp (2), and Manusela NP buffer zones (3). For exact locations, see Figure 1.

Scientific Name	N	*D*			*E*		
		1	2	3	1	2	3
Alisterus amboinensis amboinensis	18	8	0.8	9.72	0.49	0.04	0.18
Cacatua moluccensis	26	10.4	1.6	15.28	0.74	0.06	0.31
Charmosyna placentis	6	-	-	8.33	-	-	0.13
Eclectus roratus roratus	41	4.8	16.8	19.44	0.36	0.52	0.33
Eos bornea rothschildi	184	19.2	104.8	40.28	1.08	1.14	0.57
Eos semilarvata	0	-	-	-	-	-	-
Geoffroyus geoffroyi rhodops	96	9.6	40.8	45.83	0.6	1.06	0.57
Lorius domicella	3	0.8	-	2.78	0.11	-	0.05
Micropsitta bruijnii bruijnii	6	-	-	8.33	-	-	0.1
Tanygnathus megalorynchos affinis	16	0.8	7.2	8.33	0.06	0.3	0.18
Trichoglossus haematodus	134	24	63.2	34.72	0.8	1.2	0.52

The transects with the highest richness (R_2) in the Masihulan Camp area were Km 20 (R_2 = 1.6) and Illie Camp (R_2 = 1.5; Figure 2). The highest parrot richness in the Sasarata Camp was recorded in the mangrove track (R_2 = 0.9) and Wae Mual (R_2 = 0.9; Figure 2). Finally, in the buffer zone of the Manusela NP the highest parrot richness was in the shelter track (R_2 = 1.2) the and cultivation track (R_2 = 1.1; Figure 2). Our data showed that the highest parrot species diversity (H') is in secondary forest, ecotone areas, or in a transition area between primary forest and cultivation. The lowest parrot diversity was detected in the Masihulan Camp on the Hatusaka track ($H' = 0.9$) and Illie Camp ($H' = 0.7$), and in the Sasarata Camp in Negeri Pasahari ($H' = 0.9$; Figure 2).

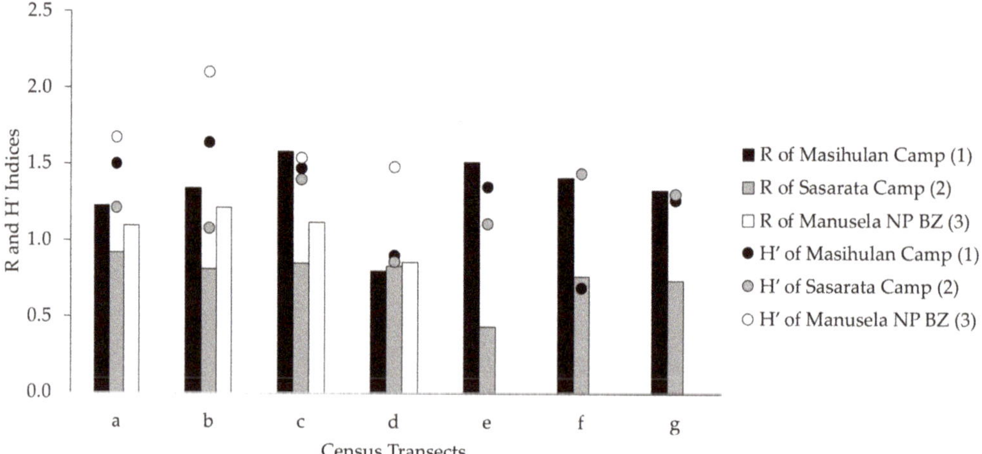

Figure 2. Comparison of species richness (R) and species diversity (H') across transects. The location of each census transect is shown in Figure 1.

Parrot species evenness (E) showed relative homogeneity in the buffer zone of the Manusela NP, because no particular species dominated this area. However, in both the Masihulan and Sasarata Camp parrot species evenness revealed higher heterogeneity. The dominating parrot species in the Masihulan Camp were *E. bornea* ($E = 1.08$), *T. haematodus* ($E = 0.8$), *C. moluccensis* ($E = 0.7$), and *G. geoffroyi* ($E = 0.6$; Table 2). The dominated species in the Sasarata Camp were *T. haematodus* ($E = 1.2$), *E. bornea* ($E = 1.1$), and *G. geoffroyi* ($E = 1.1$; Table 2), while the evenness index was very low for *C. moluccensis* ($E = 0.1$). We found similar results for both *E. bornea* and *T. haematodus* that are captured in high numbers for the wildlife trade (Table 2).

3.2. Parrot Trade and Poaching Pressure

In the past five years, BKSDA Maluku confiscated a total of 891 parrots, including 378 individuals of *E. bornea*, 216 *E. roratus*, 174 *T. haematodus*, 47 *C. moluccensis*, 45 *C. placentis*, 13 *L. domicella*, 12 *T. m. affinis*, and six *A. amboinensis*. Based on the confiscation data, the intensity of parrot poaching shows a decreasing tendency since the new government regulation (Republic of Indonesia, 2018) that protects all Indonesian parrot species by law.

The selectivity index (W) showed significant positive poaching selection for the following four species: *C. placentis*, *E. roratus*, *L. domicella*, and *E. bornea* ($p < 0.005$; Figure 3). In the case of five other species (*T. haematodus*, *T. m. affinis*, *A. amboinensis*, *M. buijnii*, and *G. geoffroyi*) we observed significant negative selection for poaching ($p < 0.005$; Figure 3). For *C. moluccensis*, we could not reject the null hypothesis that they are poached in proportion to their availability in the wild (Figure 3).

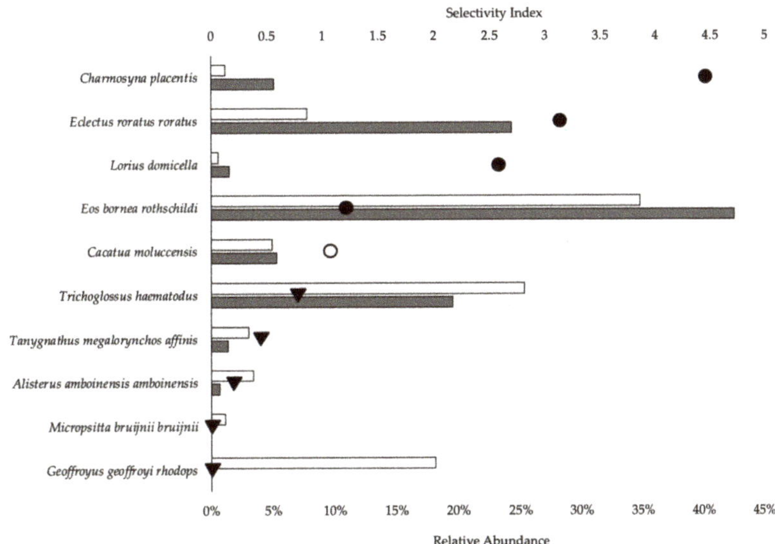

Figure 3. Relative abundance of parrot species in Seram, Indonesia as confiscated from the trade (dark gray bars) and observed in the wild (white bars), and selectivity index (W; black dots: significant positive selection; black triangles: significant negative selection; white dot: non-significant selection).

4. Discussion

4.1. Wild Parrot Populations on Seram

Seram is a highly biodiverse Indonesian island that hosts many endemic species including a high diversity of parrots. The Manusela NP is supposed to safeguard this high parrot diversity, including three species only found on this island (Table 1). However, poaching pressure for illegal trade is a significant threat to their existence both inside and outside the national park [6], where law enforcement is scarce. Illegal logging and fires also occur locally, but this forest damage covers only 10% of the area of Manusela NP [13].

The data presented in this study are from the most detailed parrot census conducted on Seram Island in recent years. *E. bornea*, *T. haematodus*, and *G. geoffroyi* were found in highest frequency and not limited to high elevation. These species are locally nomadic [25] and their populations can extend from very small to large areas, depending on the habitat carrying capacity [26]. We found *M. b. bruijnii* and *C. placentis* in the lowest frequencies and showed that both species have specific habitat and food requirements. Low evenness values in species may have been related to several limiting factors such as altitude, like in the case in *E. semilarvata* and *L. domicella*, which are difficult to find ≥700 m. Both *M. bruijnii* and *C. placentis* were closely related to the presence of certain species of trees. Species evenness was significantly associated with bird density, indicating a positive relationship with disturbed habitats as well as vertical heterogeneity with high bird density [27].

Open areas of lowland forest types, mangrove forests, savannas, and areas bordering cultivation had low species richness, though the figure may be influenced by the fruiting and flowering seasons. The majority of food sources were located inside forests [28] but parrots used cultivated and mangrove areas for foraging as well. Groups of parrots had an increasing species richness in forest types with open canopies and vertically heterogeneous vegetation structures. Frugivorous, granivorous, and nectarivorous species were closely related to disturbed areas [27]. The presence of frugivorous species is important because of their function to help plants with pollination and spreading of seeds [29,30]. The dominance of cockatoos in Masihulan Camp area is closely related to the rewilding process of confiscated birds that are concentrated in the region. The existence of PRS Masihulan

(Figure 1) possibly plays an important role in the increasing population of cockatoos in the region, and rewilded birds can be identified as they are banded.

4.2. Parrot Trade on Seram

Poaching is one of the biggest threats faced by parrots as many people buy them as pets, generating demand for illegal trapping and smuggling [4]. In Indonesia, BKSDA Maluku reported birds as the highest ranked smuggled taxon for trade accounting for 86% of species in Maluku, of which 96% were parrots [6]. Between 2016 and 2018, a total of 1135 individuals of 16 parrot species were confiscated, and about 44% came from Seram [6]. During a bird market survey in Jakarta, three out of 13 parrot species (32%) were registered as originating from Seram, including *T. haematodus*, *C. placentis*, and *E. bornea* [31].

Applying the selectivity index in this study allowed us to evaluate the differences in poaching pressure suffered by the parrot species in Seram. The highest pressure was found to be on *C. placentis*, *E. roratus*, *L. domicella*, and *E. bornea* (Figure 3), meaning that these species were found in the trade more frequently than expected given their abundance in the wild. On the other hand, the non-significant selective pressure on *C. moluccensis* does not indicate that poaching is not a risk for this species, but that their frequency on the trade corresponds to their abundancy in the wild. Moreover, the rewilding of confiscated individuals possibly skewed this index, as without these activities, there would have been fewer individuals observed in the wild, hence the index value would have been higher, i.e., the species positively selected. *C. moluccensis* is also hunted for traditional tribal ceremonies by the Huaulu, Naulu, and other tribes. Although birds account for only 6% of the wildlife consumed by the villagers in terms of the amount of protein [32], wild birds are especially important for them during certain periods.

Based on non-structural observations of communities in the area including Wahai, Air Besar, Sepa, and Masohi (Figure 1), people keep parrots as pets, such as *E. bornea*, lorikeets, and even cockatoos (D.N. unpublished data). Almost 90% of the households in Sepa keep parrots as pets (three or four individuals per family) and may sell the birds (Dr La Eddy, Pattimura University, unpublished data). *C. placentis* has also been observed for sale (20–30 individuals in small boxes) at Masohi (Figure 1). In the Manusela village, six species of parrots were caught in the forest strictly for the purpose of wildlife trade [33]. However, not every parrot species is at equal risk of being traded, and there is controversy concerning the role of demand and the opportunity-based factors driving the illicit wildlife trade [4]. The major source of income for people living in central Seram is seasonal migrant work (mainly harvesting cloves) but this income is unstable because of the fluctuation in production and the price. Hence, their dependency on wild parrots is enhanced during times of hardship caused by the decrease of their main income [32].

In order to involve local people in parrot conservation and rehabilitation, PRS Masihulan (Figure 1) was established in 2004, in the buffer zone of the Manusela NP. It has been employing former parrot trappers as caretakers who are also involved directly in the monitoring of the re-wilded birds [34]. These activities have greatly affected the decrease of direct poaching by local villagers [35]. An undercover investigation determined that trapping of cockatoos had essentially stopped in close to vicinity of PRS Masihulan [36]. For instance in 1998, the density of *C. moluccensis* was 7.9 individuals/km^2 on Seram [37]. Their density is now 20.4 individuals/km^2 near PRS Masihulan, while it remained low in other parts of Manusela NP (Figure 1, Table 2). This figure is probably due to their reproductive success and the decreased impact of direct poaching around PRS Masihulan. Sadly, the poaching of *C. moluccensis* still occurs in other parts of the island, including Manusela NP. For example, in May 2020, BKSDA Maluku confiscated 10 individuals from Tomalehu (Figure 1), and the team of the Konservasi Kakatua Indonesia tagged two birds (*C. moluccensis* and *L. domicella*) seized by BKSDA Maluku at Waypirit Harbour (Figure 1). The *L. domicella* individual had been banded before and this is the second time this bird has been trapped for the illegal parrot trade. Banding is important for recording the origin and the history of the birds and also for rewilding purposes. Other methods could include genetic tagging [38] that can aid assigning the provenance of the confiscated parrots [39].

4.3. Conservation Status of Endemic Parrot Species

The result of our current analysis of direct observations combined with recent parrot confiscation data [4,6] have important implications to the conservation status of three parrot species endemic to Seram. It is important to note that confiscations only represent a small fraction of the poached individuals, so actual figures for these species could be much higher and more worrying for their conservation status.

Cacatua moluccensis is currently considered Vulnerable (VU) on the IUCN Red List [1] and our analysis supports their uplisting to Endangered (EN). Their population size has been decreasing at least since 1994, based on previous IUCN evaluations [40]. Although its distribution previously covered some satellite islands of Seram, based on BirdLife data it has been declared extinct from Haruku, Saparua, and Nusa Laut islands (Figure 1). In Ambon, it is very difficult to find these cockatoos (D.N. pers. obs.) and they may only remain in the western part of the island. Based on abundance data in the past (7.9 individuals/km^2), their population size decreased by over 50% in some parts of the Manusela NP (currently 1.6 individuals/km^2 in Sasarata Camp), probably due to exploitation (see Section 4.2). Our selectivity analysis showed that poaching levels are consistent with their abundance in the wild (Figure 3). BKSDA Maluku and PRS Masihulan reported 47 confiscated individuals in the past five years in Maluku alone. Data from our direct field survey showed that the frequency of encounters was relatively low in Masihulan NP and its buffer zone (15.3% and 9.7% of all parrot individuals respectively) and that it was difficult to find them in the Sasarata area (relative frequency of 1.3%). These results support the IUCN criteria A2bd+3bd+4bd of EN [41].

Lorius domicella is currently considered Endangered (EN) on the IUCN Red List [1] and our analysis supports its uplisting to Critically Endangered (CR). They have a very limited distribution only in Seram in habitats above 800 m, and their population has been decreasing at least since 1994 according to IUCN evaluation [42]. Our survey showed very low relative frequency of all parrot individuals in the Masihulan NP (1.3%) and its buffer zone (1.6%). The number of mature individuals might be below 250 with fewer than 50 in each subpopulation. These results support the IUCN criteria C2a(i) of CR [41]. In addition, the threat of poaching is of concern as there were 13 recorded individuals confiscated in the past five years in Maluku and our selectivity analysis showed significant positive poaching pressure on the remining population (Figure 3). Hence, we recommend the urgent inclusion of the species to CITES Appendix I.

Eos semilarvata is currently listed as Near Threatened (NT) on the IUCN Red List [1] and our analysis supports its uplisting to Vulnerable (VU). It has a limited distribution at altitudes above 800 m in Seram, with a decreasing population size since at least 2019 based on IUCN evaluation [43]. We did not register this species during our census activity on the transects, but recorded one foraging flock of 10 individuals while in the field. Their area of occupancy is possibly less than 2000 km^2 in 10 or fewer locations in Seram with a continuing decline, and the number of mature individuals is potentially less than 10,000 with 1000 or fewer individuals in each subpopulation. These trends support the IUCN criteria B2a, B2b(ii), and C2a(i) of VU [41]. Although BKSDA Maluku confiscation data did not record this species in the past five years, 40 wild-caught birds were exported in 2019 from Soekarno Hatta airport [44], indicating that poaching activity can threaten the wild population. Hence, we also recommend the inclusion of the species to CITES Appendix I.

5. Conclusions

Manusela National Park and its buffer zone have high parrot diversity. Over the last five years confiscation data showed that nine out of the 11 parrot species of Seram Island were poached. Illegal poaching activities do not only threaten parrot populations but are also very likely to contribute to forest decline given the extremely important ecological role of parrots as seed dispersers [29,45,46]. The decline in parrot populations, as an apparent result of high demand on the wildlife market, may lead to the extinction of some species from the island. In this context, we have highlighted arguments for the uplisting on the IUCN Red List of three parrot species. We also showed that proper management and rehabilitation of confiscated parrots, and the inclusion of local communities into conservation efforts with environmental education campaigns can have positive effects on parrot conservation in the area. Our results contribute important information to local and international authorities and wildlife management programs that may help reduce the local parrot trade and inform future rewilding projects.

Author Contributions: Conceptualization, D.N., D.A. and G.O.; methodology, D.N. and D.A.; statistical analysis, D.N. and. D.A.; investigation, D.N.; writing—original draft preparation, D.N. and. D.A.; writing—review and editing, D.N., D.A., R.H. and G.O.; visualization, D.N. and G.O.; supervision, R.H. and G.O.; project administration, D.A.; funding acquisition, D.N. and D.A. All authors have read and agreed to the published version of the manuscript.

Funding: This research was funded by Critical Ecosystem Partnership Fund with Burung Indonesia as Regional Implementation Team, grant number 110662.

Institutional Review Board Statement: Not applicable.

Data Availability Statement: All data are available in this publication.

Acknowledgments: The authors would like to thank Manusela National Park and the heads of Masihulan and Sasarata, who gave the permission to carry out this study. Many thanks to the Indonesian Parrot Project (IPP), the late Stewart A. Metz, Bonnie Zimmermann (Director of IPP), and head of Manusela NP Bapak Ivan Yusfi Noor for all valuable inputs. Many thanks to all observation team from Manusela NP (Sugeng Handoyo, S. Hut, Cecep Hermawan, Ade Muliyanto, Christoperos Melaira, Sartika Sinulinga, Abdul Latif Musiin, John P. Syaranamual, Muhammad Johar Ernas, Reyn Ipakit, Nur Kholid Aloatuan, Claudia Melaira), KKI (Reki Patamanue, Mona Souhaly), and the local communities in both Masihulan (Jemy Souhaly, Federik Sapulete, Nopes Sapulete) and Sasarata (Samuel Huaulu) for their valuable assistance in the field. We also thank two anonymous reviewers, whose comments largely helped = improve this manuscript.

Conflicts of Interest: The authors declare no conflict of interests.

References

1. IUCN. *IUCN Red List of Threatened Species*; IUCN: Gland, Switzerland, 2008. Available online: http://www.iucnredlist.org (accessed on 25 August 2021).
2. Olah, G.; Butchart, S.H.M.; Symes, A.; Guzmán, I.M.; Cunningham, R.; Brightsmith, D.; Heinsohn, R. Ecological and socio-economic factors affecting extinction risk in parrots. *Biodivers. Conserv.* **2016**, *25*, 205–223. [CrossRef]
3. Republic of Indonesia. Peraturan Menteri Lingkungan Hidup Dan Kehutanan [Regulation of the Minister of Environment and Forestry]P.106/MENLHK/SETJEN/KUM.1/12/2018. 2018. Available online: https://graccess.co.id/assets/document/Permen_P106.pdf (accessed on 25 August 2021).
4. Pires, S.F.; Olah, G.; Nandika, D.; Agustina, D.; Heinsohn, R. What drives the illegal parrot trade? Applying a criminological model to market and seizure data in Indonesia. *Biol. Conserv.* **2021**, *257*, 109098. [CrossRef]
5. Olȧh, G.; Theuerkauf, J.; Legault, A.; Gula, R.; Stein, J.; Butchart, S.; O'Brien, M.; Heinsohn, R. Parrots of Oceania—A comparative study of extinction risk. *Emu-Austral Ornithol.* **2017**, *118*, 94–112. [CrossRef]
6. Setiyani, A.D.; Ahmadi, M.A. An overview of illegal parrot trade in Maluku and North Maluku Provinces. *For. Soc.* **2020**, *4*, 48–60. [CrossRef]
7. Taman Nasional Manusela. Buku Informasi Wisata Alam. Balai Taman Nasional Manusela: Seram, Indonesia, 2014; p. 32.
8. Isherwood, I.S. *Biological Surveys and Conservation Priorities in North-East Seram, Maluku, Indonesia: Final Report of Wae Bula '96*; CSḂ Conservation Publications: Cambridge, UK, 1998; ISBN 1-901399-02-8.
9. GPS Wisata Indonesia Taman Nasional Manusela Maluku Tengah Maluku. 2017. Available online: https://gpswisataindonesia.info/taman-nasional-manusela-maluku-tengah-maluku/ (accessed on 25 August 2021).
10. Edgar, P.; Lilley, R. Herpetofauna survey of Manusela National Park. In *Natural history of Seram, Maluku, Indonesia*; Edwards, I., McDonald, A., Proctor, J., Eds.; Intercept Ltd.: Andover, MA, USA, 1993; pp. 131–141.
11. FAO. *Proposed Manusela National Park, Management Plan 1982–1987. Field Report of UNDP/FAO, National Park Development Project INS/78/061. FAO Field Report No. 15*; Food and Agriculture Organization of the United Nations: Bogor, Indonesia, 1981.

12. Bowler, J.; Taylor, J. The avifauna of Seram. In *Natural history of Seram, Maluku, Indonesia*; Edwards, I., McDonald, A., Proctor, J., Eds.; Intercept Ltd.: Andover, MA, USA, 1993; pp. 143–160.
13. Bowler, J.; Taylor, J. An annotated checklist of the birds of Manusela National Park, Seram. Birds recorded on the Operation Raleigh Expedition. *Kukila* **1989**, *4*, 3–29.
14. Hutto, R.L.; Pletschet, S.M.; Hendricks, P. A fixed-radius point count method for nonbreeding and breeding season use. *Auk* **1986**, *103*, 593–602. [CrossRef]
15. Conner, R.N.; Dickson, J.G. Strip transect sampling and analysis for avian habitat studies. *Wildl. Soc. Bull. 1973-2006* **1980**, *8*, 4–10.
16. Nandika, D.; Agustina, D.; Metz, S.; Zimmermann, B. Kakatua langka abbotti dan Kepulauan Masalembu. In *Perkumpulan Konservasi Kakatua Indonesia – The Indonesian Parrot Project*; Prawiradilaga, D.M., Ed.; CV. Kurnia: Bekasi, Indonesia, 2013; p. 192.
17. Bibby, C.J.; Jones, M.; Marsden, S. *Bird Survey Field Expedition Techniques*; BirdLife International: Bogor, Indonesia, 2000; p. 179.
18. Menhinick, E.F. A comparison of some species-individuals diversity indices applied to samples of field insects. *Ecology* **1964**, *45*, 859–861. [CrossRef]
19. Pielou, E.C. Shannon's formula as a measure of specific diversity: Its use and misuse. *Am. Nat.* **1966**, *100*, 463–465. [CrossRef]
20. Spellerberg, I.F.; Fedor, P.J. A tribute to Claude Shannon (1916–2001) and a plea for more rigorous use of species richness, species diversity and the 'Shannon-Wiener' Index. *Glob. Ecol. Biogeogr.* **2003**, *12*, 177–179. [CrossRef]
21. Pielou, E.C. Species-diversity and pattern-diversity in the study of ecological succession. *J. Theor. Biol.* **1966**, *10*, 370–383. [CrossRef]
22. Engen, S. On species frequency models. *Biometrika* **1974**, *61*, 263–270. [CrossRef]
23. Manly, B.F.L.; McDonald, L.; Thomas, D.L.; McDonald, T.L.; Erickson, W.P. *Resource Selection by Animals: Statistical Design and Analysis for Field Studies*, 2nd ed.; Springer Science & Business Media: Berlin/Heidelberg, Germany, 2007; ISBN 0-306-48151-0.
24. Romero-Vidal, P.; Hiraldo, F.; Rosseto, F.; Blanco, G.; Carrete, M.; Tella, J.L. Opportunistic or non-random wildlife crime? Attractiveness rather than abundance in the wild leads to selective parrot poaching. *Diversity* **2020**, *12*, 314. [CrossRef]
25. Coates, B.J.; Bishop, K.D. *Panduan Lapangan Burung-Burung Di Kawasan Wallacea: Sulawesi, Maluku Dan Nusa Tenggara*; Dove Publications: SMK Desa Putra: Bogor, Indonesia, 2000.
26. Alikodra, H.S. *Pengelolaan Satwaliar*; Departemen Pendidikan dan Kebudayaan, Direktorat Jenderal Pendidikan Tinggi Riset Antar Universitas Ilmu Hayat IPB: Bogor, Indonesia, 2002.
27. Chettri, N.; Deb, D.C.; Sharma, E.; Jackson, R. The relationship between bird communities and habitat. *Mt. Res. Dev.* **2005**, *25*, 235–243. [CrossRef]
28. Mioduszewska, B.M.; O'Hara, M.C.; Haryoko, T.; Auersperg, A.M.I.; Huber, L.; Prawiradilaga, D.M. Notes on ecology of wild goffin's cockatoo in the late dry season with emphasis on feeding ecology. *TREUBIA* **2019**, *45*, 85–102. [CrossRef]
29. Tella, J.L.; Blanco, G.; Dénes, F.V.; Hiraldo, F. Overlooked parrot seed dispersal in Australia and South America: Insights on the evolution of dispersal syndromes and seed size in Araucaria trees. *Front. Ecol. Evol.* **2019**, *7*, 1–10. [CrossRef]
30. Partasasmita, R.; Mardiastuti, A.; Solihin, D.D.; Widjajakusuma, R.; Prijono, S.N.; Ueda, K. Komunitas Burung Pemakan Buah di Habitat Suksesi. 2009. Available online: https://repository.ipb.ac.id/bitstream/handle/123456789/58863/biosfera4.pdf?sequence=1&isAllowed=y (accessed on 25 August 2021).
31. Chng, S.C.; Eaton, J.A.; Krishnasamy, K.; Shepherd, C.R.; Nijman, V. In the Market for Extinction. 2015. Available online: http://www.parrotsdailynews.com/wp-content/uploads/2015/11/traffic_species_birds22.pdf (accessed on 25 August 2021).
32. Sasaoka, M.; Sugimura, K.; Laumonier, Y. Wildlife Conservation Compatible with Local Forest Uses on Seram Island, Eastern Indonesia: Focusing on Interrelationships between Humans and Wildlife through Indigenous Arboriculture 2011. Available online: https://www1.cifor.org/fileadmin/subsites/colupsia/documents/IALE_World_Congress_Wildlife_conservation.pdf (accessed on 25 August 2021).
33. Sasaoka, M. Customary forest resource management in Seram Island, Central Maluku: The "Seli Kaitahu" system. *Tropics* **2003**, *12*, 247–260. [CrossRef]
34. Garai, C. The Indonesian Parrot Project; Wildlife Messengers. 2019. Available online: https://wildlifemessengers.org/ipp/ (accessed on 25 August 2021).
35. Metz, S.; Zimmermann, B. *The Five Years Report on the Rehabilitation and Release of Indonesian Cockatoo and Parrots at the Kembali Bebas Avian Center, Seram Island, the Middle Molluccas*; Indonesian Parrot Project: Washington, DC, USA, 2011.
36. Metz, S.; Nursahid, R.; Trapping and smuggling of Salmon-crested Cockatoo: An undercover investigation in Seram, Indonesia. PsittaScene. 2004, 16. Available online: https://issuu.com/worldparrottrust/docs/16_4_nov_04?mode=mobile (accessed on 25 August 2021).
37. Kinnaird, M.F.; O'Brien, T.G.; Lambert, F.R.; Purmiasa, D. Density and distribution of the endemic Seram Cockatoo *Cacatua moluccensis* in relation to land use patterns. *Biol. Conserv.* **2003**, *109*, 227–235. [CrossRef]
38. Olah, G.; Heinsohn, R.G.; Brightsmith, D.J.; Espinoza, J.R.; Peakall, R. Validation of non-invasive genetic tagging in two large macaw species (*Ara macao* and *A. chloropterus*) of the Peruvian Amazon. *Conserv. Genet. Resour.* **2016**, *8*, 499–509. [CrossRef]
39. Olah, G.; Smith, B.T.; Joseph, L.; Banks, S.C.; Heinsohn, R. Advancing genetic methods in the study of parrot biology and conservation. *Diversity* **2021**, in press.
40. BirdLife International Species Factsheet: Cacatua Moluccensis. Available online: http://www.birdlife.org (accessed on 24 September 2021).

41. IUCN Standards and Petitions Committee. *Guidelines for Using the IUCN Red List Categories and Criteria. Version 14. Prepared by the Standards and Petitions Committee.* 2019, p. 101. Available online: http://www.iucnredlist.org/documents/RedListGuidelines.pdf (accessed on 24 September 2021).
42. BirdLife International Species Factsheet: Lorius Domicella. Available online: http://www.birdlife.org (accessed on 24 September 2021).
43. BirdLife International Species Factsheet: Eos Semilarvata. Available online: http://www.birdlife.org (accessed on 24 September 2021).
44. SEAIR Indonesia HS Code 1063200 Export Data with Price. Available online: https://www.indonesiatradedata.com/hs-code-1063200-export-data-indonesia (accessed on 30 July 2021).
45. Blanco, G.; Bravo, C.; Pacifico, E.C.; Chamorro, D.; Speziale, K.L.; Lambertucci, S.A.; Hiraldo, F.; Tella, J.L. Internal seed dispersal by parrots: An overview of a neglected mutualism. *PeerJ* **2016**, *4*, e1688. [CrossRef] [PubMed]
46. Hernández-Brito, D.; Romero-Vidal, P.; Hiraldo, F.; Blanco, G.; Díaz-Luque, J.; Barbosa, J.; Symes, C.; White, T.; Pacífico, E.; Sebastián-González, E.; et al. Epizoochory in parrots as an overlooked yet widespread plant–animal mutualism. *Plants* **2021**, *10*, 760. [CrossRef] [PubMed]

Article

Increasing Survival of Wild Macaw Chicks Using Foster Parents and Supplemental Feeding

Gabriela Vigo-Trauco [1,*,†], Rony Garcia-Anleu [2] and Donald J. Brightsmith [3]

[1] The Macaw Society-Sociedad Pro Guacamayos, Department of Ecology and Conservation Biology, Texas A&M University, TAMU 4467, College Station, TX 77843, USA
[2] Wildlife Conservation Society–Guatemala, Avenida 15 de Marzo, Casa No. 3. Flores, Petén CP 17001, Guatemala; rgarcia@wcs.org
[3] The Macaw Society-Sociedad Pro Guacamayos, Schubot Center for Avian Health, Department of Veterinary Pathobiology, Texas A&M University, TAMU 4467, College Station, TX 77843, USA; dbrightsmith@cvm.tamu.edu
* Correspondence: parrots@cvm.tamu.edu; Tel.: +1-979-450-9857
† Current affiliation: Schubot Center for Avian Health, TAMU 4457, College Station, TX 77843, USA.

Abstract: The use of foster parents has great potential to help the recovery of highly endangered bird species. However, few studies have shown how to successfully use these techniques in wild populations. Scarlet Macaws (*Ara macao macao*) in Perú hatch 2–4 chicks per nest but about 24% of all chicks die of starvation and on average just 1.4 of them fledge per successful nest. In this study we develop and test new techniques to increase survival of wild Scarlet Macaw chicks by reducing chick starvation. We hypothesized that using foster parents would increase the survival of chicks at risk of starvation and increase overall reproductive success. Our results show that all relocated macaw chicks were successfully accepted by their foster parents (n = 28 chicks over 3 consecutive breeding seasons) and 89% of the relocated chicks fledged. Overall, we increased fledging success per available nest from 17% (2000 to 2016 average) to 25% (2017 to 2019) and decreased chick death by starvation from 19% to 4%. These findings show that the macaw foster parents technique and post relocation supplemental feeding provide a promising management tool to aid wild parrot population recovery in areas with low reproductive success.

Keywords: foster chicks; chick starvation; chick survival; chick supplemental feeding; avian brood manipulation; wildlife management; Scarlet Macaw; Perú

Citation: Vigo-Trauco, G.; Garcia-Anleu, R.; Brightsmith, D.J. Increasing Survival of Wild Macaw Chicks Using Foster Parents and Supplemental Feeding. *Diversity* **2021**, *13*, 121. https://doi.org/10.3390/d13030121

Academic Editor: Luc Legal

Received: 31 December 2020
Accepted: 3 March 2021
Published: 12 March 2021

1. Introduction

The use of foster parents in avian population management is a technique with great potential to aid in the recovery of highly endangered species in the wild [1]. Foster parenting, the use of breeding pairs to raise young that were not part of their -own broods, is a well-known avicultural technique that has been intensively used in captive breeding and reintroduction programs over several decades [2] and also in conservation captive breeding programs to increase reproduction [1,3–5]. However, few studies have systematically studied how to successfully use this tool in the wild.

The topic of increasing productivity in parrots for conservation is not new. In the early 1990s', it was suggested that managing intensively the factors that limit species' population growth was the key to productivity maximization [6]. One of the techniques proposed at the time was to increase fledging success [7]. In psittacines, the majority of species hatch their eggs asynchronously over a period of 1 to 14 days [5,8–14] which results in a size-based hierarchy among brood members [9,15,16] which often leads to the death of younger chicks [8,9,13,14,17]. In this scenario, decreasing hatching asynchrony has been proposed as a potential management tool to increase numbers of young for harvesting for conservation purposes [7]. These harvested last and penultimate chicks could be relocated in foster nests to increase overall reproductive output. This technique has great potential for in situ conservation efforts because there is strong evidence that psittacines can be successfully used as foster parents and they are able to raise and fledge additional chicks (RG-A unpublished data [18]).

Chick fostering has been successfully used in commercial aviculture to raise finches (*Lonchura* ssp. and others) and with captive psittacines of the genera *Cyclopsitta, Alisterus, Amazona, Pionus* and *Cacatua* [4,5,8,19,20], mainly as an emergency tool when chicks were rejected by parents or fell out of the nest [19–21]. It has also been used in captive breeding programs for psittacines [4,5]. In the wild, it has proven to be useful for recovering the Puerto Rican Amazon (*Amazona vittata*) and increasing population recruitment in the Yellow-shouldered Parrot (*Amazona barbadensis*) in Venezuela [18]. It has also been used in the wild as a tool to study parent/offspring interactions in Crimson Rosellas (*Platycercus elegan*) [22] and Galahs (*Eolophus roseicapillus*) [8] in Australia. Most recently, it has been used by RG-A and collaborators from the Wildlife Conservation Society (WCS)-Guatemala as part of the efforts to recover Scarlet Macaw (*Ara macao cyanoptera*) populations around Laguna del Tigre National Park.

The Scarlet Macaw (*Ara macao*), one of the most iconic members of the Psittacidae family and an important flagship species of the tropical forest, is widely distributed in the Americas from Southeastern Mexico to Peru and Bolivia [23]. However, most populations in Central America are currently declining due to a combination of habitat loss and poaching for the local pet trade [14,24–28]. As with many other members of the family Psittacidae, the species shows brood reduction by chick starvation [11,14]. This starvation can result in the death of >22% of all hatched chicks and is the most common cause of chick death [15]. In Tambopata, Peru, an area with no nest poaching, clutches have on average three eggs, resulting in broods of about two chicks but just a mean of 1.4 chicks fledge per nest per season [29]. Overall, 27% of second chicks and all third and fourth chicks die by starvation, which results in a substantial loss of hatchlings. In areas where Scarlet Macaw populations are declining, valuable chicks that could help increase population numbers starve to death. Increasing survival of those starving chicks could provide significant numbers of young that can directly increase wild populations.

There is little information published on Scarlet Macaws as foster parents. In the late 1990s' in Carara National Park in Costa Rica one chick rescued from poachers was placed in a wild nest that had just one chick and both chicks fledged [11]. There are also reports from captivity where a Scarlet Macaw pair was used as a surrogate to raise chicks of the Blue-headed Macaw (*Primolious couloni* [30]). The most comprehensive information related to Scarlet Macaws as foster parents in the wild comes from Guatemala where the technique has been used to place 60 chicks during seven breeding seasons since 2011 [RG-A unpublished data]. In this case, foster chicks averaged 41 days old (range 12–85 days old) and 78% of them were successfully adopted and fledged [RG-A unpublished data]. The technique was used when: (1) chicks did not gain weight as expected, (2) third and fourth chicks hatched, (3) chicks lost their parents, (4) chicks hatched in the field station after eggs were rescued from nest poaching) [RG-A unpublished data].

The use of Scarlet Macaws as foster parents in the wild offers a good system not only to test the technique in situ, but also to test the main drivers of chick death by starvation. The main driver behind death by starvation appears to be brood members' age differences: first chicks apparently do not die of starvation and the chance of younger chicks starving is directly proportional to age difference in relation to the first chick of the brood [29]. In the case of second chicks, when the age difference was 3 to 4 days the probability of death was 24% but if this difference was 5 days or more, the probability of death by starvation jumped to 80% [29]. If age difference among brood members is the main reason why the younger member of the brood starved to death [29], age differences in foster broods would need to be less than 5 days to ensure that none of the brood members would perish.

The main objective of this experiment was to develop and test techniques to increase survival of Scarlet Macaw chicks in the wild by reducing chick starvation using wild foster parents. In addition, we wanted to test if the age difference among brood members was the sole driver of chick death by starvation. To do this, we tested the following main hypotheses: (1) Wild Scarlet Macaws accept chicks that are not their offspring and raise them to fledging, (2) Using foster parents increases the survival of chicks at risk of starvation and increases the overall population reproductive success, and (3) Age differences among brood members is not the only driver of chick death by starvation and (4) Wild Scarlet Macaws are able of fledge a brood of three chicks.

2. Materials and Methods

This research was conducted in the forests surrounding the Tambopata Research Center (13°8′ S, 69°36′ W), located in the Tambopata National Reserve (275,000 ha) adjacent to the Bahuaja-Sonene National Park (1,091,416) in the department of Madre de Dios, south-eastern Perú. The forest adjacent to the research station is classified as tropical moist forest (Holdridge life zones system) and is a combination of flood plain, terra firme, successional, and palm swamp forests that receives around 3200 mm of rain annually [31,32].

2.1. Background Methodology

We conducted this research from October 2016 to March 2019, during three consecutive macaw breeding seasons, as part of a program of investigation on Scarlet Macaw breeding ecology, nesting behavior, and health run by The Macaw Society -Sociedad Pro Guacamayos (www.TheMacawSociety.org and http://vetmed.tamu.edu/macawproject, accessed on 1 October 2016) [29,33–36]. This program has been monitoring macaw nests intensively since 1999. Macaw breeding season is from mid-October to mid-April, annually. Each season we monitored about 40 macaw nests (16 natural, 24 artificial) in a 5 km radius area, using single rope climbing systems [37,38]. Artificial nests were a combination of wooden boxes and PVC pipes (16″ diameter) and were hung one per tree [See [33] for a detail explanation about artificial nest used] (Figure 1). Eight artificial nests in a 3 km radius were equipped with video surveillance cameras each season. Due to the high humidity of the rainforest, video systems frequently suffered intermittent malfunctions that alter sample sizes for data reported from nest videos. Not all nests with video systems received foster chicks. All nests were checked once every 2–3 days until the first egg was found. After an egg was found, nest monitoring ceased until 26 days later and continued daily until all viable eggs had hatched. Due to this frequent monitoring, done over almost two decades, nesting macaws were habituated to human intervention and did not flee or abandon nests during nest checks. However, nesting macaws still displayed a few disturbance behaviors in the presence of climbers (i.e., alarm calls and calls they use when fighting with other macaws). They also showed aggressive behaviors towards climbers (i.e., lunging at climbers from inside nest, flying at and even hitting the climber, etc.). Despite this high level of acclimatization of the birds, we did not weigh, measure or manipulate broods until second chick hatched in each brood in order to reduce disturbance at the nest and maximize hatching success.

2.2. Chick Relocation Procedures

2.2.1. Criteria and Timing to Remove Chicks for Relocation Procedures

A total of 32 macaw chicks were removed from their original nests. Four of them perished in our nursery. Two of them died the same day they hatched probably due to the fact that they hatched underweight. Two other chicks died at <5 days old, probably due to slow digestion problems. Macaw chicks were removed from their original nests according to the criteria and timing shown in Table 1.

Figure 1. Nest check of an artificial Scarlet Macaw nest (wooden box) using a single rope climbing system. Both macaws are displaying nest defense behaviors. A stuffed leather glove attached to a metal stick is used as a tool to cover the nest entrance and prevent macaws from reentering the nest. In the picture, the researcher access door is open. Researcher: Gustavo Martinez Sovero MSc. Photo Credit: Liz Villanueva Paipay.

Table 1. Criteria and timing to remove Scarlet Macaw chicks from original nests for relocation procedures in Tambopata.

Hatch Order of Removed Chicks and Criteria for When to Remove Chicks	Timing to Remove Chick from Original Nest	# Chicks Removed
First Chicks (11 chicks removed)		
When second chick hatched > 3 days after first chick	As soon as second chick hatched	7
If chick showed signs of life-threatening botfly related infection	When infection was clearly getting worse but still localized in one area	2
To create conditions for a triple brood by adding a third chick as the younger brood member	When third chick was placed in foster nest	2

Table 1. *Cont.*

Hatch Order of Removed Chicks and Criteria for When to Remove Chicks	Timing to Remove Chick from Original Nest	# Chicks Removed
Second Chicks (7 chicks removed)		
If chick showed signs of starvation	When chick was still active and begging. Usually within 3 days of not gaining weight as expected, before they started to lose weight	5
If chick showed signs of life-threatening botfly related infection	When infection was clearly getting worse but still localized in one area	1
If chick was needed for relocation to another nest where the clutch or brood was lost due to damage to the nest	As soon as damaged nest was fixed	1
Third Chicks (12 chicks removed)		
All third chicks were removed	<24 h after hatching	12
Fourth Chick (2 chicks removed)		
All fourth chicks were removed	<24 h after hatching	2
Chicks removed from their original nests		**32**
Chicks that perished in our nursery before relocation		**4**
Total chicks relocated		**28**

2.2.2. Removed Chicks' Initial Conditions

The majority of the chicks (67%) were healthy when removed from the nest. Eleven first chicks and nine third chicks weighed as expected for their ages (see Vigo et al. [14]). Two third chicks arrived underweight (17.9 g each), one with early signs of dehydration and the other one apparently in good condition. Three second chicks were brought in as soon as they did not gain weight as expected but were still in the weight range for their ages. One of the second chicks arrived underweight for its age, showing signs of starvation (empty crop, grayish color, dry skin and prominent ocular area). Another second chick that had a congenital foot malformation was brought in as soon as it did not gain weight as expected for its age (2 days of age). The three chicks showing signs of botfly related infection did not have botflies when arrived to the nursery because they were removed in the field. Both fourth chicks arrived underweight. One arrived right after hatching at 17.1 g. The other one, was left in its original nest with a sibling 9 days older, and it was removed at age 5 days when it started to showed early signs of starvation. We removed 32 chicks in total (Season 1 = 5 chicks, season 2 = 11 chicks, season 3 = 16 chicks, Table 1).

2.2.3. Macaw Chick Rearing in the Nursery

Chicks were kept in boxes (40 cm × 40 cm × 50 cm) with three solid wood sides, a solid wood base, and a wire mesh front and top. We used a Brinsea EcoGlow Brooder as a source of heat for the chicks. Heat and humidity in each box were monitored with an off the shelf digital thermometer-hygrometer. Chicks were separated in two different boxes according to their age. Chicks under 15 days old were kept in one box, each in a separate plastic cup on wood chips and a soft piece of cloth. Chicks over 15 days old were kept together in a separate box with a woodchip substrate and a 1/2″ square mesh tray on top of substrate. Chicks were maintained at age-appropriate temperature and humidity conditions following the recommendations in Voren and Jordan [39]. In general, chicks were syringe fed Zupreem Embrace baby bird hand feeding formula prepared following the age-specific manufacture recommendations (https://www.zupreem.com/products/birds/embrace-plus/ accessed on 1 October 2016). Chicks that came in sick, weak or underweight were given custom diets used commonly in commercial aviculture (Table A1). All chicks that arrived showing signs of botfly related infection (n = 4) were treated with oral antibiotics and/or local antibiotic cream.

2.2.4. Criteria to Assign Macaw Chicks to Wild Macaw Nests

Individual chicks were assigned to foster nests with only one chick that was in the same "developmental stage" but not necessarily the same age (Table 2 and Figure 2).

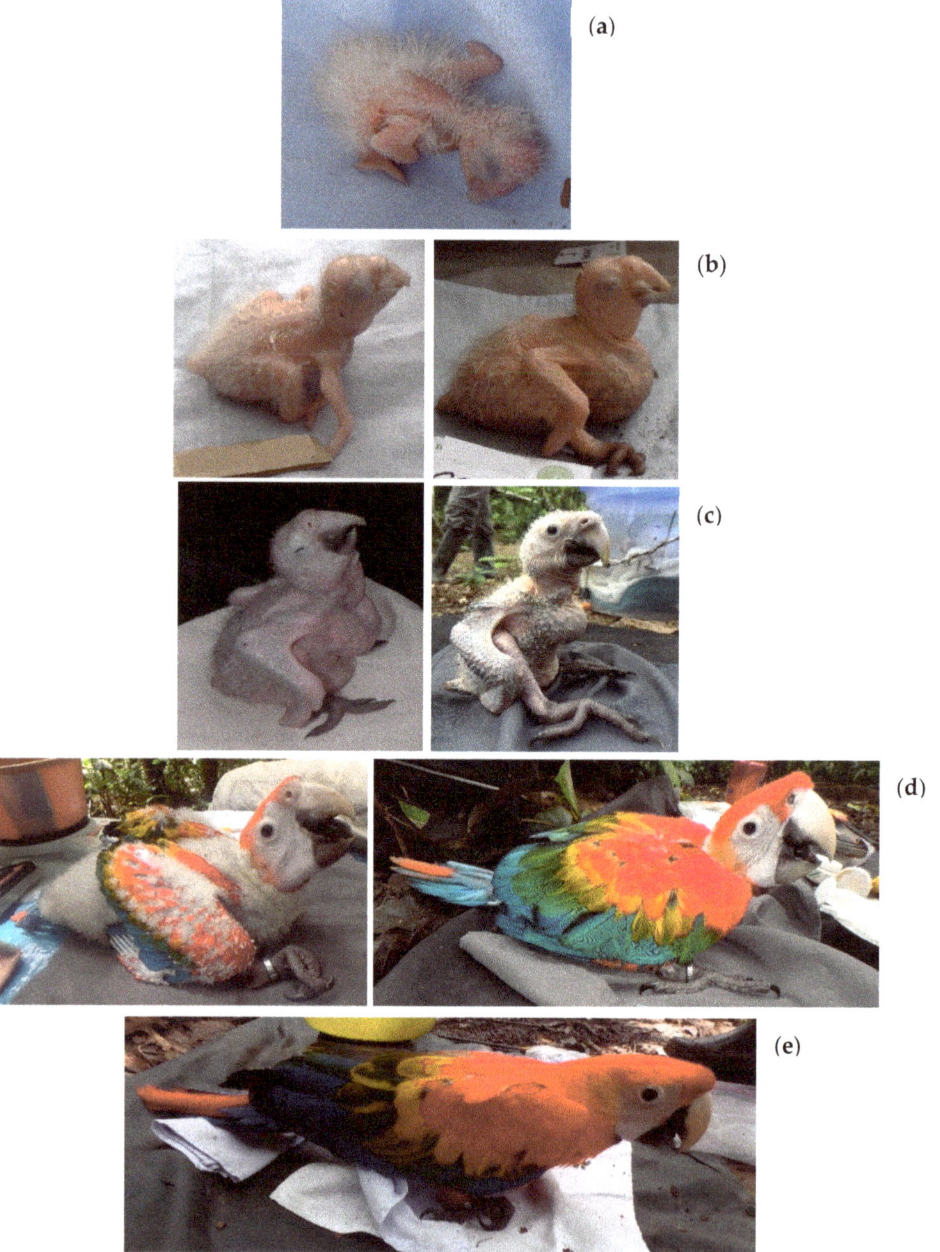

Figure 2. Developmental stages of Scarlet Macaw chicks. (a) Stage 1: 0 to 2 days old; (b) Stage 2A: from 3 to 18 day old; (c) Stage 2B: 19 to 33 days old; (d) Stage 3: 34 to 65 days old; (e) Stage 4: 66 days old to fledge. Photo credits: The Macaw Society- Sociedad Pro Guacamayos. For more details, see Table 2.

Table 2. Developmental stages in wild Scarlet macaw chicks. To define the Scarlet Macaw chick developmental stages, we used a modified version of the mass growth stages presented in Vigo et al. [14]. Scarlet macaw chicks fledge on average at 88 days old (n = 104 chicks, range = 79 to 99 day old, [15]); see Figure 2 for images of each developmental stage.

Developmental Stages in Wild Scarlet Macaw Chicks		
Stage	**Age Range**	**Description**
Stage 1	0 to 2	Hatchling
Stage 2A	3 to 18	Naked to light pinfeathers and eyes closed
Stage 2B	19 to 33	Light pinfeathers to heavy pinfeathers and eyes open
Stage 3	34 to 65	Heavy pinfeathers to mostly feathered
Stage 4	66 to fledged	Mostly feathered to fully feathered

2.2.5. Criteria to Select Foster Parents

We used the literature [18,40] and our previous knowledge of the species to create selection criteria for pairs to host foster chicks. We preferentially chose pairs with the following characteristics: (1) Pairs nesting in artificial nests; (2) Known pairs with banded individuals older than 8 years old; (3) Pairs with at least one chick that fledged in a previous season; (4) Pairs with no history of chick death by unknown causes; (5) Pairs with no history of chick death due to poor parental care in solo chicks, such as hypothermia or low daily feeding rates; and (6) Pairs with no records suggesting they have little breeding experience, such as slow growth/poor body condition in chicks. Due to a lack of suitable nesting pairs, one foster chick was placed with a nesting pair with an unknown breeding history. In total, 12 macaw pairs were used as foster parents. In 10 of them at least one individual of the pair was banded. Adult macaw genders were determined either by DNA testing of at least one individual of the pair (67% of total macaw foster pairs) or by comparing nesting behavior to known gender macaw pairs (33% of total macaw foster pairs). Seven macaw pairs were used as foster parents in more than one season and one was used in all three seasons of this experiment.

2.2.6. Foster Chick Relocation Procedures and Timing

We conducted four complementary macaw chick relocation procedures during this study. Each one was designed to test a different hypothesis related the capacity of wild macaws to foster chicks in wild conditions (Table 3). A chick was considered relocated when it was taken out of its original nest and moved to another nest, either the same day or days after. Due to the fact that we were working in natural conditions in the wild, we had little ability to set up identical conditions in each specific foster nest case. Foster nest candidates for procedures 1, 2 and 4 were checked every day from hatch to ensure nest requirements were maintained. In these procedures, the key moment to place foster chicks in foster nests was when the younger chick opened its eyes. By having both chicks with open eyes, all chicks were able to see the parents and effectively beg for food in a similar fashion. Foster nests for procedure 3 were chosen in an opportunistic way as long as foster parents fulfilled the selection criteria explained in 2.2.6. In total, we worked with 28 foster chicks in 23 foster broods.

2.2.7. Foster Chick Relocation Schedule

In all but three cases, foster chicks were placed in nests between 8:00 and 9:30 AM with a crop half-full of food. At the time of the relocation, the resident chick was pulled from the nest weighed and measured and then both chicks, the resident and foster were placed back into the nest at the same time. In the three relocations following chick/egg predation, foster chicks were placed in the nest between 11 AM and 2 PM. No chicks were relocated on rainy days as adult movements and feeding rates are lower during rain (GV-T and DJB unpublished data). Just 2 foster chicks (7% of total foster chicks) were relocated the same day they were removed from the original nest.

Table 3. Macaw foster chick relocation procedures and timing in Tambopata, Peru. Procedures outlined below were designed to evaluate the capacity of nesting wild Scarlet Macaws to accept and raise foster chicks. "Foster nest requirements" refers to the characteristics of nests needed to be eligible to foster chicks. "Foster chick age at relocation criterion" was established considering the reproductive ecology and behavior of the species [29]. In smaller font, below the number of foster chicks is the number of foster chicks in nests with video cameras. The number of foster broods is less than the number of foster chicks because some foster broods contained two foster chicks.

	Macaw Chick Relocation Procedures	Hypothesis	Foster Nest Requirements	Foster Chick Age at Relocation Criterion	Timing to Place Foster Chick	Foster Chick Age at Relocation	# Foster Broods	# Foster Chicks	# Foster Chicks that Fledged
1	Acceptance	Wild macaws will foster unrelated chicks and fledge them as their own	Nest with single resident chick	≤2 days older or younger than resident chick	Within 24 h of both foster and resident chicks, opening their eyes	~18 days old	9	11 5 w/video	9
2	Age difference	Age difference is the absolute driver of death by starvation	Nest with single resident chick	From 4 to 9 days older or younger than the resident chick.	Within 48 h of the resident chick opening its eyes	~18 days old	9	10 7 w/video	10
3	Empty nest	Wild macaws will foster and fledge chicks after losing their own brood or clutch	Nest that lost eggs due to depredation but pair was still incubating an artificial egg	Foster chick was a solo chick	At expected hatching day	<3 days old	2	2 1 w/video	4
			Nest that lost chicks due to depredation or accident (lighting)	Similar developmental stage as lost chicks	<36 h of original brood disappearance	<46 days old	2	3 No video	
4	Triple brood	Wild macaws will accept an additional chick in a brood of two chicks and fledge three chicks.	Nest that hatched three chicks	<5 days between first and third chick	Third chick swapped for first chick when eyes fully opened. First chick was placed back in foster nest 5 days after.	18 days old 24 days old	1	2 No video	2

2.2.8. Observations of Foster Parents/Foster Chick Interactions

In our first and second seasons working with macaw foster parents, video camera systems were installed in the majority of foster our nests (n = 12 foster nests with cameras of 14 foster nest total; 13 foster chicks). In the third season (n = 9 foster nests, 15 foster chicks), we did not work with video cameras in foster nest. All relocation procedures except #4 ("Triple brood") had at least one foster nest with a video camera (Table 3). At the foster nests with video cameras, an experienced observer arrived at the foster nest at about 5:00 AM on relocation day and took observations of parent/chick interactions using the nest video system until 5 PM. Observers took focal group observations of known individuals to record all contact and feeding behaviors between the parent and the chicks. The recorded behaviors were (1) feeding, (2) preening and (3) brooding. Feeding refers to when adults grasp the bill of the chick crosswise from above and bob during regurgitation. Preening refers to when an adult gently touches the chicks' body with its beak in a continuous manner. Brooding refers to when an adult positions its body in direct contact with the chick's body. We considered that the time the nesting individuals (nesting females or nesting males) entered the nest and are visible on the video camera as the moment that the adults became aware of the foster chick's presence in the nest. Nests with camera sample sizes varies slightly due to intermittent video camera malfunction during key behavioral interactions. One foster nest with video camera had problems with image (but not sound) at the time of foster chick placement and initial interactions were not recorded. Similar issues happened in another foster nest right after first interactions but before the first feeding. We had behavioral observations inside the nest of 9 different foster parent pairs.

2.2.9. Monitoring of Foster Chicks

We intensively monitored each foster nest for 10 days after each foster chick relocation. The monitoring process included (1) checking the foster chick's crop content twice per day (5 AM and 5 PM); (2) providing supplemental food to the foster chick any time we checked the nest and found its crop was more than half-empty (in this way foster chicks were fed from 0 to 2 times per day), (3) monitoring weight gain by weighing both foster and resident chicks at 5:00 AM daily; (4) monitoring interactions between the foster chicks, resident chicks, and foster parents using video cameras when available (14 foster chicks in 12 foster nests with video cameras); (5) counting feedings per day of both foster and resident chicks as seen through the video cameras both live in the field and later from video recordings when cameras were available (See Table A2 for details of hours analyzed). To count feedings, we performed focal group observations of known individuals and recorded all feedings of each particular chick in a continuous manner each time they happened.

2.2.10. Supplemental Feeding Plan after Relocation of Foster Chick

Our objective with the supplemental feeding was to allow foster parents, foster siblings, and foster chicks to learn how to interact with each other without compromising the foster chick's nutrition and overall health. We assumed that it would take time for foster parents to adjust to the new brood size and feeding requirements. We also assumed that it would take time for foster chicks that were syringe fed prior to placement, to learn how to consume chunky food regurgitated by the adult macaws. Our supplemental feeding plan had three stages: (1) Intensive supplemental feeding period. This was during the first 10 days after the foster chick was placed. The foster chick was checked twice per day and fed until the crop was 75% full. We slowly decreased the number of supplemental feedings, until by day 5 after relocation we left the chick with the crop only 50% full. This reduction in feeding was done in order to stimulate chick begging from the foster parents. In extreme cases (n = 2 chicks), foster chicks were just fed once (in the afternoon) in order to promote hunger and begging by the foster chicks. (2) Moderate supplemental feeding period. This was from day 10 after foster chick placement until the foster chick was 40 days old. In this stage, we checked the foster chick every day and provided supplemental food when its weight gain was less than 50% of the expected weight for its age on two consecutive days (expected weight was for [14] and unpublished data). (3) Passive supplemental feeding period. This was from 40 days old until the foster chick fledged. We checked the foster chick every other day and provided supplemental food if its keel was perceivable but there was moderate breast muscle development still found around it. In the 2017 breeding season, resident chicks were not supplemental fed. In the 2018 and 2019 breeding seasons, 75% of resident chicks lost weight during the 10-day adaptation period (n = 12 resident chicks). To address this, resident chicks were given supplemental food when the daily weight gain was 50% less than expected for its age. In all resident chick supplemental feeding cases, the chick was fed only until the crop was 50% full.

2.2.11. Foster Chick Acceptance Criteria

We established three levels of foster parent acceptance of foster chicks. (1) Initial acceptance: foster parents preen foster chick repeatedly and/or start attempting to feed foster chick. Attempts to feed refers to when the foster parent grabs the foster chick's beak in an attempt to start regurgitation and then releases the beak. (2) Intermediate acceptance: foster parents consistently feed foster chick (as seen by video camera) and/or foster chick shows a half-full crop on daily checkups. (3) Full acceptance: foster parents feed both foster chick and resident chick similarly. The half-full crop criterion was chosen because our previous observations on wild macaw chicks that fledge show that chicks in the 19 to 33 days old age range (Stage 2B) have on average a half-full crop (mean = 2.1 in a 0 to 4 scale, n = 515 chick crop observations, from 61 macaw chicks that fledged, during 16 breeding seasons, GV-T and DJB unpublished data). When the foster chick was the only chick in the nest, we established just one level of foster parent acceptance: relocation was considered successful when the foster chick was being fed by foster parents and was gaining weight as expected.

2.2.12. Foster Chick Acceptance Analysis

In order to better analyze the process of foster chick acceptance by their new parents we quantitatively measured acceptance using chick feeding ratios and foster chick growth.

2.2.13. Chick Feeding Ratios

In order to show how foster parents were accepting foster chicks we calculated a ratio of feedings per day (foster/resident chick) for each day in each nest during the first 10 days after relocation. We collected feeding ratio data in two different ways: from direct field observations (one season) and from video recordings (two seasons). A total of 418 h of observations were conducted live in the field by multiple observers and 573 h of video were scored by a single observer (Appendix A). Feeding ratios from video observations include 227 h of nocturnal observations. To determine if chick feeding ratios increased over the first 10 days after relocation we conducted a least squares regression with feeding ratio as the dependent variable, day post relocation as the independent variable and nest ID as a random variable.

2.2.14. Foster Chick Growth

In order to evaluate foster chick quality and acceptance we calculated the logistic growth curves for foster chicks that fledged and compared them with the growth curves of wild macaws that fledged in our study area during the previous 19 breeding seasons [14]. For this analysis, we only included chicks with 25 or more daily weight measurements and $\geq$1 measurement taken during the first week of life [14]. A total of 23 foster chicks fulfilled this criterion. The non-manipulated chicks used in this analysis were individuals that fledged from nests with no foster chicks, had no major health issues, did not receive supplemental feedings, and had fully wild parents that were not-hand raised and released (see [40] for history of releases in our site). A total of 81 wild chicks fulfilled these criteria.

To calculate the logistic growth curves, we used the chick weights and a logistic model with the equation

$$W = A/(1 + e\,(-B * (T - C))),$$

where W = weight in grams, T = age of the chick in days, A = the asymptotic body mass, B = growth rate constant, C = age in days for which the growth rate is maximal, and e = the natural constant [14]. We also compared growth parameters of foster chicks and wild chicks grouped by brood size and hatch order. Wild chick groups were as follows: single chicks (n = 17), first chicks (n = 38) and, second chicks (n = 26). Curves were fitted using Data Fit 9.1.32 (Oakdale Engineering, 2014, Oakdale, Pennsylvania, USA). To determine if growth differed between wild chicks and foster chicks, we compared the three growth parameters using a Mann Whitney U test with p-values calculated using a χ^2 approximation. The differences between groups were tested with Wilcoxon pairwise comparisons.

2.2.15. Foster Chick Influence on Breeding Success

Previous studies on breeding ecology of psittacines have used population level breeding parameters to make comparisons within and among seasons [10]. In order to measure the impact of chick relocations on overall breeding success we compared the overall breeding success for our monitored nests during the three seasons working with foster chicks and the previous 17 seasons with no foster chicks. Macaw nests were monitored from mid-October to mid-April every breeding season. We used five breeding success parameters: (1) Chicks that fledged per available nest (# chicks that fledge/# available nests), (2) Chicks that fledged per nest with eggs (# chicks that fledge/# nests with at least one egg), (3) Chicks that fledged per nest with at least one chick (# chicks that fledge/# nests with at least one chick), (4) Percentage of younger chicks that died from starvation (# of chick starved/# second chicks, third chicks and fourth chicks hatched), and (5) Percentage of chicks that fledged (# fledged chicks/# total chicks). Macaw nests included for this part of the experiment included natural and artificial cavities [33]. However, nests where total clutch size and total number hatched were not known exactly were removed from the analysis. As mentioned above, some of the adult macaws at our site were hand-raised as chicks, released and continue to consume food at the lodge [40]. Offspring of those individuals are not included in this analysis. Wild chicks that received supplemental food for any reason at some point in their lives were also excluded. To determine if breeding success differed between seasons with foster chicks and seasons without foster chicks we compared the parameters for both groups using a Mann Whitney U test with p-values calculated using a χ^2 approximation. All statistical comparisons were done using JMP Pro 15, with a confidence interval of 0.95 and $\alpha = 0.05$. All results are presented as mean $\pm$ standard deviation unless otherwise indicated.

3. Results

Twenty-eight foster Scarlet Macaw chicks were placed in nests with wild macaw foster parents. All of them were successfully accepted and 89% of them (n = 25 chicks) fledged from their foster nests. This included 23 successful foster broods (4 single foster broods, 18 double foster broods and 1 triple brood). In general, foster chicks were placed back in wild nests at 22 $\pm$ 9 days old (min = 14, max = 46 days old, n = 28 chicks). In five double foster broods, both chicks were foster chicks.

Overall, we had 15% (4 of n = 28 chicks) of foster chicks placed as solo chicks and 75% of them fledged. All 36% of foster chicks (10 of n = 28 of chicks) placed as first chicks fledged and all but two of the 46 % (13 of n = 28 chicks) placed as second chicks fledged. The one foster chick placed as a third chick successfully fledged.

The three foster chicks that did not fledge died of depredation (n = 1), lightning (n = 1) and what was probably an unknown disease (n = 1, Appendix B). We prematurely terminated one chick translocation only 3.5 h after the chick was placed in the nest because sweat bees (family *Halictidae*) from a nearby beehive started entering the nest cavity and attacking the foster chick.

In the "Acceptance" relocation procedure, 67% of foster chicks were placed as the younger member of the brood and 33% of foster chicks were placed as the older chick (n = 11 foster chicks). In this group, four chicks hatched as third chicks in their original nests, were placed as 2nd chicks (n = 3) and 1st chicks (n = 1) and were all successfully accepted.

In the "Age difference" relocation procedure, the age difference range was 4 to 9 days. All chicks in this procedure were accepted. Here, 60% of foster chicks were placed as 1st chick (n = 6 foster chicks) and 40% as 2nd chicks (n = 4 foster chicks). In four cases, chicks were true siblings that hatched in the same nests with a five-day difference which in normal conditions would have signified the death by starvation of the younger chick. Indeed, from all the procedures combined, 29% (8 of n = 28 chicks) were members of multiple broods with age differences > 4 days (4 foster chicks as first chicks, 3 foster chicks as second chicks and 1 foster chick as a third chick). In all of these eight foster broods, all chicks successfully fledged.

In the "empty nest" relocation procedure, 60% of the foster chicks were placed as solo chicks (n = 5 foster chicks). One of them was 46 days old. Two foster chicks of the same age were placed in the same foster nest but in different days. The heavier chick was placed first at age 31 and second chick was placed at age 39. All chicks in this procedure were accepted.

For the "triple brood" relocation procedure all chicks were true siblings. Here, the third chick was successfully accepted and all three chicks fledged. However, all three chicks showed inconsistent weight gain, even > 10 days after relocation of the third chick. Due to the weight gain problems, we intensively managed this brood giving them a high fat content supplemental feeding when we found them with empty crops, until the youngest chick was 45 days old. We planned to conduct more than one triple brood relocation, but were unable due to depredation events in our other chosen foster nests.

In all foster broods in the first two seasons (n = 14 broods: 12 broods with video cameras and 2 without video camera), we placed the foster chick when the nesting female was within sight of the nest (range 0 to 10 m from the nest). In two cases, the nesting female stayed inside the nest covered with a towel when we placed the foster chick. In all but two cases, the nesting male was not present. In the majority of the cases, nesting pairs were present during the precise moment when foster chicks were placed in their nest (100% females, 86% males, n = 14 nesting pairs). In these two first seasons, two double foster broods had both chicks as foster chicks.

3.1. Foster Chick Acceptance

The behavior of foster parents when seeing the foster chick for the first time followed the same pattern in all cases. In all foster broods with video cameras (n = 12 broods), the first foster parent to have physical contact with the foster chick was the nesting female. First contact behavior was usually preening (61% of the time, n = 14 chicks with video cameras) but some foster parents first attempted to feed the foster chick (39% of the time, n = 14 chicks with video cameras). None of the females showed aggression towards the chick. On average, first contact was made 4.2 min after the foster mother arrived to the nest (n = 14 chicks with video available, range = 0.8 to 14 min) and first feeding was given on average 13 min after arrival (Time range = 0.8 to 76 min, n = 13 of 14 chicks with video available).

In all nests with cameras (n = 14), the first physical contact between an adult and foster chick happened in the first 15 min after the nesting female arrived at the nest. In the two cases when females took the longest to touch the foster chick (14 min) it was because she was paying attention to the climber getting ready to repel down from the tree. After an average of 4.6 ± 3.3 days foster chicks that were members of multiple chick broods consistently had half-full crops when checked (n = 23 chicks, min = 2 days, max = 15 days).

3.2. Foster Chick-Feeding Ratios

Daily feeding ratios from observations done in the field by multiple observers and from recorded observations done by one single observer showed similar patterns. Foster chicks were initially fed less than resident chicks (daily feeding ratio of 0.37 ± 0.25 on relocation day, $n = 10$ chick pairs), but feeding ratio increased progressively until feedings were similar for both chicks 10 days after relocation (daily feeding ratio = 0.8 ± 0.4, $n = 10$ chick pairs, Figure 3). The combination of day and nest (as a random variable) explained about 22% of the variation in the data and the relationship with day post relocation was highly significant (least squares regression: $R^2 = 0.22$, df = 53, *t*-ratio 2.56, $p = 0.013$).

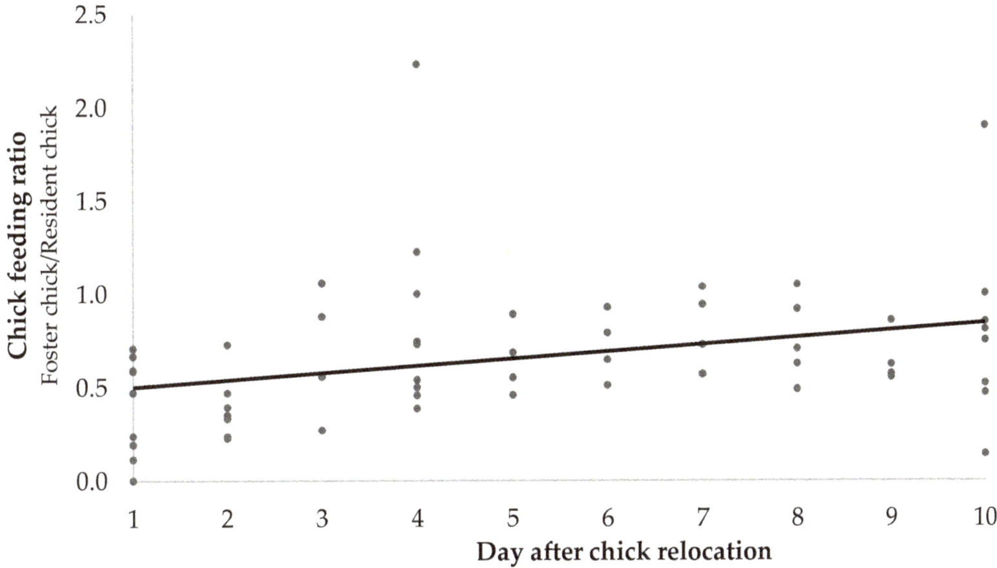

Figure 3. Acceptance of Scarlet Macaw foster chicks during the first 10 days after placement in foster nests. Acceptance of foster chicks in their new nests was measured by counting feedings per day of both foster and resident chicks and calculating daily feeding ratios ($n = 10$ chick pairs). Each point on the graph corresponds to the feeding ratio in one foster nest on one day. The solid line represents the positive linear trend observed and shows that foster chick feedings increased during the first 10 days in the foster nest. Day one on the X-axis indicates the day of relocation. These data are from the recorded observations (see Appendix A and Methods for additional descriptions of these data).

3.3. Foster Chick Growth

Foster chicks reached similar maximum weights compared to wild chicks (asymptotic size from logistic equation: all wild chicks combined: 1014.2 ± 79.7 g, $n = 81$, foster chicks 1020.3 ± 81.1 g, $n = 23$, Mann–Whitney U: $\chi^2 = 0.14$, df = 1, $p = 0.7$) and both grew at a similar rate (growth rate from logistic equation: all wild chicks combined 0.116 ± 0.016, $n = 81$, foster chicks 0.120 ± 0.014, $n = 23$, Mann–Whitney U: $\chi^2 = 0.47$ df = 1, $p = 0.5$). However, foster chicks reached maximum growth rate at a significantly younger age (Age at maximum growth rate from logistic equation: all wild chicks combined 26.3 ± 3 days, $n = 81$, foster chicks: 23.9 ± 1.7 days, Mann–Whitney U: $\chi^2 = 13.6$, $n = 23$, df = 1, $p = 0.0022$, Table 4).

Foster chicks grew significantly faster than second chicks (growth rate from logistic equation: foster chicks 0.111 ± 0.015, $n = 23$ chicks, second chicks, 0.121 ± 0.014, $n = 26$ chicks. Wilcoxon pairwise comparisons: $Z = -1.96$, $p = 0.05$). Foster chick growth was not significantly different than solo chicks or first chicks (Wilcoxon pairwise comparisons: $Z > 0.84$, $p > 0.06$, Table 4). Foster chicks reached maximum growth rate at a significantly younger age than both first and second chicks: 1.5 days younger than first chicks (first chicks: 25.7 ± 2.4 days old, $n = 38$ chicks; foster chicks: 23.9 ± 1.7 days old, $n = 23$ chicks, Wilcoxon pairwise comparisons: $Z = -2.61$, $p = 0.009$) and 3.5 days younger than second chicks (second chicks: 28.2 ± 3.2 days old, n = 26 chicks, foster chicks 23.9 ± 1.7 days old, n = 23 chicks. Wilcoxon pairwise comparisons: Z = 5.18, $p \leq 0.001$, Table 4).

Table 4. Effect of hatch order on growth parameters for Scarlet Macaw. All parameters were calculated using the logistic growth model [14,41]. "Solo chicks" are wild chicks in one-chick broods. "First chicks" are older chicks in wild two chicks' broods and "second chicks" are younger chicks in wild two chicks' broods. Within a column, values followed by a different superscript letter differed significantly using a Mann–Whitney U (DF = 1, $p < 0.05$).

Chick Type		Number of Individuals	Maximum Growth A (Mean)	Growth Rate B (Mean)	Age at Maximum Growth * C (Mean)
Wild chicks	Solo chick	17	1028 ± 83.2 ^{A}	0.126 ± 0.013 ^{B}	24.7 ± 2.3 ^{D}
	First chick	38	1022 ± 75.4 ^{A}	0.115 ± 0.014 ^{B}	25.7 ± 2.4 ^{E}
	Second chick	26	993 ± 77.7 ^{A}	0.111 ± 0.015 ^{C}	28.2 ± 3.2 ^{F}
	All combined	81	1014 ± 79.7	0.116 ± 0.016	26.3 ± 3
Foster chicks		23	1020 ± 81.1 ^{A}	0.121 ± 0.014 ^{B}	23.9 ±1.7 ^{D}

* The only growth parameter that differed significantly between wild chicks (all combined) and foster chicks is indicated by an asterisk, based on Wilcoxon pairwise comparisons (DF = 1, $p < 0.05$).

3.4. Foster Chick Influence on Breeding Success

In general, during the three seasons with chick relocations, more chicks fledged, more nests had chicks that fledged and fewer chicks died of starvation (Table 5).

Table 5. Effect of foster nests on seasonal breeding success of Scarlet Macaw. Seasons with no foster chicks (17 seasons, 2000 to 2016) were compared to seasons with chick fostering (3 seasons, 2017 to 2019). In the fostering experiments, we placed, swapped or added, foster chicks to specific nests (see text). The breeding season was mid-October to mid-April. Available nests refer to cavities (n = 40) that were not occupied by other species by the beginning of the breeding season. We calculated χ^2 and p-values using Mann–Whitney U (DF = 1).

Breeding Success Parameters	Seasons with Foster Nests (n = 3)	Seasons without Foster Nests (n = 17)	χ^2	p-Value
Chicks that fledged per available cavity	0.43 ± 0.05	0.23 ± 0.86	7.2	0.036
Chicks that fledged per nest with eggs	1.13 ± 0.15	0.56 ± 0.21	7.1	0.007
Chicks that fledged per nest where at least one chick hatched	1.5 ± 0.3	0.86 ± 0.24	7.1	0.01
Percentage of younger chicks that starved	0.06 ± 0.03	0.35 ± 0.22	5.5	0.02
Percentage of chicks that fledged	0.7 ± 0.05	0.46 ± 0.14	5.5	0.02

4. Discussion

Our use of wild Scarlet macaws as foster parents along with supplemental feeding and veterinary care was categorically successful: all foster chicks were accepted by the foster parents with no chick rejection, foster chicks were fed at rates similar to resident chicks, foster chick growth was similar to wild chick growth, and almost 90% of all foster chicks fledged. Moreover, the use of foster parents dramatically reduced chick mortality due to starvation and increased overall reproductive success in the study area.

4.1. Scarlet Macaws as Foster Parents in the Wild

There are few studies of foster parents in wild psittacines but most of them are quite complete. Their objectives varied from a management tool to increase population recruitment (RGA unpublished data, [5,18]), to a scientific technique to understand behaviors such a parent/chick recognition [8], hunger response [42] and food allocation [22]. None of these previous studies addressed the potential conflict of increasing brood size in species that show brood reduction strategies in the early stages of the nesting period nor why pairs allow their own chicks to starve at the beginning of the nesting period but then accept additional unrelated chicks later in the same nesting event.

We designed our experimental procedures to avoid placing foster chicks during the starvation risk period for the species. According to our investigations on brood reduction by chick starvation of Scarlet Macaws [29], we observed that fourth chicks are always left to starve in the first week of life and third chicks in their first two weeks. For second chicks, no death by starvation was recorded after 25 days of age. In fact, 88% of second chicks that starved were younger than 20 days old. For that reason, we consider "the starvation risk period" in Scarlet Macaw is from zero to 20 days old [29]. We did not place additional chicks in foster nests when the youngest member of the foster brood was on average younger than 22 days old. A similar strategy was used in relocating Yellow-shouldered Parrot foster chicks, where only chicks older than two weeks were used as foster chicks because mortality rates are higher in the first weeks of life [18]. Evidence from Scarlet Macaw fostering work in Guatemala support this suggestion, as foster chicks < 15 days old seemed to be rejected at higher rates than older chicks (RG-A personal observations). In foster chicks of Puerto Rican Amazons [5] and the Galahs [8] individuals as young as one week old were accepted in whole brood swaps. These two studies warn about using older foster chicks due to the evidence that adults do not recognize small chicks as individuals but they do recognize older chicks. No warnings are made about placing young foster chicks during starvation risk periods.

A difference between our foster parent experiment and previous studies with wild psittacine foster parents is the presence or absence of the nesting pair when foster chicks were placed in nests. In the two Amazon parrot studies in Puerto Rico and Venezuela and in the Galah study in Australia, foster chicks were placed in nests when parents were absent in order to minimize disturbance and possibly nest abandonment [5,8,18]. Multiple authors stated that they thought the foster parents did not detect the placement of the foster chick [5,42], but in our study that was clearly not the case. In our case, the majority of nesting pairs were present during the precise moment when foster chicks were placed in their nest. In a few cases, the female nesting individual was even inside the nest, so the argument that foster parents did not detect an additional chick is not valid in our case. It is worth emphasizing that in our experiment, we did not consider the nesting pair presence or absence at the nest as an important factor because at our study site, we have been monitoring macaw nests intensively for the last 20 years and Scarlet Macaw pairs are very accustomed to our nest checks and rarely display typical disturbance behaviors when researchers visit the nests and manipulate the chicks [5]. However, in other areas with little or no history of nest checks or chick manipulations, human presence may disturb nesting pairs and alter the results of chick fostering attempts.

4.2. Foster Chick Acceptance and Rejection

Overall acceptance of foster chicks in our investigation was excellent, as expected. There are published records of wild parrots in the genera *Cacatua, Amazona, Platycercus* and *Ara* ([8,18,42] and RG-A unpublished data) and captive *Cylopsitta, Alisterus, Amazona, Pionus, Cacatua* and *Melopsittacus* ([4,5,8,19,20] and GV-T personal observations) accepting and raising foster chicks. In all the studies done in the wild, including ours, when foster chicks were accepted, fostering manipulations caused no major disruption of adult nesting behaviors [5]. In the case of the Galah, the main disruption of adult nesting behavior happened when whole broods were swapped for unrelated broods [8]. Some pairs hesitated for several hours before first entering the nest, but once one adult entered the nest (generally the female), the other nesting adult followed. This hesitation is explained by the fact that older Galah chicks (few weeks from fledgling) reply to their own parent's calls when they arrive at the nest and foster chicks did not respond to foster parent's calls when they arrived [8]. In our case, we did not detect hesitation to enter the nest as in the Galahs. Usually Scarlet Macaw nesting females in our site were eager to come back and check on their chicks as soon as they were returned to the nest by the researchers.

High acceptance of foster chicks after chick predation or egg hatching failure was surprising but also not unexpected; mainly because it has been reported in studies with foster chicks in Amazon parrots in the wild. In Yellow-shouldered parrots, three of four foster chicks placed after full predation events were accepted [18]. In wild Puerto Rican Amazons two foster chicks placed after eggs failed to hatch were accepted, even though the foster chicks were another species: Hispaniolan Amazons (*Amazona ventralis* [5]). In our experiment in Tambopata, in the cases of foster chick acceptance after the resident brood was depredated, after chicks were killed by lightning, and after egg hatching failure, nesting pair behavior was very similar to that reported by the previous studies with Amazon parrots [5,18,24]. There was some initial hesitation, especially the very first time the foster chick was seen, but once it was fed, the nesting pair behavior fell into the normal attendance pattern according to the foster chick's age.

In all three studies, including ours, the timing in which the foster chick was placed after the nest was emptied was likely a key aspect [5,18]. In the case of replacing unviable eggs with a foster chick, the swap probably needs to be done as close to the estimated hatching date as possible. The hatching period is a very sensitive period for the nesting pair. It offers a very small window to replace eggs for foster chicks. Nesting individuals, especially nesting females, tend to decline in attentiveness a few days after the end of the normal incubation period if the eggs fail to hatch [5]. Even though the nesting pair keeps visiting the nest after egg failure, visits are likely more related to a desire to defend the nest cavity and maintain ownership GV personal observations, [43,44].

An unexpected result in our experiment was that all foster chicks were accepted. In three of four foster parent studies from the literature, a few foster chicks were rejected in each. In the Puerto Rican Amazon case [5], two older chicks that were swapped for one foster chick were rejected, even when the foster chick was an offspring of the foster parents. In the Galah study [8] 5% of foster chicks (n = 10 chicks, 3 broods) were rejected; perhaps because all of them were placed very close to fledgling time and the foster parents seemed able to recognize their own chicks either by vocalizations or by unique physical cues [8]. In the Yellow-shouldered Parrot [18], 9.3% (n = 5 chicks) were rejected. Here, rejection was attributed to a low feeding response of foster chicks and different developmental stages between foster chicks and wild offspring. In the Scarlet Macaw in Guatemala, foster chicks were occasionally rejected as well, presumably because they were too young (between 10 to 20 days old) or because brood size was increased over the maximum brood size of the species in the area (broods of three or four chicks, RG-A unpublished data). The Crimson Rosella study did not mention chick rejection at all [22].

A tentative explanation for the zero foster chick rejection found in our experiment is that we measured rejection in a different way than in the previous studies [5,8,18]. We considered that a foster chick was rejected when daily feeding ratios (foster/resident chick) were not similar and when the tendency of feeding-ratios was not positive after 10 days of foster chick relocation. In the Puerto Rican Amazon, the indicator of acceptance was also feedings but based on crop size observations after a few days and not direct observations of feeding [5]. In addition, rejection happened when whole broods were swapped but not when chicks were added to a brood. We did not swap entire broods, so these rejections are not comparable to our case. In the Yellow-shouldered Parrot, the foster chick acceptance indicator was also feedings based on crop size and observations of the absence of injuries at the end of day of relocation. Here, when foster chicks did not have large crops until the next morning after relocations, they were removed and placed in another nest [18]. Interestingly, in our experiment, the only cases in which foster chicks showed large crops in the day after relocation were when the foster chick was the only chick in the nest. In the Scarlet Macaw in Guatemala, foster chick acceptance indicator was also crop size and crop content after a maximum of three days post relocation. After that period, chicks found with crops with no macaw food content were relocated to another foster nests (RG-A unpublished data). In our case, all our foster chicks that were members of multiple broods needed on average five days to show half-full crops. They were fed by foster parents, as clearly observed in videos, but did not have large crops. Under our acceptance/rejection criteria those chicks were not considered rejected.

Zero foster chick rejection in our experiment might be due to the fact that we matched ages/developmental stages between foster chicks and resident chicks. Developmental stages in our experiment were defined based on our extensive knowledge on the nesting biology of the species [14]. Therefore, our foster chicks looked very similar to the resident chicks in nearly all cases. The importance of matching similar ages between foster chicks and resident chicks in chick additions, chick swaps, and whole brood swaps was mentioned in all the previous studies (RG-A personal observations, [5,8,18]). All investigations that worked with psittacine foster parents address the fact that foster chick acceptance and especially rejection were related to age differences among chicks involved (RG-A personal observations, [5,8,18]). In the Yellow-shouldered Amazon [18] and in the Galah [8], foster chicks and foster broods that differed in age from the resident chicks and broods were rejected. In the Puerto Rican Amazon [5] and in the Crimson Rosella [22] studies the authors considered that pairing chicks that "look similar" to be very important. In the Scarlet Macaw in Guatemala, it was considered a key aspect in order to warrant chick acceptance (RG-A unpublished data).

4.3. Foster Chick Chick-Feeding Rates

No previous studies of wild Psittacidae as foster parents have analyzed acceptance of foster chicks using chick feeding rates as an indicator of acceptance. The daily feeding ratios in our experiment showed that foster chick acceptance was a slow process that needed more than one day of post-relocation monitoring before concluding failure. In the Crimson Rosella study in Australia, feedings (food transfers) were used as a tool to: (1) quantify hunger response when broods or individual chicks were placed back in the original nest [22], and (2) to understand food allocation among brood members [42]. However, in both experiments, resident broods and chicks were placed back in their original nest three hours later, so there was no way to analyze daily feeding rates. In the Galah [8] and Puerto Rican Amazon studies [5] feedings were used in a descriptive manner, not in an analytical manner.

Other studies have addressed the first response of foster chicks to foster parents. In our experiment the results were unexpected because the foster chick reaction we observed most commonly could be considered a "distress" response. The majority of foster chicks (n = 23) were syringe fed from a few days old to 20 days of age. Because of that, they showed low or even nonexistent feeding response when approached by an adult macaw. When foster parents, grabbed foster chicks' beaks in an attempt to feed, the foster chick usually shook its head and pulled away. This pulling away behavior was consistently observed during the first days after chick relocations, even twelve hours after last supplemental feeding, when the chicks had little to no food in the crop. Our observations showed that this pulling away decreased slowly during the first 10 days post relocation as the foster chicks learned to receive food from the adult macaws. This same sequence of behaviors was also observed in Guatemala (RG-A personal observations). We consider this behavior a distress response because we have never seen any wild chicks pull away from parents in this fashion (even when their crops are nearly 100% full) in either Peru or Guatemala.

A more intense reaction was observed with the Puerto Rican Amazon in which some hand-raised chicks gave fright responses and distress vocalizations when they were placed into the wild nests and first encountered adult parrots attempting to brood and feed them [5]. The first response to foster parents in Galah Cockatoo chicks was not a distress response, instead, the chicks gave little or no response, especially for chicks in the second half of their nesting period. At that age, Galah chicks start to respond to parent calls when they arrive at the nest and foster chicks did not reply to foster parent call when they first arrived to the nest. However, after a few hours of not being fed, the tendency for rejection by the nestling was overruled by hunger. Once the nestling was hungry, begging and vocalizations attracted foster parents that proceeded to enter the nest and feed them [8].

4.4. Foster Chick Supplemental Feeding

In our experiment, chicks were given supplemental food once or twice per day during the transition period when foster chicks were learning how to be fed by wild macaws. Even though they were not responding when foster parents tried to feed them, they were not losing weight or showing signs of nutritional deficiency due to our supplemental feedings. This is similar to the fostering protocols used in Guatemala which obtained similar results (RG-A personal observations). This evidence from Peru and Guatemala leads us to conclude that supplemental feeding gave foster chicks time to learn appropriate feeding response behaviors and increased chick acceptance and the success of this technique.

Our experiment raises the question of whether or not we could have obtained similar results by just feeding the chicks in situ, without pulling them out of the nest and relocating them. Multiple lines of evidence suggest that just feeding may not have been successful. In previous breeding seasons, we provided supplemental feeding to starving younger Scarlet Macaw chicks in the nest (n = 5 chicks: 1 s chick and 4 third chicks) on average three times per day (range = 1 to 4 times per day) for an average of 4 days (range = 3 to 7 days) and this failed to prevent starvation. In addition, supplemental feeding provided to younger Green-rumped Parrotlet chicks upon hatch in the nest three times per day marginally increase the probability of survival of last-hatched chicks but they still experienced significantly higher mortality than early hatched chicks and it did not improved probability of survival of penultimate hatched chicks [45]. Moreover, our observations suggest that feeding alone may be insufficient to save younger macaw hatchlings as parents may selectively exclude them from brooding: second and third chicks that starved were excluded from brooding from 6 to 35 times more than second chicks that fledged (n = 9 macaw broods, 3 breeding seasons, 250 video hours analyzed, GV-T unpublished data). This is problematic because improper brooding of captive macaw hatchlings and young chicks can cause abrupt temperature fluctuations that may result in thermal stress and death [39].

4.5. Foster Chick Growth Rates

An interesting finding was that foster chicks in our experiment were not only accepted by foster parents, but also raised as wild chicks. Some foster chicks were in poor condition when removed from their nests; either underweight, not gaining weight as expected or sick, and they received special treatments in order to recover. Even though these individuals grew slowly when young, their overall growth rate ended up similar to wild chicks. These results fit with the compensatory growth principle that states that given adequate conditions, slow development as a result of poor nutrition is followed by accelerated growth. Growth rates become similar to nestlings that did not experience nutritional stress at all [39,46]. Besides, our hand-raising procedures in the nursery and supplemental feeding plan during the first 10 days after relocation provided enough nutrition to foster chicks, so they were able to compensate for the low caloric intake received from foster parents during the first days in foster nests, and this allowed them to catch-up and attain maximum growth rates and maximum weights similar to the wild chicks.

The fact that foster chicks reached the maximum growth rate almost two days earlier than resident chicks is likely a direct consequence of our hand-feeding procedures. Captive raised Scarlet Macaw chicks grow differently than their wild equivalents. Indeed, purely captive-raised chicks reached the maximum growth rate even sooner than foster chicks in this experiment [15]. These differences in growth might be related to differences in the consistency of macaw chicks' diets. In the wild, the diet of macaw chicks contains full seeds and even tree bark [47] so it may take more time and energy to digest than the puree like formula that is used in captivity [48]. In order to get foster chicks extra fat in preparation for the adaptation process in their new foster nest, we provided a high fat diet (formula with nuts and peanut butter added), large portions, and high feeding frequencies [29].

4.6. Testing Starvation Drivers

The main driver behind brood reduction in two chick broods appears to be the age difference between brood members [29]. This age difference effect predicts that the greater the age difference between brood members the higher the risk of starvation of younger brood members. However, work in Costa Rica and specially in Guatemala [RG-A unpublished data] suggests that macaws that allowed their own chicks to starve at the beginning of the nesting period accepted, raised, and fledged additional unrelated chicks later in the same nesting event even when age differences were >5 days (average age difference 9 days, maximum = 14 days, n = 60 chicks). In our foster parent work, we confirmed that these age difference effects on starvation did not apply when brood members were older than 20 days. We had seven foster chicks with age differences >5 days from their foster sibling. In all seven foster broods, both chick members successfully fledged. One reason why age difference might not be correlated with starvation in multiple broods with chicks older than 20 days old is that the younger the chicks, the more age specific the parental care requirements (aka: brooding and feeding). Recommendations for brooding temperatures in captivity indicate that newly hatched chicks need to be kept 2 °C warmer than 5–9 days old chicks and extreme temperature fluctuations at this time can be harmful or even fatal to the chicks [39]. When the pinfeathers of chicks start to show, around 18 days old, chicks are less affected by temperature and when they are heavily pin feathered, around 30 days old, heat requirements diminish considerably [39]. The recommended feeding frequency also varies from every hour for hatchlings to every 3 to 4 h for 5 to 9-day old chicks and are even more variable as the birds age (Appendix A, [39,49]). Hence, when chicks are older than 20 days, chicks that are >5 days apart "look similar" and their parental care requirements, brooding and feeding, are similar. The fact that foster parents were able to fledge chicks that were over 5 days apart suggests that the developmental stage in which the foster chicks were placed may have been a key factor. Since both foster siblings were at the age at which parental care requirements were very similar, even though foster parents needed time to adjust their food provisioning and foster chicks needed to learn how to be fed, death by starvation was no longer a major risk.

4.7. "Triple Brood" Chick Relocation Procedure

In our triple brood procedure, the fact that all foster chicks did not gain weight as expected suggests that the parents were unable to provide sufficient food. Even the resident chick, that regularly had the largest crop of the trio, could not consistently reach the average weight for its age. It seems like even though all chicks where being fed, the macaw foster parents were not able to feed three chicks properly. According to our work on Scarlet Macaws in southeastern Peru a maximum of 2 chicks fledge under natural circumstances. In fact, only 37% of nesting pairs managed to fledge 2 chicks and 1.3 chicks per nest is the average chick production per successful nest in the area [29]. However, there are reports of rare successful natural triple broods from Costa Rica [50] and Guatemala [51] suggesting that conditions may vary geographically. In Guatemala, RG-A's team has created a total of four triple and two quadruple foster broods across at total of three breeding seasons. From the four triple foster broods created, all chicks died in one, all chicks fledged in another and one fledged only two chicks. In the two quadruple foster broods created, only three chicks fledged from both (RG-A unpublished data).

In the Yellow-shouldered Parrot foster parent research [18] it was recommended to not create foster broods that were bigger than the optimal brood size of the species. Our results agree with this conclusion, and show that it is important to calibrate foster brood size using as a general indicator the optimal brood size of the species in the area.

5. Conclusions

Our technique of macaw foster parents and post relocation supplemental feeding was categorically successful. All relocated foster chicks were successfully accepted by their foster parents (n = 28 chicks across three seasons) and 89% of them fledged. The only three foster chick fatalities were due to unknown disease, predation, and lightning. Foster chick acceptance by foster parents was a slow process. Foster chicks were initially fed less than resident chicks, but feeding ratio increased progressively until feedings were similar for both chicks 10 days after relocation. Foster chicks needed on average about 5 days to consistently have half-full crops when checked. Growth rates of foster chicks were similar to wild chicks and both chick groups reached similar maximum weights. However, foster chicks reached maximum growth rate at a younger age. These differences were likely due to differences in diet and feeding schedule. Our foster parents technique increased the reproductive success of our studied population: fledging success per available nest increased from 23% (2000–2016) to 43% (2017–2019) and chick death by starvation decreased from 35% to 6%.

Interdisciplinary Collaboration in Parrot Conservation

Our ability to produce foster chicks that were successfully accepted and that were very similar to wild chicks by fledgling time, is the result of the integration of three different fields: parrot ecology, avian veterinary medicine, and aviculture. Psittacines have been the heart of aviculture for centuries and there are many well know breeding techniques that can be easily modified and adjusted for use in the wild [52]. In fact, the Scarlet Macaw is considered one of the most productive species of macaws in captivity (Mark Moore, co-owner of Hill Country Aviaries, USA. Personal communications, [53]). In our experiment, we used information from the aviculture literature [15,39,49,54] and worked closely with experienced psittacine breeders. We also worked with avian veterinarians that took care of chick health issues and provided insights from their experiences with captive psittacines. Lastly, we integrated our knowledge on breeding ecology and nesting behavior of the species [14,33,47,55,56]. As has been demonstrated with the Puerto Rican Parrot [5], Spix's Macaw [57,58] and our work with Scarlet Macaw, the integration of ecologists, veterinarians and aviculturists has great potential to assist management actions in the wild.

Author Contributions: Conceptualization, G.V.-T., R.G.-A. and D.J.B.; methodology, G.V.-T., R.G.-A. and D.J.B.; software, G.V.-T.; validation, G.V.-T., formal analysis, G.V.-T. and D.J.B.; investigation, G.V.-T. and D.J.B.; resources, G.V.-T.; data curation, G.V.-T.; writing—original draft preparation, G.V.-T.; writing—review and editing, G.V.-T., R.G.-A. and D.J.B.; visualization, G.V.-T.; supervision, D.J.B.; project administration, G.V.-T.; funding acquisition, G.V.-T. and D.J.B. All authors have read and agreed to the published version of the manuscript.

Funding: This research was part of G.V.-T.'s graduate studies supported by a National Science Foundation (NSF) Graduate Research Fellowship, by a Diversity fellowship from Texas A&M University and a Regents Fellowship from the College of Agriculture and Life Science from Texas A&M University. Chick rearing work was funded by generous private donations to two crowdfunding campaigns run by Experiment.com in December 2017 and December 2018 along with donations from Chris Biro from Bird Recovery International, and Janice Boyd from Amigos de las Aves USA and The Parrot Fund. Macaw chick formula was donated by ZuPreem. Discounted rates for lodging expenses and boat transportation were given by Rainforest Expeditions S.A.C. thanks to Kurt Holle. Data collection was funded in part by the Schubot Center for Avian Health thanks to Dr. Ian Tizard and funds from DJB lab.

Institutional Review Board Statement: This study follows the Animal Use Permits: AUP 2012-145 "Macaw and parrot biology in southeastern Peru" given by Texas A&M University. The work has been authorized by the Peruvian government under the following research permits: Resolución Jefatural de la Reserva Nacional Tambopata, N 16-2018-SERNAMP-JEF, Resolución Jefatural de la Reserva Nacional Tambopata, N 22-2016-SERNAMP-JEF, Resolución 33-2015-SERNAMP- JEF, Resolución 13-2014-SERNAMP-JEF, Resolución 19-2013-SERNAMP- JEF, Resolución 18-2012-SERNAMP- JEF.

Informed Consent Statement: Not applicable.

Data Availability Statement: The data presented in this study are available on request from the corresponding author. The data are not publicly available due to permit restrictions from the National Service of Protected Areas of Perú (SERNANP).

Acknowledgments: Thanks to Mark Moore, María Belen Aguirre, Sharman Hoppes, Sophie Hebert, Liz Villanueva Paipay, Gustavo Martinez, Carlos Huamaní, Shannan Courtenay, Dylan Whitaker, Roshan Tailor, George Olah, Bruce Nixon, Deysi Delgado, Jorge León, Susanne Vorbrüggen, Lauren Bazley, Patricia Gray, Simon Kiacz, Maximo Gonzales and the whole BS 2017, BS 2018 & BS 2019 "guacamayeros" volunteer teams for their invaluable hard work. Thanks to Vanesa Hilares from AIDER permit support. To Janice Boyd and Rodrigo Leon for the scientific advice. To Thomas Lacher, Ian Tizard and especially to Jane Packard for the mentoring and the corrections provided to this manuscript. To Keith Regelmann for his collaboration on our crowdfunding campaigns. Special thanks to Amanda Lucile (Mandy Lu) Brightsmith and Samantha King Brightsmith. Dedicated to Lucille Smith, Mónica Trauco and Oscar Vigo.

Conflicts of Interest: The authors declare no conflict of interest. The funders had no role in the design of the study; in the collection, analyses, or interpretation of data; in the writing of the manuscript, or in the decision to publish the results.

Appendix A

Diet details for Scarlet Macaw wild chicks in the nursery: Throughout the project macaw chicks were fed based on their age and other special circumstances as outlined here and in Table A1. Formula for neonates (<4 days old) was prepared as 1-part Zupreem formula to 4 parts water. For chicks ≥4 days old regular chick formula was 1-part Zupreem formula to 3 parts water plus peanut butter in the majority of the cases. For chicks ≥12 days old a mix of shredded raw Brazil nuts, pecans, and peanuts was added to the regular chick formula (Table A1). The majority of the time chicks were fed when their crops were empty or close to empty resulting in a feeding frequency of about once every 2.7 h when they were under 4 days of age to about once every 5 h when they were between 15 and 20 days old [modified from 39]. This protocol was followed for 21 chicks. For one chick we added shredded peanuts and peanut butter to the neonate formula starting at age 2 days and four chicks that had additional health problems received customized feeding regimes.

One underweight third chick was fed neonatal formula until it was 11 days old because its digestion was slow. From 12 days on it was fed regular chick formula. By age 15, it showed slower growth and slower development for its age but by 24 days old, its weight was as expected for its age. A similar situation happened with the underweight forth chick that arrived to the nursery right after hatching. The chick was fed neonatal emergency formula on its first day of life and neonate formula its second day of life. Subsequently it was moved up to neonate formula plus until it was 11 days old. At age 7 and age 8 chick showed early signs of slow digestion and it was given a mix of warm papaya juice and cinnamon added to its usual food until crop size increased to half crop full, once per day. From hatch, this chick showed a slower growth and slower development for its age but by age 25, it weighed as expected for its age. A second chick with signs of starvation, was given a special neonatal emergency formula (1 part Formula One by Avitech and 4 parts water; (http://www.avitec.com/Formula-One-for-hand-feeding-hatchlings-s/70.htm, accessed on 1 October 2016) and subcutaneous fluids for its first 12 h in the nursery. In these first 12 h it gained 78% of its arrival weight and after that the chick was fed according to its age.

One chick with a large botfly infection and low weight received two feedings that were a mix of neonatal emergency food and regular neonatal formula.

One first chick that was brought in as part of the acclimation process to create a triple brood, had food aspiration problems in its second day in the nursery when it was 19 days old. The chick was under antibiotics, anti-inflammatory and antifungal oral treatment for the following 20 days (15 days in the nursery, 5 days after nest relocation). This chick's weight gain was always as expected for its age. All of the remaining chicks were fed normally according to their age.

Table A1. Summary of diet of wild macaw chicks in the nursery. Food names were assigned to differentiate among five different types of food provided. Neonate food was given to younger chicks (under 4 days old) and Regular food to older ones (over 4 days old). Emergency food was given to chicks showing signs of starvation. PLUS foods contain peanut butter and EXTRA PLUS foods contain peanut butter plus shredded raw Brazil nuts, pecans, and peanuts. Chick age is given in days. Formulas used are well-known commercial formulas use to raised macaw chicks in captivity: Zupreem Embrace (https://www.zupreem.com/products/birds/embrace-plus/, accessed on 1 October 2016) and Formula One by Avitech (http://www.avitec.com/Formula-One-for-hand-feeding-hatchlings-s/70.htm, accessed on 1 October 2016). Proportion used to prepare formula was the recommended by the manufacture.

Food Name	Age Range		Ingredients					Formula/Water Proportion
	Min (days)	Max (days)	Formula Zupreem Embrace Baby Bird	Formula One Avitech	Peanut Butter	Sheered Seeds	Water	
Neonates Formula	0	9	Yes	No	No	No	Yes	1 to 4
Neonates Formula Plus	2	20	Yes	No	Yes	No	Yes	1 to 4
Regular Formula	4	43	Yes	No	No	No	Yes	1 to 3
Regular Formula Plus	4	74	Yes	No	Yes	No	Yes	1 to 3
Regular Formula Extra Plus	12	28	Yes	No	Yes	Yes	Yes	1 to 3
Special Emergency Formula	7	13	No	Yes	No	No	Yes	1 to 4

Appendix A

Table A2. Video observations of scarlet macaw behavior in foster nests. Field observations were done by a mix of 20 different assistants watching live video feeds in the field. Recorded observations were done by one experienced observer using video recordings and included recordings of both diurnal and nocturnal activity. A total of 10 chicks were observed with video, 3 in 2017 and 7 in 2018.

Type of Observation	Seasons	# Total Chicks	# Observers	Total Hours Observed	Hours Observed Per Day			
					Max	Min	Average	St Dev
Field Observations	2018	7	20	**417.9**	12.0	4.3	8.4	2.8
Recorded Observations	2017 and 2018	10	1	**573.4**	23.6	3.7	9.0	4.2

Appendix B

Foster chick fatalities. During the three years of work with wild macaws as foster parents, three foster chicks perished in their foster nests.

The first one died five days after being placed possibly because of an unknown disease. This foster chick had half-full crop by the day after relocation but just $\frac{1}{4}$ full on the following days. On those days, it was fed supplemental food. Starting on the day after relocation, this foster chick showed small red hematomas, first on the right flank, then the left flank, next to the keel with a scratch-like wound on the right leg. The foster chick was the third chick in its original nest. Both chicks in the original nest died with the same type of hematomas: the second hatched chick at 6 days of age and the first hatched at 12 days.

The second case of a foster chick death was due a combination of predation and lightning hitting the video cable systems installed in artificial nest (PVC pipe). Nest was found with cable system burned and the access door blown off. The foster chick (39 days old) and the resident chicks (41 days old) were not found inside the nest or in the surroundings. Marks of large claws were found around the door of the PVC nest box and on the tree branches from where nest was hung.

A third case of a foster chick death was due to lightning hitting the artificial nest (wooden box) that blew the base and top off the nest. The foster chick was 36 days old. Both foster and resident chick were found dead on the ground, below the nest tree, right after the thunderstorm stopped. Necropsy suggested that the foster chick's death was due to electrocution and resident chick's due to the fall.

References

1. Cade, T.J. The husbandry of falcons for return to the wild. *Int. Zoo Yearbook* **1980**, *20*, 3. [CrossRef]
2. Saint Jalme, M. Endangered avian species captive propagation: An overview of functions and techniques. *Avian Poult. Biol. Rev.* **1999**, *13*, 16. [CrossRef]
3. Fentzloff, C. Breeding, artificial incubation and release of White-tailed sea eagles. *Int. Zoo Yearb.* **1984**, *23*, 17. [CrossRef]
4. Romer, L. Management of the Double-eyed or Red-Browed fig parrot. *Zool. Soc. Lond.* **2000**, *37*, 6.
5. Snyder, N.R.R.; Wiley, J.W.; Kepler, C.B. *The Parrots of Luquillo: Natural History and Conservation of the Puerto Rican Parrot*; Western Foundation of Vertebrate Zoology: Los Angeles, CA, USA, 1987.
6. Beissinger, S.R.; Snyder, N.F.R. *New World Parrots In Crisis*; Smithsonian Institution Press: Washington, DC, USA, 1992.
7. Beissinger, S.R.; Butcher, E.H. Sustainable harvesting of parrots. In *New World Parrots in Crisis*; Beissinger, S.R., Snyder, N.F.R., Eds.; Smithsonian Institution Press: Washington, DC, USA, 1992; pp. 73–115.
8. Rowley, I. Parent-offspring Recognition in a Cockatoo, the Galah *Cacatua roseicapilla*. *Aust. J. Zool.* **1980**, *28*, 9. [CrossRef]
9. Beissinger, S.R.; Waltman, J.R. Extraordinary clutch size and hatching asynchrony of a Neotropical parrot. *Auk* **1991**, *108*, 863–871.
10. Sanz, V.; Rodriguez-Ferraro, A. Reproductive parameters and productivity of the Yellow-shouldered Parrot on Margarita Island, Venezuela: A long-term study. *Condor* **2006**, *108*, 178–192. [CrossRef]
11. Vaughan, C.; Nemeth, N.; Marineros, L. Ecology and management of natural and artificial scarlet macaw (*Ara macao*) nest cavities in Costa Rica. *Ornithol. Neotrop.* **2003**, *14*, 381–396.
12. Marineros, L.; Vaughan, C. Scarlet macaws in Carara. In *The Large Macaws: Their Care, Breeding and Conservation*; Abramson, J., Spear, B.L., Thomsen, J.B., Eds.; Raintree Publications: Ft. Bragg, CA, USA, 1995; pp. 445–468.
13. Krebs, E. Breeding biology of crimson rosellas (*Platycercus elegans*) on Black Mountain Australian Capital Territory. *Aust. J. Zool.* **1998**, *46*, 17. [CrossRef]
14. Vigo, G.; Williams, M.; Brightsmith, D.J. Growth of Scarlet Macaw (*Ara macao*) chicks in southeastern Peru. *Neotrop. Ornithol.* **2011**, *22*, 143–153.
15. Vigo Trauco, G. Crecimiento de Pichones de Guacamayo Escarlata, *Ara macao* (Linneus: 1758) en la Reserva Nacional Tambopata—Madre de Dios—Perú. Bachelor's Thesis, Universidad Nacional Agraria La Molina, Lima, Peru, 2007.
16. Smith, G.A. Systematics of parrots. *Ibis* **1975**, *117*, 99. [CrossRef]
17. Raso, T.D.; Seixas, G.H.F.; Guedes, N.M.R.; Pinto, A.A. Chlamydophila psittaci in free-living Blue-fronted Amazon parrots (Amazona aestiva) and Hyacinth macaws (*Anodorhynchus hyacinthinus*) in the Pantanal of Mato Grosso do Sul, Brazil. *Vet. Microbiol.* **2006**, *117*, 235–241. [CrossRef]
18. Sanz, V.; Rojas-Suárez, F. Los nidos nodriza como técnica para incrementar el reclutamiento de la cotorra cabeciamarilla (*Amazona barbadensis*, Aves: Psittacidae). *Vida Silv. Neotrop.* **1997**, *6*, 8–14.
19. Stoodley, A.A. Surrogate Parents. In *AFA Watchbid*; American Federation of Aviculture: Austin, TX, USA, 1986; Volume 13, p. 6. Available online: https://journals.tdl.org/watchbird/index.php/watchbird/article/view/2455 (accessed on 1 October 2016).
20. Dingle, S. Olde Tymer Bill Rattray: King of the King Parrots. In *AFA Wathcbird*; American Federation of Aviculture: Austin, TX, USA, 1998; Volume 25, Available online: https://journals.tdl.org/watchbird/index.php/watchbird/article/view/1407 (accessed on 1 October 2016).
21. Yantz, J. Notes on fostering finches. In *AFA Watchbird Magazine*; American Federationof Aviculture: Austin, TX, USA, 1986; Available online: https://journals.tdl.org/watchbird/index.php/watchbird/article/view/2299 (accessed on 1 October 2016).
22. Krebs, E.A.; Cunningham, R.B.; Donnelly, C.F. Complex patterns of food allocation in asynchronously hatching broods of Crimson Rosellas. *Anim. Behavior* **1999**, *57*, 753–763. [CrossRef] [PubMed]
23. Forshaw, J.M. *Parrots of the World*, 3rd ed.; Landsdowne Editions: Melbourne, Australia, 1989; p. 672.
24. Iñigo-Elias, E.E. Ecology and Breeding Biology of the Scarlet Macaw (*Ara macao*) in the Usumacinta Drainage Basin of Mexico and Guatemala. Ph.D. Dissertation, University of Florida, Gainesville, FL, USA, 1996.
25. Wiedenfeld, D.A. A new subspecies of Scarlet Macaw and its status and conservation. *Ornitol. Neotrop.* **1994**, *5*, 99–104.
26. Portillo Reyes, H. Distribucion actual de la guara (lapa) roja (*Ara macao*) en Honduras. *Zeledonia* **2005**, *9*, 3.
27. Tella, J.L.; Hiraldo, F. Illegal and legal parrot trade shows a long-term, cross-cultural preference for the most attractive species increasing their risk of extinction. *PLoS ONE* **2014**. [CrossRef] [PubMed]
28. Romero-Vidal, P.; Hidalgo, F.; Rosetto, F.; Blanco, G.; Carrete, M.; Tella, J.L. Opportunistic or Non-random Wildlife Crime? Attractiveness rather than Abundances in the Wild Leads to Selective Parrot Poaching. *Diversity* **2020**, *12*, 314. [CrossRef]
29. Vigo, G. Scarlet Macaw Nesting Ecology and Behavior: Implications for Conservation Management. Ph.D.Thesis, Texas A&M University, College Station, TX, USA, 2020.
30. Bentin, M.; Leyva, J.C. Model of assisted breeding of two *Primolious couloni* chicks by *Ara macao* adults. *AFA Watchbird Magazine*, 5 September 2018; Volume XLV, 23–29.
31. Brightsmith, D.J. Effects of diet, migration, and breeding on clay lick use by parrots in Southeastern Peru. In Proceedings of the Annual Convention 2004, American Federation of Aviculture, San Francisco, CA, USA, 31 March–3 April 2004; pp. 13–14.
32. Tosi, J.A. *Zonas de Vida Natural en el Perú. Memoria Explicativa Sobre el Mapa Ecológico del Perú*; Instituto Interamericano de las Ciencias Agricolas de la Organización de los Estados Americanos: Lima, Peru, 1960; p. 271.
33. Olah, G.; Vigo, G.; Heinsohn, R.; Brightsmith, D.J. Nest site selection and efficacy of artificial nests for breeding success of Scarlet Macaws (*Ara macao*) in lowland Peru. *J. Nat. Conserv.* **2014**, *22*, 176–185. [CrossRef]

34. Brightsmith, D.J.; Holle, K.M.; Stronza, A. Ecotourism, conservation biology, and volunteer tourism: A mutually beneficial triumvirate. *Biol. Conserv.* **2008**, *141*, 2832–2842. [CrossRef]
35. Gish, E. The Case for Bio-Centric Development: An Ethnographic Study of the Tambopata Macaw Project, Ecotourism and Volunteer Tourism in the Peruvian Amazon. Master's Dissertation, Wageningen University, Wageningen, The Netherlands, 2009.
36. Brightsmith, D.J. The Tambopata Macaw Project: Developing techniques to increase reproductive success of large macaws. *AFA Watchbird* **2001**, *28*, 24–32.
37. Perry, D.R.; Williams, J. The tropical rain forest canopy: A method providing total access. *Biotropica* **1981**, *13*, 283–285. [CrossRef]
38. Perry, D.R. A method of access into the crowns of emergent and canopy trees. *Biotropica* **1978**, *10*, 155–157. [CrossRef]
39. Voren, H.; Jordan, R. *Parrots: Hand Feeding & Nursery Management*; Mattachione, Silvio, & Company: Toronto, ON, Cananda, 1992.
40. Brightsmith, D.J.; Hilburn, J.; Del Campo, A.; Boyd, J.; Frisius, M.; Frisius, R.; Janik, D.; Guillén, F. The use of hand-raised Psittacines for reintroduction: A case study of Scarlet Macaws (*Ara macao*) in Peru and Costa Rica. *Biol. Conserv.* **2005**, *121*, 465–472. [CrossRef]
41. Ricklefs, R.E. Patterns of growth in birds. *Ibis* **1968**, 419–451. [CrossRef]
42. Krebs, E. Begging and food distribution in Crimson Rosella (*Platycercus elegans*) broods: Why dont hungry chicks beg more? *Behav. Ecol. Sociobiol.* **2001**, *50*, 10. [CrossRef]
43. Renton, K.; Brightsmith, D.J. Cavity use and reproductive success of nesting macaws in lowland forest of southeast Peru. *J. Field Ornithol.* **2009**, *80*, 1–8. [CrossRef]
44. Renton, K. Agonistic interactions of nesting and nonbreeding macaws. *Condor* **2004**, *106*, 354–362. [CrossRef]
45. Stoleson, S.H.; Beissinger, S.R. Hatching asynchrony, brood reduction, and food limitation in a Neotropical parrot. *Ecol. Monogr.* **1997**, *76*, 131–154. [CrossRef]
46. Schew, W.A.; Ricklefs, R.E. Developmental plasticity. In *Avian Growth and Development-Evolution within the Altricial-Precocial Spectrum*; Starck, M., Ricklefs, R.E., Eds.; Oxford University Press: Oxford, UK, 1998; pp. 288–303.
47. Brightsmith, D.J.; Matsufuji, D.; McDonald, D.; Bailey, C.A. Nutritional content of free-living Scarlet Macaw chick diets in southeastern Peru. *J. Avian Med. Surg.* **2010**, *24*, 9–23. [CrossRef]
48. Cornejo, J.; Dierenfeld, E.S.; Bailey, C.A.; Brightsmith, D.J. Nutritional and physical characteristics of commercial hand-feeding formulas for parrots. *Zoo Biol.* **2013**. [CrossRef]
49. Clubb, S.L.; Clubb, K.J.; Skidmore, D.; Wolf, S.; Phillips, A. Psittacine neonatal care and hand-feeding. In *Psittacine Aviculture: Perspectives, TECHNIQUES and Research*; Shubot, R.M., Clubb, K.J., Clubb, S.L., Eds.; Avicultural Breeding and Research Center: Loxahatchee, FL, USA, 1992.
50. Myers, M.; Vaughan, C. Movement and behavior of Scarlet Macaws (*Ara macao*) during the post-fledging dependence period: Implications for in situ versus ex situ management. *Biol. Conserv.* **2004**, *118*, 9. [CrossRef]
51. Boyd, J.; McNab, R.B. *The Scarlet Macaw in Guatemala and El Salvador: 2008 status and future possibilities. Findings and Recommendations from a Species Recovery Workshop 9–15 March 2008*; Wildlife Conservation Society-Guatemala Program: Guatemala City and Flores City, Peten, Guatemala, 2008; p. 178.
52. Clubb, S.L. The role of private avicultre in the conservation of Neotropical psittacines. In *New World Parrots in Crisis*; Snyder, N.F.R., Beissinger, S.R., Eds.; Smithsonian: Washington, DC, USA, 1992; pp. 117–132.
53. Clubb, K.J.; Clubb, S.L. Status of macaws in aviculture. In *Psittacine Aviculture: Perspectives, Techniques and Research*; Shubot, R.M., Clubb, K.J., Clubb, S.L., Eds.; Avicultural Breeding and Research Center: Loxahatchee, FL, USA, 1992.
54. Abramson, J.; Spear, B.L.; Thomsen, J.B. *The Large Macaws: Their Care, Breeding and Conservation*; Raintree Publications: Fort Bragg, CA, USA, 1995.
55. Brightsmith, D.J. Parrot nesting in southeastern Peru: Seasonal patterns and keystone trees. *Wilson Bull.* **2005**, *117*, 296–305. [CrossRef]
56. Olah, G.; Vigo, G.; Ortiz, L.; Rozsa, L.; Brightsmith, D.J. *Philornis* sp. bot fly larvae in free living Scarlet macaw nestlings and a new technique for their extraction. *Vet. Parasitol.* **2013**. [CrossRef]
57. Juniper, T. *Spix's Macaw: The Race to Save the World's Rarest Bird*; Fourth Estate: London, UK, 2003; p. 296.
58. Schischakin, N. Special Report: The Spix's Macaw Conservation Program. *AFA Watchb.* **1999**, *26*, 46–55. Available online: https://journals.tdl.org/watchbird/index.php/watchbird/article/view/1398 (accessed on 1 October 2016).

MDPI

Article

Seasonal Variation in Fecal Glucocorticoid Levels and Their Relationship to Reproductive Success in Captive Populations of an Endangered Parrot

Brian Ramos-Güivas [1,2,*], Jodie M. Jawor [2] and Timothy F. Wright [2]

[1] Puerto Rico Department of Natural and Environmental Resources, San Juan 00917, Puerto Rico
[2] Department of Biology, New Mexico State University, Las Cruces, NM 88005, USA; jjawor@nmsu.edu (J.M.J.); wright@nmsu.edu (T.F.W.)
* Correspondence: brianrg@nmsu.edu; Tel.: +1-(787)-367-5273

Abstract: Many species are threatened with extinction, and captive breeding programs are becoming more common to avoid this outcome. These programs serve to prevent extinction and produce individuals for eventual reintroduction to natural populations in historical habitat. Captive animals experience different energetic demands than those in the wild, however, and as a result may have different levels of glucocorticoid hormones. Glucocorticoids help with responses to energetically expensive and potentially stressful situations. Elevated glucocorticoid levels can also potentially alter reproduction and other key behaviors, thus complicating successful captive breeding. The Puerto Rican parrot (*Amazona vittata*) is a critically endangered parrot that currently exists in only two wild and two captive populations. Its recovery program provides a good platform to better understand how glucocorticoid levels may relate to reproductive success under captive conditions. We validated a corticosterone assay in this species and used non-invasive techniques of measuring fecal glucocorticoid metabolites of males and females from two captive populations (Rio Abajo and El Yunque) of Puerto Rican parrots over two consecutive breeding seasons, 2017 and 2018, and the pre-breeding season of 2018, which occurred just after Hurricane Maria struck Puerto Rico. Our results show that levels of fecal glucocorticoid metabolites of males measured during the breeding season of 2018 negatively correlated to the number of total eggs and fertile eggs laid by pairs. In contrast, there was a positive relationship of female fecal glucocorticoid metabolite levels during the pre-breeding season of 2018 with total eggs laid. In males from the Rio Abajo population, we found seasonal differences in fecal glucocorticoid metabolite levels, with higher levels during the pre-breeding season of 2018 compared to both 2017 and 2018 breeding seasons. There was no difference in the mean value of male fecal glucocorticoid metabolites between the 2017 breeding season and 2018 breeding season which started four months after Hurricane Maria struck Puerto Rico. We did find sex differences during the pre-breeding season of 2018 in birds from the Rio Abajo population. Adjustments in the care routine of both populations that could reduce circulating baseline glucocorticoids and avoid frequent, sudden elevations of glucocorticoids should be considered. These results provide a baseline for future comparison with reintroduced populations of this endangered species and other species with captive breeding programs.

Keywords: captive populations; endangered species; glucocorticoids; parrot; reproductive success; seasonality

Citation: Ramos-Güivas, B.; Jawor, J.M.; Wright, T.F. Seasonal Variation in Fecal Glucocorticoid Levels and Their Relationship to Reproductive Success in Captive Populations of an Endangered Parrot. *Diversity* **2021**, *13*, 617. https://doi.org/10.3390/d13120617

Academic Editors: José L. Tella, Guillermo Blanco and Martina Carrete

Received: 13 October 2021
Accepted: 21 November 2021
Published: 25 November 2021

1. Introduction

Changes in the local environment of an organism can promote responses or changes in both physiological processes and behavior, specifically an increase in glucocorticoids [1]. Environmental challenges can be either predictable or unpredictable, but in either case, can have a major impact on individual fitness and evolutionary adaptation [1]. One of the primary mechanisms to promote adaptive responses to environmental challenges in vertebrates is the elevation of glucocorticoid hormone levels by the hypothalamic-pituitary-adrenal (HPA) axis [2,3]. Glucocorticoids primary role in an organism is to aid the metabolism of protein and lipids into carbohydrates for energy consumption and multiple other functions [4,5]. During a disturbance, an animal will release glucocorticoid hormones, resulting in a process of prioritization of energy for a survival response [2,6,7] and subsequent recovery. The frequency and duration of perturbations can cause an animal to have either persistently high glucocorticoid levels, or repeated short-term elevations of glucocorticoids, sometimes termed 'allostatic overload', with potentially negative consequences [1,5]. For example, in birds, elevated levels of corticosterone (the primary avian glucocorticoid) may inhibit the production of luteinizing hormone and prolactin [8–10] and reduce affiliative behaviors, potentially affecting reproductive success [8,11,12].

Captivity is a prime example of an environment that can potentially alter normal glucocorticoid levels and reproductive success [8]. Captive breeding programs are becoming more common as human-mediated habitat changes and other anthropogenic disturbances threaten more species with population reduction and even extinction. A high production of individuals from captive breeding populations of endangered species can allow managers to develop better strategies to enhance reintroduction efforts and sustain wild populations [13,14]. But under captive conditions, animals may have altered levels of glucocorticoids compared to individuals in the wild, modifying behavior otherwise characteristic of the species [15]. In captivity, individuals may experience increased social interactions, atypical photoperiods and limited space overall [16], which can lead to an increase in glucocorticoid levels and altered behavior [17]. Captive conditions for individuals born and raised in the wild may cause a drastic increase in glucocorticoids levels, especially when these individuals spend long periods of time in captivity, with substantial effects on reproduction [17]. In black-legged kittiwakes (*Rissa tridactyla*) artificial increase of the glucocorticoid corticosterone, reduced the production of prolactin, significantly reducing reproductive success [8]. Furthermore, increased glucocorticoids could negatively impact reproductive success by altering parental behaviors, such as feeding patterns of offspring, incubation consistency in birds and the timing of nesting [8,15]. Although the endocrine system is well conserved across vertebrates, the reaction to captivity may vary depending on the species [16]. There may also be physiological differences in responses among populations and individuals [18–23]. Some species may never be able to reproduce effectively in captive conditions even after generations, while in others, artificial selection can lead to changes in reproductive physiology [16]. Therefore, studying the effects of glucocorticoid levels on parental care behavior and reproductive success in captive breeding individuals and understanding the difference among populations, sex and seasonal variation can help program managers develop improved techniques to increase reproduction.

Another factor to consider in species conservation is ongoing anthropogenic-induced climate change which has led to unpredictability in weather, potentially increasing the frequency, duration, and severity of climate events. In particular, hurricanes can be a major threat to vulnerable and threatened populations of wildlife [24]. For wild populations, direct wind effects during the hurricane can kill individuals, and those that survive must deal with limited food resources [25,26], potentially experiencing altered glucocorticoid levels. Furthermore, species that coordinate their breeding with local fruiting patterns can experience reduced reproductive success after such events [26]. Although captive populations are buffered from many weather events, hurricanes may still affect them. During and after an extreme weather event, housing facilities may suffer damage and loss of power, and caretaking staff may have limited access to facilities. Food scarcity, overcrowding and isolation from normal light cycles could alter the physiology of captive individuals in these situations. Despite these potentially important effects, there are few studies examining changes in glucocorticoid levels in captive populations after hurricanes or other natural disasters [27].

The parrots and cockatoos (Order Psittaciformes) are one group for which increased knowledge of the relation of glucocorticoid levels to captive reproduction is critical. Psittaciformes have a worldwide species distribution with 419 known species, 42.2% of which are classified from near threatened to critically endangered [28]. Major threats include habitat loss and capture for the pet trade, leading to captive breeding being increasingly used as a conservation tool to protect these species [29–33]. One endangered species in which captive breeding programs play a critical role is the Puerto Rican parrot (*Amazona vittata*). This species is endemic to the island of Puerto Rico and underwent a drastic decline during the 20th century due primarily to habitat loss as the island's native forest was converted to agriculture [34]. The periodic threat of hurricanes combined with low population numbers has stalled population growth in the wild [35,36] and potentially perpetuates the genetic bottleneck observed in the species [37].

The Puerto Rican Parrot Recovery Program is an excellent model in which to test the relationship of glucocorticoids to parental care and reproductive success in captivity. The program currently consists of two captive populations (Rio Abajo and El Yunque) in different locations on the island that are closely monitored with good record-keeping practices. Climate and husbandry methods differ between each population. The urgent need for the establishment of wild populations in historical habitat demands high productivity of individuals from the captive populations for release. Birds are reintroduced at both wild populations' sites at least once a year, with the number of released birds varying among releases depending on the yearly production of the captive populations. Captive breeding is an essential part of the plan to save this species, but no study to date has investigated the relationship between glucocorticoid levels and reproductive success in this, or any other, captive breeding program for parrots.

In this study, we validate a commercial corticosterone assay and use it to examine fecal glucocorticoid metabolite levels in both captive populations of Puerto Rican parrot in 2017 and 2018 and relate these measures to reproductive success of captive individuals. Assessment of fecal glucocorticoid metabolites is non-invasive and can provide a broader picture of general circulating levels of glucocorticoids in animals [38]. We predicted that high glucocorticoid metabolite levels during the pre-breeding season and the beginning of the breeding season would have a negative relationship to reproductive success. If so, we expected to see females with lower levels of fecal glucocorticoid metabolites during the pre-breeding season produce more fertile eggs and males with lower levels of fecal glucocorticoid metabolites during the breeding season produce more chicks and fledglings. Additionally, we were provided with an opportunity to test for variation of glucocorticoid metabolites of captive individuals before and after a major environmental perturbance in the form of a hurricane. We also explore the effects of Hurricane Maria, a Category 5 hurricane that caused extensive damage to local forests and impacted the captive breeding facilities and their staff, on fecal glucocorticoid metabolites levels before and during the following breeding season, while also testing for differences among season, populations, and sexes.

2. Materials and Methods

The Puerto Rican parrot has a monogamous mating system with biparental care in which only the females incubate, and the male provides food to the female and the chicks. Chicks hatch asynchronously, and once the youngest can thermoregulate (at approximately 14 days post-hatch), the female joins the male in foraging for food for the chicks. In captivity, first eggs are laid from the end of January to the beginning of February. Hatching time is on average 26 days after egg laying and a range of 55–75 days until chicks fledge from the nest [34]. Clutch size and the number of chicks that fledge from nests vary among mated pairs, but typically 3 eggs are laid per pair each year in a single clutch [34]. The production of fledglings varies with population. Fledgling production from 2015 to 2018 in the Rio Abajo captive population was 1.02 per pair, and in the Rio Abajo wild population was 1.29 per pair. In the El Yunque captive population fledgling production was 0.71 per pair. The El Yunque wild population was extirpated by Hurricane Maria in 2017 but releases have been ongoing since summer 2019. The birds produced in captivity are used for eventual reintroductions into the wild or as new breeding individuals in the captive populations.

Each captive population houses around 220 individuals in outside cages with yearly fluctuations due to births, deaths, translocations, and reintroductions. Each year there are at least 80 to 120 individuals that form mated pairs at each captive population. Housing conditions and daily care routines differ somewhat between the two captive populations. At Rio Abajo, the captive breeding season starts between January 15–25 when the breeding pairs are each placed in their own breeding cages and ends when the last chicks fledge in July-August. During the breeding season at Rio Abajo, personnel enters the breeding areas in the morning to feed the birds and in the afternoon to collect food dishes. All nests are inspected on Monday mornings by personnel, with secondary nest checks done on some nests on Thursday or Friday mornings. After all the chicks have fledged from the nest, mated pairs are placed in retention cages in which some pairs are maintained together in the same cage and other pairs are separated with males and females placed in individual cages next to each other. All pair members at Rio Abajo remain in these cages until the beginning of the next breeding season.

At El Yunque, the captive breeding season also starts between January 15–25 and ends when the last chick fledges from the nest in July-August. El Yunque personnel feed the birds in the morning and nest checks are done throughout the week at variable times, with personnel potentially entering breeding areas multiple times during a day. At the end of the breeding season, the nest entrance is closed until the initiation of the next breeding season and pairs remain in these breeding cages year-round except for a week-long period, during which their cage is washed and prepared for the next breeding season. The activity of cleaning the breeding cages can occur at any moment between August to December.

2.1. Ethical Considerations

The Puerto Rican parrot is classified as critically endangered by the IUCN Red List [28]. This status imposes particular ethical considerations regarding the type of handling, sampling, and experimental manipulations that might be conducted in a study of this type. For example, frequent bleeding for measurement of circulating hormone levels or experimental manipulation of the HPA axis via adrenocorticotropic hormone or dexamethasone challenges, are considered overly invasive and are not permitted by the two managing authorities for this species (Puerto Rico Department of Natural and Environmental Resources and United States Fish and Wildlife Service). To minimize disruption to reproductive efforts in this species we primarily used non-invasive sampling of fecal glucocorticoids metabolites collected as part of the regular daily care routine already established at the two breeding facilities. This study was conducted under NMSU IACUC protocols 2014-030 and 2021-014, approved by the Puerto Rican Parrot Recovery Program Interagency Operational Team, supported by the management under the Puerto Rico Department of Natural and Environmental Resources and conducted under the United States Fish and Wildlife Service Permit TE125521-4.

2.2. Reproductive Information

We collected reproductive information for the pairs that were included in our study from captive population managers. This information included the production of total eggs, fertile eggs, first egg fertility, chicks, and fledglings.

2.3. Pair Selection and Breeding Stages

For this study, we randomly selected 46 pairs from all the breeding pairs in the two captive populations as follows: 21 pairs for the 2017 breeding season (Rio Abajo = 12 and El Yunque = 9), 12 for 2018 breeding season (Rio Abajo) and 13 for the intervening pre-breeding season of 2018 (Rio Abajo). We resampled 10 pairs from 2017 in the 2018 breeding seasons. At the Rio Abajo captive population, the breeding areas are clusters of breeding pairs, and each area is located in a different section of the facility. To minimize the potential effect of breeding area the pairs selected were placed in Area I. At the El Yunque captive population, all breeding cages are in close proximity compared to the structure used at the Rio Abajo captive population, for this reason, there was no potential effect of area. We define the breeding season for the captive population as the period of the year from January when pairs are placed in their breeding cage (Rio Abajo) or when the nest box is opened (El Yunque) up to the date the last chick in the population fledged. We define pre-breeding as the time before pairs are placed in their respective breeding cages (November to December in both populations); during this period, it is common to see birds showing pair-bonding behaviors (synchronized flights, duets, allopreening, copulations, allofeeding) and territorial displays.

2.4. Assay Validation

We used the DetectX Corticosterone Enzyme Immunoassay Kit (Arbor Assays, Ann Arbor MI, USA) to measure fecal glucocorticoids. To validate this assay for the measurement of corticosterone in plasma and fecal glucocorticoids in the Puerto Rican parrot, we conducted a small study with a limited subset of birds and under supervision of a veterinarian. We employed two different manipulations that were each anticipated to be stressful to the birds and collected blood and fecal samples at regular intervals after each to determine whether we could detect a rise in plasma corticosterone and fecal glucocorticoids following these stressful events. The first manipulation was to capture birds from a group flight cage and transfer them to individual cages, with subsequent collection of fecal samples to monitor fecal glucocorticoids. The second manipulation was capture and immediate blood collection followed by 30 min of restraint and then a second blood collection. Individuals were returned to cages and additional fecal samples were collected following blood collection. This manipulation allowed us to compare baseline and stress response circulating corticosterone to fecal glucocorticoid metabolites measured before and after the stressful event. Assessment of circulating corticosterone and levels to which it elevates with handling were key to assessing later secretion of glucocorticoids into fecal material and to confirm that general handling procedures are seen as stressful to this species.

To conduct our validation, we selected twelve birds from the captive population at Rio Abajo, six males and six females. All birds were two years old at the time of the study. For the first manipulation, we captured individuals with a large butterfly net from their standard group housing in a large flight cage and placed them into the smaller cages for fecal sampling. This method of capture, while standard in the facility, is thought to be stressful for the birds as it involves chasing them with the net, during which time they often produce alarm calls. We captured birds at 7:00 and then collected fecal samples every 2–3 h on the day of capture and the following 4 days to assess changes in fecal glucocorticoid levels after this putatively stressful event. For this study, we only used the day of capture and the following day after capture. If this event caused an increase in circulating glucocorticoids, as anticipated, then we also expected to see a transitory rise in fecal glucocorticoids later on the day of capture compared to samples collected in following days.

For the second part of the validation, we collected blood samples and performed a capture restraint on the same 12 birds after two weeks of individual housing in the same cages. Approximately 0.3 cc of whole blood was collected from the jugular vein from each bird between 8:00 am and 9:30 am. Samples were collected in under 3 min from approaching each cage, and again after 30 min of restraint in a small wooden box to measure baseline and stress-response circulating corticosterone. The blood samples were centrifuged, and plasma extracted and frozen immediately. We collected fecal samples for our baseline values before the restraint protocol was conducted. We then continued to collect fecal samples from birds for the remainder of the day after blood samples were collected. For the stress response analysis, we used from each individual the fecal sample with the highest value after restraint collected between 13:30 to 18:00.

2.5. Fecal Sampling and Analysis

For our main study, we collected fecal samples from males in both the breeding and pre-breeding seasons and from females during the pre-breeding season only. Pre-breeding (November to January) samples were collected only at Rio Abajo where the members of each pair are separated, which allowed us to distinguish which individual produced each sample. Breeding season samples were collected from males in both populations due to their presence outside nesting cavities while females incubate. To collect the fecal samples, we placed a clean PVC plank under a perch where individuals roosted at night after sunset between 7:30 to 8:00 PM and collected the plank before the following sunrise between 5:30 to 6:00 AM. The droppings collected at sunrise were thus collected no more than 10 h after defecation and when no rainfall occurred overnight. A study examining the effects of environmental changes on fecal glucocorticoid metabolites found no effect of room temperature for 12h on fecal glucocorticoid metabolite concentration but it did find effects of rainfall on the concentration of fecal glucocorticoid metabolites [39]. Samples were collected in a 2ml microcentrifuge tube and then stored at −20 °C until drying and analysis.

Fecal samples were dried using a gravity convection oven (Fisher Scientific, No: 3511FS) preheated to 90 °C, the samples remained in the oven for at least 2 h until fully desiccated. A study examining the effect of this procedure on glucocorticoid metabolites levels found that samples that are frozen and then dried in conventional ovens do not change significantly in glucocorticoid metabolite concentration [40]. After drying the samples were pulverized and centrifuged, then stored at −20 °C until analysis. Rainfall and small amounts of fecal material in the samples of some individuals, lead to some individuals having low sample collection during the pre-breeding season of 2018. We randomly selected 3 samples per individual during the pre-breeding and a range of 3 to 15 samples per individual during the breeding seasons that had ≥0.2 g of dry feces to measure glucocorticoids metabolites. We extracted glucocorticoid metabolites from fecal samples following the DetectX Steroid Extraction Protocol (Arbor Assay), after drying, 2 mL of ethanol was added to the solid to initiate the extraction of the glucocorticoids. We did minor adjustments to the centrifugation step during which extracted samples were centrifuged at 4000 rpm for 20 min. We used a SpeedVac (Eppendorf) centrifuge to dry down extracted samples, after which samples were stored at −20° until they were assayed (within 24 h), sample recovery was assumed to be 100% in subsequent analyses. We adjusted the kit recommended reconstitution procedure slightly and dissolved extracted samples in 50 µL of ethanol followed by 600 µL of Assay Buffer. Fecal glucocorticoid metabolite levels were then analyzed using the DetectX® Corticosterone Enzyme Immunoassay kit (EIA, Arbor Assays, Inc., #K014-H5). This kit has high affinity for corticosterone (100%), with much lower cross-reactivity for other glucocorticoids and their metabolites (Desoxycorticosterone—12.30%, Tetrahydrocorticosterone—0.76%, Aldosterone—0.62%, Cortisol—0.38%, Progesterone—0.24%, Dexamethasone—0.12%, Corticosterone-21-Hemisuccinate—<0.1%, Cortisone—<0.08%, Estradiol—<0.08%); since in fecal samples there is low presence of corticosterone, we report on glucocorticoid metabolites as opposed to corticosterone.

2.6. Statistical Analysis

Since we analyzed samples in multiple plates and years we corrected our glucocorticoid estimates using a correction factor applied to all samples to account for inter-assay variation that could be linked to variation in kits or lab conditions as glucocorticoid analyses spanned across years [as in 41]. We used the estimated levels of glucocorticoid metabolites from the fourth point of the standard curve as a basis for our correction factor. We used the grand mean for all point 4 data points from all standard curves and divided this by the mean point 4 standard curve point for each individual plate to develop a correction factor [41–47]. We then multiplied each hormone value on each plate by the corresponding correction factor. Inter-assay variation was measured as the coefficient of variation (N = 21, Mean = 1241.44, SD = 131.78, CV = 0.106 equaling an inter-assay variation of 10.6%, intra-assay variation range = 3.22–9.09%). The adjusted glucocorticoid values were then log-transformed for subsequent analyses. For all analyses, we used the mean glucocorticoid values obtained for each individual across all measures within a season.

From the full set of data available, we used a subset to address different questions depending on the data available. We used twelve males for the 2017 and 2018 breeding seasons, and ten for the 2018 pre-breeding from the Rio Abajo captive population. From the Rio Abajo captive population, nine males had data across all seasons, two males had data only for the 2017 breeding season, only one male had data for pre-breeding and breeding season of 2018. We had data from nine males for 2017 breeding season from the El Yunque captive population. For females, we had data from only eleven females from the Rio Abajo captive population from the 2018-pre-breeding season. After a Shapiro Wilk Test we Log transform our glucocorticoid values for all the analyses. For the determination of the relation of reproductive success measures and glucocorticoid metabolite levels at each season, we used only males (see Table 1) from the Rio Abajo captive population and performed a Spearman correlation analysis. To compare glucocorticoid metabolite levels between the pre-breeding and two breeding seasons we used an ANOVA followed by Tukey post hoc tests using only the males (14) from the Rio Abajo captive population because this was the only population sampled over all three seasons. Since the breeding season of 2017 occurred before Hurricane Maria and the breeding season of 2018 occurred after, we used the results from the previous analysis to compare levels before and after Hurricane Maria. We performed a paired t-test to investigate differences among males (10) and females (11) from the Rio Abajo captive population during the pre-breeding season. For population differences in glucocorticoid metabolite levels, we compared the data from the breeding season of 2017 of only males (21) from each population using a paired t-test. We used a logistic regression to determine the relationship of glucocorticoids metabolite levels during the pre-breeding to first egg fertility. Statistical analysis was performed using the JMP statistical software package, version 14.0.0 (SAS Institute, Inc. 2018, Cary, NC, USA).

Table 1. Relationship between male fecal glucocorticoid metabolite levels and reproductive success variables †.

Reproductive Success	Season	n	r	p
Total eggs	Breeding 2017	12	0.2587	0.4169
	Pre-breeding 2018	10	0.4411	0.2019
	Breeding 2018	12	−0.7255	**0.0076**
Fertile eggs	Breeding 2017	12	−0.0431	0.8942
	Pre-breeding 2018	10	−0.0187	0.9591
	Breeding 2018	12	−0.6538	**0.0211**
Chicks	Breeding 2017	12	−0.4139	0.1810
	Pre-breeding 2018	10	0.0781	0.8301
	Breeding 2018	12	−0.3897	0.2105
Fledglings	Breeding 2017	12	−0.4193	0.1810
	Pre-breeding 2018	10	0.0855	0.8143
	Breeding 2018	12	−0.3528	0.2607

† All tests Spearman correlations.

3. Results

3.1. Validation of Corticosterone Assay

In the first potentially stressful event of our validation study (capture and placement in cages), we saw a sharp but transitory rise in estimated fecal glucocorticoid metabolite levels in the 12 birds following initial capture from group housing and movement to individual housing. Although we were not able to obtain fecal samples from all individuals immediately upon capture, all samples collected in the 6 h following capture had estimated fecal glucocorticoid metabolite levels under 20,000 pg/mg (Figure 1a). After 6 h there was a general rise in CORT levels that persisted for 30 h after capture, at which time levels decreased (Figure 1a). In our second stressful event (capture and restraint) we saw a significant rise in circulating plasma CORT from baseline to stress-response levels following 30 min of a standard capture restraint protocol for the whole group (Figure 1b, paired *t*-test, df = 20.12, t = 5.92, p = 0.001). When we compared levels based on sex, we found that at time zero there was no difference in plasma glucocorticoid levels between the sexes (paired *t*-test, df = 7.41, t = 0.02, p = 0.987) but after restraint males showed higher plasma glucocorticoid levels than females (paired *t*-test, df = 9.77, t = 2.76, p = 0.021). Both sexes demonstrated a restraint response with an increase of plasma glucocorticoids (paired *t*-test males, df = 11.04, t = 5.59, p = 0.0002 and for females df = 4.93, t = 2.86, p = 0.036). There was a concomitant rise in estimated fecal glucocorticoid levels in fecal samples collected on the same day before and after the capture restraint (Figure 1c, paired *t*-test, df = 13.09, t = 4.76, p = 0.0004). Comparing the sexes, both sexes demonstrated a significant increment of fecal glucocorticoids after the restraint protocol (paired *t*-test for males, df = 5.89, t = 3.48, p = 0.0136 and for females df = 5.47, t = 6.59, p = 0.0009). Before the restraint protocol females had higher fecal glucocorticoid levels than males (paired *t*-test, df = 10.00, t = −3.82, p = 0.003) but there was no significant difference between the sexes after the restraint protocol (paired *t*-test, df = 6.83, t = −1.46, p = 0.1998). As a final assessment concerning the ability of this assay to detect fecal glucocorticoids in this species, we used extra, mixed fecal samples from individuals that we then measured as split into two aliquots (n = 10). These split aliquots were analyzed on the same plate with final calculated levels of fecal glucocorticoid metabolites compared for similarity. Levels from split samples were highly correlated, supporting the reproducibility of this assay and its ability to detect fecal glucocorticoid metabolites for this species (Pearson Correlation, r = 0.963, n = 10, p < 0.0001). These results validate that the Arbor Assay Detect-X ELISA Assay is effective in detecting both CORT in plasma and feces of this species. No birds were harmed during this procedure and veterinary monitoring detected no adverse impacts on their health.

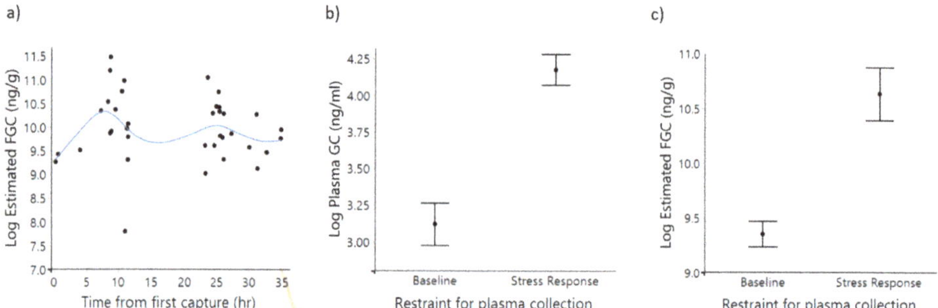

Figure 1. Glucocorticoids levels in fecal and plasma samples. A total of 12 individuals were used for this test with 6 males and 6 females. (**a**) Fecal glucocorticoids fluctuation same day of capture up to 40 h after first capture. Multiple samples per individual, with a total of 37 samples. (**b**) Plasma glucocorticoids levels before and after bleeding, n = 12, 6 males and 6 females. Figure shows means ± SE. (**c**) Fecal glucocorticoids levels before and after restraining for blood capture, n = 12, 6 males and 6 females. Figure shows means ± SE.

3.2. Glucocorticoid Metabolite Levels and Measures of Reproductive Success

We assessed how male fecal glucocorticoid metabolite levels in different seasons related to reproductive success. For the 2018 breeding season we found a significant relationship between fecal glucocorticoid metabolite levels in males and measures of reproductive success in the egg stage (total eggs and fertile eggs), but not at the post-egg stage (chicks and fledglings) (Table 1). Fecal glucocorticoid metabolite levels had a negative relationship with both the number of eggs laid and the number of fertile eggs (Figure 2). In contrast, we did not find any relationship between male fecal glucocorticoid metabolite levels during either the 2017 breeding season or the pre-breeding season of 2018 with measurements of reproductive success at the egg stage (total eggs laid, number of fertile eggs) or chick stage (number of chicks hatched, number of fledglings).

In females, fecal glucocorticoid metabolite levels during the pre-breeding stage of 2018 were positively related to the total number of eggs laid but were not related to the number of fertile eggs, chicks produced, or fledglings (Table 2, Figure 2). There was no significant relationship between fecal glucocorticoids metabolite levels of either males or females during the pre-breeding stage in 2018 to the fertility of the first egg (Female df = 1, $r^2 = 0.1748$, $X^2 = 2.1361$, $p = 0.1439$; Male, df = 1, $r^2 = 0.0727$, $X^2 = 0.6927$, $p = 0.4052$).

Table 2. Relationship between fecal glucocorticoid metabolite levels in females during the pre-breeding season and reproductive success variables †.

Reproductive Success	*n*	r	*p*
Total eggs	11	0.6404	**0.0338**
Fertile eggs	11	0.3202	0.3371
Chicks	11	0.2741	0.4147
Fledglings	11	0.2513	0.4561

† All tests Spearman correlations.

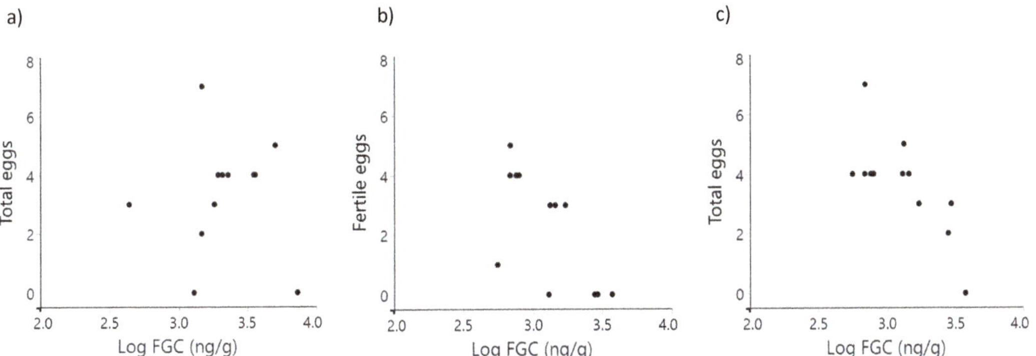

Figure 2. Significant relationships of fecal glucocorticoid metabolites to reproductive success measures. (**a**) Females showed a positive relationship of fecal glucocorticoid metabolites during the pre-breeding to the total eggs laid by female in the following breeding season (2018). Males have a negative relationship of fecal glucocorticoid metabolites to (**b**) total eggs and (**c**) total fertile eggs laid in the following breeding season (2018).

3.3. Hurricane Maria and Seasonal Variation

For the Rio Abajo captive population, we did not identify differences between the 2017 and 2018 breeding seasons (d.f = 16.53, t = −1.018, $p = 0.323$), which represented pre- and post-Hurricane Maria, respectively (Figure 3). We found a difference in male fecal glucocorticoid metabolite levels between seasons (df = 2, F = 7.015, $p = 0.003$; Figure 3). Here, male birds had higher fecal glucocorticoid metabolite levels during the pre-breeding season of 2018 (N = 10, mean ± SE = 3.83 ± 0.15 ng/g) compared to the breeding season of 2017 (N = 12, mean ± SE = 3.28 ± 0.13 ng/mL) and the breeding season of 2018 (N = 12, mean ± SE = 3.108 ± 0.13 ng/mL). After a Tukey HSD post-hoc comparison, pre-breeding levels in 2018 were significantly different from the 2017 breeding season (df = 12.821, t = −3.795, $p = 0.0023$), as well as the 2018 breeding season (df = 19.144, t = 2.354, $p = 0.029$).

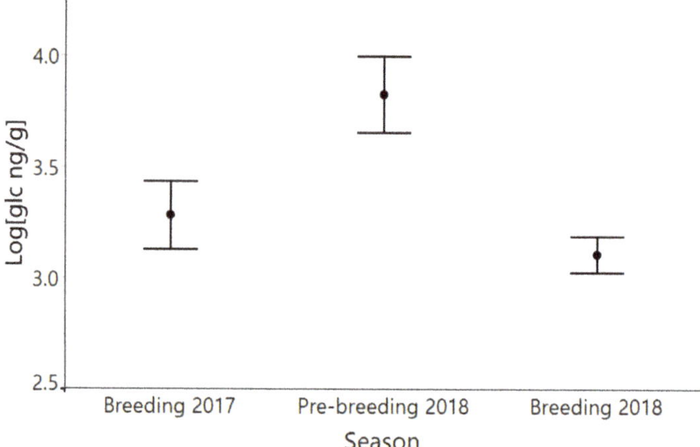

Figure 3. Seasonal variation in fecal glucocorticoid metabolite of males at Rio Abajo. During pre-breeding 2018 fecal glucocorticoid metabolites were higher relative to both the 2017 and 2018 breeding seasons, which did not differ from each other. The 2017 breeding season occurred prior to Hurricane Maria while the pre-breeding and breeding season of 2018 occurred after Hurricane Maria. Graphs show means ± SE.

3.4. Glucocorticoid Metabolite Levels Differed between Sexes and Populations

We found that the fecal glucocorticoid metabolites ranged from 10.988 to 133.750 ng/g for males and 14.057 to 40.606 ng/g for females. We compared pre-breeding season fecal glucocorticoid metabolite levels in males and females and found that males had higher levels (mean ± SE = 3.827 ± 0.17 log ng/g) than females (mean ± SE = 3.284 ± 0.86 log ng/g; df = 1, t = 2.836, p = 0.00138; Figure 4). In addition, we found a trend towards males' glucocorticoid levels differing between populations during the 2017 breeding season (Figure 5). Males at the Rio Abajo captive population had lower glucocorticoid metabolite levels (mean ± SE = 3.285 ± 0.15 log ng/g) than males at the El Yunque captive population (mean ± SE = 3.647 ± 0.11 log ng/g), but this trend was not statistically significant (df = 1, t = −1.895, p = 0.0737).

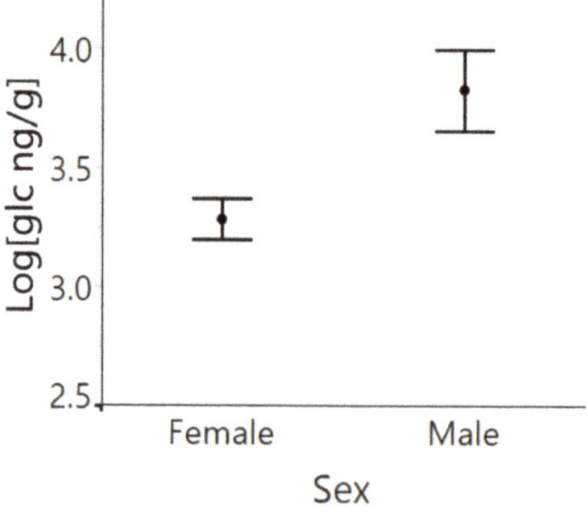

Figure 4. Sex difference in fecal glucocorticoid metabolites during the pre-breeding 2017–2018. Males had higher fecal glucocorticoid metabolite levels than females during the pre-breeding season. These samples include individuals from Rio Abajo captive populations (Rio Abajo males = 10, Rio Abajo females = 11). Graphs shows means ± SE.

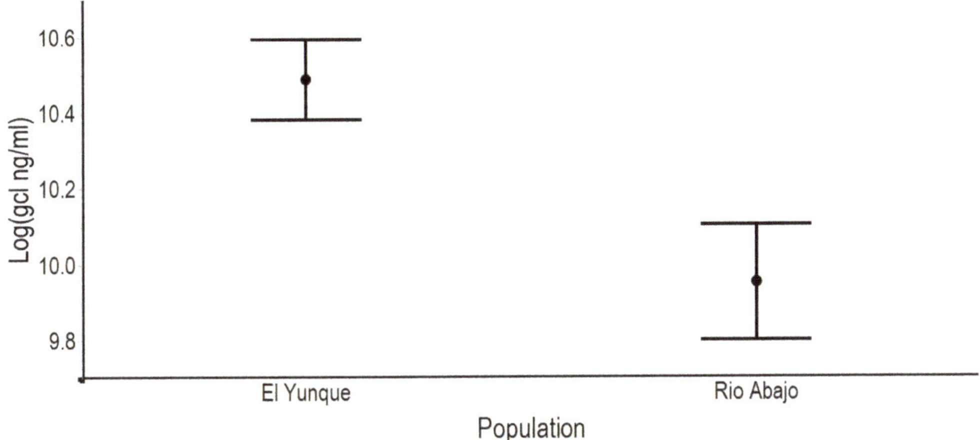

Figure 5. Populations comparison in fecal glucocorticoid metabolites between males in the Rio Abajo (12) and El Yunque (9) captive populations during the 2017 breeding season. Graphs show means ± SE.

4. Discussion

Captive breeding can be an important tool for many conservation efforts. Understanding how a threatened species is affected by captive conditions is critical for conservation success [16]. We evaluated fecal glucocorticoid metabolite levels of captive populations of Puerto Rican parrots using a newly validated commercial corticosterone assay and related this to reproductive success over two breeding seasons and the intervening non-breeding season. We expected to see a negative relationship between the reproductive success of captive breeding pairs and fecal glucocorticoid metabolite levels of males and females during pre-breeding, and glucocorticoid metabolite levels of males in the breeding seasons. Glucocorticoids have been shown to negatively impact reproductive behavior and success in other avian species [8]. Our results showed a negative relationship of male fecal glucocorticoid metabolite levels to total eggs and fertile eggs during the breeding season of 2018 (Figure 2), while female fecal glucocorticoid metabolite levels during the pre-breeding season of 2018 showed a positive relationship to total eggs laid. We also found differences in fecal glucocorticoid metabolite levels between the sexes and seasonal variation in males. Below we discuss the implications of these results and put them in the context of similar work on the physiology of captive populations.

4.1. Fecal Glucocorticoid Metabolite Levels and Reproductive Success

Previous research has found that high levels of glucocorticoids can downregulate reproductive efforts [8]. In a previous study of captive breeding Puerto Rican parrots [48] visual inspection of adrenal gland and testes showed that in some males the adrenal gland was hypertrophied and the testes were small. It was suggested that these findings could be due to high levels of stress [48], however, circulating or excreted glucocorticoids were not assessed. Here we showed that a standard capture restraint protocol produced a rise in circulating corticosterone that was mirrored by a rise in fecal glucocorticoid metabolites. Using this validated assay, we found that male fecal glucocorticoid metabolites during the breeding season of 2018 had a negative relationship to total eggs and fertility of eggs providing support to the previous study's findings related to the testes [48]. For females during the pre-breeding season of 2018, we found a positive relationship of fecal glucocorticoid metabolites levels with total eggs laid. However, our results, for either sex, do not show any relation of fecal glucocorticoid metabolites to the number of chicks and fledglings produced.

In males, increases in glucocorticoids have been documented to suppress production of testosterone [49–51] and luteinizing hormone [49]. Suppression of testosterone and luteinizing hormone can negatively affect reproductive behaviors. On the other hand, in a study of wild populations of house wrens (*Troglodytes aedon*), females whose glucocorticoids levels were artificially elevated before breeding produced heavier eggs and had higher nestling feeding rates, significantly increasing chick body mass [52]. Elevated glucocorticoids levels in females may mobilize energy to prepare for the breeding season. Together, these results suggest that, in some species, males and females are differently impacted by changes in glucocorticoid levels.

The relationship between reproductive success and glucocorticoids may also be affected by a species' natural history [1,53]. Short live species may have mechanisms that suppress the effects of high glucocorticoids while long-lived species may experience selection for survival over reproduction [53]. Like many parrots, the Puerto Rican parrot is a long-lived species and may live for up to 37 years in captivity [54]. Under natural conditions, in long-lived species, individual survival may be prioritized before reproduction and males, in particular, may not have evolved the capacity to modulate the effects of frequent energetic challenges experienced in captivity. On the other hand, females may evolve mechanisms to modulate any negative effects of elevated glucocorticoid just prior to egg-laying [55] and this may be held over into captivity. Additionally, egg production is an energetically demanding process, in which the female must place energetically rich resources into the production of eggs and may use glucocorticoids for the mobilization of that energy. This may explain the positive relationship observed between fecal glucocorticoid metabolite levels in the pre-breeding season and egg production in females seen in this study.

4.2. Seasonal Variation

Many animals have seasonal variation in hormone levels in the wild [53,56–59] but it remains unclear whether that variation is retained under captive conditions. Because of the outdoor housing and seasonal breeding cycle of the Puerto Rican parrot in the captive populations assessed here, we expected to see variation among the breeding and pre-breeding seasons in fecal glucocorticoid metabolite levels. We found that male fecal glucocorticoid metabolite levels in the Puerto Rican parrot were higher during the pre-breeding than in the breeding season.

Seasonal variation has also been documented in captive red-tailed parrots (*Amazona brasiliensis*), with mean fecal glucocorticoid metabolite concentrations highest just before breeding [60]. Research in wild populations of northern cardinals (*Cardinalis cardinalis*) has shown that females and males have higher plasma glucocorticoid corticosterone levels in the pre-breeding compared to the breeding season [47]. In northern cardinals, it was suggested that high glucocorticoid corticosterone levels during the non-breeding season could be an adaptation for energy needed to survive cold conditions. In the Puerto Rican parrot, each pair displays territorial behaviors year-round defending their nesting cavity with a peak in the pre-breeding season which could explain higher fecal glucocorticoid metabolite levels in males at that time [34], elevated plasma glucocorticoids could provide the energy needed for expensive territorial activities in this species.

Puerto Rican parrots at Rio Abajo are housed outdoors and exposed to normal daylight changes, temperature changes and rainfall, and social activity of wild parrots, all of which may affect glucocorticoids levels between seasons [1]. In European starlings (*Sturnus vulgaris*) individuals of both sexes placed in indoor housing experienced delayed reproductive behaviors while outdoor-housed individuals initiated breeding earlier [61]. Pairs of European starlings housed indoors had significantly lower sex steroids and higher expression of gonadotropin inhibitory hormone (GnIH). It is possible that the year-round outdoor housing experienced by these birds results in individuals having hormonal seasonal variation similar to what would be seen in wild individuals.

Alternatively, birds at Rio Abajo are housed in different cages during the pre-breeding versus the breeding season; these conditions may prevent the expression of the full range of natural behaviors during this period and may in turn alter glucocorticoid levels [16,61]. The cages used in the pre-breeding period are smaller than breeding cages and they are placed in closer proximity to other individuals. The proximity of the cages provokes frequent territorial displays (B. Ramos-Guivas, pers obs), potentially at a higher rate than when birds are in their breeding cages. This could result in males having higher glucocorticoid levels during the pre-breeding than during the breeding season. In addition, closer proximity to human activity and capturing for medical antiparasitic treatments (twice over two weeks) during the pre-breeding season may contribute to the seasonal glucocorticoid variations observed.

4.3. Glucocorticoid Levels Do Not Differ between Breeding Seasons, despite Hurricane Maria

The breeding seasons of 2017 and 2018 occurred pre and post-Hurricane Maria. At the Rio Abajo captive population, we found no significant differences in fecal glucocorticoid metabolite levels between the two seasons. Although initially surprising, it is important to note that captive conditions were relatively stable compared to those experienced by wild populations. Many wild populations of animals suffer increased mortality both during a hurricane (due to climatic conditions), and afterward due to increased exposure to predators and reduced food availability [26,62]. Indeed, the wild population of the Puerto Rican parrot at El Yunque, which was only 31 birds before Hurricane Maria, was totally extirpated by the hurricane. New releases started in 2019 to re-establish this wild population. The wild population at Rio Abajo fared somewhat better but is thought to have declined from 134 to 110 birds as a result of direct effect of the hurricane. Secondary effects as scarcity of food and exposure to predators reduced the population even more in the following months after the hurricane. In contrast, the Rio Abajo captive population was sheltered from the direct impacts of the storm in a hurricane shelter and had no changes in food availability.

4.4. Males Show Higher Fecal Glucocorticoid Metabolite Levels than Females

Differences between the physiology of males and females are common, therefore we expected to see a difference in fecal glucocorticoid metabolite levels between males and females during the pre-breeding season even when faced with similar captivity stimuli (variation linked to preparation for breeding). We found that males had significantly higher fecal glucocorticoid metabolite levels than females during the pre-breeding. It could be that lower levels of glucocorticoids in females compared to males represent suppression of glucocorticoid levels during the pre-breeding in preparation for egg-laying [60]. Alternatively, higher energetic demands for males than females during the pre-breeding may be due to increases in courtship or territorial behavior shown by males, which may increase glucocorticoid levels [63,64].

4.5. Differences between Captive Populations

For our study, we collected fecal samples from both extant captive populations during the 2017 breeding season. Because of the difference in location and management techniques we expected to see differences in the glucocorticoid levels between the populations. We found that fecal glucocorticoid metabolite levels among males were somewhat higher at the El Yunque population than at the Rio Abajo population, although the difference was not quite statistically significant. Housing conditions, climate and daily routines differ between the two captive populations. At the El Yunque captive population, the breeding cages are more exposed to personnel activity, so recurrent perturbations by people walking by may stimulate continuous energetic challenges [65]. In addition, the nest checks and daily feeding are not performed in a strict schedule at El Yunque captive population adding an element of unpredictability for individuals. Recurrent intrusion to breeding areas can maintain glucocorticoids levels above baseline, eventually impacting reproductive success [65]. At the Rio Abajo captive population, the breeding cages are located behind vegetation that may serve as a screen from personnel, potentially reducing glucocorticoid levels. Another difference between the populations is the daily feeding routine. At El Yunque, feeding time lasts longer than at Rio Abajo. The longer presence of personnel inside the breeding areas may increase baseline glucocorticoids levels. High levels of glucocorticoids during the breeding season could negatively affect reproduction.

5. Conclusions

Fecal glucocorticoid metabolite analysis provides an opportunity to study species in a minimally invasive fashion. Our results provide insight into the relationship of fecal glucocorticoids metabolites to reproductive success under normal daily routines in captivity. Although it has been reported in previous studies that captive individuals have a more attenuated response to energetic challenges than wild individuals [17,66,67], captive individuals can still be affected by high glucocorticoids levels [68–70]. The negative relationship of fecal glucocorticoid metabolite levels in males to total eggs laid and total fertile eggs may indicate a susceptibility to anthropogenic activities. For endangered species programs, strategies should be developed to reduce perturbations and regularize their occurrence during key periods of the year to minimize the potential negative effects on reproduction. The differences between populations in glucocorticoid levels suggest that managers of captive breeding species should carefully evaluate factors that may be driving these differences (e.g., housing placement and the frequency and predictability of anthropogenic activities during critical stages of the breeding season). Future research should examine the role of glucocorticoids in reproduction of captive populations of other endangered species, and where possible, compared to wild populations of these species.

Author Contributions: Conceptualization, B.R.-G. and T.F.W.; methodology, B.R.-G., T.F.W. and J.M.J.; validation, B.R.-G., J.M.J.; formal analysis, B.R.-G. and J.M.J.; investigation, B.R.-G., J.M.J.; resources, B.R.-G., T.F.W. and J.M.J.; data curation, B.R.-G. and J.M.J.; writing—original draft preparation, B.R.-G. and T.F.W.; writing—review and editing, T.F.W. and J.M.J.; visualization, B.R.-G.; supervision, T.F.W.; project administration, B.R.-G. and T.F.W.; funding acquisition, B.R.-G. and T.F.W. All authors have read and agreed to the published version of the manuscript.

Funding: This work was supported by the World Parrot Trust (grant date 2015, 2016 and 2018).

Institutional Review Board Statement: All research activities and animal handling were carried out in full compliance and accordance with the NMSU IACUC protocols 2014-030 and 2021-014, the Puerto Rican Parrot Recovery Program Interagency Operational Team, supported by the management under the Puerto Rico Department of Natural and Environmental Resources and conducted under the United States Fish and Wildlife Service Permit TE125521-4.

Informed Consent Statement: Not applicable.

Data Availability Statement: Data availability in this study are available upon request from corresponding author.

Acknowledgments: We thank Marisel López, Jafet Vélez, Iris Rodriguez-Carmona, Gabriel Benitez-Soto, Limary Ramírez-Esquilín, Arelys Johnson-Camacho from the USFWS for their cooperation in collecting the samples at the Iguaca facilities at El Yunque National Forest. We thank Miguel A. García-Bermudez, Ricardo Valentín de la Rosa, Tanya Martínez-Ramirez, Jong Piel Banchs-Plaza and Roseanne Medina-Miranda from the PR DNER for their cooperation during the study. We thank Elisa Boyd for the help provided during the preparation of the samples, Dominique Hellmich, Angela Medina and Grace Smith Vidaure for sharing their knowledge all members of the Wright Lab at the New Mexico State University for the help provided during the analysis process. We thank Antonio Rivera-Guzman, Tomás Medina-Cortés, Magaly Massanet, Valeria Perez-Marrufo, Natasha L. Perez-Nieto, Mariana Travieso-Diffoot, Gabriela Fonseca-Cordero, Adriana Pérez-Vega, Alondra Villalba, Echo Ruff for their labor during the validation.

Conflicts of Interest: The authors declare no conflict of interest.

Permits and Approval Statement: The study was approved by the Puerto Rico Department of Natural and Environmental Resources Puerto Rican Parrot Program coordinator leadership and by the US Fish and Wildlife Puerto Rican Parrot program coordinator.

References

1. Romero, L.M.; Wingfield, J.C. *Tempest, Poxes, and Predators, and People: Stress in Wild Animals and How They Cope*; Oxford University Press: Oxford, UK, 2016; Volume 1.
2. Boonstra, R. Reality as the leading cause of stress: Rethinking the impact of chronic stress in nature. *Funct. Ecol.* **2013**, *27*, 11–23. [CrossRef]
3. Romero, L.M.; Reed, J.M.; Wingfield, J.C. Effects of weather on corticosterone responses in wild free-living passerine birds. *Gen. Comp. Endocrinol.* **2000**, *118*, 113–122. [CrossRef]
4. Busch, D.S.; Hayward, L.S. Stress in a conservation context: A discussion of glucocorticoid actions and how levels change with conservation-relevant variables. *Biol. Conserv.* **2009**, *142*, 2844–2853. [CrossRef]
5. McEwen, B.S.; Wingfield, J.C. The concept of allostasis in biology and biomedicine. *Horm. Behav.* **2003**, *43*, 2–15. [CrossRef]
6. Sapolsky, R.M.; Romero, L.M.; Munck, A.U. How do glucocorticoids influence stress responses? Integrating permissive, suppressive, stimulatory, and preparative actions. *Endocr. Rev.* **2000**, *21*, 55–89. [CrossRef]
7. Thiel, D.; Jenni-Eiermann, S.; Palme, R.; Jenni, L. Winter tourism increases stress hormone levels in the Capercaillie Tetrao urogallus. *Ibis* **2011**, *153*, 122–133. [CrossRef]
8. Angelier, F.; Clément-Chastel, C.; Welcker, J.; Gabrielsen, G.W.; Chastel, O. How does corticosterone affect parental behaviour and reproductive success? A study of prolactin in black-legged kittiwakes. *Funct. Ecol.* **2009**, *23*, 784–793. [CrossRef]
9. Love, O.P.; Breuner, C.W.; Vézina, F.; Williams, T.D. Mediation of a corticosterone-induced reproductive conflict. *Horm. Behav.* **2004**, *46*, 59–65. [CrossRef] [PubMed]
10. Bowers, E.K.; Thompson, C.F.; Bowden, R.M.; Sakaluk, S.K. Posthatching Parental Care and Offspring Growth Vary with Maternal Corticosterone Level in a Wild Bird Population. *Physiol. Biochem. Zool.* **2019**, *92*, 496–504. [CrossRef]
11. Lin, H.; Decuypere, E.; Buyse, J. Oxidative stress induced by corticosterone administration in broiler chickens (Gallus gallus domesticus): 1. Chronic exposure. *Comp. Biochem. Physiol. B Biochem. Mol. Biol.* **2004**, *139*, 737–744. [CrossRef]
12. Shi, L.; Zhang, J.; Lai, Z.; Tian, Y.; Fang, L.; Wu, M.; Xiong, J.; Qin, X.; Luo, A.; Wang, S. Long-term moderate oxidative stress decreased ovarian reproductive function by reducing follicle quality and progesterone production. *PLoS ONE* **2016**, *11*, e0162194. [CrossRef] [PubMed]
13. Earnhardt, J.M. The Role of Captive Populations in Reintroduction Programs. In *Wild Mammals in Captivity: Principles and Techniques for Zoo Management*; The University of Chicago Press: Chicago, IL, USA, 2010; pp. 268–280.
14. Earnhardt, J.; Vélez-Valentín, J.; Valentin, R.; Long, S.; Lynch, C.; Schowe, K. The puerto rican parrot reintroduction program: Sustainable management of the aviary population. *Zoo Biol.* **2014**, *33*, 89–98. [CrossRef] [PubMed]
15. Gilby, A.J.; Mainwaring, M.C.; Rollins, L.A.; Griffith, S.C. Parental care in wild and captive zebra finches: Measuring food delivery to quantify parental effort. *Anim. Behav.* **2011**, *81*, 289–295. [CrossRef]
16. Mason, G.J. Species differences in responses to captivity: Stress, welfare and the comparative method. *Trends Ecol. Evol.* **2010**, *25*, 713–721. [CrossRef] [PubMed]
17. Cabezas, S.; Carrete, M.; Tella, J.L.; Marchant, T.A.; Bortolotti, G.R. Differences in acute stress responses between wild-caught and captive-bred birds: A physiological mechanism contributing to current avian invasions? *Biol. Invasions* **2013**, *15*, 521–527. [CrossRef]
18. Adkins-Regan, E. Do hormonal control systems produce evolutionary inertia? *Philos. Trans. R. Soc. B Biol. Sci.* **2008**, *363*, 1599–1609. [CrossRef]
19. Ball, G.F.; Balthazart, J. Individual variation and the endocrine regulation of behaviour and physiology in birds: A cellular/molecular perspective. *Philos. Trans. R. Soc. B Biol. Sci.* **2008**, *363*, 1699–1710. [CrossRef]
20. Dawson, A. Control of the annual cycle in birds: Endocrine constraints and plasticity in response to ecological variability. *Philos. Trans. R. Soc. B Biol. Sci.* **2008**, *363*, 1621–1633. [CrossRef]
21. Nelson, B.F.; Daunt, F.; Monaghan, P.; Wanless, S.; Butler, A.; Heidinger, B.J.; Newell, M.; Dawson, A. Protracted treatment with corticosterone reduces breeding success in a long-lived bird. *Gen. Comp. Endocrinol.* **2015**, *210*, 38–45. [CrossRef]
22. Wingfield, J.C. The concept of allostasis: Coping with a capricious environment. *J. Mammal.* **2005**, *86*, 248–254. [CrossRef]
23. Wingfield, J.C. Environmental endocrinology: Insights into the diversity of regulatory mechanisms in life cycles. *Integr. Comp. Biol.* **2018**, *58*, 790–799. [CrossRef] [PubMed]
24. Lance, V.A.; Elsey, R.M.; Butterstein, G.; Trosclair, P.L.; Merchant, M. The effects of hurricane Rita and subsequent drought on alligators in Southwest Louisiana. *J. Exp. Zool. Part A Ecol. Genet. Physiol.* **2010**, *313*, 106–113. [CrossRef]
25. Waide, R.B. The effect of hurricane Hugo on bird populations in the Luquillo Experimental Forest, Puerto Rico. *Biotropica* **1991**, *23*, 475–480. [CrossRef]
26. Wunderle, J.M. Pre- and Post-Hurricane fruit availability: Implications for Puerto Rican Parrots in the Luquillo Mountains. *Caribb. J. Sci.* **1999**, *35*, 249–264.
27. Anestis, S.F. Urinary cortisol responses to unusual events in captive chimpanzees (Pan troglodytes). *Stress* **2009**, *12*, 49–57. [CrossRef] [PubMed]

28. IUCN. The IUCN Red List of Threatened Species. Available online: https://www.iucnredlist.org/search/stats?taxonomies=22672853&searchType=species (accessed on 10 August 2020).
29. Berkunsky, I.; Quillfeldt, P.; Brightsmith, D.J.; Abbud, M.C.; Aguilar, J.M.R.E.; Alemán-Zelaya, U.; Aramburú, R.M.; Arce Arias, A.; Balas McNab, R.; Balsby, T.J.S.; et al. Current threats faced by Neotropical parrot populations. *Biol. Conserv.* **2017**, *214*, 278–287. [CrossRef]
30. Dahlin, C.R.; Blake, C.; Rising, J.; Wright, T.F. Long-term monitoring of Yellow-naped Amazons (Amazona auropalliata) in Costa Rica: Breeding biology, duetting, and the negative impact of poaching. *J. Field Ornithol.* **2018**, *89*, 1–10. [CrossRef]
31. Olah, G.; Butchart, S.H.M.; Symes, A.; Guzmán, I.M.; Cunningham, R.; Brightsmith, D.J.; Heinsohn, R. Ecological and socio-economic factors affecting extinction risk in parrots. *Biodivers. Conserv.* **2016**, *25*, 205–223. [CrossRef]
32. Grajal, A. The Neotropics (Americas). In *Parrots. Status Surveys and Conservation Action Plan 2000–2004*; IUCN: Gland, Switzerland; Cambridge, UK, 2000; pp. 98–151, ISBN 2831705045.
33. Wright, T.F.; Toft, C.A.; Enkerlin-Hoeflich, E.; Gonzalez-Elizondo, J.; Albornoz, M.; Rodríguez-Ferraro, A.; Rojas-Suárez, F.; Sanz, V.; Trujillo, A.; Beissinger, S.R.; et al. Nest poaching in Neotropical parrots. *Conserv. Biol.* **2001**, *15*, 710–720. [CrossRef]
34. Snyder, N.F.R.; Wiley, J.W.; Kepler, C.B. *The Parrots of Luquillo: Natural History and Conservation of the Puerto Rican Parrot*; Western Foundation of Vertebrate Zoology: Los Angeles, CA, USA, 1987.
35. Beissinger, S.R.; Wunderle, J.M.; Meyers, J.M.; Saether, B.E.; Engen, S. Anatomy of a bottleneck: Diagnosing factors limiting population growth in the puerto rican parrot. *Ecol. Monogr.* **2008**, *78*, 185–203. [CrossRef]
36. Snyder, N.F.R.; Derrickson, S.R.; Beissinger, S.R.; Wiley, J.W.; Smith, T.B.; Toone, W.D.; Miller, B. Limitations of captive breeding in endangered species recovery. *Conserv. Biol.* **1996**, *10*, 338–348. [CrossRef]
37. Wilson, M.H.; Kepler, C.B.; Snyder, N.F.R.; Derrickson, S.R.; Josh, F., Wiley, J.W.; Wunderle, J.M.; Lugo, A.E.; Graham, D.L.; Toone, D. Puerto Rican Parrots of the Potential Approach Metapopulation Species Conservation. *Conserv. Biol.* **1994**, *8*, 114–123. [CrossRef]
38. Mohlman, J.L.; Navara, K.J.; Sheriff, M.J.; Terhune, T.M.; Martin, J.A. Validation of a noninvasive technique to quantify stress in northern bobwhite (Colinus virginianus). *Conserv. Physiol.* **2020**, *8*, 1–10. [CrossRef] [PubMed]
39. Washburn, B.E.; Millspaugh, J.J. Effects of simulated environmental conditions on glucocorticoid metabolite measurements in white-tailed deer feces. *Gen. Comp. Endocrinol.* **2002**, *127*, 217–222. [CrossRef]
40. Terio, K.A.; Brown, J.L.; Moreland, R.; Munson, L. Comparison of different drying and storage methods on quantifiable concentrations of fecal steroids in the cheetah. *Zoo Biol.* **2002**, *21*, 215–222. [CrossRef]
41. McGlothlin, J.W.; Jawor, J.M.; Greives, T.J.; Casto, J.M.; Phillips, J.L.; Ketterson, E.D. Hormones and honest signals: Males with larger ornaments elevate testosterone more when challenged. *J. Evol. Biol.* **2008**, *21*, 39–48. [CrossRef]
42. Jawor, J.M.; McGlothlin, J.W.; Casto, J.M.; Greives, T.J.; Snajdr, E.A.; Bentley, G.E.; Ketterson, E.D. Seasonal and individual variation in response to GnRH challenge in male dark-eyed juncos (Junco hyemalis). *Gen. Comp. Endocrinol.* **2006**, *149*, 182–189. [CrossRef]
43. McGlothlin, J.W.; Jawor, J.M.; Ketterson, E.D. Natural variation in a testosterone-mediated trade-off between mating effort and parental effort. *Am. Nat.* **2007**, *170*, 864–875. [CrossRef]
44. DeVries, M.S.; Winters, C.P.; Jawor, J.M. Testosterone elevation and response to gonadotropin-releasing hormone challenge by male Northern Cardinals (Cardinalis cardinalis) following aggressive behavior. *Horm. Behav.* **2012**, *62*, 99–105. [CrossRef]
45. DeVries, M.S.; Jawor, J.M. Natural variation in circulating testosterone does not predict nestling provisioning rates in the northern cardinal, Cardinalis cardinalis. *Anim. Behav.* **2013**, *85*, 957–965. [CrossRef]
46. Susan Devries, M.; Holbrook, A.L.; Winters, C.P.; Jawor, J.M. Non-breeding gonadal testosterone production of male and female Northern Cardinals (Cardinalis cardinalis) following GnRH challenge. *Gen. Comp. Endocrinol.* **2011**, *174*, 370–378. [CrossRef]
47. Duckworth, B.M.; Jawor, J.M. Corticosterone profiles in northern cardinals (Cardinalis cardinalis): Do levels vary through life history stages? *Gen. Comp. Endocrinol.* **2018**, *263*, 1–6. [CrossRef] [PubMed]
48. Clubb, S.; Velez, J.; Garner, M.M.; Zaias, J.; Cray, C. Health and Reproductive Assessment of Selected Puerto Rican Parrots (Amazona vittata) in Captivity. *J. Avian Med. Surg.* **2015**, *29*, 313–325. [CrossRef]
49. Johnson, B.H.; Welsh, T.H.; Juniewicz, P.E. Suppression of luteinizing hormone and testosterone secretion in bulls following adrenocorticotropin hormone treatment. *Biol. Reprod.* **1982**, *26*, 305–310. [CrossRef] [PubMed]
50. Orr, T.E.; Mann, D.R. Role of glucocorticoids in the stress-induced suppression of testicular steroidogenesis in adult male rats. *Horm. Behav.* **1992**, *26*, 350–363. [CrossRef]
51. Deviche, P.; Gao, S.; Davies, S.; Sharp, P.J.; Dawson, A. Rapid stress-induced inhibition of plasma testosterone in free-ranging male rufous-winged sparrows, Peucaea carpalis: Characterization, time course, and recovery. *Gen. Comp. Endocrinol.* **2012**, *177*, 1–8. [CrossRef]
52. Bowers, E.K.; Bowden, R.M.; Sakaluk, S.K.; Thompson, C.F. Immune activation generates corticosterone-mediated terminal reproductive investment in a wild bird. *Am. Nat.* **2015**, *185*, 769–783. [CrossRef]
53. Breuner, C.W. Stress and Reproduction in Birds. In *Hormones and Reproduction of Vertebrates*; Academic Press: Cambridge, MA, USA, 2011; Volume 4, pp. 129–151, ISBN 9780123749291.
54. Young, A.M.; Hobson, E.A.; Lackey, L.B.; Wright, T.F. Survival on the ark: Life-history trends in captive parrots. *Anim. Conserv.* **2012**, *15*, 28–43. [CrossRef] [PubMed]

55. Lattin, C.R.; Romero, L.M. Seasonal variation in corticosterone receptor binding in brain, hippocampus, and gonads in House Sparrows (Passer domesticus). *Auk* **2013**, *130*, 591–598. [CrossRef]
56. Li, D.; Wang, G.; Wingfield, J.C.; Zhang, Z.; Ding, C.; Lei, F. Seasonal changes in adrenocortical responses to acute stress in Eurasian tree sparrow (Passer montanus) on the Tibetan Plateau: Comparison with house sparrow (P. domesticus) in North America and with the migratory P. domesticus in Qinghai Province. *Gen. Comp. Endocrinol.* **2008**, *158*, 47–53. [CrossRef]
57. Romero, L.M. Seasonal changes in plasma glucocorticoid concentrations in free-living vertebrates. *Gen. Comp. Endocrinol.* **2002**, *128*, 1–24. [CrossRef]
58. Romero, L.M.; Cyr, N.E.; Romero, R.C. Corticosterone responses change seasonally in free-living house sparrows (Passer domesticus). *Gen. Comp. Endocrinol.* **2006**, *149*, 58–65. [CrossRef] [PubMed]
59. Romero, L.M.; Meister, C.J.; Cyr, N.E.; Kenagy, G.J.; Wingfield, J.C. Seasonal glucocorticoid responses to capture in wild free-living mammals. *Am. J. Physiol.-Regul. Integr. Comp. Physiol.* **2008**, *294*, 614–622. [CrossRef] [PubMed]
60. Popp, L.G.; Serafini, P.P.; Reghelin, A.L.S.; Spercoski, K.M.; Roper, J.J.; Morais, R.N. Annual pattern of fecal corticoid excretion in captive Red-tailed parrots (Amazona brasiliensis). *J. Comp. Physiol. B Biochem. Syst. Environ. Physiol.* **2008**, *178*, 487–493. [CrossRef] [PubMed]
61. Dickens, M.J.; Bentley, G.E. Stress, captivity, and reproduction in a wild bird species. *Horm. Behav.* **2014**, *66*, 685–693. [CrossRef]
62. White, T.H.; Collazo, J.A.; Vilella, F.J.; Guerrero, S.A. Effects of Hurricane Georges on habitat use by captive-reared Hispaniolan Parrots (Amazona ventralis) released in the Dominican Republic. *Ornitol. Neotrop.* **2005**, *16*, 405–417.
63. Helfenstein, F.; Wagner, R.H.; Danchin, E.; Rossi, J.M. Functions of courtship feeding in black-legged kittiwakes: Natural and sexual selection. *Anim. Behav.* **2003**, *65*, 1027–1033. [CrossRef]
64. Seymour, R.M.; Sozou, P.D. Duration of courtship effort as a costly signal. *J. Theor. Biol.* **2009**, *256*, 1–13. [CrossRef]
65. MacLeod, K.J.; Sheriff, M.J.; Ensminger, D.C.; Owen, D.A.S.; Langkilde, T. Survival and reproductive costs of repeated acute glucocorticoid elevations in a captive, wild animal. *Gen. Comp. Endocrinol.* **2018**, *268*, 1–6. [CrossRef] [PubMed]
66. Cyr, N.E.; Romero, L.M. Fecal glucocorticoid metabolites of experimentally stressed captive and free-living starlings: Implications for conservation research. *Gen. Comp. Endocrinol.* **2008**, *158*, 20–28. [CrossRef]
67. Vidal, A.C.; Roldan, M.; Christofoletti, M.D.; Tanaka, Y.; Galindo, D.J.; Duarte, J.M.B. Stress in captive Blue-fronted parrots (Amazona aestiva): The animalists'tale. *Conserv. Physiol.* **2019**, *7*, 1–11. [CrossRef] [PubMed]
68. Costa, P.; Macchi, E.; Valle, E.; De Marco, M.; Nucera, D.M.; Gasco, L.; Schiavone, A. An association between feather damaging behavior and corticosterone metabolite excretion in captive African grey parrots (Psittacus erithacus). *PeerJ* **2016**, *2016*, 1–14. [CrossRef] [PubMed]
69. Jill Heatley, J.; Oliver, J.W.; Hosgood, G.; Columbini, S.; Tully, T.N. Serum corticosterone concentrations in response to restraint, anesthesia, and skin testing in hispaniolan amazon parrots (Amazona ventralis). *J. Avian Med. Surg.* **2000**, *14*, 172–176. [CrossRef]
70. Owen, D.J.; Lane, J.M. High levels of corticosterone in feather-plucking parrots (Psittacus erithacus). *Vet. Rec.* **2006**, *158*, 804–805. [CrossRef] [PubMed]

 *diversity*

MDPI

Communication

Reintroduction of the Golden Conure (*Guaruba guarouba*) in Northern Brazil: Establishing a Population in a Protected Area

Marcelo Rodrigues Vilarta [1,2], William Wittkoff [1], Crisomar Lobato [3], Rubens de Aquino Oliveira [3], Nívia Gláucia Pinto Pereira [3] and Luís Fábio Silveira [1,2,*]

1 Lymington Foundation, Estrada Dias Gomes 2.200, Juquitiba 06950-970, SP, Brazil; marcelovilarta@hotmail.com (M.R.V.); fund.lymington@gmail.com (W.W.)
2 Museu de Zoologia da Universidade de São Paulo, MZUSP, Avenida Nazare 481, Ipiranga 04263-000, SP, Brazil
3 Instituto de Desenvolvimento Florestal e da Biodiversidade do Estado do Pará, IDEFLOR-Bio, Avenida João Paulo II, Utinga, Belém 66610-770, PA, Brazil; crisomarlobato@yahoo.com.br (C.L.); rubens.aquino25@gmail.com (R.d.A.O.); niviabio@gmail.com (N.G.P.P.)
* Correspondence: lfs@usp.br

Abstract: Brazil has the highest number of parrots in the world and the greatest number of threatened species. The Golden Conure is endemic to the Brazilian Amazon forest and it is currently considered as threatened by extinction, although it is fairly common in captivity. Here we report the first reintroduction of this species. The birds were released in an urban park in Belem, capital of Para State, where the species was extinct more than a century ago. Birds were trained to recognize and consume local food and to avoid predators. After the soft-release, with food supplementation and using nest boxes, we recorded breeding activity in the wild. The main challenges before the release were the territorial disputes within the aviary and the predation by boa snakes. During the post-release monitoring the difficulties were the fast dispersion of some individuals and the dangers posed by anthropic elements such as power lines that caused some fatalities. Released birds were very successful at finding and consuming native foods, evading predators, and one pair reproduced successfully. Monitoring continues and further releases are programmed to establish an ecologically viable population.

Keywords: reintroduction; soft-release; acclimatization; monitoring; Amazon; dispersion

Citation: Vilarta, M.R.; Wittkoff, W.; Lobato, C.; Oliveira, R.d.A.; Pereira, N.G.P.; Silveira, L.F. Reintroduction of the Golden Conure (*Guaruba guarouba*) in Northern Brazil: Establishing a Population in a Protected Area. *Diversity* **2021**, *13*, 198. https://doi.org/10.3390/d13050198

Academic Editors: Guillermo Blanco, Martina Carrete and Luc Legal

Received: 31 March 2021
Accepted: 29 April 2021
Published: 8 May 2021

1. Introduction

Parrots are among the most endangered birds in the world, having over 29% of their 402 extant species at risk of extinction [1]. This vulnerability is mostly due to their charismatic nature that leads to a high demand in the illegal pet trade, which, consequently, reduces wild populations. In turn, this situation attracts many conservation actions on their behalf [2,3].

The reintroduction of captive-bred wildlife is an important conservation tool, being increasingly used to compensate human impacts on populations and ecosystems [4,5]. In extreme cases, it has already been used to recover species that were on the brink of extinction such as *Petroica traversi* and is currently the only option to return extinct in the wild species such as *Pauxi mitu* and *Cyanopsitta spixii* [6–9].

Brazil is known for having both the greatest richness of parrot species and the largest number of endangered birds in the world [10,11]. However, reintroduction programs are still rare in the country, while indiscriminate release actions, carried out without technical rigor and monitoring are common [12]. This seems to be a rule in most of South America [13], but there are successful examples with parrots that highlight the potential for their reintroduction [14,15].

The Golden Conure (*Guaruba guarouba*) is an endemic species of the Brazilian Amazon forest. Given their exuberant appearance, these conures have suffered a dramatic population decline due to the illegal pet trade, which in addition to severe habitat loss, has led them to be vulnerable to extinction [16,17]. Although it is estimated that their wild population is small, with less than 10,000 remaining individuals, Golden Conures are prolific and quite common in captivity [17].

Considering the Amazon's progressive habitat destruction, and how the occurrence area of this species has drastically shrunk over the years, the long-term survival of this species in the wild is seriously jeopardized, and actions for its conservation are urgent [17–20].

In such an unfavorable scenario, the reintroduction of captive-bred Golden Conures may be a viable way to restore the species to key areas, reduce the risk of extinction and raise awareness of their importance as a conservation symbol to the local population. However, there are no records of previous attempts to reintroduce this species. Additionally, they remain poorly studied, with many aspects of their behavior and natural history not yet clarified [16,21].

Lymington Foundation, a Brazilian non-profitable organization located in Juquitiba, São Paulo state, Brazil, has successfully bred this species over the last 20 years, and in 2017, teamed up with IDEFLOR-Bio, from Belém, Pará state, Brazil, to start the first attempt to reintroduce the Golden Conure in the wild. Here we report the preliminary results of this collaboration.

2. Study Area and Methods

Identify and remove the causes or the main threats that lead the targeted taxa to be extinct in a given area are of paramount importance for a successful reintroduction [22]. Hence, we chose a recently created protected area in Belem, capital of the state of Para, where the last credible sighting of the Golden Conure dates from over 150 years ago [23].

The Utinga State Park, located in Belém, Pará state, has 1393 km^2 mostly represented by lowland rainforest, connected to the continuous forested area to the east, the Guamá River to the south, and the city by the west and north (Figure 1a) [24]. The area is constantly under surveillance by both private security and the public environmental police, so the main causes for the Golden Conures' extirpation, which were habitat loss and capture, are controlled in this area [24].

Figure 1. (**a**) The Utinga State Park (dark green), located inside the environmental protected area (light green). (**b**) The acclimatization aviary is composed of two modules connected.

We built the acclimatization aviaries in the intended release site at the center of the park. They consisted of a maintenance module of 6 m × 2 m × 2 m, connected to a larger one with 5 m × 5 m × 5 m. Two nest boxes were available in the maintenance module for the birds to spend the night, the second module had a tree inside, and both with enough space for flight (Figure 1b). We chose a previously opened site with sparse secondary vegetation, known to be preferred by the species for roosting and nesting [18]. The site was rich with *Byrsonima* sp. Trees, an important resource in the post-fledgling period [17] and provided good visualization during the monitoring. The adjacent patches were composed by primary forest fragment and a lake, providing various environmental options.

Individuals selected for the reintroduction were bred at Lymington Foundation, where the installations are surrounded by Atlantic Forest, with visual and acoustic exposure to local wildlife. We selected birds both hand-raised and raised by their parents. The group was composed of birds genetically diverse and kept at Lymington Foundation in aviaries similar to the one built in the reintroduction area. Individuals had numbered metallic bands to allow for identification up close. Males and females had their bands placed on the right or the left leg respectively, for quick identification. Identification of individuals outside of close range relied on physical traits such as plumage patterns, eye coloration, and beak marks.

After performing the standard health evaluations (including PCR exams for circovirus, bornavirus, and herpesvirus, among other important diseases), and the results being negative in all cases, two groups of birds were sent to the release site. Both groups were composed of mostly young individuals of two to three years old, in equal proportion of males and females. The first group was sent to Belem in August 2017, consisting of 14 birds, and the second group of 10 individuals was sent in May 2018.

The birds were moved to the aviary, where they stayed for five months for adaptation to the local conditions. The diet was gradually adapted to native fruits until a total replacement was achieved. Native food was served attached to the branches of the respective plants and placed on hard-to-reach spots, so the birds would have to practice the recognition and foraging techniques. Predator recognition training was performed using live boa snakes placed in the proximity of the enclosure. Birds of prey were naturally present in the area and could occasionally be seen diving for lizards on the ground and at the top of the aviary. During these occasions we evaluated the behavioral response of the Golden Conures by scanning their reactions after the predators were sighted and approached the enclosure. The group was considered apt when all individuals reacted together, demonstrating alertness and emitting alarm vocalizations when potential predators approached. We made daily ad libitum observations to record social interactions in search of aggressive behaviors and couple formation, in order to identify which individuals were more inclined to fights and/or reproduction.

To provide experience to gain and increase site fidelity before the group release, we selected three males that showed the strongest attachment to females for prereleases. These males were released individually and on different days in the morning (Figure 2a), allowing them to explore the surroundings during the day, but still being aware of the group vocalizations in the aviary. Before the afternoon, they were attracted back to the aviary by food offering (Figure 2b).

(a)

(b)

Figure 2. (**a**) A single male being prereleased separately from the group. (**b**) The single male returning to the aviary at the end of the day.

For the reintroduction, we used the soft release method, which consists of opening the windows of the aviary and allowing them to decide when to leave and to return if they desire [25]. Supplementary food was offered daily on the top of the aviary and in two suspended feeders, distributed around the release site. In an additional effort to promote site fidelity, we installed nest boxes on the top of the enclosure, using the same model that the birds were already familiar with. A total of 20 individuals were released.

3. Results and Discussion

Golden Conure is a social species, usually living in groups varying from 3 to 30 birds [17,26]. However, in the acclimatization aviary we recorded territorial fights, with two females being killed by a very aggressive female. This bird was separated from the main group, returning a few days before the release. The fact that females were involved in aggressions, and not males, contradicts anecdotical reports from breeders, which indicated that males were the most aggressive sex.

During the prerelease we considered a group of 10 individuals as the optimal size for release. The aggression events recorded in the first group but not in the second, suggest that the birds should be released before the breeding season. The breeding season of the Golden Conure starts in October, lasting to March, and the casualties recorded were related with a female defending a nest site. It is relevant to report that the second group showed no negative interactions when we avoided the breeding season and, after release, no territorial disputes in the wild were recorded.

We recorded a random dispersion of the conures after the release, with a few birds staying near the enclosure. In the first group, 72% of the Conures dispersed in one week, while in the second event, 50%. The birds that showed site fidelity spent most of the time over the enclosure, but without re-entering, as reported by other psittacine releases [27,28]. From these dispersed groups, three birds were located 10 km away from the releasing area. On multiple occasions, other sightings of flying birds were reported in different sites in urban areas of Belem. Although we cannot confirm if those were the same three individuals or the others from the group, we had evidence of a quite long-distance dispersion for some individuals. The second group showed a stronger site fidelity over 12 months after release, but afterward they moved to the border between the protected area with the city, 4 km away from the release site, leaving it vacant (Figure 3).

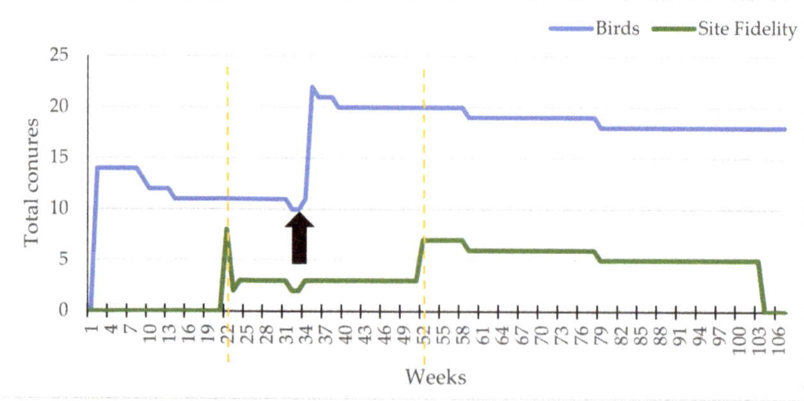

Figure 3. Number of golden conures potentially alive over time (blue) and those that remained near the release site (green); every reduction in the blue line represents a recorded death. The space between the blue and green lines represents the individuals with uncertain fate during a given time. Dashed orange lines mark the first and second release, the arrow points to the first breeding in the wild, in which one individual is added to the total population.

The dispersion of the Golden Conures after release can be associated with fluctuations of food availability in the area, which included fragmented landscapes and pristine forests. Since the supplementary food was constant throughout the monitoring, we believe there may be other causes for the higher dispersion and nomadism. Groups of other psittacines, mostly *Amazona amazonica*, were often seen flying from northeast to southwest during mornings and the way back in the afternoons. Even though we did not see any interaction with the conures, they may have influenced these movements, since the five conures were found in the northeast of the release site, in the same path that these parrots use daily. It is also possible that they were attracted to the numerous plantations of *Euterpe oleracea*, that are concentrated in the area where the dispersed birds moved to, since this is a favorite food item with a nutritious pulp. The uncertainty of the fate of some individuals that dispersed was the main problem to attest the level of success of this reintroduction. Dispersal and difficulty of monitoring has already been reported in the reintroduction of the Thick-billed Parrot, when it deeply affected the survival rates [29]. The recent release of the Orange-bellied Parrot (*Neophema chrysogaster*) in Tasmania also showed an elevated dispersion, with 38–46% of birds flying over the monitoring range, despite the training to instill site fidelity [30]. These cases suggest that elevated dispersal is still one of the main challenges for monitoring in psittacine reintroductions. Even so, this trait can be positive for a population, given the possibility of finding more suitable habitats and being able to expand their range.

We recorded a successful reproductive event only one month after the first release (Figure 4) by the same aggressive female that caused two deaths during prerelease. We did not register any behavior of nest helpers as we expected from previous reports [18,26]. Despite the first breeding attempt being successful, the two following attempts failed, even with the female incubating to the eggs apparently in the same way. It was not possible to evaluate if either the eggs were unfertilized or if they suffered any mechanical damage, but the latter might be the most probable cause, since individuals from the second group were sharing the nest box at night, and broken eggs were later found.

Figure 4. Released Golden Conure parents positioned in the extremities guiding the wild-born young at the center.

Reintroductions often present pronounced mortality in the first month of release, and predation is often considered the main factor leading to loss of individuals [3]. In our study, after the release, one individual was predated but the main threat was electrocution in powerlines, which caused two deaths. Predation by boa snakes, *Boa constrictor*, was a major problem in the prerelease. Despite the conures showed an aversive response to predator exposure and being alert near their presence, three individuals were preyed upon inside their nest boxes during the night, two before the release. We addressed this problem by moving the nest boxes to trees that had no contact with the rest of the canopy and protecting their base with metallic belts, avoiding snake access. Other reintroductions also reported psittacines preyed mostly inside their nests [25,28], therefore the selection of sites for the installation of artificial nests must be carefully evaluated to avoid predation.

In previous studies with psittacines, the absence of antipredator training led to high losses to raptors, especially with hand-reared birds [29,31]. In our case, the golden conures were alert in the presence of bigger birds of prey like *Heterospizias meridionalis*, evading their attacks on multiple occasions. *Milvago chimachima* a smaller falcon that does not pose a risk to them was abundant in the area, and in that case, the conures did not show any concern and even shared perches occasionally. By the end of the monitoring, no losses to avian predators were recorded.

Toucans are known to be nest predators of conures [26], and many individuals or groups of *Pteroglossus aracari* and *Ramphastos vitellinus* were often seen around the release site. However, no interaction between them was recorded during the breeding period or in the rest of the monitoring. Two individuals of black tamarins, *Saguinus niger,* tried to access the nest box once but were rapidly fended off by the breeding couple. Given the conures´ positive reactions to conspicuous predators after going through exposure in captivity and given that captive parrots tend to lose antipredator behavior without exposure [32], we reinforce the importance of these conditions during the prerelease period.

As for post-release feeding, we did not register any individual suffering from starvation during monitoring. On multiple occasions, individuals dispersed for weeks or months and were found later without apparent signs of food deprivation, meaning they were able to forage and survive without supplementary food. Thus, we attest to the success of the food recognition training and recommend it before any release. The supplementation of food was important in the maintenance of site fidelity for the individuals that did not disperse far, given they were routinely present in the site at the exact time of the food exchange. This importance has also been attested in the reintroduction of *Ara macao* [15]. Similarly, reintroduced Conures switched supplementary feeding for natural foraging gradually over time, despite the offer remaining constant.

4. Concluding Remarks

Even while this reintroduction is still in its first steps, important milestones of success were already achieved such as an early reproduction in the wild, conures developing natural foraging skills, and avoiding avian predators. However, we still lack accurate data on post-release dispersal that is vital to understand how this species will move and occupy the region. We aim to address this matter with the use of telemetry, since these questions rely on it. We also seek to standardize the methods to better record feeding habits and group interactions. With the learning acquired in this stage, we are confident that future releases will be more successful, and, in the future, an ecologically viable population of Golden Conures will settle in this protected area.

Author Contributions: Conceptualization, L.F.S. and M.R.V.; methodology, L.F.S., M.R.V.; formal analysis, M.R.V. and L.F.S.; resources, W.W., C.L., N.G.P.P. and R.d.A.O.; data curation, M.R.V. and L.F.S.; writing—original draft preparation, M.R.V. and L.F.S.; writing—review and editing, M.R.V., L.F.S., W.W., C.L., N.G.P.P. and R.d.A.O.; supervision, L.F.S.; project administration, N.G.P.P., R.d.A.O., C.L., L.F.S. and W.W.; funding acquisition, N.G.P.P., R.d.A.O., C.L., L.F.S. and W.W. All authors have read and agreed to the published version of the manuscript.

Funding: This research was funded by Instituto de Desenvolvimento Florestal e da Biodiversidade do Estado do Pará, IDEFLOR-Bio.

Institutional Review Board Statement: The study was approved by the Ethics Committee of the Instituto de Biociências da Universidade de São Paulo (protocol code 227/2015, approved in 30/06/2015) and has the Federal license issued by ICMBio number 56616-11.

Informed Consent Statement: Not applicable.

Data Availability Statement: Not applicable.

Acknowledgments: This work was (and is) possible due the collaboration of many people and institutions: Luiz Carlos Oliveira and Nancy Fragnan Oliveira, Maria Fernanda Gondim (Zooparque Itatiba), Instituto Viva Cerrado, Roberto Reis Veloso, Priscilla Amaral, Juliana Anaya Sinhorini, Thaís Tamamoto, Glaucia Bio, Marina Somenzari, Rosemary Low, Roelant Jonker, Bruno Ehlers, IBAMA, ICMBio, Soraya Tatiana Macedo Alves, Ana Claudia Aranha C. Moreira, Carlos Boelter, José Aluísio Fernandes Junior, Ana Paula Vitoria C. Rodrigues, José Leonardo de Lima Magalhães, Neusa Renata Lima Emin, Aina Gorayeb, Marina Benarrós, and the staff of IDEFLOR-Bio and Lymington Foundation, for their constant support. To the anonymous reviewers and to Guillermo Blanco and Martina Carrete for the comments t improved the manuscript. This article is dedicated to the memory of Linda Wittkoff, for her inspiration and contribution to the conservation of Brazilian parrots. Linda was the creator of this reintroduction project, and we are honored by the opportunity to learn from her.

Conflicts of Interest: The authors declare no conflict of interest.

References

1. IUCN. The IUCN Red List of Threatened Species. Available online: https://www.iucnredlist.org (accessed on 24 October 2020).
2. Latas, P.J. Rescue, rehabilitation, and release of psittacines: An international survey of wildlife rehabilitators. *J. Wildl. Rehabil.* **2019**, *39*, 15–22.
3. White, T.H., Jr.; Collar, N.J.; Moorhouse, R.J.; Sans, V.; Stolen, E.D.; Brightsmith, D.J. Psittacine reintroductions: Common denominators of success. *Biol. Conserv.* **2012**, *148*, 106–115. [CrossRef]
4. Armstrong, D.P.; Seddon, P.J. Directions in reintroduction biology. *Trends Ecol. Evol.* **2008**, *23*, 20–25. [CrossRef]
5. Pérez, I.; Anadón, J.D.; Díaz, M.; Nicola, G.G.; Tella, J.L.; Giménez, A. What is wrong with current translocations? A review and a decision-making proposal. *Front. Ecol. Environ.* **2012**, *10*, 494–501. [CrossRef]
6. Sutherland, W.J.; Armstrong, D.; Butchart, S.H.M.; Earnhardt, J.M.; Ewen, J.; Jamieson, I.; Jones, C.G.; Lee, R.; Newbery, P.; Nichols, J.D.; et al. Standards for documenting and monitoring bird reintroduction projects. *Conserv. Lett.* **2010**, *3*, 229–235. [CrossRef]
7. Massaro, M.; Chick, A.; Whitsed, R.; Kennedy, E.S. Post-reintroduction distribution and habitat preferences of a spatially limited island bird species. *Anim. Conserv.* **2018**, *21*, 54–64. [CrossRef]
8. Plano de Ação Nacional para Conservação da Ararinha-Azul. Available online: https://www.icmbio.gov.br/portal/faunabrasileira/plano-de-acao-nacional-lista/2752-plano-de-acao-nacional-para-conservacao-da-ararinha-azul (accessed on 15 March 2021).
9. Francisco, M.R.; Costa, M.C.; Azeredo, R.M.A.; Simpson, J.G.P.; Dias, T.D.C.; Fonseca, A.; Pinto, F.J.M.; Silveira, L.F. Recovered after an extreme bottleneck and saved by ex situ management: Lessons from the Alagoas curassow (Pauxi mitu [Linnaeus, 1766]; Aves, Galliformes, Cracidae). *Zoo Biol.* **2021**, *40*, 76–78. [CrossRef] [PubMed]
10. BirdLife International Country Profile: Brazil. Available online: http://www.birdlife.org/datazone/country/brazil (accessed on 8 August 2020).
11. Forshaw, J.M. *Parrots of the World*; De Gruyter: Berlin, Germany, 2010; p. 573.
12. Wanjtal, A.; Silveira, L.F. A soltura de aves contribui para a sua conservação? *Atualidades Ornitológicas* **2000**, *98*, 7.
13. Galetti, M.; Root-Bernstein, M.; Svenning, J.-C. Challenges and opportunities for rewilding South American landscapes. *Perspect. Ecol. Conserv.* **2017**, *15*, 245–247. [CrossRef]
14. Sanz, V.; Grajal, A. Successful Reintroduction of Captive-Raised Yellow-Shouldered Amazon Parrots on Margarita Island, Venezuela. *Conserv. Biol.* **1998**, *12*, 430–441. [CrossRef]
15. Brightsmith, D.; Hilburn, J.; Campo, A.; Boyd, J.; Frisius, M.; Janik, D.; Guilen, F. The use of hand-raised psittacines for rein-troduction: A case study of scarlet macaws (Ara macao) in Peru and Costa Rica. *Biol. Conserv.* **2005**, *121*, 465–472. [CrossRef]
16. Birdlife International. Scientists Discover an "Invisible Barrier" that Holds the Answer to One of Nature's Little Mysteries. *Birdlife News Posts, 16 March 2012*. Available online: http://www.birdlife.org (accessed on 14 June 2012).
17. Laranjeiras, T.O. Biology and population size of the Golden Parakeet (Guaruba guarouba) in western Pará, Brazil, with recommendations for conservation. *Rev. Bras. Ornitol.* **2011**, *19*, 303–314.
18. Silveira, L.F.; Belmonte, F.J. Comportamento reprodutivo e hábitos da Ararajuba, Guarouba guarouba, no município de Tailândia, Pará. *Ararajuba* **2005**, *13*, 89–93.
19. Laranjeiras, T.O.; Cohn-Haft, M. Where is the symbol of Brazilian Ornithology? The geographic distribution of the Golden Parakeet (Guarouba guarouba—Psittacidae). *Rev. Bras. Ornitol.* **2009**, *17*, 1–19.

20. Soares-Filho, B.S.; Nepstad, D.C.; Curran, L.M.; Cerqueira, G.C.; Garcia, R.A.; Ramos, C.A.; Voll, E.; McDonald, A.; Lefebvre, P.; Schlesinger, P. Modelling conservation in the Amazon basin. *Nat. Cell Biol.* **2006**, *440*, 520–523. [CrossRef]
21. Juniper, T.; Parr, M. *Parrots—A Guide to Parrots of the World*; Yale University Press: New Haven, CT, USA, 1998; p. 436.
22. Guidelines for Reintroductions and Other Conservation Translocations. Available online: https://www.iucn.org/content/guidelines-reintroductions-and-other-conservation-translocations (accessed on 20 December 2020).
23. Moura, N.G.; Lees, A.C.; Aleixo, A.; Barlow, J.; Dantas, S.M.; Ferreira, J.; Lima, M.D.F.C.; Gardner, T.A. Two Hundred Years of Local Avian Extinctions in Eastern Amazonia. *Conserv. Biol.* **2014**, *28*, 1271–1281. [CrossRef] [PubMed]
24. Plano de Manejo do Parque Estadual do Utinga. Available online: http://ideflorbio.pa.gov.br/utinga/wp-content/uploads/2018/03/PMUtinga_26out2013.pdf (accessed on 23 December 2020).
25. White, T.H., Jr.; Abreu, W.; Benitez, G.; Jhonson, A.; Lopez, M.; Ramirez, L.; Rodriguez, I.; Toledo, M.; Torres, P.; Velez, J. Minimizing Potential Allee Effects in Psittacine Reintroductions: An Example from Puerto Rico. *Diversity* **2021**, *13*, 13. [CrossRef]
26. Oren, D.C.; Novaes, F.C. Observations on the golden parakeet Aratinga guarouba in Northern Brazil. *Biol. Conserv.* **1986**, *36*, 329–337. [CrossRef]
27. Lopes, A.R.S.; Rocha, M.S.; Mesquita, W.U.; Drumond, T.; Ferreira, N.F.; Camargos, R.A.L.; Vilela, D.A.R.; Azevedo, C.S. Translocation and Post-Release Monitoring of Captive-Raised Blue-fronted AmazonsAmazona aestiva. *Acta Ornithol.* **2018**, *53*, 37–48. [CrossRef]
28. Seixas, S.G.H.F.; Mourão, G.M. Assesment of restocking of blue-fronted Amazon (Amazona aestiva) in the Pantanal of Brazil. *Rev. Bras. Ornitol.* **2000**, *8*, 73–78.
29. Koenig, S.E.; Koschmann, J.; Snyder, H.A.; Johnson, T.B. Thick-Billed Parrot Releases in Arizona. *Condor* **1994**, *96*, 845–862. [CrossRef]
30. Save the Orange-Bellied Parrot. Mainland Release Trial 2020 Four-Week Update. Available online: https://orangebelliedparrot1.blogspot.com/2020/06/mainland-release-trial-2020-four-week.html?m=0 (accessed on 21 December 2020).
31. White, T.H., Jr.; Collazo, J.A.; Vilella, F.J. Survival of captive-reared Puerto Rican Parrots released in the Caribbean National Forest. *Condor* **2005**, *107*, 424–432. [CrossRef]
32. Carrete, M.; Tella, J.L. Rapid loss of antipredatory behaviour in captive-bred birds is linked to current avian invasions. *Sci. Rep.* **2015**, *5*, 18274. [CrossRef] [PubMed]

MDPI

Article

Minimizing Potential Allee Effects in Psittacine Reintroductions: An Example from Puerto Rico

Thomas H. White Jr. *, Wilfredo Abreu †, Gabriel Benitez, Arelis Jhonson, Marisel Lopez, Limary Ramirez, Iris Rodriguez, Miguel Toledo ‡, Pablo Torres and Jafet Velez

United States Fish and Wildlife Service—Puerto Rican Parrot Recovery Program, Box 1600, Rio Grande, PR 00745, USA; wilfredo_abreu@fws.gov (W.A.); gabriel_benitez@fws.gov (G.B.); arelisjohnson@hotmail.com (A.J.); marisel_lopez@fws.gov (M.L.); limary_ramirez@fws.gov (L.R.); iris_rodriguez@fws.gov (I.R.); miguel_toledo@fws.gov (M.T.); pablo_torres@fws.gov (P.T.); jafet_velez@fws.gov (J.V.)

* Correspondence: thomas_white@fws.gov

† Current address: PR-909, Barrio Mariana, Humacao, PR 00791, USA.

‡ Current address: PR-9988, Barrio Rio Chiquito, Luquillo, PR 00773, USA.

Abstract: The family Psittacidae is comprised of over 400 species, an ever-increasing number of which are considered threatened with extinction. In recent decades, conservation strategies for these species have increasingly employed reintroduction as a technique for reestablishing populations in previously extirpated areas. Because most Psittacines are highly social and flocking species, reintroduction efforts may face the numerical and methodological challenge of overcoming initial Allee effects during the critical establishment phase of the reintroduction. These Allee effects can result from failures to achieve adequate site fidelity, survival and flock cohesion of released individuals, thus jeopardizing the success of the reintroduction. Over the past 20 years, efforts to reestablish and augment populations of the critically endangered Puerto Rican parrot (*Amazona vittata*) have periodically faced the challenge of apparent Allee effects. These challenges have been mitigated via a novel release strategy designed to promote site fidelity, flock cohesion and rapid reproduction of released parrots. Efforts to date have resulted in not only the reestablishment of an additional wild population in Puerto Rico, but also the reestablishment of the species in the El Yunque National Forest following its extirpation there by the Category 5 hurricane Maria in 2017. This promising release strategy has potential applicability in reintroductions of other psittacines and highly social species in general.

Keywords: Psittacidae; reintroduction; Allee effect; population; survival; reproduction; site fidelity; flock cohesion

Citation: White, T.H., Jr.; Abreu, W.; Benitez, G.; Jhonson, A.; Lopez, M.; Ramirez, L.; Rodriguez, I.; Toledo, M.; Torres, P.; Velez, J. Minimizing Potential Allee Effects in Psittacine Reintroductions: An Example from Puerto Rico. *Diversity* **2021**, *13*, 13. https://doi.org/10.3390/d13010013

Received: 21 November 2020
Accepted: 29 December 2020
Published: 2 January 2021

1. Introduction

The family Psittacidae is comprised of over 400 species, an ever-increasing number of which are considered threatened or endangered [1,2]. In recent decades, conservation strategies for these species have increasingly employed reintroduction as a technique for reestablishing populations in previously extirpated areas [2–5]. However, reintroductions in general face substantial biological and methodological challenges, and many are ultimately unsuccessful [2,3,6,7]. Among these challenges are inherent—but often overlooked—Allee effects associated with small populations [7–10]. Allee effects are generally considered as consisting of either component effects (i.e., those which affect a component of individual fitness), or demographic effects, which affect per capita growth rates at the population level [11]. Examples of Allee effects include increased per capita predation risk, reduced foraging efficiency, and reduced pair-formation and reproductive effort, all of which contribute to reduced population growth [7–9]. These effects are particularly notable in group-living social species [9,11]. For example, the viability of African wild dog (*Lycaon pictus*) populations declines markedly once group size falls below a critical threshold, as also occurs with schools of bluefin tuna (*Thunnus thynnus*) and social groups of suricates (*Suricata suricatta*) [9]. Although Allee effects typically are considered as affecting populations as they decline from previously robust levels, in recent years, component and demographic Allee effects have increasingly been recognized during the establishment phase of reintroduced populations, before population size has achieved a robust, sustainable level [9,12–14]. Because most psittacines are highly social and flocking species, reintroduction efforts may face the numerical, behavioral and methodological challenges of minimizing initial Allee effects during the critical establishment phase of the reintroduction [9,14–16]. For example, although Snyder et al. [3] did not explicitly identify a component Allee effect as affecting releases of Thick-billed parrots (*Rhynchopsitta pachyrhyncha*) in Arizona, USA, they clearly implied such by stating that there appeared to be a "critical mass" of group size that conferred greater protection to released birds from avian predators. Further, Brightsmith et al. [4] reported that post-release survival of hand-reared Scarlet macaws (*Ara macao*) increased with increasing cohort sizes, and that macaws established at release sites facilitated survival of subsequent releases, also suggestive of a potential component Allee effect. Common challenges of reintroducing psittacines include, but are not limited to: (1) excessive or premature dispersal from the release area, (2) maintaining flock cohesion, (3) maximizing survival, and (4) obtaining reproduction rapidly following release [3,14,17–20]. If these challenges are not recognized and adequately addressed or ameliorated, they can result in failed efforts and wasted resources [2,3,9,13,15,16]. Management efforts for meeting these challenges typically fall into three general categories: (1) managing release group size and composition, (2) reducing post-release dispersal and mortality, and (3) direct management of Allee effects (e.g., predator control, supplemental feeding) [9,16].

The Puerto Rican parrot (*Amazona vittata*) is a critically endangered psittacine endemic to the island of Puerto Rico, for which an ongoing species conservation and recovery program has existed since the early 1970s [17,18]. Like most psittacines, Puerto Rican parrots are primarily frugivorous canopy-dwellers, secondary cavity nesters, and also exhibit marked natal philopatry and nest-site fidelity [17,18]. The total wild population has remained precariously low throughout the recovery program, ranging from a low of 13 to nearly 200 individuals over the period 1973–2017 [17,18] (USFWS, unpubl. data). Since 2000, captive-reared parrots have been released under a variety of scenarios in order to augment the sole relict wild population of the species in the El Yunque National Forest (hereafter EYNF), and to reestablish the species at an additional location on the island (i.e., Rio Abajo Commonwealth Forest) [17,20–23]. Moreover, future releases to reestablish the species at yet a third location (i.e., Maricao Commonwealth Forest) are anticipated. Here, we examine these scenarios, specifically those related to the relict population in the EYNF, in terms of how and why specific release strategies have achieved the desired objectives and potentially minimized some of the inherent initial Allee effects often associated with reintroduced populations [9,12,14,16]. We believe our findings have direct relevance to reintroductions of not only psittacines, but also other highly social or group-living species. We use the term "reintroduction" herein in its broadest sense, inclusive of all its recognized variants [2,24].

2. Materials and Methods

We examined and compared four (4) distinct captive release strategies in terms of their efficacy in establishing a resident, breeding population of Puerto Rican parrots. These strategies included: (1) Soft release of individual groups of captive-reared parrots translocated to a wild release site (hereafter "traditional release"); (2) Hard release of small numbers of captive-reared parrots translocated directly to a wild release site (hereafter "precision release"); (3) Soft release of multiple groups of captive-reared parrots translocated to a wild release site with conspecifics held on-site briefly following release (hereafter "soft release type A"; (4) Soft release of captive-reared parrots released on-site at a captive-breeding facility (hereafter "soft release type B"). In the context of this study, the term "soft release" refers to those methods that include on-site acclimation at the release site, and post-release support or supplementation. "Hard release" refers to methods involving no on-site acclimation, and no post-release support [8]. All parrots were released in the EYNF of northeastern Puerto Rico (approx. 18.32° N; 65.78° W, Figure 1), a mountainous forest reserve consisting of 19,650 ha of subtropical rainforest at elevations ranging from 200 to 1074 m ASL [20]. However, all releases occurred at elevations ranging from 500 to 700 m ASL, in the Tabonuco and Palo Colorado forest types [18,20], within the subtropical montane rainforest life zone [18,20]. Detailed descriptions of all forest types in the EYNF are found in Snyder et al. [18].

Figure 1. Location of Puerto Rico within the Caribbean Basin, and locations of the El Yunque National Forest and Rio Abajo Commonwealth Forest (sites of current wild Puerto Rican parrot populations) and the Maricao Commonwealth Forest, where future releases of Puerto Rican parrots are planned for reestablishing the species at a third location.

2.1. General Release Methodologies

2.1.1. Traditional Release

"Traditional release" is the most common release method for captive-reared animals in the reintroduction literature, e.g., [3,8,19,20]. For the Puerto Rican parrot, this consisted of rearing a group (15–20) of parrots together to desired age of release (1–4 years of age), and providing pre-release flight training, wild foods (>50 species) and predator aversion training for a minimum of 6 months in large (approx. 9 m × 8 m × 5 m) outdoor flight cages. Because the parrots were to be monitored post-release using radio-telemetry, replica "dummy" radio transmitters were attached to all release candidates during pre-release training to accustom them to flying and foraging with the device prior to release [19–21]. Flight cages were equipped with both stable and non-stable perches comprised of natural materials. Natural wild foods were offered in the same fashion as parrots would encounter in the wild. Complete fruiting branches, racemes, etc., were suspended from perches and cage roofs and sides to accustom birds to identifying and manipulating these foods in the wild. Following the initial training period, parrots were transported to a release cage for on-site acclimation (30–40 days) at a release site occupied by wild conspecifics [20]. All parrots were equipped with a functioning radio-transmitter (Holohil®, Ottawa, ON, Canada, SI-2C model) approximately 5–7 days prior to release to allow monitoring of their movements and survival post-release. Immediately following on-site acclimation, all parrots were allowed to exit the release cage at will. All releases occurred at dawn. Supplemental food sources were provided daily at the release site for a period of 2–4 weeks following release. Following each release, parrots were radio-tracked 3–5 days/week for the duration of transmitter life (approx. 10–16 months), or parrot mortality. The traditional releases occurred in EYNF at the end of the wild parrot nesting season (June) during 2000–2002 and 2004 [17,20]. Because of numerous predations of released parrots and wild parrot fledglings by Red-Tailed Hawks (Buteo jamaicensis) during 2000–2002, active predator control was implemented within 1.5 km of the release site, beginning in 2003, and continued throughout all subsequent years for all releases [17,20].

2.1.2. Precision Release

"Precision release" was an experimental methodology aimed at the fostering rapid integration of limited numbers of captive-reared parrots into an existing population. Like traditional releases, all release candidates were provided with at least six months of extensive pre-release training. However, unlike traditional releases, only two individuals were released during any given release event. All parrots released ranged from 1–2 years in age. Moreover, each release occurred within 100 m of an active wild parrot nest site, immediately (1–2 days) following fledging of the last nestling from the nest (typically May–June [17,18]). Releases were thus timed to take advantage of the early post-fledging phase, during which wild Puerto Rican parrot family groups remain relatively sedentary for several days, and with greatly diminished nest site territoriality [18]. The objective was to promote greater and more rapid interaction between wild and released parrots than had been observed following traditional releases [20]. All parrots so released were transported directly from the captive-rearing facility and released immediately at the wild nest site shortly after dawn, with no post-release supplemental food sources provided. Each parrot was also equipped with a radio-transmitter to allow post-release monitoring. Precision releases occurred in EYNF in alternate years from 2008 to 2010, and then yearly thereafter until 2014.

2.1.3. Soft Release Type A

"Soft release type A" was also an experimental release methodology. The objective was to promote increased site fidelity and flock cohesion of the release cohorts, as well as social interactions between the relict wild population and the released captive-reared birds. As with both the traditional releases and precision releases, all release candidates underwent at least 6 months of extensive pre-release training prior to transport to the release site. Parrots released ranged from 1 to 3 years of age. During pre-release training, candidates were closely observed for signs of potential pair bonds developing (e.g., allopreening, allofeeding). The use of unique color and shape-coded tags facilitated identification of individual parrots during training. Soft release type A also involved a 30–40 day on-site acclimation period at the release site (as with traditional releases), following which the parrots were released in two (2) groups over a period of 6–8 days. The release cage was divided into two equal-size segments to allow release group separation. Group 1 consisted of the males of any apparent pairs, together with a mix of unpaired males and females, and was released first. Group 2 consisted of the females of apparent pairs, together with a mix of other unpaired males and females, and released 6–8 days following Group 1. All releases occurred at dawn, and parrots were allowed to exit cages at will. Once all parrots had vacated both release cages, the following morning an additional group of 6–8 captive-reared parrots was placed in one of the release cages for a period of 2–3 weeks to serve as an additional "social attractant" for the newly released parrots. Supplemental food sources were provided and replenished daily at the release site for at least one year following release. All parrots were equipped with radio-transmitters to allow post-release monitoring. The soft release type A releases occurred in EYNF at the end of the wild parrot nesting season (June) during 2015–2017.

2.1.4. Soft Release Type B

"Soft release type B" releases were conducted on the grounds of the two captive-rearing facilities for the species, one of which is in the Rio Abajo Commonwealth Forest (hereafter, RAF) in northcentral Puerto Rico and the other in the EYNF [17,22,23]. At each of these facilities, a large number (currently 225–275) of captive-reared Puerto Rican parrots are housed and maintained in outdoor cages in a natural setting. Because the objective of these releases was to reestablish a free-flying wild population in an area from whence it had previously been extirpated, these releases were true "reintroductions" as defined by IUCN guidelines [24]. As with the other release strategies for the species, all release candidates received at least 6 months of extensive prerelease training, with some individuals receiving up to one year of training. As with soft release type A, parrots for release in EYNF ranged from 1 to 3 years of age, with the exception of one individual of six years of age. The 6-year-old parrot was a parrot previously released in 2015, which returned to the captive rearing facility following hurricane Maria in 2017. However, unlike soft release type A, all birds of soft release type B were reared and trained at the actual release site, and released directly from the prerelease training cage, instead of being transported to a separate release site. As with other releases, all parrots were equipped with radio transmitters to allow post-release monitoring. Upon release, supplemental food sources were provided daily near (10–30 m) the release site, and maintained continuously following each release. Several (8–10) artificial nest cavities [25] were also strategically placed within the release area in order to provide immediate nesting opportunities for the parrots following release. Soft release type B releases occurred during November 2006 at RAF [23] and January/February 2020 at EYNF, immediately prior to the species' normal nesting season (February–June).

2.2. Data Analyses and Reporting

We report and discuss the results of each type of release in EYNF in terms of four key parameters we considered important for successful reestablishment of psittacine populations: (1) survival, (2) site fidelity, (3) flock cohesion, and (4) prompt reproduction. We define "prompt reproduction" as successful reproduction by parrots within 18 months post-release, expressed as the proportion of surviving breeding age ($\geq$3 years) parrots that successfully nested during this period. We choose 18 months as a temporal benchmark as it allows all release cohorts time to adapt to the release environment and experience at least one complete nesting season following release, independent of their actual month of release. We considered prompt reproduction important in the context of establishing a resident population. This is because wild Puerto Rican parrots exhibit an annual philopatry of 87.5% to previously successful nesting sites [18,26]. Thus, prompt reproduction may more quickly and effectively "anchor" released individuals in the release area. We define "site fidelity" as the percentage of released parrots that established a stable activity area within 1.5 km of the release site, excluding any temporary longer distance forays. We choose 1.5 km as a spatial delineator because it corresponds to the radius of the primary area utilized by the relict wild Puerto Rican parrot population in EYNF during the last two decades, prior to their extirpation by the Category 5 hurricane Maria in 2017 (USFWS, unpublished data). We define "flock cohesion" as the percentage of surviving individuals that directly interacted (e.g., flying, foraging, roosting) as a group within the release area. With the exception of soft release type B, this also includes any direct interactions with, or integration into, groups of wild conspecifics at or near the release site. We report "survival" as the percentage of released individuals that survived for at least one year post-release [17,20]. Additionally, because most post-release mortalities of captive-reared psittacines occur within the first three months (90 days) after release ([3,17,19,20,27]; see [4,28]), we also report initial 3-month post-release survival for each release type. We estimated weekly, 3-month and annual survival of soft release types A and B in EYNF using the Kaplan–Meier product-limit estimator, in order to directly compare with published survival estimates for "traditional" captive releases of this and similar species [17,19,20], and because there were occasional censored observations (i.e., missing individuals and/or transmitter failures) during these releases. Censored observations were more frequent following soft release type B in EYNF due to personnel limitations and restrictions associated with the ongoing COVID-19 global pandemic, which resulted in reduced monitoring intensity compared to that of previous releases. For precision releases, because of the very low sample sizes (n = 2 individuals/release) for each of the multiple releases of this type, we report the overall range and average known survival pooled across all releases, as these releases did not meet the sample size requirements for Kaplan–Meier methods [19]. Finally, because of the distinct differences in release area habitat and microclimate between RAF and EYNF [17,18], we also report and discuss the first-year survival results of RAF soft release type B [23] for comparative purposes only. We did this in order to eliminate an additional and unquantifiable source of variability in the overall results, and facilitate a more accurate and direct comparison of the actual release strategies without the confounding effects of habitat or environmental differences. All percentages reported were rounded to nearest percent for simplicity. Kaplan–Meier first-year survival estimates are reported with associated 95% confidence intervals (CI), and compared using a log-rank test [29]. Differences in survival trajectories were considered significant at alpha < 0.05.

3. Results

3.1. Traditional Releases

3.1.1. Survival

A total of 39 captive-reared parrots were released in EYNF using traditional releases from 2000 to 2004 [17]. As previously reported see [17,20], overall first-year survival for traditional releases was 41% (CI: 22–61%), whereas survival at three months post-release was 74% (see [20], Figure 2). Similar traditional releases of captive-reared Hispaniolan parrots (*A. ventralis*) in the Dominican Republic resulted in a first-year survival of 30%, with a 3-month post-release survival of 60% [19]. In the EYNF, raptor predation was responsible for at least 53% (9/17) of the documented mortalities [17]. The causes of the remaining mortalities could not be determined, although additional raptor predations were possible [20]. Interestingly, most (67%) of the raptor predations occurred following increased dispersal of individual parrots from the release area approximately 6–8 weeks post-release (see [17] (pp. 22, 49)).

3.1.2. Site Fidelity

According to White et al. [20], individual parrots began dispersing from the immediate release area approximately two months following each release. Dispersing parrots often travelled up to 6–8 km from the release site (USFWS, unpublished data). All parrots which so dispersed did not return to the release area (i.e., 1.5 km radius of release site), and most were later recorded as mortalities or censored observations [20]. On one occasion, a pair (male, female) of released parrots returned to the captive-rearing facility, approximately 3 km from the release site, 11 months post-release. These parrots also did not return to the release area. Overall, site fidelity for traditional releases was low, with approximately 30–40% of released parrots remaining within the release area one year post-release.

3.1.3. Flock Cohesion

As with site fidelity, flock cohesion of traditional release cohorts was low. Few (approximately 20–25%) of the surviving parrots that remained within the release area for up to one year post-release were observed engaging in typical flocking behaviors with either wild or other released captive-reared parrots.

3.1.4. Prompt Reproduction

There was no successful reproduction (or attempts at such) by traditionally released parrots within 18 months post-release, despite the fact that several parrots (44%) were released at or entering breeding age within said period [20]. According to White et al. [20], there were documented nesting attempts by only three traditionally released parrots. These attempts first occurred in 2004 and consisted of a pair of captive-reared parrots released in 2002 at the ages of one and two years. This attempt was unsuccessful, and no subsequent nesting attempts by this pair were documented. The other was a captive-reared male released in 2001 at the age of one year, which successfully nested with a wild female and fledged two chicks in 2004 [20]. This pair continued to successfully nest each year thereafter until the disappearance of the male in 2009.

3.2. Precision Releases

3.2.1. Survival

A total of 36 captive-reared parrots were released in 18 separate release events in EYNF during six different years. Overall first-year survival of precision releases averaged 59%, although it ranged widely from 25% to 75% annually. Survival at three months post-release averaged 76%, while also ranging annually from 50% to 87%. Although raptor predation was confirmed as a cause of mortality in at least six (40%) cases, the cause of most mortalities remained unknown due to their occurrence in inaccessible areas, which precluded recovery of transmitters or parrot remains. In such cases, mortality was presumed when parrot movements ceased and transmitters remained stationary thereafter. We based this presumption on past experience with radio-tracking parrots in this environment [17,20].

3.2.2. Site Fidelity

Site fidelity of precision releases was markedly low. Released parrots typically remained near (<200 m) the release site for 2–5 days post-release, and then rapidly engaged in extensive movements both within and outside the immediate release area (USFWS, unpublished data). These movements also include the longest distances documented by captive-reared parrots released in EYNF, many of which resulted in parrot locations within suburban and urban areas up to 23 km from the EYNF release site. On four separate occasions, a precision released parrot returned to the captive-rearing facility, and did not return to the release area. Indeed, of 21 parrots known to have survived for one year, only 8–10 (38–48%) were subsequently observed within the release area. Thus, only 22–28% of all precision released parrots during 2008–2014 remained within the release area after one year.

3.2.3. Flock Cohesion

Flock cohesion of precision released parrots was relatively low. Although most released parrots exhibited some initial vocal interactions with both wild and previously released parrots (T. White, pers. observation), there were few instances (4–6) of their long-term integration into existing groups of resident birds. Nevertheless, it was notable that there were no agonistic interactions witnessed between released parrots and wild parrots, despite the fact that parrots were released in close proximity (<100 m) to family groups of newly fledged wild parrots (T. White, pers. Observation). Released parrots were also occasionally seen flying with or towards wild parrots during the initial days post-release.

3.2.4. Prompt Reproduction

Not surprisingly, given the very low site fidelity and flock cohesion, there were likewise no cases of prompt reproduction by precision released parrots with either wild parrots or other captive-reared parrots. Notwithstanding, during 2014, a male parrot that had been precision-released in 2012 was observed nesting with a wild female (USFWS, unpublished data). This pair successfully fledged three chicks from an artificial nest cavity that year. Unfortunately, that was the only year this particular male was observed nesting, and his subsequent fate remains unknown.

3.3. *Soft Release Type A*

3.3.1. Survival

A total of 65 captive-reared parrots were released during 2015–2017; with 20, 24 and 21 being released in 2015, 2016, and 2017, respectively. Overall first-year survival of soft release type A releases (2015, 2016) was 64% (CI: 50–79%; Figure 2), while initial 3-month survival averaged 85% (all three cohorts). No first-year survival data exist for the year 2017 release cohort, as virtually the entire wild population in EYNF was extirpated by hurricane Maria approximately 12 weeks following the 2017 release. However, at the time of the hurricane (20 September 2017), survival of the 2017 cohort was 95% (USFWS, unpublished data).

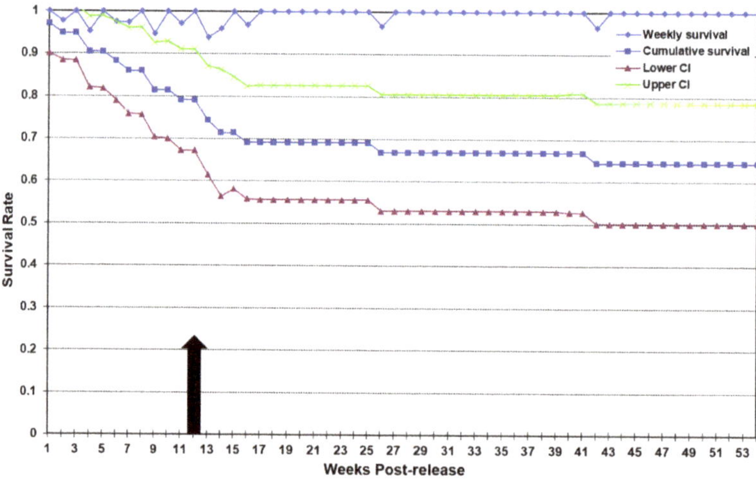

Figure 2. Kaplan–Meier weekly and cumulative survival estimates for 44 captive-reared Puerto Rican parrots released during soft release type A in the El Yunque National Forest, Puerto Rico, 2015–2016. Black arrow denotes 3-months post-release.

3.3.2. Site Fidelity

Site fidelity of soft release type A parrots was comparatively high. All surviving parrots remained within the release area following release, despite occasional longer distance forays by some individuals lasting from 2 to 4 days. All parrots who engaged in such forays later returned to the release area. Thus, the locations and status of most released parrots were known each day of monitoring, and there were few censored observations except for a single week in 2015, when a predation attempt at the release site by a Red-tailed hawk resulted in the rapid dispersal of over 50% of the released birds from the area. However, all dispersed parrots returned to the release area within five days.

3.3.3. Flock Cohesion

As with site fidelity, flock cohesion by soft release type A parrots was very high. All surviving parrots of each release event interacted on a daily basis with not only members of their release cohort, but also parrots of other release cohorts. Moreover, approximately four months post-release, beginning in 2015, released parrots and wild parrots were observed and video-recorded directly interacting at the release site (location of supplemental feeders) and within the surrounding area. These interactions consisted of vocal exchanges as well as flying and foraging together on wild foods. Interestingly, wild parrots were never observed approaching or utilizing supplemental feeders, despite the fact that they (wild birds) would often perch in the canopy immediately above supplemental feeders being used by released parrots.

3.3.4. Prompt Reproduction

Reproduction by soft release type A parrots began the first breeding season (<1 year) following release. A pair of captive-reared parrots released in June 2015 nested and fledged two chicks during both the 2016 and 2017 breeding seasons, in addition to a pair released in 2016 that also nested and fledged three chicks during the subsequent 2017 season. Thus, there were three successful nesting attempts during the first 18 months post-release. For parrots released in 2015 and 2016, this represented 18% and 17%, respectively, of the total breeding-age birds released each year. Two additional pairs of captive-reared parrots were observed engaging in stereotypical nesting behavior (e.g., allofeeding, defending and entering nest cavities) during the 2017 season, but did not actually nest.

3.4. *Soft Release Type B*

3.4.1. Survival

A total of 30 captive-reared parrots were released in EYNF during late January-early February 2020. As with soft release type A, captive-reared parrots were released in two groups; one group of 15 birds on January 30 followed by a second group of 15 birds on February 6. To date (i.e., 10 months post-release), the survival estimate is 68% (CI: 47–87%, Figure 3). Three-month (12-weeks) post-release survival of this cohort was 94% (Figure 3), very similar to the 92% survival reported by Estrada [5] for Scarlet macaws released in Mexico. The greatest declines in overall survival occurred 24–26 weeks and 41–42 weeks post-release (Figure 3), when at least six parrots were lost to raptor predations during two separate episodes. Survival of type B releases was higher than that of traditional releases ($\chi^2 = 8.779$, df = 1, $p = 0.003$), but not significantly greater than type A releases ($\chi^2 = 0.792$, df = 1, $p = 0.373$). However, early post-release survival of type B releases was much higher than that of type A during the ensuing breeding season over the 20 weeks immediately following the type B release ($\chi^2 = 7.647$, df = 1, $p = 0.006$; Figure 4).

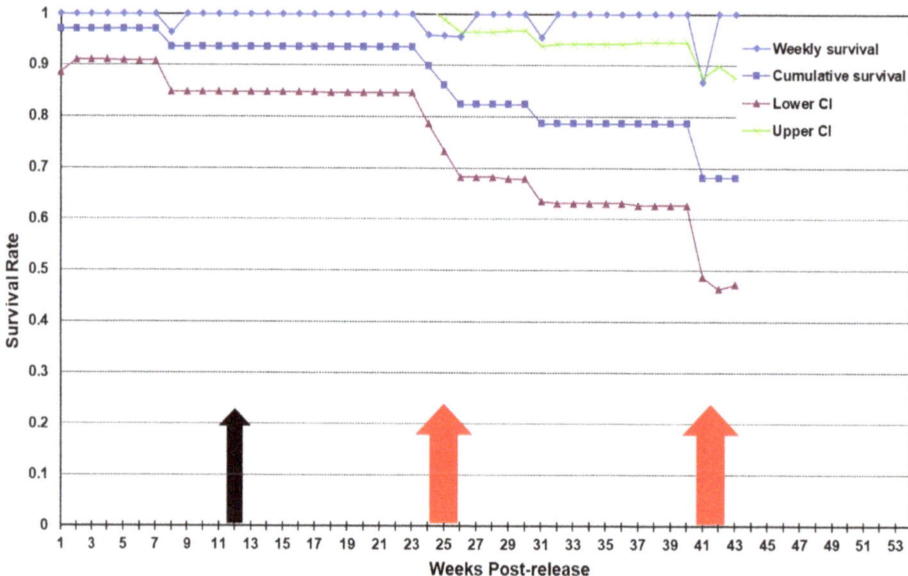

Figure 3. Kaplan–Meier weekly and cumulative survival estimates for 30 captive-reared Puerto Rican parrots released during a soft release type B in the El Yunque National Forest, Puerto Rico, 2020. Black arrow denotes 3-months post-release. Red arrows denote 24–26 weeks and 41–42 weeks post-release, corresponding to two episodes of raptor predation of several released parrots near release site, which accounted for 75% of all documented post-release mortalities.

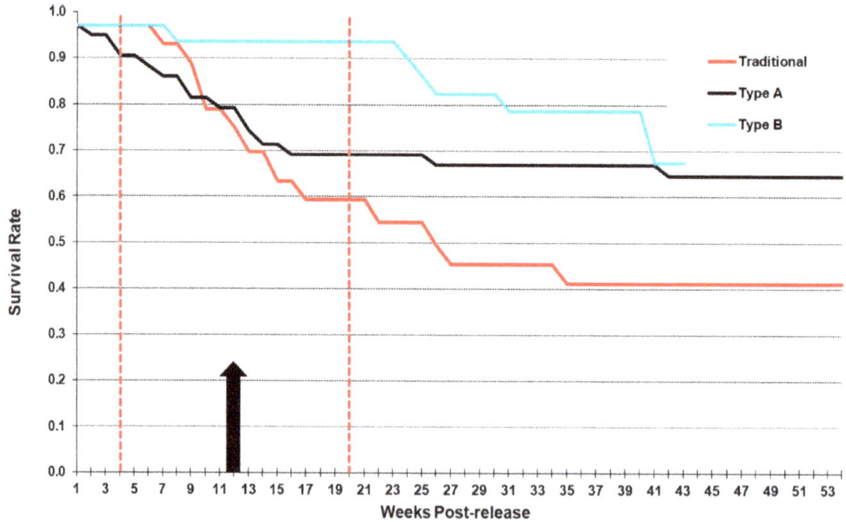

Figure 4. Comparison of survival trajectories based on Kaplan–Meier survival estimates for traditional, type A, and type B releases of captive-reared Puerto Rican parrots in the El Yunque National Forest, Puerto Rico, 2000–2020. Survival trajectory of traditional releases adapted from White et al. [17,20] and used with permission. Vertical red lines delineate temporal span (approx. 16 weeks) of the species' reproductive season. Black arrow denotes 3-months post-release.

3.4.2. Site Fidelity

Site fidelity of released parrots was moderately high. Of the 30 birds released, 26 (87%) remained within the release area until 24–26 weeks post-release, when a series of raptor attacks resulted in the temporary dispersal of several individuals from the area. Although most of the dispersed parrots later returned to the release area, two did not and their current locations are unknown. Approximately 10 months (40 weeks) post-release, 20 (67%) of the released parrots remained within the release area and immediate vicinity of the release site, until a second episode of raptor attacks during weeks 41–42 resulted in the dispersal of several parrots and additional censored observations.

3.4.3. Flock Cohesion

As with site fidelity, flock cohesion of soft release type B parrots was high. Virtually all (95%) of the surviving parrots remained together as a flock within the release area. Moreover, parrots were observed daily flying and foraging together, and engaging in group antipredator behaviors (e.g., posting "sentinels" while foraging, coordinated flights to confuse raptors) [30].

3.4.4. Prompt Reproduction

Reproduction by soft release type B parrots was very rapid, with two pairs of released parrots initiating nesting activities within two months of release. This represented 25% of the total number (n = 16) of breeding-age birds released (Table 1). Interestingly, the male of one breeding pair was a 6-year-old parrot that was, at the time of release, the sole surviving individual of the former wild population prior to hurricane Maria (Figure 5). Both pairs successfully fledged chicks, with one pair fledging three chicks and the other two. Indeed, this was the most rapid reproduction of captive-released parrots ever documented for this species. As occurred with the soft release type A releases, there were two additional pairs that engaged in stereotypical nesting behaviors post-release, but failed to actually nest. In both of these cases, at least one member of the pair was only two years of age, and thus unlikely to be sexually mature [18]. A graphical comparison of summary statistics for all four release types is presented in Figure 6.

Table 1. Summary statistics for four different types of release for captive-reared Puerto Rican parrots in the El Yunque National Forest, Puerto Rico, 2000–2020. Site fidelity refers to percentage of released parrots that remained within 1.5 km of release site; Flock cohesion refers to percentage of surviving parrots that remained in release area interacting as a group; Prompt reproduction refers to percentage of reproductive age parrots released that successfully nested within 18 months post-release.

Release Type	Survival 1-Year	Site Fidelity	Flock Cohesion	Prompt Reproduction
Traditional	41%	30–40%	20–25%	0
Precision	59%	22–28%	11–17%	0
Soft Release A	64%	65%	100%	18%
Soft Release B	68% [1]	67% [1]	95% [1]	25%

[1] 10 months post-release (November 2020).

Figure 5. Pair of captive-reared Puerto Rican parrots (male–upper; female–lower) released during a soft release type B in the El Yunque National Forest, 30 January 2020. The pair began nesting at this artificial nest cavity 29 February 2020 (30 days post-release) and subsequently fledged three chicks from this nest. This was the first active nest of free-flying parrots in the El Yunque National Forest following hurricane Maria in 2017. Photograph taken by Thomas White.

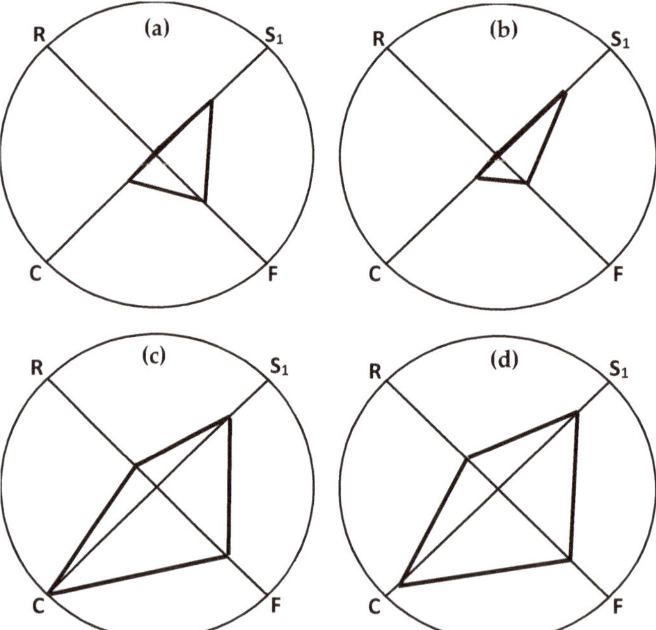

Figure 6. Graphical representation of the relative efficacy of four different strategies for establishing a resident breeding population of Puerto Rican parrots from captive-reared individuals in the El Yunque National Forest, Puerto Rico, 2000–2020. R-axis is prompt reproduction, S_1-axis is first-year survival, F-axis is site fidelity, C-axis is flock cohesion. Circumference of circle represents 100%, while intersect at center represents zero for associated axial parameter values. Area of interior polygons indicates degree of maximization of the four component parameters. Release types depicted: (**a**) Traditional release, (**b**) Precision release, (**c**) Soft release type A, (**d**) Soft release type B.

4. Discussion

We examined the results of four different strategies for the release of captive-reared Puerto Rican parrots during 26 distinct release events totaling 170 parrots from 2000 to 2020 in the El Yunque National Forest in Puerto Rico. Because the fundamental pre-release training of all released parrots was the same, we were able to compare the actual release strategies and methodologies in terms of their effectiveness at promoting survival, site fidelity, flock cohesion and prompt reproduction by released parrots. We believed these parameters to be important in mitigating or reducing potential Allee effects associated with small populations, as commonly occurs during the initial establishment phase of reintroductions [9,16]. As such, our study adds to the findings of White et al. [2] regarding factors influencing the success of psittacine reintroductions. Although survival is the single most commonly reported parameter of reintroduction attempts (e.g., [2,4,19,20,28]), site fidelity and flock cohesion are seldom addressed explicitly, as reported in this study. However, these parameters are all inextricably interrelated when reintroducing highly social species [14,16].

For captive-reared Puerto Rican parrots released in the EYNF, post-release survival tended to increase not only with the size of release cohorts, but also with numbers of conspecifics at or near the release site. This was particularly apparent in the case of soft release type B in EYNF (Figure 4), in which the largest release of parrots (n = 30) occurred, and in the presence of approximately 250 captive conspecifics held in outdoor cages at the release site. Indeed, of the four release methods, soft release type B resulted in improved post-release survival (Table 1, Figures 3 and 4), with a 3-month survival of 94%. In comparison, Llerandi-Román et al. [23] reported a first-year survival of 48% for a similar soft release type B in the RAF. Although post-release survival of type B eventually (approx. 10 months post-release) approximated that of type A releases, the initial survival during the critical early establishment phase and concomitant reproductive season was very high (>90%) throughout the season (Figure 4). In a previous study of factors associated with success of psittacine reintroductions, White et al. [2] did not find these factors (i.e., numbers released, conspecifics present) to be significant, perhaps due to the high variability in this parameter among those reintroductions examined. Nevertheless, many other studies have found positive relationships between numbers released and establishment success ([6,31,32], but see [16] for some caveats). In this study, we believe the presence of a large number of conspecifics held on-site aided newly released parrots in terms of more effective predator detection and avoidance. This was because, having been held in outdoor cages, all captive parrots had substantial prior exposure to avian predators, and most had even witnessed raptor predations of other avian species (e.g., *Zenaida* spp.) at the captive-rearing facility (T. White, pers. observations; see also [28]). These numerous captive "sentinels" quickly sounded alarm calls that alerted released parrots to impending dangers (T. White, pers. observation), thereby increasing the "effective flock size" of released parrots in terms of predator detection and avoidance [29]. We believe this increase in effective flock size helped to mitigate or reduce per capita risk associated with a potential predator-driven Allee effect, as suggested by White et al. [17] and demonstrated theoretically by Gascoigne and Lipsius [33] and empirically by Angulo et al. [34]. This is very important because the primary source of mortality for parrots in the EYNF has historically been raptor predation ([17,18,20], this study). Following soft release type B, survival trajectories (weekly, cumulative) were characterized by high survival for extended periods, punctuated by two separate episodes of raptor predations that resulted in multiple mortalities (Figures 3 and 4). This is in contrast to the temporal pattern of raptor predations following traditional releases, when at least 23% of all released parrots were lost to raptors during the first 27 weeks post-release [17], p. 22. Following traditional releases, raptor predations occurred concomitantly with the exodus of individual parrots from the release group and area, and the dispersing individuals were also the predominant victims of predation [17,20]. This is consistent with a predator-driven Allee effect increasing per capita predation risk with decreasing group size. Angulo et al. [34] also reported a similar predator-driven component Allee effect related to population size in reintroduced island foxes (*Urocyon littoralis*). In that study [34], larger group sizes resulted in lower per capita predation risk from Golden eagles (*Aguila chrysaetos*). Indeed, Llerandi-Román et al. [23] reported that survival of subsequent establishment releases of captive-reared Puerto Rican parrots increased annually following an initial soft release type B at RAF, and attributed this to a steadily increasing group size of resident survivors. During the RAF reintroduction, there were survival benefits that accrued to successive release cohorts due to cultural transmission of acquired survival skills by survivors of previous releases. Similarly, the initial 3-month survival (95%) of the third (2017) soft release type A in EYNF was also higher than that of the previous two such releases, as parrots were also released into a larger group of resident survivors of previous cohorts.

Although attaining adequate survival is paramount in any reintroduction, how surviving individuals distribute themselves within the release landscape is likewise critical, especially in the case of social species. Accordingly, we were encouraged by the comparatively high site fidelity exhibited by parrots released during both the type A and type B releases (Figure 6). Site fidelity not only promotes increased social interactions among individuals of a given cohort, but also—in conjunction with high survival—promotes increased survival and integration of subsequent release cohorts [4,23]. Conversely, low site fidelity can result in greater post-release mortality of individuals dispersing into areas with few, if any, conspecifics, and attendant increased per capita predation risk [17,20], as occurred following traditional and precision releases. This "dilution" and reduction in release group size via low site fidelity can contribute directly to initial Allee effects [9,12,14,16]. In the case of the Puerto Rican parrot, the presence of conspecifics held on-site following releases of individual cohorts likely aided in reducing excessive or premature dispersal. For example, for both the type A and type B releases, we believe that our technique of releasing only one member (male) of potential breeding pairs in a partial cohort release, followed by a second release soon after consisting of the other members of the cohort (including females of pairs), further promoted site fidelity. Because Puerto Rican parrots—like most psittacines—form lifelong monogamous pairs, the strength of this bond may be harnessed in order to retain initially released individuals on-site long enough for them to locate supplemental feeders and begin the adaptation process to the release area. In all such cases, we observed males initially released visiting their mates still held in the release cage, and even attempting to feed them through the cage sides and roof. Upon later release, the other members of the cohort immediately integrated into the previously released group. Moreover, in the case of release type B, we believe the presence of over 250 conspecifics held in captivity at the release site constituted yet another significant factor that aided in promoting site fidelity (Table 1) via increased conspecific attraction (see also [4,5,28]). The long-term presence of on-site supplemental feeders following soft releases type A and type B was also a likely factor in maintaining site fidelity, as also reported by Brightsmith et al. [4] and White et al. [2]. The continual presence of supplemental food sources not only promoted fidelity to the release area, but also greatly aided in monitoring the behaviors and fates of parrots post-release (Figure 7).

Figure 7. A group of captive-reared Puerto Rican parrots converging at a post-release supplemental feeder during March 2016. These parrots had been released nine months earlier (June 2015) during a soft release type A in the El Yunque National Forest of Puerto Rico. Photograph taken by Dailos Hernández-Brito.

Closely related to site fidelity was flock cohesion. For both the soft release type A and type B releases, post-release social cohesion and interactions of released cohorts was extremely high (Table 1, Figure 6), in contrast to traditional and precision releases, in which group cohesion was markedly low. The use of larger and mixed-age release cohorts, combined with long-term supplemental feeding, were the most probable reasons for these findings. The presence of on-site supplemental feeders facilitated daily social interactions by released parrots (Figure 7), which strengthened post-release social bonds and attendant group cohesion, as also reported by Brightsmith et al. [4]. Indeed, the lowest flock cohesion occurred with precision releases (Table 1, Figure 6), in which we released only two individuals during any given release event, and with no post-release support. Moreover, parrots released during precision releases were also released into a very small and low-density wild population, with attendant low conspecific attraction.

Among the parameters recognized as indicative of success in psittacine reintroductions, there are two—survival and subsequent reproduction—that are the most characteristic metrics of success [2]. Of the four types of captive releases of Puerto Rican parrots, only soft releases type A and type B resulted in prompt successful reproduction by released birds (Table 1, Figures 5 and 6). Indeed, in the case of EYNF soft release type B, the successful nesting by two pairs of captive-reared parrots within two months of release—in only 30 days in one case—was unprecedented (Figure 4). Reasons for the more rapid reproduction most likely relate to increased social interactions of all parrots, both released and captive, resulting from larger group sizes associated with type B releases. Importantly, the higher early post-release survival following type B releases also maximized the number of potential breeding individuals during the ensuing reproductive season (Figures 4 and 5), thereby increasing the likelihood of spatial anchoring and establishment of the incipient population. Increased pair formation and breeding efforts in response to larger group sizes have been similarly documented in several social species, including Royal penguins (*Eudyptes schlegeli*) [35], African wild dogs [14], Kakapo (*Strigops habroptila*) [15], and Flamingos (*Phoenicopterus* spp.) [36]. Indeed, for many such species there appears to be a critical group size threshold, below which social interactions such as pair formation and reproduction are disrupted or inhibited by component Allee effects [17,36,37]. In the case of the Puerto Rican parrot, White et al. [17] hypothesized such a flock size threshold, and believed it to be at least >50 individuals, based on past breeding performance of both wild and captive-released populations. For example, although the number of nesting pairs in the small relict population in EYNF had never exceeded six pairs/year in over 50 years, the breeding population reintroduced at RAF consisted of ten pairs within only six years of initial release [17]. However, since the 1960s, the relict population in EYNF had also never exceeded 50 individuals [17,18] until shortly before hurricane Maria in 2017, when it briefly reached 53–56 individuals (USFWS, unpublished data). In contrast, at RAF the combined presence of both the released and captive parrots resulted in an "effective social population" of 150–200 birds during the critical establishment phase of the reintroduction [17]. Following the initial establishment releases, the RAF wild population—and number of breeding pairs—increased rapidly [17,23]. Similarly, for the soft release type B at EYNF, this "effective social population" approached 300 individuals, and the celerity of pair-formation and post-release reproduction was unprecedented. Thus, in both cases (RAF, EYNF), we hypothesize that soft release type B surpassed a species-specific social threshold and demographic Allee effect inherent to the smaller populations associated with both the traditional and precision releases.

5. Conclusions

Our findings highlight the importance of developing effective strategies for achieving high survival, site fidelity, flock cohesion and prompt reproduction during psittacine reintroductions. Maximizing these parameters may aid in reducing the inherent vulnerability of such reintroductions to potential Allee effects, as described by Deredec and Courchamp [9] and recommended by Armstrong and Wittmer [16].

Consequently, we highly recommend use of soft release type A and B strategies, appropriately adapted to local and species-specific conditions and requirements. For captive-reared Puerto Rican parrots, these strategies have resulted in higher post-release survival, site fidelity and flock cohesion than either the traditional or precision releases. Most importantly, type A and B releases were the only methods that also resulted in prompt post-release reproduction by released parrots (Figures 5 and 6), and associated establishment of a resident breeding population. For those reintroductions in which a substantial numbers of conspecifics are available to be held on-site both before and after any initial releases, soft release type B would be the favored strategy—particularly if releases of reproductive-age individuals can occur shortly before or at the onset of the species reproductive season (sensu Figure 4). Examples include reintroductions at existing captive-rearing or rehabilitation facilities, or the a priori establishment of small captive populations of conspecifics on-site at proposed reintroduction locations [38]. Nevertheless, we recognize that for many reintroduction efforts, the required resources—both biological and financial—may be insufficient to effectively employ this particular strategy. In such cases, a potentially viable option would be the soft release type A strategy. Regardless of the specific strategy employed, diligent efforts directed at minimizing potential Allee effects should be incorporated into the overall reintroduction plan. For psittacines, we believe that both of our recent release strategies (types A, B) have clearly demonstrated the potential to achieve this critical reintroduction goal.

Author Contributions: Conceptualization, T.H.W.J. and M.L.; methodology, T.H.W.J.; software, T.H.W.J., J.V.; validation, T.H.W.J.; formal analysis, T.H.W.J.; investigation, T.H.W.J., W.A., G.B., A.J., L.R., I.R., M.T., P.T., J.V.; resources, M.L., P.T., J.V.; data curation, T.H.W.J.; writing—original draft preparation, T.H.W.J.; writing—review and editing, T.H.W.J.; visualization, T.H.W.J.; supervision, M.L. and T.H.W.J.; project administration, M.L.; funding acquisition, M.L. All authors have read and agreed to the published version of the manuscript.

Funding: This research received no external funding.

Institutional Review Board Statement: All research activities and animal handling procedures were carried out in full compliance and accordance with USFWS Permit PRT-697819 and USFWS Endangered Species Authorization SA-00-03, MBTA Permits MB182071-0; MB087748-0, and Puerto Rico DNER Permit 08-IC-002 (including all subsequent renewals).

Informed Consent Statement: Not applicable.

Data Availability Statement: Data presented in this study are available upon request from the corresponding author. Data are not publicly available due to security concerns regarding detailed locations of a critically endangered species.

Acknowledgments: We are grateful to the United States Fish and Wildlife Service, United States Forest Service–El Yunque National Forest, and the Puerto Rico Department of Natural and Environmental Resources for financial and logistical support of this study. Three anonymous reviewers provided helpful and insightful comments that improved an earlier version of the manuscript. The findings and conclusions in this article are those of the authors and do not necessarily represent the views of the U.S. Fish and Wildlife Service. Use of trade names in this article does not imply endorsement by the United States government.

Conflicts of Interest: The authors declare no conflict of interest.

References

1. IUCN Red List of Threatened Species. Version 2020. Available online: http://www.iucnredlist.org/apps/redlist (accessed on 26 October 2020).
2. White, T.H., Jr.; Collar, N.J.; Moorhouse, R.J.; Sanz, V.; Stolen, E.D.; Brightsmith, D.J. Psittacine reintroductions: Common denominators of success. *Biol. Conserv.* **2012**, *148*, 106–115. [CrossRef]
3. Snyder, N.F.R.; Koenig, S.E.; Koschmann, J.; Snyder, H.A.; Johnson, T.B. Thick-billed Parrot releases in Arizona. *Condor* **1994**, *96*, 845–862.
4. Brightsmith, D.J.; Hilburn, J.; del Campo, A.; Boyd, J.; Frisius, M.; Frisius, R.; Janik, D.; Guillén, F. The use of hand-raised psittacines for reintroduction: A case study of scarlet macaws (*Ara macao*) in Peru and Costa Rica. *Biol. Conserv.* **2005**, *121*, 465–472. [CrossRef]
5. Estrada, A. Reintroduction of the scarlet macaw (*Ara macao cyanoptera*) in the tropical rainforests of Palenque, Mexico: Project design and first-year progress. *Trop. Conserv. Sci.* **2014**, *7*, 342–364. [CrossRef]
6. Wolf, C.M.; Garland, T., Jr.; Griffith, B. Predictors of avian and mammalian translocation success: Reanalysis with phylogenetically independent contrasts. *Biol. Conserv.* **1998**, *86*, 243–255. [CrossRef]
7. Moseby, K.E.; Read, J.L.; Paton, D.C.; Copley, P.; Hill, B.M.; Crisp, H.A. Predation determines the outcome of 10 reintroduction attempts in arid South Australia. *Biol. Conserv.* **2011**, *144*, 2863–2872. [CrossRef]
8. Ewen, J.G.; Armstrong, D.P.; Parker, K.A.; Seddon, P.J. (Eds.) *Reintroduction Biology: Integrating Science and Management*; John Wiley and Sons: West Sussex, UK, 2012.
9. Deredec, A.; Courchamp, F. Importance of the Allee effect for reintroductions. *Écoscience* **2007**, *14*, 440–451. [CrossRef]
10. Dennis, B. Allee effects in stochastic populations. *Oikos* **2002**, *96*, 389–401. [CrossRef]
11. Stephens, P.A.; Sutherland, W.J.; Freckleton, R.P. What is the Allee effect? *Oikos* **1999**, *87*, 185–190. [CrossRef]
12. Stephens, P.A.; Sutherland, W.J. Consequences of the Allee effect for behaviour, ecology and conservation. *Trends Ecol. Evol.* **1999**, *14*, 401–405. [CrossRef]
13. Collazo, J.A.; Fackler, P.F.; Pacific, K.; White, T.H., Jr.; Llerandi-Roman, I.; Dinsmore, S.J. Optimal allocation of captive-reared Puerto Rican Parrots: Decisions when divergent dynamics characterize managed populations. *J. Wildl. Manag.* **2013**, *77*, 1124–1134. [CrossRef]
14. Somers, M.J.; Graf, J.A.; Szkman, M.; Slotow, R.; Gusset, M. Dynamics of a small re-introduced population of wild dogs over 25 years: Allee effects and the implications of sociality for endangered species' recovery. *Oecologia* **2008**, *158*, 239–247. [CrossRef] [PubMed]
15. Courchamp, F.; Clutton-Brock, T.; Grenfell, B. Inverse density dependence and the Allee effect. *Trends Ecol. Evol.* **1999**, *14*, 405–410. [CrossRef]
16. Armstrong, D.P.; Wittmer, H.U. Incorporating Allee effects into reintroduction strategies. *Ecol. Res.* **2011**, *26*, 687–695. [CrossRef]
17. White, T.H., Jr.; Collazo, J.A.; Dinsmore, S.J.; Llerandi-Román, I. Niche restriction and conservatism in a neotropical psittacine: The case of the Puerto Rican parrot. In *Habitat Loss: Causes, Effects on Biodiversity and Reduction Strategies*; Devore, B., Ed.; Nova Science Publishers: New York, NY, USA, 2014; pp. 1–83. Available online: https://novapublishers.com/wp-content/uploads/2019/05/978-1-63117-231-1_ch1.pdf (accessed on 14 November 2020).
18. Snyder, N.F.R.; Wiley, J.W.; Kepler, C.B. *The Parrots of Luquillo: Natural History and Conservation of the Puerto Rican Parrot*; Western Foundation of Vertebrate Zoology: Los Angeles, CA, USA, 1987.
19. Collazo, J.A.; White, T.H., Jr.; Vilella, F.J.; Guerrero, S.A. Survival of captive-reared Hispaniolan Parrots released in Parque Nacional del Este, Dominican Republic. *Condor* **2003**, *105*, 198–207. [CrossRef]

20. White, T.H., Jr.; Collazo, J.A.; Vilella, F.J. Survival of captive-reared Puerto Rican Parrots released in the Caribbean National Forest. *Condor* **2005**, *107*, 426–434. [CrossRef]
21. White, T.H., Jr.; Abreu-González, W. Dummy transmitters for pre-release acclimation of captive-reared birds. *Reintro. News* **2007**, *26*, 28–30.
22. USFWS. *Recovery Plan for the Puerto Rican Parrot (Amazona vittata)*; United States Fish and Wildlife Service: Region 4, Atlanta, GA, USA, 2009.
23. Llerandi-Román, I.; White, T.H., Jr.; Garcia-Bermúdez, M.A. Successful reintroduction of Puerto Rican parrots in the moist karst region of northcentral Puerto Rico. Unpubl. Abstract. In Proceedings of the 8th Caribbean Biodiversity Congress, Universidad Autónoma de Santo Domingo, Santo Domingo, Dominican Republic, 28 January–1 February 2014.
24. The 2012 IUCN Guidelines for Reintroductions and Other Conservation Translocations. Available online: http://www.iucnsscrsg.org (accessed on 2 November 2020).
25. White, T.H., Jr.; Abreu-González, W.; Toledo-González, M.; Torres-Báez, P. From the field: Artificial nest cavities for *Amazona* parrots. *Wildl. Soc. Bull.* **2005**, *33*, 756–760. [CrossRef]
26. White, T.H., Jr.; Brown, G.G., Jr.; Collazo, J.A. Artificial cavities and nest site selection by Puerto Rican Parrots: A multiscale assessment. *Avian Conserv. Ecol.* **2006**, *1*, 5. [CrossRef]
27. Oehler, D.A.; Boodoo, D.; Plair, B.; Kuchinski, K.; Campbell, M.; Lutchmendial, G.; Ramsubage, S.; Maruska, E.J.; Malowski, S. Translocation of Blue and Gold Macaw *Ara ararauna* into its historical range on Trinidad. *Bird Conserv. Int.* **2001**, *11*, 29–141. [CrossRef]
28. Sanz, V.; Grajal, A. Successful reintroduction of captive-raised Yellow-shouldered Amazon Parrots on Margarita Island, Venezuela. *Conserv. Biol.* **1998**, *12*, 430–441. [CrossRef]
29. Harrington, D.P.; Fleming, T.R. A class of rank test procedures for censored survival data. *Biometrika* **1982**, *69*, 553–566. [CrossRef]
30. Caro, T. *Antipredator Defenses in Birds and Mammals*; Univ. of Chicago Press: Chicago, IL, USA, 2005.
31. Griffith, B.; Scott, J.M.; Carpenter, J.W.; Reed, C. Translocation as a species conservation tool. *Science* **1989**, *245*, 477–480. [CrossRef] [PubMed]
32. Green, R.E. The influence of numbers released on the outcome of attempts to introduce exotic bird species to New Zealand. *J. Animal Ecol.* **1997**, *66*, 25–35. [CrossRef]
33. Gascoigne, J.C.; Lipsius, R.N. Allee effects driven by predation. *J. Appl. Ecol.* **2004**, *41*, 801–810. [CrossRef]
34. Angulo, E.; Roemer, G.W.; Berek, L.; Gasco, J.; Courchamp, F. Double Allee effects and extinction in the Island fox. *Conserv. Biol.* **2007**, *21*, 1082–1091. [CrossRef]
35. Waas, J.R.; Caulfield, M.; Colgan, P.W.; Boag, P.T. Colony sound facilitates sexual and agonistic activities in royal penguins. *Animal Behav.* **2000**, *60*, 77–84. [CrossRef]
36. Pickering, S.; Creighton, E.; Stephens-Woods, B. Flock size and breeding success in Flamingos. *Zoo Biol.* **1992**, *11*, 229–234. [CrossRef]
37. McCarthy, M.A. The Allee effect, finding mates and theoretical models. *Ecol. Model.* **1997**, *103*, 99–102. [CrossRef]
38. Instituto Chico Mendes de Conservação da Biodiversidade. *Executive Summary of the National Action Plan for the Spix's Macaw Conservation*; Instituto Chico Mendes de Conservação da Biodiversidade: Brasilia, Brasil, 2012.

MDPI

Article

Parrot Free-Flight as a Conservation Tool

Constance Woodman [1,*], Chris Biro [2] and Donald J. Brightsmith [1]

1 Schubot Avian Health Center, Department of Veterinary Pathobiology, Texas A&M University, College Station, TX 77843, USA; brightsmith1@tamu.edu
2 Liberty Wings Freeflight Training, P.O. Box 169, McNeal, AZ 85617, USA; director@birdrecoveryinternational.com
* Correspondence: gryphus@gmail.com; Tel.: +1-424-262-4743

Abstract: The release of captive-raised parrots to create or supplement wild populations has been critiqued due to variable survival rates and unreliable flocking behavior. Private bird owners free-fly their parrots in outdoor environments and utilize techniques that could address the needs of conservation breed and release projects. We present methods and results of a free-flight training technique used for 3 parrot flocks: A large-bodied (8 macaws of 3 species and 2 hybrids), small-bodied (25 individuals of 4 species), and a Sun Parakeet flock (4 individuals of 1 species). Obtained as chicks, the birds were hand-reared in an enriched environment. As juveniles, the birds were systematically exposed to increasingly complex wildland environments, mirroring the learning process of wild birds developing skills. The criteria we evaluated for each flock were predation rates, antipredator behavior, landscape navigation, and foraging. No parrots were lost to predation or disorientation during over 500 months of free-flight time, and all birds demonstrated effective flocking, desirable landscape navigation, and wild food usage. The authors conclude that this free-flight method may be directly applicable for conservation releases, similar to the use of falconry methods for raptor conservation.

Keywords: psittaciformes; macaw; conure; parakeet; reintroduction techniques; hand-rearing; pioneer flock; training; survival; flocking; predator evasion

Citation: Woodman, C.; Biro, C.; Brightsmith, D.J. Parrot Free-Flight as a Conservation Tool. *Diversity* **2021**, *13*, 254. https://doi.org/10.3390/d13060254

Academic Editors: Luc Legal, José L. Tella, Martina Carrete and Guillermo Blanco

Received: 30 March 2021
Accepted: 31 May 2021
Published: 8 June 2021

1. Introduction

Reintroduction is often a necessary conservation strategy in the face of rapid environmental change and anthropogenic impacts [1]. However, the successful release of captive-raised parrots has been limited due to a variety of problems, including predation, loss of fear of humans, inadequate foraging skills, poor landscape navigation skills, and inappropriate socialization [2–4]. In terms of best practices, released parrots often do better when added to established flocks [1,3,4]. However, there are not always appropriate flocks available, and creating a wild parrot flock de novo from captive-reared birds is a challenge [3,5,6].

For parrots, prerelease training can be a key factor in project success. Prerelease training is broadly defined and can encompass a wide variety of behavior-developing or -modifying techniques. Techniques include the birds observing predation events, keepers providing experience with wild foods to encourage food plant recognition, or operant conditioning training to recall to a protective aviary [4,7,8]. Researchers have been successful in encouraging birds to recognize wild foods, remain near the release site, interact in group settings, increase stamina, and recognize predators [2,4,9–14]. However, many of these methods could be improved, and methods for creating other key survival skills, including effective flocking, landscape navigation, and coordinated response to predators, remain undocumented.

During raptor conservation activities, many key elements of breed and release projects are developed using or modifying established practices of falconry [15], including captive breeding, rearing, physical conditioning, and release methods. Falconry methods applied to conservation have traditionally outperformed newly developed techniques [16,17], allowing these practices to speed species recovery using predeveloped, field-proven methods. For raptors, release success can be impressive: The long-term survival of captive-reared kestrels can match that of wild-bred individuals using falconry techniques and falconer staff participation [18].

Similar to falconry, there is a system for flying parrots outdoors called free-flight [19]. However, unlike falconry, parrot owners and breeders have historically had less participation in conservation actions [9]. Current parrot free-flight includes the sport flying of pet parrots, outdoor educational bird shows, and parrot keeping, where parrots fly in and out of building windows, similar to an indoor-outdoor pet door. Free-flight tends to utilize internet groups, classes, and in-person seminars to disseminate this practice [19,20]. This paper focuses on a popular method developed by the author Chris Biro (C.B.), heretofore referred to as the free-flight method. Since 1999, C.B. has trained over 400 students in using this method.

This system starts with the trainer creating a strong human-animal bond and site fidelity through the attendance, nurturance, and comforting of chicks during early development. Certain behaviors, including recalling to a trainer, flying point to point between trainers, getting off objects on command, and becoming wary when humans warn of danger are developed using an operant conditioning approach [21]. Once these basic behaviors are established, the trainer takes the birds outside and allows them to interact with the environment, then recalls them back into the safety of captivity. The trainer systematically exposes the birds to more and more complex and dangerous environments. Shortly after fledging and without the need for an operant conditioning protocol, the birds develop skills in flocking, aerial maneuvers, alertness, predator evasion, landscape navigation, wild food consumption, and utilization of information from heterospecifics. These behaviors appear to be generated through animal-environment interaction.

By comparison, most parrots in breed and release projects are provided normal captive care in cages and aviaries in breeding facilities and release sites [6,13,14,22–24]. Unfortunately, these conditions do not allow the animals to develop many of the "instinctive" behaviors that are needed for survival in the wild [3]. In wild individuals, these survival skills normally emerge as a product of the animals' interactions with the environment during their development [25], and in young parrots, these interactions often occur under the guidance and protection of their parents or other conspecifics [26]. In this way, the animals' survival behaviors are calibrated to the environment in which they are raised. Even in domestic animals, such as dogs and mice, that are carefully bred for consistent temperament and behavior, variations in postnatal experiences can have significant lifelong effects [27].

Studies of predator recognition and avoidance in birds have shown how inappropriate escape behaviors can form. Development in a captive environment may create less functional responses than wild development [28]. This is likely because each bird undergoes a threat learning process and has an individualized set of responses to the world formed during early development [29]. Without the necessary experiences during growth and development, the brain circuits that underlie normal behaviors may not form [30]. As a result, captive-raised birds are often considered poor candidates for release into the wild [3,7].

The free-flight method outlined here attempts to overcome this issue by providing the needed experiences during postnatal and juvenile development. Using the broad definition of prerelease training, free-flight "training" can be thought of as developing key survival techniques in release candidates through a combination of limited formal operant conditioning training in early development followed by intentional and sequential exposure to carefully selected and increasingly complex environments. Simply, free-flight is allowing birds' developmental processes to spontaneously fulfill their function by providing the opportunity at the correct age.

The objective of this paper is to introduce a community-based method that that has the potential for use in conservation science. To communicate this method, we document the mortalities and behavioral outcomes of 37 parrots of 7 species and 2 hybrids trained by C.B. using this method in 3 different flocks over a total period of 17 years.

2. Materials and Methods

C.B. began experimenting with free-flight techniques in 1993 and began using the specific method reported here in 1997. The activities reported on in this paper were conducted between 1997 and 2016. Author Constance Woodman, C.W., and C.B. began their collaboration in 1999. Starting in 2008, C.B. and C.W. worked together to document the method in writing [31] and compiled training and behavioral records of all the birds flown by C.B. In 2010, C.B. began formally teaching his expanded version of these methods to pet owners.

In 2016, we conducted a more formal research project to document the methods used and the results during the first year of creating a new free-flying flock. This process was reviewed by the Texas A&M University Institutional Animal Care and Use Committee (IACUC), College Station, TX, USA, and determined to be exempt from Animal Use Protocol on 3 February 2016, as the study utilized recording the outcomes when using pre-existing methods of private individuals (C.B.) from outside the university.

The training process used in this study begins with unweaned, pre-fledge birds and trains them in a series of more and more complicated physical and ecological systems. The guiding principle of this process is that, when placed in the appropriate environments, the birds' behaviors are shaped by interaction with the environment and other animals [31]. The method relies on the birds' natural responses to wild environments during juvenile development as opposed to behaviors shaped one at a time through interactions with a human trainer. Through this process, normal parrot survival skills develop by mimicking what happens in the natural rearing process of parrots raised in the wild by their parents.

The birds learn in 6 distinct environment levels (heretofore referred to as training levels, Figure 1). As the birds' abilities improve, they progress from simple environments (level 0: Indoors in a room) to highly complex environments (level 5: Forests and landscapes with major elevation changes inhabited by dangerous avian and mammalian predators with potentially dangerous weather conditions).

As a note, some verbiage in this paper disagrees with the language used by the free-flight community. We have attempted to codify and explicitly define some activities that are part of the culture of free-flight. For example, a trainer will verbally warn their birds of impending threats intuitively and not define communication as a part of training. Here, we refer to the "human alarm call" that alerts birds to threats, even though that specific term is not a part of the practice of free-flight.

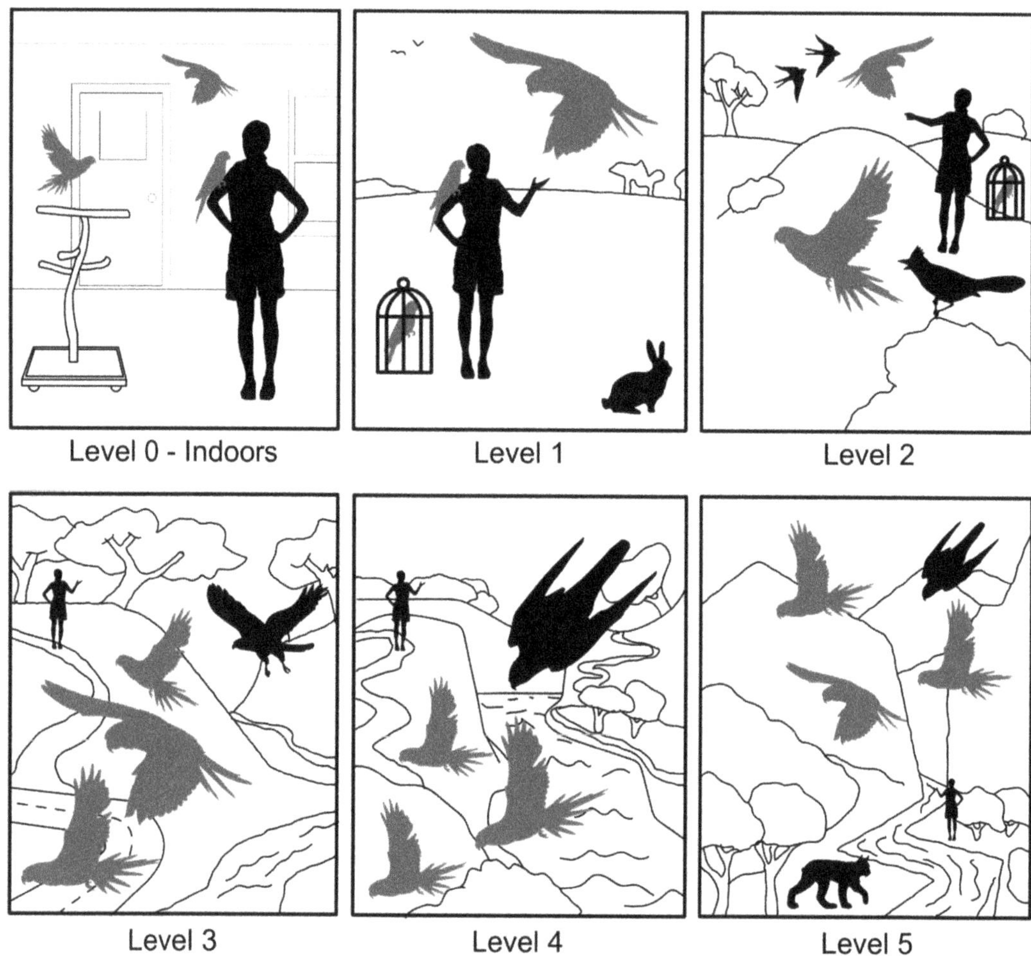

Figure 1. Schematic diagram showing the physical and ecological complexity of sites used for training parrots in this study. Loss of line of sight and landscape feature complexity increases with level. Key level elements include the presence of wild animals: Harmless at level 1, harassing to level 2, casual predator investigations in level 3, occasional determined predator at level 4, and immediate predation threat at level 5. Each image in the sequence shows how landscape features influence the ease of retrieving birds by vehicle or by foot, from contained birds indoors at level zero; to retrieval not being possible at level 5. Note the caged member of the social group (the "anchor bird") in levels 1 and 2 whose contact calls help keep other birds near the training site.

2.1. Flock Descriptions

For this study, we report on the raising and training of 37 individual birds. These birds were flown in 3 flocks, a large-bodied mixed-species macaw flock, a small-bodied mixed-species parrot flock, and a Sun Parakeet flock. All birds were reared and trained similarly except as noted below.

The small-bodied flock included a total of 25 different birds: Sun Parakeets (*Aratinga solstitialis*) $n = 16$, Mitred Parakeets (*Psittacara mitratus*) $n = 5$, Senegal Parrots (*Poicephalus senegalus*) $n = 3$, and a Burrowing Parakeet, (*Cyanoliseus patagonus*) $n = 1$. This group was active for 16 years (1997–2013). Not all birds were intended to be made fully independent, as C.B. focused on a subgroup of show flyers and others were less intensively trained.

The large-bodied flock included a total of 8 different birds: Hybrid "Calico" macaws (*Ara chloroptera* x *Ara militaris*) $n = 3$, Blue-Throated Macaws (*Ara glaucogularis*) $n = 2$, a Scarlet Macaw (*Ara macao*) $n = 1$, a Blue-and-Yellow Macaw (*Ara ararauna*) $n = 1$, and a hybrid "Shamrock" Macaw (*Ara macao* x *Ara militaris*) $n = 1$. This group was active for 13 years (2000–2013). This flock was trained to be maximally independent.

The Sun Parakeet flock included a total of 4 birds, all Sun Parakeets $n = 4$. This group was active for 1 year (2015–2016). This flock was used only for documentation of the early rearing process, and transition from indoor to outdoor flying and training was only conducted at levels 1–3.

When outdoor nesting attempts occurred in mature birds, the birds were not allowed to progress to wild reproduction to avoid creating naturalized populations.

2.2. Locations

The large and small-bodied flocks primarily flew in a rural area outside of Moab, UT, USA. The average temperature during the study period was 14.2 °C, with an extreme maximum of 43.9 °C and an extreme minimum of −21.1 °C. Average annual rainfall was 233 mm [32].

The birds were also transported by C.B. and flown in multiple locations in the Western United States, including locations in Washington State, California, and Oregon. The Sun Parakeet flock was fledged in College Station and primarily flown outdoors in Dripping Springs, TX, USA. The average temperature during the study period was 20.1 °C, with an extreme maximum of 39.4 °C and an extreme minimum of −15.6 °C. Average annual rainfall was 1189 mm [32].

Each group of birds added to the flocks had a different set of location experiences. The sites utilized for level 1, for example, comprised about 20 sites utilized across all 3 flocks. Some birds were trained in only 1 level and 1 area, others were trained in multiple level1 locations. For the 2 longer-term flocks, the small and large-bodied flocks, new level 2, 3, and 4 locations were frequently identified and utilized. Site identification included casual recognition of a site while traveling, where birds might only be flown once with permission of a property owner.

The 3 free-flight flocks varied in their range size based on training. The Sun Parakeet flock was not trained to travel between locations, while the 2 longer-term flocks were. The large-bodied flock was encouraged to follow a vehicle over multi-kilometer trips, further than what was done for the small-bodied flock.

2.3. Data Types and Collection

Data on the large-bodied and small-bodied flocks were drawn from C.B.'s archives and C.W.'s photography and notes. The archives consisted of dated emails, SMS text messages, content and meta-data of digital photographs, and content and meta-data from videos. The data included each birds' name, species, age at first outdoor training, date of each bird's entry into their flock, duration of participation, the reason the bird left the flock, a maximum level reached, and total time spent flying outdoors. To record the Sun Parakeet flock rearing process, a video camera with a time-lapse recording function was mounted above the playpen to record the chicks and monitor how they utilized the space. Records for the Sun Parakeet flock consisted of content and meta-data from normal and time-lapse video, content and meta-data data from photographs, and contemporaneous notes taken by C.W.

Total time flying in a natural environment was estimated based on 12 h of daily flying when not working at seasonal educational shows. Hours flying were calculated per bird, meaning if a group of 10 birds flew for 4 h, there would be 40 h of flying time recorded. To understand how outdoor flight mortality outcomes compare to conservation outcomes of similar outdoor duration, a "flight months" metric was created. The hours of outdoor flying are converted to "flight months", consisting of 30 counts of 12 h outdoors. Mortality outcomes were analyzed using the Mayfield method [33], calculating the risk of death for 1 year. All data are presented as mean ± standard deviation unless otherwise noted.

2.4. Level 0

2.4.1. Goals

The skills the birds gained at level 0 were skills for socialization, weaning, and fledging. Meeting all the criteria in Table 1 were needed for the bird to move to a level 1 environment.

Table 1. Level p environmental characteristics and mastery criteria for parrot free-flight training. The birds in this study completed level p criteria between the time of fledge and weaning, ~70 d for Sun Parakeets, ~100 d for macaws.

Environmental Features	Mastery Criteria
• Handfeeding location. • Enclosed spaces such as a living room or outdoor aviary • No wild species.	• Trainer linked with the consistent meeting of care needs through associative learning. • Accepts food and water from the trainer. • Accepts interaction from trainer including snuggles and toy play readily. • Steps up on the trainer. • Approaches trainer on foot or wing when separated. • Returns to the trainer with recall cue. • Leaves perch with "get off" cue. • Lands on difficult to reach perches. • Flies throughout the entire space. • Orients to other birds in flight ("tagging," "chasing"). • Aerial maneuvers (i.e., "jinking" sudden turn in the air).

2.4.2. Acquisition

To document the general early rearing process for all flocks, 4 captive-bred, hand-reared, incubator-hatched Sun Parakeets from different clutches were purchased from a commercial bird breeding facility and assembled into an aggregated group of young. The hatch dates of the birds were unknown, but the developmental stages were roughly estimated as 33 days old (n = 1) and 40 days old (n = 3). When acquired, the chicks were able to walk between locations, thermoregulate, and possessed adequate stamina and coordination to climb up and over Carefresh-brand bedding (http://www.carefresh.com/ accessed on 22 February 2021) substrate and return to the nest box after play periods. At the time of acquisition, the chicks were not yet human-socialized. Gaping, swaying, and cowering in the presence of human beings were observed.

2.4.3. An Enriched Rearing Environment

For all 3 flocks, the rearing setup was intended to maximize opportunities for interaction with the environment. The environment, built as a playpen, was roughly 1 m × 1 m with 0.5 m-high cardboard walls (Figure 2). A small box with a paper towel flap provided a cavity for the birds. The box and playpen were routinely refilled with clean bedding. Various objects, climbing opportunities, and foods were placed on the bedding, including toys and soft comfort "cuddle" items. A lamp on a timer provided 12 h per day of direct lighting. The chicks were old enough to thermoregulate so they could be safely reared without a temperature-controlled brooder. This general setup focused on free-choice activity, where chicks could remain inside a dark box or leave the box and engage with multiple activities in the environment. Additional rearing and housing details are similar to those described by Speer [34]. Additional parrot developmental complexity, a topic much too complex for this methodology, has been described by Bond and Daimon [35].

Figure 2. Stages during free-flight training: (**A**) Sun parakeets flock at the time of acquisition, 33–40 days old. Chicks showed a lack of human socialization through gaping and swaying as well as cowering; (**B**,**C**) Playpen rearing area. 1. Feeding access door. 2. Wire cored rope climbing coil. 3. Box with paper towel entry flap. 4. Overhang to prevent climbing out. 5. Carefresh-brand bedding on the floor and in brooder box; (**D**) Author Chris Biro at a level 1 area appropriate for small birds, an open area of about 3 hectares. Note the transport carrier and anchor birds' cage; (**E**) Author Chris Biro at a level 1 area appropriate large-bodied birds, utilizing a much larger open area of about 16 hectares. Note the portable perch for back and forth flying; (**F**) Complex landscape navigation training (levels 3–5). Trainers on either side of a canyon and cliff complex recall the birds at the safest crossing points to train landscape navigation; (**G**) The large and small-bodied flock escape from a hawk (arrow) at the home base.

2.4.4. Feeding and Training

All chicks in all 3 flocks were hand-fed using plastic syringes and commercial Kaytee-brand hand-feeding formula (https://www.kaytee.com accessed on 22 February 2021). The objective of syringe feeding was to enable normal use of the beak and tongue as opposed to feeding by gavage needle where the mouth is bypassed. Solid food, including apple slices, breakfast cereals, and Zupreem parrot pellets (https://zupreem.com accessed on 22 February 2021), were provided daily to enable a smooth transition to weaning and maximize options for chick activities.

Feedings broadly followed the manufacturer's recommendations and varied based on individuals' ingested amount per feeding, digestion speed, and age. Body condition scoring, a common veterinary technique, was utilized to monitor health [36].

The introduction of the recall cue was paired with feeding times. During feeding times, the chicks ran to the syringe and followed the human hand to different areas of the playpen while the recall cue was presented. The cue was a verbal "here birds" or the bird's specific name to train for individual recall versus full group. Over time, the birds came to the cue whether or not the syringe was present.

The Sun Parakeets weaned at approximately 60 days of age. To check that the wean was complete, the birds were weighed at the time of cessation of hand feeding and 1 week later. Weight losses of <5% indicated birds were maintaining body condition and the wean was successful.

2.4.5. Handling

To ensure that the chicks became comfortable interacting with the researchers, chicks were handled several times a day. Handling consisted of petting, holding, carrying, and interacting. Chicks approached human hands spontaneously 3 days after acquisition for the Sun Parakeets flock. The chicks were taken on 30-min trips to indoor or outdoor spaces away from the playpen roughly every 2 days.

2.4.6. Fledging

Once chicks had well-developed wing feathers, at approximately 50 days of age for the Sun Parakeets, they began spontaneously climbing to higher perches and intensely flapping their wings. By the time the chicks fledged, they were already responsive to the recall cue, having run to the hand while being called during feedings. As the birds became proficient at hopping to the trainer, hops were regularly practiced until they became short flights. The goal was to produce maximum flight skills available within this contained environment and to establish a behavior routine of flying to the human on cue, building the recall behavior prior to fledge.

The playpen environment was modified for the fledge by adding a second rope perch, with a loop extending above the playpen. A perch "tree" was set up near the playpen for flight practice. By day 60 of age, the Sun Parakeets spontaneously flew to the researcher and areas around the rearing area. The Sun Parakeets flew frequently throughout the day. As weaning occurs after fledging, parakeets who flew to the researcher for food were fed first, creating a competitive situation that rewarded fast response to the recall cue.

To develop a "get off" behavior, birds were spoken to sharply immediately upon landing in an unsafe location. The harsh volume and tone of voice resulted in them flying off. The birds appeared to become more sensitive to the sharp "get off" cue and readily responded more immediately as time went on. When birds could fly as a group, engage in aerial acrobatics, be individually or jointly recalled, and responded to a "get off" cue, they were ready to transition to a level 1 environment. The birds were called over for food, touch, or play, then shooed back to the perch or placed on the perch. Then, they were recalled again and given more attention or hand-feeding as a reward. The flying away and back to the trainer repeatedly led to a habit of back-and-forth flying to nearby approved objects, called "point-to-point" or "A-to-B" flight. Nonapproved landing sites were identified through the get off cue.

2.4.7. Human Alarm Call

When chicks were observed engaging in a problematic activity, such as climbing an object that would fall over, the trainer would warn the chick in a louder, stern tone. As chicks frequently had such problems, the chicks learned to associate the tone with coming danger, a precursor to the training creating increased outdoor wariness in later levels.

2.4.8. Move to Outdoor Caging

After confirming that weaning was complete, the Sun Parakeet flock was moved full-time to a tall outdoor aviary that was approximately 5 m by 3.5 m by 2.7 m tall in Dripping Springs, TX, USA. The aviary allowed for nearly constant, unmonitored flying, and physiological adaptation to the mild early summer outdoor environment. The large-bodied and small-bodied flocks were split across similar aviary buildings when not out flying. These outdoor aviaries were at the home base site. Back-and-forth flying was developed from the food- and comfort-seeking flights to the trainer. Large, portable perches were introduced into the outdoor aviary and utilized for back-and-forth flying practice.

2.5. Level 1

2.5.1. Landscape Setting

The landscape features of these sites were all similar and can be summarized as large, flat areas with few trees or shrubs similar to prairies or agricultural fields. There were limited opportunities for biotic interactions and only mild weather (Table 2, Figure 1). The transport vehicle was parked adjacent to the flying area to train the birds to return to this easily discernable landmark.

Table 2. Level 1 environmental characteristics and mastery criteria for parrot free-flight training. Training occurred as close to fledging as possible, and older individuals were observed to be more likely to panic fly or not bond with the group. Birds in this study gained mastery within about 3 weeks of flying. All criteria were mastered before birds were moved to the next-level environment.

Environmental Features	Mastery Criteria
• Open field. • Light wind. • No precipitation. • Distant wildlife. • Simple retrieval by foot or vehicle.	• All previous criteria. • Repeated practice flying at low and high altitudes. • Fly with and against the wind. • Demonstrate endurance through multi-minute continuous flapping flight. • Introduced to flocking outdoors with others. • Fly low the majority of the time (high flight is associated with nervous behavior, indicating the bird is unready for more complexity). • Tendency to stay near rally point vehicle between flights. • Develop complex movements initiated during aerial play. • Utter alarm and contact calls. • Respond appropriately to flockmate's contact and alarm calls through increased wariness, reply calling, and approaching calling flockmate.

2.5.2. Goals

The skills the birds gained at level 1 were foundational skills for flying in an outdoor space and returning to the trainer. Meeting all the criteria in Table 2 was needed for the bird to move to a level 2 environment.

2.5.3. Point-to-Point Flying

Before the training sessions, portable perching stored in the rally vehicle was set up adjacent to the rally vehicle. The bird(s) were taken from the carrier by hand and placed onto a portable perch. The trainer walked a few meters away and began the "point-to-point" back-and-forth perch flight routine developed during level 0. This back-and-forth routine was utilized to acclimate the birds to the new conditions in level 1 through a familiar routine. During the first outdoor flights, 1 bird at a time practiced point to point.

2.5.4. Rally Vehicle and Anchor Bird

During initial training, not all birds were taken out to fly at once. Birds not being trained were placed in a cage on the top of the rally vehicle, as shown in Figure 2. These caged bird(s) were able to contact call with the bird(s) being trained, forming an "anchor." During training sessions, the birds rarely flew outside of the contact call range of these anchor birds to which they were socially bonded, which helped them remain near the rally vehicle.

2.5.5. Recall Cue

The recall cue developed at level 0 was put into practice at level 1. Recall practice began with the back-and-forth flying routine and continued each time the bird flew off the perch and explored the area. When multiple trainers were available, birds could be recalled between trainers to practice distance flying and build stamina. The constant presence of the vehicle and anchor bird(s) during recall, as shown in Figure 2 helped reinforce the vehicle as the return point.

2.5.6. "Get Off" Cue

The "get off" cue, developed at level 0, was utilized at the level 1 outdoor location. Birds were cued to "get off" when they entered dangerous situations such as approaching powerlines or landing on a vehicle that was not the rally vehicle.

2.5.7. Human Alarm Call

Using warning tones while speaking in a louder voice, the trainers could verbally increase the birds' awareness. For example, if another vehicle approached but the birds were oblivious, the trainer would speak in a louder, warning tone, and the birds would increase their attention to the environment and notice the oncoming car. Through practice, the birds learned that the warning tone signaled a need for increased vigilance.

2.5.8. Flying in a Group

Chicks initially flew 1 at a time. Other socially bonded birds were held back in a cage on the rally vehicle. As the birds explored, they were praised for exploratory flights and increasingly complex aerial maneuvers. Once each bird was competent in outdoor point-to-point flying, the birds would be placed as a group on the portable perches and to fly point to point as a group until they became confident enough to explore the area and expand beyond point-to-point flights. Confidence was judged by a lack of fear-associated behaviors. Fear-associated behaviors included high flight, increased respiration, raised hackle feathers for moderate fear, completely smooth feathering for strong fear, dilated pupils, panting, tight gripping of the perch or arm, alarm vocalizations, and distress vocalizations [37].

2.5.9. Feeding on Plants

Feeding on plants was limited in level 1 except when birds were flown near lone trees or shrubs present in the landscape. Utilization of sparse trees of shrubs for practicing recall coming down from trees and flying up into them was conducted occasionally. Upon contacting a tree or shrub, the parrot inevitably began chewing on buds, seeds, shoots, and leaves. The "get off" cue was utilized to discourage chewing on plants that the trainer felt were inappropriate or might have been toxic.

2.5.10. Situations Special to Level 1

If the trainer felt that 1 or more birds were fearful, the birds were placed back in their carriers to allow them to acclimate to the site and watch their socially bonded fellows fly. When startled, some birds occasionally flew up very high (>40 m). When this happened, the anchor bird usually initiated back-andforth contact calls. The high-flying bird would circle the anchor bird and the trainer, eventually tiring and circling and gliding back to the anchor bird and trainer at the rally vehicle. The recall cue was utilized during the high flying to encourage the bird to return.

Circling flights, increased speed, and increased distance away from the trainer occurred. Eventually, all birds engaged in sudden movements using their tail to maneuver, called "jinking," recreating the aerial play patterns seen at level 0. This initial pattern of behavior was similar for all flocks.

Uncontrolled flights associated with strong fear states were called panic flights. In a panic flight, there was no response to the recall cue. Panic flights were rare. A prolonged panic flight was observed on a single occasion in 2014 when C.B was building a new macaw flock. The event is worst-case and is noteworthy enough to include even though the bird was not from the 3 studied flocks. A straight-line panic flight away from the rally vehicle was observed by C.W. when C.B. was flying a macaw. After 13 min of flight, the bird tired, lost altitude, and landed. The bird was not observed to engage in another panic flight over subsequent weeks. As the bird was being flown in an appropriately wide, agricultural field complex, the bird never left the line of sight or entered a forested area. Nervous flying at unusually high altitudes was only observed at level 1.

2.6. Level 2

2.6.1. Landscape Setting

Level 2 landscapes consisted of various shrubby fields, gentle hills, and sparsely treed areas (Table 3). Flying through trees introduced the birds to territorial songbirds, while flying in the vicinity of bodies of water provided harassing, curious gulls. Level 2 landscapes did not contain known dangerous predators except as aerial silhouettes on the horizon. Retrieval of birds was possible by off-road vehicle.

Table 3. Level 2 environmental characteristics and mastery criteria for parrot free-flight training. The average time to master level 2 was 3 weeks. Mastery time could be extended depending on the exact location and wildlife presence. The frequency of wildlife interactions was a limiting factor.

Environmental Features	Mastery Criteria
• Hills, shrubs, and small or isolated trees. • Breezy or gusting wind. • Mist or drizzle. • Non-dangerous wild species that follow or harass. • Retrieval by foot or vehicle relatively easy	• All previous criteria. • Recalls to the trainer from shrubs and trees. • Chooses perches for easy takeoff. • Startle response to strange species. • Joins flock in flight. • Coordinated group escape from curious or harassing wildlife initiated by any flock member. • Recalls after a momentary loss of sight of the trainer. • Returns to and follows rally vehicle over short distances.

2.6.2. Goals

The primary goals of level 2 were to encourage brief, independent navigation when the line of sight is broken, build stronger flocking skills, introduce interaction with shrubs and trees, and interact with wildlife to begin the development of antipredation behaviors (Table 3).

2.6.3. Rally Vehicle and Anchor Bird

During level 2, 1 bird was typically an anchor bird while the others were flying. During level 2, birds required less individual monitoring of behavior as panic flights and confusion were less frequent than during initial level 1 experiences.

The rally vehicle was parked close to the trainer, continuing to build an association of returning to the vehicle after periods of activity. The vehicle was often driven a short distance during training, changing the location of both the trainer and the vehicle. These alterations in location made it possible to train the flock to follow the vehicle and orient to a changing rally point.

2.6.4. Point-to-Point Flying

Similar to level 1, back-and-forth flying was utilized to adapt the birds to the new environment until they became comfortable with exploring. Birds were let out individually for training or as a group.

2.6.5. Recall Cue

The recall was practiced throughout the 1- to 6-h sessions, with significant focus on coming down from trees and shrubs. The birds followed the trainer around single trees of isolated forest fragments and learned to follow and recall even when visibility was blocked by trees and hills.

2.6.6. "Get Off" Cue

Birds were cued to "get off" when they entered potentially dangerous situations or attempted to consume unsafe items. Observed uses included interrupting perching on a stump near to the ground, landing on dangerous cacti, and landing on powerlines.

The "get off" cue was utilized to direct the birds to safely utilize perching in trees and shrubs. Members from all 3 flocks were not permitted to rest in dense tree cover or other locations where the birds could not see approaching predators. Inexperienced birds would initially perch close to the trunk of a tree and would be discouraged from doing so using the "get off" cue. Using the "get off" cue led to permanent behavior of perching on outer branches where emergency takeoffs were unobstructed by dense branches.

2.6.7. Human Alarm Call

The human alarm vocalizations initially developed in level 1 were utilized in subsequent levels. By increasing alertness in the flock, the trainer selectively sensitized the birds to dangerous situations. The level of volume and harshness of tone were commiserated with the danger. Birds were alerted to be wary at the approach of harassing wildlife. Bird wariness was increased selectively, such as for a dangerous hawk's silhouette flying far away. However, bird wariness was intentionally not increased for a harmless vulture silhouette at the same distance, building recognition of predators before close encounters.

2.6.8. Flying in a Group

Birds from the same cohort were permitted to fly together when each individual showed competence in recalling from trees or shrubs and when there was a break in the line of sight to the trainer. Birds were flown as individuals or in subgroups of the full flock to focus on skill development in specific members.

Birds from all flocks tended to group in response to the approach of harassing wild animals. When available, more experienced birds were added to level 2 birds in training once the newly flying birds showed competency in recalling from trees and broken line of sight. When flying with more experienced birds from outside the study, the Sun Parakeets flock learned to respond to the alarm calls and escape flights of the macaws and cockatoos. Sometimes, the Sun Parakeet flock would follow and perch next to the larger birds C.B. brought out to go flying, apparently gaining information about how to use the landscape from the more experienced flyers.

2.6.9. Feeding on Plants

Birds would almost always chew spontaneously on the nearest plant parts whenever they landed in foliage. The "get off" cue was utilized to discourage landing on spiny plants or chewing on undesirable plants.

2.6.10. Situations Special to Level 2

Northern mockingbirds, (*Mimus polyglottos*), blue jays, (*Cyanocitta cristata*), grackles (*Quiscalus* spp.), and various gulls (genus *Larus*) were observed to chase and threaten the parrots. Interactions with aggressive, non-dangerous birds like these allowed the free-flight flocks to practice grouping and responding to threats. The flocks spontaneously grouped up and fled or stood their ground in response to harassment. For example, the Sun Parakeet flock would occasionally group and chatter or chase harassing wildlife, beginning the development of mobbing behavior.

Through repetition, the flocks learned what stimuli indicated real danger. Initial inappropriate hypersensitivity to certain kinds of harmless events, such as a vulture high and far away on the horizon, became appropriate after multiple repetitions. Eventually, the birds learned to accept a distant vulture while still reacting to approaching raptors.

2.7. Level 3 through 5

2.7.1. Landscape Progression and Training Activities

Level 0 developed a bond between the trainer and birds and established many basic flight skills within a contained space, while levels 1 and 2 focused on expanding early skills to unconfined but open spaces. Basic outdoor flight skills, including beginner-level navigation, flocking strengthened recall, and avoidance of harassing wildlife, were achieved in levels 1 and 2. Increased 3-dimensional flying and brief loss of sight to the handler were achieved in level 2. The next levels were incremental increases toward fully independent function in the landscape. The environmental complexity increased from level 3 through 5 (Table 4), matching the trainer's evaluation of behavior mastery. Level 4 conditions were frequently similar to level 5, and only differed based on landscape access for the trainer. The trainer's ability to access a disoriented, injured, or struggling bird was an important factor in choosing a level 4 versus 5 locations. At level 5, there was no ability for recovery or rescue, emphasizing the need for fully independently functioning birds.

During level 3 through 5 training, the tools developed in earlier levels were utilized to encourage more complex behavior. The human alarm call was used to sensitize birds to new dangers in their environment without deleterious trial and error. The rally vehicle was parked out in the open as much as was possible to keep the return point visible to the birds in the increasingly hilly and forested terrain. Anchor birds were mainly utilized during the initial visits to new sites. An anchor bird was carried on the trainer's hand to encourage other birds to follow while on a hike, teaching routes and moving birds to desired training locations. Hand-carried anchor birds were also used to encourage reluctant birds to fly down from a tree or cliff or enter an area with novel features. Birds were recalled while the rally vehicle was in motion. The vehicle drove along access roads between sites, guiding the small and large-bodied flocks to fly between sites.

Back-and-forth flying practice was utilized to encourage the birds to safely interact with complex landscape features (Figure 2). A second trainer was often present to recall the birds to a location where the birds were unlikely to fly alone. Examples included canyon and cliff navigation, selection of safest crossing points over water or forested terrain, and selection of cliff diving sites to develop skill in diving escape behaviors.

Table 4. Level 3 through 5 environmental characteristics and mastery criteria for parrot free-flight training. The time for new flocks to reach levels 4 and 5 was normally within 1 year and before 2 years of age. Not all birds reached level 5, as intentionally flying the birds without the ability to retrieve or under immediate predator threat was not necessary for developing skills.

Level 3 Environmental Features	Mastery Criteria
• Substantial elevation variation. • Open forest. • Small ponds/small streams. • Windy, light precipitation. • Investigative pursuit by aerial predators. • Retrieval by foot and off-road vehicle.	• All previous criteria. • Exploration and learning of landscape, circling and exploration patterns. • Consistent routes between features, and preferred, safe, perching areas. • Habituation to weather and precipitation, respond by appropriate sheltering instead of anxiety behaviors. • Ability to fly during wind gusts. • Some mobbing or intimidating behaviors toward harassing wildlife. • Complex aerial escape maneuvers. • Recall after 2–3 min of loss of sight of the trainer.
Level 4 environmental features	**Mastery criteria**
• Water basins or major streams. • Windy, heavy precipitation. • Chance of pursuit by a determined aerial predator. • Retrieval is possible only by foot or specialty vehicle due to limited vehicle access.	• All previous criteria. • Fly up and down cliffs. • Complex diving and escape maneuvers. • Habituation to heavy precipitation. • Strong flight negotiating wind gusts. • Strong flock mobbing, escape, and predator confusion behaviors. • Recall readily after 5–10 min out of sight of the trainer. • Intelligent disobedience, refuse cues if there are hazards present.
Level 5 environmental features	**Mastery criteria**
• Extreme elevation changes and landforms. • Low visibility due to precipitation. • Large bodies of water or swift-moving water. • Immediate threat from determined predators. • Retrieval not possible due to landscape or lack of specialty vehicles.	• All previous criteria. • Function completely independently between sporadic recall cues. • Safely negotiate immediate and serious predator threats.

2.7.2. Goals

Level 3 through 5 training developed familiarity and appropriate responses to a variety of landforms, predators, local food plants, and weather conditions. Most bird activities consisted of the birds experiencing and reacting to biotic and abiotic environmental factors in the human selected environment, with guidance to move the birds through the landscape where certain experiences in the environment was provided by the trainers. As the birds functioned more independently, they were expected to engage in "intelligent disobedience," a concept most often encountered in service dog training [38]. The animal should be aware enough of the environment to refuse a trainer's cues that increase risk until the risk passes.

3. Results

A total of 37 parrots across 3 free-flight flocks logged a total of 501.2 flight months during this study. Total combined mortality during outdoor flying was six birds or 16%. The causes of outdoor flying mortality were human environmental hazards (pesticides

$n = 2$, powerline $n = 1$, wind turbine $n = 1$) and weather associated with flying birds in cold climates ($n = 2$). A total of 20 birds were retired either before or at the end of the study.

The large-bodied flock was flown over 13 years. The members of the large-bodied flock logged 147.3 flight months total (18 ± 3.2 months per individual, $n = 8$ individuals). The longest membership was 25.5 flight months over 9 years for a scarlet macaw, who was retired, the shortest membership was 15.3 flight months over 7 years for a blue-throated macaw, who was also retired (Table 5).

Table 5. Outcomes for three free-flight parrot flocks from 1997–2016 flown in the continental United States. Of 37 birds, 6 died due to abiotic hazards in the environment and 11 died due to husbandry-related issues. LB is large-bodied flock, SB is small-bodied flock, S is Sun Parakeet flock. Flight months are defined as 30 twelve-hour days flying in wildland spaces. Age level 1 is the age, in months, when a bird began flying outside. The level attained is the highest level on the free-flight Biro system of 0–5 environmental complexity.

Species	Flock	Age Level 1	Start Level 1	End Training	Membership Months	Flight Months	Level Attained	Fate
Blue & Yellow Macaw	LB	3	Apr-00	Apr-07	84	21	4	Wind turbine mortality
Scarlet Macaw	LB	3	Oct-04	Mar-13	102	25.5	5	Retired
macaw hybrid	LB	3	Jul-06	Mar-13	78	19.5	5	Retired
macaw hybrid	LB	3	Jul-06	Jan-12	66	16.5	5	Aviary fight mortality
Blue-Throated Macaw	LB	3	Jul-06	Mar-13	66	16.5	5	Retired
Blue-Throated Macaw	LB	3	Dec-06	Mar-13	61	15.25	5	Retired
macaw hybrid	LB	12	Oct-07	Mar-13	66	16.5	5	Retired
macaw hybrid	LB	3	Oct-07	Mar-13	66	16.5	5	Retired
Mean ± SD		4.1 ± 3.0				18.4 ± 3.2	4.9	
Patagonian Parrot	SB	3	Jun-97	Mar-13	154	38.5	5	Retired
Mitred Parakeet	SB	3	Jun-97	Aug-06	99	24.75	4	Electrical line mortality
Mitred Parakeet	SB	3	Jun-97	Aug-04	75	18.75	5	Pesticide mortality
Mitred Parakeet	SB	3	Jun-98	Jul-07	87	21.75	4	Aviary fight mortality
Mitred Parakeet	SB	3	Jun-98	Aug-04	63	15.75	4	Pesticide mortality
Sun Parakeet	SB	3	Apr-99	Mar-07	94	23.5	4	Aviary fight mortality
Sun Parakeets	SB	3	Nov-04	Mar-13	101	25.25	5	Retired
Sun Parakeets	SB	3	Nov-04	Nov-06	24	6	4	Husbandry issue
Sun Parakeets	SB	3	Nov-04	Nov-06	24	6	4	Husbandry issue
Sun Parakeets	SB	3	Nov-04	Nov-06	24	6	4	Husbandry issue
Mitred Parakeet	SB	4	Feb-08	Mar-13	61	15.25	5	Retired
Sun Parakeet	SB	3	Mar-08	Mar-13	60	15	4	Retired
Sun Parakeet	SB	3	Mar-08	Mar-13	60	15	4	Retired
Sun Parakeet	SB	3	Mar-08	Mar-13	60	15	4	Retired
Sun Parakeet	SB	3	Mar-08	Mar-13	60	15	4	Retired
Sun Parakeet	SB	3	Mar-08	Aug-10	30	7.5	4	Natural death
Sun Parakeet	SB	3	Mar-08	Feb-13	60	15	4	Weather mortality
Sun Parakeet	SB	3	Mar-08	Feb-13	60	15	4	Weather mortality
Sun Parakeet	SB	3	Mar-08	Mar-13	60	15	4	Husbandry issue
Sun Parakeet	SB	3	Nov-08	Nov-08	0.1	0	0	Husbandry issue
Senegal Parrot	SB	3	Mar-08	Mar-11	36	9	3	Husbandry issue
Sun Parakeet	SB	3	Mar-10	Mar-13	36	9	4	Retired
Sun Parakeet	SB	3	Mar-10	Mar-13	36	9	4	Retired
Senegal Parrot	SB	5	Mar-10	Mar-13	34	8.5	5	Retired
Senegal Parrot	SB	5	Mar-10	Mar-10	0.1	0	0	Husbandry issue
		3.8 ± 0.57				15.2 ± 7.6	4.2	
Sun Parakeet	S	3	Jul-15	Jul-16	12	1.1	3	Retired
Sun Parakeet	S	3	Jul-15	Jul-16	12	1.1	3	Retired
Sun Parakeet	S	3	Jul-15	Jul-16	12	1.1	3	Retired
Sun Parakeet	S	3	Jul-15	Jul-16	12	1.1	3	Retired
		3.0 ± 0				1.1 ± 0	3.0	

The small-bodied flock was flown over 16 years. The members of the small-bodied flock logged 349.5 flight months total (15.2 ± 7.6 months per individual, $n = 25$ individuals). The longest membership in the small-bodied flock was 38.5 flight months over a 16-year span for a burrowing parrot, who was retired, and the shortest membership was 0 flight months for a Senegal parrot and Sun Parakeet that were not yet bonded to a human trainer, escaped before starting outdoor training, and were subsequently unrecovered (Table 5). These two birds' zero values of outdoor training duration were omitted to calculate means and standard deviations.

The Sun Parakeet flock was flown for 1 year, and the total flight months were 4.4 (1.1 ± 0 flight months per individual, $n = 4$ individuals). All birds from the Sun Parakeet flock were retired after 1 year, with no early exits from the flock.

3.1. Predation and Husbandry-Related Mortality

No birds were killed by predators even though they flew in predator-rich environments. Predators seen at the Moab, Utah location included *Accipiter* hawks, *Buteo* hawks, peregrine falcons (*Falco peregrinus*), golden eagles (*Aquila chrysaetos*), coyotes (*Canis latrans*), fox species (genus *Vulpes*), and bobcats (*Lynx rufus*). The two long-term flocks, small-bodied and large-bodied, were primarily flown in a hawk migration area. The largest observed migration was a kettle of 197 hawks. The predators seen at the Dripping Springs, TX, USA location included *Buteo* hawks, *Accipiter* hawks, feral domesticated cats, fox species, and coyotes. Mortality was primarily due to husbandry issues (Table 5). Of the 37 birds studied, 11 died during captive management. These deaths occurred unrelated to outdoor training and included dying naturally during sleep or accidental escape of a young bird before any training began. The death during husbandry and training combined translates into a mortality rate of about 45%.

To understand this mortality in terms of risk over time in outdoor environments, the Mayfield method [33] was utilized. The calculation did not include the two fledged chicks that escaped before the start of outdoor training. During birds' first year of flight months, there was 100% annualized daily survival probability during outdoor training. During the first year, six birds were considered husbandry-related moralities, creating a 59% annualized daily survival probability related to handling and care. After the first year, annualized survival probability during training decreased to 77%. Annualized post-first year captivity and husbandry survival probability were 60%.

Flocking and Responses to Predator Threats

During level 0 training, hand-fed chicks flew as a group to be fed when formula was presented, practicing the fundamentals of group flight. The birds also tended to follow one another around the human home while expanding their activity area from the playpen. Social play during flight consisted of chasing, following, and pouncing, such as landing by grabbing the tail of a flockmate. During level 1, the groups became more cohesive, with birds increasingly seeking to remain with the group. During level 2, defensive flocking was developed through repeated interactions with harassing wild birds. Sometimes, flocking coordination was developed from a single, prolonged set of interactions with a particularly tenacious wild bird, such as a black vulture (*Coragyps atratus*), that followed the Sun Parakeet flock for an hour. In other cases, interactions with multiple wild birds formed the basis of a predator response. After each iteration or harassment, flocking behavior became more cohesive, forming coordinated vigilance, escape, and mobbing behaviors as seen in wild birds. A gull or a jay that might initially scatter the birds during early interactions would face a coordinated, alarm calling group during subsequent interactions.

Once birds gained level 2 mastery, flocking behavior was highly developed and consistent in all three flocks, with birds seldom leaving the line of sight of the group. Coordinated alarm calling and escape developed at that time. Level 3 training developed birds' discrimination between non-dangerous wildlife and animals that posed a predation threat. Early mobbing behaviors of agitated chatter and approach of predators by the flock, observed in earlier levels, grew to be aggressive in rare circumstances, unrelenting mobbing in level 4 conditions.

Interactions with predators were primarily with avian predators. It is estimated that over 100 aerial predation attempts were observed across the 3 flocks, primarily hunting attempts by bird-hunting *Buteo* and *Accipiter* hawks. When a predator was observed, one bird would typically alarm call and launch into flight, and its fellows would immediately launch as well. All three flocks responded to predator observation with a pattern of identification, alarm calling, launching, forming tight flying groups, predator avoidance, effective perching for escape, and exhibiting wariness. All three flocks utilized loud, continuous vocalizations in the presence of predators. If the birds were already airborne when a predator was observed, an initial bird would alarm call and the birds would form into a tighter group while already in the air.

The large and small-bodied flocks were observed, in some cases, mobbing predators and strange animals that approached the flock. Mobbing was a spectrum of behavior, ranging from tentatively approaching the target while the group alarm-called to the extreme of chasing and biting. Typically, the flocks alarm-called and stood their ground, facing the target as a group. C.W. observed one instance in the large-bodied flock and one instance in the small-bodied flock where flock members aggressively chased a target. In one instance of extended mobbing, the large-bodied flock drove a golden eagle that approached the flock out of a valley and up over the cliff rim about 2km away for approximately 10 min before breaking off pursuit. The small-bodied flock showed high aggression when they chased a pet parrot of a species that was not a part of their flock that had been accidentally let loose and flew into their midst. The flock surrounded the bird in the air, physically pushed the offending bird to the ground, and forced it to land, where a trainer broke up the skirmish.

3.2. Behavioral Outcomes

Behavioral outcomes are summarized in Figure 3. The two long-term flocks were outdoors regularly for long durations. The large- and small-bodied flocks were most regularly free-flown in the area around the home base, ranging up to 2 km normally. The two flocks occasionally flew further away when at the home base, but excursions were difficult to verify due to the lack of telemetry. The conditions at the Utah home base ranged from level 2–4 based on predator presence and weather. The normal flying day was approximately 12 h a day of flight time, varying depending on seasonal day length. Outdoor flight time for the small- and large-bodied flocks involved periods of no supervision, estimated to be up to 2 h, while trainers were in a nearby building. There were almost always more experienced birds present at the home base when new juvenile birds were let out to free-fly. Occasionally, birds would not recall at the end of the day and would outdoors overnight, but the frequency of these overnights was not recorded.

The Sun Parakeet flock home base in Dripping Springs, TX, USA was adjacent to a heavily forested area ranging from level 3 to 4, requiring the development of level 3 skills before flying at the home base. Their free-flight sessions were up to 6 h a day. Experienced free-flight trained birds from outside the Sun Parakeet flock were less often present at the home base when the Sun Parakeet flock was free-flown due to the difficulty of casual tracking of birds in among the dense trees.

Figure 3. Survival behaviors in free-flight trained parrots. (**A**) Large-bodied flock coordinated during an escape launch. (**B**) Blue-throat macaw and hybrid macaw evade a hawk. (**C**) Sun parakeet foraging on ocotillo (*Fouquieria splendens*) flowers. (**D**) Hybrid macaw foraging on juniper (Genus *Juniperus*) berries. (**E**) Scarlet and hybrid macaws forage alongside a wild turkey, *Meleagris gallopavo*. (**F**) Multispecies flocking in response to a predator. (**G**) Large-bodied flock engaging in long-distance navigation.

3.2.1. Landscape Navigation

No birds permanently left the home base site during training. Failed site fidelity was seen only in two fledged birds that escaped from the small-bodied flock prior to the start of formal outdoor training, a Sun Parakeet and a Senegal parrot, that had 0 h of outdoor training (Table 5).

Physical fitness was developed early in training, starting at level 1. The birds in all three flocks made extended flights as a form of social or individual play. Play flying was indicated by nonaggressive aerial dogfighting and jinking. Aerial circling in response to novel situations or wildlife presence was common, with investigative flights greater than 10 min of length regularly observed.

The free-flight training occurred in multiple spatially disparate landscapes. Through experience, the birds learned to navigate in novel landscapes. Once familiar with areas, the birds spontaneously went to nearby locations which were visible from the air, apparently recognizing landmarks and flying between them. The large- and small-bodied flock sometimes returned home or flew to the next location in known training routes spontaneously. The large-bodied flock executed the longest spontaneous navigation recorded: The group flew 11 km to return to the home base after training. The birds were also trained on routes through repetition. Travel involved repeating the route of the rally vehicle or the foot route of the trainer. Experienced birds would fly ahead of the trainer through a complex canyon or drainage system, having learned how the group would travel through the area based on earlier experiences. The large-bodied flock followed the rally vehicle for the longest duration, more than 3 km.

Practice within the landscape focused on navigating cliffs, canyons, hills, trees, and other landscape features at each level, emphasizing staying up high and enabling maximum line of sight for the three flocks. Birds flew over and not through heavily treed areas when navigating between locations, stayed above narrow canyons, and perched at the highest point of landforms whenever possible. The only flock that did not go between identified training locations spontaneously was the Sun Parakeet flock, as they were in a semirural residential area where it was not possible to fly between areas without disturbing property owners. Skill gains were an obvious progression. For example, macaws would dive off a 4m bluff with a hiking trail at a level 3 location while a trainer above and below used point-to-point flying to encourage diving. For level 4, those same macaws dove and rode the air currents down a landscape-sized, steep cirque, where the trainer had less access and ability to interact. At level 5, macaws fully and independently navigated major canyons and were not accessible to the trainer.

3.2.2. Foraging on Wild Foods

All three flocks were observed feeding on local plants (Figure 3). In all three flocks, all the birds routinely consumed the berries of junipers (Genus *Juniperus*), and specific individuals occasionally ate maple (*Acer* ssp.) seeds. The birds of all flocks daily chewed on leaf buds, seeds, any present fruits, and catkins of local plants.

The three parrot flocks were joined daily by other wild birds that foraged nearby on the ground, in adjacent trees, or in the same tree as the parrots. The parrots and wild birds appeared to form temporary foraging assemblages, where the parrots could receive information and copy behaviors of the wild birds. At least one time, the large-bodied flock was observed dropping to the ground to search for food in the grass with a single wild turkey (*Meleagris gallopavo*) despite the flock's training to stay off the ground (Figure 3). The turkey and macaws foraged safely within this novel multispecies complex, and the macaws' non-wary behavior suggested that this event had previously occurred. When wild birds, which were most often doves and songbirds, alarm-called or flushed, the parrot flocks increased wariness or launched into flight, demonstrating learning of heterospecific signals and behavioral cues.

4. Discussion

These hand-raised parrots trained with free-flight methods successfully developed skills in flocking, predator evasion, navigation of complex landscapes, and wild food use. These successes align well with the key goals of parrot prerelease training [4,6] and show that our methodology can avoid skill deficiency and aberrant behavior associated with many hand-raised parrots [2,7,26]. Whereas the level of human effort for free-flight training is high, it is comparable to other intensive bird management schemes utilizing hand-rearing, wild nest management, cross-fostering, and intensive soft release [12,39–42]. As a result, we feel that this free-flight method of human-guided learning has great potential for use in conservation releases.

4.1. Flocking, Predation, and Mortality

Captive-bred parrots commonly lack vital survival skills, such as being able to form a cohesive flock, and are often considered unsuitable for release due to lack of antipredator behaviors [2,7]. As a result, predation is a major cause of failure in parrot releases. A review of 100 releases for 10 species showed that high predator presence was the main predictor of release program failure, and that predator training was a predictor of post-release survival [4]. Fortunately, all birds in our study demonstrated appropriate antipredator behaviors, including identification of predators, flocking, increased vigilance, mobbing, and evasive landscape use. As a result, there were zero predation events in the studied flocks despite multiple observed interactions with predators. This contrasts with projects that have shown major losses of released birds and failure to establish a second generation due to predation [2,5]. Current antipredator training techniques teach release candidates to associate a predator with a fear state. For example, training used with Puerto Rican Amazons includes the following steps: (1) A silhouette of a hawk is passed over the cage while playing a hawk call, (2) a captive hawk attacks the aviary, and (3) a captive hawk attacks an armored Hispaniola parrot (*Amazona ventralis*) in full sight of the caged birds [12]. Whereas White et al. utilized a captive raptor, this free-flight training utilized naturally occurring encounters with non-dangerous harassing birds present in the environment to build early individual and group skills, then utilized increasingly dangerous predator interactions in the field to train aversion to specific species. Although both techniques increase wariness and vigilance, only free-flight training improves coordinated group responses to predators including flocking, evasive maneuvers, and mobbing. As a result, the use of free-flight training may help further reduce predation rates in hand-raised and released Psittacines.

4.2. Landscape Use

None of our birds permanently left the home base area or got lost in the landscape during this study. This stands in stark contrast to soft-release projects with a variety of macaws and parrot species where birds permanently left the release area, reducing the success of the projects [2,6,7,43,44]. The panic flights that have caused problems for these other reintroduction projects only occurred during our level 1 flying, which was always conducted in areas with few places to perch where it was relatively easy to recover the bird once it flew until exhaustion. Our success in preventing flyoffs was likely due to the gradual way that our training introduced birds to navigation in the landscape and our use of the anchor bird during the early stages of free-flying. Our use of anchor birds resembles the widely adopted practice of using caged conspecifics as an attractant to help keep released Psittacines near the release site [9,10,14,45].

During our study, we trained the two long-term flocks to navigate among major landmarks and find high-quality patches in a semi-arid and marginal landscape in Utah. The birds flown near springs and streams with fruiting trees and shrubs knew how to travel among high-quality patches. This skill is likely beneficial during conservation release projects, as multiple studies have shown that released Amazon parrots that ranged farther had higher survival rates presumably because they could exploit resources over a wider geographic area [10,11]. In addition, many native populations of large parrots move across hundreds of kilometers, ostensibly tracking food resources [26,45–47], and this may be key to long-term survival. To date, no parrot reintroduction projects have reported methods to train individuals to navigate to distant points in the landscape. Long-distance training may have benefits, where trainers fill a role typically provided by conspecifics. Random exploration by newly released blue-and-yellow macaws without an established flock were associated with higher mortality than newly released macaws following an established flock's pattern of landscape use [44]. Our results suggest that free-flight training may be useful in simultaneously reducing unwanted abandonment of the release area and teaching birds how to navigate among food sources, habitat patches, and other distant resources in the landscape.

4.3. Foraging

Teaching birds to forage on their own was not a focus of our free-flight training. However, through their natural habit of chewing plants they encountered, the parrot flocks all learned to consume wild foods. One of the unique aspects of this free-flight method is that interactions with naturally occurring native wild birds occurred during flying and foraging. These interactions appear to have led to the unintended benefit of learning to forage with a mix of wild species. Members of all three free-flight flocks foraged alongside other native species. These types of mixed-species foraging groups likely have multiple benefits as multispecies bird flocks are more likely to successfully utilize novel food sources [48]. Multispecies flocking might help naïve released birds utilize food sources and is an area in need of further study. In addition, all three flocks increased wariness and scanning in response to alarm calls or flushing of other birds. Eavesdropping on the signals of other animals, even when not participating in a multispecies flock, can also confer survival benefits as information transfer among different taxa likely improves predator avoidance [49,50]. As a result, free-flight training that includes interaction with native species can produce birds that can both forage more effectively and benefit from interactions with native species.

4.4. Human-Guided Learning

Our ability to recall birds allowed us to move them in and out of captivity and move them among training sites with different sets of resources, risks, and physical features. However, the use of recall in conservation releases of parrots is not unique. Release methodologies used with echo parakeets (*Psittacula eques*) included a recall cue to bring birds to a home aviary, where supplemental food was provided [8]. These parakeets were then given increasing exposure to the environment around the release site, through longer and longer outdoor periods between recall, until they were free-living. In our free-flight training, using complete recall back to cages provided us with the ability to transport our birds to new locations and expose them to sequentially more complex and dangerous sites and ecosystems throughout the training process. In this way, we shaped the flocks' landscape use through human knowledge and intent. Effectively, the human trainer determined the landscape usage patterns the birds normally learn from their parents and other conspecifics.

The flexible, "plastic" development of young parrots is not spontaneous, as behaviors develop from extended environmental and social interaction [26,51,52]. This plastic developmental process allows parrots to adapt to widely varying circumstances.

Naturalized populations of parrots can adapt to environments strikingly dissimilar to their ancestral range, such as *Amazona* parrots in Germany [53] and other locations throughout the world [54]. In the wild, behavioral flexibility allows wild parrots to adapt to human-altered environments [52,55] and transmit behavior socially among individuals [26]. When carefully planned, hand-rearing has the potential to magnify the ability of a parrot to adapt to its environment. When animals are raised by human caregivers, their behavioral repertoire may increase through the introduction to novel food types, foraging behaviors, and habitats unused by their ancestors [56]. Using the free-flight technique presented here, trainers should be able to customize the birds' landscape knowledge to the exact locations and resources that the trainers want them to exploit, even if those resources are quite distant. This should allow researchers to customize birds for specific release areas by imparting knowledge of the landscape that is not a part of traditional soft-release techniques.

The long-term effects of hand-rearing parrots for release are not well understood. Captive breeding programs regularly utilize hand-rearing with successful reproduction by hand-reared birds [57], suggesting sexual imprinting is not a major problem in this clade. Concerns about human-socialized animals being easily poached or engaging in human-wildlife conflict could be reduced if human-socialized birds were recalled after functioning as a core flock for the release of non-socialized birds. It is not yet known if the natural dispersal phase when subadult birds leave their family group would disrupt the parrot-human bond.

4.5. Potential Use of Free-Flight Training in Conservation

The 0 to 5 level system presented here is a useful way to compare animal survival skills to the complexity and dangers of potential release sites, even when the skill-building and learning processes are different among projects. For example, using the level system to analyze the classic thick-billed parrot releases in Arizona [2], the birds for release lacked coordinated flocking responses, which are required for mastery of a level 2 environment in the free-flight methodology we present here. By comparison, the environment and predator presence suggest that the Arizona release site was appropriate only for birds that had mastered skills equivalent to level 4 or 5 training. From such an analysis, the thick-billed parrots would not have been considered ready for release at this site, and plans for additional training or an alternative release site could have been considered.

In some instances, getting the first released parrots established at sites without conspecifics can be challenging, especially in areas with high predation rates [2,5,6,44]. We propose that, using the techniques we outline here, projects should be able to create pioneer flocks of birds that can (1) be recalled to captivity as needed, (2) have highly advanced flocking and predator avoidance skills, (3) safely forage on a wide diversity of natural foodstuffs, (4) utilize other native bird species as information sources and foraging partners, and (5) navigate safely and effectively among resource patches in the landscape surrounding the release area. Though, unlike a permanent pioneer flock, these birds would form a kernel for new additions but would not be intended to remain in the wild. Once the kernel flock is established, additional young birds could be raised with minimal human socialization and be released shortly after fledging into this pioneer flock using techniques similar to traditional soft-release methods. This mixture of human-socialized and non-human-socialized birds could then remain together in the wild for months without being recalled until the non-human-socialized birds have learned the core survival skills and landscape navigation. This overlap period could be similar to the amount of time wild parrot chicks stay with their parents post-fledging. Once this overlap period is complete, the human socialized birds could be recalled and removed from the environment, leaving only a core flock of non-human-socialized birds. After recall, this trained flock could then be used to establish new flocks in other areas or to help raise additional release candidates. In this way, parrot free-flight training could help jumpstart parrot reintroduction efforts similar to the ways that falconry has revolutionized raptor reintroduction science.

Author Contributions: Conceptualization C.B., C.W.; data Collection, C.B., C.W.; analysis, C.W., D.J.B.; writing, C.W., D.J.B.; editing, C.W., D.J.B., C.B. All authors have read and agreed to the published version of the manuscript.

Funding: Funding provided by the National Science Foundation IGERT program, award #0654377, and the Texas A&M Applied Biodiversity Sciences Program.

Institutional Review Board Statement: The project was reviewed by the Texas A&M University Institutional Animal Care and Use Committee (IACUC) and determined to be exempt from Animal Use Protocol on 3 February 2016, as the study utilized recording pre-existing methods of private individuals from outside the university.

Informed Consent Statement: Not applicable.

Data Availability Statement: Video and photographs of early rearing and bird flight are available upon request. These videos and photos include private individuals, their property, and their homes, containing personal details. As such, the videos cannot be posted publicly.

Acknowledgments: Texas A&M IACUC for oversight and long conversations to understand how to best record this practice, Hill Country Aviaries, LLC. in Dripping Springs, Texas for being the home base of the Sun Parakeet flock and housing of C.B. and C.W. during training.

Conflicts of Interest: The authors declare no conflict of interest. For transparency, we note that C.B.s free-flight classes and seminars are fee-based.

References

1. Seddon, P.J. From reintroduction to assisted colonization: Moving along the conservation translocation spectrum. *Restor. Ecol.* **2010**, *18*, 796–802. [CrossRef]
2. Snyder, N.F.; Koenig, S.E.; Koschmann, J.; Snyder, H.A.; Johnson, T.B. Thick-billed Parrot releases in Arizona. *Condor* **1994**, *96*, 845–862.
3. Snyder, N.F.; Derrickson, S.R.; Beissinger, S.R.; Wiley, J.W.; Smith, T.B.; Toone, W.D.; Miller, B. Limitations of captive breeding in endangered species recovery. *Conserv. Biol.* **1996**, *10*, 338–348. [CrossRef]
4. White, T.H.; Collar, N.J.; Moorhouse, R.J.; Sanz, V.; Stolen, E.D.; Brightsmith, D.J. Psittacine reintroductions: Common denominators of success. *Biol. Conserv.* **2012**, *148*, 106–115. [CrossRef]
5. Ziembicki, M.; Raust, P.; Blanvillain, C. Drastic decline in the translocated ultramarine lorikeet population on Fatu Iva, Marquesas Islands, French Polynesia. *Reintrod. News* **2003**, *23*, 17–18.
6. Brightsmith, D.; Hilburn, J.; Del Campo, A.; Boyd, J.; Frisius, M.; Frisius, R.; Janik, D.; Guillen, F. The use of hand-raised psittacines for reintroduction: A case study of scarlet macaws (*Ara macao*) in Peru and Costa Rica. *Biol. Conserv.* **2005**, *121*, 465–472. [CrossRef]
7. Lopes, A.R.S.; Rocha, M.S.; Mesquita, W.U.; Drumond, T.; Ferreira, N.F.; Camargos, R.A.L.; Vilela, D.A.R.; Azevedo, C.S. Translocation and Post-Release Monitoring of Captive-Raised Blue-fronted Amazons *Amazona aestiva*. *Acta Ornithol.* **2018**, *12*, 37–48. [CrossRef]
8. Woolaver, L.; Jones, C.; Swinnerton, K.; Murray, K.; Lalinde, A.; Birch, D.; Ridgeway, E. The release of captive bred echo parakeets to the wild, Mauritius. *Reintrod. News* **2000**, *19*, 12–15.
9. Clubb, K.J.; Clubb, S.L. Reintroduction of Military Macaws in Guatemala. In *Psittacine Aviculture: Perspectives, Techniques and Research*; Shubot, R.M., Clubb, K.J., Clubb, S.L., Eds.; Avicultural Breeding and Research Center: Loxahatchee, FL, USA, 1992; Chapter 23.
10. Sanz, V.; Grajal, A. Successful reintroduction of captive-raised yellow-shouldered Amazon parrots on Margarita Island, Venezuela. *Conserv. Biol.* **1998**, *12*, 430–441. [CrossRef]
11. Collazo, J.A.; White, T.H.; Vilella, F.J.; Guerrero, S. Survival of captive-reared Hispaniolan Parrots released in Parque Nacional del Este, Dominican Republic. *Condor* **2003**, *105*, 198–207. [CrossRef]
12. White, T.H.; Collazo, J.A.; Vilella, F.J. Survival of captive-reared Puerto Rican parrots released in the Caribbean National Forest. *Condor* **2005**, *107*, 424–432. [CrossRef]
13. Estrada, A. Reintroduction of the Scarlet Macaw (Ara Macao Cyanoptera) in the tropical rainforests of Palenque, Mexico: Project design and first year progress. *Trop. Conserv. Sci.* **2014**, *7*, 342–364. [CrossRef]
14. White, T.H.; Abreu, W.; Benitez, G.; Jhonson, A.; Lopez, M.; Ramirez, L.; Rodriguez, I.; Toledo, M.; Torres, P.; Velez, J. Minimizing potential allee effects in psittacine reintroductions: An example from Puerto Rico. *Diversity* **2021**, *13*, 13. [CrossRef]
15. Bolton, M. Birds of prey and modern falconry. In *Conservation and the Use of Wildlife Resources*; Springer: Berlin/Heidelberg, Germany, 1997; pp. 153–170.
16. Kenward, R.E. Conservation Values from Falconry. In Recreational Hunting, Conservation and Rural Livelihoods: Science and Practice. 2009. Available online: https://onlinelibrary.wiley.com/doi/abs/10.1002/9781444303179.ch11 (accessed on 22 February 2021). [CrossRef]
17. Weaver, J.D.; Cade, T.J. *Falcon Propagation*; The Peregrine Fund: Boise, ID, USA, 1991.
18. Nicoll, M.; Jones, C.G.; Norris, K. Comparison of survival rates of captive-reared and wild-bred Mauritius kestrels (*Falco punctatus*) in a re-introduced population. *Biol. Conserv.* **2004**, *118*, 539–548. [CrossRef]
19. Moser, D. Parrots and Flight: Flight and the Companion Parrot. *AFA Watchbird* **2004**, *31*, 50–55.
20. Biro, C. Pirates'n parrots (or free-flying birds) stage II: Behind the scenes training. *AFA Watchbird* **2000**, *27*, 60–63.
21. Ramirez, K. *Animal Training: Successful Animal Management through Positive Reinforcement*; Shedd Aquarium: Chicago, IL, USA, 1999.
22. Clubb, S.L. Thick-billed Parrots: Homecoming for our native parrot. In *Psittacine Aviculture: Perspectives, Techniques and Research*; Shubot, R.M., Clubb, K.J., Clubb, S.L., Eds.; Avicultural Breeding and Research Center: Loxahatchee, FL, USA, 1992; Chapter 22.
23. Groffen, H.; Watson, R.; Hammer, S.; Raidal, S.R. Analysis of growth rate variables and postfeeding regurgitation in hand-reared Spix's Macaw (*Cyanopsitta spixii*) chicks. *J. Avian Med. Surg.* **2008**, *22*, 189–198. [CrossRef] [PubMed]
24. Earnhardt, J.; Vélez-Valentín, J.; Valentin, R.; Long, S.; Lynch, C.; Schowe, K. The Puerto Rican parrot reintroduction program: Sustainable management of the aviary population. *Zoo Biol.* **2014**, *33*, 89–98. [CrossRef] [PubMed]
25. Griffin, A.S.; Blumstein, D.T.; Evans, C.S. Training captive-bred or translocated animals to avoid predators. *Conserv. Biol.* **2000**, *14*, 1317–1326. [CrossRef]
26. Collar, N.J. Family psittacidae. In *Handbook of the Birds of the World*; Hoyo, J.D., Elliott, A., Sargatal, J., Eds.; Lynx Edicions: Barcelona, Spain, 1997; pp. 280–479.

27. Foyer, P.; Wilsson, E.; Jensen, P. Levels of maternal care in dogs affect adult offspring temperament. *Sci. Rep.* **2016**, *6*, 19253. [CrossRef] [PubMed]
28. Belin, L.; Formanek, L.; Heyraud, C.; Hausberger, M.; Henry, L. Influence of early experience on processing 2D threatening pictures by European starlings (*Sturnus vulgaris*). *Anim. Cogn.* **2018**, *21*, 749–758. [CrossRef] [PubMed]
29. Schleidt, W.; Shalter, M.D.; Moura-Neto, H. The hawk/goose story: The classical ethological experiments of Lorenz and Tinbergen, revisited. *J. Comp. Psychol.* **2011**, *125*, 121. [CrossRef] [PubMed]
30. Galván, A. Neural plasticity of development and learning. *Hum. Brain Mapp.* **2010**, *31*, 879–890. [CrossRef] [PubMed]
31. Biro, C.; Woodman, C. Expanding freeflight using outdoor environments: Safer bird shows and incredible experiences. In Proceedings of the International Association of Avian Trainers and Editors, Cincinnati, OH, USA, 2–5 May 2009.
32. NOAA, National Centers of Environmental Information Climate Data Online Search. 2020. Available online: https://www.ncdc.noaa.gov/cdo-web/search (accessed on 22 February 2021).
33. Mayfield, H. Nesting success calculated from exposure. *Wilson Bull.* **1961**, *73*, 255–261.
34. Speer, B. *Current Therapy in Avian Medicine and Surgery*; Elsevier: Amsterdam, The Netherlands, 2015.
35. Diamond, J.M. Mixed-species foraging groups. *Nature* **1981**, *292*, 408–409. [CrossRef]
36. Burton, E.; Newnham, R.; Bailey, S.; Alexander, L. Evaluation of a fast, objective tool for assessing body condition of budgerigars (*Melopsittacus undulatus*). *J. Anim. Physiol. Anim. Nutr.* **2014**, *98*, 223–227. [CrossRef] [PubMed]
37. Luescher, A. *Manual of Parrot Behavior*; John Wiley & Sons: Hoboken, NJ, USA, 2008.
38. Eames, E.; Eames, T.; Gingold, A. A Guide to Guide Dog Schools. Baruch College Guide Dog Book Fund. 1986. Available online: https://books.google.com/books?id=_EFUAAAAYAAJ (accessed on 22 February 2021).
39. Ewen, J.G.; Armstrong, D.P.; Parker, K.A.; Seddon, P.J. *Reintroduction Biology: Integrating Science and Management*; John Wiley & Sons: Hoboken, NJ, USA, 2012.
40. Fund, K.; Fund, B.; Plans, S.; Taiao, T.T.; Grants, S.T.; Scheme, N.R.M. Cross-fostering of the Chatham Island black robin. *N. Z. J. Ecol.* **1983**, *6*, 156–157.
41. Lloyd, B.; Powlesland, R. The decline of kakapo *Strigops habroptilus* and attempts at conservation by translocation. *Biol. Conserv.* **1994**, *69*, 75–85. [CrossRef]
42. Vigo-Trauco, G.; Garcia-Anleu, R.; Brightsmith, D.J. Increasing survival of wild macaw chicks using foster parents and Supplemental feeding. *Diversity* **2021**, *13*, 121. [CrossRef]
43. Willie, C. Military macaws in Guatemala. *Am. Birds* **1992**, *46*, 25–31.
44. Plair, B.L.; Kuchinski, K.; Ryan, J.; Warren, S.; Pilgrim, K.; Boodoo, D.; Mohammed, N. Behavioral monitoring of Blue-and-yellow Macaws (*Ara ararauna*) reintroduced to the Nariva Swamp, Trinidad. *Ornitol. Neotrop* **2008**, *19*, 113–122.
45. Salinas-Melgoza, A. Dinámica Espacio-Temporal de Individuos Juveniles del loro Corona lila (*Amazona finschi*) en el Bosque seco de la Costa de Jalisco. Master's Thesis, Instituto de Biologia, Universidad Nacional Autónoma de México, Mexico City, Mexico, 2003; p. 59.
46. Bjork, R.D. Delineating Pattern and Process in Tropical Lowland: Mealy Parrot Migration Dynamics as a Guide for Regional Conservation Planning. Ph.D. Thesis, Oregon State University, Corvalis, OR, USA, 2004.
47. Brightsmith, D.J.; Boyd, J.D.; Hobson, E.A.; Randel, C.J. Satellite telemetry reveals complex movement patterns of large macaws in the western Amazon basin. *Avian Conserv. Ecol. Écologie Conserv. Oiseaux* **2021**, in press. [CrossRef]
48. Freeberg, T.M.; Eppert, S.K.; Sieving, K.E.; Lucas, J.R. Diversity in mixed species groups improves success in a novel feeder test in a wild songbird community. *Sci. Rep.* **2017**, *7*, 43014. [CrossRef] [PubMed]
49. Schmidt, K.A.; Lee, E.; Ostfeld, R.S.; Sieving, K. Eastern chipmunks increase their perception of predation risk in response to titmouse alarm calls. *Behav. Ecol.* **2008**, *19*, 759–763. [CrossRef]
50. Fallow, P.M.; Magrath, R.D. Eavesdropping on other species: Mutual interspecific understanding of urgency information in avian alarm calls. *Anim. Behav.* **2010**, *79*, 411–417. [CrossRef]
51. Mason, G.; Burn, C.C.; Dallaire, J.A.; Kroshko, J.; Kinkaid, H.M.; Jeschke, J.M. Plastic animals in cages: Behavioural flexibility and responses to captivity. *Anim. Behav.* **2013**, *85*, 1113–1126. [CrossRef]
52. Renton, K.; Salinas-Melgoza, A.; De Labra-Hernández, M.Á.; de la Parra-Martínez, S.M. Resource requirements of parrots: Nest site selectivity and dietary plasticity of Psittaciformes. *J. Ornithol.* **2015**, *156*, 73–90. [CrossRef]
53. Martens, J.M.; Woog, F. Nest cavity characteristics, reproductive output and population trend of naturalised Amazon parrots in Germany. *J. Ornithol.* **2017**, *158*, 823–832. [CrossRef]
54. Jones, S. *Naturalized Parrots of the World: Distribution, Ecology, and Impacts of the World's Most Colorful Colonizers*; Princeton University Press: Princeton, NJ, USA, 2021.
55. Salinas-Melgoza, A.; Salinas-Melgoza, V.; Wright, T.F. Behavioral plasticity of a threatened parrot in human-modified landscapes. *Biol. Conserv.* **2013**, *159*, 303–312. [CrossRef]
56. Dinets, V. Can interrupting parent-offspring cultural transmission be beneficial? The case of Whooping Crane reintroduction. *Condor* **2015**, *117*, 624–628. [CrossRef]
57. Jordan, R.; Moore, M.; Woodman, C.J.; Donnelly, R. *A Guide to Macaws as Pet and Aviary Birds*; ABK Publications: Burleigh Bc, QS, Australia, 2015.

MDPI

Review

The Number and Distribution of Introduced and Naturalized Parrots

Carlos E. Calzada Preston and Stephen Pruett-Jones *

Department of Ecology and Evolution, University of Chicago, 1101 East 57th St., Chicago, IL 60637, USA; carlosc@uchicago.edu
* Correspondence: pruett-jones@uchicago.edu; Tel.: +1-630-248-2360

Abstract: Parrots have been transported and traded by humans for at least the last 2000 years and this trade continues unabated today. This transport of species has involved the majority of recognized parrot species (300+ of 382 species). Inevitably, some alien species either escape captivity or are released and may establish breeding populations in the novel area. With respect to parrots, established but alien populations are becoming common in many parts of the world. In this review, we attempt to estimate the total number of parrot species introduced into the wild in non-native areas and assess how many of these have self-sustaining breeding populations. Based the public databases GAVIA, eBird, and iNaturalist, 166 species of Psittaciformes have been introduced (seen in the wild) into 120 countries or territories outside of the native range. Of these, 60 species are naturalized, and an additional 11 species are breeding in at least one country outside of their native range (86 countries or territories total). The Rose-ringed Parakeet (*Psittacula krameri*) and Monk Parakeet (*Myiopsitta monachus*) are the most widely distributed and successful of the introduced parrots, being naturalized in 47 and 26 countries or territories, respectively. Far and away, the United States and its territories support the greatest number of naturalized parrots, with 28 different species found in either the continental US, or Hawaii or Puerto Rico. Naturalized species as well as urbanized native species of parrots are likely to continue increasing in numbers and geographical range, and detailed studies are needed to both confirm species richness in each area as well mitigate potential ecological impacts and conflicts with humans.

Keywords: naturalized parrots; introduced species; invasive species; world parrot trade; invasion biology

Citation: Calzada Preston, C.E.; Pruett-Jones, S. The Number and Distribution of Introduced and Naturalized Parrots. *Diversity* **2021**, *13*, 412. https://doi.org/10.3390/d13090412

Academic Editors: Michael Wink, José L. Tella, Guillermo Blanco and Martina Carrete

Received: 15 June 2021
Accepted: 26 August 2021
Published: 29 August 2021

1. Introduction

Parrots have been transported and traded by humans for at least the last 2000 years and this trade continues today [1]. Cardador et al. [2] summarized trade data available through the Convention on International Trade in Endangered Species of Wild Fauna and Flora (CITES [3]) and documented that during the 20-year period 1975 to 2015, more than 19 million individual parrots of 336 species were legally traded among countries. This involved an average of more than half a million birds each year, with the parrot trade representing approximately 25% of all legal bird trade [2].

Inevitably, some individuals of introduced alien species either escape captivity and/or are accidentally or purposefully released and may begin breeding in the wild in the novel area [4,5]. Parrots are no exception and released or escaped parrots are often quite successful at surviving in the wild in new areas. Over time, if a successful breeding population is established, the species would be considered naturalized in that area. In some cases, the new populations can expand rapidly and grow exponentially in size [6–10]. If the species extends its naturalized range and establishes additional populations, it may become invasive.

Naturalized and invasive species are increasing worldwide, and parrots represent an increasingly large proportion of the naturalized bird species [11,12]. Although the invasive nature of established foreign parrot species is debated [13–15], naturalized parrot populations are increasing in distribution and size. Additionally, their interactions with humans are also increasing and becoming more complex and involve both positive and negative aspects [16–20]. This interaction with humans also includes control of some populations. In many cities around the world two common introduced parrots, the Rose-ringed Parakeet (*Psittacula krameri*) and Monk Parakeet (*Myiopsitta monachus*) are being controlled due to real or perceived problems with human activity. This is also true for some species in their native distribution [21].

The wildlife trade that ultimately gives rise to naturalized populations of parrots can also directly and negatively impact populations of species in their native ranges [22]. In many cases, this trade is causing species to be endangered in their native area, while at the same time inadvertently creating the possible situation where a population may establish itself in a novel and foreign area. In addition, the established populations can have impacts on local and native species [13]. It seems critical, therefore, to know exactly how many parrot species have established breeding populations in novel areas outside of their natural distribution. Such information is critical for monitoring introduced populations, informing management priorities, and understanding how introduced population may relate to the conservation of endangered populations in the native range of species [20]. That is the purpose of this review. We summarize multiple databases and attempt to arrive at an estimate for the number of parrot species both introduced and naturalized in the world. Our effort includes providing a database combining information from separate sources for use by other researchers.

Efforts to estimate the number of naturalized parrots have been made for almost two decades, and a comparison of the results highlights that the number and distribution of naturalized parrots is increasing. In one of the first efforts at counting naturalized parrots, Lever [23]; see also [24] reported that 34 species of parrots established naturalized populations. Two years later, Runde et al. [25] reported that there were 39 naturalized parrot species. Subsequently, Menchetti and Mori [13] reported about 60 parrot species were breeding outside their native distribution, and Avery and Shiels [26] reported 54 species have been introduced into foreign areas and 38 of these have become established. Most recently, Royle and Donner [24] examined records in the Global Avian Invasion Atlas (GAVIA) database [27] from 1993–2012 and documented records of 129 species of parrots observed in 106 countries. From these records, Royle and Donner concluded that there were at least 47 species of parrots in 21 genera that are naturalized in at least one country outside their native range. Lastly, a recent estimate of the geographical range of naturalized parrots is that of Mori and Menchetti [15] in which they conclude that species are found in 47 countries and all continents except Antarctica [28–30]. The variation in recent estimates is due in part to the sources of the information reviewed, and the time frame considered. Although our study is also subject to the same limitations, our review represents the first attempt to estimate the number of naturalized species based on a combination of the available data sets that have previously been analyzed separately. Additionally, for the United States, we compare the data from the public data sets with detailed reviews and field observations to examine the consistency and accuracy in the public data sets.

2. Methods

Our examination of world parrot species follows the taxonomy of the International Ornithological Congress (IOC) [31]. According to that taxonomy, there are 399 recognized species of parrots, including 17 taxa now extinct, and 382 extant species.

2.1. Terminology

There are many terms now used in the literature on introduced and naturalized alien species [32]. We use the terminology of Blackburn et al. [33,34] and Richardson et al. [32] as follows: (a) introduced species–a non-native/alien species that has been transported outside of its native range by human means and for which individuals have been observed in the wild in the new and novel area; (b) Breeding-non-native/alien species for which there is evidence of breeding in the wild; (c) Naturalized-non-native/alien species that has established a self-sustaining population; (d) Invasive–non-native/alien species that has established a self-sustaining populations at multiple sites across a range of habitats.

2.2. Databases

We were focused on identifying populations of species of Psittaciformes that occur in areas outside of their natural ranges. Thus, records of sub-species were subsumed under their corresponding species. To assess the status of each species, we summarized all records in the Global Avian Invasions Atlas (GAVIA) database [27]. GAVIA is a spatial and temporal database that summarizes published literature on naturalized birds and classifies the occurrence of species into various categories based on published findings. The GAVIA database consists of 27,723 records of observations and/or data on alien birds, representing 971 species and spanning the period 6000 BCE–2014 CE. Each record details an introduced species' status within a country, as referenced by a particular publication. For our analysis, the GAVIA dataset was filtered to only include records of Psittaciformes. Furthermore, 76 records of introductions (corresponding to 22 species) for conservation purposes or reintroductions back into a species' native range (known or presumed) were excluded. The final GAVIA dataset we examined consisted of 3422 records of 127 species introduced into 109 different countries and administrative regions. Of these 127 species, 96 were also present in the eBird database (see below), whereas 31 were unique to the GAVIA database.

In the GAVIA database, the status of species is classified into one of six categories: Breeding = a species that is known to be breeding or to have bred in the area of introduction, but for which the population is not self-sustaining; Established = a species that has formed self-sustaining populations in the area of introduction; Unsuccessful = an introduced species that has been seen in the wild but has not been able to establish a breeding population; Died Out = a species that was once established in the area of introduction, but has become extinct (by non-human means); Extirpated = a species that was once established in an area, but has subsequently been exterminated by humans; and lastly Unknown = a species that is observed in the wild in the area of introduction but whose status is unknown relative to the other categories [27].

We sorted these records by species and country and collapsed the six categories to four: Introduced (Unsuccessful or Unknown status in GAVIA), Breeding, Naturalized (Established status in GAVIA), and Historic (Extirpated or Died Out status in GAVIA).

We complimented the above data from GAVIA with citizen science records from eBird [35] and the Alien Parrots Observatory project in iNaturalist [36]. These are spatial and temporal databases of species' observations as reported by citizen scientists. All eBird records (whether from checklists or individual observations) between 1960–2017 were downloaded and filtered to include only extant species of Psittaciformes (N = 2,342,926). We then mapped these observations onto a high-resolution world map (from the R packages rworldmaps and rworldxtra [37–39]) to identify the country/territory where the observation was made. Observations of a given species were excluded if they were made within that species' native range, as based on distribution maps available from BirdLife International [40]. Furthermore, observations within 250 km of the native range were also excluded under the assumption that these observations likely represent extralimital sightings rather than observations of introduced birds. If there were at least three observations of individuals that occurred on different days and were more than 250 km outside of their native range, we considered those observations to represent an introduced population.

The final data set of eBird observations comprised 215,699 records of 135 species. Observations in iNaturalist were handled similarly to those from eBird, and the resulting data comprised 12,760 observations of 34 species from 1960–2017. All 34 species present in the iNaturalist data set were also in the eBird database.

Although the records from GAVIA provide information on the status of introduced parrots (breeding, etc.) the records from eBird and iNaturalist generally do not, at least in terms of the occurrence data that we summarized. In our data set (Tables S1–S3), we scored a species as 'Observed' if the records came from eBird or iNaturalist. The category Observed is thus the same as Introduced (from GAVIA) but these are listed separately in the database to indicate where those data came from. In cases where the GAVIA database indicated a status of 'Historic' for a species, but there were also records in eBird and iNaturalist, the status was listed as 'Historic/Observed'.

Lastly, using the eBird and iNaturalist records, we determined each species' area of occupancy (AOO) using the R package redlistr [41] to quantify the area (in km^2) occupied by each species outside its native distribution. The AOO analysis examines a species distribution based on 2 $\times$ 2 km grids, and the total AOO for a given species is the sum of the area for the total number of grids in which that species has been recorded outside is native distribution subject to the criteria listed above (three observations at least 250 km distant). For the six species of introduced parrots with the largest values for AOO, we also map their worldwide distribution, using the R packages rworldmaps and rworldxtra [37–39].

For both the identification of introduced populations and the mapping of sightings for calculation of the AOO, consistent nomenclature between the various databases is critical. Although we used the IOC checklist [31], we also compared that taxonomy with that of the BirdLife/Handbook of Birds of the World checklist [42] to identify cases where taxonomic changes have been proposed. We identified 21 cases/taxa where the taxonomy has changed (Table S4) and that would impact our examination of introduced populations. First, as regards the GAVIA database and the identification of naturalized species, the newly recognized species are not present as distinct taxa in the GAVIA database. Thus, the GAVIA records could conceivably refer to naturalized populations of any one of the species listed in association with a particular species in the original taxonomy. This possible error is unavoidable until new research is done on these newly designated taxa. Similarly, there are no labeled sightings of the newly recognized species identified in either eBird or iNaturalist. When calculating the AOO for introduced species, we could only make the calculations for the species taxa listed in the 'original' taxonomy in Table S4. To calculate the AOO for each such species, we filtered out sightings within the distribution of the newly recognized taxa, as such observations would have artificially increased the AOO of the taxa under consideration.

Separate from examining the records in the above databases, we examine in detail the parrot species present in the United Sates. Several recent, and in-depth analyses and reviews of introduced parrots in the United States and its territories [43–46], permit comparisons between various data sets.

2.3. Political Designations

The GAVIA database, and records on eBird and iNaturalist, are only as geographically widespread as the publications or actual observations themselves. Thus, there are not records or observations for every country or geographical area. In our summary, we designated the country of observation as that location on the observation or reference publication.

Many countries administer political territories. When there were data on introduced parrots in territories, these data were summarized for the specific territory as separate from the country itself (Tables S1–S3). For example, Puerto Rico and US Virgin Islands are listed separately from the United States.

3. Results

Based on the GAVIA, eBird, and iNaturalist databases (hereafter referred to as the combined database), there are records of 166 species of Psittaciformes having been introduced (seen in the wild) in 120 countries or territories outside of the native range (Figure 1; Tables S1 and S2). These species comprise 43% (166 of 382) of all known species of Psittaciformes and approximately half (49.4%, 166 of 336) of the species of parrots identified in the international parrot trade [2]. Of these 166 species, 60 species have been recorded or are now known to be naturalized and an additional 11 species are breeding in at least one country outside of their native range, being present in a total 86 countries or territories.

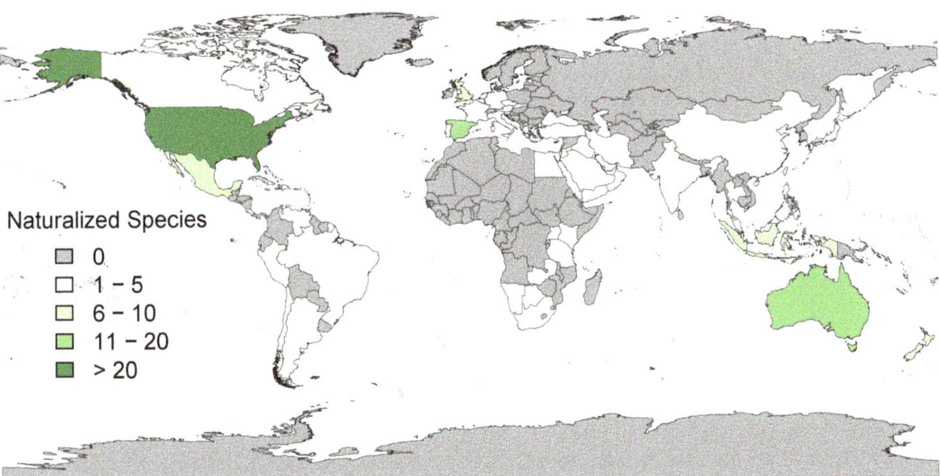

Figure 1. The distribution of naturalized and breeding species of parrots (Psittaciformes), according to the GAVIA (Dyer et al., 2017) dataset. The map depicts how many species of introduced Psittaciformes are naturalized or breeding per country. See text for definitions.

For the 71 species either breeding or naturalized, the mean number of countries (or territories) in which these species occur is 3.8 with a wide range of 1–51 (Figure 2). Almost half (30) of these species are recorded as either breeding or having a naturalized population in just one country. The six most widely distributed naturalized parrots, in terms of countries occupied are: Rose-ringed Parakeet, naturalized in 47 countries or territories; Monk Parakeet, naturalized in 26 countries or territories; Budgerigar (*Melopsittacus undulatus*), naturalized in 12 countries or territories; Alexandrine Parakeet (*Psittacula eupatria*) naturalized in 12 countries or territories; Brown-throated Parakeet (*Eupsittula pertinax*), naturalized in eight countries or territories; and Grey-headed Lovebird (*Agapornis canus*), naturalized in six countries or territories (Table S1).

Countries vary enormously in size, and the area of occupancy (AOO) is a more objective measure of the geographical distribution of introduced populations than number of countries occupied. For introduced parrots (species observed in the wild outside their native range), the AOO varied widely. The mean AOO was 714.3 km^2 (n = 135; range = 4–21,944 km^2; SD = 2595.7; Figure 3). Above, the six most widely distributed parrots are listed in terms of countries occupied. This list changes when considering AOO. The six species with the largest AOO of introduced populations are: Monk Parakeet (21,944 km^2), Rose-ringed Parakeet (18,812 km^2), Eastern Rosella (*Platycercus eximus*, 5976 km^2), Nanday Parakeet (*Aratinga nenday*, 4840 km^2), Red-crowned Amazon (Amazona viridigenalis, 3376 km^2) and Budgerigar (3172 km^2). Only the Monk Parakeet and Rose-ringed Parakeets overlap in these two ranked lists. Figures S1–S6 illustrate the global distributions of the sightings of these six species outside their native ranges.

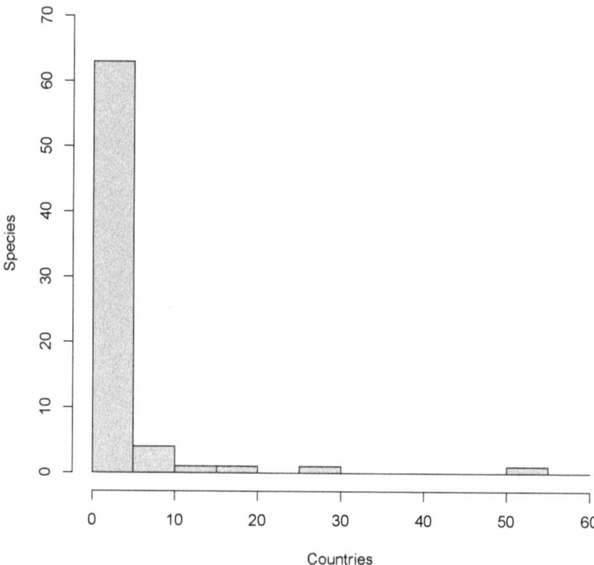

Figure 2. The frequency distribution of introduced and naturalized or breeding species of parrots (Psittaciformes) across countries.

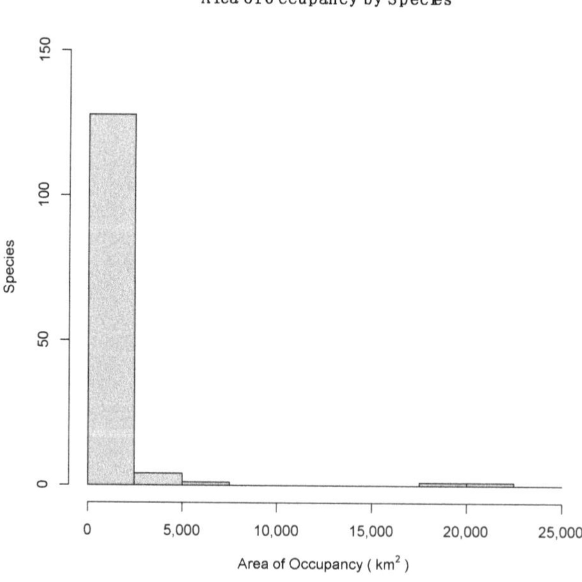

Figure 3. The frequency distribution of introduced and naturalized or breeding species of parrots (Psittaciformes) by their AOO (Area of Occupancy). The AOO only refers to introduced populations.

Despite the difference between countries as an indicator of geographical spread and AOO, there was a significant correlation between the number of countries a species was introduced in and the AOO (Figure 4; Spearman $R_s = 0.732$, $p < 0.001$).

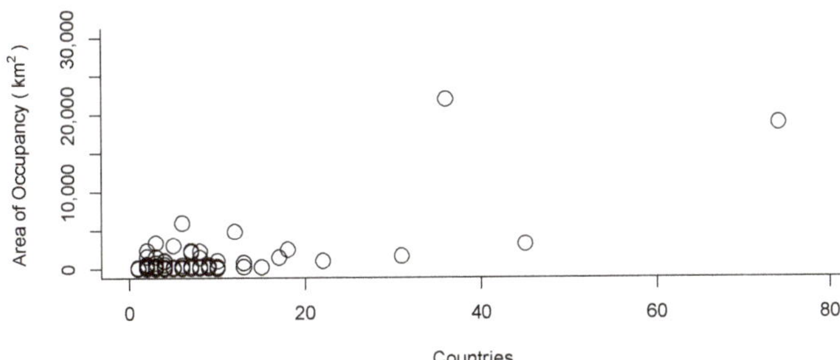

Figure 4. Relationship between area of occupancy and the number of countries that introduced parrots have been seen in the wild or are naturalized or breeding.

In terms of countries supporting naturalized parrots, and based on the combined database, the six countries or territories with the largest number of naturalized or breeding species are: United States (40 species), Australia, Spain, and Puerto Rico each with 14 species, Taiwan (9 species), and Singapore (8 species). This order is different if we consider records for all introduced species combined. That list is: United States (87 species), Brazil and Spain with 52 species, Australia and Puerto Rico with 35 species, and Mexico (20 species) (Table S1).

The records for Australia of 13 naturalized species (Table S1) illustrate the complexity of the parrot trade and the current distribution of introduced species. Currently in Australia, there is only one introduced species likely to be naturalized at present, the Rose-ringed Parakeet [47]. The other naturalized species in Australia isted in Table S1 are native to Australia but introduced in areas outside of their native range on the continent [47]. Thus, these species fall within the definition of transported, introduced, and naturalized used by authors, but the species' novel distributions are still within their native country Australia.

For the continental United States, there are records of 85 species of parrots introduced, breeding, or naturalized (Tables S1 and S2). At least two of these records are suspected to be in error or are inaccurate (that of Kuhl's Lorikeet *Vini kuhlii* and Kakapo *Strigops habroptila*), leaving 83 species. In comparison, the work by Uehling et al. [43,44], focusing on the continental United States during the 15-year period from 2002–2016, documented records of 56 species of parrots either introduced or naturalized. These two lists (the combined database (Table S1) and Uehling et al. [44,45]) overlap considerably when only considering naturalized species, but less so when considering all species. Thus, of the 25 naturalized species listed in [44], all but three are listed as naturalized or breeding in the combined database. Similarly, of the 22 species listed as naturalized in the combined database, 16 species are also listed as naturalized by Uehling et al. [44]. There is even greater overlap for the data in Hawaii and Puerto Rico. Of the five species of parrots listed by VanderWerf and Kalodimos [46] as naturalized in Hawaii, each of those species is listed as naturalized in the combined database (Table S3). For Puerto Rico, of the 12 naturalized species identified by Falcón and Tremblay [43], all of these species are listed as either introduced, breeding, or naturalized in the combined database (Table S3). Despite this considerable overlap when considering currently known naturalized species, the combined database (Table S1) also contains records of many species that have not been recently confirmed or verified. Thus, for the continental US, the combined database contains records of 28 introduced and eight breeding or naturalized species not confirmed by Uehling et al. [44,45].

Combining the lists of the recent studies [43–46], 28 species of Psittaciformes are naturalized in either the continental US, Hawaii, or Puerto Rico, and an additional 15 species are breeding there (43 species total). If we ask the same question of the combined database, there are records of 26 species as naturalized in either the continental US, Hawaii, or Puerto Rico, and an additional 14 species are breeding there (40 species total).

4. Discussion

Parrots are one of the most endangered groups of birds in the world, and in part this is because of the global trade driven primarily by the pet trade. As a result of this international trade, parrots as introduced and naturalized species are also among the most widely distributed groups of birds in the world, although much of this distribution is in novel areas outside of species' native ranges. It was our goal in this review to attempt to estimate the number of naturalized species of parrots in the world. This effort updates past estimates [13,23,25,26] and provides a combined database of parrot specific records from GAVIA, eBird, and iNaturalist available for use by other researchers. While previous efforts have utilized separate data sets, by combining data sets our goal was to a reliable, current estimate for introduced parrots around the world.

Of the 382 extant species of Psittaciformes, the majority of these (336) have been transported around the world through the global pet trade [2]. Our review indicates that almost half of these species (49.4%, 166 of 336) have escaped captivity or been released in novel areas and observed in the wild in no less than 120 countries or territories. Not surprisingly, introduction in a new area does not guarantee establishment success, but nevertheless at least 71 species are known to have established breeding or naturalized populations in 86 different countries or territories. Considering past estimates of the number of naturalized species [13,23,25,26] it is obvious that the number of naturalized parrots has increased over time. Part of this increase is related to a general increase in parrot trade around the globe [2], although this trade has changed drastically in some areas due to bans on trade that been imposed by the governments in some areas, e.g., the United States and the European Union (2,4,11,45). Some of the increase in naturalized parrots is likely also related to increased numbers of escapes or releases of individuals already present in a locality as the result of past trade activity.

There are necessary qualifications to the data that we summarized as well as our methods of analysis. Citizen science data are increasingly used to examine distributional patterns of species worldwide including introduced parrots [24,43,48–50]. Nevertheless, issues concerning species identification and spatial and temporal biases in sampling must be considered in analysis and interpretation [49–52]. Our combined database (Table S1) is subject to these considerations, and our conclusions about the numbers of introduced and naturalized species should be viewed as our best attempt to conservatively review the data available in public databases.

The GAVIA database is an important resource as a starting point, but given that it is not being updated with respect to changes in taxonomy and current research on the distribution and status of individual species, the importance of this database will likely decrease over time. It is also the case for many species, the status as based on publications listed in the GAVIA database needs confirmation from more recent sources. Similarly, our use of a 250 km distance as a filter for observations from eBird affects our conclusion about the number of introduced species. Without such a filter, every extralimital observation of a species would have been included but, in our opinion, would not have improved our understanding of the number or distribution of naturalized parrots. If a transported species establishes a new population on a new island or in a far-distant country, it is clearly a novel naturalized population. However, if an extralimital population establishes itself close to the native population, it is simply a matter of judgement or semantics whether that population is considered naturalized or just an example of a range expansion. This is particularly true in some countries, e.g., Australia, where the majority of naturalized parrot species are also species native to Australia. It is also the case that for some poorly studied species, the actual native distribution may not be fully known. In these cases, observations listed in eBird and iNaturalized that meet our distance criteria of 250 km may, in fact, simply be observations of birds in the native range but misclassified as representing introduced populations.

For any geographical area, combining citizen science records with detailed field observations by knowledgeable researchers will ultimately yield the most accurate and reliable records for distribution of introduced parrots, as exemplified by [46]. We hope that by providing the combined database (Table S1) other researchers can use these data as the starting point for such field observations. Our comparison of the combined database with recent publications on parrots in the United States illustrates one method of checking for consistency and accuracy. This comparison showed general but not exact agreement for species either breeding or naturalized, but less so for all introduced species. Considerable overlap was expected given that both Uehling et al. [44] and this study made use of eBird data. However, Uehling et al. [44] reported species for which there was at least one observation recorded in eBird, whereas we used a minimum of three observations. Clearly, any conclusion we or other researchers reach is dependent on the exact data set examined. Although not summarized specifically here, comparison of the combined database with recent surveys of introduced parrots in England [53], Europe [54], Spain and Portugal [55], and South Africa [56] also show general agreement with respect to naturalized and breeding species.

Calculation of the area of occupancy (AOO) for introduced species allows for a more objective analysis of a species' spread than just comparing the number of countries a species is recorded in. The number of countries a species has colonized as a naturalized species is important, but we expect that any examination of life-history correlates of success would be more likely to identify significant factors if such analyses focused on AOO. A comparison of the data for the two most common introduced species, the Rose-ringed Parakeet and Monk Parakeet, highlight the value of examining both measures of success. The Rose-ringed Parakeet is now naturalized in 47 countries, whereas naturalized Monk Parakeets are found in 26 countries. In contrast, the AOO of Monk Parakeets is ~15% larger than that of Rose-ringed Parakeets (21,944 km^2 compared to 18,812 km^2; Table S1). One possible explanation for this difference is that the Rose-ringed Parakeet is more widely traded worldwide in the pet trade than is the Monk Parakeet, leading to Rose-rings establishing themselves in more countries. In contrast, Monk Parakeets are highly adaptable and successful in areas where they establish themselves [57], leading to population increases and range expansions that would be observed through calculation of the AOO. We encourage consideration of both the AOO and countries occupied in future studies of the spread and success of introduced parrots.

Naturalized parrots are increasingly common in some areas and can present a host of both positive and negative interactions with humans. As Kiacz and Brightsmith [20] review, naturalized parrots offer timely and significant opportunities for conservation, research, and human society. The potential negative impacts of naturalized parrots, thoroughly reviewed by Mori and Menchetti [15] and Brightsmith and Kiacz [14] can be significant in some situations, as with damage to electrical infrastructure by Monk Parakeets or localized agriculture by some species. Nevertheless, overall, Brightsmith and Kiacz [14] conclude that these impacts are minor and do not in general justify the widespread and indiscriminate control of naturalized parrot species.

Given that populations of naturalized parrots are expanding, becoming urbanized in many cities, and generally representing larger fractions of local avifaunas, a greater understanding of their population biology, behavior, and interactions with humans is needed. We encourage regular local and regional surveys for species presence and abundance as well as large scale reviews of global patterns. Accurate data on the species richness and diversity of naturalized parrots will be critical for understanding the role of parrots as introduced and possibly invasive species, conservation efforts of threatened or endangered species, any management efforts when needed, and increasing the public knowledge and understanding of this important group of birds.

Supplementary Materials: The following are available online at https://www.mdpi.com/article/10.3390/d13090412/s1. Table S1: Status of all species of Psittaciformes identified in the GAVIA, eBird, and iNaturalist databases as having been introduced (seen in the wild) in a country or territory outside of their native range. Introduced = a non-native/alien species that has been transported outside of its native range by human means and for which individuals have been observed in the wild in the new and novel area. Breeding = a non-native/alien species for which there is evidence of breeding activity in the wild. Naturalized = non-native/alien species that has established a self-sustaining population. Historic = a non-native/alien species that was previously recorded as breeding but which was extirpated or it died out. Observed = a non-native/alien species that has been transported outside of its native range by human means and for which individuals have been observed in the wild in the new and novel area, according to observations in eBird and iNaturalist. This designation is thus the same as 'Introduced', but we separate the terms to indicate which database the record came from. All instances of 'Introduced' are from the GAVIA database. All instances of 'Observed' are from the eBird or iNaturalist database. Table S2: A summary of the numbers of species of parrots (Psittaciformes) outside of their native range according to the GAVIA database, eBird and iNaturalist. Here the category Introduced includes the category Observed from Table S1. See text for definitions of each category. Note that the categories in this table are mutually exclusive. Thus, species counted in the Naturalized column are not counted in the Breeding column. Table S3: Status of parrots (Psittaciformes) in the USA recorded in the GAVIA, eBird, and iNaturalist databases (This Study* in table) compared with those in recent and in-depth studies and reviews. Species arranged in alphabetical order of scientific name. Table S4: Recent taxonomic changes [31,41] in parrots that were considered when calculating the Area of Occupancy (AOO) of introduced populations. Figure S1: Distribution of sightings of Monk Parakeets (*Myiopsitta monachus)* outside of their native range based on records in eBird and iNaturalist. See text for explanation. Figure S2: Distribution of sightings of Rose-ringed Parakeets (*Psittacula krameri)* outside of their native range based on records in eBird and iNaturalist. See text for explanation. Figure S3: Distribution of sightings of Eastern Rosellas (*Platycercus eximus)* outside of their native range based on records in eBird and iNaturalist. See text for explanation. Figure S4: Distribution of sightings of Nanday Parakeets (*Aratinga nenday*) outside of their native range based on records in eBird and iNaturalist. See text for explanation. Figure S5: Distribution of sightings of Red-crowned Amazons (*Amazona viridigenalis*) outside of their native range based on records in eBird and iNaturalist. See text for explanation. Figure S6: Distribution of sightings of Budgerigar (*Melopsittacus undulatus*) outside of their native range based on records in eBird and iNaturalist. See text for explanation.

Author Contributions: Both authors contributed equally to the conceptualization, methods, analysis, and writing of this paper. Both authors have read and agreed to the published version of the manuscript.

Funding: This research received no external funding.

Institutional Review Board Statement: Not applicable.

Informed Consent Statement: Not applicable.

Data Availability Statement: The data summarized in this paper are available at the publicy archived websites for GAVIA [27], eBird [35], iNaturalist [36] well as in references [43–46].

Acknowledgments: Thanks to José Tella for the invitation to contribute to this special issue of *Diversity* and for extensive comments, to Emma Li for editorial assistance, and an anonymous reviewer for comments.

Conflicts of Interest: The authors declare no conflict of interest.

References

1. Scheffers, B.R.; Oliveira, B.F.; Lamb, L.; Edwards, D.P. Global wildlife trade across the tree of life. *Science* **2019**, *366*, 71–76. [CrossRef]
2. Cardador, L.; Abellán, P.; Anadón, J.D.; Carrete, M.; Tella, J.L. The world parrot trade. In *Naturalized Parrots of the World: Distribution, Ecology, and Impacts of the World's Most Colorful Colonizers*; Pruett-Jones, S., Ed.; Princeton University Press: Princeton, NJ, USA, 2021; pp. 13–21.
3. CITES Trade Database. 2015. Available online: http://trade.cites.org/ (accessed on 22 April 2020).
4. Carrete, M.; Tella, J. Wild-bird trade and exotic invasions: A new link of conservation concern? *Front. Ecol. Environ.* **2008**, *6*, 207–211. [CrossRef]
5. Vall-llosera, M.; Cassey, P. Leaky doors: Private captivity as a prominent source of bird introductions in Australia. *PLoS ONE* **2017**, *12*, e0172851. [CrossRef] [PubMed]
6. van Bael, S.; Pruett-Jones, S. Exponential population growth of monk parakeets in the United States. *Wilson Bull.* **1996**, *108*, 584–588.

7. Hobson, E.A.; Smith-Vidaurre, G.; Salinas-Melgoza, A. History of nonnative Monk Parakeets in Mexico. *PLoS ONE* **2017**, *12*, e0184771. [CrossRef] [PubMed]
8. Postigo, J.L.; Shwartz, A.; Strubbe, D.; Muñoz, A.R. Unrelenting spread of the alien monk parakeet in Israel. Is it time to sound the alarm? *Pest Manag. Sci.* **2017**, *73*, 349–353. [CrossRef]
9. Postigo, J.L.; Strubbe, D.; Mori, E.; Ancillotto, L.; Carneiro, I.; Latsoudis, P.; Menchetti, M.; Pârâu, L.G.; Parrott, D.; Rein, L.; et al. Mediterranean versus Atlantic monk parakeets *Myiopsitta monachus*: Towards differentiated management at the European scale. *Pest Manag. Sci.* **2019**, *75*, 915–922. [CrossRef] [PubMed]
10. Jackson, H.A. Global invasion success of the rose-ringed parakeet. In *Naturalized Parrots of the World: Distribution, Ecology, and Impacts of the World's Most Colorful Colonizers*; Pruett-Jones, S., Ed.; Princeton University Press: Princeton, NJ, USA, 2021; pp. 159–172.
11. Cardador, L.; Lattuada, M.; Strubbe, D.; Tella, J.L.; Reino, L.; Figueira, R.; Carrete, M. Regional bans on wild-bird trade modify invasion risks at a global scale. *Conservat. Lett.* **2017**, *10*, 717–725. [CrossRef]
12. Aloysius, S.; Yong, D.; Lee, J.; Jain, A. Flying into extinction: Understanding the role of Singapore's international parrot trade in growing domestic demand. *Bird Conserv. Internat.* **2020**, *30*, 139–155. [CrossRef]
13. Menchetti, M.; Mori, E. Worldwide impact of alien parrots (Aves Psittaciformes) on native biodiversity and environment: A review. *Ethol. Ecol. Evol.* **2014**, *26*, 172–194. [CrossRef]
14. Brightsmith, D.J.; Kiacz, S. Are naturalized parrots priority invasive species? In *Naturalized Parrots of the World: Distribution, Ecology, and Impacts of the World's Most Colorful Colonizers*; Pruett-Jones, S., Ed.; Princeton University Press: Princeton, NJ, USA, 2021; pp. 133–155.
15. Mori, E.; Menchetti, M. The ecological impact of introduced parrots. In *Naturalized Parrots of the World: Distribution, Ecology, and Impacts of the World's Most Colorful Colonizers*; Pruett-Jones, S., Ed.; Princeton University Press: Princeton, NJ, USA, 2021; pp. 87–101.
16. Senar, J.C.; Domènech, J.; Arroyo, L.; Torre, I.; Gordo, O. An evaluation of monk parakeet damage to crops in the metropolitan area of Barcelona. *Anim. Biodiver. Conserv.* **2016**, *39*, 141–145. [CrossRef]
17. Crowley, S.L. Human dimensions of naturalized parrots. In *Naturalized Parrots of the World: Distribution, Ecology, and Impacts of the World's Most Colorful Colonizers*; Pruett-Jones, S., Ed.; Princeton University Press: Princeton, NJ, USA, 2021; pp. 41–53.
18. Crowley, S.L.; Hinchliffe, S.; McDonald, R.A. Conflict in invasive species management. *Front. Ecol. Environ* **2017**, *15*, 133–141. [CrossRef]
19. Crowley, S.L.; Hinchliffe, S.; McDonald, R.A. The parakeet protectors: Understanding opposition to introduced species management. *J. Environ. Manag.* **2019**, *229*, 120–132. [CrossRef]
20. Kiacz, S.; Brightsmith, D.J. Naturalized parrots: Conservation and Research Opportunites. In *Naturalized Parrots of the World: Distribution, Ecology, and Impacts of the World's Most Colorful Colonizers*; Pruett-Jones, S., Ed.; Princeton University Press: Princeton, NJ, USA, 2021; pp. 71–86.
21. Bucher, E.H. Management of human-parrot conflicts: The South American Experience. In *Naturalized Parrots of the World: Distribution, Ecology, and Impacts of the World's Most Colorful Colonizers*; Pruett-Jones, S., Ed.; Princeton University Press: Princeton, NJ, USA, 2021; pp. 123–132.
22. Tella, J.L.; Hiraldo, F. Illegal and legal parrot trade shows a long-term, cross-cultural preference for the most attractive species increasing their risk of extinction. *PLoS ONE* **2014**, *9*, e107546.
23. Lever, C. *Naturalised Birds of the World*; A&C Black: London, UK, 2005.
24. Royle, K.; Donner, W.B. The distribution of naturalized parrot populations. In *Naturalized Parrots of the World: Distribution, Ecology, and Impacts of the World's Most Colorful Colonizers*; Pruett-Jones, S., Ed.; Princeton University Press: Princeton, NJ, USA, 2021; pp. 22–40.
25. Runde, D.E.; Pitt, W.C.; Foster, J. Population ecology and some potential impacts of emerging populations of exotic parrots. In *Managing Vertebrate Invasive Species: Proceedings of an International Symposium*; Witmer, G.W., Pitt, W.C., Flagerstone, K.A., Eds.; USDA/APHIS Wildlife Services, National Wildlife Center: Fort Collins, CO, USA, 2007; pp. 338–360.
26. Avery, M.L.; Shiels, A.B. Monk and rose-ringed parakeets. In *Ecology and Management of Terrestrial Vertebrate Invasive Species in the United States*; Pitt, W.C., Beasley, J.C., Witmer, G.W., Eds.; CRC Press: Boca Raton, FL, USA, 2018; pp. 333–357.
27. Dyer, E.E.; Redding, D.W.; Blackburn, T.M. The Global Avian Invasions Atlas: A database of alien bird distributions worldwide. *Sci. Data* **2017**, *4*, 170041. [CrossRef] [PubMed]
28. Ancillotto, L.; Strubbe, D.; Menchetti, M.; Mori, E. An overlooked invader? Ecological niche, invasion success and range dynamics of the Alexandrine parakeet in the invaded range. *Biol. Invas.* **2015**, *18*, 583–595. [CrossRef]
29. Menchetti, M.; Mori, E.; Angelici, F.M. Effects of the recent world invasion by ring–necked parakeets Psittacula krameri. In *Problematic Wildlife. A Cross-Disciplinary Approach*; Angelici, F.M., Ed.; Springer: New York, NY, USA, 2016; pp. 253–266.
30. iNaturalist Alien Parrots Observatory. 2019. Available online: https://www.inaturalist.org/projects/alien-parrots-observatory?tab=species (accessed on 5 December 2019).
31. Gill, F.; Donsker, D.; Rasumussen, P. (Eds.) IOC World Bird List (v 11.1). 2021. Available online: http://www.worldbirdnames.org/ (accessed on 19 August 2021).

32. Richardson, D.M.; Petr Pyšek, P.; Carlton, J.T. A compendium of essential concepts and terminology in invasion ecology. In *Fifty Years of Invasion Ecology: The Legacy of Charles Elton*, 1st ed.; Richardson, D.M., Ed.; Wiley-Blackwell Publishing Ltd.: West Sussex, UK, 2011; pp. 409–420.
33. Blackburn, T.M.; Lockwood, J.L.; Cassey, P. *Avian Invasions: The Ecology and Evolution of Exotic Birds*; Oxford University Press: Oxford, UK, 2009.
34. Blackburn, T.M.; Pyšek, P.; Bacher, S.; Carlton, J.T.; Duncan, R.P.; Jarošík, V.; Wilson, J.R.U.; Richardson, D.M. A proposed unified framework for biological invasions. *Trends Ecol. Evol.* **2011**, *26*, 333–339. [CrossRef] [PubMed]
35. *eBird: An Online Database of Bird Distribution and Abundance [Web Application]*; eBird; Cornell Lab of Ornithology: Ithaca, NY, USA, 2020; Available online: http://www.ebird.org (accessed on 1 October 2020).
36. iNaturalist Alien Parrots Observatory. 2020. Available online: https://www.inaturalist.org/projects/alienparrots-observatory (accessed on 9 November 2020).
37. R Core Team. *R: A Language and Environment for Statistical Computing*; R Foundation for Statistical Computing: Vienna, Austria, 2018. Available online: https://www.R-project.org/ (accessed on 9 November 2020).
38. South, A. Rworldxtra: Country Boundaries at High Resolution. R Package Version 1.01. 2012. Available online: https://CRAN.R-project.org/package=rworldxtra (accessed on 18 August 2021).
39. Brunsdon, C.; Chen, H. GISTools: Some Further GIS Capabilities for R. R Package Version 0.7-4. 2014. Available online: https://CRAN.R-project.org/package=GISTools (accessed on 18 August 2021).
40. BirdLife International and Handbook of the Birds of the World. Bird Species Distribution Maps of the World. Version 2019.1. 2019. Available online: http://datazone.birdlife.org/species/requestdis (accessed on 1 May 2021).
41. Handbook of the Birds of the World and BirdLife International Digital Checklist of the Birds of the World. Version 5. 2020. Available online: http://datazone.birdlife.org/userfiles/file/Species/Taxonomy/HBWBirdLifeChecklistv5Dec20.zip (accessed on 19 August 2021).
42. Lee, C.; Murray, N. redlistr: Tools for the IUCN Red List of Ecosystems and Species. R Package Version 1.0.3. 2019. Available online: https://CRAN.R-project.org/package=redlistr (accessed on 18 August 2021).
43. Falcón, W.; Tremblay, R.L. From the cage to the wild: Introductions of Psittaciformes to Puerto Rico. *PeerJ* **2018**, *6*, e5669. [CrossRef]
44. Uehling, J.; Tallant, J.; Pruett-Jones, S. Status of naturalized parrots in the United States. *J. Ornith.* **2019**, *160*, 907–921. [CrossRef]
45. Uehling, J.; Tallant, J.; Pruett-Jones, S. Introduced and naturalized parrots in the Continental United States. In *Naturalized Parrots of the World: Distribution*; Pruett-Jones, S., Ed.; Princeton University Press: Princeton, NJ, USA, 2021; pp. 193–210.
46. VanderWerf, E.; Kalodimos, N.P. Statis of naturalized parrots in the Hawaiian Islands. In *Naturalized Parrots of the World: Distribution, Ecology, and Impacts of the World's Most Colorful Colonizers*; Pruett-Jones, S., Ed.; Princeton University Press: Princeton, NJ, USA, 2021; pp. 211–226.
47. Rogers, A.M.; Kark, S. Australia's urban cavity nesters and introduced parrots: Patterns, processes, and impact. In *Naturalized Parrots of the World: Distribution, Ecology, and Impacts of the World's Most Colorful Colonizers*; Pruett-Jones, S., Ed.; Princeton University Press: Princeton, NJ, USA, 2021; pp. 277–292.
48. Bonter, D.N.; Zuckerberg, B.Z.; Dickinson, J.L. Invasive birds in a novel landscape: Habitat associations and effects on established species. *Ecography* **2010**, *33*, 494–502. [CrossRef]
49. Dickinson, J.L.; Zuckerberg, B.; Bonter, D. Citizen science as an ecological research tool: Challenges and benefits. *Annu. Rev. Ecol. Syst.* **2010**, *41*, 149–172. [CrossRef]
50. Minor, E.S.; Appelt, C.W.; Grabiner, S.; Ward, L.; Moreno, A.; Pruett-Jones, S. Distribution of exotic monk parakeets across an urban landscape. *Urban Ecosys.* **2012**, *15*, 979–991. [CrossRef]
51. Ratnieks, F.L.; Schrell, F.; Sheppard, R.C.; Brown, E.; Bristow, O.E.; Garbuzov, M. Data reliability in citizen science: Learning curve and the effects of training method, volunteer background and experience on identification accuracy of insects visiting ivy flowers. *Methods Ecol. Evol.* **2016**, *7*, 1226–1235. [CrossRef]
52. Robinson, O.J.; Ruiz-Gutierrez, V.; Fink, D. Correcting for bias in distribution modelling for rare species using citizen science data. *Diver. Distribut.* **2018**, *24*, 460–472. [CrossRef]
53. Butler, C.J. Naturalized parrots in the United Kingdom. In *Naturalized Parrots of the World: Distribution, Ecology, and Impacts of the World's Most Colorful Colonizers*; Pruett-Jones, S., Ed.; Princeton University Press: Princeton, NJ, USA, 2021; pp. 249–259.
54. Braun, M.P. Introduced and naturalized parrots in Europe. In *Naturalized Parrots of the World: Distribution, Ecology, and Impacts of the World's Most Colorful Colonizers*; Pruett-Jones, S., Ed.; Princeton University Press: Princeton, NJ, USA, 2021; pp. 227–239.
55. Carrete, M.; Abellán, P.; Cardador, L.; Anadón, J.D.; Tella, J.L. The fate of multistage parrot invasions in Spain and Portugal. In *Naturalized Parrots of the World: Distribution, Ecology, and Impacts of the World's Most Colorful Colonizers*; Pruett-Jones, S., Ed.; Princeton University Press: Princeton, NJ, USA, 2021; pp. 240–248.
56. Symes, C.T.; Ivanova, I.M.; Howes, C.G.; Martin, R.O. Introduced and naturalized parrots of South Africa: Colonization and the wildlife trade. In *Naturalized Parrots of the World: Distribution, Ecology, and Impacts of the World's Most Colorful Colonizers*; Pruett-Jones, S., Ed.; Princeton University Press: Princeton, NJ, USA, 2021; pp. 260–276.
57. Calzada Preston, C.E.; Pruett-Jones, S.; Eberhard, J.R. Monk parakeets as a globally naturalized species. In *Naturalized Parrots of the World: Distribution, Ecology, and Impacts of the World's Most Colorful Colonizers*; Pruett-Jones, S., Ed.; Princeton University Press: Princeton, NJ, USA, 2021; pp. 173–192.

 diversity

MDPI

Article

Sex and Age Effects on Monk Parakeet Home-Range Variation in the Urban Habitat

Juan Carlos Senar [1,*], Aura Moyà [1], Jorge Pujol [1], Xavier Tomas [1] and Ben J. Hatchwell [2]

1 Museu de Ciències Naturals de Barcelona, Pº Picasso s/n, 08003 Barcelona, Spain; aura.moya.4@gmail.com (A.M.); jpujoldelrio@gmail.com (J.P.); fxato@hotmail.com (X.T.)
2 Ecology, Evolution and Behaviour, School of Biosciences, University of Sheffield, Western Bank, Sheffield S10 2TN, UK; b.hatchwell@sheffield.ac.uk
* Correspondence: jcsenar@bcn.cat

Abstract: Home-range size is a key aspect of space-use, and variation in home-range size and structure may have profound consequences for the potential impact of damage and control strategies for invasive species. However, knowledge on home-range structure of naturalized parrot species is very limited. The aim of this study was to quantify patterns of home-range variation according to sex and age of the monk parakeet *Myiopsitta monachus*, an invasive parakeet in Europe. Mean kernel home-range size was 12.4 ± 1.22 ha (range 1.7–74.1 ha; $N = 73$ birds). Juveniles had a larger home-range size than adults, but sexes did not differ in kernel home-range size. The mean maximum distance moved by monk parakeets was 727 ± 37.0 m (range: 150–1581 m), and it was not dependent on either the sex or age of the birds. Having a small home range is one of the conditions for the feasible eradication of an invasive species; hence, the small home range of urban monk parakeets that we report here is good news for pest managers. However, this small home-range size can limit the effectiveness of culling operations with traps or feeders with contraceptives or poison, and other alternatives, such as funnel nets or traps, should be used.

Keywords: *Myiopsitta monachus*; home range; sex; age; urbanization; invasive alien species

Citation: Senar, J.C.; Moyà, A.; Pujol, J.; Tomas, X.; Hatchwell, B.J. Sex and Age Effects on Monk Parakeet Home-Range Variation in the Urban Habitat. *Diversity* **2021**, *13*, 648. https://doi.org/10.3390/d13120648

Academic Editor: Luc Legal

Received: 25 September 2021
Accepted: 3 December 2021
Published: 6 December 2021

1. Introduction

Animal movement and space-use is a key topic in ecology [1]. Early work mostly focused on describing movement patterns and their links with external factors (e.g., the environment), neglecting the individual causal drivers of this movement [2–4]. More recently, research effort has focused primarily on understanding the reasons for consistent intraspecific variation among individuals, investigating how morphological, behavioral, sexual, or age variation affect movement patterns [5–7].

Knowledge on the reasons and pattern of individual movements is especially important in the management of pests, to ensure that pest control actions are undertaken at a scale relevant to the species. Knowledge on home-range size and use is critical, for instance, to determine number and density of traps, their placement, and timing of trapping operations [8–11]. The same scale problem is applicable to other control methods, such as contraceptives or poison baiting [12,13]. Simulation models to manage the population dynamics and spread of pest species also need estimates of home-range parameters [13,14]. Sex and age are two main individual causal drivers of variation in home range [7], and because of that, any pest control plan has to scale actions having taken into account these two key variables [14].

Psittaciformes (parrots) is one of the most endangered bird orders in the world [15], and, at the same time, this group also contains some of the most invasive and damaging alien species [16,17]. However, knowledge on movement patterns of parrot species, including invasive ones, is very limited. Current available information is based on the study of a few species and provides data on just a few radio-tagged individuals [18–22], which reduces generalization of results found. Given that many parrot species are generally monomorphic [23], current available work does not provide any data on sexual differences in home-range use. Invasive parrot species are typically linked to urbanized habitats [24], but apart from the ring-necked parakeet *Psittacula krameri* [19], no information is available on home range-use in urban habitats.

The limited knowledge on movement patterns of parrots is probably due to the fact that parrots are difficult to capture and that their strong beaks easily destroy most devices that allow individual identification of the birds without having to recapture them [20]. However, we recently designed a metal tag attached to the bird neck with a collar that has proved very useful for the long distance identification of marked parakeets [25]. The method is simple and cheap, which allows marking a high number of individuals in an economically feasible way.

The aim of this paper is to take advantage of this marking method to study monk parakeet *Myiopsitta monachus* variation in home-range size and movements according to the sex and age of the birds. The monk parakeet is a highly successful invasive species in Europe and North America [24,26]. The marking device has been successfully used to determine dispersal patterns of the species in an urban environment [27]. However, similar to other parrot species, no detailed data is currently available on its home-range size. The monk parakeet is sexually monomorphic, but we use specifically developed genetic methods [28] to sex our birds and to analyze for sex specific home-range patterns. Data on home-range variation and movements is later used to delineate strategies for the control of the species.

2. Materials and Methods

The study was performed in the city of Barcelona, located on the northeastern coast of the Iberian Peninsula. Extending over 102.16 km^2, Barcelona is structured in 10 districts, with approximately 73% of the city built up. We concentrated our sampling efforts in and around the Parc de la Ciutadella area (625 ha) (Figure 1), which holds the highest monk parakeet density in the city [29]. Previous information (based on a smaller sample size) on the movement of monk parakeets showed that maximum home-range distances (between the center and the maximum extremes of the range) moved by juveniles were 1113 $\pm$ 103 m and 496 $\pm$ 122 m for adults [30]. This made us confident that the size of the area sampled was adequate to locate most daily home-range movements of the birds. Monk parakeets were captured using a modified Yunick trap at the Natural Sciences Museum of Barcelona, and marked with metal rings and a medal attached to a collar [25]. These unique identification tags allowed the identification of the birds without having to recapture them. The marking of birds has been carried out for two 6-week sampling periods (winter and summer) every year since 2002 [31].

On capture, we determined the age of the birds (juveniles or adults) based on molt patterns and capture history [32]. We also obtained a blood sample that allowed us to sex the birds molecularly (see Dawson-Pell et al. [28]).

The sampling of parakeets for the study of home range was carried out from 15 January to 15 July of 2016 and 2017 by direct observation and georeferencing of the marked individuals (visual recapture) in the Ciutadella Park area (Figure 1). This period of sampling was chosen because natal dispersal by juveniles in their first year, if present, has already taken place and the birds are ready to breed, building or maintaining nests, or actively breeding [30]. Monk parakeets are sedentary, long-lived, and use the nests year-round for roosting and breeding for multiple years [27,31]. The species is also colonial, building large compound nests [27]. About 80% of birds breed in the same nest in successive years, and dispersing adult birds move an average distance of just 37 m [27], removing any bias of using two years. Data from juvenile birds includes only resightings within the year in which they were aged as juveniles, because in the second year they were, by definition, adult birds, but also because some juveniles may disperse in their second year. In Barcelona, only about 50% of birds breed in their first year, with the others showing delayed breeding [27,33]. A transect was made on foot every weekday (on average 3 h per day, avoiding days with poor weather). In practice, the use of transects means that there was only one location per individual per day. Four different transects, with a total length of 15.4 km, were established through the study area. The transects visited all the green areas and locations where previous information indicated there could be parakeets breeding, resting, or foraging. Note that the area has been intensively surveyed since 2001 during a succession of different studies [27,29–31,33,34]. Data on individual identification, location, date, and time was stored in the field using app *iNaturalist* (https://www.inaturalist.org; accessed on 10 November 2021). The transects were carried out at different times of the day at random in order to avoid any bias due to the activity of the birds (observations made between 800 and 1400 h and between 1400 and 2000 h). We additionally included observations of birds seen at the trap at the Museum, since this is a regular feeding point for many birds at the Ciutadella Park (Figure 1). We additionally visited, in an opportunistic way, some green areas out of the study area to confirm that no marked birds were present.

Figure 1. Map of the study area. The yellow box includes the study area (625 ha) where the transects were carried out. The blue circle indicates the location of the Museum trap. The green areas in the city are colored green.

We used the program Ranges 9 [35] (http://www.anatrack.com; accessed on 10 July 2017) to analyze home-range size of Monk parakeets. We used the kernel estimation of home-range size, which determines the probability of use of space, creating nuclei according to the density of observations of the animal at different locations [36]. This is a measure of the area most used by the individual, calculated using 95% of the estimated total area [37]. Ranges 9 also computes the activity center as the location at which the Gaussian kernel estimator indicates highest density [36]. The kernel method, however, excludes movements outside of "normal activities". Hence, and in order to have information on movement potential away from the "normal" home range, we also computed maximum distances moved by an individual, which are computed by Ranges 9 as the maximum distances from locations to the range center. Incremental area analysis, also implemented in Ranges 9, was used to determine the minimum number of locations to be used per individual. Consecutive areas, which tend to increase initially as the animal is observed using different parts of its range, were plotted against number of locations until there was evidence of stability, indicating that adding further locations did not improve the home-range estimate [35]. The number of locations needed to reach the asymptote was estimated to be 21, and so analyses were conducted only on individuals located ≥21 times (N = 73). Kernel home-range size did not follow a normal distribution, the distribution being skewed to the right, and did not follow homogeneity of variances (Shapiro–Wilk W = 0.70, $p < 0.001$; Levene's test for homogeneity of variances $F_{3,69} = 6.34$, $p < 0.001$; skewness = 3.43 ± 0.28). Hence, to test for the relationship between home-range size and sex and age, we used a general linear model (GLM) on the logarithmic transformation. Logarithmic transformed data fit to a normal distribution and showed homogeneity of variances (Shapiro–Wilk W = 0.99, $p = 0.57$; Levene's test $F_{3,69} = 1.55$, $p = 0.21$; skewness = −0.05 ± 0.28). We used the kernel home-range size as a dependent variable, and we included sex and age (juvenile or adult) of the birds as categorical fixed factors. Mean maximum distances moved, again, did not follow a normal distribution (Shapiro–Wilk W = 0.93, $p < 0.001$; skewness = 0.76 ± 0.28; Levene's test $F_{3,69} = 0.97$, $p = 0.41$). Logarithm transformation overcorrected the data, which still did not fit to a normal distribution and was skewed to the left (Shapiro–Wilk W = 0.94, $p = 0.01$; Levene's test $F_{3,69} = 0.03$, $p = 0.99$; skewness = −0.73 ± 0.28). In such cases, square root transformation is advised [38,39]. The squared root transformed data did not fit to a normal distribution (Shapiro–Wilk W = 0.96, $p = 0.02$), but showed homogeneity of variances ($F_{3,69} = 0.27$, $p = 0.85$) and reduced skewness (0.07 ± 0.28), and thus this was the transformation we used in analyses. We used a GLM with the maximum distance as a dependent variable, and sex and age (juvenile or adult) of the birds as categorical fixed factors.

3. Results

We recorded 471 different marked individuals, with 4807 visual "recaptures". From these, we selected the subsample of birds recorded ≥21 times (N = 73 birds; average N observations per individual: 30 ± 1.12 SE, range 21–62; see methods). Mean kernel home-range size was 12.4 ± 1.22 ha (N = 73), with a median value of 10.1 ha (range 1.7–74.1 ha). Home-range size was negatively correlated to the number of observations per individual (r = −0.25, $p = 0.03$, N = 73). Some individuals showed a compact home range while some other individuals showed a multinuclear home-range area with two or three main activity areas (Figure 2).

Juveniles had larger home-range sizes than adults (juveniles: 16.0 ± 1.96 SE ha, N = 32; adults: 10.4 ± 1.66 ha, N = 41) (Figure 3), and sexes did not differ in kernel home-range size (Table 1).

The mean maximum distance moved by monk parakeets, from the center of the range, was 727 ± 37.0 m (range: 150–1581 m). Maximum distance moved was not correlated to the number of observations per individual (r = −0.001, $p = 0.99$, N = 73), but it correlated positively to kernel home range (r = 0.40, $p < 0.001$, N = 73). Maximum distance moved was not dependent on either the sex or age of the birds (Table 2).

Figure 2. Study area displaying all the visual observations of the 73 marked monk parakeets and three examples of home-range area (kernel method). Home ranges display 50, 75, and 95% contours. The home range of individual I76 is compact, while that of E44 and E71 is multinuclear, with two and three activity areas, respectively. The activity center, computed by Ranges 9 as the location at which the Gaussian kernel estimator indicates highest density, is marked with the sign +.

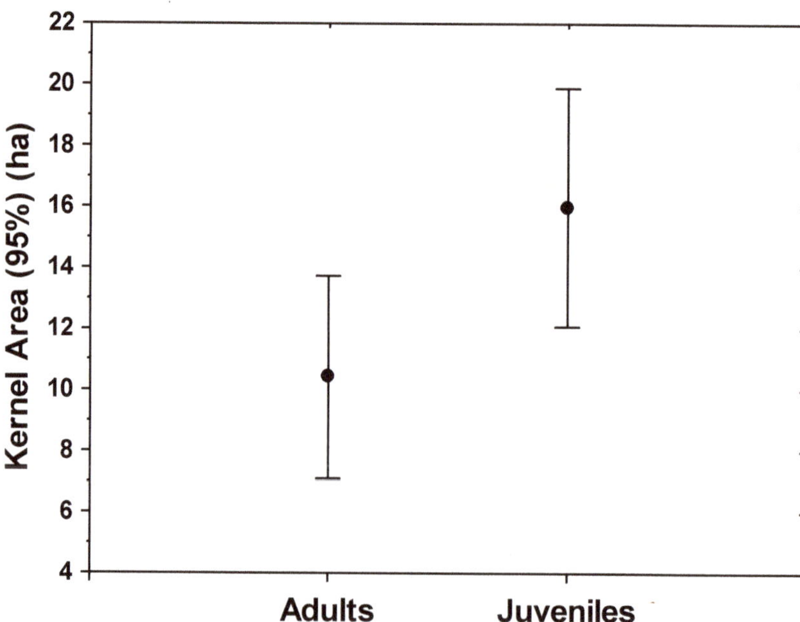

Figure 3. Variation in monk parakeet home-range size (ha ± SE) (kernel 95%) in Barcelona urban area according to the age of the birds ($N = 73$). Test of effects in Table 1.

Table 1. Results from the general linear model analysis (GLM) of the variation in monk parakeet kernel home-range size (ha) according to the sex and age (juveniles and adults) of the birds.

	F	Df	P
Sex	0.005	1.69	0.95
Age	4.50	1.69	0.04
Sex x Age	0.008	1.69	0.93

Table 2. Results from the general linear model analysis (GLM) of the variation in monk parakeet maximum distances moved (m) according to the sex and age (juveniles and adults) of the birds.

	F	Df	P
Sex	0.65	1.69	0.42
Age	1.12	1.69	0.29
Sex x Age	0.11	1.69	0.74

4. Discussion

Juvenile monk parakeets had larger home ranges than adults. Although there is a lot of interspecific variation, this is typical of many species [7,40] and may be related to the fact that in Barcelona monk parakeets, about 50% of juveniles do not breed in their first year [33], and hence are not so strongly tied to the nest as adult breeding birds. Juvenile birds, therefore, may wander more widely, returning to the nest only for roosting. Similarly to other species, the nest in adult birds may therefore act as a central place which limits movements [41,42]. In contrast, we found that extreme movements did not differ between sexes and ages, which suggests that occasional forays away from the home range are carried out by all the individuals irrespective of sex and age. However, we have to acknowledge that detailed GPS data on movement patterns is needed to quantify the distance and frequency of forays more precisely. This GPS data can also help to determine whether visual data can underestimate home-range size estimation.

Monk parakeet home ranges in our study area are surprisingly small, with a median value of 10 ha. This contrasts with data for other similarly sized parrot species. Bahama parrots *Amazona leucocephala* had average home ranges of 16,000 ha (95% kernel) [22], maroon-fronted parrots *Rhynchopsitta terrisi* of 4000 to 12,000 ha (50% kernel) [21], and the mean home ranges of Hispaniolan parrots *Amazona ventralis* was 864 ha [43]. The small home ranges of our monk parakeets could be the result of extremely high resource loads in the urban environment, in part because of the public providing food to parakeets [44]. In their native range, monk parakeets normally have home ranges of 3–5 km in diameter [45], and have been found to travel as far as 16 km to feed on grain crops [46]. Although anecdotal, this was also the explanation for the small home range of an urban blue-crowned parakeet *Aratinga acuticaudata* [47]. This seems also to be the case for other urban-dwelling species with access to anthropogenic food supplies, such as foxes *Vulpes vulpes* [48] or raccoons *Procyon lotor* [49]. Ring-necked parakeets studied in an urban habitat with anthropogenic food supplies also showed a reduced home-range size (86 ha, 95% kernel) compared to parrots in the wild [19], although this home range was still substantially higher than that of the smaller monk parakeet.

The monk parakeet is considered an invasive alien species of high concern, and calls are being raised to control their populations [26,50]. Having a small home range is one of the conditions for the feasible eradication of an invasive species [51,52], and hence the small home range of urban monk parakeets that we report here is good news for pest managers. The method to be used for the control, however, is dependent on the demography and home range of the focal species. Population dynamic models have suggested that the culling of adult birds is twice as effective as efforts to suppress breeding [31]. From the different culling methods available for monk parakeets, we can generally distinguish between those methods in which the animals have to move to the control source, as in the case of traps or feeders with contraceptives or poison, and methods in which the control operation moves to locate the parakeets, as in approaches such as shooting or the use of funnel nets at nests [50]. The first group of methods can be difficult when the home range of the focal species is small, because the number of traps or feeders to be used increases inversely with the home-range size of the species under control [8,53]. In the case of monk parakeets, the small home range of the species in urban settings makes the costs of setting enough traps or feeders to cover the whole urban area of a city such as Barcelona prohibitively high, which advises against its use [50]. Hence, the alternative methods in which the control operation moves to locate the parakeets are necessary. Given that monk parakeets roost at their nests, showing high nest-site fidelity [27], focusing on the capture of birds at the nest at night with funnel nets or funnel traps could be a good alternative [50].

Author Contributions: J.C.S. and B.J.H. designed the study. J.C.S., A.M., J.P. and X.T. conducted fieldwork. J.C.S. analyzed the data. J.C.S. wrote the paper with input from co-authors. J.C.S. and B.J.H. provided financial support. All authors read and approved the final manuscript.

Funding: Laboratory work was conducted at the University of Sheffield, UK and funded by a NERC studentship to Francesca Dawson Pell through the ACCE Doctoral Training Partnership and by a Natural Environment Research Council (NERC) Biomolecular Analysis Facility (NBAF) grant (NBAF1078). The fieldwork was supported by CGL-2020 PID2020-114907GB-C21 research project to JCS from Ministry of Science and Innovation.

Institutional Review Board Statement: Birds were handled, and blood samples taken with special permission EPI 7/2015 (01529/1498/2015) from Direcció General del Medi Natural i Biodiversitat, Generalitat de Catalunya, following Catalan regional ethical guidelines for the handling of birds. J.C.S. received special authorization (001501-0402.2009) for the handling of animals in research from Servei de Protecció de la Fauna, Flora i Animals de Companyia, according to Decree 214/1997/30.07.

Data Availability Statement: Data supporting reported results will be found at ResearchGate https://www.researchgate.net/publication/356784880_Data_from_paper_Sex_and_Age_Effects_on_Monk_Parakeet_Home-Range_Variation_in_the_Urban_Habitat_published_in_Diversity_2021.

Acknowledgments: We thank Francesc Uribe for drawing maps and Lluïsa Arroyo and Alba Ortega-Segalerva for field assistance. We are also most grateful to Institut de Parcs i Jardins from the Barcelona City Council for their support and facilities during field work, and to Francesca Dawson Pell for conducting the lab work, and to NBAF lab at Sheffield for facilities and training.

Conflicts of Interest: The authors declare no conflict of interest.

References

1. Nathan, R.; Getz, W.M.; Revilla, E.; Holyoak, M.; Kadmon, R.; Saltz, D.; Smouse, P.E. A movement ecology paradigm for unifying organismal movement research. *Proc. Natl. Acad. Sci. USA* **2008**, *105*, 19052–19059. [CrossRef]
2. Holyoak, M.; Casagrandi, R.; Nathan, R.; Revilla, E.; Spiegel, O. Trends and missing parts in the study of movement ecology. *Proc. Natl. Acad. Sci. USA* **2008**, *105*, 19060–19065. [CrossRef] [PubMed]
3. Ellison, N.; Hatchwell, B.J.; Biddiscombe, S.J.; Napper, C.J.; Potts, J.R. Mechanistic home range analysis reveals drivers of space use patterns for a non-territorial passerine. *J. Anim. Ecol.* **2020**, *89*, 2763–2776. [CrossRef] [PubMed]
4. Börger, L.; Dalziel, B.D.; Fryxell, J.M. Are there general mechanisms of animal home range behaviour? A review and prospects for future research. *Ecol. Lett.* **2008**, *11*, 637–650. [CrossRef] [PubMed]
5. Clobert, J.; Danchin, E.; Dhondt, A.A.; Nichols, J. *Dispersal*; Oxford University Press: Oxford, UK, 2001.
6. Hansson, L.-A.; Åkesson, S. *Animal Movement Across Scales*, 1st ed.; Oxford University Press: Oxford, UK, 2014; ISBN 9780199677184.
7. Rolando, A. On the Ecology of Home Range in Birds. *Revue d'écologie* **2002**, *57*, 53–73.
8. Adams, A.L.; Recio, M.R.; Robertson, B.C.; Dickinson, K.J.M.; van Heezik, Y. Understanding home range behaviour and resource selection of invasive common brushtail possums (Trichosurus vulpecula) in urban environments. *Biol. Invasions* **2014**, *16*, 1791–1804. [CrossRef]
9. Cameron, B.G.; van Heezik, Y.; Maloney, R.F.; Seddon, P.J.; Harraway, J.A. Improving predator capture rates: Analysis of river margin trap site data in the Waitaki Basin, New Zealand. *N. Z. J. Ecol.* **2005**, *29*, 117–128.
10. Norbury, G.L.; Norbury, D.C.; Heyward, R.P. Space use and denning behaviour of wild ferrets (Mustela furo) and cats (Felis catus). *N. Z. J. Ecol.* **1998**, *22*, 149–159.
11. Bengsen, A.J.; Butler, J.A.; Masters, P. Applying home-range and landscape-use data to design effective feral-cat control programs. *Wildl. Res.* **2012**, *39*, 258. [CrossRef]
12. Moseby, K.E.; Stott, J.; Crisp, H. Movement patterns of feral predators in an arid environment–implications for control through poison baiting. *Wildl. Res.* **2009**, *36*, 422. [CrossRef]
13. Croft, S.; Aegerter, J.N.; Beatham, S.; Coats, J.; Massei, G. A spatially explicit population model to compare management using culling and fertility control to reduce numbers of grey squirrels. *Ecol. Model.* **2021**, *440*, 109386. [CrossRef]
14. Suppo, C.; Naulin, J.M.; Langlais, M.; Artois, M. A modelling approach to vaccination and contraception programmes for rabies control in fox populations. *Proc. Biol. Sci.* **2000**, *267*, 1575–1582. [CrossRef]
15. Olah, G.; Butchart, S.H.M.; Symes, A.; Guzmán, I.M.; Cunningham, R.; Brightsmith, D.J.; Heinsohn, R. Ecological and socio-economic factors affecting extinction risk in parrots. *Biodivers. Conserv.* **2016**, *25*, 205–223. [CrossRef]
16. Bucher, E.H. Neotropical parrots as agricultural pests. In *New World Parrots in Crisis, Solutions from Conservation Biology*; Beissinger, S.R., Snyder, N., Eds.; Smithsonian Intitution Press: Washington, DC, USA, 1992; pp. 201–219.
17. Menchetti, M.; Mori, E. Worldwide impact of alien parrots (Aves Psittaciformes) on native biodiversity and environment: A review. *Ethol. Ecol. Evol.* **2014**, *26*, 172–194. [CrossRef]
18. Lindsey, G.D.; Arendt, W.J.; Kalina, J.; Pnedelton, G.W. Home ranges and movements of juvenile Puerto Rican Parrots. *J. Wildl. Manag.* **1991**, *55*, 318–322. [CrossRef]
19. Strubbe, D.; Matthysen, E. A radiotelemetry study of habitat use by the exotic Ring-necked Parakeet Psittacula krameri in Belgium. *Ibis* **2011**, *153*, 180–184. [CrossRef]
20. Brightsmith, D.J.; Boyd, J.D.; Hobson, E.A.; Randel, C.J. Satellite telemetry reveals complex migratory movement patterns of two large macaw species in the western Amazon basin. *Avian Conserv. Ecol.* **2021**, *16*, 14. [CrossRef]
21. Ortiz-Maciel, S.G.; Hori-Ochoa, C.; Enkerlin-Hoeflich, E. Maroon-Fronted Parrot (Rhynchopsitta terrisi) Breeding Home Range and Habitat Selection in the Northern Sierra Madre Oriental, Mexico. *Wilson J. Ornithol.* **2010**, *122*, 513–517. [CrossRef]
22. Stahala, C. Seasonal movements of the Bahama parrot (Amazona leucocephala bahamensis) between pine and hardwood forests: Implications for habitat conservation. *Ornitol. Neotrop.* **2008**, *19*, 165–171.
23. Forshaw, J.M. *Parrots of the World*; Princeton University Press: Princeton, NJ, USA, 2010.
24. Pruett-Jones, S. *Naturalized Parrots of the World*; Princeton University Press: Princeton, NJ, USA, 2021.
25. Senar, J.C.; Carrillo-Ortiz, J.; Arroyo, L. Numbered neck collars for long-distance identification of parakeets. *J. Field Ornithol.* **2012**, *83*, 180–185. [CrossRef]

26. Postigo, J.L.; Strubbe, D.; Mori, E.; Ancillotto, L.; Carneiro, I.; Latsoudis, P.; Menchetti, M.; Pârâu, L.G.; Parrott, D.; Reino, L.; et al. Mediterranean versus Atlantic monk parakeets *Myiopsitta monachus*: Towards differentiated management at the European scale. *Pest Manag. Sci.* **2019**, *75*, 915–922. [CrossRef] [PubMed]
27. Dawson-Pell, F.S.E.; Senar, J.C.; Hatchwell, B.J. Fine-scale genetic structure reflects limited and coordinated dispersal in the colonial monk parakeet, *Myiopsitta monachus*. *Mol. Ecol.* **2021**, *30*, 1531–1544. [CrossRef] [PubMed]
28. Dawson-Pell, F.S.E.; Hatchwell, B.J.; Ortega-Segalerva, A.; Dawson, D.A.; Horsburgh, G.J.; Senar, J.C. Microsatellite characterisation and sex-typing in two invasive parakeet species, the monk parakeet *Myiopsitta monachus* and ring-necked parakeet Psittacula krameri. *Mol. Biol. Rep.* **2020**, *47*, 1543–1550. [CrossRef] [PubMed]
29. Rodríguez-Pastor, R.; Senar, J.C.; Ortega, A.; Faus, J.; Uribe, F.; Montalvo, T. Distribution patterns of invasive Monk parakeets (*Myiopsitta monachus*) in an urban habitat. *Anim. Biodiv. Conserv.* **2012**, *35*, 107–117. [CrossRef]
30. Carrillo-Ortiz, J. Dinámica de poblaciones de la cotorra de pecho gris (*Myiopsitta monachus*) en la ciudad de Barcelona. Ph.D. Thesis, University of Barcelona, Barcelona, Spain, 2009.
31. Conroy, M.J.; Senar, J.C. Integration of demographic analyses and decision modeling in support of management of invasive monk parakeets, an urban and agricultural pest. *Environ. Ecol. Stat.* **2009**, *3*, 491–510.
32. Navarro, J.L.; Martín, L.F.; Bucher, E.H. El uso de la muda de remiges para determinar clases de edad en la cotorra (*Myiopsitta monachus*). *El Hornero* **1992**, *13*, 261–262.
33. Senar, J.C.; Carrillo-Ortiz, J.; Ortega-Segalerva, A.; Dawson-Pell, F.S.E.; Pascual, J.; Arroyo, L.; Mazzoni, D.; Montalvo, T.; Hatchwell, B.J. The reproductive capacity of monk parakeets *Myiopsitta monachus* is higher in their invasive range. *Bird Study* **2019**, *66*, 136–140. [CrossRef]
34. Mori, E.; Pascual, J.; Fattorini, N.; Menchetti, M.; Montalvo, T.; Senar, J.C. Ectoparasite sharing among native and invasive birds in a metropolitan area. *Parasitol. Res.* **2019**, *118*, 399–409. [CrossRef]
35. Kenward, R.E.; Casey, N.M.; Walls, S.S.; South, A.B. *Ranges 9: For the Analysis of Tracking and Location Data, Online Manual*; Anatrack Ltd.: Wareham, UK, 2014.
36. Worton, B.J. Kernel methods for estimating the utilization distribution in home-range studies. *Ecology* **1989**, *70*, 164–168. [CrossRef]
37. White, G.C.; Garrott, R.A. *Analysis of Wildlife Radio-Tracking Data*; Elsevier: New York, NY, USA, 1990.
38. Osborne, J.W. *Best Practices in Data Cleaning: A Complete Guide to Everything You Need to Do before and after Collecting Your Data*; SAGE: Thousand Oaks, CA, USA, 2013; ISBN 1412988012.
39. Erickson, B.H.; Nosanchuk, T. *Understanding Data*, 2nd ed.; Open University Press: Buckingham, UK, 2009; ISBN 9780335096626.
40. Warnock, S.E.; Takekawa, J.Y. Wintering site fidelity and movement patterns of Western Sandpipers Calidris mauri in the San Francisco Bay estuary. *Ibis* **1996**, *138*, 160–167. [CrossRef]
41. Adams, E.S. Approaches to the study of territory size and shape. *Annu. Rev. Ecol. Syst.* **2001**, *32*, 277–303. [CrossRef]
42. Lameris, T.K.; Brown, J.S.; Kleyheeg, E.; Jansen, P.A.; van Langevelde, F. Nest defensibility decreases home-range size in central place foragers. *Behav. Ecol.* **2018**, *29*, 1038–1045. [CrossRef]
43. White, T.H.; Collazo, J.A.; Vilella, F.J.; Guerrer, S.A. Effects of Hurricane Georges on habitat use by captive-reared Hispaniolan Parrots (*Amazona ventralis*) released in the Dominican Republic. *Ornitol. Neotrop.* **2005**, *16*, 405–417.
44. Borray-Escalante, N.A.; Mazzoni, D.; Ortega-Segalerva, A.; Arroyo, L.; Morera-Pujol, V.; González-Solís, J.; Senar, J.C. Diet assessments as a tool to control invasive species: Comparison between Monk and Rose-ringed parakeets with stable isotopes. *J. Urban Ecol.* **2020**, *6*, juaa005. [CrossRef]
45. Spreyer, M.F.; Bucher, E.H. Monk Parakeet (*Myiopsitta monachus*). In *The Birds of North America*; Poole, A., Gill, F., Eds.; Cornell Lab of Ornithology: Ithaca, NY, USA, 1998; pp. 1–23.
46. Larson, G.E. The Monk Parakeet in Illinois: New Views of Alarm. *Ill. Audubon Bull.* **1973**, *166*, 29–30.
47. Brooks, D.M. Behavioral ecology of a blue-crowned parakeet (*Aratinga acuticaudata*) in a subtropical urban landscape far from its natural range. *Bull. Texas Ornithol. Soc.* **2009**, *42*, 78–82.
48. White, J.G.; Gubiani, R.; Smallman, N.; Snell, K.; Morton, A. Home range, habitat selection and diet of foxes (*Vulpes vulpes*) in a semi-urban riparian environment. *Wildl. Res.* **2006**, *33*, 175. [CrossRef]
49. Prange, S.; Gehrt, S.D.; Wiggers, E.P. Influences of anthropogenic resources on raccoon (Procyon lotor) movements and spatial distribution. *J. Mammal.* **2004**, *85*, 483–490. [CrossRef]
50. Senar, J.C.; Conroy, M.J.; Montalvo, T. Decision-making Models and Management of the Monk Parakeet. In *Naturalized Parrots of the World*; Pruett-Jones, S., Ed.; Princeton University Press: Princeton, NJ, USA, 2021; pp. 102–122.
51. Bomford, M.; O'Brien, P. Eradication or control for vertebrate pests? *Wildl. Soc. Bull.* **1995**, *23*, 249–255.
52. Rainbolt, R.E.; Coblentz, B.E. A different perspective on eradication of vertebrate pests. *Wildl. Soc. Bull.* **1997**, *25*, 189–191.
53. Murton, R.K.; Thearle, R.J.P.; Thompson, J. Ecological studies of the feral pigeon Columba livia var.: I. population, breeding biology and methods of control. *J. Appl. Ecol.* **1972**, *9*, 835–874. [CrossRef]

MDPI
St. Alban-Anlage 66
4052 Basel
Switzerland
Tel. +41 61 683 77 34
Fax +41 61 302 89 18
www.mdpi.com

Diversity Editorial Office
E-mail: diversity@mdpi.com
www.mdpi.com/journal/diversity

www.ingramcontent.com/pod-product-compliance
Lightning Source LLC
LaVergne TN
LVHW071618170726
843515LV00009B/2428

* 9 7 8 3 0 3 6 5 5 0 0 9 1 *